Encyclopedia of South American Aquatic Insects

Illustrated Keys to Known Families, Genera, and Species in South America

Series editor
Charles W. Heckman, Olympia, Washington, USA

This series is designed to provide books permitting entomologists in all special fields of biology to identify the specimens of aquatic insects they acquire in South America, and, if the species has not yet been described, to determine its family and usually also its genus, and to facilitate its description as a species new to science. Volumes in this series provide keys to families, genera, and species of both aquatic and terrestrial insects. All books contain richly illustrated keys and encompass adults of all genera, larvae and pupae.

More information about this series at http://www.springer.com/series/15144

Charles W. Heckman

Neuroptera (Including Megaloptera)

Springer

Charles W. Heckman
Olympia, WA, USA

Encyclopedia of South American Aquatic Insects
ISBN 978-3-319-35124-7 ISBN 978-3-319-35125-4 (eBook)
DOI 10.1007/978-3-319-35125-4

Library of Congress Control Number: 2017932552

Printed on acid-free paper

This Springer imprint is published by Springer Nature
The registered company is Springer International Publishing AG
The registered company address is: Gewerbestrasse 11, 6330 Cham, Switzerland

Contents

Introduction

This work was begun to provide keys to the aquatic insect species known from Brazil. The original goal was to include all genera known from South America and all species from Brazil, but for most groups, the scope was expanded to encompass all species in South America and, in some cases, to include terrestrial species of orders that encompass both terrestrial and aquatic taxa. In no case is a taxonomic revision of any group undertaken, although recommendations for such revisions are included where appropriate, and the probable synonymy of nominal species still treated as valid in the literature is noted.

Two different approaches will be employed according to the taxon being treated. For phylogenetic groups encompassing overwhelmingly or exclusively aquatic species, such as the orders Plecoptera and Ephemeroptera or the families Dytiscidae and Culicidae, keys are provided to distinguish all genera and species known to occur in South America. An effort has been made to include every identifiable species so that the user of the key can determine with reasonable certainty whether or not his specimen belongs to a species that has already been described or whether it is one that is not yet known to science. Where feasible, complete keys will be prepared for groups containing both aquatic and terrestrial species, if they do not encompass an extraordinarily large number of species. This has already been done for the order Collembola.

The second approach will be used for aquatic species belonging to predominantly terrestrial taxa, such as the order Lepidoptera or the families Curculionidae and Muscidae. In such cases, the number of terrestrial species involved is too great to deal with conveniently. For example, a work confined to aquatic insects cannot include a key to all of the approximately 45,000 weevil species (O'Brien and Wibmer 1978) or even the portion of its enormous fauna inhabiting South America because only a small percentage of them live in or on aquatic plants. However, without such a comprehensive key, it becomes difficult for a person not specializing in the taxonomy of this group to be sure in which genus or subfamily his specimen belongs. For aquatic species in such groups, a descriptive approach is used. It is attempted to describe the morphology completely enough for the nonspecialist to recognize his aquatic species. At the same time, descriptions or keys to higher taxa within these groups are provided as necessary so that the aquatic species can be distinguished from the terrestrial ones and phylogenetic affinities of the specimens can be recognized.

The geographical limitations of this work are not strictly maintained throughout. For some genera, keys are provided that include mention or descriptions of species still known only from Central America or Caribbean islands, while for others, the keys are strictly limited to the South American fauna. The choice of comprehensiveness was dictated mainly by convenience. For genera encompassing few species and those that have been revised by competent taxonomists, keys to most or all Neotropical species could easily be provided. In the cases of other genera encompassing poorly described species, those which have not been reported south of Panama and the Antilles were not included in the keys.

In general, the ranges of the aquatic insect species are very poorly known. The reported occurrences indicate more the locations at which entomologists have worked than the actual distribution of the species. The type specimens of a great many South American species were collected during expeditions to individual regions of the continent, and systematic surveys of the fauna have only been undertaken for a few groups. Among the prominent early collectors was Charles Darwin. Larger expeditions were undertaken during the late nineteenth and early twentieth century. A considerable number of species have been described after examination of the specimens collected in Patagonia and South Chile during the early part of the twentieth century. In Brazil, Rio de Janeiro and Santa Catarina appear very frequently in the distribution reports due to the extensive collection in those states by F. Plaumann, while many other species were described from Amazonas because of the presence of a large research station, the Instituto Nacional de Pesquisas da Amazônia, at Manaus. Naturally, a great many species were described from specimens collected at São Paulo and Rio de Janeiro, where most of the Brazilian institutes of research are located. Reports from other Brazilian states are generally much rarer, so the ranges of Brazilian species can generally be expected to be much more extensive than indicated by the published reports. Notable research efforts in other parts of South America have resulted in large numbers of species being recorded for Surinam and the Amazon region of Peru. Knowledge of the fauna in other regions varies considerably from taxon to taxon, depending upon the efforts of talented individuals specializing in individual orders or families. Generally, the insect taxa with the greatest impact on public health and agriculture are the best known.

For some rarer species in neglected taxa, the ranges are scarcely defined at all. Specimens described during the nineteenth century were sometimes reported from "Brazil" or even "South America" without any more exact collecting data. The fact that the descriptions of such species are generally sketchy, at best, makes it doubtful whether they can be recognized again. There is mention in many of the review papers cited in this work of the confusion caused by falsely labeled specimens, incomplete collecting data, and doubt among the collectors about the names of the places from which their specimens originated. Names such as "Chapada" in Brazil frequently appear without mention of which Chapada is meant. This problem will continue to cause confusion in the future as it has in the past.

The problems encountered in systematic biology worldwide are especially pronounced in the Neotropical region. This is due to the dearth of detailed taxonomic studies on the regional fauna. Theoretically, the scientist who needs to know the correct name of a specimen can accomplish this by a procedure that assures a high probability of success, although a good deal of time and expense may be required. The first step in this procedure is the use of a comprehensive key treating the higher taxon to which the specimen belongs. This process is much easier for a specialist familiar with the group than for a person who needs to know the identity of species being used in physiological experiments or surveyed in synecological studies. In most cases, after the specimens have been identified using the keys, original or revised descriptions listed by the author of the key should be consulted to confirm the identifications. If the available descriptions are so poor that the identity of the species remains uncertain, a comprehensive taxonomic review of the taxon to which it belongs must be consulted. If such a review has not yet been published, the specimen to be identified must be compared with type specimens of each species belonging to the higher taxon to which the specimen belongs. These should be found in museum collections.

According to taxonomic convention, the author of a published original description of a new species should designate one of his specimens as a holotype. This type then becomes the standard on which the identifications of all specimens collected in the future are based. The species to which the holotype belongs is then referred to by the name proposed by the author, at least when the same name has not previously been applied to another species.

In addition to the holotype, other specimens believed to be of the same species by the author are designated as paratypes. If a holotype is not available, a paratype can be examined to confirm identifications. If the holotype has been lost or extensively damaged, a specialist revising the taxonomy of the group may designate a lectotype apparently belonging to the species in question. This then replaces the holotype as a standard for recognizing the species, at least until the lost holotype or paratype is found.

Anyone wishing to take the trouble to identify a specimen beyond the shadow of a reasonable doubt should be able to follow this procedure. If his specimen is not the same species as any of the holotypes preserved in museums, he should prepare a description of his specimen and have it published with his proposed name for the new species or give it to a specialist who is interested in doing this.

Unfortunately, the procedures described above often fail to work, especially when South American insects are involved. First of all, keys to identify the species are seldom available, and when they are, they are generally out of date or incomplete for the region being studied. This leaves the researcher with the chore of collecting a large number of original descriptions to match with his specimens. These descriptions are sometimes very sketchy, leaving the reader with no reliable way of identifying his specimen from the available literature. Comparisons with type specimens may not only be difficult in many cases because the types were deposited in museums on other continents, but they also are often impossible because some authors have failed to mention the name of the museum in which they intended to place their type specimens or because the specimen cannot be found again due to war damage or sloppiness by the curators of the museum in question. Even when the type can be found, it often proves to be in very poor condition, and relatively few paratypes of South American species have been designated that can be examined in the absence of a usable holotype.

Using established procedures for identifying South American insects is made even more problematic by the practice of some authors of describing new species on the basis of only one life stage. Not only have many species been described by taxonomists who examined only adults, some are known only from larvae. Furthermore, countless species are known only from adults of one sex. While the description of an adult may provide a legitimate basis for establishing a new species, it is difficult to find justification for naming a species based only on larvae. Some authors have designated larvae as the types of species congeneric with others known only as adults. These individuals seemed to be in such a hurry to publish that they deliberately left the arduous task of matching larvae with adults to other researchers.

In practice, convention should dictate which gender and life stage should be chosen as the holotype. For example, adult male chironomids are presently those on which species descriptions should be based. Earlier descriptions of female adults are generally useless for determining a species unless someone has taken the trouble to match the female to a described male. In an ideal system of nomenclature, descriptions of adults of both sexes and of the larval stages should be provided. However, the state of the art still requires specimens to be identified according to partial descriptions of one stage and often one sex. Therefore, before fully workable systems for identifying South American species will be possible, an enormous number of revisions and supplemental descriptions will be necessary. This work is meant as a first step in the process: providing keys as reliable as the available publications permit.

Finally, keys to taxa higher than families are provided only to delimit groups of aquatic species from terrestrial ones. It must be noted that obsolete names of insect taxa are defined or noted in the keys whenever users of these books are likely to encounter the names in old publications. They are neither endorsed nor suppressed but merely defined. Keys to insect orders can be found in any number of textbooks on entomology or invertebrate zoology, and most specimens can be quickly assigned to the correct higher taxon by referring to the illustrations in this series. With a little experience, anyone can learn to recognize the order of most insects almost at a glance.

An Appeal for Quality in Taxonomic Work

This appeal is addressed to two groups, the first consisting of those responsible for deciding who obtains what portion of the available research funds and the second being the taxonomists themselves.

It has long been recognized by experienced ecologists that identification of the species present in a community is an absolute necessity for thorough ecological research. The biota of any water body, for example, is more than a quantity of "biomass" or a "pathway for energy." Many ecologists, especially those beginning their research careers, have the greatest difficulty in identifying the species present, and the quality of their work is limited by this difficulty. The ecologist may indeed investigate systems using methods very different from those engaged in descriptive biology, but he is nevertheless dependent on a sound basis of taxonomic information for the proper reporting of his results. Briefly stated, ten may be equal to ten, but ten oranges are not equal to ten cows, and ten of one dragonfly species are not equal to ten of another. The "emergency measure" of identifying a species only as far as the genus or family is not satisfactory because two congeneric species may have completely different habitat preferences, feeding habits, and seasonal activity cycles. In short, for one ecologist to properly compare his work with that of another, he must know the name of the species he has encountered and of those that have influenced the biotic communities he has been studying.

In the past, ecologists usually relied on taxonomic specialists to identify their specimens. Since the specimens invariably belonged to a wide variety of taxa, they were distributed among a number of taxonomists, each of whom was an expert only for his own small phylogenetic group. This practice has ceased to function in recent years because of the failure of those in charge of distributing research funds to support scientists who wish to engage themselves in taxonomic work professionally. As the taxonomists reached retirement age, they left active research without being able to pass their knowledge on to a successor. As time went on, more and more personal knowledge and skill that is difficult or impossible to get from books was permanently lost.

The disappearance of the best taxonomists was accompanied by a decline in the quality of ecological work. This may not be considered totally undesirable by persons in government and business who are constantly confronted by the demands of environmentalists. As ecology moves increasingly out of the laboratory and into the courtroom, the ground rules change, and where principles remain unclear, the lawyers have more room to maneuver, using their skills of persuasion rather than being forced to confront research results already proven beyond the shadow of a reasonable doubt. Furthermore, laws passed to protect endangered species can be better circumvented if no people can be found who are capable of identifying such species.

Although there may well be incentives to suppress taxonomic research, the virtual demise of taxonomy as a field of biology has been instigated mainly by the scientific community itself. Ambition and the desire for quick success have motivated the staffs of many institutions to seek persons promising to achieve great breakthroughs, and no place was left for routine systematic work, which is by nature slow and methodical. Often, it was the number of publications rather than their quality or length that determined whether or not a person could succeed in finding gainful employment in science. Recently, science citation indices have become the criterion for judging success, prompting young researchers to investigate whatever everyone else is investigating in order to assure themselves of a maximum number of other scientists who would have reason to cite their work. Pressure was placed on the young scientist to come up with something clever very quickly, leaving no time to learn in depth about the organisms he was encountering. It is especially unfortunate that the worldwide elimination of taxonomist positions came just prior to the development of computer equipment that makes it very easy to establish data banks. Thus, science was robbed of the personnel competent to systematize and disseminate the great body of information that had accumulated during two centuries of intensive work by systematic biologists in all parts of the world.

A long time went by without any crisis becoming evident because the taxonomists and systematic biologists already working in the museums and universities continued donating their time to help identify specimens collected or used in the laboratory by ecologists, physiologists, geneticists, and biochemists. Even after their retirements, many continued contributing time to help younger colleagues in other fields of biology. Moreover, a number of excellent specialists were employed in other fields of work rather than in the biological sciences and made their contributions to the knowledge of the world's flora and fauna as a hobby.

With the gradual elimination of the competent taxonomists and systematic biologists through death or disability, however, the problems of defining community structure, determining species diversity, and identifying the organisms causing some problem in the field have been intensifying. For many taxa, there are no longer experts who can be consulted, or those experts who are still active are hopelessly overworked.

The result is the loss of a vast body of information about the flora and fauna in all parts of the world. Without knowledge of the species that have been encountered, the ecologist is faced with a dilemma. He must learn to identify a vast array of species belonging to the widest variety of phylogenetic groups, or he must limit his research to superficial phenomena. Hence, there has been an increasing tendency to use numbers instead of names and to try to equate quantities of unlike and undefined substances or objects, although this cannot yield any reliable results, as mentioned above.

The second aspect of the problem must be dealt with by the taxonomists themselves, who have in the past often been guilty of laxity in quality control of their work. The ambition to gain status by naming new species has certainly been a motivating factor for many of these scientists, and for this purpose, it is easier to write short papers with sketchy descriptions of aberrant specimens than to produce comprehensive reviews compiling the available information on genera or families together with thorough, well-illustrated descriptions of new taxa. As mentioned above, it is evident that the examination of type specimens for positive identification of South American insects is often impossible because none were designated, or if they were, they have been lost or greatly damaged. For better or worse, the published literature frequently provides the only criteria for identifying species.

While compiling this key, the enormously broad spectrum of quality in the taxonomic publications became evident. Some papers are of very high quality, such as the works of Belle on the Odonata. These publications provide descriptions of all essential morphological characters, the exact locations of type specimens, and precise collecting data. Any morphological details lacking in the text can be seen in the comprehensive illustrations. They make it clear that the species described are actually distinct from all species previously described.

In general, more recent publications tend to include better descriptions than older ones, but this is not the case with all taxa. Many works more than a century old are still useful for identifying species, and some of the information they provide cannot be found in any more recent publications. On the other hand, many recent publications are extremely poor, providing either very sketchy descriptions or showing a very poor comprehension of the concept of species as a closed phylogenetic grouping of individuals, which may possess a considerable degree of individual variability (Mayr 1963).

Many of the better taxonomic publications are rather long, but length is no guarantee of quality. There are some excellent concise works, which can be used with great reliability for the identification of species. Some very long works, including incredible amounts of detail, on the other hand, are very poorly organized and omit much essential information while including vast amounts of trivia. Authors who are more collectors than biologists have used very keen, practiced observation abilities to find the most minor differences among single populations to produce massive numbers of nominal species that are indistinguishable to less skilled observers. Among the European fauna, long lists of synonyms attest to the vast amount of printer's ink that has been squandered on the description and suppression of spurious taxa. Unfortunately, revisions of many South American taxa are still lacking, and many of the nominal species included in the keys will certainly prove to be junior synonyms when more is known about the variability of natural populations.

In the following sections, minimum criteria will be suggested for describing new taxa. All of the features an author should describe to define a new species or subspecies within a particular taxonomic group will be mentioned. It is further suggested, as a general rule, to base descriptions on a holotype and several paratypes. New species described from only one or two specimens should be regarded as questionable, particularly when the morphological differences between them and specimens of other nominal species are very slight. The practice of choosing type specimens that are missing various structures, such as legs, abdomens, or antennae, should be discouraged. With modern means of mass transportation available, there should be no serious obstacle to collecting more material from the locations at which the specimens of the alleged new species were encountered, and funds should be provided for such collecting for the reasons outlined above.

It would probably be justified to declare many names of species poorly described and without known type specimens as nomena inquerendae and omit them from the key. However, whenever distinctive features were found in the original description that can be used to distinguish a species from all other known species in the genus, the name is included in the keys. Names from the keys must therefore be used with caution because one or more undescribed species may also have the characteristics in question. Where such situations are apparent, it is hoped that specialists for the groups in question, if any are presently active, will designate lectotypes for the species most likely to have been referred to by the earlier authors. Holotypes, paratypes, and original descriptions should be provided for any similar, congeneric species that were hitherto undescribed.

Scope of the Work

The families recognized are mainly those found in the key of Brues et al. (1954), and some of the new families that have been described by various authors after the appearance of that key are omitted because they have not been generally accepted, are too poorly defined, are too similar to other families to recognize any general definitive characteristics for use in the key, or seem for another reason to be of doubtful validity.

Geographically, the work is limited to continental South America and islands very close to the coast, such as Tierra del Fuego (Map 1). Islands far enough offshore to be zoogeographically distinctive, such as the Galapagos Islands, are only included in some of the volumes, even though politically, they are parts of South American countries. Case by case decisions to include them depend upon the availability and reliability of published surveys of the local fauna. Except in the case of Brazil, the distribution reported in the keys usually refers only to the country in which the species have been found. When they have been reported in the literature, the Brazilian states are shown, as depicted in Map 2. However, some inaccuracies may have been inadvertently introduced in the cases of states which have changed their boundaries since the insects were described. For example, Mato Grosso formerly encompassed the present states of Mato Grosso, Mato Grosso do Sul, and Rondônia, so species reported in older literature from Mato Grosso may actually be confined to any one of these three present states.

Acknowledgments The preparation of these keys required the search of a large amount of literature. A great deal of help was provided in obtaining many obscure or archaic papers by the staff of the Library of the Zoologisches Institut und Zoologisches Museum of the University of Hamburg and through the Timberland Public Library System in Washington State, USA.

Map 1 The geographical scope of this work is limited to continental South America and the off-shore islands shown on this map

Map 2 The Brazilian states where the species were found are generally reported as shown on this map. Older reports, however, may not reflect the modern political boundaries, so it often requires knowledge of the travels of entomologists to determine the precise locations of known occurrence. D.F. designates the Federal District of Brasilia

Literature

Brues CT, Melander AL, Cartenter FM (1954) Classification of insects. Keys to the living and extinct families of other terrestrial arthropods. Bulletin of the Museum of Comparative Zoology at Harvard College 108:1–797

Mayr E (1963) Animal species and evolution. Harvard University Press, Cambridge, MA, 797 pp

O'Brien CW, Wibmer GJ (1978) Genera and species of Curculionidae (Coleoptera). Entomological News 89(2–3):89–92

Part I
Order 7: Neuroptera (*sensu lato*)

Chapter 1
General Information

Abstract The general characteristics of the order Neuroptera, tentatively including the species in the nominal order Megaloptera, are described, including the morphology, larval development, ecology, and zoogeography. The methods of collecting, preserving, and examining specimens are discussed, and some needs for future research are mentioned. It is shown that many species names, which are still technically valid, may never be correctly used again unless a type specimen is discovered or unless it is possible to provide an up-to-date description and lectotype.

Keywords Morphology • Preservation • Ecology • Zoogeography • Taxonomy

Until the early part of the twentieth century, the order Neuroptera was understood to encompass a large number of insects with two pairs of similar wings, often supported by many veins and with transparent or tinted membranes. Species belonging to Odonata, Plecoptera, Ephemeroptera, Trichoptera, and smaller orders were included by many taxonomists within Neuroptera. As the fundamental differences between some of the major taxa became evident, many insects formerly considered to be neuropterans were placed in orders now recognized as distinct, leaving behind a rather heterogeneous group of families in the order. The variety of common names for the individual groups reflects their diversity in morphology, ecology, and distribution.

Those species regarded as members of Neuroptera at the present time are holometabolic; that is, they have complete metamorphosis, distinguishing them from many of the taxa with which they were formerly grouped. The adults display many similarities, usually including a moderate to complex network of veins in the wing, from which the name Neuroptera is derived. The wings lack scales and are typically folded posteriad to cover the abdomen while at rest. Exceptions to this are owlflies belonging to the family Ascalaphidae, which typically rest with the abdomen pointed dorsad, almost perpendicular to the longitudinal axis of the head and thorax, and thread-winged lacewings in the family Crocidae, which usually point the forewings obliquely dorsad when at rest.

Neuropteran antennae are typically long and often thread-like, but they are clubbed in some members of the Ascalaphidae and Myrmeleontidae, the owlflies and antlions, respectively.

C.W. Heckman, *Neuroptera (Including Megaloptera)*, Encyclopedia of South American Aquatic Insects, DOI 10.1007/978-3-319-35125-4_1

The mouths of most larval and adult neuropterans are modified for chewing, but the jaws of spongillafly larvae, in the family Sisyridae, appear to have adaptations suitable for piercing and sucking. However, they may actually be used for pipetting microscopic prey. Larvae in the family Osmylidae, usually found at the water's edge or under stones and tree bark, also have long, narrow mouthparts, as do larvae in the family Berothidae, which inhabit termite nests and presumably feed chiefly on the termites. Unlike sisyrid larvae, which inhabit freshwater sponges and possibly also unspecified ectoprocts, those in the families Osmylidae and Berothidae lack external gills (Tauber et al. 2003). Sisyridae and Osmylidae are the families of the most conspicuously aquatic species belonging to Planipennia, the taxon also encompassing all of the terrestrial species in this book.

Sialidae, the family of the alder flies, and Corydalidae, the dobsonflies and fishflies, constitute distinctive taxa of aquatic insects. They were originally classified with the Neuroptera, but for many years, they were believed to constitute a group deserving the status of an independent order, which was named Megaloptera. Thus, in early literature, alder flies were considered to be members of the Neuroptera, but during the middle decades of the twentieth century, almost all publications included Megaloptera as an independent order, even though it had long been known that the differences between the adult megalopterans and neuropterans did not justify assigning them to separate orders. Only differences in the larvae seemed to make them distinctive. For the past several decades, doubt has been expressed about whether alder flies, dobsonflies, and their relatives are really different enough phylogenetically to merit their separation from the other groups still classified as subdivisions of Neuroptera. Because the major differences between alder, dobson, and fishflies and members of the Neuroptera *sensu stricto* are observed only during the larval stages, many authors pointed out that this by no means makes the taxon fundamentally different from other groups of neuropterans. All major taxa of true neuropterans are characterized by distinctive morphological modifications during the larval stage, which are necessary for the specialized ways of life they have adopted. As a result, there has been a trend to reintegrate the Megaloptera into the order Neuroptera. However, to avoid controversy, some authors avoid taking sides in the debate by using the term "neuropteroid insects" for all controversial groups. To make the relationship clearer, an additional stratum of classification was introduced, and Neuroptera and Megaloptera have sometimes been maintained as independent orders in a superorder called Neuropteroidea (Aspöck 1986) or Neuropterida (Aspöck et al. 2001; Aspöck 2002).

Pending additional research, there is still no general agreement on the taxonomy of the Neuroptera, *sensu lato*. Some authors treat the taxon as an order encompassing three suborders: Megaloptera, Raphidioptera, and Planipennia. Presently, this seems to be the predominant classification system and is the one used in this volume. Others have continued to treat these three taxa as orders within the superorder Neuropteroidea (Berland and Grass 1951) or Neuropterida (Tauber et al. 2003). At times, the name Sialoidea has been substituted for Megaloptera as a suborder. Those who continue to use the older classification system utilize the name Neuroptera for the Planipennia alone, treating the other two names as those of independent orders.

Of the three main subdivisions of Neuroptera, treated herein as the suborders Megaloptera, Raphidioptera, and Planipennia, only two are positively known from South America. Species of Raphidioptera are not known south of northern Central America, and Aspöck and Aspöck (1969) reported none south of Mexico. While the suborder Megaloptera encompasses two families characterized by aquatic larvae, Planipennia includes mainly families of typically terrestrial insects, although, as already mentioned, distinct groups of species with aquatic or littoral larvae are also included in this suborder.

Neuropteran larvae display a wide variety of modifications for distinctive modes of life. They are found in terrestrial, aquatic, and wetland habitats. The locations at which adults are encountered can be indicative of the habitats in which the larvae develop, even though a few adults are strong flyers and may be found far from the sites at which they will mate and deposit their eggs. Usually, the best places to seek adults of aquatic species are along streams and rivers, as well as near lakes. When more is known about the larvae of South American species, it is likely that wetlands and floating islands of plants will also prove to be particularly rich in aquatic neuropterans. Most larvae and pupae in temperate zones seem to show seasonal growth cycles related to temperature or the seasonal length of the day. In the tropics, seasonal rainfall patterns or changes in the water level of lakes and rivers are apparently determinants of the seasonal reproductive patterns of the species. Unfortunately, reliable information on the seasonal development of South American species is still fragmentary, so our present state of knowledge does not provide a firm basis for more than speculation about the life cycles of most species.

A genus of apparently flightless species in South America is of controversial phylogenetic origin, and after its discovery by Navás (1927a), it was maintained in its own family, Brucheiseridae, which was included in the Megaloptera. More recently, the family was reduced to the status of a subfamily, Brucheiserinae, within the family Coniopterygidae (Riek 1975), which is now assigned to the suborder Planipennia (New 1989). The larvae remained unknown for more than half a century, after which the first specimens were found in one of the dryest regions of the world, the Atacama Desert of Chile, at about 1400 m above sea level. These larvae have been described by Sziráki and Flint (2007). To date, four species in two genera have been discovered.

Most other terrestrial species of Planipennia overpower and chew up small to moderately large insects as both larvae and adults. The larvae of antlions, members of the Myrmeleontidae sometimes called doodlebugs, are often encountered in dry sand. Some species dig cone-shaped ant traps. When an ant stumbles into one, it usually descends gradually to the bottom as the sand shifts under its feet. The larval antlion waits buried in the sand at the lowest point of the inverted cone to devour it. If the ant seems to be succeeding in climbing out of the trap, the larva throws sand at it to make it fall to the bottom. Some genera in this family, however, encompass species which do not construct such traps. Their larvae are encountered under pieces of wood or among leaf litter, where they stalk or ambush their prey.

A small family, Ascalaphidae, encompasses only a few relatively large and conspicuous species called owlflies. The habits of the known species in the Americas include predatory feeding, mainly on small arthropods, and larval development among leaf litter or on trees. The family is considered to be closely related to Myrmeleontidae.

Crocidae is a small family related to Myrmeleontidae and Ascalaphidae. The larvae of crocid species often inhabit hiding places in caves or buildings, where they ambush or stalk prey. They are remarkable for their elongated collars between the head and thorax. Adults have thread-like hind wings and seem to feed on pollen or nectar from flowers (Tauber et al. 2003).

Another remarkable family is Mantispidae, which includes species resembling the preying mantis in both superficial appearance and feeding habits, although they are not closely related to their orthopteran look-alikes phylogenetically. A few species mimic the shape and coloration of wasps. None of the species are known to colonize aquatic or semi-aquatic habitats. Most fly well and seek out locations where relatively large prey can be found, grasped in fore-legs suitably modified and armed with spines, and then devoured. Mantispid larvae have been observed living in the egg cases of spiders to feed on the eggs (Biraben 1960).

The remaining families encompass fully terrestrial species with common names containing the words lacewing or duskywing. Species in the family Dilaridae, known as pleasing lacewings, have larvae differing from those of other lacewings in developing in rotting wood rather than on plants, as well as by transforming to pupae within cocoons. The larvae have been encountered in burrows bored by other insects through wood (Tauber et al. 2003). The dilarids are believed to be primitive relatives of mantisflies in the family Mantispidae.

Many and perhaps most of the South American species in the remaining families prey on small insects, including aphids, which feed on sap sucked from fruits, stems, or leaves of plants, including many valuable cultivated crops. The stomach contents of a few of the lacewings examined, however, included only pollen or other material from plants. A considerable number of lacewing species have been investigated as possible agents for biologically controlling agricultural pests (Souza et al. 1990; Souza and Ciociola 1995, 1997; New 2001; Albuquerque et al. 2001; Frietas and Penny 2001; Anad et al. 2003). Species of Chrysopidae, the green lacewings, are especially promising for facilitating orgainic farming in the tropics by feeding during both their larval and adult stages on large numbers of small, harmful, herbivorous insects. Also rich in species are the brown lacewings in the family Hemerobiidae and the duskywings in the family Coniopterygidae. The giant lacewings in the family Polystoechotidae are poorly known, in spite of their large size. There are only two species known from South America, both from Chile. A third species is found in North and Central America, as far south as Panama (Oswald 1998).

Neuroptera is an ancient insect order, dating at least from the Permian. Fossil specimens identified as belonging to the Megaloptera and Planipennia date from this geological period. They seem to belong among the oldest known extant taxa of insects with complete metamorphosis (Carpenter 1930).

1.1 Morphology

The species within the individual subdivisions of Neuroptera possess many common features as adults, even though their larvae are quite distinct in morphology and preferred habitats. The distribution patterns of many species are limited by ecological factors affecting the larvae. In the case of the Megaloptera, the morphology and ecological requirements of the larvae were formerly considered distinctive enough from all other neuropteran larvae to merit giving the group the rank of an independent order. As already explained, the general opinion of neuropterologists has swung slowly toward the consensus that the anatomy of the adults should be given more consideration, and the Neuroptera and Megaloptera should be rejoined at or slightly above the level of order.

Naturally, the larval morphology of the individual families will have to be considered on an individual basis. Adults, on the other hand, have many features shared by the entire order, although extreme individual modifications of some of the body parts make each family distinctive and sometimes bizarre in appearance.

The morphological features most frequently used to distinguish the species include the genitalia, especially those males, as well as the shape and venation of the wings, the color pattern, and morphometric characters of various structures and appendages. Unfortunately, a general overview of a typical neuropteran cannot be conveniently provided because the morphology of most families is so distinctive from most others. Needless to say, such morphological differences make identification of specimens easier.

Although the majority of species within this order are terrestrial, keys to all South American taxa as complete as current knowledge permits will be provided because the order is small and can best be handled as an intact group in one volume. Most of the major taxa can be readily distinguished from each other, usually at a glance, by many morphological features, including general shape and size, the form of the head and antennae, the mouthparts, the wing venation, and even the coloration of the body and wings (Fig. 1.1). General features of adult specimens which should be noted at the beginning of any examination include whether the mouth is directed anteriad or ventrad, whether or not the antenna is clubbed, and modifications of the fore-legs, especially those indicating whether these legs are raptorial. The larvae are even more distinctive than the adults, and many can be recognized immediately from their habitats and habits. For example, the antlions in the family Myrmeleontidae are terrestrial, and most larvae live on dry ground, sometimes at the base conical pits, while spongillafly larvae in the family Sisyridae usually inhabit freshwater sponges and possibly also colonies of ectoprocts.

The general morphology of most adults in the order Neuroptera appears to be relatively unspecialized and conforms to a hypothetical body plan of a primitive insect. The **mesothoracic** and **metathoracic wings**, that is, the fore and hind wings, respectively, appear similar in size and venation. However, the **metathoracic wings** of most species tend to be slightly smaller, lack a few veins that are present in the fore-wing, and often have fewer markings. A few species, however, such as the

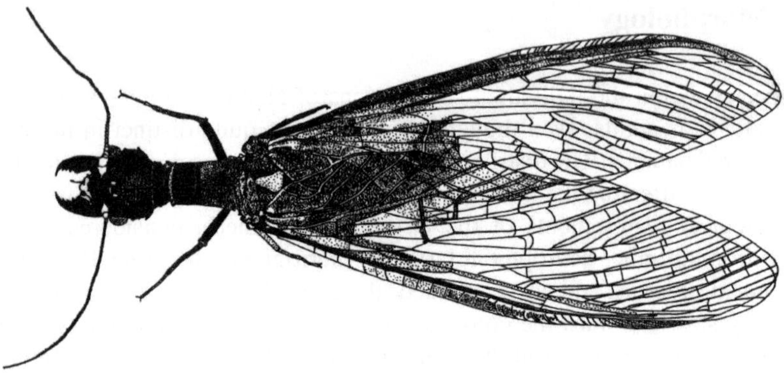

Fig. 1.1 Habitus of a male member of the suborder Megaloptera, *Corydalus wanningeri,* in dorsal view. Based on Contreras-Ramos and von der Dunk (2010).

threadwinged lacewings of the family Crocidae, have highly modified hind wings, as the common name of these insects implies.

The legs of most species are not notably modified. Typically, the **prothoracic legs** are similar to, although smaller than the mesothoracic and metathoracic legs. A notable exception to this is found in members of the mantisflies of the family Mantispidae, which are characterized by greatly enlarged prothoracic legs modified for seizing and holding prey. The **mesothoracic** and **metathoracic legs** are modified for walking rapidly and stabilizing the body of the mantisfly while it holds relatively large, struggling prey in its modified fore-legs. In the keys, the legs will be referred to by the less technical terms: fore, middle, and hind legs. The resemblance of the legs of these insects to those of the preying mantis, belonging to the very distantly related order Orthoptera, accounts for the common name of this family.

As shown in Fig. 1.1, adults of the Megaloptera have no notable modifications of the thorax and abdomen. Only the head displays some eye-catching modifications in some species (Fig. 1.2). Although males and females of many Neuroptera species are generally similar in morphology and coloration, with differences being confined to the genitalia, there are cases of sexual dimorphism or dichromatism affecting other parts of the body. The mouthparts are obviously adapted for grasping and chewing in all species of Megaloptera. However, the **mandibles** of males in some species of the family Corydalidae are greatly enlarged and elongated. The results of investigations of this modification suggest that both males and females of some species always possess short, robust **mandibles** (Fig. 1.3), while the long, forceps-like **mandibles** (Fig. 1.2) are characteristic of all males of other species (Contreras-Ramos 1998). A few species, however, encompass both males with elongated mandibles and those with short ones, like those of females.

Bowles et al. (2007) noted that the males of some South American species, including *Corydalus arpi, C. cephalotes, C. hecate,* and *C. ignotus,* have only short mandibles similar to those of the females, while *C. colombianus* develops somewhat elongated mandibles of a length the authors designated as "transitional"

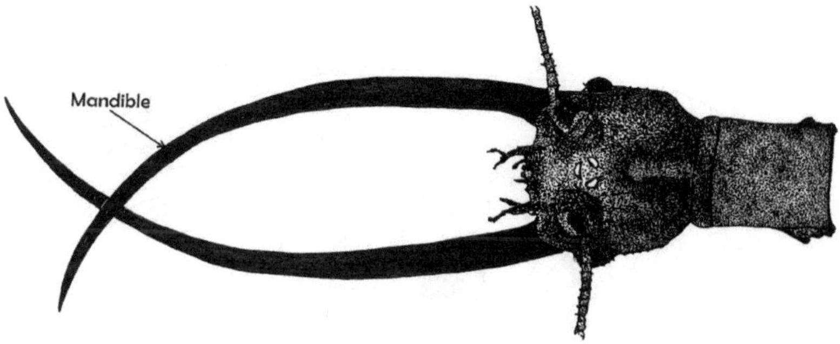

Fig. 1.2 The head and prothorax of a male *Corydalus imperiosus*, showing its elongate mandibles. Based on Contreras-Ramos (1998).

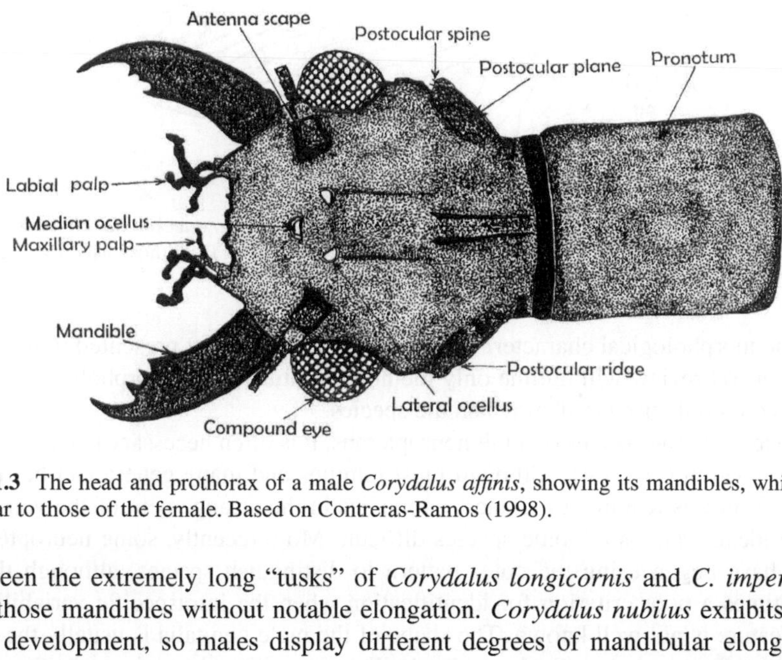

Fig. 1.3 The head and prothorax of a male *Corydalus affinis*, showing its mandibles, which are similar to those of the female. Based on Contreras-Ramos (1998).

between the extremely long "tusks" of *Corydalus longicornis* and *C. imperiosus* and those mandibles without notable elongation. *Corydalus nubilus* exhibits variable development, so males display different degrees of mandibular elongation. Subsequent investigations will be necessary to reveal the genetic and ecological implications of this phenomenon, especially the use to which the long mandibles are put in each species.

The heads of adult planipennians show a much greater spectrum of modfication than those of megalopterans. Most have mouthparts modified for grasping and chewing prey. However, those of species in the family Sisyridae are elongated and appear to be suitable for piercing and sucking, although the modification seems more suitable to permit the insects to "pipette" particularly small prey.

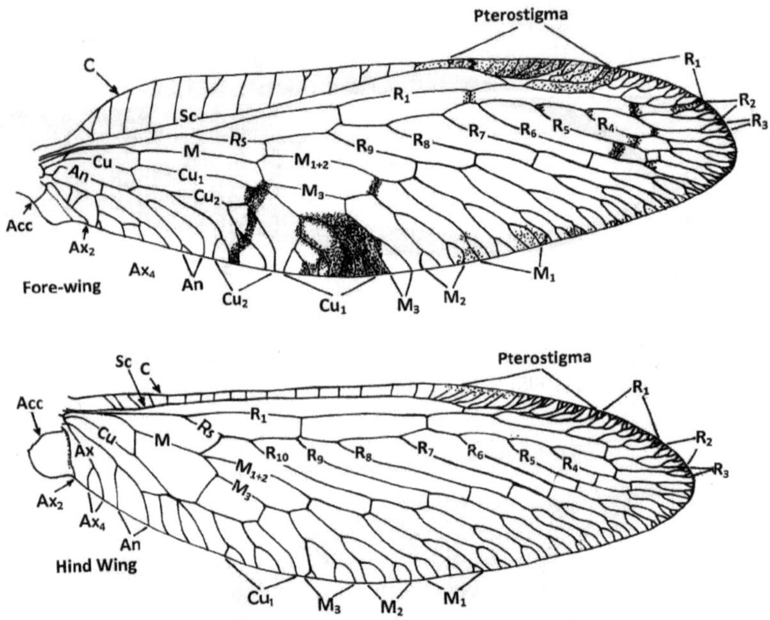

Fig. 1.4 Commonly used abbreviations for the wing veins of *Gerstaeckerella gigantea,* in the family Mantispidae and suborder Planipennia, modified somewhat from those used by Enderlein (1910) to conform to English names. Based on Enderlein (1910).

The morphological characteristics of each family are best presented in the keys. The general review will outline only the nomenclature of the morphological structures commonly used to distinguish the species.

To identify the species of adult neuropterans, it is often necessary to examine the wing venation (Fig. 1.4). Although most families and many genera can be recognized from this feature at a glance, anomalies in the arrangement of the veins can make identifications of some species difficult. More recently, some neuropterologists have begun using the color pattern to distinguish species, although this is sometimes a poor criterion for identification when the intraspecific variability in this feature is not well known. The shape of the male genitalia is usually the most definitive feature, but, as the keys clearly illustrate, either the male or the female of many species remain undescribed, forcing the researcher to seek other distinguishing characters. The disadvantage of relying on almost any other character for identification when specimens of only one sex have been examined is the lack of certainty whether differences may exist due to sexual dimorphism or dichromatism. The size of the adult, measured either as total body length or wing length, is also a poorly defined feature of neuropterans, especially when descriptions are based on only one or a few specimens. Size variability within this insect order seems to be considerable in comparison to that of most other insect orders, but little can be said with certainty until greater numbers of specimens have been examined.

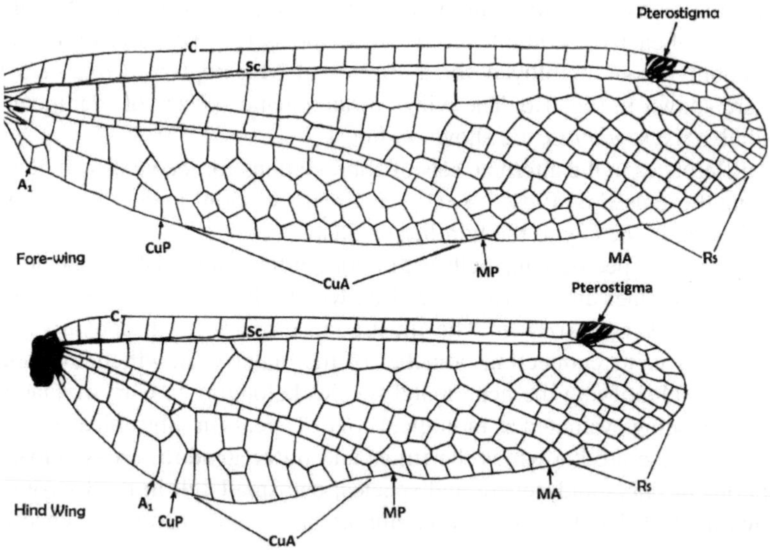

Fig. 1.5 The nomenclature of the wing veins of *Fillus amazonicus,* a species in the family Ascalaphidae using a somewhat different set of abbreviations from that employed in Fig. 1.4. Based on Machado and Rafael (2011).

Fossils of neuropterans are not uncommon, and they show that the order is ancient. Modifications from the primitive body plan of winged insects are remarkable in only a few taxa. However, most families have distinctive characteristics that facilitate their identification almost at a glance. The arrangement of the wing veins is usually one such character.

The nomenclature of the wing veins has not deviated too much in the literature, although fundamental inconsistencies persist. For example, some authors use the abbreviations M_1, M_2, Cu_1, Cu_2, 1A, 2A, and 3A for the same veins designated by others as MA, MP, CuA, CuP, A_1, A_2, and A_3, respectively. This discrepancy also exists in the literature on the other orders, which had been grouped in the Neuroptera during the nineteenth century. The commonly used alternative abbreviations for the wing veins are shown for a species of Ascalaphidae (Fig. 1.5), which has a network of wing veins that is roughly equal in complexity to that shown in Fig. 1.4.

The names of the wing veins followed by the abbreviations used for them in parentheses are the **costa** (C), **subcosta** (Sc), **radial sector** (Rs), **radial** (R), **median** (M), **cubital** (Cu), and **anal** (A) **veins**. The abbrevations sometimes also include designations for the positions of the veins, either with A and P, standing for anterior and posterior, respectively, or as a number written as a subscript, noting the first to final branch of the vein, counting clockwise on a dorsal view of the right wing. Thus, R_1, R_2, R_3, R_4, R_5, R_6, R_7, R_8, and R_9, shown on the fore-wing in Fig. 1.4, are the first branch of the **radial vein** to reach the wing margin through the ninth to do so. These are followed by the first through third branches of the **median vein** to reach the margin, abbreviated as M_1, M_2, and M_3, respectively. In species with multiple anal veins, the numbering may be 1A, 2A, and 3A, presumably allowing the branches of each to be numbered by subscript, as $1A_1$, $1A_2$, and so forth.

The abbreviations employed for a species of Megaloptera, *Corydalus wanningeri*, are shown in Fig. 1.6. The wings of an aquatic species of Planipennia, the spongillafly, *Sisyra minuta,* are shown in Fig. 1.7.

The differences in the alternate sets of abbreviations for the veins can be seen by comparing those used on Figs. 1.4 and 1.5. Great variations in degrees of complexity can be seen in the key to the families in Chap. 2, p. 47.

In some cases, specific wing cells are important for distinguishing species. Some of the important ones are shown in Fig. 1.8., which illustrates wings of a chrysopid species, *Titanochrysa ferreirai*. The illustration also shows the positions of two series of cross veins connecting branches of the radial vein, which are especially important for identifying the genera and species belonging to Chrysopidae, which are presently being evaluated as biological control agents in subtropical and tropical agriculture. These are known as the **inner** and **outer gradate cross veins.** In the descriptions of chrysopid genera and species, one small cell in the fore-wing, the **intramedian cell,** has become a very important. Its presence or absence and its shape are defining features of certain genera.

Other features commonly used to distinguish species are morphometric, based on the relationships between individual dimensions of the body structures. Some of the more common ones include size relationships between the length and width of the antennal segments, pronotum, thorax, femora, and tarsal segments. They may also include the relationships between the lengths of the femora, tibiae, and tarsi to one another; the lengths of certain setae relative to other structures; and the

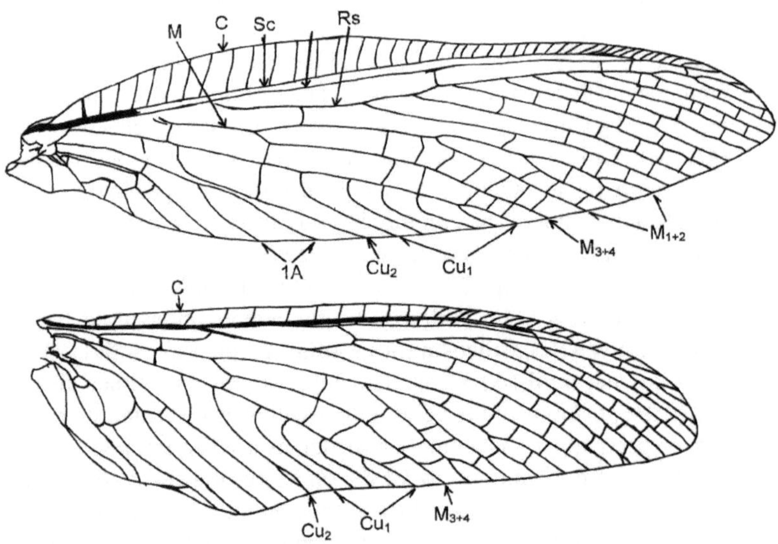

Fig. 1.6 The wing veins of a megalopteran in the family Corydalidae, *Corydalus wanningeri,* showing their nomenclature. Based on Contreras-Ramos and von der Dunk (2010).

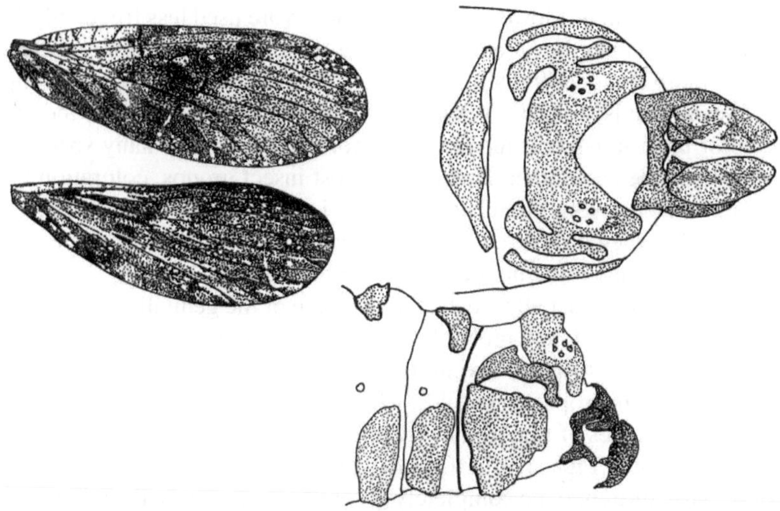

Fig. 1.7 *Sisyra minuta* male: fore and hind wing (left) and apex of the abdomen in dorsal (upper left) and lateral view (lower center). Based on Parfin and Gurney (1956).

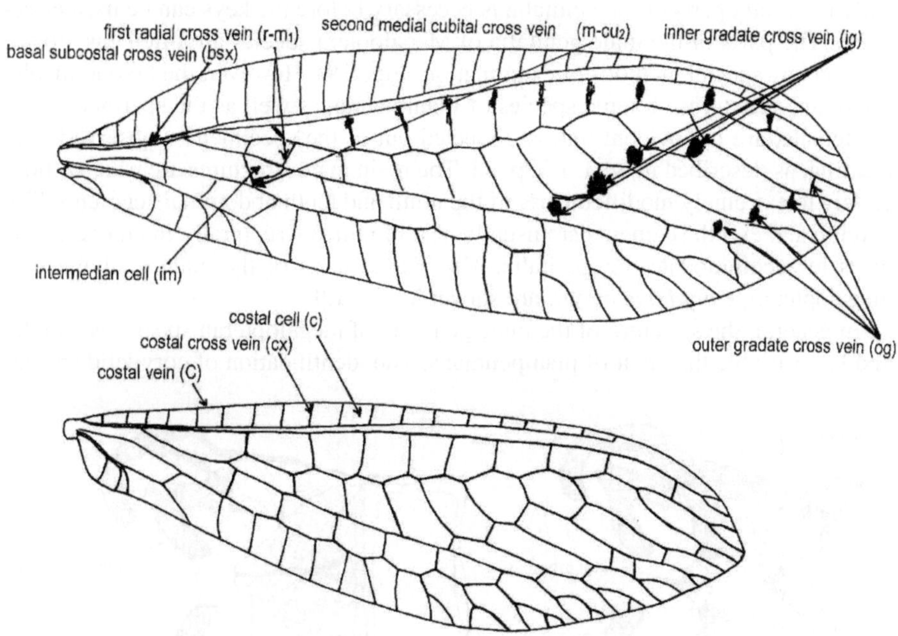

Fig. 1.8 Fore and hind wing of *Titanochrysa ferreirai,* showing the names of veins and wing cells particularly important for the identification of species. The wing membrane adjacent to the gradate cross veins of the fore-wing is darkened. The numbers of inner and outer gradate cross veins on one or both wings are sometimes important characters used to identify species. The intermedian cell, often abbreviated "im," has become a particularly important character for the identification of chloropids. The presence or absence of this cell and its shape are characteristic of certain genera. Based on Sosa and de Freitas (2012).

dimensions of the wings. However, these characters are used less frequently for species of Neuroptera than they are for most other insect orders. Within certain genera, species are more frequently distinguished on the basis of whether or not the forewing is falcate, that is, concave along the posterior margin proximal to the wing tip.

The color pattern has become important for distinguishing many species which otherwise resemble each other closely. For most insect groups, coloration is a particularly unreliable taxonomic character. This is especially true for poorly known species. Clearly, when only a few specimens are available for study, almost nothing can be determined about the natural intraspecific variability.

As in the case of almost all insects, the structure of the genitalia, especially those of males, is an important, and probably the most important morphological character for distinguishing the species. Taxonomists have developed a nomenclature for the genital structure of neuropteran species, which includes names for structures unique to the order. The names of important structures characteristic for males from several families are shown in Figs. 1.9, 1.10, and 1.11. It is important to keep in mind that some of the structures may be completely absent from species believed to be closely related to others which possess the structure, a situation which will be encountered when using the keys to the species.

Learning the vocabulary used by neuropterologists to distinguish the taxonomically important parts of the genitalia is necessary before the keys can be used effectively. The parts of the male genitalia of Megaloptera species are often easy to see without any special methods of examination (Fig. 1.9). However, the taxonomically important structures of some species of Planipennia, as well as the spermatheca of the female, are not evident unless dissected out of the specimen or observed after clearing, as described in Sect. 1.3, p. 33. The main male structures are usually moderately to extremely modified parts of the ninth and tenth abdominal segments. The seventh and eighth segments are usually also modified structurally in characteristic ways to accommodate the genitalia. The visible parts of the male genitalia of a megalopteran, *Corydalus mayri,* are shown in Fig. 1.9.

In general, the structure of the male genitalia of megalopteran species is simpler and less variable than that of planipennians. The identification of corydalid species

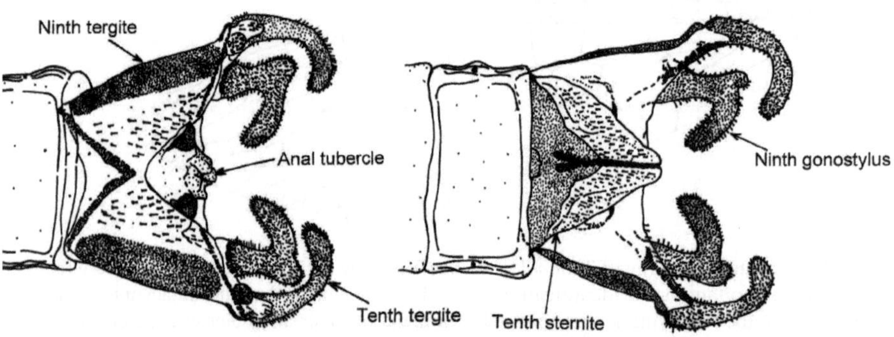

Fig. 1.9 The apex of the male abdomen of a species in the suborder Megaloptera, *Corydalus mayri,* shown in dorsal (left) and ventral view (right). Based on Contreras-Ramos (2002).

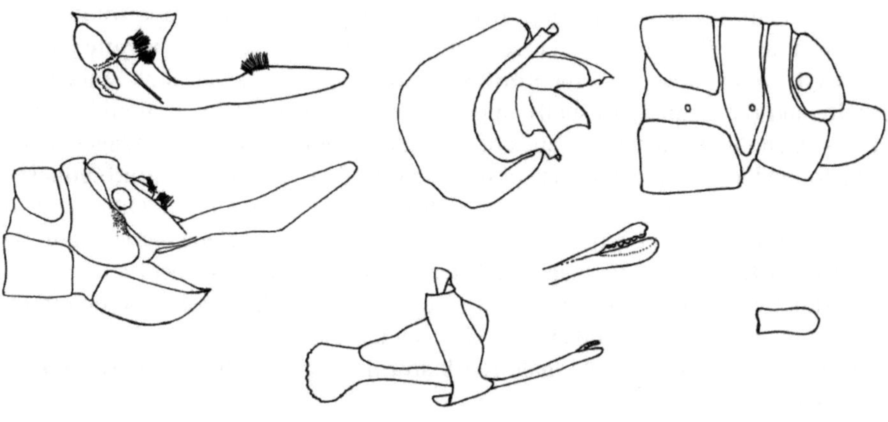

Fig. 1.10 *Biramus aggregatus,* a member of the Hemerobiidae, showing the locations of parts of the genitalia: apex of the male abdomen in lateral view (lower left), ectoproct of the male in dorsal view (upper left), gonarcus and mediuncus of the male in lateral view (upper center), parabaculum and supraparabacular process of the male in lateral view (lower center), terminal lobe at the apex of the parabaculum in dorsolateral view (right of center), apex of the female abdomen in lateral view (upper right), and subgenitale in ventral view (lower right). Based on Oswald (2004).

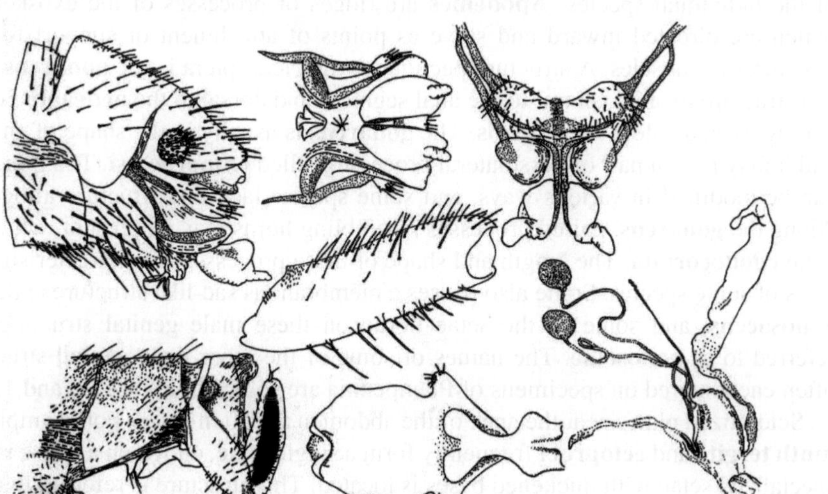

Fig. 1.11 The genitalia of a male and a female *Phymatosmylus caprorum,* a member of the family Osmylidae, and the left fore-coxa of a female (*center*); the morphological parts of the insect are designated as follows: the apices of the male (*upper left*) and female abdomen in lateral view (*lower left*); the male gonarcus and gonocoxite in ventral (*upper center*) and posterior view (*upper right*), the female eighth sternite and gonocoxite in ventral view (*lower center*), and the female genital system as revealed in a cleared specimen (*lower right*). (Based on Adams (1969)).

is based on the relative shapes and sizes of two appendages at the apex of the male abdomen, as well as by the relative shapes and sizes of the **tenth tergite** and the **gonostyli,** which arise from the **ninth sternite**. The lateral extensions of the **tenth tergite** and the **gonostyli** usually appear to be two pairs of long, digitiform processes extending from near the apex of the male abdomen. These appendages appear to be formed as lateral extensions of the sclerotized plates, with the **ninth gonostylus** usually extending dorsad and the **tenth tergite** curving ventrad. Other parts of the male terminalia frequently used to distinguish species include the **lobes on the tenth sternite,** which have shapes and sizes distinctive for most species.

Distinctive features of the male genitalia are generally less useful for identifying species in the family Sialidae than the genital structures of corydalid species. However, the structure of the male **tenth sternite** seems to be distinctive in most sialid species.

The structure of the male genitalia of planipennians is illustrated, exemplified by a member of the Hemerobiidae, *Biramus aggregatus* (Fig. 1.10) and a species of Osmylidae, *Phymatosmylus caprorum* (Fig. 1.11). Figures showing the male genitalia of the other families are found in the appropriate keys. The male genital structures are of great importance for identifying the species. Generally, the presence, location, and shapes of **apodemes** on the male genital structures are characteristic of the individual species. **Apodemes** are ridges or processes of the exoskeleton, which are directed inward and serve as points of attachment or support for soft structures or muscles. A structure peculiar to the Neuroptera is the **gonarcus.** This is a structure located ventral to the anal segment and dorsal to the **aedeagus**, which is roughly equivalent to the **penis**. The **gonarcus** is usually in the shape of an arch, and it may have a pair of dorsolateral processes called **entoprocessi**. The **gonarcus** can be modified in various ways, and some species lack this structure altogether. Along the **gonarcus**, paired processes resembling horns may be present, which are called **gonocornua**. The length and shape of these processes are characteristic features of some species. Some also posses a membranous sac-like structure called the **gonosaccus,** and some of the setae borne on these male genital structures are referred to as **gonosetae**. The names of some of the other male genital structures often encountered on specimens of Planipennia are shown in Figs. 1.10 and 1.11.

Sclerotized plates near the apex of the abdomen are often fused. For example, the **ninth tergite** and **ectoproct** frequently form a single plate, on which a dense ring of specialized setae with thickened bases is located. This structure is referred to as the **callus cercus.** The **eighth** and **ninth sternites** may also be fused. Some species have structures formed by fusion of the **parameres** with a structure called the **mediuncus**. This structure is important for identifying a few species, especially members of the Berothidae.

The external female genital structures of Neuroptera species are usually much simpler in structure than those of the males. However, taxonomic specialists who have studied the Neuroptera during recent decades have generally dissected or cleared female specimens and selected the **sclerotized ducts** of the **spermatheca** as important characters for distinguishing the species. Clearing the specimens is preferable to dissection, which may damage the organs examined, reducing the

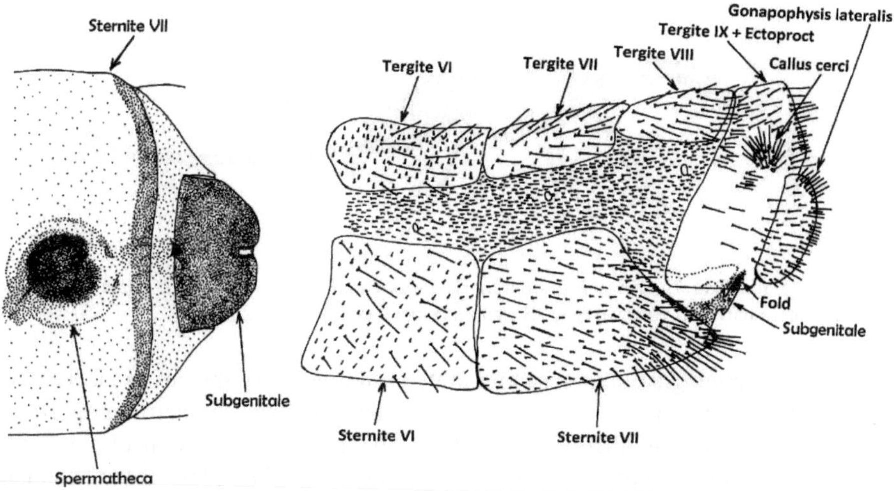

Fig. 1.12 The genitalia of a female *Vieira brooksi,* showing the nomenclature of characteristic structures. Based on Tauber (2006), who described the species under the name *Berchmansus brooksi*.

usefulness of the type specimens for future examination. Dissection or clearing by the methods described in Sect. 1.3 may be unavoidable for taxonomic revisions, but dissection of specimens designated as holotypes or lectotypes needs to be discouraged, even if it is not prohibited by the institutions in which the collections are kept. Too many of the rare type specimens on which descriptions prior to about 1950 are based have lost major structures, greatly impeding systematic revision based on published descriptions.

Typical parts of the female planipennian genitalia useful for identification are shown in Figs. 1.10, 1.11, and 1.12. In addition to the **spermatheca**, modifications of the apical segments of the abdomen and the shape of the **subgenitale**, the ventral plate usually located at or near the apex of the abdomen, are most often used to identify female specimens.

Many illustrations of both the male and female genitalia of Planipennia species have been made after clearing the tissues to make the sclerotized structures inside visible without dissection. The methods available are summarized in Sect. 1.3, p. 33. For the nomenclature of the genital structures of Planipennia species, Brooks and Bernard (1990) provided a particularly thorough list of the terms used to designate the individual parts of the male and female terminalia and a few of the other important parts of the body. This terminology is based on that provided by Tjeder (1966), but it was somewhat modified. Although the group to which the publication by Brooks and Bernard (1990) pertains encompasses only green lacewings in the family Chrysopidae, the list of names is fairly complete for the entire order, some of which have much simpler genitalia and hence fewer structures to identify. It is useful to have this list at hand because different authors sometimes use different names for the same structure.

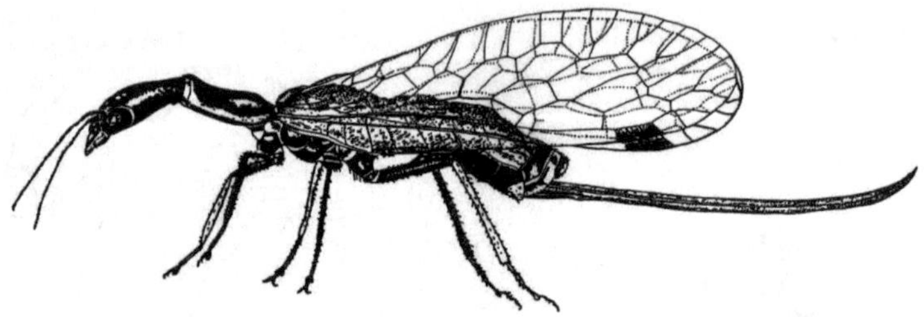

Fig. 1.13 Habitus of *Turcoraphidia acerba* H. Aspöck and U. Aspöck, 1966, a member of the Raphidioptera native to Western Asia. Based on Aspöck and Aspöck (2003).

Other morphological features commonly used to identify species of Neuroptera include the structure and coloration of the antennae and morphometric proportions of its segments. The number of segments is also reported for many species, but this may be a feature difficult to rely upon. Many specimens lose parts of the antennae during capture, preservation, and examination.

1.1.1 Morphology of Adult Megaloptera

Species belonging to Megaloptera can be distinguished immediately from species of Planipennia by the orientation of their mouths, which are directed anteriad rather than ventrad. This trait is shared with members of the third neuropteran suborder, Raphidioptera, which has not been found in South America. Species of Raphidioptera are easily distinguished from megalopterans because their pronota are elongate, and their heads articulate with great freedom of movement (Fig. 1.13). The pronota of megalopterans are flattened dorsally and shorter than the remainder of the thorax (Figs. 1.1, 1.2, and 1.3).

Typically, the wings of members of the family Sialidae are held in a tectiform position when at rest, while those of species of Corydalidae are held in a nearly horizontal position, parallel to or slightly divergent from the body axis over the abdomen (Fig. 1.1). This permits these two families to be distinguished almost at a glance. In addition, most members of the Corydalidae are considerably larger than most sialids. The position of the wings, relatively short pronotum, and long antennae consisting of many segments are the general characteristics of Megaloptera. The presence or absence of three **ocelli** on the dorsal surface of the head is diagnostic for the two families in South America. They are evident on the heads of species of Corydalidae (Fig. 1.1) but absent from the species of Sialidae. However, the considerable difference in size typically permits the two families to be distinguished at a glance. South American sialids rarely have wings as long as 20 mm, while the wings of corydalids are much longer.

While the shape of the individual parts of the male terminalia are usually diagnostic for megalopteran species (Fig. 1.9), color patterns on the wings sometimes facilitate recognition of certain species. In addition, the lengths of the **mandibles** of male specimens is helpful in recognizing corydalid species, but this feature is less reliable for making identifications because a considerable natural variability in this length has been observed in a few species. Many more specimens will have to be examined to find out when and why the variations in mandibular length occur before the degree of natural variability can be determined for all species.

The female genital structures are not as distinctive as those of the male, so other features should be relied upon to recognize species. Generally, the species of female specimens can be recognized without difficulty from the size, general coloration, and wing markings. It should be kept in mind that teneral specimens may not be well suited for identification, so insects newly emerged from the pupal stage should be maintained alive for at least several days before attempting to determine their species.

1.1.2 Morphology of Adult Planipennia

From their heads, adults belonging to the Planipennia can be distinguished easily from adult megalopterans, as well as from species of Rhaphidoptera, none of which have yet been found in South America. Planipennian mouthparts are obviously directed ventrad (Fig. 1.14). They are typically modified for grasping and chewing prey, except in a few planipennian species, including most of the typically aquatic ones, which have narrow, elongate mouthparts. The adults of relatively few species are characterized by a head shape greatly deviating from a nearly equilateral triangle. One of the exceptions is found in certain species of Crocidae (Fig. 1.15). The most obviously variable organ on the heads of most confamilial and congeneric species is the antenna, which often differs significantly in length and in the number, shape, and color of the individual segments from species to species. Unfortunately, using features of the antenna to reliably distinguish individual species in most families is made impossible due to the frequent loss of all or part of it from many specimens, including the types.

The position of the wings while the adults are at rest is most frequently tectiform. However, a few tend to hold the wings in positions reminiscent of species of Odonata. Many antlion species somewhat resemble damselflies in flight. There are too many variations from family to family to generalize about planipennian morphology, but the keys describe some of the best morphological characteristics on which to base identifications of specimens.

A few members of the order Neuroptera possess **stridulatory structures**, first discovered by Adams (1962) on the North American species, *Meleoma schwarzi* (Banks, 1903). Since then, additional chrysopid species belonging to other genera, including two represented in South America, were discovered by Brooks (1987) to have stridulatory structures on their legs and abdomens. They include *Meleoma* and *Chrysoperla*. It is not at all unlikely that careful examination of chrysopid

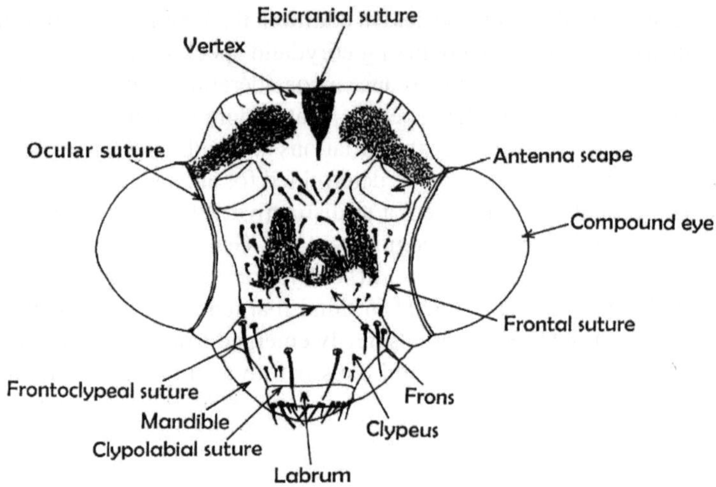

Fig. 1.14 Head of *Ameromyia nigriventris* in anterior view. Based on Stange (1994).

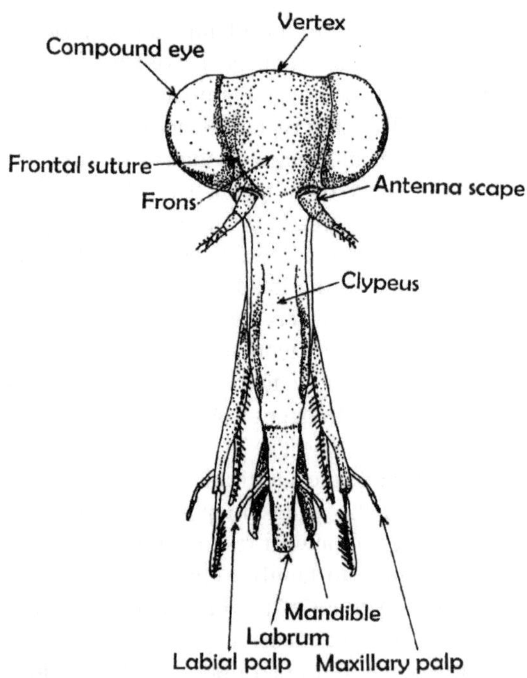

Fig. 1.15 Anterior view of the head of *Amerocroce boliviana,* a member of the Crocidae, showing the names of its parts. Based on Mansell (1983).

specimens from South America belonging to these and other genera would lead to the discovery of more species capable of stridulation.

Relatively little is known about the structure and function of the diverse organs observed on or in the bodies of neuropteran species. The functions of many sensory organs are not well known. A structure named the Eltringham organ has been observed on antlions, and results of studies on this organ were provided by Elofsson and Löfqvist (1974), who also discovered a thoracic gland. They believed that these structures were part of a system for producing and detecting pheromones.

Preliminary studies of other morphological characters of species native to other continents have shown that members of the Planipennia and Megaloptera possess many similar cuticular structures, made visible using scanning electron microscopy. **Tibial spurs,** for example, are present on almost all species examined, with a few exceptions in various families. However, the structures formerly classified as spurs on the hind legs of members of the Sisyridae are **socketed spines** rather than true spurs (Vshivkova and Makarkin 2010).

The sex of Neuroptera species that have been studied is determined according to the presence of sex chromosomes, with females having two X chromosomes and males, one X and one Y. However, an unusual feature of these chromosomes is that they display distance pairing (Tauber et al. 2003). Only recently has DNA sequencing been used to determine phylogenetic relationships between families (Winterton and Makarkin 2010).

The production of eggs by some owlfly species, in the family Ascalaphidae, is accompanied by the production of structures resembling eggs, called repagula (New 1971). These structures act to take the attention of predators away from the eggs. After hatching, the larvae remain together for defensive purposes, threatening would-be predators with an array of open mandibles (Fig. 1.16).

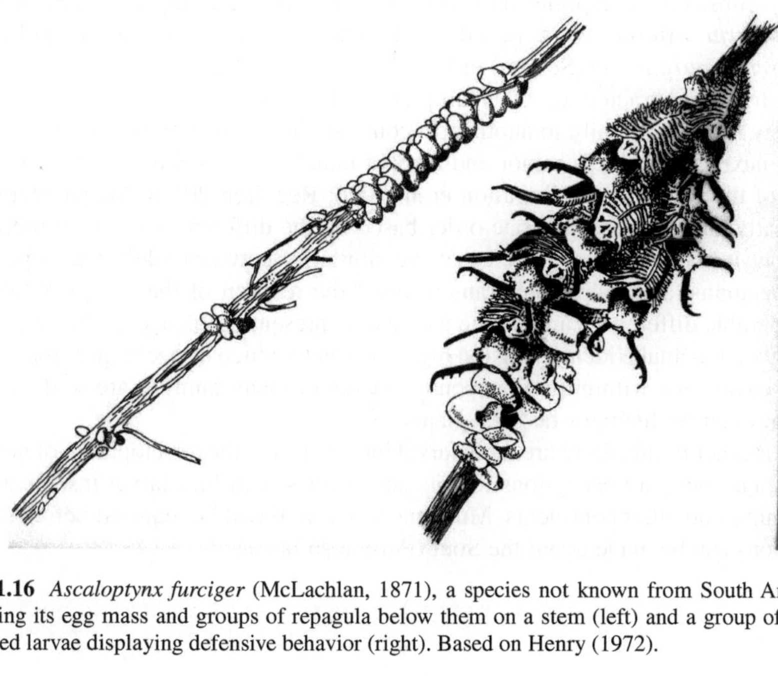

Fig. 1.16 *Ascaloptynx furciger* (McLachlan, 1871), a species not known from South America, showing its egg mass and groups of repagula below them on a stem (left) and a group of newly hatched larvae displaying defensive behavior (right). Based on Henry (1972).

Most other morphological features vary greatly from family to family, making it advisable to introduce them only in the keys. As already mentioned, the differences in the morphology of adults are generally much less pronounced than those of the larvae.

1.1.3 Changes During Larval Development

Important morphological features for identifying the species of larvae include the presence or absence of **external gills** and, if present, their locations on the body; morphometric features; location of setae or groups thereof; structure of the mouthparts; and color pattern. It is still not possible to identify the species of the great majority of South American neuropterans from larvae. For identifying the family of a larva, however, the general habitus is often sufficient. Nevertheless, it is possible that the features now used to identify the genera of known South American larvae may not be adequate to distinguish some of the many larvae not yet described.

Unfortuntely, knowledge of the larvae of South American Neuroptera species is fragmentary, although several thorough studies have been made of a few species. Therefore, at the present time, knowledge of the larval development of many families in South America is based mainly or exclusively on studies of only a tiny minority of the species. Generalizations are often made based only on the knowledge of species native to other continents.

It has been reported that all larvae are predatory, even if the adults of the same species subsist only on pollen and nectar from flowers. Recently, however, larvae of *Chrysoperla externa* were raised to adulthood exclusively on the pollen of *Pennisetum purpureum* (Schum) in Brazil (Oliveira et al. 2010).

A striking characteristic of Neuroptera is the extreme variation of larval morphology from one family to another. In contrast, the changes from one larval instar to the next are relatively minor and involve mainly increases in size and development of the appendages (Reguilón et al. 2006; Reguilón 2010). Megaloptera was originally elevated to a separate order based on the differences in larval morphology, but it was later recognized that the similarity between adult megalopterans, planipennians, and rhaphidopterans justified the reunion of these taxa. While the considerable differences among the larvae may present problems for the systematic biologist, they make identifying the higher taxon to which a specimen belongs relatively easy. Even within the Planipennia, larvae of many families are so distinctive that they can be distinguished at a glance.

As a general rule, there are three larval instars during the development of neuropterans, but there are exceptions to this, and species with four larval instars are not uncommon on other continents. Much more research will be required before generalizations can be made about the South American fauna.

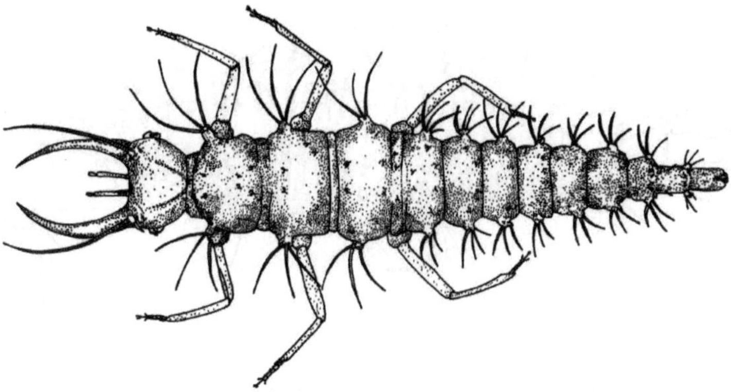

Fig. 1.17 The first larval instar of *Chrysoperla argentina* in dorsal view. Based on Reguilón et al. (2006).

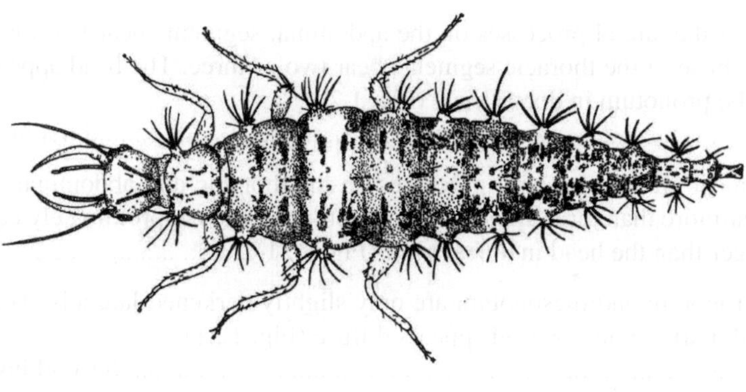

Fig. 1.18 The second larval instar of *Chrysoperla argentina* in dorsal view. Based on Reguilón et al. (2006).

Because of the great interest in members of the Chrysopidae for controlling pests in tropical agriculture, a few studies of larval morphology have been undertaken on species in that family. Three larval instars of *Chyrsoperla argentina* have been identified and illustrated by Reguilón et al. (2006). The first instar is shown in Fig. 1.17; the second, in Fig. 1.18; the third, in Fig. 1.19. Metamorphosis to the pupal stage of this species occurs after the third larval stage is completed. A key to the larval instars of chrysopids was prepared by Reguilón (2010), based on the species *Ungla binaria*.

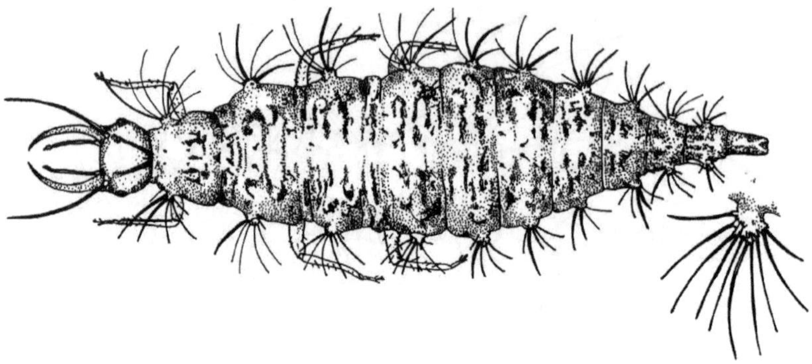

Fig. 1.19 The third larval instar of *Chrysoperla argentina* in dorsal view and an enlarged lateral tubercle from the first abdominal segment to the lower right. Based on Reguil6n et al. (2006).

Key to the Three Larval Instars of *Ungla binaria*, Family Chrysopidae

The key provided here is based, in part, on that of Reguil6n (2010).

1. Most of the lateral processes on the abdominal segments bear two long spines, while those of the thoracic segments bear two or three. The head appears larger than the pronotum in dorsal view (Fig. 1.20).
   ....First instar larva
   – Almost all of the lateral processes on the thoracic and abdominal segments bear more than three spines. The pronotum appears approximately equal to or larger than the head in dorsal view (Fig. 1.21)... 2
2. The pronotum and mesonotum are only slightly darkened laterally. The anteromedial marking on the head appears diffuse (Fig. 1.21).
   ....Second instar larva
   – The pronotum and mesonotum are sculptured and darkly clouded laterally. The anteromedial marking on the head is sharply defined (Fig. 1.22).
   ....Third instar larva

1.1.4 Morphology of the Pupa

The dearth of information about the pupae of South American neuropterans makes it impossible to provide accurate and comprehensive information about the pupal morphology of any family. It is possible to characterize the pupae of most families based on examination of species inhabiting other continents, but how similar their general morphology is to the confamilial species in South America remains unobserved in most cases.

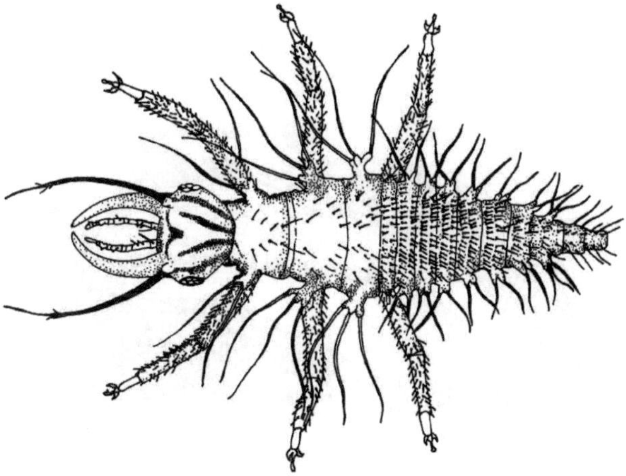

Fig. 1.20 First larval instar of *Ungla binaria* in dorsal view. Based on Reguilón (2010).

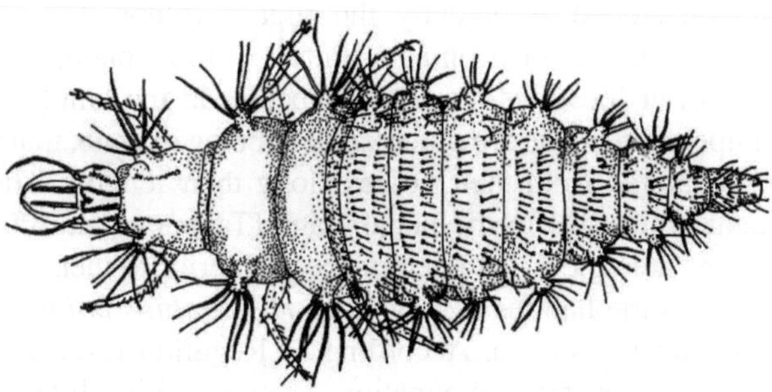

Fig. 1.21 Second larval instar of *Ungla binaria* in dorsal view. Based on Reguilón (2010).

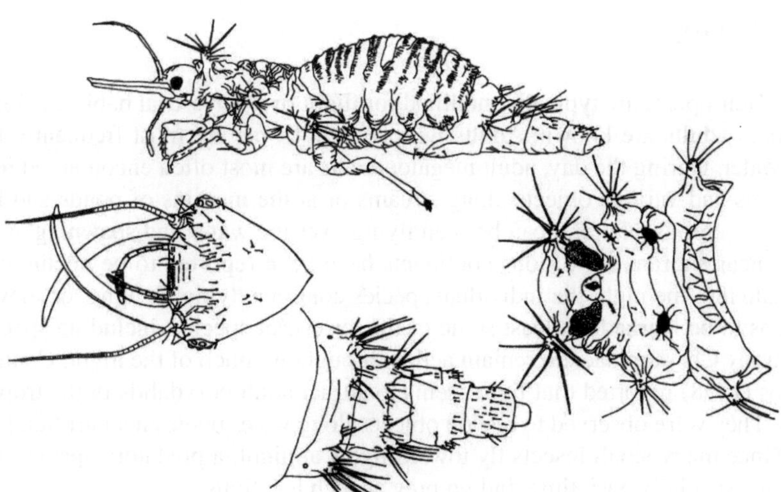

Fig. 1.22 *Ungla binaria* third and final larval instar: habitus in lateral view (above), head in dorsal view (middle left), thorax in dorsal view (right), and the apex of the abdomen in dorsal view (below). Based on Monserrat and de Freitas (2005).

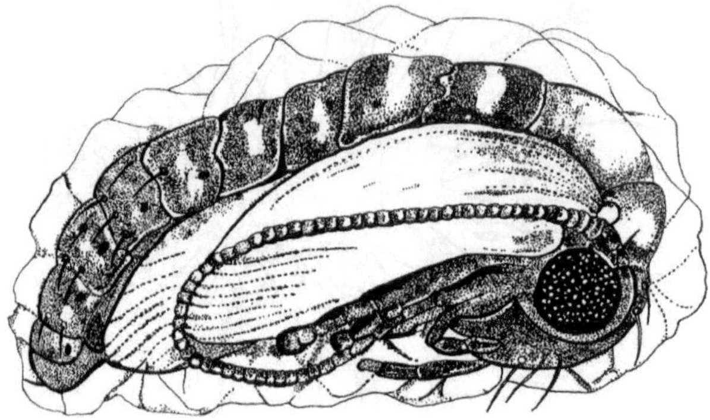

Fig. 1.23 A pupa of *Hemerobius bolivari* in its cocoon. Based on Reguilón (2002).

Cocoons of megalopterans are usually spun at the edge of streams or lakes in which the larvae developed. Apparently, the pupa develops above the local water level but remains in a humid environment. Pupae of most planipennians spin their silken cocoons in drier locations, except for species in the families Sisyridae and Osmylidae. Planipennian pupae are remarkable in being exarate, that is, having their wings and legs separated from their bodies along their lengths. This allows them limited movement while still in the pupal stage (Tauber et al. 2003). The strong mandibles of a pupa are able to cut open the cocoon prior to metamorphosis.

Figure 1.23 shows the habitus of a pupa of *Hemerobius bolivari,* in the family Hemerobiidae, within its cocoon. According to Reguilón (2002), there are three larval stages of this species prior to pupation. The cocoon is white.

1.2 Ecology

Adult Neuroptera are typically modified for flight over terrestrial habitats, although flightless adults are known. Adults with aquatic larvae are most frequently found near water. During the day, adult megalopterans are most often encountered resting on plants and various objects along streams or at the margins of ponds and lakes. Many are crepuscular and can be seen flying over the water and spawning at dusk. Some neuropterans on various continents have been reported to be nocturnal, but how late into the night the individual species continue flying, feeding, or spawning remains to be learned. At least some of the terrestrial species, including species in the family Chrysopidae, do remain active throughout much of the night. Contreras-Ramos (1998) reported that light seems to attract adult corydalids in the tropics at night. They were observed to rest on objects along water bodies near artificial lighting. Since many small insects fly toward lights at night, a predatory species would have an especially easy time finding prey at such locations.

Relatively little has been reported concerning the activities of adult neuropterans in South America. In contrast to various other insect groups but like the dragonflies, many adult neuropterans seem to remain alive and active for relatively long periods of time. They are adept at capturing and feeding on prey, and most seem to be capable of survival without depending upon nutrients stored during the larval stage.

Relatively little attentions has been given to the role of aquatic neuropteran larvae in controlling certain insect species known to be vectors of the microorganisms causing serious diseases in man and domestic animals. Studies by Gorayeb and Pinger (1978) indicate that the aquatic larvae of megalopterans feed on the larvae of *Simulium* species, which suck the blood of vertebrates and spread diseases. Much more attention has been given to the role of terrestrial planipennians, which can consume prodigious numbers of aphids and other pests, thereby protecting valuable plants, especially those grown in tropical and subtropical agriculture.

Terrestrial neuropterans, especially the lacewing flies, are probably attracted to plantations of food crops, which are usually frequented by aphid species, providing the predators with an ample food supply. The rapid decimation of aphids and other phytophagous insects by both adult and larval lacewings has attracted much attention to them as agents for the biological control of pests in tropical agriculture. First noted on other continents (New 1975), the consumption of small, rapidly reproducing homopterans and similar pests common in cultivated fields and orchards by lacewings of several families is proving to be an effective means of controlling damage to crops in the tropics without the use of agricultural chemicals. de Freitas and Penny (2001) provided a long list of green lacewing species, which have been found on specific crops in Brazil, to which they were probably attracted by large numbers of their prey.

The larvae of neuropterans live in a variety of terrestrial and aquatic habitats. The terrestrial species, including antlions, lacewings, and other predators, are considered in this section in a cursory manner because the feeding habits of few South American species have been determined. However, examinations of the stomach contents of adults from agricultural areas suggest that some adults consume nectar or pollen rather than insects or other invertebrates. Studies of adult chrysopids and other lacewings being considered for employment as pest control agents, however, suggest that most adults are predatory rather than herbivorous (de Freitas and Penny 2001). However, even the larvae of at least one species can survive exclusively on pollen (Oliveira et al. 2010). Giant lacewings in the family Polystoechotidae may be entirely phytophagous.

All known megalopteran larvae are aquatic, epibenthic predators or scavengers. A few larvae of Planipennia species are also aquatic, developing in freshwater sponges or living as predators in wetlands. The known aquatic planipennian species in South American belong to Osmyloidea. The association of sisyrid larvae with sponges has been investigated only superficially, and it is not certain whether the sponge benefits from their presence or whether the neuropterans simply feed on the host. It is still not fully certain how the sisyrid larvae develop inside the sponges and feed with their narrow, sucking mouthparts. It has been assumed by some entomologists that cells of the sponges are consumed, but others have speculated that these larvae are commensals and do not harm the host. Similarly, it is not certain whether the larvae found in colonies of ectoprocts actually develop in them. They might also have been in nearby sponges but were forced to leave and take shelter in the ectoprocts after the host was damaged.

Not enough is known about the natural enemies of Neuroptera species, especially those in South America, to describe the bionomics of each. However, a few important reports about species which consume neuropterans have been published, showing that some species are specialized for utilizing them as a food source. Clancy (1946) studied the insect parasites which develop in three chrysopid species in North America and found parasitic species belonging to one family of Coleoptera and eight families of Hymenoptera. His review of the literature on parasites developing in species in the families Chrysopidae and Hemerobiidae revealed that 61 such parasitic species had been found, and all developmental stages of the lacewings were attacked. The only dipteran among the species described as parasites was a species in the family Chironomidae.

Although relatively few studies on insect parasites of Neuroptera species have been conducted in South America, it is already clear that hymenopteran species are among the chief enemies of certain neuropterans on that continent, as well. As studies continue, the list of these parasites is certain to become considerably longer.

In Chile, the antlion species, *Elicura litigator,* is the host of a hymenopteran hyperparasite, the ichneumonid, *Itamuton stangei* Porter, 1989. The eggs of the ichneumonid are laid inside the pupae of antlions, where the larvae feed on the internal organs of their individual hosts until they are ready to eat their way out and transform into pupae themselves (Porter 1989). A member of the hymenopteran family Scelionidae, *Telenomus chrysoperlae* Loiácono, Lanati, and Neila, 2006, was discovered in Argentina parasitizing the eggs of the chrysopid, *Chrysoperla asoralis* (Banks, 1915), which seems to be its specific host (Loiácono et al. 2006). The interest in the destruction of chrysopids by hymenopteran parasitoids is economic as well as scientific because of the value of certain neuropterans as predators of agricultural pests.

Although information about the habitat preferences of the terrestrial larvae of neuropterans is still fragmentary for most families in South America, those few studies which have been conducted indicate that all but the antlions show little specialization, either for food items consumed or for substrate. Souza et al. (2008) tested the preferences of several species of chrysopid larvae for leaf-litter substrate and could find no preference at all for either sizes or sources of the materials provided. That would tend to make such insects poor candidates for indicator species. At the same time, it makes them seem particularly useful for agents of biological control for relatively small agricultural pests. As long as they go where the best food supplies can be found without regard for the physical environment, they could be expected to come to the rescue whenever a farmer finds his crops covered with rapidly reproducing populations of aphids, "mealy bugs," or other destructive arthropods. As soon as adults of certain chrysopids or other lacewings arrive on the affected crops, they begin feeding and also reproducing so that their larvae can assist in destroying the pests. Indeed, studies of chrysopids encountered in tropical plantations have shown that prodigious numbers of many known species take full advantage of pest abundance on cultivated crops (de Freitas and Penny 2001). Similarly, certain coniopterygids and hemerobiids have also been investigated as potential biological control agents (New 2001).

It has been noted that some plants, which protect themselves by producing their own insecticides, may reduce the populations of useful predators as well as the herbivorous insect pests that feed on the plants. The toxicity of one such plant, *Schinus molle* Linnaeus, a species in the family Anacardiaceae, on four insect species was determined by Iannacone and Alvariño (2010). The eggs and larvae of two of the predatory species belonging to the family Chrysopidae were tested, while an adult hemipteran and an adult hymenopteran were selected as the other two species. The eggs and larvae of *Ceraeochrysa cincta* turned out to be the most sensitive of the four species to the insecticide from the plant, while eggs and larvae of *Chrysoperla asoralis* were considerably more resistant. Its LC_{50}, the concentration at which 50 % of the insects are killed, was greater than that of the hemipteran but less than that of the hymenopteran tested.

1.2.1 Ecology of Megaloptera

Adult megalopterans in the family Sialidae, called alder flies, are often encountered on plants and other objects not far from water bodies. They are generally predators, feeding on small insects and other invertebrates. Generally, the activity of the adults is seasonal, but they may persist for moderately long periods of time, after a much longer total period of larval and pupal development. Eggs are typically laid on plants projecting out over the water, so that the larvae can drop into the water after hatching. Larvae are common epibenthic predators, creeping among detritus at the bottom of the water body or on submerged plants. They breathe through prominent, filamentous gills attached to the abdominal and sometimes to the thoracic segments. The larvae leave the water shortly before pupation, which occurs below the soil or leaf litter layer.

Alder flies usually develop in lentic or slowly flowing water. They do not seem to be among the species usually encountered in fast or moderately flowing streams. However, the larval development of these species in South America has rarely been studied, and exceptions to the general course of larval development by other members of the family cannot be ruled out.

Those megalopterans in the family Corydalidae, called dobsonflies, seem to share the general preference for proximity to water bodies and crepuscular or nocturnal hunting as adults with the sialids. Naturally, the size of the preferred prey can be expected to vary in proportion to the size of the adult predator. Of course, this must be confirmed on a species to species basis in South America, a process that has already begun but not yet advanced enough to make the keys to most groups of larvae better than tentative, at best.

The larvae of corydalid species, however, may differ more in habits from the sialids than the adults. Personal observations indicate that the large larvae of corydalid species are not uncommon under rocks in fast-flowing streams. Sialids, on the other hand, are found among various sediments in standing or slowly flowing water bodies, as well as in streams.

Little is known about the natural enemies of Megaloptera species. It is presumed that fishes would consume the larvae, while birds would consume adults. It has been reported that a fly in the family Chloropidae, *Pseudogaurax ideogenes* Wheeler, 2009, feeds on the eggs of one or more species of *Corydalus* in southern Brazil. Other species in this chloropid genus are known to consume the eggs of spiders (Melo and Wheeler 2009). Two species of *Corydalus, C. diasi* and *C. hecate,* were observed in the same watershed in which chloropid larvae had consumed eggs, but only *C. diasi* was collected at the site of attacks by the flies on the eggs. In addition, adult *Chloronia corripiens* were observed where the eggs were consumed, but the authors suggested that the egg masses belonged to *C. diasi* and/or *C. hecate* because those consumed were white as are those of the two *Corydalus* species, the eggs of which were surmised to be the food of the fly. The eggs of *Chloronia hieroglyphica* and perhaps those of congeneric species are brown.

The general bionomics of both sialids and corydalids on other continents are very similar except for the relative sizes of the prey and the speed of the water movement in the habitat. The larvae of all species seem to be general predators that stalk and consume species within a size range that they can overpower with their strong mandibles. The usefulness of corydalid larvae to control larvae of the black fly, *Simulium fulvinotum* Cerqueira and Mello, 1968, has been suggested by Gorayeb and Pinger (1978). Adult megalopterans are good flyers, even if somewhat slow, and can actively seek suitable hunting grounds.

As already stated, it is still to be determined whether South American species conform to the behavior patterns observed for related species on other continents. It is likely that deviations from these general patterns and special adaptations to tropical and subtropical conditions will be found during future studies of native South American megalopterans.

1.2.2 Ecology of Planipennia

Most species in this sub-order are terrestrial as larvae and as adults. However, species in the family Sisyridae are aquatic as larvae. Those in the subfamily Sisyrinae are known as spongillaflies and develop as larvae inside of freshwater sponges and possibly also in colonies of ectoprocts. Their relationship with the host sponge is still not fully understood. The other subfamily of Sisyridae is the Climaciinae, which encompasses species that develop as larvae in or beside both lentic and lotic water bodies (Monserrat 2005).

A second family of Planipennia also encompasses species which may be considered aquatic as larvae. The Osmylidae are associated with lotic water courses, along which their larvae develop. The ecology of the South American species requires more thorough study to provide additional details of their littoral habits.

Other members of Planipennia occupy the widest variety of terrestrial ecological niches. What seem to be common to the known Neotropical species are predatory feeding habits, both as larvae and adults. However, exceptions are known. Too little is known about the larvae of this group to make sweeping generalizations about anything. Other exceptional species are almost certain to be found, so the discussion must be confined to those few species with larval stages which have been studied by entomologists.

Adults of most planipennian species are commonly seen flying, often at night, although some species seem to display diurnal or crepuscular activity patterns. Their slow, fluttering mode of flight, as well as the large size of many of them, could easily attract the attention of observers. When at rest, they are frequently observed on plants or other objects, where they also capture their prey, mainly smaller insects. Their efficient capture and consumption of common herbivorous insects have made many of them prime candidates for use in biological control of agricultural pests, particularly in tropical and subtropical regions, as already discussed.

The most remarkable and probably best known of the planipennian larvae are the trap-building antlions, members of the family Myrmeleontidae, which catch the attention of even casual observers while they trap and consume ants. Typically, traps are constructed as conical depressions in dry, loose sand. The larva remains buried and flings sand out of the depression until it takes on an even, conical shape. After the trap is complete, the larva remains buried below the inverted tip of the cone. If excess sand or other material falls into the trap, the larva cleans it out by flinging it upward and outward.

Passing ants which slip over the edge of the cone find themselves sliding downward toward the bottom. Some ants may regain their footing and continue upward to leave the trap. When the antlion larva observes this, it flings sand grains at the ant to make the loose sand under its feet slide downward, carrying the ant toward the predatory larva at the bottom of the pit. When the prey comes close enough, it is seized by the large jaws of the antlion, injected with venom, and devoured. During this capture, the antlion remains buried and invisible to human observers above the trap. Once in the jaws of the antlion, the ant is usually dragged beneath the sand before it is eaten. Apparently, some or all antlions are capable of quickly tranquilizing the ants with the venomous substances injected when the prey is seized.

Antlion larvae can be preyed upon by certain hymenopteran parasitoid larvae, which develop inside the antlions without consuming any of the vital organs until ready to transform to a pupa. An example of such a hyperparasite is *Itamuton* sp., a member of the Ichneumonidae, which was reared in the larvae of *Elicura litigator* by Porter (1989).

Not all myrmeleontids build traps. Some live under objects on the ground and move about in search of their prey. These species bear a physical resemblance to the trap builders, and all of them can be recognized easily as myrmeleontids by their general appearance (Fig. 1.24).

A small but remarkable family of terrestrial planipennians is Ascalaphidae, the owlflies. Like antlions, they are fully terrestrial predators, both as larvae and as adults. As the morphology of the Ascalaphidae and Myrmeleontidae suggests, their habits are very similar, both as larvae and adults. Larval owlflies are aggressive predators, and like antlions, the species that have been studied inject venom into their prey to disable it. The ascalaphids which have been studied lay eggs around a

Fig. 1.24 Habitus of a larval antlion, *Sical peralinus,* in dorsal view. Based on Stange (1994).

plant stem, and after the larvae hatch, they act together to ward off potential predators (Fig. 1.16). Eventually, they fall to the ground, where they begin to search for prey. There is still much to learn about the habits of the South American species, and most of the information about their behavior was obtained through observation of North American and European species, which may turn out to be significantly different from their South American relatives.

The pleasing lacewings in the family Dilaridae are small, and little specific information is available on their habits. They are presumed to be predatory. Larvae of species outside of South America have been found under bark, where they presumably search for insect prey. The adults are small, usually with densely hairy wings, which are often spotted.

Dilaridae is now considered to encompass primitive relatives of the mantisflies in the family Mantispidae. These species resemble the preying mantis, a member of the Orthoptera, not closely related to Neuroptera. At least one species mimics a wasp. Mantisflies are fully terrestrial predators, recognized at a glance by their powerful fore-legs, modified for seizing and holding prey while it is devoured.

As both larvae and adults, brown lacewing species, in the family Hemerobiidae, prey on insects that are small enough be seized and devoured. The predators patrol the surfaces of plants seeking suitable prey to consume. Their eggs are laid on surfaces, presumably near potential food supplies for the larvae after they hatch. Small herbivorous insects frequently become abundant enough to provide large food supplies for the predators, which then reproduce fast enough to greatly reduce the numbers of prey. For this reason, some hemerobiid species have attracted attention as possible biological control agents for agricultural pests in the tropics (New 1975, 2001; Lara and de Freitas 2003; Lara and Perioto 2003; Lara et al. 2010). After food supplies are diminished, the adults can fly to seek out new food sources at other locations.

The green lacewings of the family Chrysopidae have attracted even more attention as potential biological control agents than the brown lacewings. The family presently encompasses many recognized species, although taxonomic revisions may reduce the number in South America, as they already have on other continents. Their habits as general predators of smaller insects are somewhat similar to those of the brown lacewings. Both larvae and adults share predatory habits. Adults of those species which have been studied seek out new food supplies by flying to new locations, often during twilight hours. Many of the green lacewing species have already been found in agricultural areas of Brazil (de Freitas and Penny 2001), and their importance as biological control agents for pests can hardly be overestimated (New 1975, 2001; Costa et al. 2010). They can reproduce at rates which almost match those of the herbivorous insects they consume, and they can devour large numbers of prey throughout their larval and adult stages. However, their large populations also provide attractive food sources for insects that prey on them. Special attention has been given to certain species of Hymenoptera, which develop as parasitoids within larval chrysopids. A species of *Telenomus*, a hymenopteran in the family Scelionidae, has been identified as one such parasitoid (Loiácono et al. 2006).

There is still a great deal to discover about the feeding habits of most planipennians. Many of the lacewings have been observed feeding on insects, both as larvae and as adults. However, the stomach contents of other, apparently related species, have revealed no insect remains at all. The stomach contents of a few specimens have been limited to pollen grains. This may indicate that the insect had not succeeded in capturing prey recently, but some of these insects are thought to feed exclusively on nectar or pollen.

At least one South American hemerobiid, *Gayomyia falcata*, has been shown to have a color pattern excellently camouflaging it on a common plant in the region where it occurs, *Ribes magellanicum* Poiret. This suggests that a symbiotic relationship exists between the plant and the insect (Faúndez 2005). Because the color pattern of the insect protects members of this species on only one kind of plant, it would provide no benefit for the insect to remain on any other, where its color pattern would catch the eye of predators. Thus, the insect finds protection from natural enemies on one specific plant, which also attracts herbivorous insects on which the hemerobiid feeds. The plant is protected from the insects that come to feed on it by the hemerobiid, so each of the symbionts gains great benefit from the relationship.

1.3 Preservation and Examination

Traditionally, adult neuropterans have been pinned and dried. Many species have very long, thin antennae, and this demands that great care be taken with the collections of dried material. To avoid losses of appendages useful for distinguishing different species, collectors began preserving specimens of Neuroptera in liquids relatively early. Sealed in jars, the detached appendages are at least retained for examination along with the rest of the insect, even after being broken off.

Specimens can be preserved very well in ethyl alcohol, preferably with an amount of glycerine not to exceed 10 % added. Preserving the color is especially difficult with strongly patterned species, and it is still not known how to keep lighter colors from changing dramatically almost as soon as they are introduced into a preservative fluid. When working with such insects, it is strongly recommended to take color photographs of living specimens and keep detailed records permitting both living and preserved insects to be matched with their photographs in collections. In this way, color artifacts related to specific preservatives can be recognized as such.

Because taxonomists of Neuroptera have selected certain internal structures of the insects as characters for recognizing species, it is advisable to clear parts of the insects for examination rather than trying to remove the structures by dissection. The genitalia and accessory organs are located within the terminal segments of the abdomen. Clearing typically reveals only the sclerotized parts of the organs, but those are the ones taxonomists mention and illustrate in their descriptions of the species. Since clearing may destroy certain tissues that might in the future be needed to confirm identifications or for analyses of various kinds, it is recommended that a sufficient number of each species be collected to assure that some undamaged type specimens remain.

Clearing as currently practiced by neuropterologists is usually destructive to the soft tissues and sometimes causes the sclerotized plates to become somewhat deformed. Less radical methods of clearing employ lactic acid or Berlese fluid, which are effective on relatively soft-bodied invertebrates. However, strongly pigmented sclerites of large neuropterans may still obscure the view of the inner organs after less drastic clearing, and the pigmentation and dense setae of smaller species sometimes make the organs difficult to see.

The radical methods of completely clearing the apex of the abdomen to view the genital structures usually entail heating part of the specimen in a solution of potassium hydroxide (KOH) or sodium hydroxide (NaOH). Delicate specimens are best cleared at room temperature rather than at or near boiling to avoid their complete dissolution. The tissues being cleared at or near the boiling temperature of water have to be watched carefully to make sure that they do not completely disappear. NaOH is usually somewhat milder than KOH and may do less damage to delicate structures. However, the process employing NaOH may take longer.

After clearing, almost all soft materials and even lightly sclerotized structures have been dissolved by the solution, leaving parts made of chitin transparent. Illustrations of the genitalia of many species in this volume showing both outer and internal chitinous structures are based on specimens cleared in this way.

After clearing, permanent slides should be made of the preparations. This should be done with the goal of keeping the relative form and positions of the organs intact. The specimens can be carefully transferred to distilled water for washing out the caustic substance and then to baths of ethanol-distilled water solutions containing increasing percentages of ethanol until a suitable solution for permanent storage is reached (Gurney et al. 1964). To make permanent slides containing specimens, they may be passed along such a series of baths until a 100 % ethanol bath extracts the last amounts of water from the specimen on the slide. It can then be transferred to a solution of 50 % absolute ethanol mixed with 50 % xylene, then to 100 % xylene, and finally to a solution of Canada balsam or a similar imbedding material. This method has grown unpopular because of the health risks of working with xylene, and a suitable hood and protective garments are required to perform this process safely.

Less drastic methods that are becoming increasing popular as chemicals with superior properties are discovered, such as clearing with lactic acid. The solution dissolves fats and protein but leaves even relatively delicate sclerotized structures visible. After spending time in storage, however, the specimens may become obscured.

Another method of clearing employs Berlese fluid, which contains chloral hydrate and gummi arabicum, allowing the specimen to be embedded on a slide. The formula was first provided by Swan (1936), although it was apparently not that for the fluid personally used by Berlese. While the preparation hardens, the specimen is cleared by the fluid, and the internal sclerotized structures become clearly visible under the microscope. This method is easy, produces relatively stable, semi-permanent slides, and does not require exposure of the entomologist to caustic substance and dangerous xylene fumes during preparation of the mounts, although chloral hydrate is toxic and and must also be used in a laboratory with suitable safety features. Over the long term, however, mounts using this and similar clearing fluids deteriorate due to the continual action of the chemicals on the tissues. Therefore, it is recommended to not to prepare mounts from holotypes or lectotypes of any species being described, following the advice of Upton (1993).

The selection of fluids for clearing and mounting should be made based on the purposes the mounts are to be used for and the robustness of the specimens to be cleared. Clearing with a hydroxide and mounting in balsam produces the most durable preparations, but the method is time-consuming and must be conducted with considerable caution to prevent the destruction of the specimens. Lactic acid is easier to use, but preparations tend to become cloudy with time. The methods using Berlese and similar mixtures of chloral hydrate and gummi arabicum are easy to employ and usually produce much more natural-looking preparations of chitinous structures, but they may deteriorate over the long-term. If the specimen being prepared is not a type specimen, the best solution is usually to make a preparation using Berlese fluid and photograph the structures as soon as they become sufficiently clear.

An optimal method for the study of neuropteran life cycles is rearing them in the laboratory. For Megaloptera and members of the planipennian families Sisyridae and Osmylidae, this entails the production of artificial lentic and lotic environments in which the larval stages can develop. A zone along the edge of each provides a habitat in which the final larval stage can seek a suitable site for metamorphosis to the pupal stage. This method not only provides specimens of each life stage for morphological studies, it also gives entomologists opportunities to observe the behavior of previously identified species throughout their life cycles.

For terrestrial planipennians, providing suitable artificial environments may be easy or, in the case of such families as Myrmeleontidae and Berothidae, they may not always be necessary. In the case of Myrmeleontidae, observations of antlions in the field may provide much useful information, although creating suitable habitats in a laboratory may facilitate far more detailed studies (Missirian et al. 2006). Members of the Berothidae could not be raised under natural conditions in a laboratory at all, unless a termite nest could be established there first. Even if this were possible, observing the development and behavior of the berothid larvae inside the tunnels of the termite colony would be a problem requiring much electronic endoscopic equipment, which would also function in the field.

1.4 Zoogeography

Species of Neuroptera belong to an ancient order in geological terms. The earliest fossils recognizable as Neuropterans date from the Permian, and fossils belonging to extant families date from the Mesozoic, when a great variety of species were found throughout the world. By the Cretaceous, many of the families known today had already appeared. Especially well-preserved specimens have been found in Baltic amber (Tauber et al. 2003; Makarkin et al. 2012). It appears from the distribution of many extant species that they are relicts of a declining insect order. Most modern neuropteran families seem to have had a cosmopolitan distribution during earlier geological periods but are somewhat more limited geographically today. However, many families and even genera still display cosmopolitan distribution patterns. Some of the larger and more oddly modified species belong to small families found at isolated sites around the world, confirming through their fossil record that they are the relicts of a much more extensive prehistoric fauna.

Although several higher taxa are widespread, individual genera and species show clear associations with well defined zoogeographical regions as well as preferred kinds of habitat. For example, a considerable number of genera are confined to the Andes or the relatively cool southern parts of South America. More than a few of the higher taxa known from the cool regions of southern South America range to Australia and New Zealand, giving them a classical circum-Antarctic distribution pattern. Other genera are typically Neotropical and range widely from North America throughout the warmer parts of Central and South America.

Earlier, many of the genera were considered cosmopolitan, absent only from the coldest parts of the world. A considerable number of these, however, have since been redefined according to consistent morphological differences, by which the Old World taxa can be distinguished from the New World ones. In a few cases, however, genera with representatives on several continents in widely separated parts of the world have continued to be recognized.

Neuroptera taxa encountered frequently as fossils are often similar to and sometimes close relatives of extant species. Although beyond the scope of this volume, it can be said with certainty that studies on the affinities of the fossils to living species will provide a considerable amount of insight into the zoogeography of the Neuroptera in general.

It is not hard to predict that comparisons of fossil forms with living species will demonstrate that neuropterans developed into their respective orders in relatively early geological periods. This accounts for the individual families having representatives in most parts of the world. There is no known family of Neuroptera endemic to South America. Many genera, however, as presently defined, are limited to the Neotropics, and several are endemic to the temperate regions toward the southern part of the continent and the upper elevations of the Andes Mountains farther north. Morrone (2001) has been mapping discrete biogeographical regions and provinces in South America, and his works can be referred to for more specific locations of distinct climatological environments.

Morrone (2000) divided the Subantarctic Subregion into discrete provinces based on their fauna. Several species of Neuroptera were used in the analysis. Generally speaking, the ranges of neuropteran species are confined to one of the large subregions of the continent, and the precise limitations of the ranges seem to depend on the general climatology. There is a rich fauna of Neuropteran species in the colder regions of Argentina and Chile, and many of the species native to this part of the continent extend their ranges northward at suitable elevations in the Andes. On the other hand, most species in the warm and moist tropical and subtropical lowlands to the east of the Andes range widely within the neotropics but are absent from the colder regions to the south and west. The distinctiveness of the cold temperate regions of South America was appreciated by early taxonomists, who confined a large number of early publications on neuropteran species to this zoogeographical area. A few such older publications and a few newer ones include Navás (1910; 1911a, 1918, 1921, 1922, 1924a, 1924b, 1926a, 1930, 1933a, b, 1934, 1936), Porter (1923), Gazulla and Ruiz (1928), Handschin (1955), Flint (1973), Gonzales Olazo (1981), Figueroa et al. (2003), and Monserrat and de Freitas (2005). Some of these studies also encompassed species inhabiting the colder regions of Argentina, especially Patagonia. A compilation of the species known from Chile has been provided more recently by Peña (1987), and the entire aquatic macrobenthos in the cold regions of southern Chile were surveyed by Figueroa et al. (2003).

A large percentage of South American neuropteran species inhabit the warm and moist tropical and subtropical lowlands east of the Andes. Central America seems to provide an effective boundary between the species native to North and Central America and those endemic to South America. Surprisingly many species are known from range only to the north or south of Panama. However, this may due either to a poorly understood zoogeophical barrier to these species at the southern end of Panama or to insufficient surveys of the fauna. A general review of the global diversity of Megaloptera and aquatic and semi-aquatic Planipennia families was provided by Cover and Resh (2008).

Recently, attention has been given to identifying plant species preferred as substrata for neuropterans in order to explain their geographical distribution in South America (Rebolledo et al. 2009). It is important to establish the relationship of predatory species to specific plant species providing the preferred substrate, not just because much scientific knowledge about the habits of the insects can be gained, such as the use of camouflage to escape detection by both potential predators and prey, but also because the species that prefer to live in association with cultivated crops may prove to be of great economic value for the control of agricultural pests. The presence of plant species on which certain neuropteran species are excellently camouflaged is, of course, very important as a zoogeographical factor of great significance for the distribution of the insect species. Although neither species depends upon the other for food, the absence of the plant species would make the survival of the insect very difficult. Conversely, crops being introduced to new continents could well be protected by suitable planipennian symbionts from the continents of their origin, which have long been naturally associated with the plant species due to a color pattern providing camouflage when the insect is resting on that particular species of plant.

1.5 Taxonomic Problems

The most obvious problem encountered when reviewing taxonomic literature on South American Neuroptera is the dearth of information about the larval and pupal stages. With notable exceptions, species descriptions refer only to adult specimens, sometimes only of one sex. It is therefore seldom possible to identify a larva to species, and even genera of South American species may not be identifiable, not only because distinguishing characteristics of most species are unknown but also because it has never been determined whether those features used to distinguish genera in North and Central America are possessed by all congeneric species in South America.

Much recent taxonomic work has been good to excellent, and most authors have taken the trouble to write comprehensive monographs rather than many short publications, each describing one or a few species. Illustrations have almost always been provided in publications after about 1940, permitting specimens to be identified with a reasonable degree of confidence.

Taxonomic and systematic work on South American Neuroptera is still at an early stage. Prior to the 1940s, many species were described inadequately and without illustrations. In the case of Megaloptera, most of these species have been re-examined and re-described. Many terrestrial species of Planipennia, however, still cannot be satisfactorily identified because type specimens were damaged, or only one sex has been described. Morphological features presently used to distinguish genera and species were not described in earlier literature, making all identifications of the species that have not been re-examined tentative. The state of confusion in the taxonomy of the family Chrysopidae is particularly frustrating in view of the fact that some of the species apparently have great importance in tropical agriculture as predators of pests. This confusion is particularly difficult to overcome because, until the 1980s, most species of green lacewing were placed in the genus *Chrysopa,* which is presently believed to be completely absent from South America. Most of the old descriptions of the species do not include illustrations, and it is impossible even to determine in which genera the species belong, the type specimens having been badly damaged, destroyed, or lost. Many good reviews on the Chrysopidae have been written based on re-examinations of the type specimens of the *"Chrysopa"* species, but a considerable number of them still remain to be placed in a currently recognized South American genus, and a few may forever remain as *species inquerendae.* No complete list of all Planipennia species that have not yet been correctly assigned to a genus or placed in synonymy is available yet, and the names of unassigned South American species listed in older publications without adequate descriptions are to be considered questionable until the types are located and revised descriptions have been published.

Many species have been described before enough specimens were collected to establish natural variability within the populations sampled. Many authors of systematic works have had problems finding undamaged type specimens of some species, leaving enough remains to identify a species again but not enough to determine the genus in which it should be placed. A large part of South America has still not been visited enough by collectors of neuropterans to provide more than cursory accounts of the fauna, and regional knowledge of some groups is based on expeditions conducted many decades ago.

As already mentioned, until the latter half of the twentieth century, large numbers of species were described in a cursory manner, and the descriptions were scattered in short publications in journals which are presently difficult to obtain. As a result of a lack of comprehensive publications, some common species were described many times and given a large number of different names. This is reflected in the long lists of synonyms found after individual valid species in the keys. In reviewing the chrysopid fauna, Adams (1978a) noted that almost 700 species formerly thought to inhabit the New World actually belong to only about 350 species. That means that approximately half of the specific names, based only on examinations of specimens from the Nearctic and Neotropical Regions, were invalid because the species had already been described and named in earlier publications. It is likely that some of the nominal species described on the basis of one or a few specimens belonging to only one sex will turn out to be conspecific after more is learned about their sexual dimorphism or dichromatism. Further work will undoubtedly reduce many of currently accepted names to synonyms, just as it will add new names for specimens of previously undiscovered species. Fortunately, most taxonomists have used the good judgment to refrain from naming larval specimens as new species without first having specimens of conspecific adults.

One of the first tasks that must be accomplished is to match larvae correctly with conspecific adults. Pupae of almost all species remain to be described. Naturally, rearing the species will provide the necessary information, and the usefulness of lacewings for the biological control of agricultural pests almost guarantees that rearing them will become an important commercial enterprise.

Because the aquatic larvae of neuropteran species have not yet found a place in agriculture or public health and sanitation, rearing only for scientific purposes might not be undertaken in the near future. However, larvae of megalopterans may well prove to exert some control on disease-carrying aquatic insects or other undesirable species after their habits are better known.

1.6 Suggestions for Continued Research

Deficiencies in descriptions, illustrations, and interpretations of the available information are not common problems encountered in the recent literature about species of Neuroptera. Of course, such dificiences are apparent when comparing publications from the nineteenth and early twentieth century with more recent literature. They were a major cause of the frequently encountered problem of twentieth century authors, who made false identifications of their specimens due to poor original descriptions and later provided detailed and accurate ones, but under the names that turned out to be junior synonyms of the older ones.

Incomplete lists of publications prior to about 1982 were compiled for the Neotropical Region by Flint (1977) and Penny (1981, 1982). It is still impossible to account for all of the specific names in the literature published prior to about 1950. Some nominal species may never be accounted for again. In most cases, descriptions are based on one or a few specimens, the sex of which was not reported. Attempts to find many of the type specimens have been unsuccessful, or, in some cases, only fragments of the holotype could be found. Lists of valid species often include names for which only sketchy descriptions and no type specimens are available. Some of these may have junior synonyms that will never be recognized as such. The Spanish Civil War and World War II took their tolls on museum collections in which the type specimens had been placed.

A continuing problem is encountered when the type specimens are sought for re-examination. Usually, newly described species were kept in the collection of the taxonomists who described them. When they were later placed in museum collections, the names of the museum in which they were deposited were not published anywhere. Only after taxonomists and systematic entomologists catalogued the species in the collections of individual museums were the locations of many type specimens reported. These publications are extremely helpful in saving time and effort to locate a type specimen of an individual species. However, copies of many of these publications are still difficult to find, in spite of the existence of Internet literature data banks. The following publications can be recommended when seeking the institutes in which type specimens of South American neuropterans were originally placed and were still available at the time the lists were compiled: Walker (1853) for the British Museum (Natural History), van der Weele (1908, 1910) for the collection of Baron Edmond de Selys Longchamps, Enderlein (1910) for the German Entomological Institute in Berlin-Dahlem, Navás (1911b) for the Museo de Munich, Navás (1912a, 1913a) for the Zoologische Staatssammlung in Munich, Navás (1912b) for mantispids in the St. Petersburg Museum of Zoology of the Imperial Academy of Sciences of St. Petersburg, Navás (1914a) for the Museum of Naples, Navás (1914band 1914c) for the chrysopid specimens in the Museum of London, Navás (1926b, 1927b) for the Museum of Paris, Navás (1926c) for the German Entomological Institute in Berlin-Dahlem, Navás (1928a, 1929a, 1934) for the "Museo de Hamburgo," Navás (1928b) for the "Museo de Estocolmo" (Stockholm Museum,) Navás (1929b) for the Civic Museum of Natural History of Genoa, Navás for the Zoological Museum of the Royal Academy of Turin (1932), Navás (1933c) for the German Entomological Institute in Berlin-Dahlem, Nakahara (1965) for the United States National Museum of Natural History, Weidner (1972) for the Zoological Institute and Zoological Museum of the University of Hamburg, Gonzales Olazo (1996) for entomological collections in Argentina, Flint (2002) for the United States National Museum of Natural History, von Ellenrieder (2009) for the Museo de Ciencias Naturales de Salta, Argentina. The names of the museums are reported as stated in the publications, and where the collections might be today are not always clear.

In spite of the problems in recognizing species described in a cursory manner almost or more than a century ago, recent revisions have resulted in much synonymy being revealed and other mistakes of the past being corrected. Recent work has still failed to provide corresponding descriptions of the larvae and pupae of species already described, with a few notable exceptions. This is the most pressing problem for neuropterologists, but finding and describing the still unknown South American species will hardly be less challenging.

A special problem involves South American chrysopid species described under the name *Chrysopa*. At present, no species in this genus is known to inhabit South America. Revisions of the chrysopid genera have moved a large number of South American species, formerly considered members of *Chrysopa*, into new genera. However, many "*Chrysopa*" species were not described adequately enough to reveal in which of the more recently described genera they belong. Unfortunately, not all of the type specimens have been found, and some of those that have been are missing important morphological structures necessary to correctly place them in any of more recently described genera. Especially because of the potential economic importance of chrysopids for the biological control of agricultural pests, a completion of the revision of Chrysopidae in order to place many of the dubious *"Chrysopa"* species in their correct genera should be given special attention. An extensive list of these species was provided by Penny (1977). Since this list was published, the classification of many species has been revised, while the systematic relationships of others remain unresolved, preventing their inclusion in the keys.

Obviously, attention should also be given to describing the larval and pupal stages and both sexes of the adults of all South American species. Eventually, faunal surveys and ecological studies will require identifications of both sexes and all stages to species, which is not yet possible in the case of most genera. As information on the morphology and morphological variability of both sexes increases, so will the number of names added to the lists of junior synonyms. This reduction in the number of species known from South America will certainly be compensated for by the discovery of new species in extreme and unusual habitats not yet visited by neuropterologists.

Many of the most recent taxonomic works on families and genera of both Megaloptera and Planipennia are exemplary, especially with regard to the many illustrations. Hopefully, this standard will be maintained.

Literature

Adams PA (1962) A stridulatory structure in Chrysopidae (Neuroptera). Pan-Pacific Entomologist 38(3):178–180

Adams PA (1969) A new genus and species of Osmylidae [Neuroptera] from Chile and Argentina, with a discussion of planipennian genitalic homologies. Postilla 141:1–11

Adams PA, (1978) Zoogeography of New World Chrysopidae, a progress report. Folia Entomologica Mexicana 39–40:210–211

Aguilar F, Lamas C (1980) Apuntes sobre el control biologico y el control integrado de las plagas agrícolas en el Perú. Revista peruana de Entomología 23(1):83–110

Albuquerque GS, Tauber CA, Tauber MJ (2001) *Chrysoperla externa* and *Ceraeochrysa* spp.: potential for biological control in the New World tropics and subtropics. In: McEwen PK, New TR, Wittington AE (eds) Lacewings in the crop environment. Cambridge University Press, Cambridge, pp 408–423

Anad AM, de Freitas S, Barbosa LR (2003) Potencial de alimentação de *Chrysoperla externa* (Hagen, 1861) (Neuroptera, Chrysopidae) em diferentes densidades de *Uroleucon ambrosiae* (Thomas, 1878) (Hemiptera, Aphididae). Revista Brasileira de Entomologia 47(1):15–18

Aspöck U (1986) The present state of knowledge of the family Berothidae (Neuropteroidea: Planipennia). In: Gepp J, Aspöck H, Hölzel H (eds) Recent research in neuropterology. Proceedings of the 2nd international symposium on neuropterology, Graz, pp 87–101

Aspöck U (2002) Phylogeny of the Neuropterida (Insecta: Holometabola). Zoologica Scripta 31(1):51–55

Aspöck U, Aspöck H (1969) Zur Frage des Vorkommens der Ordnung Raphidiodea auf der südlichen Hemisphaere. Beiträge zur Entomologie 19:45–47

Aspöck U, Aspöck H (2003) Raphidioptera (Snakeflies). In: Resh VH, Conde RT (eds) Encyclopedia of insects. Academic, Amsterdam, pp 973–975

Aspöck H, Hoelzel H, Aspöck U (2001) Kommentierte Katalog der Neuropterida (Insecta: Raphidioptera, Megaloptera, Neuroptera) der Westpalaearktis. Denisia 2:1–606

Banks N (1915 [1914]) New neuropteroid insects, native and exotic. Proceedings of the Academy of Natural Sciences of Philadelphia 66:608–632 + pl. 27

Berland L, Grasse P-P (1951) Superordre des Neuropteroides. In: Grasse P-P (ed) Traite de Zoologie, vol 10. Masson, Paris, pp 3–69

Biraben M (1960) *Mantispa* (Neuroptera) parasita en el cocon de *Metepeira* (Araneae). Neotrópica (La Plata) 6(29):61–64

Bowles DC, Contreras-Ramos A, Sites RW (2007) Gynomorphic mandible morphology in the dobsonfly, *Corydalus cornutus*. Journal of Insect Science 7(23):1–5

Brooks SJ (1987) Stridulatory structures in three green lacewings (Neuroptera: Chrysopidae). International Journal of Insect Morphology and Embryology 16(3–4):237–244

Brooks SJ, Bernard PC (1990) The green lacewings of the world: a generic review (Neuroptera: Chrysopidae). Bulletin of the British Museum, Natural History (Entomology) 59(2):117–286

Carpenter FM (1930) A review of our present knowledge of the geological history of the insects. Psyche 37:15–34

Clancy DW (1946) The insect parasites of the Chrysopidae (Neuroptera). University of California Publications in Entomology 7(13):403–496

Contreras-Ramos A (1998) Systematics of the dobsonfly genus *Corydalis* (Megaloptera: Corydalidae). Thomas Say Publications in entomology (Monographs), Entomological Society of America, Lanham, Maryland, 360 pp

Contreras-Ramos A, von der Dunk K (2010) A new species of *Corydalus* Latreille from Venezuela (Megaloptera, Corydalidae). ZooKeys 67:11–19

Costa RIF, Souza B, de Freitas S (2010) Dinâmica espaço-temporal de taxocenoses de crispídeos (Neuroptera: Chrysopidae) em ecosistemas naturais. Neotropical Entomology 39(4):470–475

Cover MR, Resh VH (2008) Global diversity of dobsonflies, fishflies, and alderflies (Megaloptera; Insecta) and spongillaflies, nevrorthids, and osmylids (Neuroptera; Insecta) in freshwater. Hydrobiologia 595:409–417

de Freitas S, Penny ND (2001) The green lacewings (Neuroptera: Chrysopidae) of Brazilian agroecosystems. Proceedings of the California Academy of Sciences 52(19):245–395

de Oliveira SA, Sousa B, Auad AM, Carvalho CA (2010) Can larval lacewings *Chrysoperla externa* (Hagen) (Neuroptera, Chrysopidae) be reared on pollen? Revista Brasileira de Entomologia 54(4):697–700

Elofsson R, Löfqvist J (1974) The Eltringham organ and a new thoracic gland: ultrastructure and presumed pheromone functions (Insecta, Myrmeleontidae). Zoologica Scripta 3:31–40

Enderlein G (1910) Eine neue *Sialis* aus Columbien. Stettiner Entomologische Zeitung 71:380–381

Faúndez E (2005) Asociación críptica de *Gayomyia falcata* (Blanchard, 1851) (Neuroptera: Hemerobiidae) con la Zarzaparrilla *Ribes magellanicum* Poiret, 1812 (Saxifragaceae) en la región de Magallanes. Anales Instituto Patagonia (Chile) 33:63–64

Figueroa R, Valdovinos C, Araya E, Parra O (2003) Macroinvertebrados bentónicos como indicadores de calidad de agua de ríos del sur de Chile. Revista Chilena de Historia Natural 76:275–285

Flint OS Jr (1973) The Megaloptera of Chile (Neuroptera). Revista Chilena de Entomologia 7:31–45

Flint OS Jr (2002) List of the name-bearing Neuropterida types in the collection of the National Museum of Natural History, Smithsonian Institution, USA. Acta Zoologica Academiae Scientarum Hungaricae 48(Suppl 2):89–98

González Olazo EV (1981) El genero *Megalomus* Rambur (Neurop.-Hemerobiidae) en Argentina y Chile. Acta Zoológica Lilloana 36:97–113

González Olazo EV (1996) Catálogo de los tipos de Neuroptera (Insecta: Holometabola) depositados en colecciones argentinas. Acta Zoológica Lilloana 43(2):373–391

Gorayeb I, Pinger RR (1978) Detecção de predadores naturais das larvas de *Simulium fulvinotum* Cerq. e Mello, 1968 (Diptera: Nematocera). Acta Amazonica 8(4):629–637

Gurney AB, Kramer J, Steyskal GC (1964) Some techniques for the preparation, study, and storage in microvials of insect genitalia. Annals of the Entomological Society of America 57:240–242

Handschin E (1955) Los insectos de la Islas Juan Fernandez. 15. Neuroptera. Revista Chilena de Entomologia 4:3–20

Henry CS (1972) Eggs and rapagula of *Ululodes* and *Ascaloptynx* (Neuroptera: Ascalaphidae): a comparative study. Psyche 79(1–2):1–22

Iannacone J, Alvariño L (2010) Toxicidad de *Schinus molle* L. (Anacardiaceae) a cuatro controladores biológicos de plagas agrícolas en el Perú. Acta Zoológica Mexicana (ns) 26(3):603–615

Lara RIR, de Freitas S (2003) Caracterização morphológica de espécies de *Hemerobius* Linnaeus, 1758 (Neuroptera, Hemerobiidae) associadas a cultivos de café (*Coffea aribica* L.), milho (*Zea mays* L.) e erva-mate (*Ilex paraguariensis* St. Hill.). Revista Brasileira de Entomologia 47(3):427–434

Lara RIR, Perioto NW (2003) Bioecologia de hemerobiídeos (Neuroptera, Hemerobiidae). Arquivo do Instituto de Biologia, São Paulo 70(4):517–523

Lara RIR, Perioto NW, de Freitas S (2010) Diversidade de hemerobiídios (Neuroptera) e suas associações com presas em cafeeiros. Pesquisa Agropecuária Brasileira, Brasilia 45(2):115–123

Loiácono MS, Lanati SJ, Neila C (2006) Una especie nueva del género *Telenomus* (Hymenoptera: Scelionidae) parasitoide de posturas de Chrysopidae (Neuroptera) en Mendoza, Argentina. Revista de la Sociedad Entomologica Argentina 65(1–2):23–26

Machado RJP, Rafael JA (2011) A new species of *Fillus* Navás, 1919 (Neuroptera: Ascalaphidae) from the Brazilian Amazon Basin. Zootaxa 2907:22–28

Makarkin VN, Wedmann S, Weiterschan T (2012) First record of a fossil larva of Hemerobiidae (Neuroptera) from Baltic amber. Zootaxa 3417:53–63

Mansell MW (1983) New Crocinae (Neuroptera: Nemopteridae) from South America, with descriptions of larvae. Journal of the Entomological Society of Southern Africa 44:245–257

Melo AS, Wheeler TA (2009) A new species of *Pseudogaurax* Malloch (Diptera: Chloropidae) reared from dobsonfly egg masses (Megaloptera: Corydalidae) in Brazil. Zootaxa 1972:53–58

Missirian GLB, Ochoa-Fernandes MA, Fischer E (2006) Development of *Myrmeleon brasiliensis* (Navás) (Neuroptera, Myrmeleontidae), in laboratory, with different natural diets. Revista Brasileira de Zoologia 23(4):1044–1050

Monserrat VJ (2005) Nuevos datos sobre angunas pequeñas familias de neurópteros (Insecta: Neuroptera: Nevrothidae, Osmylidae, Sisyridae, Dilaridae). Heteropterus Revista de Entomologia 5:1–26

Monserrat VJ, de Freitas S (2005) Contribución al conocimiento de los crisópidos de Coquimbo, Patagonis y Tierra del Fuego (Argentina, Chile) (Insecta, Neuroptera, Chrysopidae). Graellsia 61:163–179

Morrone JJ (2000) Biogeographic delimitation of the subantartic subregion and its provinces. Revista de la Museo Argentino de las Ciencias Naturales, ns 2(1):1–15

Morrone JJ (2001) A formal definition of the Paramo-Punan Biogeographic Subregion and its provinces, based mainly on animal taxa. Revista de la Museo Argentino de las Ciencias Naturales, ns 3(1):1–12

Navás L (1910) Neurópteros chilenos (1ª Serie). Revista Chilena de Historia Natural 14(4–6):235–241

Navás L (1911a) Neurópteros chilenos (Series 2a.). Revista Chilena de Historia Natural 15:121–128

Navás L (1911b) Notas sobre Neurópteros del Museo de Munich. I. Mitteilungen der Münchener Entomologischen Gesellschaft 2:22–28

Navás L (1912a) Bemerkungen über die Neuropteren der Zoologischen Staatssammlung in München IV. Mitteilungen der Münchener Entomologischen Gesellschaft 3:89–95

Navás L (1912b [1911]) Quelques Mantispides (Insectes Névroptères) du Musée Zoologique de l'Académie Impèriale des Sciences de St.-Pétersbourg. Annuaire du Musée Zoologique de l'Academie Impèriale des Sciences de St.-Pétersbourg 16:535–538

Navás L (1914a) Algunos Neurópteros exóticos del Museo de Nápoles con 3 incisioni. Annuario del Museo Zoologico della R Università di Napoli (Nuova Serie) 4:1–4

Navás L (1914b) Les Chrysopides (Ins. Névr.) du Musée de Londres. Annales de la Société scientifique de Bruxelles 37(2):292–330

Navás L (1914c) Les Chrysopides (Ins. Névr.) du Musée de Londres. Conclusion. Annales de la Société scientifique de Bruxelles 38(1):73–114

Navás L (1918) Insectos Chilenos. Boletin Sociedad Aragonesa de Ciencias Naturales 17:212–230

Navás L (1921) Névroptéres des Islas Juan Fernandez et de L'Isle de Pâques. In: Skottsberg C (ed) The natural history of Juan Fernandez and Easter island, 3rd edn. Almqvist & Wiksells Boktryckeri, Uppsala, pp 125–128

Navás L (1922 [1921]) Algunos insectos de Chile. Revista Chilena de Historia Natural 25:443–445

Navás L (1923 [1922]) Algunos insectos del Museo de París. Revista de la Academia de Ciencias de Zaragoza 7:15–51

Navás L (1924a [1922]) Insectos de la Argentina y Chile. Estudios (Buenos Aires) 22:358–368

Navás L (1924a) Algunos insectos de Chile. II. Revista Chilena de Historia Natural 28:2–16

Navás L (1926a) Insectos de la Argentina y Chile. Secunda Serie. Estudios (Buenos Aires) 31:103–111

Navás L (1926b) Trichoptera, Megaloptera und Neuroptera aus dem Deutsch. Entomolog. Institut (Berlin-Dahlem). II Serie. Entomologische Mitteilungen 15(1):57–63+Tafel I

Navás L 1926c (1925) Algunos insectos del Museo de Paris. 3ª Serie. Brotéria (Série Zoológica) 23:95–115

Navás L (1927a) Vienticinco formas neuvas de insectos. Boletin de la Sociedad Ibérica de Ciencias Naturales 26:48–75

Navás L (1927b) Insectos des Museo de Paris (4ª Serie). Brotéria, Série Zoológica 24:5–33

Navás L (1928a) Insectos del Museo de Estocolmo. Revista de la Real Academia de Ciencias Exactas Fisicas y Naturales de Madrid 24:28–39

Navás L (1928b) Insectos del Museo de Hamburgo. Primera Serie. Boletin de la Sociedad Entomológica de España 11:59–67, 90–100, 121–138
Navás L (1929a) Insectos Neurópteros del Museo de Hamburgo. Memorias de la Real Sociedad España de Historia Natrual 15:315–322
Navás L (1929b) Insectos exóticos neurópteros y afines del Museo Civico di Storia Naturale di Génova. Annali del Museo Civico di Storia Naturale di Génova 53:354–389
Navás L (1930 [1929]) Algunos insectos de Chile. 4.ª Serie (1). Revista Chilena de Historia Natural 34:350–366
Navás L (1932) Alcuni insetti del Museo de Zoologia della R. Università de Torino. Bollettino dei Musei de Zoologia e di Anatomia Comparata della R Universita de Torino Serie 3 42(26):1–38
Navás L (1933a [1932]) Insectos de la Argentina y Chile. 3.ª serie. Revista de la Sociedad Entomológica Argentina 5:79–86
Navás L (1933b) Algunos insectos de Chile. 4. Serie. Revista Chilena de Historia Natural 37:230–234
Navás L (1933c) Neurotteri e tricotteri del "Deutsches Entomologisches Institut" di Berlino – Dahlem. Bollettino della Società Entomologica Italiana 65(5):105–113
Navas L (1934) Algunos insectos de Chile 5.a Serie. Revista Chilena de Historia Natural 38:9–12
Navás L (1936) Insectos neurópteros de Chile poco conocidos. Revista Chilena de Historia Natural 40:179–181
New TR (1971) Ovariolar dimorphism and repagular formation in some American Ascalaphidae (Neuroptera). Journal of Entomology (A) 46(1):73–77
New TR (1975) The biology of Chrysopidae and Hemerobiidae (Neuroptera), with reference to their usage as biological control agents: a review. Transactions of the Entomological Society of London 127(2):115–140
New TR (1989) Planipennia. In: Fischer M (ed) Handbuch der Zoologie Band 4, Arthropoda: Insecta. Lieferung 30. De Gruyter, Berlin, p 132
New TR (2001) Introduction to systematics and distribution of Coniopterygidae, Hemerobiidae, and Chrysopidae used in pest Management. In: McEwen PK, New TR, Wittington AE (eds) Lacewings in the crop environment. Cambridge University Press, Cambridge, pp 6–28
Oswald JD (1998) Rediscovery of *Polystoechotes gazullai* (Neuroptera: Polystoechotidae). Proceedings of the Entomological Society of Washington 100(3):389–394
Oswald JD (2004) Review of the brown lacewing genus *Biramus* (Neuroptera: Hemerobiinae), with the description of a new species from Costa Rica and Panama. Tijdschrift voor Entomologie 147:41–47
Parfin SI, Gurney AB (1956) The spongilla-flies with special reference to those in the Western Hemisphere (Sisyridae, Neuroptera). Proceedings of the U. S. National Museum 105(3360):421–529 + 3 pl
Peña LE (1987) Introducción a los Insectos de Chile. Editorial Universitaria, Santiago, Chile. 256 pp
Penny ND (1978 [1977]) Lista de Megaloptera, Neuroptera e Raphidioptera do Mexico, America Central, ilhas Caraibas e America do Sul. Acta Amazonica, Supplement 7(4):1–61
Porter CE (1923) Los estudios sobre neuropteros chilenos (Ápuntes históricos y bibliográficos). Revista Chilena de Historia Natural 27:168–181
Porter CC (1989) New Chilean *Itamuton* (Hymenoptera: Ichneumonidae: Mesostenini) reared from *Elicura litigator* (Neuroptera: Myrmeleontidae). Florida Entomologist 72(4):660–664
Rebolledo R, Pinot P, Ojeda N, Aguilera A (2009) Distribución geográphica y plantas substratos de *Semidalis kolbei* Enderlein, 1906 (Neuroptera: Coniopterygidae) en la región de la Araucanía, Chile. IDESIA 27(1):41–46
Reguilón C (2002) Morfología de los estadios inmaduros de *Hemerobius bolivari* (Neuroptera: Hemerobiidae). Revista de la Sociedad Entomológica Argentina 61(1–2):63–68
Reguilón C (2010) Morfología de los estadios inmaduros y cicla biológico de *Ungla binaria* (Neuroptera: Chrysopidae). Acta Zoológica Lilloana 54(1–2):78–86

Reguilón C, González Olazo EV, Nuñez Campero SR (2006) Morfología de los estadios inmaduros de *Chrysoperla argentina* (Neuroptera: Chrysopidae). Acta Zoológica Lilloana 50(1–2):31–39

Riek EF (1975) On the phylogenetic position of *Brucheiser argentinus* Navás 1927 and description of a second species from Chile (Insecta: Neuroptera). Studies on the Neotropical Fauna 10(2):117–126

Sosa F, de Freitas S (2012) A new genus of Neotropical Chrysopini (Neuroptera: Chrysopidae). Zootaxa 3351:1–14

Souza B, Ciociola AI (1995) Aspectos comportamentais de *Nusalala uruguaya* (Navás) (Neuroptera; Hemerobiidae), em laboratório. Anais da Sociedade Entomológica do Brasil 24(1):173–175

Souza B, Ciociola AI (1997) Efeito de diferentes dietas sobre la fecundidade e longevidade de *Nusalala uruguaya* (Navás, 1923) (Neuroptera; Hemerobiidae). Pesquisa Agropecuária Brasileira 32(1):27–32

Souza B, Matioli JC, Ciociola AI (1990) Biologia comparada de *Nusalala uruguaya* (Navás, 1923) (Neuroptera; Hemerobiidae) alimentada com diferentes espécies de afídeos. I. Fase de larva. Anais de Escola Superior de Agricultura "Luíz de Queiroz" 47(2):283–300

Souza B, Costa RIF, Louzada JNC (2008) Influência do tamanho e de la forma de fragmentos florestais na composição da taxocenose de crisopídeos (Neuroptera: Chrysopidae). Arquivo do Instituto de Biologia, São Paulo 75(3):351–358

Stange LA (1994) Reclassification of the New World antlion genera formerly included in the tribe Brachynemurini (Neuroptera: Myrmeleontidae). Insecta Mundi 8:67–119

Swan DC (1936) Berlese's fluid: remarks upon its preparation and use as a mounting medium. Bulletin of Entomological Research 27(3):389–391

Sziráki G, Flint OS Jr (2007 [2005]) Larva of *Brucheiser penai* Riek, 1975 (Neuroptera Coniopterygidae). Annali del Museo Civico di Storia Naturale di Ferrara 8:45–48

Tauber CA (2006) A new species of *Berchmansus* (Neuroptera: Chrysopidae) and nomelclatural notes on *B. elegans* (Guérin Méneville). Proceedings of the Entomological Society of Washington 108(4):830–841

Tauber CA, Tauber MJ, Albuquerque GS (2003) Neuroptera (Lacewings, Antlions). In: Resh VH, Conde RT (eds) Encyclopedia of insects. Academic, Amsterdam, pp 785–798

Tjeder B (1966) Neuroptera – Planipennia 5. Family Chrysopidae. South African Animal Life 12:228–534

Upton MC (1993) Aqueous gum-chloral slide mounting media: an historical review. Bulletin of Entomological Research 83(2):267–274

van der Weele HW (1908) Ascalaphiden. In: Collections Zoologiques du Baron Edm. de Selys Longchamps, vol 8. HAYEZ, Brussels, 326 pp

van der Weele HW (1910) Megaloptera (Latrelle) Monographic Revision. In: Collections Zoologiques du Baron Edm. de Selys Longchamps 5(1):1–93 + 4 pl

von Ellenrieder N (2009) Type specimens of Insecta housed at the Museo de Ciencias Naturales de Salta, Argentina. Revista de la Sociedad Entomologica Argentina 68(3–4):253–262

Vshivkova TS, Makarkin VN (2010) Ultrastructural morphology of leg cuticle derivatives useful for phylogenetic study of Neuropterida (Insecta: Megaloptera, Neuroptera): preliminary report. In: Devetak D, Lipovšek S, Arnett AE (eds) Proceedings of the tenth international symposium on neuropterology. Univerza v Mariboru, Faculty of Natural Sciences and Mathematics, Maribor, pp 287–300

Walker F (1853) List of the specimens of neuropterous insects in the collection of the British Museum. Part II (Sialidae – Nemopterides). Trustees, London, pp 193–476

Weidner H (1972) Die entomologische Sammlungen des Zoologischen Instituts und Zoologischen Museums der Universität Hamburg VIII, Teil Insecta V. Mitteilungen aus dem Hamburgischen Zoologischen Museums und Institut 68:107–134

Winterton S, Makarkin VN (2010) Phylogeny of moth lacewings and giant lacewings (Neuroptera: Ithonidae, Polystoechotidae) using DNA sequence data, morphology, and fossils. Annals of the Entomological Society of America 103(4):511–522

Chapter 2
Keys to the Neuroptera Families in South America

Abstract Keys are provided to assign South American specimens of adult and larval Neuroptera to the correct suborder and family. Where there are major gaps in the knowledge of South American neuropteran fauna, such as lacks of specimens of the larvae and knowledge of their development and metamorphosis, these are pointed out in the keys. The user of these keys is directed to the chapters in which keys to genera and species are provided.

Keywords Key to suborders • Key to families • Adults • Pupae • Larvae

2.1 Key to the Suborder of Adults

Information for the key was provided by Stitz (1931), Berland and Grass (1951), Brues et al. (1954), Imms (1957), Oswald and Penny (1991), and Miller and Stange (2014).

1. The head is hypognathous; that is, the mouth is directed ventrad (Fig. 2.1).
   ....**Part III**..................................Suborder Planipennia..p. 133

   – The head is prognathous; that is, the mouth is directed anteriad (Fig. 2.2) ... 2

2. The pronotum is cylindrical and greatly elongated (Fig. 1.13).
   ....Suborder Raphidioptera
   This suborder is not known from South America and is not further considered.

   – The pronotum is shorter than the remainder of the thorax (Fig. 2.2)................
   ....**Part II**.. Suborder Megaloptera..p. 74

C.W. Heckman, *Neuroptera (Including Megaloptera)*, Encyclopedia of South American Aquatic Insects, DOI 10.1007/978-3-319-35125-4_2

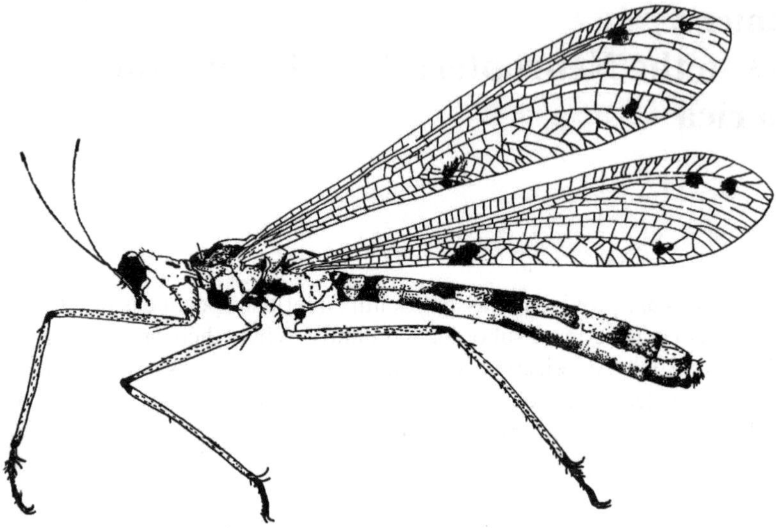

Fig. 2.1 Habitus of *Elachyleon punctipennis*. Based on Stange (2002).

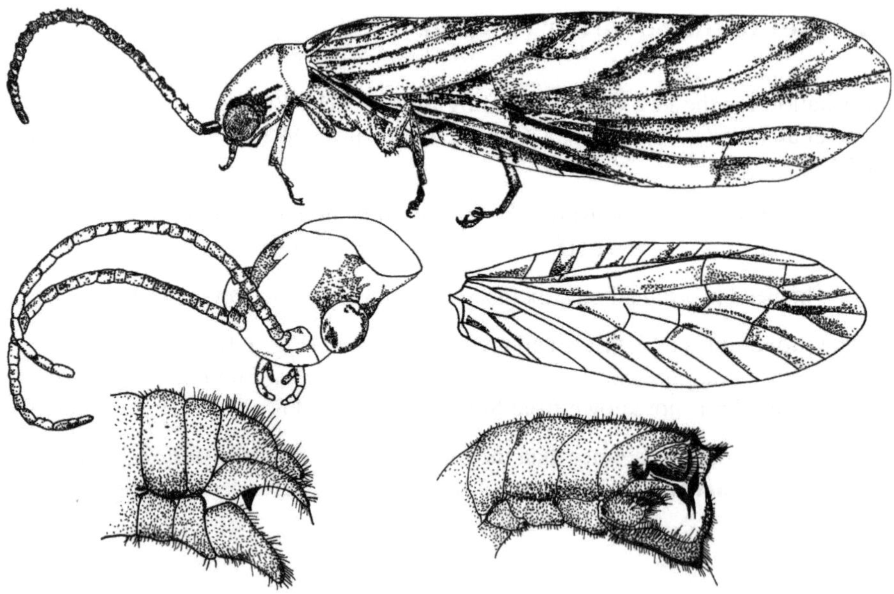

Fig. 2.2 *Protosialis flammata* male: habitus in lateral view (above), head in oblique dorsolateral view (middle left), fore-wing (middle right), and the apex of the abdomen in lateral (lower left) and oblique posterior view (lower right). Based on Penny (1981a).

Key to the Suborder of Larvae

Information for the key was taken from Berland and Grass (1951), Brues et al. (1954), Chandler (1956), and Riek (1975).

1. The larvae are fully aquatic and have grasping and chewing mouthparts (Fig.2.3)
   .... Suborder Megaloptera..p. 74
   – The larvae are fully terrestrial or semi-aquatic with various mouthparts or aquatic with a mouth that appears to be modified for piercing and sucking, although it may be used for pipetting minute prey (Fig. 2.4)........................ 2
2. The larvae are usually considerably different from the adults, and if the larvae are relatively slender, the head typically bears elongate, slender mandibles and maxillae, apparently modified for sucking (Fig. 2.4). Maxillary palps are absent. Usually, the tarsi each consist of one segment ending in two claws.
   ....Suborder Planipennia..p. 122
   – The larvae are terrestrial, with elongate and flattened heads and bodies. The head has strong, biting mouthparts. The head and thorax are sclerotized, while the abdomen is soft (Fig. 2.5).

   ....Suborder Raphidioptera
   This suborder is not known from South America and is not further considered.

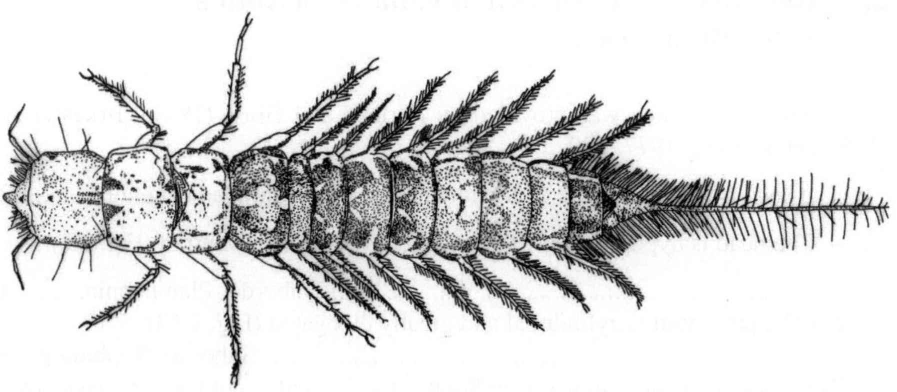

Fig. 2.3 *Protosialis chilensis* larva: habitus in dorsal view. Based on Flint (1973).

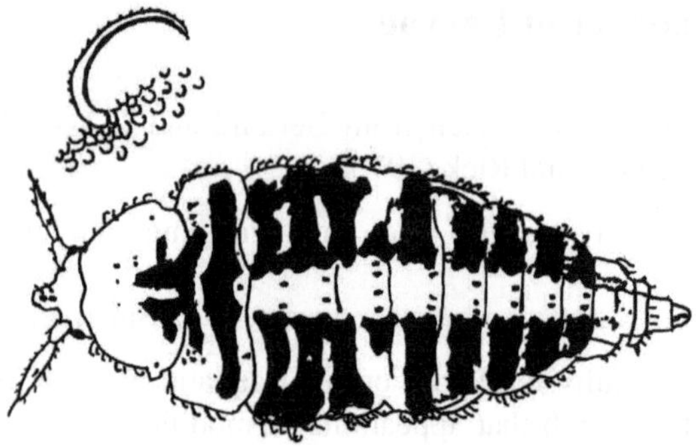

Fig. 2.4 *Semidalis kolbei* larva: habitus in dorsal view with enlarged papillae and a typical seta (upper left). Based on Monserrat (2005).

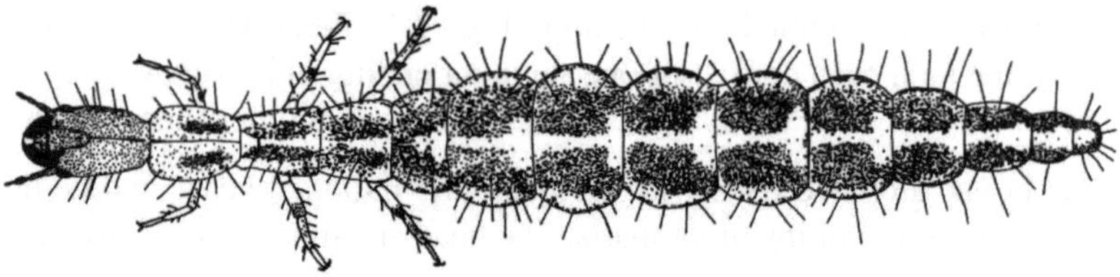

Fig. 2.5 Habitus of an *Indianoinocellia mayana* Aspöck et al., 1992, larva native to Mexico. Based on U. Aspöck and H. Aspöck (2003).

2.2 Keys to the Suborders and Families of Adults in South America

Information for the key was provided by Berland and Grass (1951), Brues et al. (1954), and Penny (1977, 2002a).

1. The head is prognathous; i.e., the mouth is directed anteriad (Fig. 2.6)........... 2

 – The head is hypognathous; i.e., the mouth is directed ventrad (Fig. 2.7).

 ....Suborder Planipennia...........4

2. (1) The pronotum is cylindrical and greatly elongated (Fig. 1.13).
 ....Suborder Raphidioptera
 This suborder is not known from South America and is not further considered.

 – The pronotum is shorter than the remainder of the thorax (Fig. 2.6)

 .... Suborder Megaloptera...3

3. (2) Ocelli are absent. The fourth tarsal segment is enlarged and with two lobes (Fig. 2.8). The wing is generally less than 20 mm long.
 ....**Chapter 3**..Sialidae..p. 75

 – Three ocelli are present. The fourth tarsal segment is not modified (Figs. 1.2, 1.3, and 2.6). The wing length is about 20 mm or more, usually much more.
 ....**Chapter 4** ....Corydalidae..p. 81

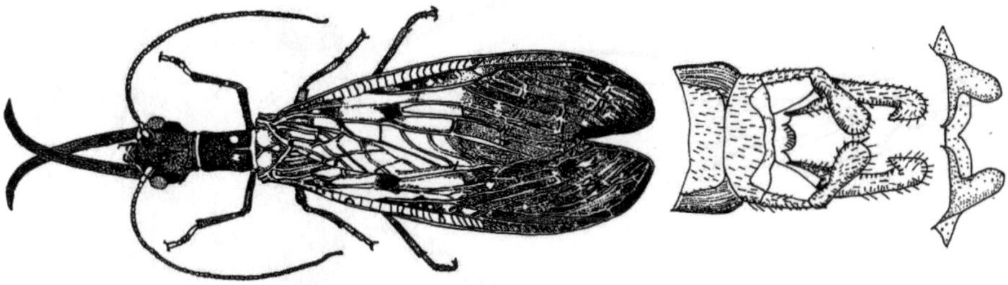

Fig. 2.6 *Corydalus batesi* male (left to right): habitus in dorsal view, apex of the abdomen, and the tenth gonostyli. Based on Penny (1982a).

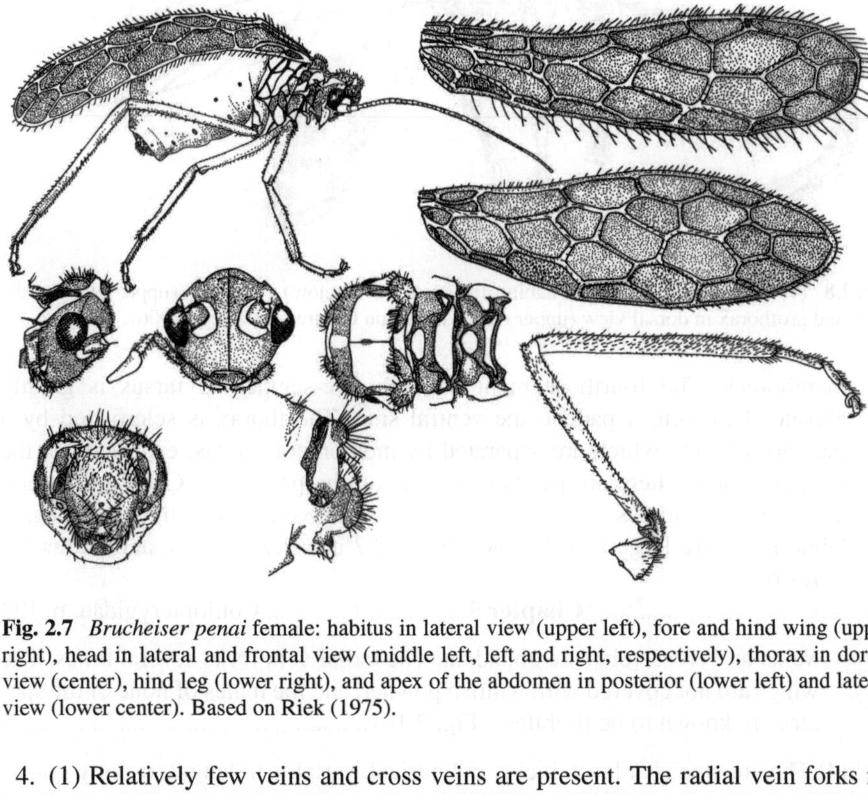

Fig. 2.7 *Brucheiser penai* female: habitus in lateral view (upper left), fore and hind wing (upper right), head in lateral and frontal view (middle left, left and right, respectively), thorax in dorsal view (center), hind leg (lower right), and apex of the abdomen in posterior (lower left) and lateral view (lower center). Based on Riek (1975).

4. (1) Relatively few veins and cross veins are present. The radial vein forks no more than once. The wings are covered with whitish powder. The species are very small, slender, and pale, measuring only 3–10 mm from wing tip to wing tip, and usually no more than 3 mm in length. A few species, which are apparently incapable of flight, belong to this family. In these brachypterous species, a short distance from the base of the wing, vein R fuses with the costal margin. Both the fore and hind wings are brachypterous, heavily sclerotized along the entire margin, without trichosors, and with microtrichia covering the

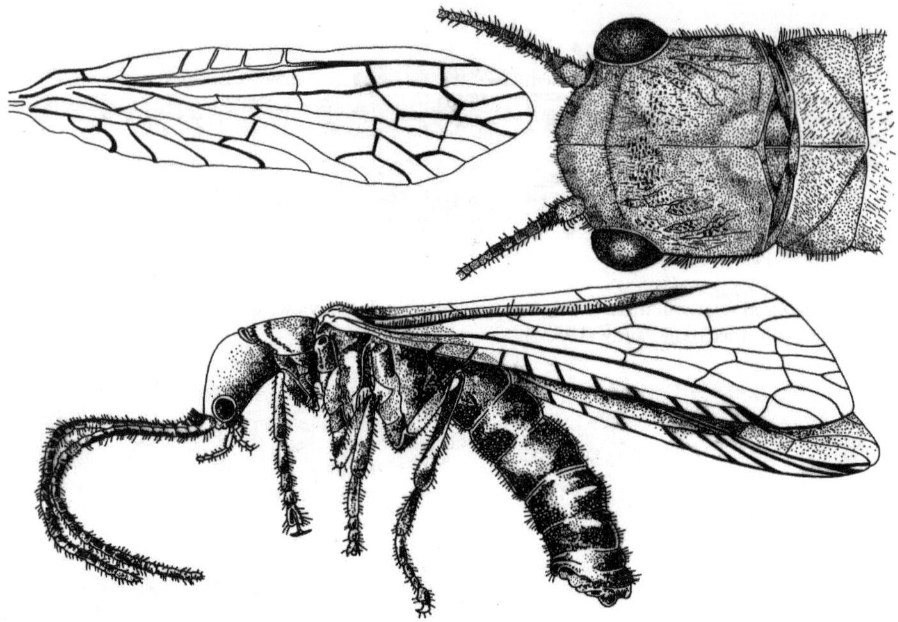

Fig. 2.8 *Protosialis ranchograndis:* habitus in lateral view (below), fore-wing (upper left), and the head and prothorax in dorsal view (upper right). Based on Contreras-Ramos (2006).

membranes. The fourth segment of each five-segmented tarsus is greatly expanded to form a pad on the ventral side. The thorax is sclerotized by a network of bars, which are separated by membranous areas, especially in the pleural region. There are patches of setae on the pronotum. Cerci are absent. The apical segments of the five-segmented maxillary and three-segmented labial palps are large and flattened (Figs. 2.7 and 2.9). Larvae and adults are terrestrial.
....................................**Chapter 5**..............................Coniopterygidae..p. 123

– Veins and cross veins are abundant. The radial vein forks several times. The wings are not covered with whitish powder, and the males of none of the species are known to be flightless (Fig. 2.10) 5

5. (4) The antenna is at least thickly cylindrical, variable in length, usually gradually enlarged toward the tip or filiform with a clavate apex. At least the discal portion of the wing is densely reticulate. Veins Sc and R_1 are fused apically. Cu ends in the apical part of the wing and often has a long, straight branch behind Cu_1 (Fig. 2.10).. 6

– The antenna is never enlarged toward the apex; it is moniliform or filiform and rarely pectinate. Cu usually ends at or proximal to the middle of the wing. There is no straight longitudinal branch behind Cu_1 (Fig. 2.11) 7

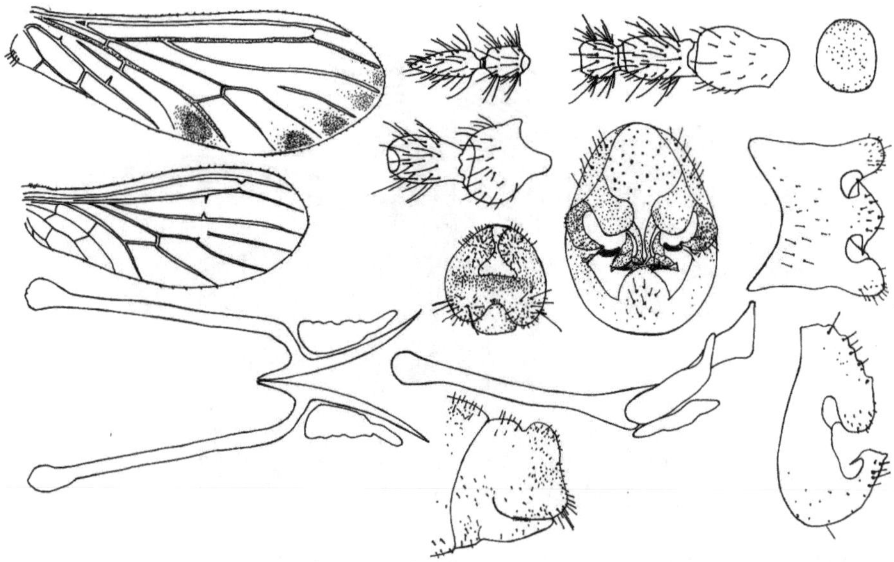

Fig. 2.9 *Semidalis isabelae:* fore and hind wing (upper and middle left), apical (upper left center) and basal antenna segments of a male in dorsal view (upper right center), scape and pedicel of the antenna in lateral view (below apical segments), male genitalia in posterior view (right center), male ninth sternite with the hypandrium and ectoproct in ventral (middle right) and lateral view (lower right), apex of the female abdomen in posterior (center) and lateral view (lower center), male parameres in dorsal (lower left) and lateral view (left center to right center, between views of the female abdomen), and an egg (upper right). Based on Monserrat (1981).

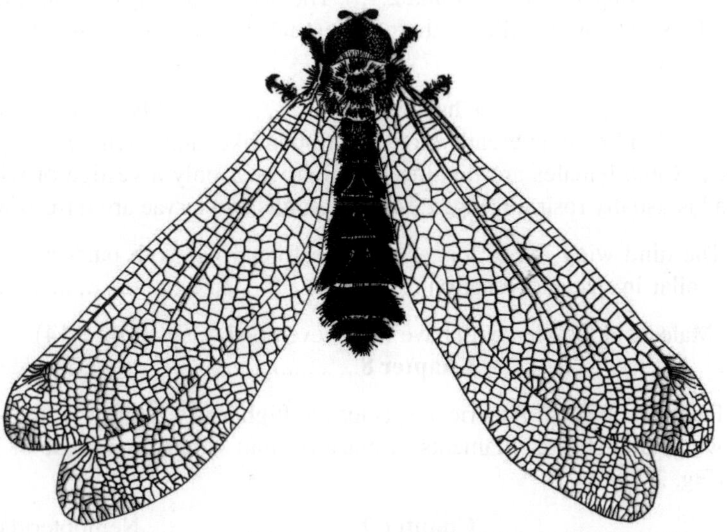

Fig. 2.10 Habitus of *Albardia furcata* in dorsal view. Based on Penny (1983).

Fig. 2.11 Habitus of *Nallachius limai*. Based on Henry et al. (1992).

6. (5) The antenna is long, slender, and strongly clavate at the apex. The eyes are usually divided into two parts by a groove. No elongate hypostigmatic cell is formed (Figs. 2.10 and 2.12). The larvae and adults are terrestrial.
   ....**Chapter 6**..................................Ascalaphidae..p. 183

   – The antenna is weakly clubbed or flattened at the apex. The hypostigmatic cell is elongate (Figs. 2.1 and 2.13). The body and wings are pubescent. The adults are weak fliers, their terrestrial larvae have earned the name "antlion."

   ....**Chapter 7**............................Myrmeleontidae..p. 225
7. (5) The hind wing is greatly elongate, ribbon like, and often with a spoon-like apex. Some females are flightless and may bear only a vestige of wings. The head is usually rostrate (Fig. 2.14). The adults and larvae are terrestrial 8

   – The hind wing is not longer than the fore-wing; both pairs are somewhat similar in size and venation (Fig. 2.11) 9

8. (7) Males and females both have fully developed wings (Fig. 2.14).
   ....**Chapter 8**..Crocidae..p. 285

   – Females of South American species are flightless with tiny vestiges of fore-wings and shorter filaments in place of hind wings than those of the male (Fig. 2.15).

   ....**Chapter 9**................................Nemopteridae..p. 293

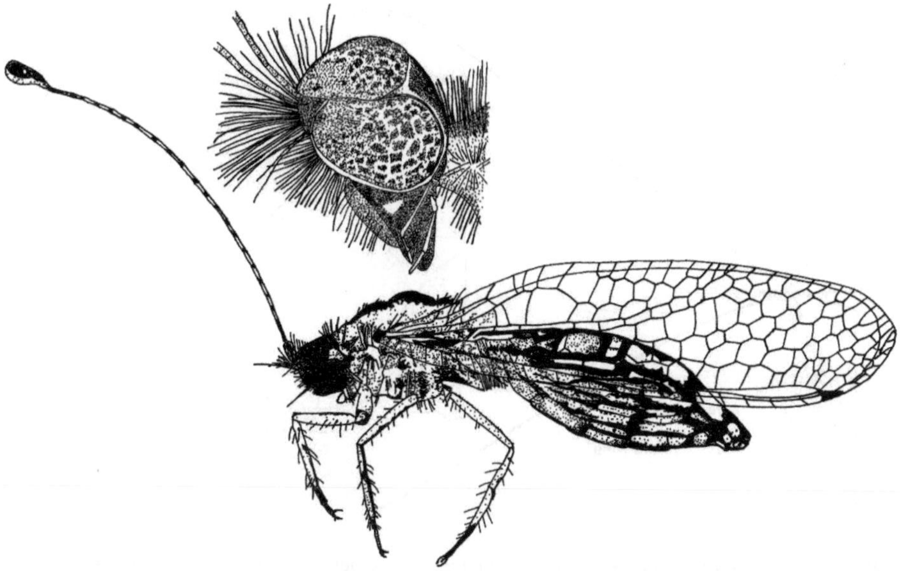

Fig. 2.12 Habitus of *Ululodes macleayanus* in lateral view with a lateral view of the head above it. Based on Penny (2002b).

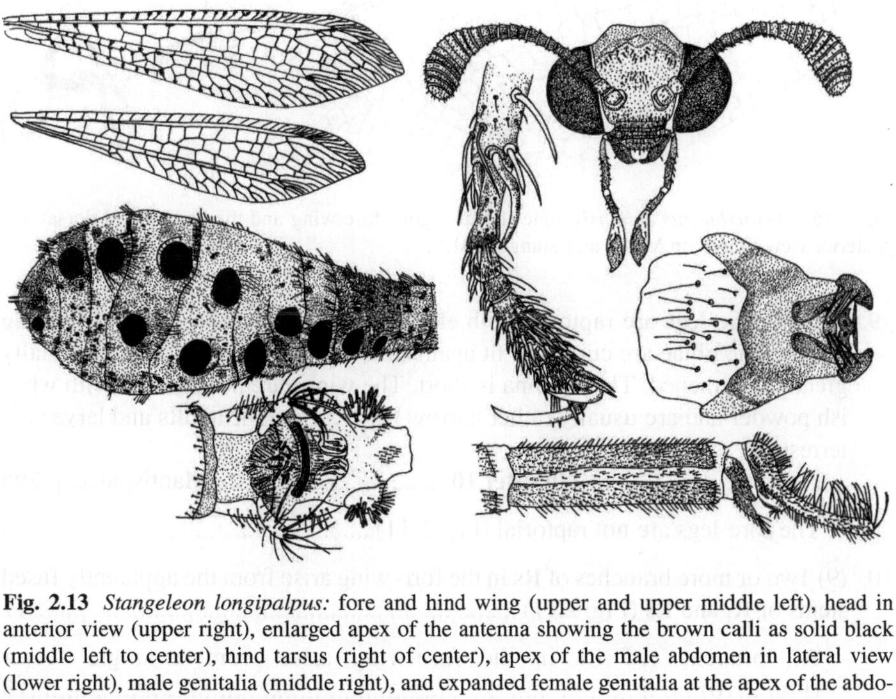

Fig. 2.13 *Stangeleon longipalpus:* fore and hind wing (upper and upper middle left), head in anterior view (upper right), enlarged apex of the antenna showing the brown calli as solid black (middle left to center), hind tarsus (right of center), apex of the male abdomen in lateral view (lower right), male genitalia (middle right), and expanded female genitalia at the apex of the abdomen (lower left center). Based on Miller (2008).

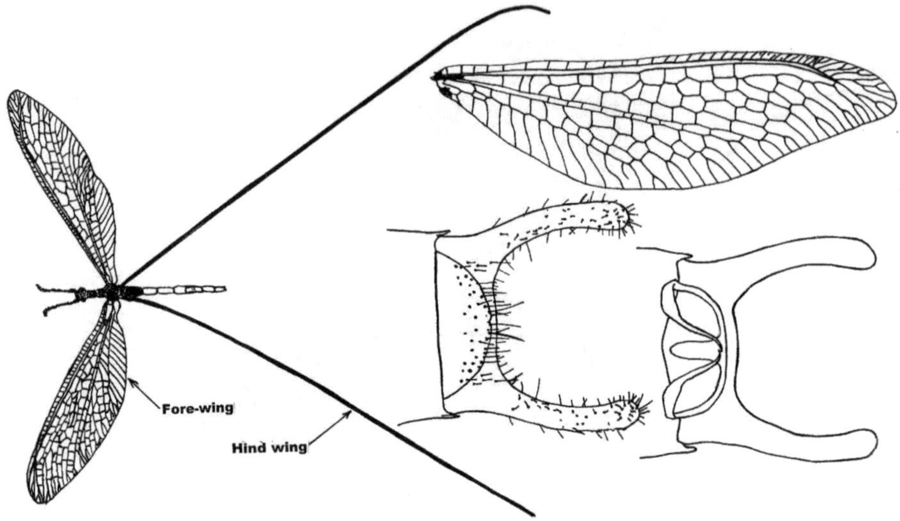

Fig. 2.14 *Pastranaia riojana* male: habitus in dorsal view (left), fore-wing (upper right), and apex of the abdomen in dorsal (center) and ventral view (lower right). Based on Orfila (1955).

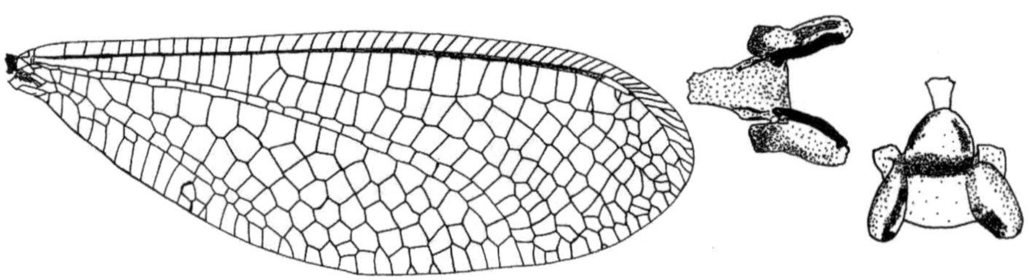

Fig. 2.15 *Stenorrhachus chilensis* male (left to right): fore-wing and the genitalia in dorsal and posterior view. Based on Miller and Stange (2012).

9. (7) The fore-legs are raptorial with elongate coxae and robust femora bearing spines. The tibiae are curved to fit against the femora. The prothorax is usually greatly lengthened. The antenna is short. The wings are not covered with whitish powder and are usually rather narrow (Fig. 2.16). The adults and larvae are terrestrial.
   ....**Chapter 10**..................................Mantispidae..p.295
   – The fore-legs are not raptorial (Fig. 2.11) 10
10. (9) Two or more branches of Rs in the fore-wing arise from the apparently fused stems of R_1 and Rs (Fig. 2.11) 11
   – All branches of Rs in the fore-wing arise from a single sector (Fig. 2.16)...13

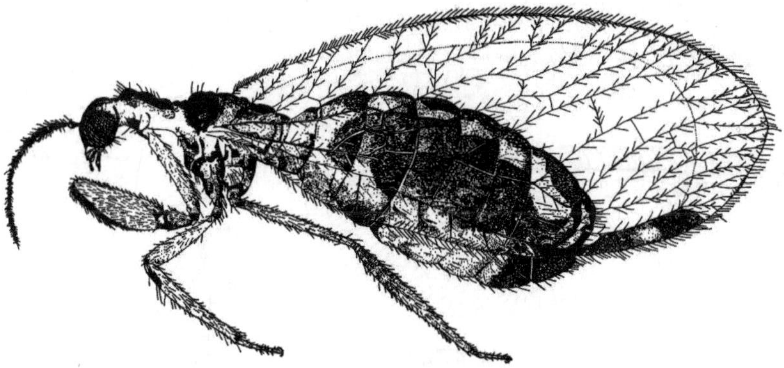

Fig. 2.16 Habitus of *Trichoscelia iridella*. Based on Penny (1982b).

11. (10) The antenna of the male is coarsely pectinate. The ovipositor is exserted. The vertex bears three prominent tubercles resembling ocellae. Cross veins are numerous (Fig. 2.11). The adults and larvae are terrestrial. ....**Chapter 11**....................................Dilaridae..p. 343

   – The antenna is moniliform in both sexes. The ovipositor is not exserted. Cross veins are few. Ocellae or similar structures are absent (Fig. 2.13).........12

12. (11) Three or more branches of the radius are present in the fore-wing. Veins R_4 and R_5 arise separately (Fig. 2.19). Adults and larvae are terrestrial. ....**Chapter 12**..Hemerobiidae..p.355

   – The fore-wing appears to have two radii, one R_{2+3} and the other R_{4+5} (Fig. 2.18). The adults and larvae are terrestrial.

   ....**Chapter 13**.............Osmylidae, Subfamily Sympherobiinae p.413

13. (10) Ocelli are present. The discal area of the wings bears many cross veins. The marginal area lacks cross veins but has many forked veinlets (Fig. 2.17). The length of the species is moderate to long, and the body is slender. The species are found along streams, in or beside which the larvae develop. ....**Chapter 13**...Osmylidae pars..p. 413

   – Ocelli are absent (Fig. 2.20) 14

14. (13) The humeral cross vein forms a recurrent vein at the humeral angle. The discal area of the wing displays a simple gradate series of cross veins and is distinct from the costal and marginal areas, which have many forked veinlets. Sc and R_1 are fused near the wing tip (Fig. 2.20). The vertex is convex. The antenna is moderate in length. Adults are large and nocturnal with wing spans from 40 to 75 mm, tip to tip. The adults and larvae are terrestrial. ....**Chapter 14**...Polystoechotidae..p. 425

   – There are no recurrent veins at the humeral angle. The discal and marginal areas are not clearly distinguished (Fig. 1.7). The antenna is longer than the head and thorax (Fig. 2.21). The adult is small 15

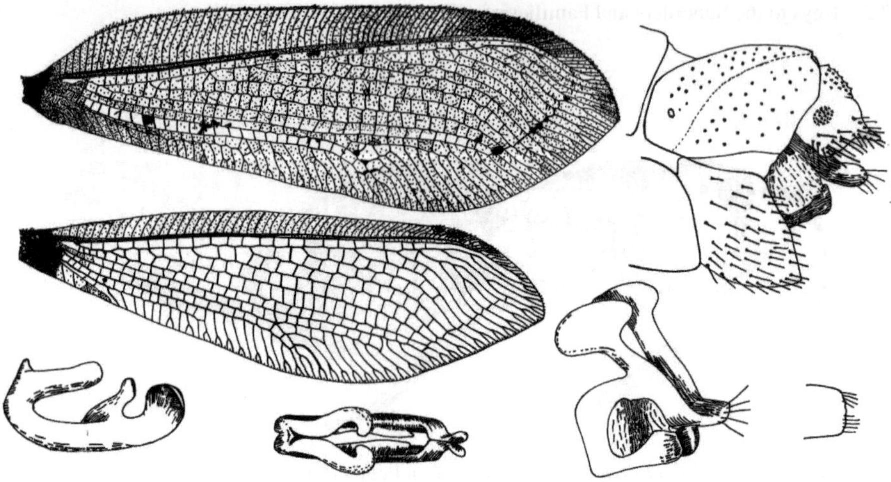

Fig. 2.17 *Isostenosmylus fusciceps* male: fore and hind wing (upper and middle left), apex of the abdomen in lateral view (upper right), paramere in lateral (lower left) and dorsal view (lower left center), tenth sternite in lateral view (lower right center) with its apex in dorsal view (lower right). Based on Kimmins (1940).

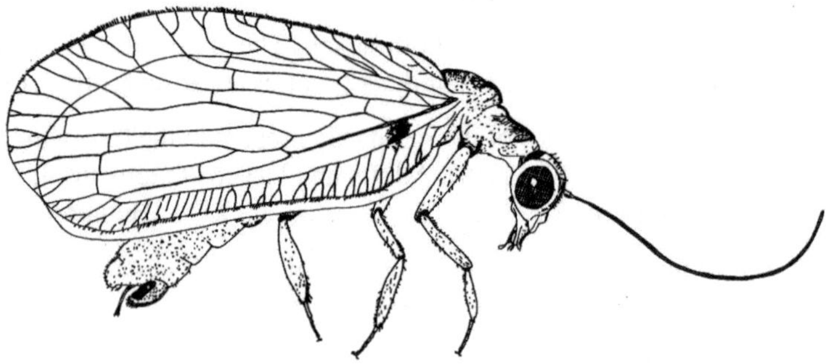

Fig. 2.18 Habitus of *Notiobiella paddiae*. Based on Penny and Monserrat (1983).

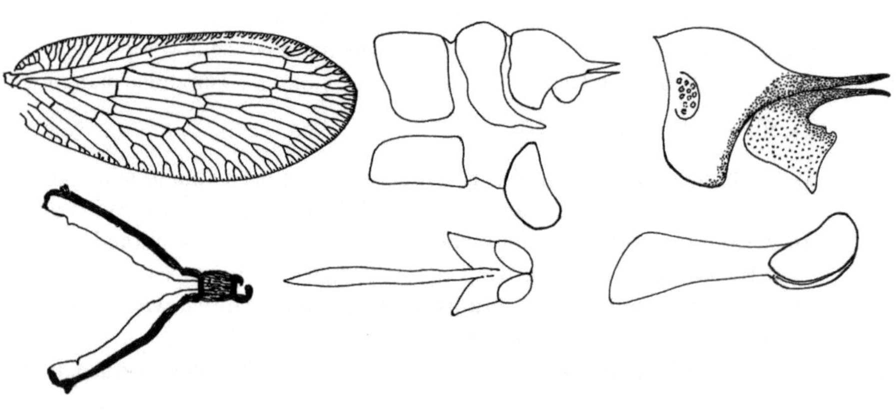

Fig. 2.19 *Sympherobius gayi* male (above, left to right): fore-wing, the apical sclerites of the abdomen in lateral view, a lateral view of the anal plate, and (below, left to right): the tenth sternite in dorsal view, and the parameres in dorsal and lateral view. Based on Nakahara (1965), who referred to the species by its synonym: *Sympherobius maculipennis*).

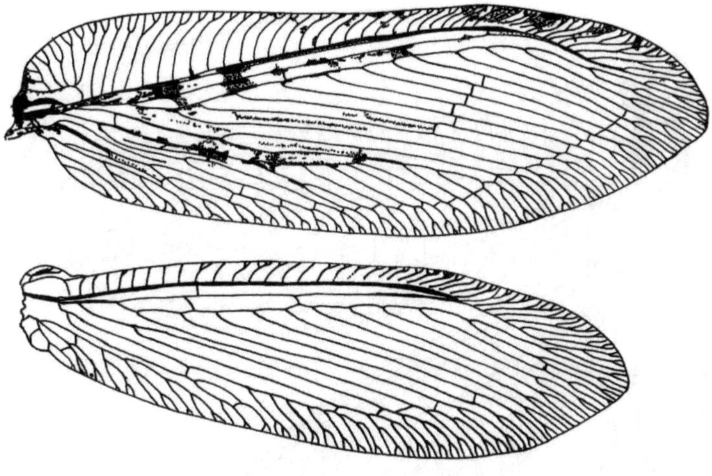

Fig. 2.20 Fore and hind wing of *Polystoechotes gazullai*. Based on Oswald (1998).

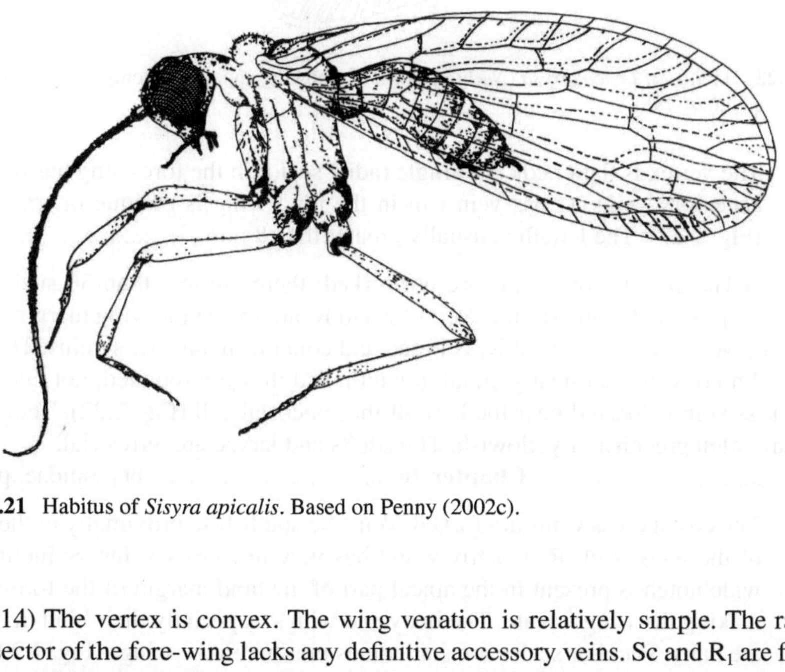

Fig. 2.21 Habitus of *Sisyra apicalis*. Based on Penny (2002c).

15. (14) The vertex is convex. The wing venation is relatively simple. The radial sector of the fore-wing lacks any definitive accessory veins. Sc and R_1 are fused near the wing tip. The costal cross veins are not forked. Cross vein r-m is in the axis of the hind wing (Fig. 2.21). Length: 6–8 mm. The larvae in one subfamily, Sisyrinae, develop in freshwater sponges and possibly also ectoproct colonies. Those in the other subfamily, Climaciinae, are found in or beside lentic and lotic water bodies.
....**Chapter 15**....................................Sisyridae..p. 429

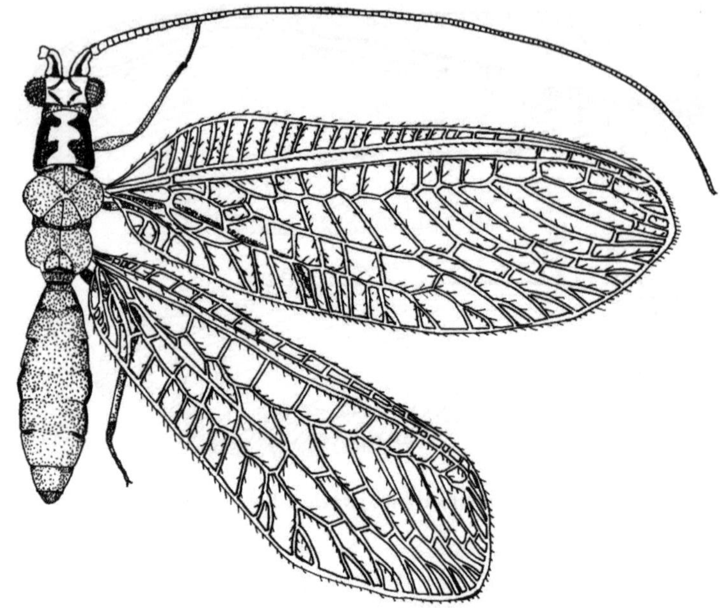

Fig. 2.22 Habitus of *Leucochrysa (Nodita) cruentata* showing appendages only on the right side. Based on Núñez (1988).

– The vertex is flattened. The single radial sector in the fore-wing has obvious accessory veins. Cross vein r-m in the hind wing is oblique or transverse (Fig. 2.22). The length is usually greater than 8 mm................................. 16

16. (15) The costal cross veins are not forked; there are less than 30 such cross veins proximal to the stigma. Veins Sc and R_1 are free at the wing margins. Vein Rs diverges from R_1. Cell R_1 is broad and contains many cross veins. The fore and hind wings are nearly equal in width, and they are rounded, not falcate. A cross vein is located near the base of the subcostal cell (Fig. 2.22). The adults are often greenish or yellowish. The adults and larvae are terrestrial.
....**Chapter 16**...................................Chrysopidae..p. 447

– The costal cross veins are forked. Veins Sc and R fuse proximally to the apex of the wing. Cell R_1 is narrow and has very few cross veins. Sometimes a wide notch is present in the apical part of the hind margin of the fore-wing, making the wing falcate. The body and wings, especially their hind margins, are hairy. Peculiar scales are sometimes present on the wings (Fig. 2.23). All known South American species have cursorial fore-legs, although some African species have raptorial fore-legs and a straight free stem of vein MA in the hind wing.

....**Chapter 17**.................................Berothidae..p. 595

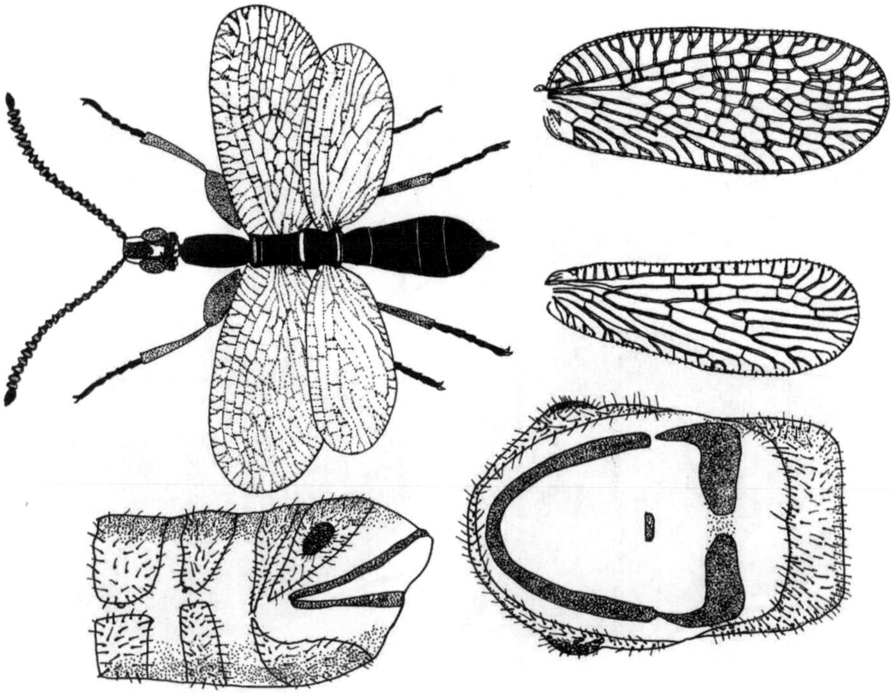

Fig. 2.23 *Ormiscocerus nitidipennis:* dorsal view of habitus (upper left), fore (upper right) and hind wing (middle right), and the apex of the abdomen in lateral (lower left) and posterior view (lower right). Based on Penny and Winterton (2007).

Key to the Suborders and Known Families in South America with notes of Pupae

Information for the key was provided by Berland and Grass (1951), Gurney and Parfin (1959), Chandler (1956), Miller and Stange (2012), Penny (1982a), and Bachmann (1995). The larvae of all South American genera have not yet been described, and there is no description available for the late larval instars of any South American species in the family Polystoechotidae. Grebennikov (2004) speculated that they might be grub-like, which would place them in a different position in the key. The key should not be used for species known only from other continents.

1. The aquatic larvae possess robust grasping and chewing mouthparts. There are gills on the first seven or eight abdominal segments (Fig. 2.24). ....Suborder Megaloptera...
   - The larvae are fully terrestrial, semi-aquatic, or aquatic. If they are aquatic, their mouths are narrow and elongate, with the mouthparts modified to stylets (Fig. 2.25)..3

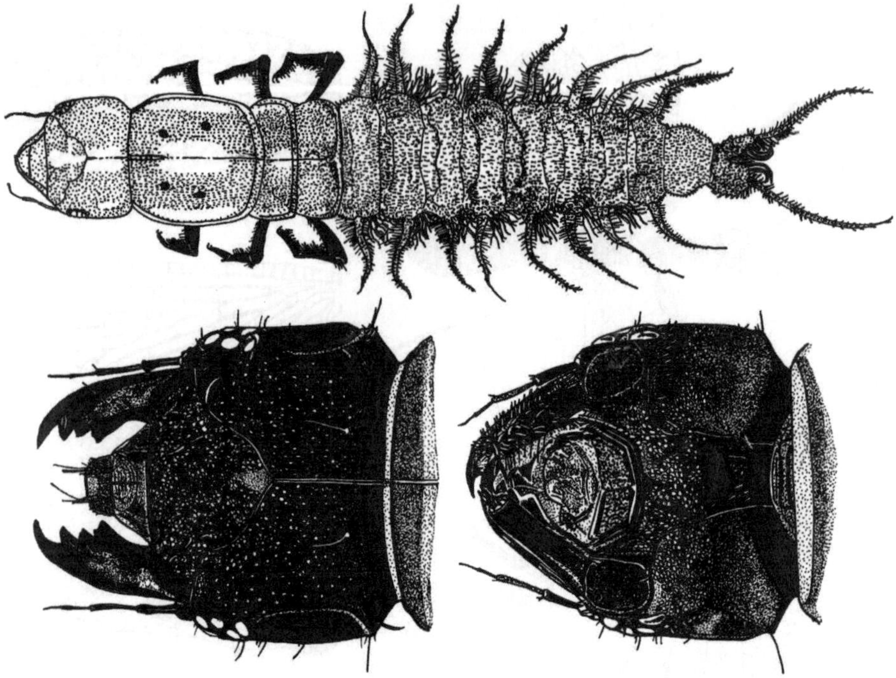

Fig. 2.24 *Chloronia hieroglyphica* larva: habitus in dorsal view (above) and the head in dorsal (lower left) and ventral view (lower right). Based on Penny and Flint (1982).

2. (1) A long median filament extends from the last segment, which lacks lateral hooks. Only the first seven abdominal segments bear lateral gills (Fig. 1.3). .... Sialidae..p. 75

   – The last segment lacks a median filament but possesses a pair of lateral hooks. The first eight abdominal segments bear lateral filaments (Fig. 2.24), ... Corydalidae..p. 81

3. (1) The larvae are aquatic and inhabit freshwater sponges or ectoprocts, or they live in wetlands, at the edge of water bodies, under stones and tree bark, or in termite colonies. Their mouthparts are modified to form long stylets, which may be used for piercing and feeding on termites or may appear to be modified for pipetting tiny prey, although the latter have also been described as modified for piercing and sucking sponges, in which some larvae live (Fig. 2.2)..............4

   – The larvae are semi-aquatic or terrestrial and have grasping and chewing mouthparts (Fig. 2.25) 6

4. (3) The known larvae inhabit termite nests and feed on the termites, which they immobilize with toxic substances. The larvae are long and thin with relatively small heads. Their legs are less than twice as long as the width of a thoracic segment (Fig. 2.26). .... Berothidae..p. 595

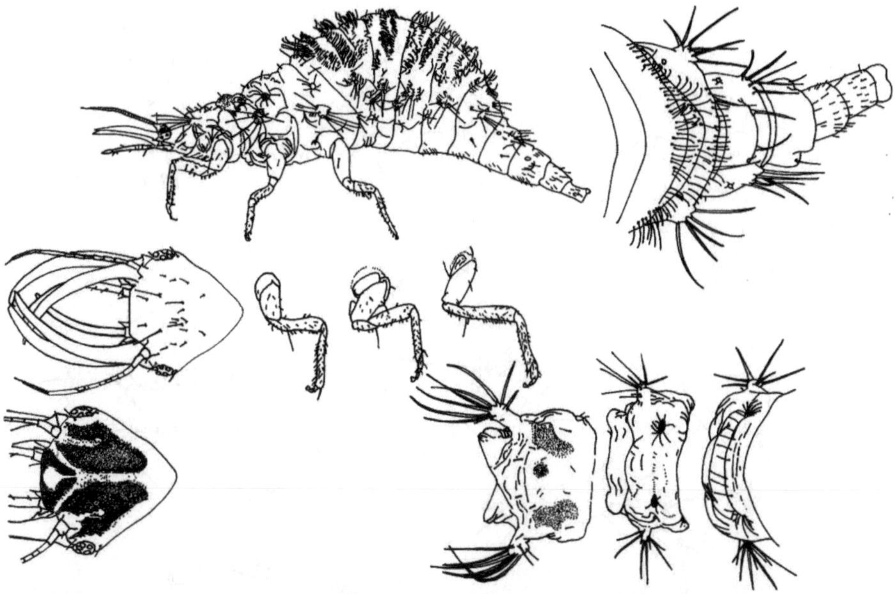

Fig. 2.25 *Ungla argentina* final stage larva: habitus in lateral view (upper left); head with appendages in dorsal view not showing the color pattern to reveal the locations of setae (middle left) and the head without the appendages to show the color pattern (lower left); the fore, middle, and hind leg (center, left to right, respectively); the prothorax, mesothorax, and metathorax in dorsal view (lower right); and the apex of the abdomen in dorsal view (upper right). Based on Monserrat and de Freitas (2005).

– The larvae do not inhabit termite nests, and most are aquatic or semi-aquatic. Their mouthparts are modified to form long stylets, which appear to be modified for pipetting tiny prey, although they have also been described as modified for piercing and sucking sponges, in which some larvae live. The length of a leg is much more that twice the width of a thoracic segment (Fig. 2.27) 5

5. (4) The head is small with very long and slender stylets directed anteriad. There is a pair of gills on each of the abdominal sternites. Each tarsus bears only a single claw (Fig. 2.27).
.... Sisyridae..p. 429

– The head is large with thick stylets, which are somewhat curved and directed outward. Gills are lacking. The last segment of the abdomen bears a pair of large, spiny tubercles directed outward. Each tarsus bears two claws (Fig. 2.28).
.... Osmylidae..p. 413

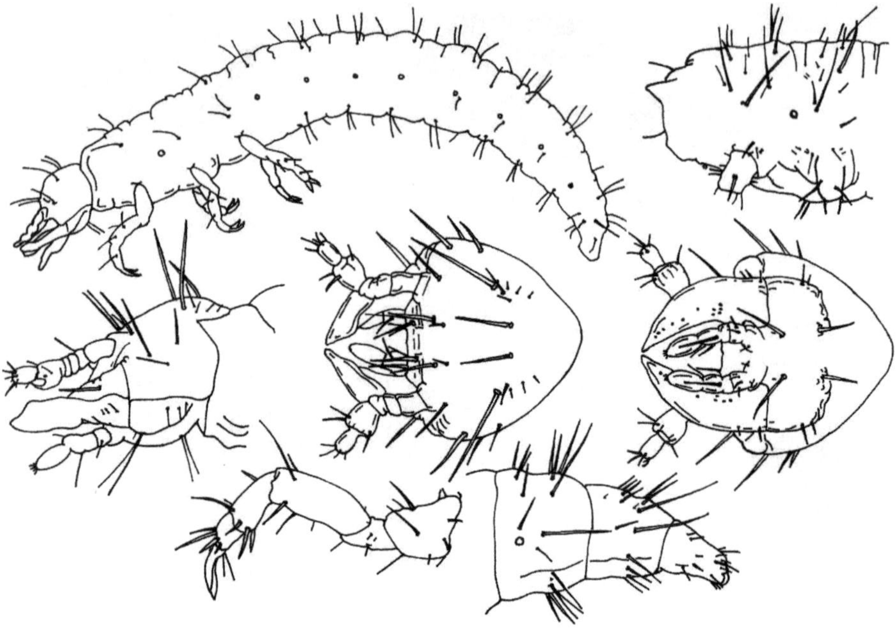

Fig. 2.26 An unidentified larva in the family Berothidae: habitus in lateral view (upper left to center); prothorax and mesothorax in lateral view (upper right); head in lateral, dorsal, and ventral view (middle row, left to right); fore-leg (lower left); apex of the abdomen in lateral view (lower right). Based on Grebennikov (2004).

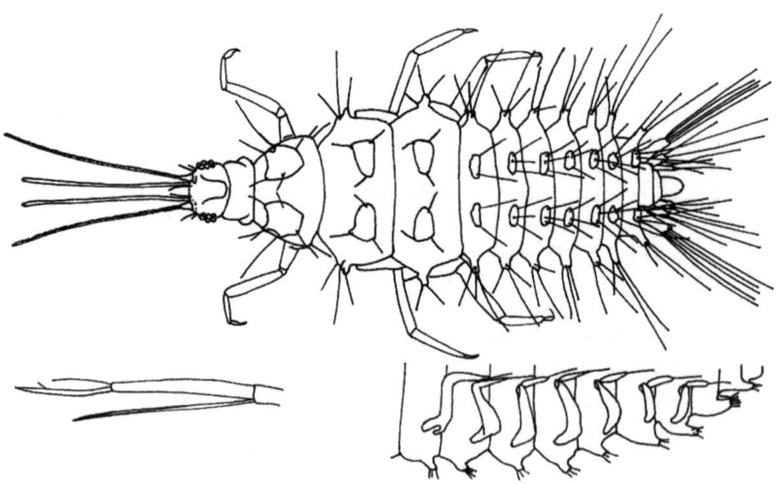

Fig. 2.27 *Sisyra vicaria* (Walker, 1853) larva, a North American species: habitus in dorsal view (above), apical antenna segments (lower left), and ventral view of one side of the abdomen showing the tracheal gills. Based on Parfin and Gurney (1956).

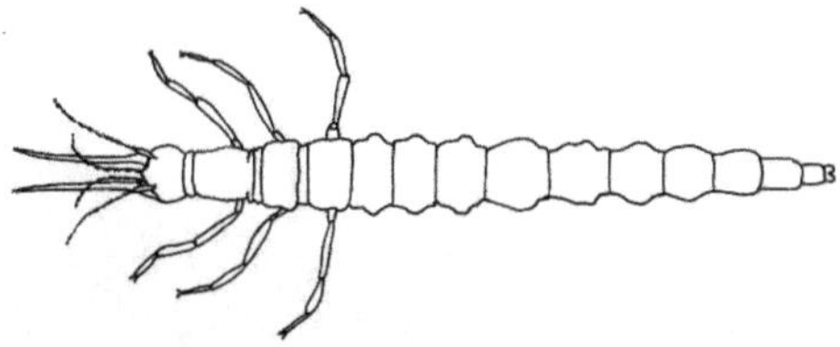

Fig. 2.28 Habitus of a larva of an unidentified species of Osmylidae, which is not known from South America. Based on New (1992).

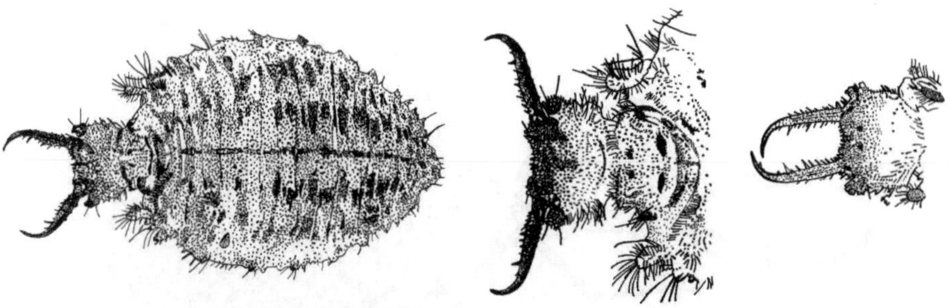

Fig. 2.29 *Stangeleon longipalpus* larva (left to right): habitus in dorsal view, enlarged head and anterior part of thorax in dorsal view, and ventral view of the head from an exuvia. Based on Miller (2008).

6. (3) The larvae are terrestrial and have robust, curved mandibles, usually much longer than the head and forming a powerful forcep (Fig. 2.29). The body is short and fairly broad, generally widest at or near the junction of the thorax and abdomen 7

   – Larvae may have thin, straight, or slightly curved mandibles (Fig. 2.30). If the mandibles are strongly curved and longer than the head, then the body is fusiform with the widest part usually at or near the third or fourth abdominal segment 8

7. (6) The known larvae prey on insects, frequently ants, which they actively hunt, ambush, or trap in conical depressions they dig in dry soils. The abdominal segments may have tufts of setae on their lateral margins, but these tufts are not borne on relatively large, tubercle-like processes (Fig. 2.29). .... Myrmeleontidae..p. 225

   – The larvae hide on the ground or in trees to ambush insects. They can widely open the jaws while waiting. The lateral margins of the abdominal segments of the known South American species bear relatively large, tubercle-like processes, which are usually covered by strong setae (Fig. 2.31). .... Ascalaphidae..p. 183

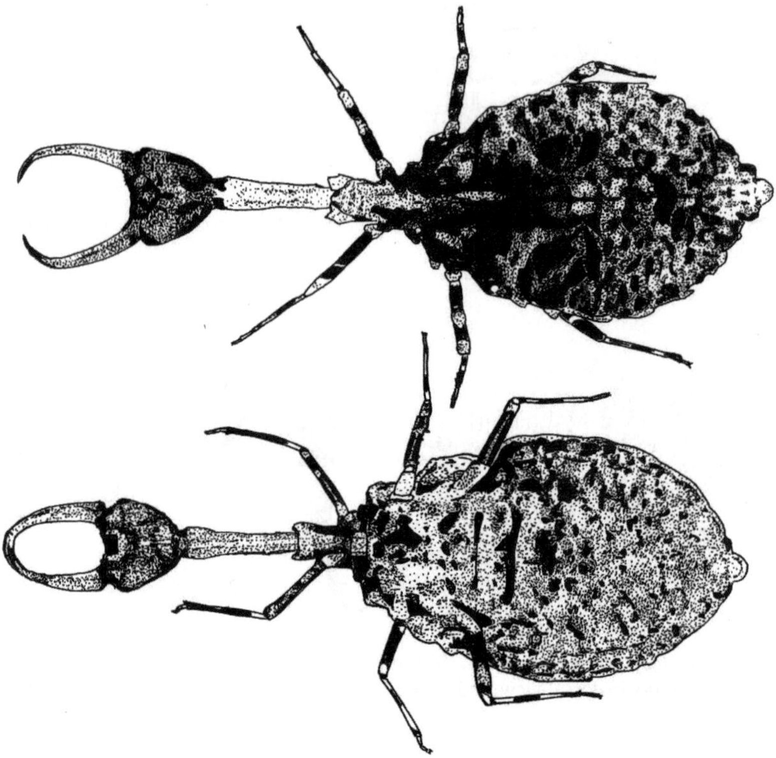

Fig. 2.30 Habitus of a *Moranida manselli* larva in dorsal (above) and ventral view (below). Based on Miller and Stange (1989).

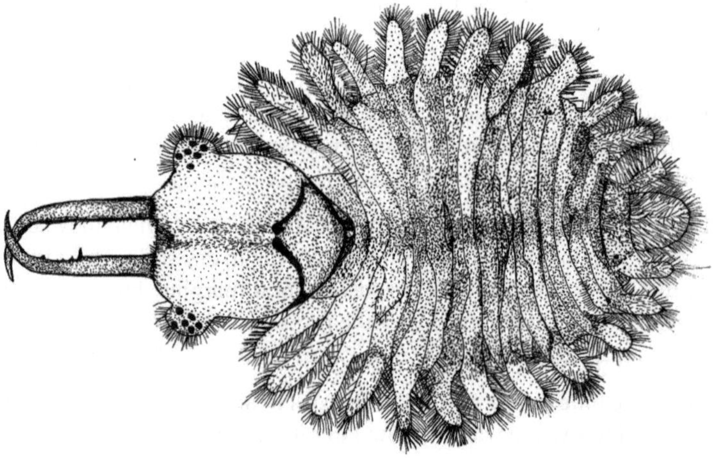

Fig. 2.31 The larva of an unidentified South American species in the family Ascalaphidae. Based on Penny (1981b).

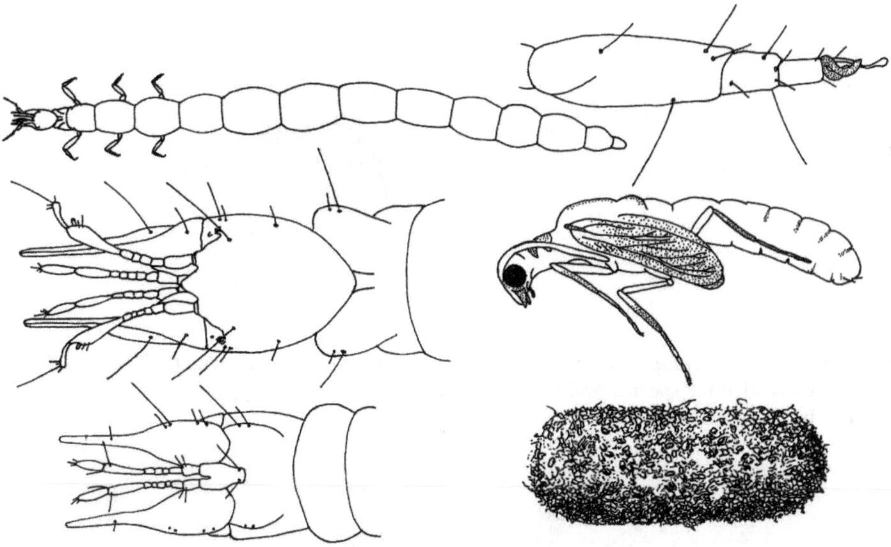

Fig. 2.32 *Nallachius americanus* (McLachlan, 1881) larva from North America: habitus (upper left), fore-leg (upper right), head in dorsal and ventral view (middle and lower left, respectively), pupa (middle right), and cocoon (lower right). Based on Gurney (1947).

8. (6) The combined length of the collar and pronotum is greater than the length of the head capsule, which is much wider than the collar and the pronotum. The anterior part of the abdomen is usually wider than the thorax. The mandibles are longer than the head capsule and usually evenly curved (Fig. 2.30). .... Crocidae..p. 285
   – The collar is inconspicuous, and the pronotum is shorter or wider than the head capsule. The mandibles are usually subequal to or shorter than the head capsule. If longer, the larvae are usually fusiform and widest between the junction of the thorax and abdomen and the third of fourth abdominal segments (Fig. 2.32) 9
9. (8) The larvae are long and very narrow with the segments of the thorax and abdomen varying little in width. The head capsule is narrower than the pronotum. There are two simple eyes on each side of the head. The mandibles are straight and subequal to the length of the head capsule (Fig. 2.32). .... Dilaridae..p. 343
   – The larvae are grub-like or fusiform, often notably narrow (Fig. 2.33) 10
10. (9) The larvae are eyeless, fusiform, and burrow through sand with their fossorial legs. A captured specimen seems to have been eating a beetle larva. The mandibles are strongly curved but shorter than the head capsule (Fig. 2.34). .... Nemopteridae..p. 293
   – Simple eyes are present, or, if not, the bodies are not fusiform, or the legs are not fossorial (Fig. 2.35) 11

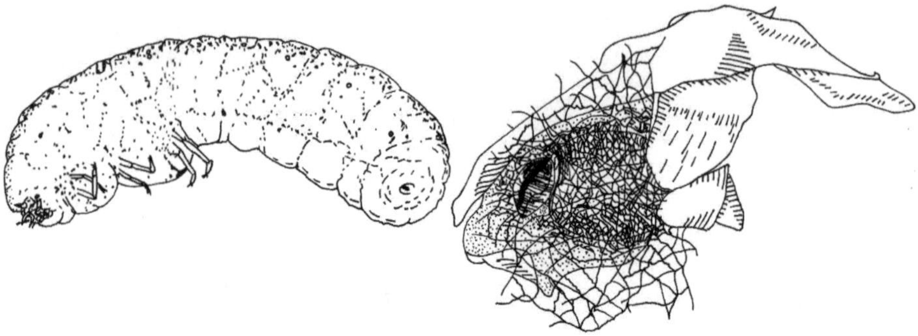

Fig. 2.33 *Nacarina valida:* Larva (left) and cocoon containing a pupa (right). Based on Weber (1942), who called the species *Nadiva valida,* a junior synonym. The specimen was identified at the time by Nathan Banks.

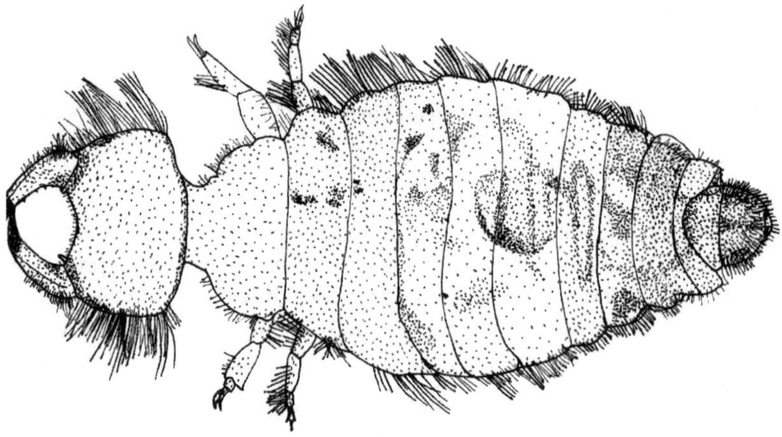

Fig. 2.34 Habitus of a larval *Stenorrhachus walkeri* in dorsal view. Based on Miller and Stange (2014).

11. (10) The first larval instar of an unidentified species from Mexico is fusiform. There is a gula on the ventral side of the cranium. The mandibles are large and wide at their bases (Fig. 2.35). No specimens of larvae from South America have been described. Descriptions of the final larval instars are needed to determine whether they become grub-like and whether they are herbivorous. ....Polystoechotidae..p. 425

    – There is no gula on the ventral side of the cranium. The mandibles are not notably large or wide at their bases (Fig. 2.36) 12

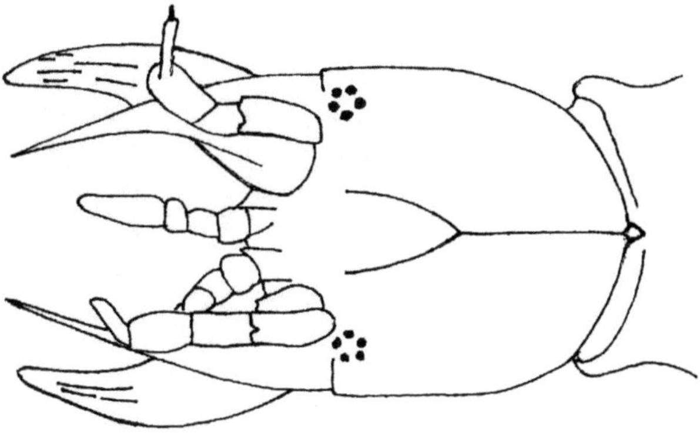

Fig. 2.35 Head of an unidentified member of the family Polystoechotidae from a specimen not from South America. Based on Tauber (1991).

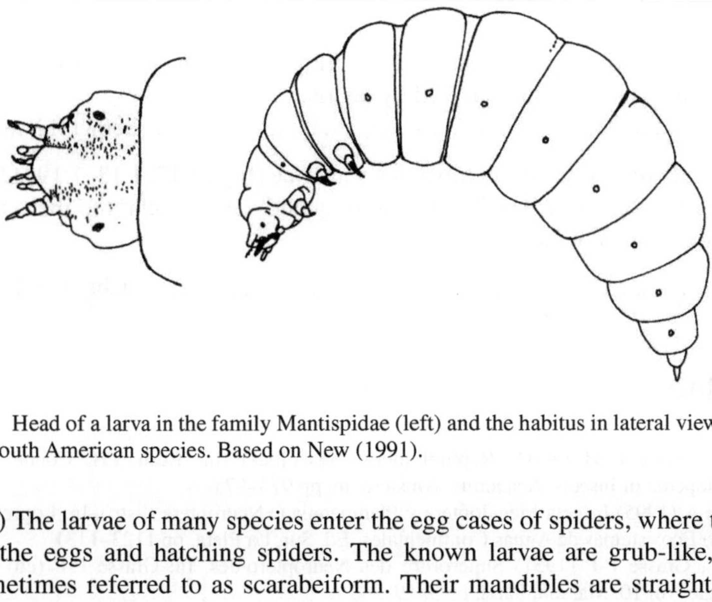

Fig. 2.36 Head of a larva in the family Mantispidae (left) and the habitus in lateral view (right). It is not a South American species. Based on New (1991).

12. (11) The larvae of many species enter the egg cases of spiders, where they feed on the eggs and hatching spiders. The known larvae are grub-like, a shape sometimes referred to as scarabeiform. Their mandibles are straight, narrow, and not notably long (Fig. 2.36). The larvae of few South American species have been described.
 .... Mantispidae..p.295
 – The larvae do not live in spider egg cases. They are fusiform, and most walk relatively quickly on plants to hunt for insects, such as aphids (Fig. 2.37) 13
13. (12) There are five ocelli on each side of the head. The mandibles are small and narrow (Fig. 2.28).
 ....Coniopterygidae..p. 123
 – The mandibles are strong and sometimes longer than the head capsule (Fig. 2.37) 14

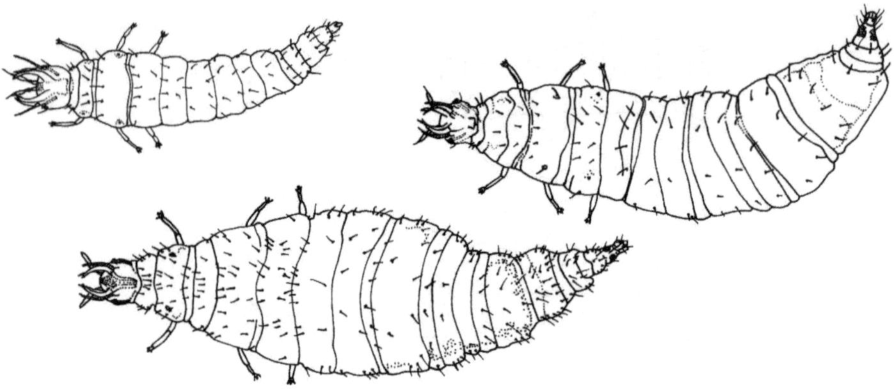

Fig. 2.37 *Sympherobius gayi* larvae: first (upper left), second (upper right), and third larval instar (below). Based on Reguilón and Nuñez Campero (2005), who used the synonym *Sympherobius marmoratipennis*.

14. (13). The body surface lacks tubercles (Fig. 2.37). The body surface bears few setae and is usually not covered by debris.
 ....Hemerobiidae..p. 355
 – The body surface bears tubercles and setae (Figs.1.17, 1.18, 1.19, 1.20, 1.21, 1.22, 2.22, and 2.33). The larvae frequently attach debris to their body surfaces as camouflage.
 ....Chrysopidae..p. 447

Literature

Aspöck U, Aspöck H (2003) Raphidioptera (Snakeflies). In: Resh VH, Conde RT (eds) Encyclopedia of insects. Academic, Amsterdam, pp 973–975

Bachmann A (1995) Insecta Megaloptera y Planipennia (=Neuroptera s. str.). In: Lopretto E, Tell G (eds) Ecosistemas de Aguas Continentales. Ed. Sur, La Plata, pp 1123–1131

Berland L, Grasse P-P (1951) Superordre des Neuropteroides. In: Grasse P-P (ed) Traite de Zoologie, vol 10. Masson, Paris, pp 3–69

Brues CT, Melander AL, Cartenter FM (1954) Classification of insects. Keys to the living and extinct families of other terrestrial arthropods. Bulletin of the Museum of Comparative Zoology at Harvard College 108:1–792

Chandler HP (1956) Megaloptera. In: Usinger RL (ed) Aquatic insects of California. University of California Press, Los Angeles, pp 229–236

Contreras-Ramos A (2006) *Protosialis ranchograndis*, a new species of alderfly from Venezuela, with a redescription of *P. brasiliensis* Navás (Megaloptera: Sialidae). Proceedings of the Entomological Society of Washington 108(4):977–984

Flint OS Jr (1973) The Megaloptera of Chile (Neuroptera). Revista Chilena de Entomologia 7:31–45

Grebennikov VV (2004) Grub-like larvae of Neuroptera (Insecta): a morphological review of the families Ithonidae and Polysteochotidae and a description of *Oliarces clara*. European Journal of Entomology 101:409–412

Gurney AB (1947) Notes on Dilaridae and Berothidae, with special references to immature stages of the Nearctic genera (Neuroptera). Psyche 54:145–169

Gurney AB, Parfin S (1959) Neuroptera. In: Edmondson WT (ed) Fresh-water biology. Wiley, New York, pp 973–980, 1248 pp

Henry CS, Penny ND, Adams PA (1992) The neuropteroid orders of Central America (Neuroptera and Megaloptera). In: Quintero D, Aiello A (eds) Selected studies: insects of Panama and Mesoamerica. Oxford University Press, Oxford, pp 432–458

Imms AD (1957) A general textbook of entomology, 9th edn. Methuen, London, 886 pp

Kimmins DE (1940) New genera and species of Hemerobiidae (Neuroptera). Annals and Magazine of Natural History 11(6):222–236

Miller RB (2008) A new genus and species of Brachynemurini from Venezuela (Neuroptera: Myrmeleontidae). Insecta Mundi 0059:1–5

Miller RB, Stange LA (1989) A new species of *Moranida* Mansell from Venezuela (Neuroptera: Nemopteridae). Insecta Mundi 3(1):65–70

Miller RB, Stange LA (2012) A new species of *Stenorrhachus* McLachlan from Chile (Neuroptera: Nemopteridae) with biological notes. Insecta Mundi 0226:1–8

Miller RB, Stange LA (2014) A revision of the genus *Purenleon* Stange (Neuroptera: Myrmeleontidae: Nemoleontini). Insecta Mundi 0384:1–62

Monserrat VJ (1981) (1980): Dos nuevos coniopterigidos de Paraguay (Insecta, Neuroptera, Coniopterygidae). Eos 55–56:155–163

Monserrat VJ (1983) Contribución al conocimiento de los coniopterígidos de Venezuela (Neuroptera, Planipennia, Coniopterygidae). Boletín de Entomología Venezolana, NS 2(19):137–156

Monserrat VJ (2005) New data on the dusty wings from Coquimbo, Patagonia and Tierra del Fuego (Neuroptera, Coniopterygidae). Entomofauna, Zeitschrift für Entomologie 26(25):421–436

Monserrat VJ, de Freitas S (2005) Contribución al conocimiento de los crisópidos de Coquimbo, Patagonis y Tierra del Fuego (Argentina, Chile) (Insecta, Neuroptera, Chrysopidae). Graellsia 61:163–179

Nakahara W (1965) Neotropical Hemerobiidae in the United States National Museum. Proceedings of the United States National Museum 117(3509):107–122 + 2 pl

New TR (1991) Neuroptera. In: The insects of Australia, 2nd edn, vol 1. University Press, Carlton, pp 525–542, xvii - 1137 - vi pp

New TR (1992) The lacewings (Insecta, Neuroptera) of Tasmania. Papers and Proceedings of the Royal Society of Tasmania 126:29–45

Núñez E (1988) Chrysopidae (Neuroptera) del Perú y sus especies más comunes. Revista Peruana de Entomologia 31:69–75

Orfila RN (1955) Un nuevo Nemopteridae (Neuropt.) americano con una sinopsis de la familia. Revista de la Sociedad Entomológica Argentina 17:29–32

Oswald JD (1998) Rediscovery of *Polystoechotes gazullai* (Neuroptera: Polystoechotidae). Proceedings of the Entomological Society of Washington 100(3):389–394

Oswald JD, Penny ND (1991) Genus-group names of the Neuroptera, Megaloptera and Raphidioptera of the world. Occasional Papers of the California Academy of Sciences 147:1–94

Parfin SI, Gurney AB (1956) The spongilla-flies with special reference to those in the western hemisphere (Sisyridae, Neuroptera). Proceedings of the U. S. National Museum 105(3360):421–529 + 3 pl

Penny ND (1977) (1978): Lista de Megaloptera, Neuroptera e Raphidioptera do Mexico, America Central, ilhas Caraibas e America do Sul. Acta Amazonica, Supplement 7(4):1–61

Penny ND (1981a) Neuroptera of the Amazon basin, part 4 Sialidae. Acta Amazonica 11(4):843–846

Penny ND (1981b) Neuroptera of the Amazon basin, part 3 Ascalaphidae. Acta Amazonica 11(3):605–651

Penny ND (1982a) Neuroptera of the Amazon basin. Part 2. – Corydalidae. Acta Amazonica 12:825–832
Penny ND (1982b) Review of the generic level classification of new world Mantispidae (Neuroptera). Acta Amazonica 12(2):209–223
Penny ND (1983) Neuroptera of the Amazon basin. Part 9. – Albardinae. Acta Amazonica 13(3–4):697–699
Penny ND (1984) Recent progress in the taxonomy of South American Neuroptera. In: Gepp J, Aspöck H, Hölzel H (eds) Progress in world's neuropterology. – Proceedings of the 1st international symposium on neuropterology. Graz, Austria, pp 75–77
Penny ND (ed) (2002a) A guide to the lacewings (Neuroptera) of Costa Rica. Proceedings of the California Academy of Sciences 53(12):161–452
Penny ND (2002b) Family Ascalaphidae. In: Penny ND (ed) A guide to the lacewings (Neuroptera) of Costa Rica. Proceedings of the California Academy of Sciences 53(12):176–186
Penny ND (2002c) Family Dilaridae. In: Penny ND (ed) A guide to the lacewings (Neuroptera) of Costa Rica. Proceedings of the California Academy of Sciences 53(12):236–232
Penny ND, Flint OS, Jr. (1982) A revision of the genus *Chloronia* (Neuroptera: Corydalidae). Smithsonian Contributions to Zoology No. 348, 27 pp
Penny ND, Monserrat VJ (1985) (1983): Neuroptera of the Amazon basin. Part 10. - Hemerobiidae. Acta Amazonica 13(5–6):879–909
Penny ND, Winterton SL (2007) Rediscovery of the unusual genus *Ormiscocerus* (Neuroptera: Berothidae: Cyrenoberothinae). Proceedings of the California Academy of Sciences, Fourth Ser 58(1):1–6
Reguilón C, Nuñez Campero SR (2005) Morfología de los estadios larvales de *Sympherobius marmoratipennis* (Neuroptera: Hemerobiidae). Revista de la Sociedad Entomologica Argentina 64(3):81–85
Riek EF (1975) On the phylogenetic position of *Brucheiser argentinus* Navás 1927 and description of a second species from Chile (Insecta: Neuroptera). Studies on the Neotropical Fauna 10(2):117–126
Stange LA (2002) Family Myrmeleontidae. In: PennyND (ed) A guide to the lacewings (Neuroptera) of Costa Rica. Proceedings of the California Academy of Sciences 52(12):275–289
Stitz J (1931) Planipennia. Biologie der Thiere Deutschlands 35:67–304
Tauber CA (1991) Neuroptera. In: Stehr FW (ed) Immature insects, vol 2. Kendall-Hunt, Dubuque, pp 126–143
Walker F (1853) List of the specimens of neuropterous insects in the collection of the British museum (Sialidae – Nemopterides), 2:193–476
Weber NA (1942) A neuropterous myrmecophile, *Nadiva valida* Erichs. Psyche 49(1–2):1–3

Part II
Megaloptera

Chapter 3
Sialidae

Abstract To date, all specimens of sialids known to inhabit South America are assigned to the genus *Protosialis*. Keys are provided to identify the species of adult specimens in this genus known to occur in South America. The countries or states in which specimens of each species have been found and invalid synonyms are reported.

Keywords *Protosialis* • Key to species • Specimens • Synonyms • Neotropical

The only Neotropical genus in this family is *Protosialis* van der Weele, 1909. The imagos are known as alder flies.

Key to the Species of Adult *Protosialis* Inhabiting South America

Information for the key was provided by Walker (1853), Davis (1903), van der Weele (1910), Enderlein (1910), Stitz (1914), Navás (1933, 1936), Banks (1943), Penny (1977, 1981), Archangelsky (2004), and Contreras-Ramos (2006, 2007, 2008). The identity of several species is doubtful because type specimens were lost or damaged, as noted for *Protosialis brasiliensis* in the key. The species assumed to be *Protosialis flavicollis* has not yet been formally removed from the genus *Sialis* due to a lack of specimens to examine. New specimens fitting the descriptions of those lacking illustrations in the key are needed for examination. The identification of larvae is still not possible because specimens and descriptions of most are lacking.

1. The head, including the clypeus and labrum, is uniformly orange brown. The color is predominantly orange brown. The antenna is brown with a somewhat paler scape; it is pilose and consists of about 35 segments. The tenth sternite of the male forms a double sclerotized plate, which appears fusiform in posterior view (Fig. 2.8). Fore-wing length: c. 8.4–10.4 mm. ....*Protosialis ranchograndis* Contreras-Ramos, 2006 (Venezuela). In the description of this species, Contreras-Ramos (2006) discusses the liklihood of its synonymy with other species of this genus.

C.W. Heckman, *Neuroptera (Including Megaloptera)*, Encyclopedia of South American Aquatic Insects, DOI 10.1007/978-3-319-35125-4_3

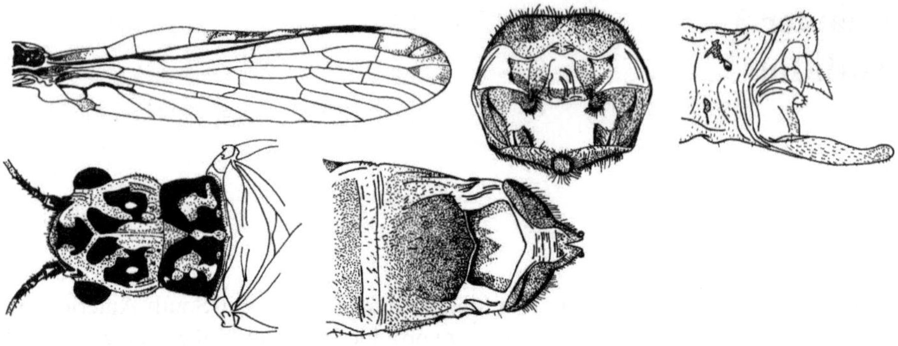

Fig. 3.1 *Protosialis chilensis* (above, left to right): fore-wing, apex of the male abdomen in posterior and lateral view, and (below, left and right): head and anterior part of the thorax and the apex of the female abdomen in ventral view. Based on Contreras-Ramos (2008).

– The head is generally dark, sometimes with ferrugineous markings, or it is yellowish or orange brown with a distinct fuscous stripe or other prominent dark markings (Fig. 3.1) 2

2. (1) The head and pronotum are yellowish without a broad, longitudinal stripe on the head but with a black spot posterior to the base of each antenna and a black trifid marking posterior to each compound eye. The general coloration of the rest of the insect is black. The wings are fumose. The length was reported to be 15 mm to the apices of the folded wings.
.... *Protosialis bimaculata* Banks, 1920 (Bolivia).

– The head is generally dark or, if it is orange or yellowish, a longitudinal fuscous stripe is present, or it may have a black marking with four prongs posterior to each compound eye, or it may have only a bifid marking there or no dark marking there at all (Fig. 3.1).. 3

3. (2) The fore-wing usually exceeds 11 mm..4

– The fore-wing is shorter than 11 mm..6

4. (3) The head is pale orange with a broad, longitudinal, fuscous stripe, which is wider at the anterior end and becomes poorly defined toward the posterior part of the head, where there are two narrow, paler bands. The labrum and clypeus are fuscous, and the antenna is black with paler apical segments. The dark orange pronotum has small brownish spots, and the metanotum is almost black. The wings are narrow (Fig. 3.2). Fore-wing length of male: c. 12 mm; female: c. 13 mm.
...............................*Protosialis hauseri* Contreras, Florentin, and Urakami, 2005 (Rio Grande do Sul).

– The predominant color is dark with paler markings (Fig. 3.1)........................5

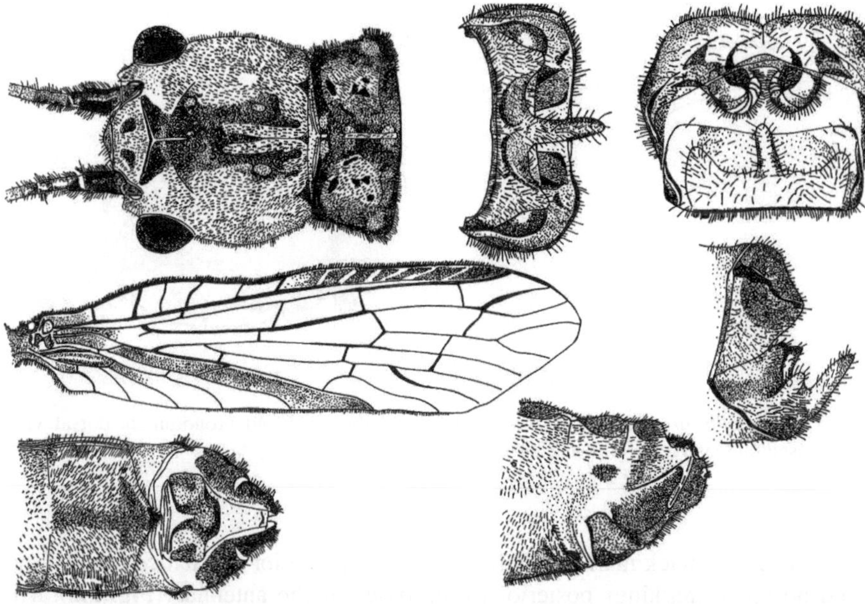

Fig. 3.2 *Protosialis hauseri:* fore-wing (middle left), head with basal antenna segments and anterior part of the thorax (upper left); apex of the male abdomen in ventral (upper center), posterior (upper right), and lateral view (middle right); apex of the female abdomen in ventral (lower left) and lateral view (lower right). Based on Contreras-Ramos et al. (2005).

5. (4) The ground color of the entire body, antennae, and palps is black, except for a pale brownish yellow pronotum and a brownich clypeus and labrum. The legs are blackish brown. The antenna consists of c. 34 segments. Fore-wing length of male: c. 14.5 mm; female: c. 13 mm.
   ....*Protosialis flavicollis* (Enderlein, 1910) (Colombia). Syn: *Sialis flavicollis* Enderlein, 1910.
   – The head is reddish or orange with extensive black markings and completely black antennae. There is no fuscous trifid marking posterior to each compound eye, but there is a fuscous spot beneath each eye and fuscous infuscations on the frons and along the median line on the head (Fig. 3.1). The fore-wing is conspicuously narrow. Fore-wing length of male: 14–17 mm; female: 16–19.5 mm.

   ....*Protosialis chilensis* (McLachlan, 1870) (Chile, Argentina). Syn: *Sialis chilensis* McLachlan, 1870.

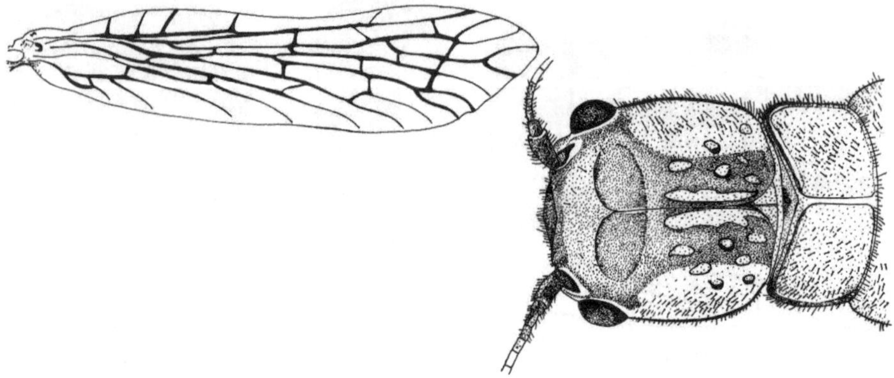

Fig. 3.3 *Protosialis brasiliensis:* fore-wing (left) and the head and pronotum in dorsal view (right). Based on Contreras-Ramos (2006).

6. (3) There is a black marking with four prongs posterior to each compound eye and no black markings posterior to the bases of the antennae (Fig. 2.2). The wings are fuscous. Fore-wing length of male: c. 7–8 mm.
   ....*Protosialis flammata* Penny, 1981 (Peru, Amazonas).
   – A black, four-pronged marking on a lighter background is not present posterior to each compound eye (Fig. 3.3)..7

7. (6) The head is pilose, predominantly dark brown, and there are paler posterolateral areas and pale coloration narrowly surrounding the compound eyes (Fig. 3.3). Fore-wing length: 9.5–10 mm.
   ....*Protosialis brasiliensis* Navás, 1936 (São Paulo). This species lacks an adequate discription. Contreras-Ramos (2006) examined the only specimen, which he supposes to be a male. The abdomen, antenna, one fore-wing, both hind wings, and hind legs are missing, so even its sex cannot be determined with certainly.
   – The head is mainly black with ferrugineous markings bordering the eyes and black pilosity. Fore-wing length: c. 10 mm; hind wing: c. 8.3 mm.

   ....*Protosialis nubila* Navás, 1933 (Mato Grosso). The holotype was destroyed, so identifications are tentative.

Literature

Archangelsky M (2004) Nuevas citas de Coleoptera acuaticos y Megaloptera para la provincia de Cubut (Argentina). Revista de la Sociedad Entomologica Argentina 63(3–4):66–68
Banks N (1920) New neuropteroid insects. Bulletin of the Museum of Comparative Zoology 64(3):297–362+pl. 1–7
Banks N (1943) Neuroptera of northern South America. Part I. Boletin de Entomologia Venezolana 2:59–66
Contreras-Ramos A (2005 [2007]) Recent accounts on the systematics and biogeography of Neotropical Megaloptera (Corydalidae, Sialidae). Annali del Museo Civico di Storia Naturale di Ferrara 8:67–72
Contreras-Ramos A (2006) *Protosialis ranchograndis*, a new species of alderfly from Venezuela, with a redescription of *P. brasiliensis* Navás (Megaloptera: Sialidae). Proceedings of the Entomological Society of Washington 108(4):977–984
Contreras-Ramos A (2008) Notes on some Neotropical alderflies (Sialidae: Megaloptera). Annals of the Entomological Society of America 101(5):808–814
Contreras-Ramos A, Florentin GL, Urakami Y (2005) A new species of alderfly (Megaloptera: Sialidae) from Rio Grande do Sul, Brazil. Amazoniana 18(3–4):267–272
Davis KC (1903) Sialididae (sic) of North and South America. Bulletin of the New York State Museum 68:442–486, 499+pl. 51–52
Enderlein G (1910) Eine neue *Sialis* aus Columbien. Stettiner Entomologische Zeitung 71:380–381
McLachlan R (1870) On the occurrence of the neuropterous genus *Sialis* in Chili. Entomologists' Monthly Magazine 7:145–146
Navás L (1933 [1932]) Décadas de insectos nuevos. Decada 23. Brotéria 29(1):34–44
Navás L (1936) (1931): Insectos del Brasil. 5ª Serie. Revista do Museu Paulista 20:721–734
Penny ND (1977 [1978]) Lista de Megaloptera, Neuroptera e Raphidioptera do Mexico, America Central, ilhas Caraibas e America do Sul. Acta Amazonica, Supplement 7(4):1–61
Penny ND (1981) Neuroptera of the Amazon Basin, part 4. Sialidae. Acta Amazonica 11(4):843–846
Stitz J (1914) Sialiden der Sammlung des Berliner Museums. Sitzungsberichte der Gesellschaft naturforschender Freunde zu Berlin 1914(5):191–205+pl. 6–7
van der Weele HW (1909) New genera and species of Megaloptera Latr. Notes from the Leyden Museum 30:249–253
van der Weele HW (1910) Megaloptera (Latrelle) monographic revision. In: Collections Zoologiques du Baron Edm. de Selys Longchamps 5(1):1–93+4 pl
Walker F (1853) List of the specimens of neuropterous insects in the collection of the british museum. Part II (Sialidae – Nemopterides). Trustees, London, pp 193–476

Chapter 4
Corydalidae

Abstract Illustrated keys are provided for identification of the genera of adult corydalids known to occur in South America and for those larvae that have been examined and described. Illustrated keys are also provided to identify the species of each genus known to occur in South America. The countries or states in which specimens of each species have been found and invalid synonyms are reported.

Keywords *Corydalus* • *Chloronia* • Dobsonfly • Fishfly • Key to species

The imagos in this family are relatively large. English common names for the species include dobsonfly and fishfly. Adults are terrestrial and rest on trees and various objects not far from streams or lakes. The larvae are usually benthic inhabitants of streams or lakes with continual water currents, where they feed on prey small enough for them to overpower. They are often found beneath large, flat rocks where water currents are at least fairly rapid. The habits of most South American species have yet to be studied in detail. Gills are the organs of respiration for the larvae.

4.1 Key to the Genera of Adults in the Family Corydalidae in South America

Information for the key is taken from Banks (1908); Penny (1982, 1999); Penny and Lee (1996); New and Theischinger (1993); Romero (2003).

1. Gonostyli are present on the ninth abdominal segment. The antennae are never pectinate (Fig. 4.1)............................Subfamily Corydalinae......................... 2
   – The ninth abdominal segment lacks gonostyli. The antenna is sometimes pectinate or serrate (Fig. 4.2)......................Subfamily Chauliodinae................ 3
2. (1) The last vein of the radial sector (Rs) is branched. The color is usually yellow with dark markings. The ninth gonostylus of the male usually bears a small apical tooth (Fig. 4.1).
   .... *Chloronia* Banks, 1908..p. 85

C.W. Heckman, *Neuroptera (Including Megaloptera)*, Encyclopedia of South American Aquatic Insects, DOI 10.1007/978-3-319-35125-4_4

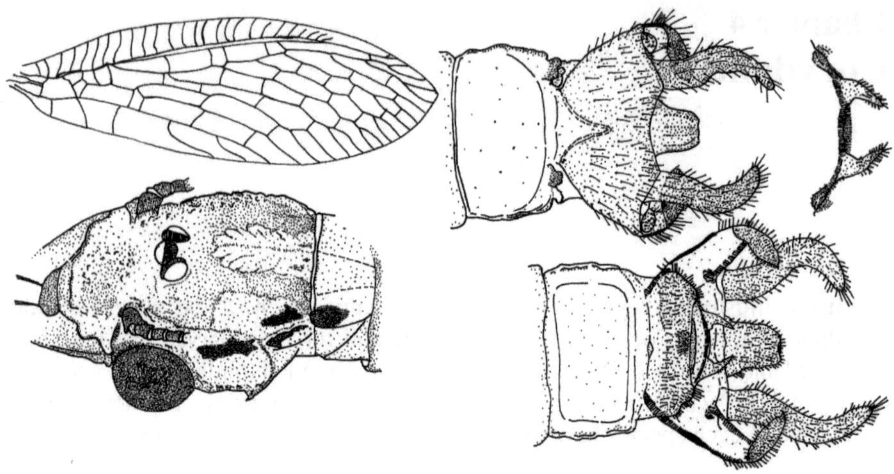

Fig. 4.1 *Chloronia gaianii* male: fore-wing (*upper left*), head in dorsolateral view with only a few basal segments of the antennae shown (*lower left*), apex of the abdomen in dorsal (*upper center*) and ventral view (*lower right*), and the tenth sternite (*upper left*). Based on Contreras-Ramos (2002).

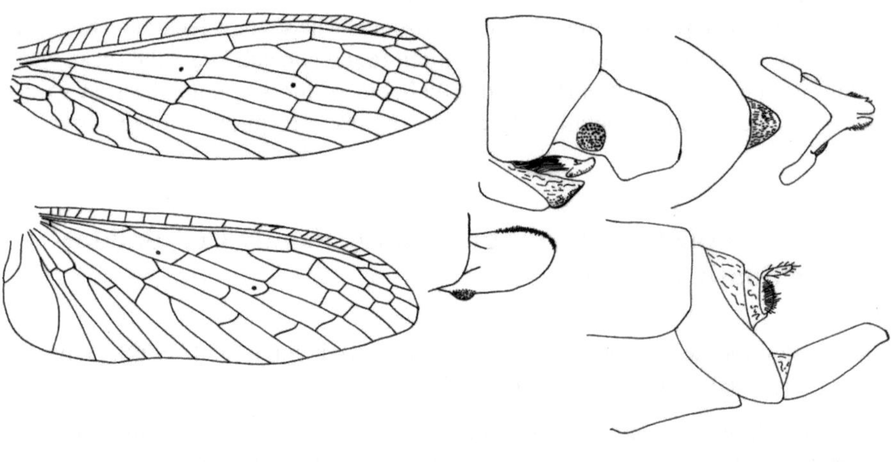

Fig. 4.2 *Archichauliodes chilensis:* fore and hind wing of a male (*left*, *above* and *below*, respectively), apex of the male abdomen in lateral view (*upper center*), ninth sternite in ventral view (*upper right center*), aedeagus in dorsal view (*upper right*), left anal plate in dorsal view (*lower center*), and the apex of the female abdomen in lateral view (*lower right*). Based on Kimmins (1954).

– The last vein of the radial sector (Rs) is not branched. The color is usually brown or black with lighter markings. The ninth gonostylus of the male lacks an apical tooth (Fig. 4.3).
....*Corydalus* Latreille, 1802..p. 93

3. (2) In the fore-wing, the anterior branch of vein 2A remains separate from vein 1A, so that cell A_1 is closed at its distal end by a cross vein between veins 1A and 2A (Fig. 4.2). The antennae are filiform.
   ....*Archichauliodes* van der Weele, 1909..p. 114

   – In the fore-wing, the anterior branch of vein 2A is joined to vein 1A but separates from it distally. It is the anterior fork of 2A which closes cell A_1 (Fig. 4.4)..4

4. (3) Vein R_4 in both wings is forked, and there is only occasionally a cross vein joining the two branches of the fork. Each segment of the antenna is free from whorls of erect setae, which distinguishes the genus from those found in other parts of the world. The genitalia are shown in Fig. 4.5.
   ....*Protochauliodes* van der Weele, 1909..p. 115

   – Vein R_4 in both wings is unforked (Fig. 4.4).
   ....*Nothochauliodes* Flint, 1983
   The only South American species in this genus is *Nothochauliodes penai* Flint, 1983, from Chile.

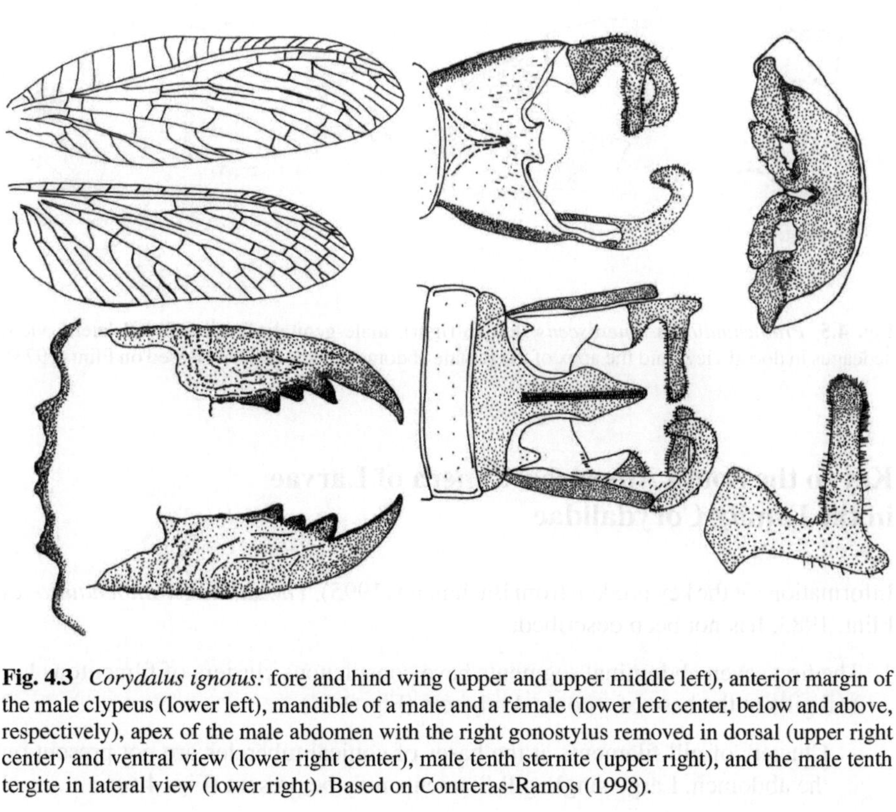

Fig. 4.3 *Corydalus ignotus:* fore and hind wing (upper and upper middle left), anterior margin of the male clypeus (lower left), mandible of a male and a female (lower left center, below and above, respectively), apex of the male abdomen with the right gonostylus removed in dorsal (upper right center) and ventral view (lower right center), male tenth sternite (upper right), and the male tenth tergite in lateral view (lower right). Based on Contreras-Ramos (1998).

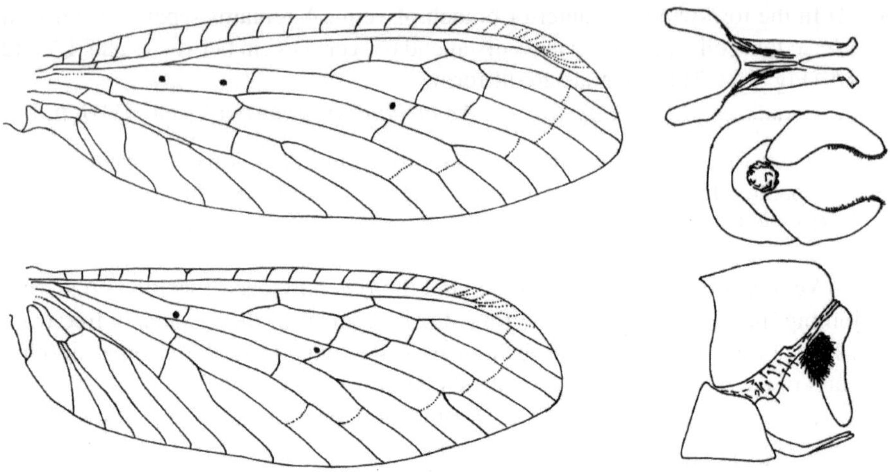

Fig. 4.4 *Nothochauliodes penai* male: fore and hind wing (upper and lower left), dorsal view of the penis (upper right), ninth and tenth tergite in posterior view with the dorsal part facing left (middle right), and the male genitalia in lateral view (lower right). Based on Flint (1983).

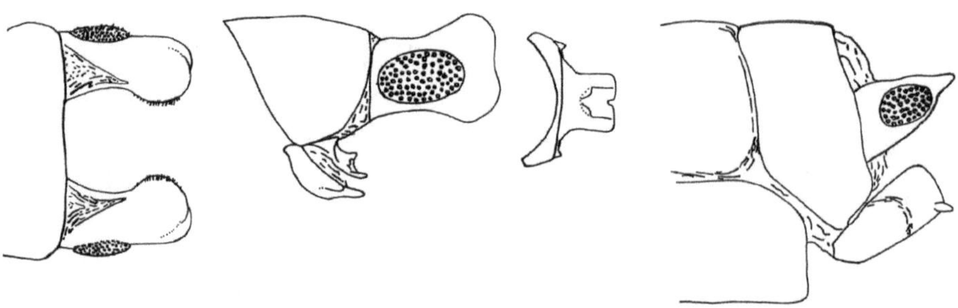

Fig. 4.5 *Protochauliodes cinerascens* (left to right): male genitalia in dorsal and lateral view, aedeagus in dorsal view, and the apex of the female abdomen in lateral view. Based on Flint (1973).

Key to the South American Genera of Larvae in the Family Corydalidae

Information for the key is taken from Bachmann (1995). The larva of *Nothochauliodes* Flint, 1983, has not been described.

1. The first seven abdominal segments have conspicuous clusters of filmentous lateral gills at the base of a conical tubercle (Fig. 2.24) 2
   – Clusters of gill filaments at the bases of conical tubercles are not present on the abdomen. Large single gill filaments may be present (Fig. 4.6) 3

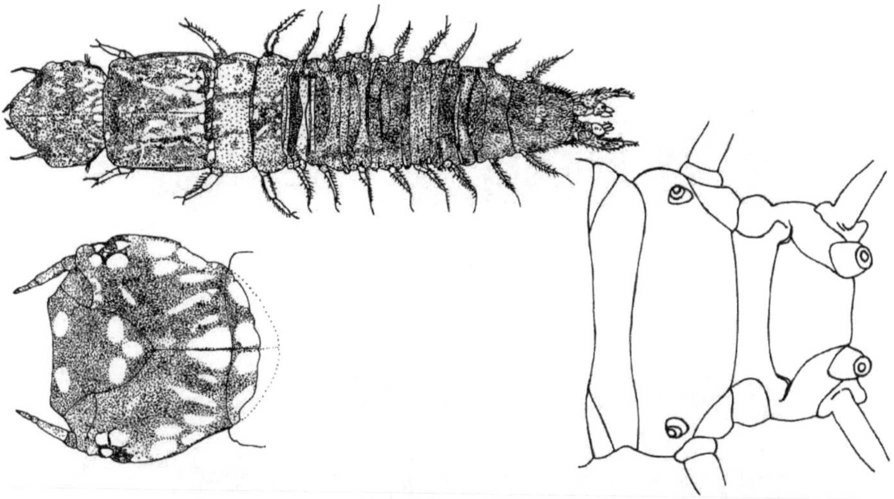

Fig. 4.6 Habitus of a larva of *Archichauliodes* sp., presumably from Chile (upper left to center), head in dorsal view (lower left), and the seventh and eighth segments of the abdomen in dorsal view (right). Based on Flint (1973).

2. (1) The head and pronotum are nearly uniform in color dorsally, but the pronotum may also have a few well defined dark markings (Fig. 2.24) .... *Chloronia* Banks, 1908..p. 85
   – The head and pronotum have irregular dark markings dorsally, which are almost always poorly defined (Fig. 4.7). .... *Corydalus* Latreille, 1802..p. 93
3. (1) The head is rounded laterally. The muscle scars on the head are pale and distinct. The abdominal spiracles are large (Fig. 4.6). .... *Archichauliodes* van der Weele, 1909..p. 114
   – The head has nearly straight lateral margins (Fig. 4.8). The muscle scars are dark. The spiracles are small. .... *Protochauliodes* van der Weele, 1909..p. 115

4.2 Subfamily Corydalinae

Key to the Species of Adult Male *Chloronia* in South America

The key was prepared with information found in Navás (1925, 1928a, 1934a), Penny and Flint (1982) and Contreras-Ramos (1995, 1999, 2000, 2002, 2004a, 2006b), Thouvenot (2008).

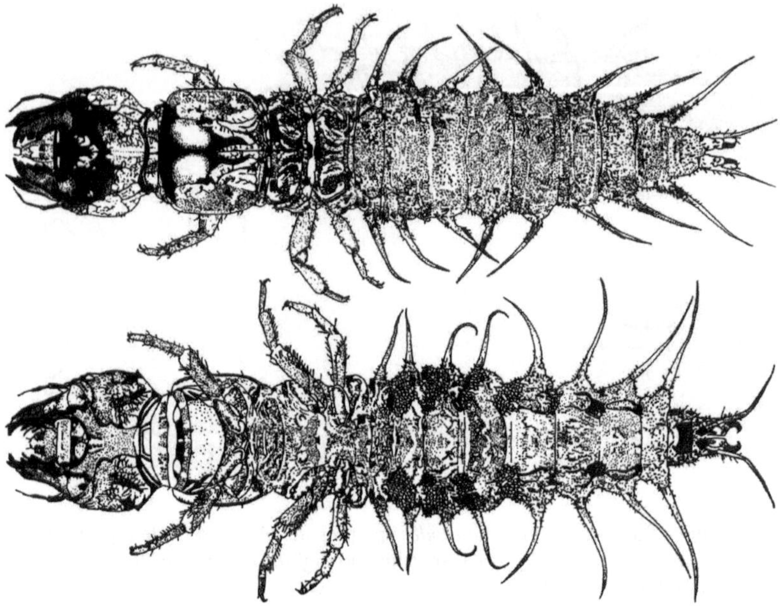

Fig. 4.7 Habitus of a larva of *Corydalus ignotus* in dorsal (above) and ventral view (below). Based on Azevêdo and Hamada (2007).

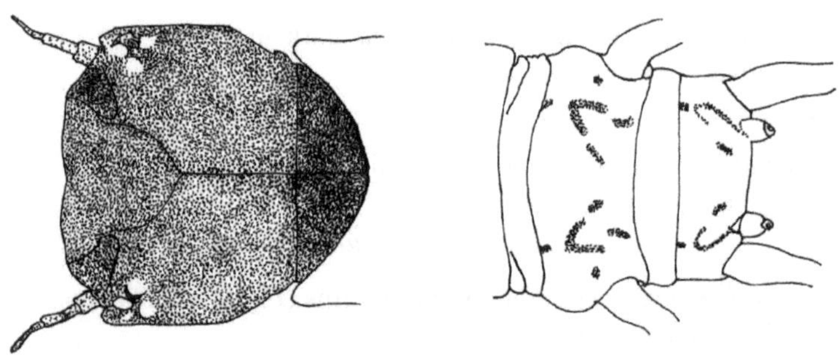

Fig. 4.8 *Protochauliodes* sp. larva, presumably from Chile: head in dorsal view (left), and seventh and eighth segments of the abdomen in dorsal view (right). Based on Flint (1973).

1. The head is uniformly colored or has infuscated lateral margins. There are lateral membranous pouches between the eighth and ninth segments of the male. The ninth tergite is divided into two broad, triangular plates. The ninth sternite of the male is almost rectangular (Fig. 4.9). The length is at least 32 mm 2
   – The posterior part of the head has one or two pairs of fuscous spots (Fig. 4.10) 3

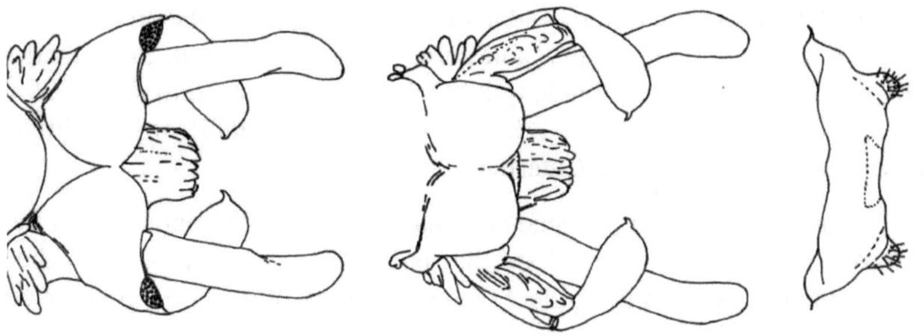

Fig. 4.9 *Chloronia plaumanni* male (left to right): apex of the abdomen in dorsal (center) and ventral view, and tenth gonocoxites and gonostyli in ventral view. Based on Penny and Flint (1982).

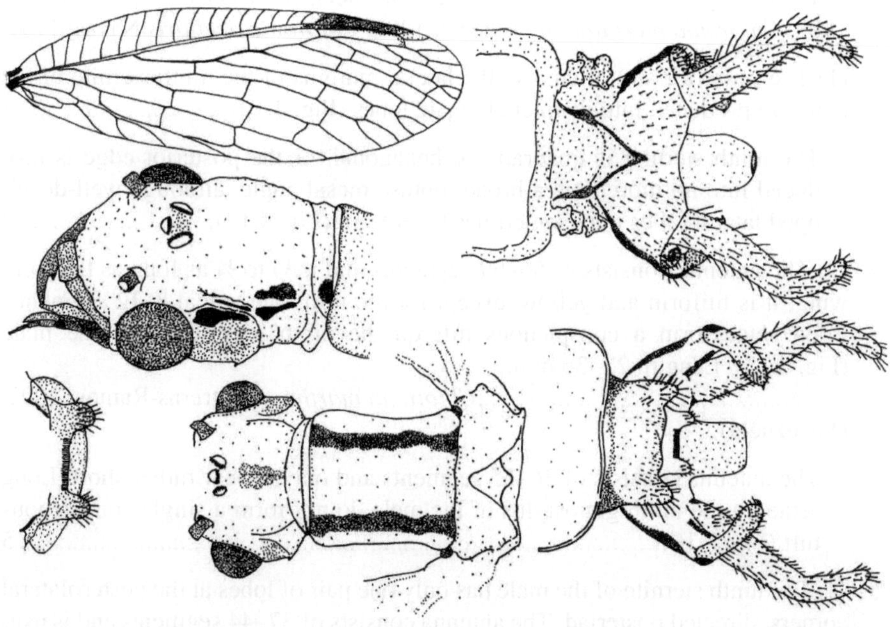

Fig. 4.10 *Chloronia pennyi* male: fore-wing (upper left), head in dorsolateral view (middle left), posterior part of head and anterior part of the thorax in dorsal view (lower center), tenth sternite in ventral view (lower left), and apex of the abdomen in dorsal (upper right) and ventral view (lower right). Based on Contreras-Ramos (2000).

2. (1) The lateral margin of the head is fuscous. The lateral lobe on the tenth gonostylus is small, rounded, hairy, and borne on the very broad, fused tenth gonocoxites (Fig. 4.9). Fore-wing length of male: 35–42 mm; female: 37–39 mm. ....*Chloronia plaumanni* Penny and Flint, 1982 (Rio Grande do Sul, Santa Catarina).

   – The lateral margin of the head is unmarked. The tenth gonostylus of the male has a small, hairy lateral lobe that is barely differentiated from the fused tenth gonocoxites (Fig. 4.11). Fore-wing length of male: 32–39 mm; female: 36–50 mm.
   ....*Chloronia corripiens* (Walker, 1858) (Santa Catarina, Paraná, Espirito Santo, Rio de Janeiro, Sao Paulo, Minas Gerais). Syn: *Hermes corripiens* Walker, 1858; *Corydalis livida* Hagen, 1861 *nomen nudum; Neuromus corripiens* (Walker, 1858) MacLachlan, 1869; *Corydalis affinis* Hagen, 1861 (nec Burmeister); *Neuromus winthemi* Davis, 1903, *Chloronia meridionalis* Wheele, 1909; *Chloronia ocellaris* Navás, 1934.

3. (1) The posterior and often also the lateral margins of the ninth sternite of the male are produced into distinct elongate lobes (Fig. 4.10) 4

   – The ninth sternite is quadratic or hexagonal, or the posterior edge is produced into no more than a broad, obtuse mesal angle, although well-developed lateral lobes may sometimes be present (Fig. 4.12)............................ 8

4. (3) The antenna consists of 56–69 segments and is ½ to ¾ as long as the forewing; it is filiform and yellow, except for the infuscate apical 8–12 segments. Long setae form a conspicuous tuft on the ninth gonostylus of the male (Fig. 4.13). Length: 27–36 mm.
   ....*Chloronia marthae* Contreras-Ramos, 2002 (Venezuela).

   – The antenna consists of 36–55 segments and may appear rather short. Long setae on the ninth gonostylus of the male do not form a single conspicuous tuft (Fig. 4.10).. 5

5. (4) The ninth sternite of the male has only one pair of lobes at the posterolateral corners, directed posteriad. The antenna consists of 37–44 segments and is usually entirely yellow except for the infuscate terminal segment. The pronotum is marked with two fuscous, longitudinal stripes (Fig. 4.10). Fore-wing length of male: 27.4–29.4 mm; female: 34.3–34.7 mm.
   ....*Chloronia pennyi* Contreras-Ramos, 2000 (Rondônia, Amazonas).

   – The ninth sternite has a pair of lateral lobes and sometimes also a pair of posterior lobes (Fig. 4.14).. 6

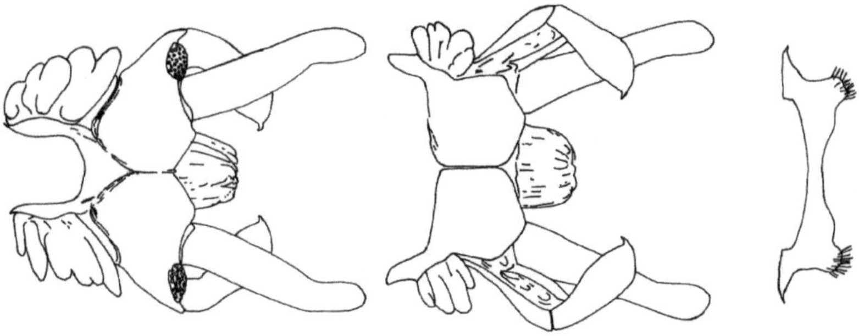

Fig. 4.11 *Chloronia corripiens* male (left to right): apex of the abdomen in dorsal (center) and ventral view, and tenth gonocoxites and gonostyli in ventral view. Based on Penny and Flint (1982).

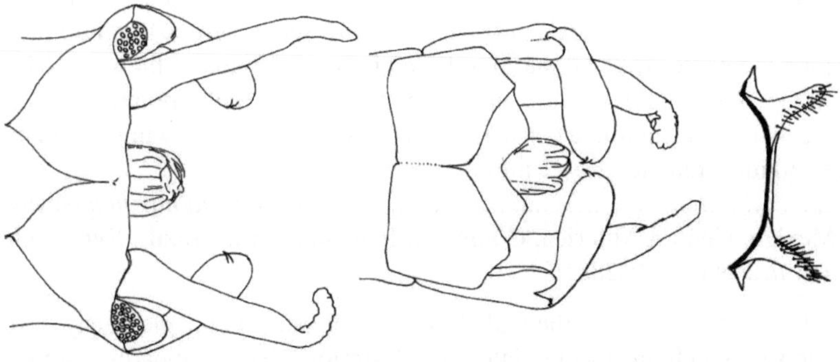

Fig. 4.12 *Chloronia banksiana* male (left to right): apex of the abdomen in dorsal (center) and ventral view, and tenth gonocoxites and gonostyli in ventral view. Based on Penny and Flint (1982).

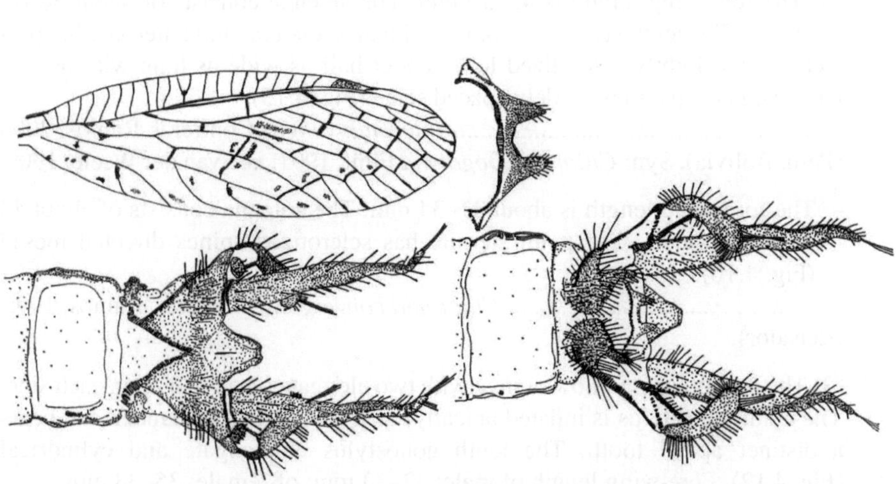

Fig. 4.13 *Chloronia marthae* male: fore-wing (upper left), tenth sternite (upper center), and the apex of the abdomen in dorsal (lower left) and ventral view (right). Based on Contreras-Ramos (2002).

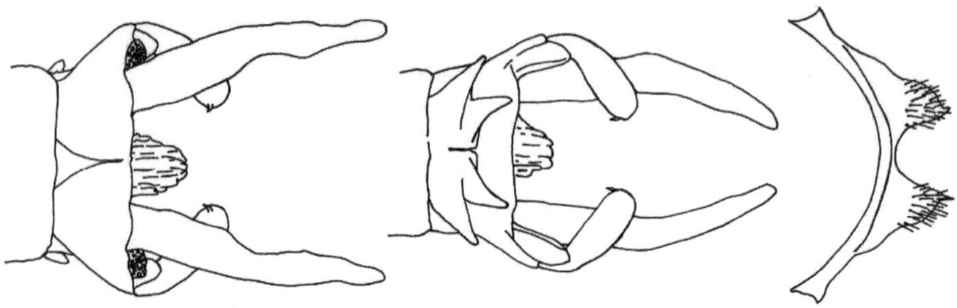

Fig. 4.14 *Chloronia mirifica* male (left to right): apex of the abdomen in dorsal (center) and ventral view, and tenth gonocoxites and gonostyli in ventral view. Based on Penny and Flint (1982).

6. (5) The ninth sternite of the male has one pair of posterior and one pair of lateral lobes, both of which are conspicuous. Long setae form a longitudinal fringe along the posteroventral margin of the ninth gonostylus of the male. The lobes on the male tenth sternite are triangular and acutely pointed at the apex (Fig. 4.14). The antenna consists of 36–55 segments. Fore-wing length of male: 29–40 mm; female: 25–42 mm.
   ....*Chloronia mirifica* Navás, 1925 (Mexico, Central America, Colombia, Ecuador, Peru, Brazil). Syn.: *Chloronia hieratica* Navás, 1928a.

   – The ninth sternite of the male has only one pair of conspicuously developed lobes, which are located laterally. Posterior lobes, if apparent, are no more than moderately developed (Fig. 4.15) 7

7. (6) The fore-wing length is 42–50 mm. The antenna consists of about 38–45 segments. The tenth sternite is convex with anterolateral and anteromedial projections and lightly sclerotized lobes about half as wide as long with narrow bases and conspicuous, widely spaced setae (Fig. 4.15).
   ....*Chloronia yungas* Contreras-Ramos, 2006 (Peru, Bolivia). Syn: *Chloronia bogotana* Flint, 1991, *nec* van der Weele, 1909.

   – The fore-wing length is about 31–34 mm. The antenna consists of about 45 or 46 segments. The tenth sternite has sclerotized spines directed mesad (Fig. 4.16).
   ....*Chloronia convergens* Contreras-Ramos, 1995 (Ecuador).

8. (3) The pronotum is yellow orange with two elongate black marks on each side. The ninth gonostylus is inflated apically; its apex is broadly rounded and bears a distinct apical tooth. The tenth gonostylus is elongate and cylindrical (Fig. 4.12). Fore-wing length of male: 32–33 mm; of female: 35–38 mm.
   ....*Chloronia banksiana* Penny and Flint, 1982 (Venezuela). Syn: *Chloronia bogotana* Banks, 1943 *nec* van der Weele.

   – The pronotum is pale yellow with two anterior and two posterior fuscous spots (Fig. 4.17)..9

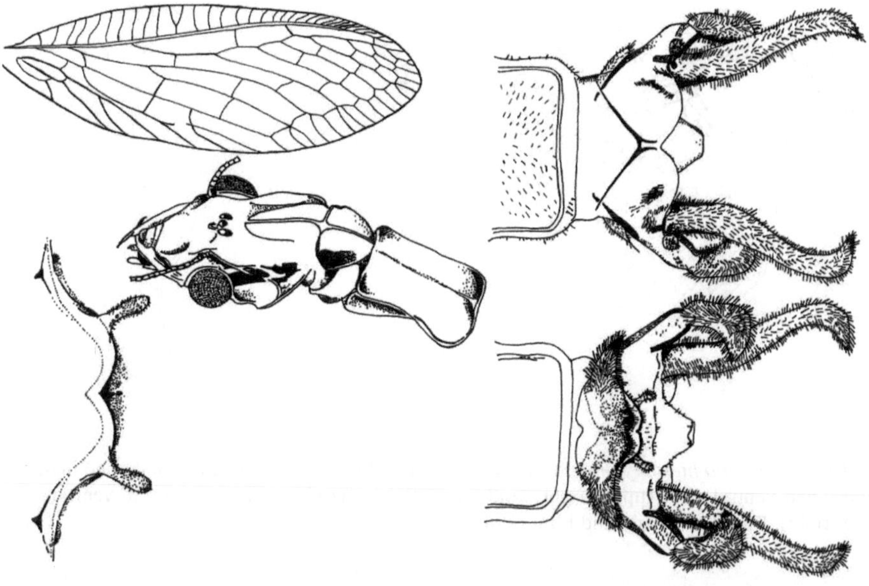

Fig. 4.15 *Chloronia yungas* male: fore-wing (upper left), head and prothorax in dorsolateral view (left center), apex of the abdomen in dorsal (upper right) and ventral view (lower right), and the tenth abdominal sternite (lower left). Based on Contreras-Ramos (2006b).

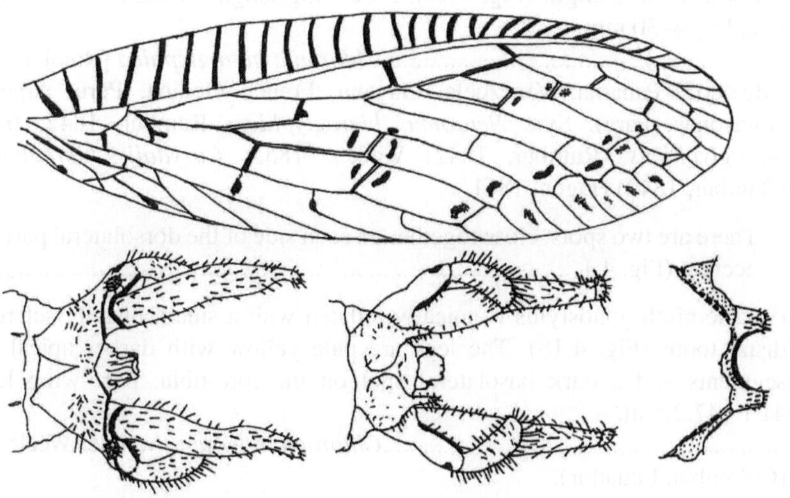

Fig. 4.16 *Chloronia convergens* male: fore-wing (above), and (below, left to right): apex of the male abdomen in dorsal and ventral view and the tenth sternite. Based on Contreras-Ramos (1995).

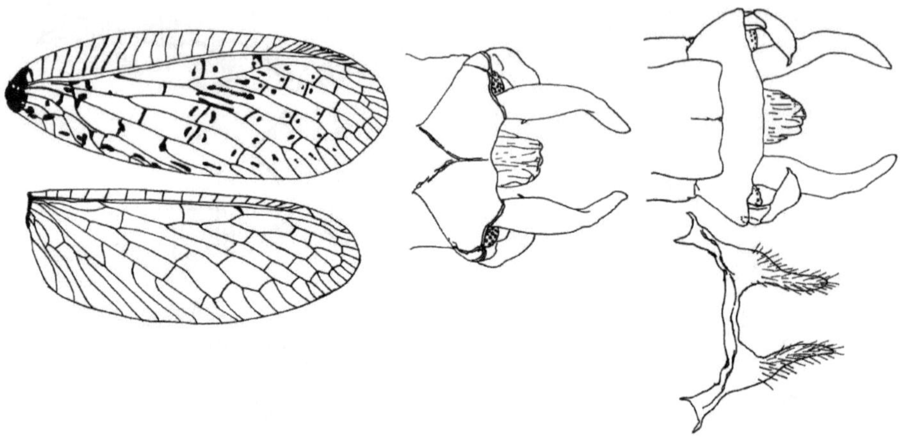

Fig. 4.17 *Chloronia hieroglyphica:* fore and hind wing (left), apex of the male abdomen in dorsal (center) and ventral view (upper right), and the tenth gonocoxites and gonostyli in ventral view (lower right). Based on Penny and Flint (1982).

9. (8) There is one dark spot on each side of the dorsolateral part of the occiput. The pronotum is marked with four fuscous spots, an anterior and a posterior pair. The legs are uniform pale yellow. The ninth gonostylus is distinctly narrowed and produced into a sharp apical tooth. The tenth gonostylus is shouldered at its midlength (Fig. 4.17). Fore-wing length of male: 23–25 mm; of female: 24–30 mm.
   ....*Chloronia hieroglyphica* (Rambur, 1842) (Mexico?, Panama, Venezuela, Guyana, French Guiana, Peru, Argentina, Amazonas, Pará). Syn: *Neuromus hieroglyphicus* Rambur, 1842; *Hermes hieroglyphicus* (Rambur, 1842) Walker, 1853; *Corydalis hieroglyphicus* (Rambur, 1842) Hagen, 1861.
   - There are two spots, close together on each side of the dorsolateral part of the occiput (Fig. 4.1.. 10
10. (9) The ninth gonostylus is apically inflated with a small, heavily sclerotized distal tooth (Fig. 4.18). The legs are pale yellow with darker apical tarsal segments and a dark basolateral spot on the fore-tibia. Fore-wing length: 41.1 –47.2 mm.
   ....*Chloronia bogotana* van der Weele, 1909 (Colombia, Ecuador).
   - The ninth gonostylus is fusiform and curves mesad; it has a convex anterior margin and a sharp apical point (Fig. 4.1). Fore-wing length of male: 28.2–32.3 mm; female: 32.5–38.3 mm.
   ....*Chloronia gaianii* Contreras-Ramos, 2002 (Venezuela).

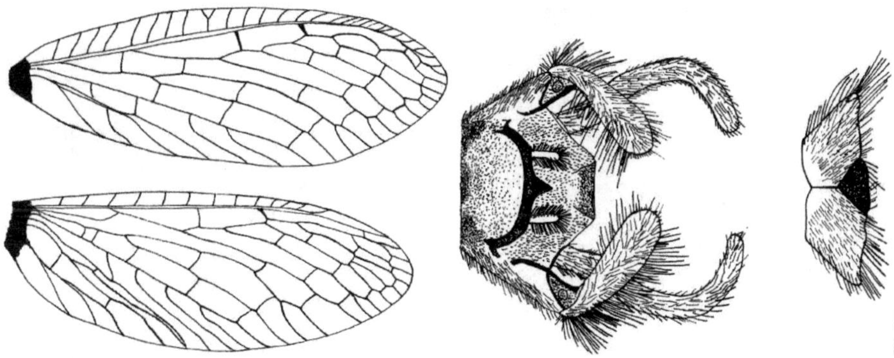

Fig. 4.18 *Chloronia bogotana* male (left to right): fore (above) and hind wing (below), apex of the abdomen in ventral view, and the ninth abdominal tergites in ventral view. Based on Penny and Flint (1982).

Also fitting into the last coupet of the key are four species not yet reported from South America: *C. mexicana* Stitz, 1914, and *C. pallida* (Davis, 1903) from Mexico; *C. gloriosoi* Penny and Flint, 1982, from northern Panama; *C. antilliensis* Flint, 1970, from Dominica and perhaps Guadeloupe. Penny and Flint (1982) suggested that *C. pallida* may be a color variety of *C. mexicana*.

Key to the Species of Adult *Corydalus* Known from South America

The key was prepared with information found in Navás (1915a, 1920, 1928b, 1934b, 1936), Penny (1982), Contreras-Ramos (1998, 2002, 2004a, b, 2011), Thouvenot (2008), and Contreras-Ramos and von der Dunk (2010). The features in the key are usually those of the male. Two names lack descriptions and have not yet been matched with species later described: *Corydalus illota* Hagen, 1861, *nomen nudum*; *Corydalus armigera* Hagen, 1861, *nomen nudum*. They appear in lists but have no taxonomic significance because Hagen failed to provide descriptions.

1. The head of the female is not flattened but robust. It is uniform in color, except for infuscated lateral stripes along sclerotized dorsolateral ridges. The antenna is slender, pale yellow, and infuscate at the apex. There is a well-developed sternal pouch visible between the sixth and seventh abdominal sternites (Fig. 4.19). Fore-wing length of female: 55.6–59.2 mm. The male has not been described.
....*Corydalus amazonas* Contreras-Ramos, 1998 (Amazonas, Rondônia).

– The head is flattened (Fig. 4.20) 2

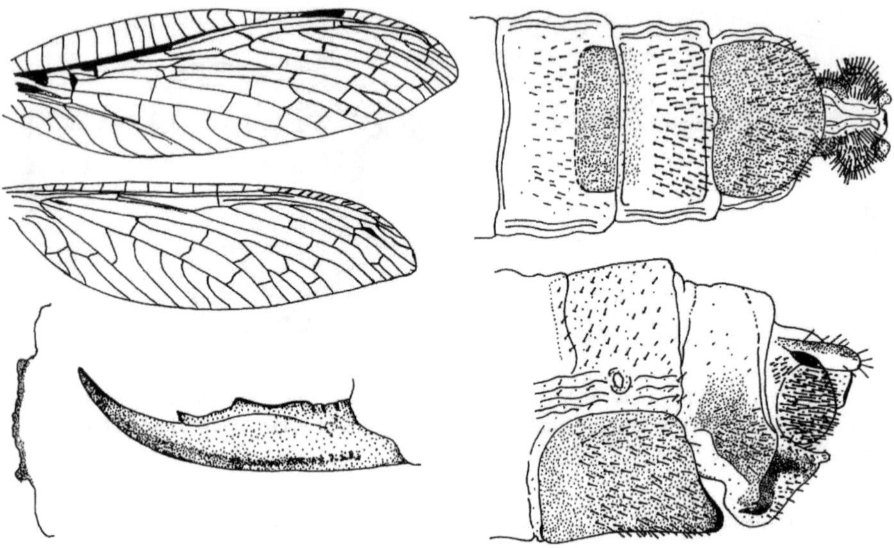

Fig. 4.19 *Corydalus amazonas* female: fore and hind wing (upper and middle left), anterior margin of the clypeus (lower left), mandible (lower left center), and apex of the abdomen in ventral (upper right) and lateral view (lower right). Based on Contreras-Ramos (1998).

2. (1) The tenth tergite of the male is either short and broad, or it is tubular and bent at an angle of about 90°. The lobes of the tenth sternite are reduced and inconspicuous, or flattened and fused at their apical ends to the tenth sternite (Fig. 4.20 3

   – The tenth tergite may be elongate, but it is never sharply bent. The lobes of the tenth sternite are digitiform or papillate, but they are never plate-like or fused apically with the sternite (Fig. 4.21) 7

3. (2) The head has a conspicuous brown and yellow pattern. There are large basal protrusions on the mandibles. There is only a trace of a postocular spine. The fore-wing is strongly spotted on a translucent background, but the stigma is not distinct. Black costal cross veins make the area proximal to the stigma resemble a stigma. The ninth gonostyli are subclavate, and the internal apodeme shows no outgrowth. The tenth sternite is broadly conical; it bears a small ventral protrusion at the base but lacks dorsal processes (Fig. 4.20). Forewing length: 43.6–44.8 mm.
   .... *Corydalus flinti* Contreras-Ramos, 1998 (Venezuela).

   – The head is almost uniform in color (Fig. 4.22) 4

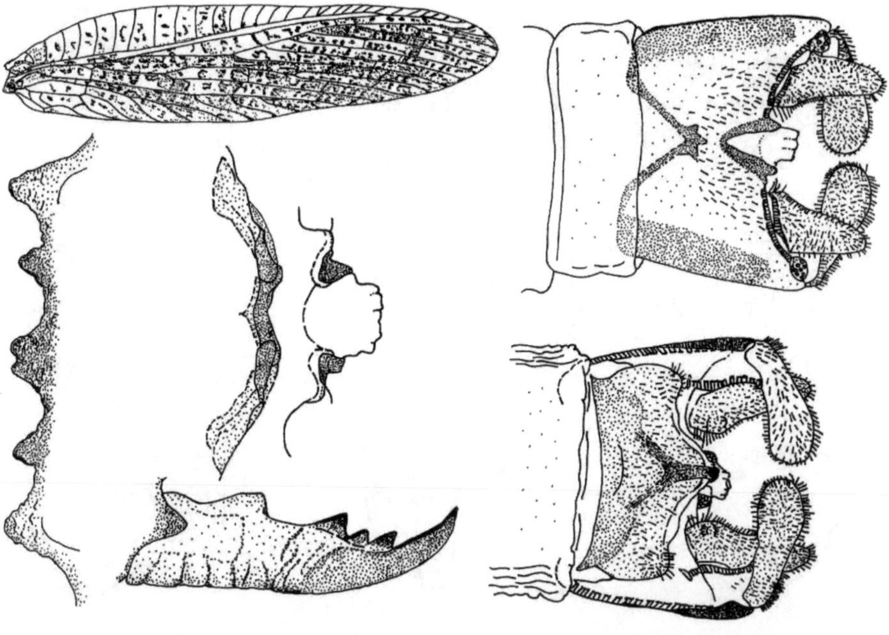

Fig. 4.20 *Corydalus flinti* (male): fore-wing (upper left); anterior margin of the clypeus (lower left); mandible (lower left center); tenth sternite with tubercle and the eleventh sternite, which is ring-like, without showing the lateral connections between them (left center); apex of the abdomen in dorsal (upper right) and ventral view (lower right). Based on Contreras-Ramos (1998).

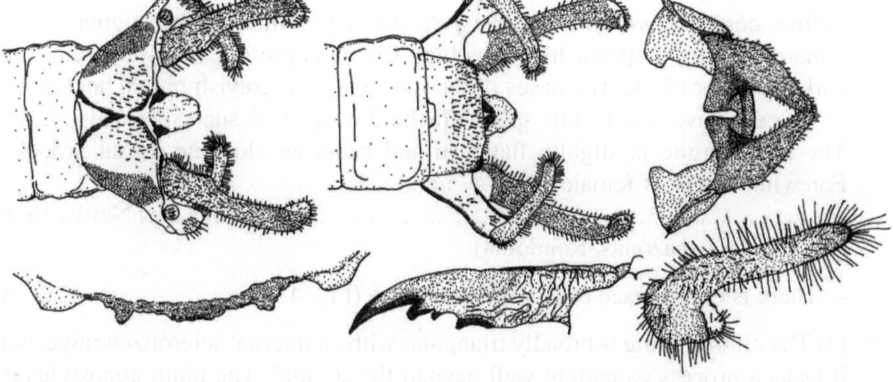

Fig. 4.21 *Corydalus crossi* (above, left to right): apex of the male abdomen in dorsal and ventral view, tenth sternite of a male, and (below, left to right): the outer margin of the clypeus of a male with front facing downward, the mandible of a female viewed from the right, and the ninth gonostylus of the male in posterior view. Based on Contreras-Ramos (2002).

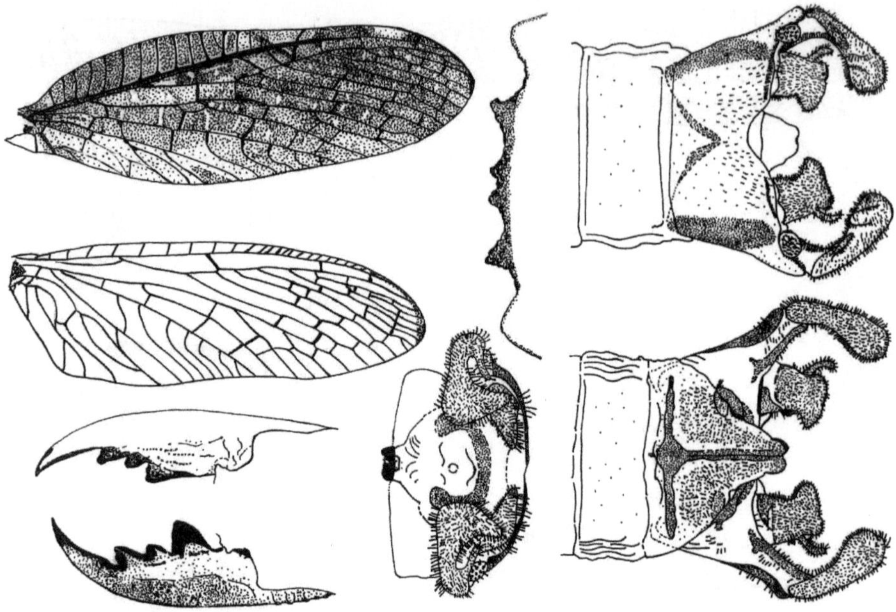

Fig. 4.22 *Corydalus arpi:* fore and hind wing of a female (upper and middle left), mandible of a male (lower middle left) and a female (lower left), anterior margin of the clypeus (upper center), male genitalia in posterior view (lower center), and the apex of the male abdomen in dorsal (upper right) and ventral view (lower right). Based on Penny 1982 and Contreras-Ramos (1998).

4. (3) There is a well-developed postocular spine with its apex clearly separated from the margin of the head. There is no basal protrusion on either mandible. Yellow costal cross veins form a pale patch proximal to the stigma of the female, making it appear like a doubled stigma is present. All other veins are dark brown or black. The bases of the fore-wing are grayish brown, and many of the cells have small white spots. The hind wings lack such spots (Fig. 4.22). The tenth tergite is slightly flattened and bears an elongate dorsal process. Forewing length of female: 37.5–47 mm. .... *Corydalus arpi* Navás, 1936 (Venezuela, Amazonas, Rondônia).

   – There is only a trace of a postocular spine (Fig. 4.23) 5

5. (4) The ninth sternite is broadly triangular with an internal sclerotized ridge, but it lacks a process extending well beyond the sternite. The ninth gonostylus is tubular. The tenth tergite is strongly curved and has a broad process directed ventrad (Figs. 1.9, and 4.23). Fore-wing length of male: 42.4–47.5 mm. .... *Corydalus mayri* Contreras-Ramos, 2002 (Venezuela).

   – The ninth sternite bears a long medial process at its apex, which extends between the gonostyli (Fig. 4.24) 6

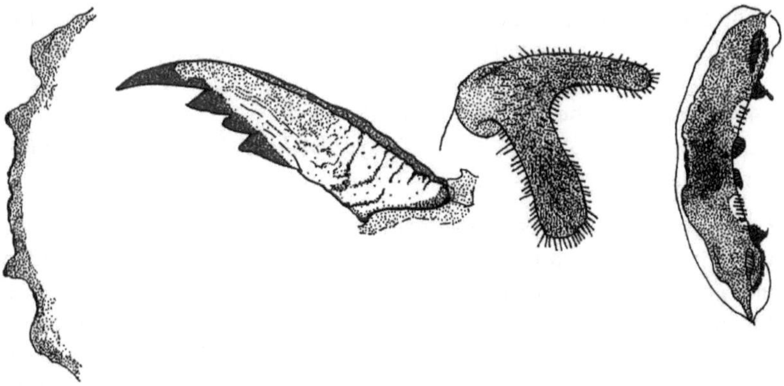

Fig. 4.23 *Corydalus mayri* male (left to right): anterior margin of the clypeus, right mandible, tenth tergite in dorsoposterior view, and the tenth sternite. Based on Contreras-Ramos (2002).

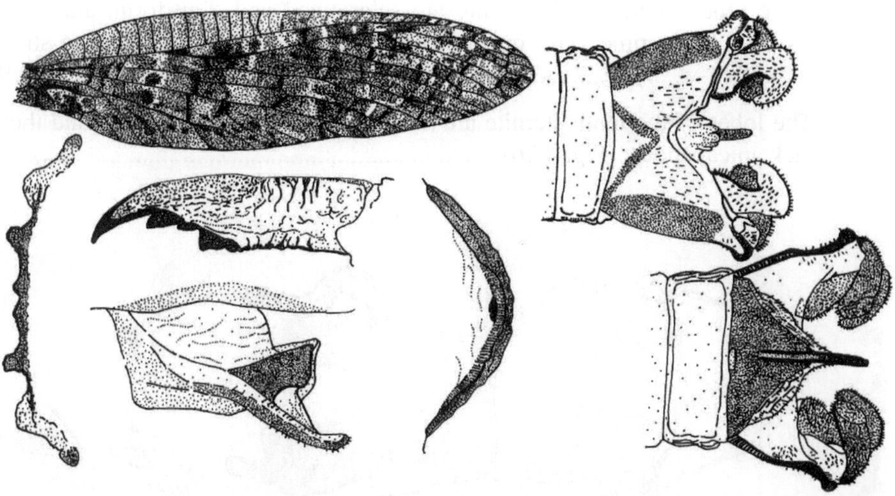

Fig. 4.24 *Corydalus hayashii* male: fore-wing (upper left), anterior margin of the clypeus (lower right), mandible (left of center), ninth sternite in lateral view (lower left center), tenth sternite (lower center), and the apex of the abdomen in dorsal (upper right) and ventral view (lower right). Based on Contreras-Ramos (2002).

6. (5) The long medial process of the ninth sternite broadens beyond its midlength to form a rhomboid apex. The tenth tergite forms a distinct angle of about 90°. The lobes of the tenth sternite are flattened, directed ventrad, and fused to the sternite at their apices (Fig. 4.3). Fore-wing length: 37.7–41.3 mm. ....*Corydalus ignotus* Contreras-Ramos, 1998 (French Guiana, Brazil).

– The long medial process on the ninth sternite is narrow and spine-like. The tenth tergites curve sharply anterodorsad. The fore-wing is subelliptical and has two areas resembling pterostigmas (Fig. 7.24). Fore-wing length: c. 45.6 mm. Only the male has been described.
....*Corydalus hayashii* Contreras-Ramos, 2002 (Venezuela).

7. (2) The sclerotized lobes on the tenth sternite of the male are elongate tranguloid with the apical parts bent mediad so that the apices contact each other. The ninth gonostylus is shorter and narrower than the tenth tergite (Fig. 4.21). The filiform antenna consists of about 72 segments. The length of the fore-wing is about 57 mm. The general color is dark reddish brown with orange markings.
....*Corydalus crossi* Contreras-Ramos, 2002 (Venezuela).

– The sclerotized lobes on the tenth sternite of the male are not elongate tranguloid, or their apices are not held in contact (Fig. 4.25).............................. 8

8. (7) The lobes of the tenth sternite are sclerotized and digitiform, and each bears an apical spine. The mandibles of the male are always rather short (Fig. 4.25)... 9

– The lobes of the tenth sternite are short and broad or papilliform, and they lack apical spines (Fig. 4.26)... 10

Fig. 4.25 *Corydalus cephalotes:* anterior margin of the male clypeus (left), mandible of a male (lower left center) and those of two females (lower center and right center), apex of the male abdomen in dorsal (upper left center) and ventral view (upper right center), the male tenth sternite (upper right) and one lobe of the tenth sternite showing the small apical spine (lower right). Based on Contreras-Ramos (1998).

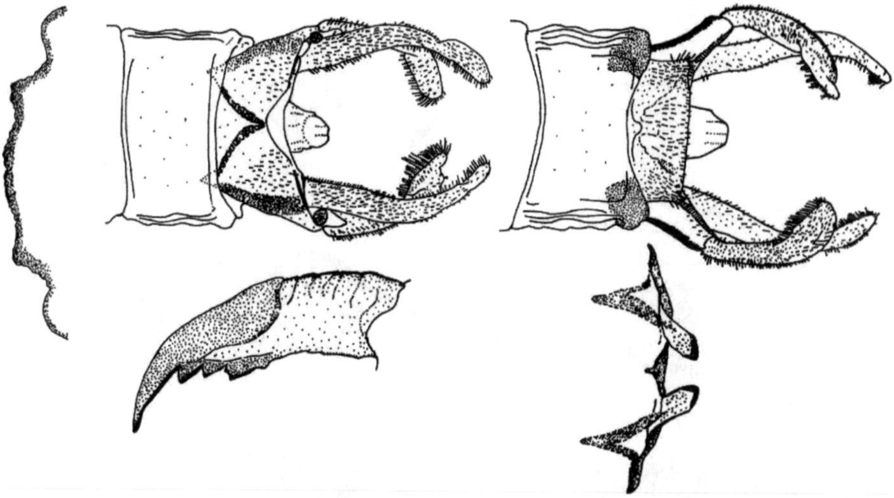

Fig. 4.26 *Corydalus affinis:* anterior margin of the clypeus of a male (left), mandible of a female (lower left center), apex of the male abdomen in dorsal (upper left center) and ventral view (upper right), and tenth sternite (lower right). Based on Contreras-Ramos (1998).

9. (8) In the fore-wing, vein M_{1+2} has three branches, with only its anterior branch forked. The posterior margin of the hind wing is almost straight, and it tapers into a rather narrow apex. There are small white spots in the wing cells, which are indistinct in the fore-wing. A distinct incision is evident in the pterostigmal region of the fore-wing. The antennae are yellow or brown. The dorsovental surface of the tenth tergite is obviously flattened and has a base considerably wider than the inner margin (Fig. 4.25). Fore-wing length of male: 52–57 mm; female 58–68 mm. Color: dark brown with pale brown legs.
....*Corydalus cephalotes* Rambur, 1842 (Colombia, Venezuela, Rio de Janeiro, Mato Grosso). Syn: *Corydalis affinis* Walker, 1853 nec Burmeister, 1839 pars? (also see *Chloronia corripiens*); *Corydalis lutea* Hagen, 1861, *nomen nudum; Neuromus cephalotes* (Rambur, 1842) Davis, 1903; *Corydalus intricatus* Navás, 1921.
   - In the fore-wing, vein M_{1+2} has four branches. The posterior margin of the hind wing is slightly sinuate, and it tapers into a moderately broad apex. The dorsovental surface of the tenth tergite is slightly flattened and has a base that is tubular and parallel-sided (Fig. 4.27). Forewing length: 65–75 mm.
   ....*Corydalus hecate* (MacLachlan, 1866) (Venezuela, Peru, Brasilia DF, Espirito Santo, Minas Gerais, São Paulo). Syn: *Corydalis hecate* MacLachlan, 1866; *Neuromus cephalotes* Davis, 1903 *nec* Rambur; *Corydalus cephalotes* van der Weele, 1910 *nec* Rambur, 1842, pars; *Corydalus raymundoi* Navás, 1920; *Corydalus sallei* Navás, 1920.

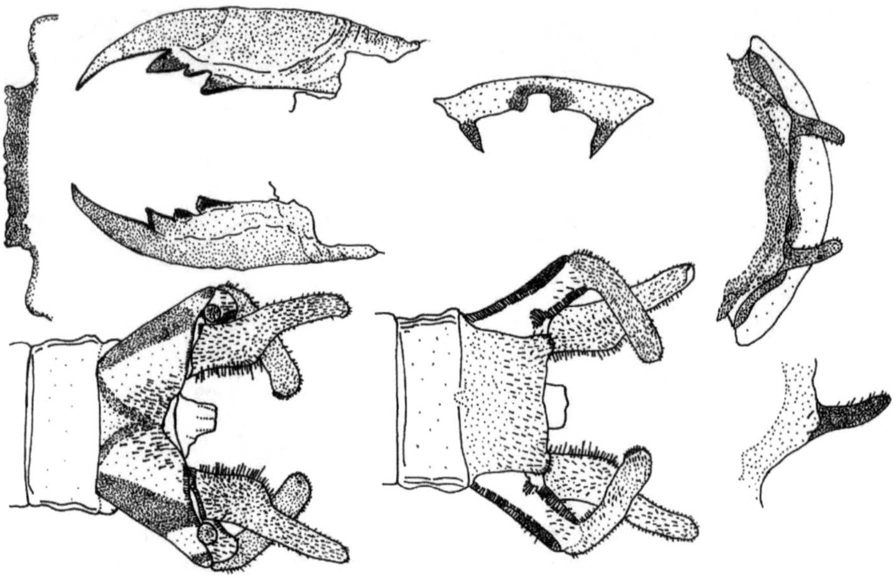

Fig. 4.27 *Corydalus hecate:* anterior margin of the clypeus of a male (left), mandible of a male and female (left center, above and below, respectively), apex of the male abdomen in dorsal (lower left) and ventral view (lower center), and the male tenth sternite in posterior (upper right center) and ventral view (upper right) with one lobe enlarged (lower right). Based on Contreras-Ramos (1998).

10. (8) The tenth tergite and ninth gonostyli are subequal in length; they resemble each other and both have an elongate, tubular shape (Fig. 4.26) 11

 – The tenth tergite is noticeably longer than the ninth gonostyli. The tenth tergite is tubular, and the ninth gonostyli are unguiform or subclavate (Fig. 4.28) 13

11. (10) The head is pale brown with dark brown lateral stripes. The teeth on the long mandibles of the male are reduced in size and number. The tenth tergite and ninth gonostyli are slender and elongate, densely covered by setae, and subequal in shape and length (Figs. 1.3, and 4.26). Length of fore-wing: 37–62 mm.
 .... *Corydalus affinis* (Burmeister, 1839) (Colombia, Ecuador, Peru, Venezuela, French Guiana, Guyana, Argentina, Bolivia, Paraguay, Acre, Rondônia, Amapá, Amazonas, Roraima, Pará, Bahía, Espirito Santo, São Paulo, Santa Catarina, Rio Grande do Sul, Mato Grosso). Syn: *Corydalis affinis* Burmeister, 1839: *Corydalis ancilla* Hagen, 1861 *nomen nudum*; *Corydalus affinis* Hagen, 1861 *nomen nudum; Corydalis nubilus* van der Weele, 1910 *nec* Erichson 1848 auctt.; *Corydalus* sp. Glorioso, 1981; *Corydalus* spec. nov. Geijskes, 1984.

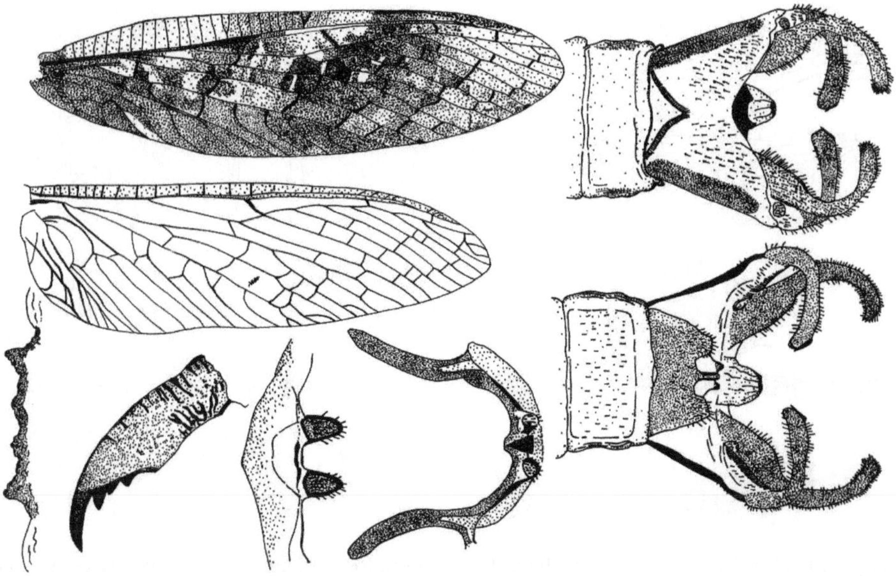

Fig. 4.28 *Corydalus clavijoi:* fore and hind wing of a female (upper and middle left), apex of the male abdomen in dorsal view (upper right), and (below, left to right): anterior margin of the clypeus, mandible of a female, tenth sternite of a male in posterior view with the tubercle retracted and in ventral view, and the apex of the male abdomen in ventral view. Based on Contreras-Ramos (2002).

– The head is dark brown with a faint or a strong pattern of markings but without lateral stripes. The teeth on the long mandibles of the male are distinctly developed. The tenth tergite and ninth gonostyli are different in shape, with the tergite slender, elongate, and narrowing toward the apex, and the gonostyli are slender, elongate, and sub-cylindrical (Fig. 4.27) 12

12. (11) The color pattern on the head is faint to moderate. The antennae are moderately long and thick. The mandibles of the male are long, and each has a conspicuous basal protrusion. The apices of the ninth gonostyli are small and claw-like. The tenth sternite has a narrow medial band and wide, subquadrate, and strongly divergent lobes (Fig. 4.29). Fore-wing length of male: 46.5–47.9 mm; female: 48.6–54.0 mm.
....*Corydalus colombianus* Contreras-Ramos, 1998 (Colombia). Syn: *Corydalus ecuadorianus* Glorioso, 1981 *nec* Banks, 1948; *Corydalus* sp. Contreras-Ramos, 1993.

– The color pattern on the head is strong and distinct on a mainly blackish background. The antennae are short and narrow. The mandibles of the male are short and resemble those of the female. The apices of the ninth gonostyli are not claw-like. The tenth sternite has plate-like lobes, which are subequal

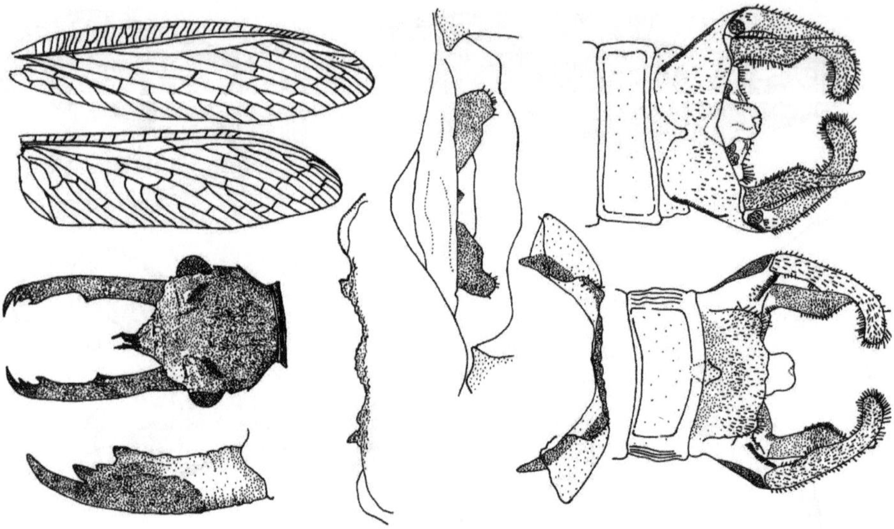

Fig. 4.29 *Corydalus colombianus* (above, left to right): fore and hind wing (upper left, above and below, respectively), head of a male in dorsal view (lower middle left), mandible of a female (lower left), anterior clypeal margin (lower left center), tenth sternite in dorsal (upper center) and ventral view (lower right center), and apex of the male abdomen in dorsal (upper right) and ventral view (lower right). Based on Contreras-Ramos (1998).

in length and width, papilliform, and only slightly divergent (Fig. 4.30). Fore-wing length of male: c. 44.0 mm; female: c. 62.8 mm.
.... *Corydalus ecuadorianus* Banks, 1948 (Ecuador).

13. (10) The ninth sternite of the male has a posteromedial thumb-like projection almost as long at the sternite itself, and internal, V-shaped sclerotized ridges. The ninth gonostyli are elongate and bear well-developed apical or subapical claws (Fig. 4.28) 14

– The ninth sternite of the male is subquadrate, sometimes with a moderately sclerotized posterior margin but always lacking a long projection or internal ridges. The ninth gonostyli are subclavate and not unguiform; they do not bear apical claws (Fig. 4.31) 15

14. (13) The ninth gonostylus is elongate, unguiform, and bears a well-developed apical claw (Fig. 4.32). Fore-wing length: 44.2–62.0 mm.
.... *Corydalus tesselatus* Stitz, 1914 (Colombia, Venezuela). Syn: *Corydalus batesi tesselatus* Stitz, 1914; *Corydalis nubila* Hagen, 1861 *nec* Erichson, 1843; *Corydalus bolivari* Banks, 1943.

– The ninth gonostylus is rounded at the apex and has a preapical claw (Fig. 4.28). Fore-wing length: c. 50.7 mm.
.... *Corydalus clavijoi* Contreras-Ramos, 2002 (Venezuela).

Fig. 4.30 *Corydalus ecuadorianus* male (left to right): anterior clypeal margin, mandible (below) and apex of the male abdomen in dorsal (above) and ventral view, and the tenth sternite. Based on Contreras-Ramos (1998).

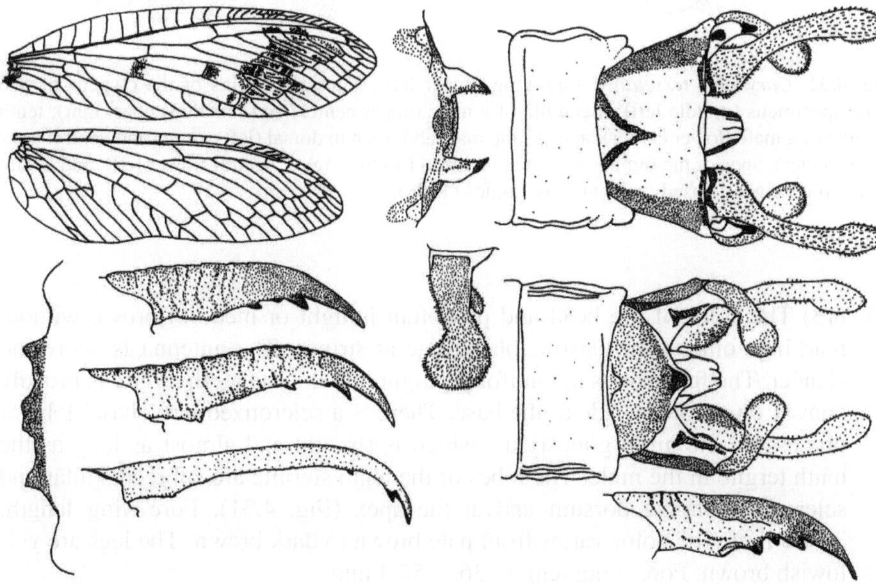

Fig. 4.31 *Corydalus nubilus:* fore and hind wing (upper left, above and below, respectively), anterior margin of the clypeus (lower left), mandibles from three different male specimens (lower left center) and one female (lower right), apex of the male abdomen in dorsal (upper right) and ventral view (middle right), tenth sternite (upper right center), and ninth gonostylus in lateral view (center). Based on Contreras-Ramos (1998).

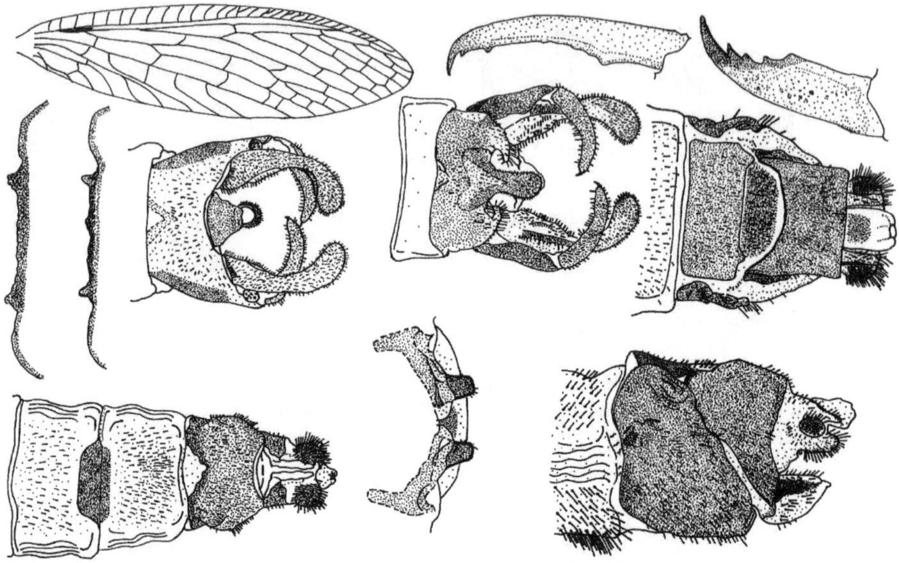

Fig. 4.32 *Corydalus tesselatus:* fore-wing (upper left), anterior margins of the clypeus of two male specimens (middle left); mandible of a male (upper center) and female (upper right); tenth sternite of a male (lower center); apex of the male abdomen in dorsal (left center) and ventral view (right center); apex of the female abdomen in dorsal (middle right), ventral (lower left), and lateral view (lower right). Based on Contreras-Ramos (1998).

15. (13) The color of the head and pronotum is light or medium brown without markings other than narrow, black, lateral stripes. The antenna is short and slender. The fore-wings are uniformly light brown. The tenth tergite is broadly convex on the inner side at the base. There is a sclerotized dorsobasal lobular process on the ninth gonostylus, which is straight and almost as long as the tenth tergite in the male. The lobes of the tenth sternite are subrectangular and sclerotized on the dorsum and at the apex (Fig. 4.31). Fore-wing length: 39–65 mm. The color varies from pale brown to dark brown. The legs are yellowish brown. Fore-wing length: 36.8–57.3 mm. .... *Corydalus nubilus* (Erichson, 1848) (Guyana, French Guiana, Venezuela, Colombia, Peru, Amazonas, Roraima, Pará, Mato Grosso, Rondônia, Acre). Syn: *Corydalis nubila* Erichson, 1848; *Corydalus titschacki* Navás, 1928b; *Corydalus nevermanni* Penny, 1982 *nec* Navás, 1934b.

– The head and pronotum may have markings, but they lack narrow black stripes. The antenna is thick and moderately to very long. The base of the tenth tergite has semiparallel or angled sides. The ninth gonostylus lacks a dorsal process. The lobes on the tenth sternite are papilliform (Fig. 4.33) 16

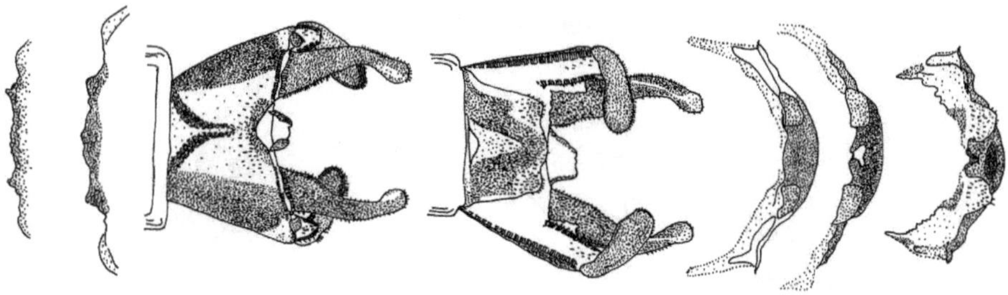

Fig. 4.33 *Corydalus tridentatus* male (left to right): anterior margin of the clypeus of two specimens, apex of the abdomen in dorsal and ventral view, and tenth sternite of three specimens. Based on Contreras-Ramos (1998).

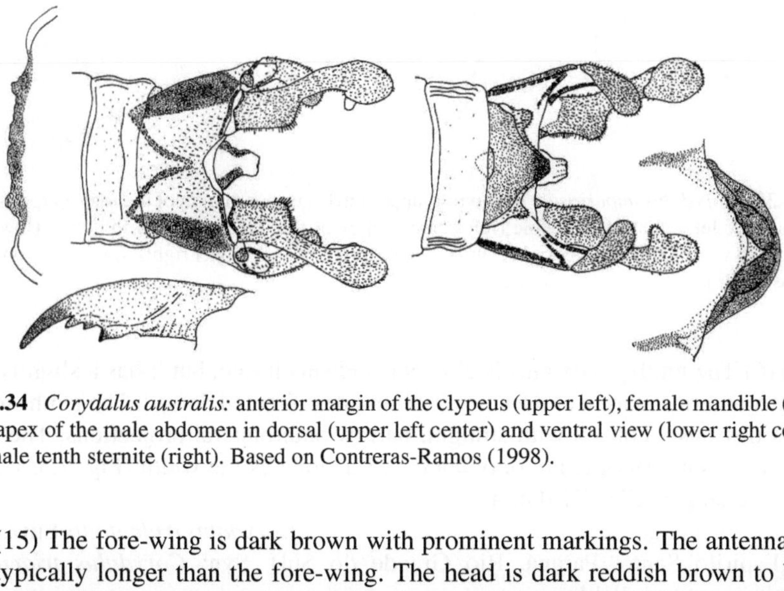

Fig. 4.34 *Corydalus australis:* anterior margin of the clypeus (upper left), female mandible (lower left), apex of the male abdomen in dorsal (upper left center) and ventral view (lower right center), and male tenth sternite (right). Based on Contreras-Ramos (1998).

16. (15) The fore-wing is dark brown with prominent markings. The antennae are typically longer than the fore-wing. The head is dark reddish brown to black with uniform black mandibles. There is no apical invagination of the tenth tergite. The tenth sternite has wide, unsclerotized lobes (Fig. 4.33) 17

    – The wings are medium brown with only weak markings. The antennae are only moderately long. The head and mandibles are typically dark brown. There is an apical invagination on the tenth tergite. The tenth sternite has somewhat elongate lobes, which are only moderately sclerotized (Fig. 4.34). 18

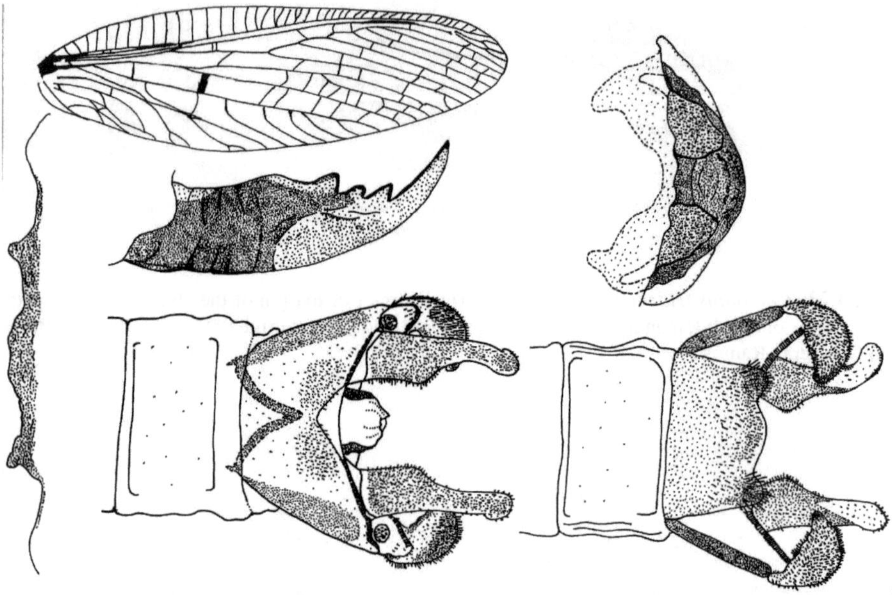

Fig. 4.35 *Corydalus imperiosus:* fore-wing (upper left), anterior margin of the male clypeus (middle to lower left), female mandible (left center), apex of the male abdomen in dorsal (lower left center) and ventral view (lower right), and the male tenth sternite (upper right). Based on Contreras-Ramos (1998).

17. (16) The ninth gonostylus is elongate and subclavate, but it has a slightly convex anterior margin and a narrow. blunt apex directed dorsad. The tenth sternite is obviously convex with straight, narrow anterolateral projections. The lobes on the tenth sternite are only about 2/3 as long as the width (Fig. 7.33). Fore-wing length: 57.6–71.0 mm.
    ....*Corydalus tridentatus* Stitz, 1914 (Espirito Santo, Paraná, Rio Grande do Sul). Syn: *Corydalus tridentatus nigripes* Stitz, 1914.

    – The ninth gonostylus is subclavate with a crenulated anterior margin and a slightly convex posterior border. The tenth sternite is usually strongly convex with a slight dorsoventral prominence. The anterolateral projections are short and wide. The lobes on the tenth sternite have an irregular, subtriangular shape (Figs. 1.2, and 4.35). Fore-wing length: 50–67 mm.
    ....*Corydalus imperiosus* Contreras-Ramos, 1998 (Argentina). Syn: *Corydalus tridentatus* Glorioso, 1981 *nec* Stitz, 1914.

18. (16) The ninth sternite is somewhat produced posteromedially and well sclerotized. The tenth sternite is strongly convex, and its lobes are longer than wide and obviously directed mesad (Fig. 4.34).. 19

    – The ninth sternite is subquadrate and moderately sclerotized. The tenth sternite is weakly or moderately convex, and its lobes are about as long as wide and directed mesad, posteriad, or outward (Fig. 4.36) 20

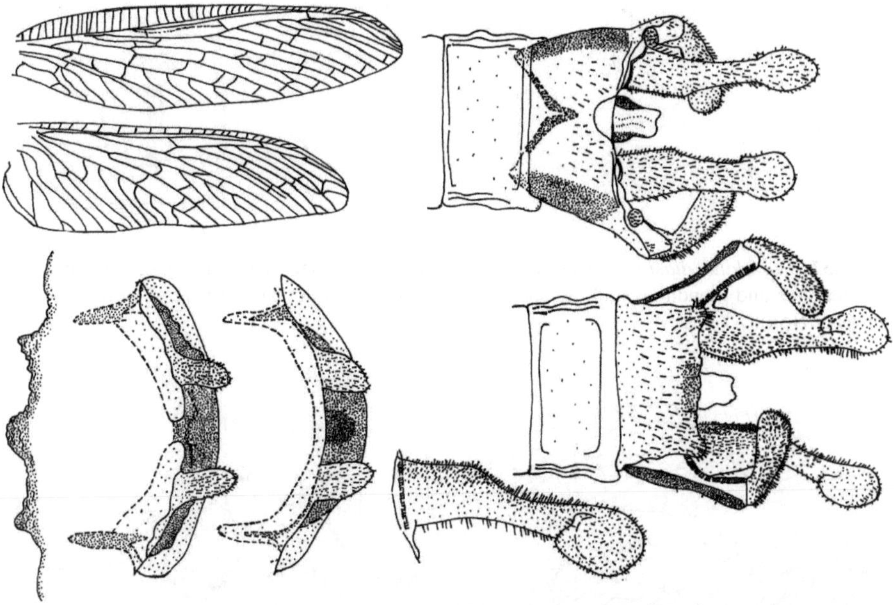

Fig. 4.36 *Corydalus longicornis* (male): fore and hind wing (upper left), anterior margin of the clypeus (lower left), apex of the abdomen in dorsal (upper right) and ventral view (lower right), tenth sternites of two specimens (lower center and left center), and the right tenth tergite in lateral view (lower right center). Based on Contreras-Ramos (1998).

19. (18) The tenth tergite is tubular without forming obvious angles but enlarged dorsoventrally toward the apex; its inner surface at the apex drops into the cavity of the tergite to make it appear like a forceps (Fig. 4.37). Fore-wing length: 45.3 to 66.0 mm.
....*Corydalus diasi* Navás, 1915
(Argentina, Paraguay, Ceará, Bahía, Goiás, Minas Gerais, São Paulo, Rio Grande do Sul). Syn: *Corydalis armigera* Hagen, 1861, *nomen nudum; Corydalis vetula* Hagen, 1861, *nomen nudum*; *Corydalus finoti* Navás, 1921.

   – The tenth tergite forms an obvious angle and curves mesad toward the apex; its inner surface does not appear like a forceps (Fig. 4.34). Length of forewing: 46.8–70.0 mm.
....*Corydalus australis* Contreras-Ramos, 1998
(Argentina, Uruguay, Minas Gerais, Santa Catarina, Rio Grande do Sul). Syn: *Corydalus affinis* van der Weele, 1910 *pars;* Penny, 1977 *pars;* Glorioso, 1981.

20. (18) The tenth tergite abruptly narrows to one third or half of its basal width; the distal 2/3 of its lateral margins diverge outward, and their apices are capitate (Fig. 4.36).. 21

   – The tenth tergite remains tubular without conspicuous narrowing, although it may gradually narrow somewhat, and it does not diverge abruptly toward the apex, which is not capitate (Fig. 4.38).. 23

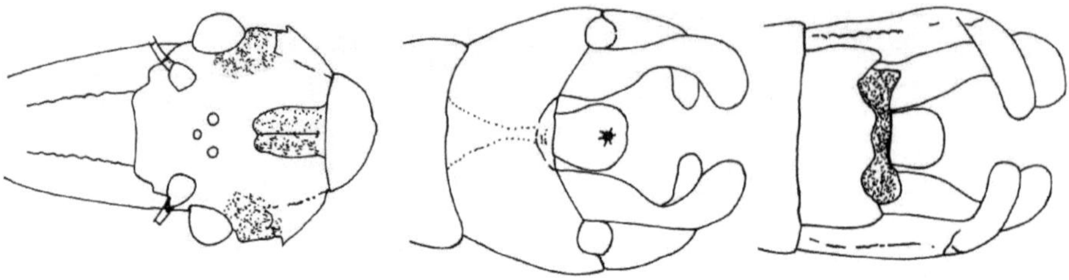

Fig. 4.37 *Corydalus diasi* male (left to right): the head in dorsal view showing only the bases of the antennae and mandibles and the apex of the abdomen in dorsal and ventral view. Based on Navás (1915a).

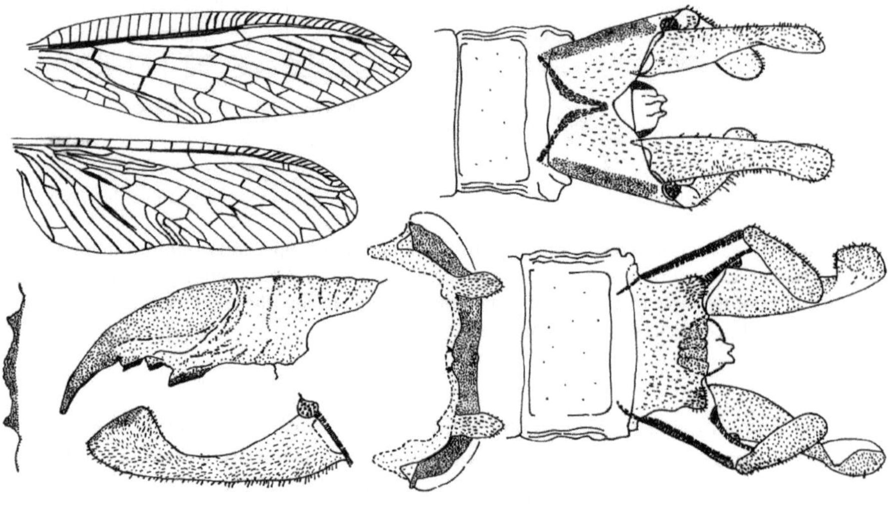

Fig. 4.38 *Corydalus neblinensis:* fore and hind wing (upper and middle left), anterior margin of the clypeus of a male (lower left), mandible of a female (below hind wing), apex of the male abdomen in dorsal (upper right) and ventral view (lower right), male tenth sternite (lower center), and male tenth tergite in lateral view (lower left center). Based on Contreras-Ramos (1998).

21. (20) The fore-wings are semi-translucent grayish or reddish brown, with their membrane evenly covered by small white spots ringed with dark brown or black. The antenna is brown. The mandibles of the male may be short or may be twice as long as the head. The apex of the tenth tergite is enlarged dorsoventrally and is incurved for a short distance. The length and width of the lobes on the tenth sternite are subequal; the lobes are broadly rounded (Fig. 4.39). Fore-wing length: 41.2–64.7 mm.
....*Corydalus primitivus* van der Weele, 1909 (Peru, Bolivia, Paraguay, Argentina, Minas Gerais, Mato Grosso).

– The pattern on the wings consists of large markings and not small white spots ringed with dark brown or black. The apex of the tenth tergite is

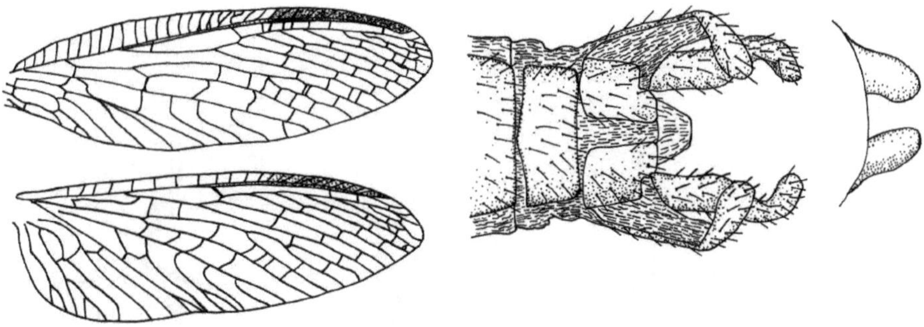

Fig. 4.39 *Corydalus primitivus* male (left to right): venation on the fore and hind wing, genitalia, and gonostyli. Based on Penny (1982).

concave. The lobes on the tenth sternite are slightly narrower than long; they taper bluntly at the apex (Fig. 4.36) 22

22. (21) The head lacks longitudinal stripes. The fore-wings are a nearly uniform pale yellowish brown or dark greenish brown (Fig. 4.36). The mandibles of the male are usually elongate. Fore-wing length: 62.1–74.0 mm. ....*Corydalus longicornis* Contreras-Ramos, 1998 (Ecuador, Bolivia, Argentina).

 – The head has lateral postocular stripes but otherwise lacks a pattern. The fore-wing has a strongly contrasting pattern of brown with irregular pale areas on the distal part of the wing and pale areas in the cells of R_1 and in the adjacent R cell (Fig. 4.40). Length of fore-wing: 47.7–63.0 mm. ....*Corydalus parvus* Stitz, 1914 (Ecuador, Peru). Syn: *Corydalus armatus* Glorioso 1981, *nec* Hagen, 1961.

23. (21) The head is smooth and uniformly dark reddish brown. The antenna is pale yellow with an apical infuscation. The fore-wings are clear and nearly translucent with very dark reddish ground color covering the distal half, parts of the coastal area, the R_1 and adjacent R cell, and the area of the fork of M_{1+2} and M_{3+4}. The ninth gonostylus of the male is incurved, swollen apically, and only half as long as the tenth tergite (Fig. 2.30). Length of fore-wing: 45.8–67.6 mm. ....*Corydalus batesi* (MacLachlan, 1868) (Colombia, Venezuela, Surinam, French Guiana, Guyana, Ecuador, Peru, Bolivia, Amazonas, Pará, Rondônia). Syn: *Corydalis batesii* MacLachlan, 1868.

 – The head is pale brown to dark greenish or grayish brown, sometimes with a pattern or sculpturing. The fore-wings lack the pattern of translucent and dark reddish brown as described in the first part of the couplet 24

24. (23) The tenth tergite lacks an apical concavity. The lobes of the tenth sternite are well separated (Fig. 4.38). The color of the head is pale to dark reddish brown 25

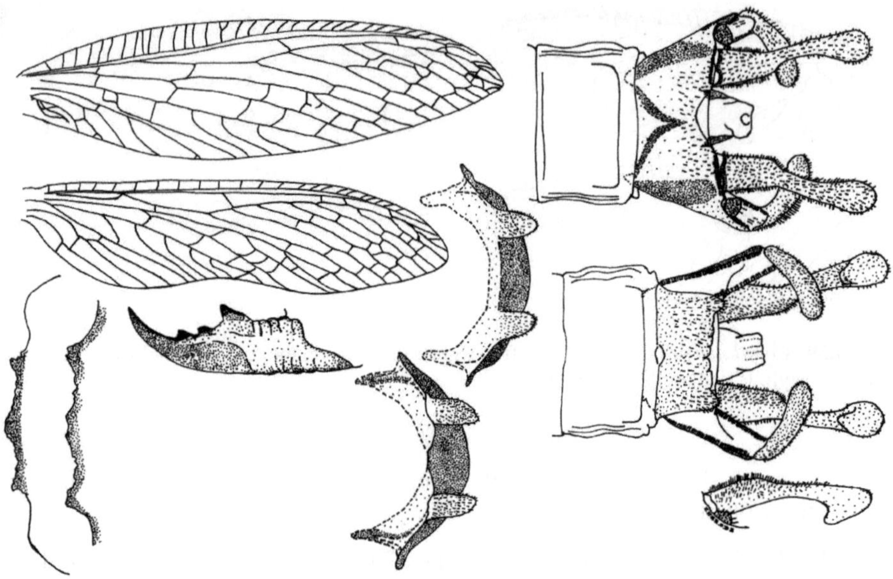

Fig. 4.40 *Corydalus parvus:* fore and hind wing (upper and middle left), anterior margins of the clypeus of two male specimens (lower left), female mandible (below hind wing), apex of the male abdomen in dorsal (upper right) and ventral view (lower middle right), tenth male tergite in lateral view (lower right), and the tenth sternites of two male specimens (center and lower center). Based on Contreras-Ramos (1998).

– The tenth tergite is concave along the apical margin. The lobes of the tenth sternite are somewhat close together (Fig. 4.41). The head is pale brown, dark greenish brown, or dark grayish brown 27

25. (24) The head is smooth, and it has only a weak color pattern, if any. The antenna is mainly pale yellow or pale brown with an infuscated apex. The forewing is pale and translucent with a few white spots and cell R_1 and the basal R cell pale brown. The ninth gonostylus lacks any notable modifications. The lobes on the tenth sternite are twice as long as wide and directed posteriad (Fig. 4.38). Fore-wing length: 46.7–59.6 mm. ....*Corydalus neblinensis* Contreras-Ramos, 1998 (Venezuela, Guyana, French Guiana, Roraima, Amazonas, Pará).

– The head is finely to moderately sculptured (Fig. 4.42) 26

26. (25). The head is finely sculptured and has a conspicuous color pattern. The antenna is pale reddish brown with the distal 1/3 to 1/4 infuscated. The forewings are various shades of reddish brown with scattered small white spots, translucent, and darker on the distal half. The length and width of the lobes on the tenth sternite are subequal and slightly divergent, or the width may be slightly greater than the length (Fig. 4.42). Fore-wing length of male: 51.4–60.8 mm; female: 63.1–66.0 mm. ....*Corydalus holzenthali* Contreras-Ramos, 1998 (Peru, Bolivia).

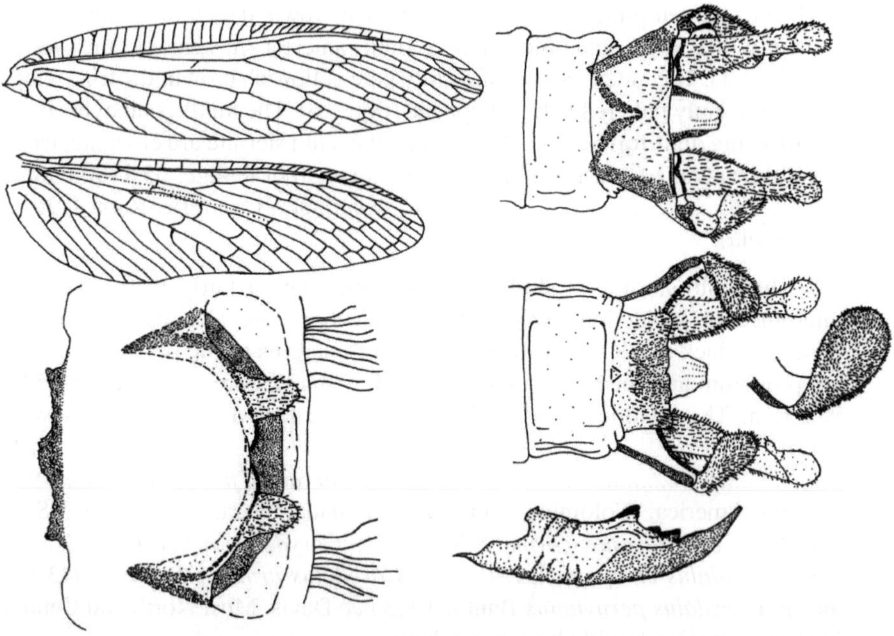

Fig. 4.41 *Corydalus clauseni:* fore and hind wing of a male (upper left), anterior margin of the clypeus of a male (lower left), mandible of a female (lower right), male tenth sternite (lower left center), and the apex of the male abdomen in dorsal (upper right) and ventral view (right of center) with an enlargement of its ninth gonostylus (middle right). Based on Contreras-Ramos (1998).

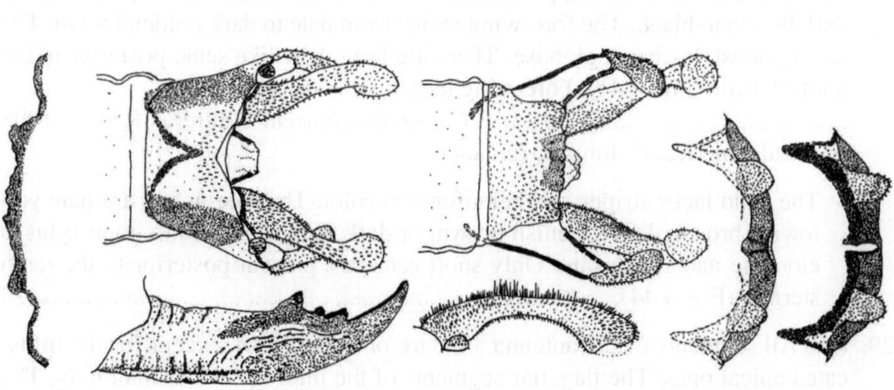

Fig. 4.42 *Corydalus holzenthali:* anterior margin of the male clypeus (left), female mandible (lower left center), apex of the male abdomen in dorsal (upper left center) and ventral view (upper right center), male tenth tergite in lateral view (lower right center), and the tenth sternites of two specimens (right), that on the left from Bolivia, and that on the right from Peru. Based on Contreras-Ramos (1998).

– The head is sculptured and is mostly reddish brown, darkening anteriad with mainly blackish mandibles and a dark anterior margin of the vertex (Figs. 1.1). The antennae are uniformly pale yellowish brown; the fore-wings are uniformly pale reddish and opaque (Fig. 1.6). The ninth gonostylus has a narrowing digitiform apex. The lobes on the tenth sternite are elongate, trianguloid, and close to each other. Fore-wing length of female: 65.1–80.6 mm. ....*Corydalus wanningeri* Contreras-Ramos, 2010 (Venezuela).

27. (24) The tenth sternite is slightly convex and bears a fairly well developed anteromedian projection. Anterolateral projections are also usually visible. Wing cells lack small white spots. The lateral margins of the pale brown head and pronotum are lined with a dark stripe (Fig. 4.43). Fore-wing length: 35.8–59.1 mm. The abdomen is dark brown, and the legs are entirely yellowish brown. ....*Corydalus flavicornis* Stitz, 1914 (Central America, Colombia, Venezuela, Ecuador, Peru, Amazonas). Syn: *Corydalus armatus flavicornis* Stitz, 1914; *Corydalus nevermanni* Navás, 1934; *Corydalus camposi* Navás, 1935; *Corydalus armatus* Banks, 1943 *nec* Hagen; *Corydalus peruvianus* Banks, 1943 *nec* Davis. Most North and Central American species would also key out here.

– The tenth sternite is convex but lacks an anteromedial projection. Its anterolateral projections are well developed (Fig. 4.44). 28

28. (27) There are faint lateral postocular stripes on the head. The antenna is shiny dark brown to black. The fore-wing varies from pale to dark golden brown. The ninth gonostylus is subglobose. There are long, hair-like setae posterior to the tenth sternite (Fig. 4.41). Fore-wing length: 48.8–67.4 mm. ....*Corydalus clauseni* Contreras-Ramos, 1998 (Central America, Colombia, Ecuador).

– The head lacks stripes and is uniform in color. The fore-wings are pale yellowish brown, dark greenish brown, or dark brown. The ninth gonostylus is elongate and subclavate. Only short setae are present posterior to the tenth sternite (Fig. 4.44). 29

29. (28) All segments of the antenna are dark or light brown, except for the infuscated apical ones. The flagellar segments of the male are wider than long. The basal part of the tibia is black, and the apical part, pale. The antenna of the female is black. The length and width of the lobes on the tenth sternite are subequal, and less than half of each lobe extends beyond the posterior margin of the sternite. Pregenital sacs are evident (Fig. 4.45). Fore-wing length: 44.7–82.2 mm. Color: entirely dark brown. ....*Corydalus armatus* (Hagen, 1861) (Colombia, Venezuela, Ecuador, Peru, Bolivia, Argentina, Brazil). Syn: *Corydalis armata* Hagen, 1861, *Corydalis armatus* Hagen, 1861; *Corydalis cornuta* Rambur, 1842 *nec* Linnaeus, *Corydalus quadrispinosus* Stitz, 1914, *Corydalis peruvianus* Banks, 1943 *nec* Davis, 1903 *pars*; *Corydalus ormatus* (sic) Navás, 1929.

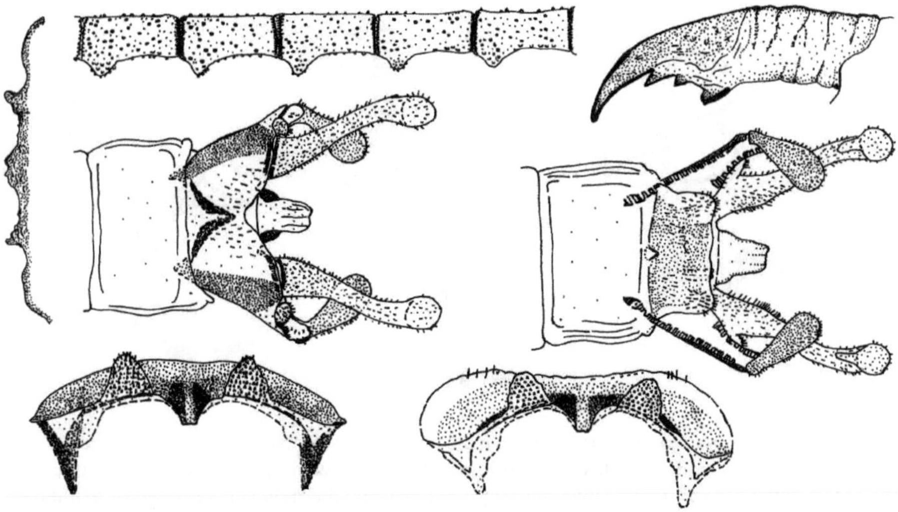

Fig. 4.43 *Corydalus flavicornis:* anterior margin of the male clypeus (upper left), segments of the antennal flagellum of a male (upper left center to center), female mandible (upper right), apex of the male abdomen in dorsal (left center) and ventral view (middle right), and the tenth sternite of a specimen from Venezuela (lower left) and one from Costa Rica, each with the posterior part oriented upward. Based on Contreras-Ramos (1998).

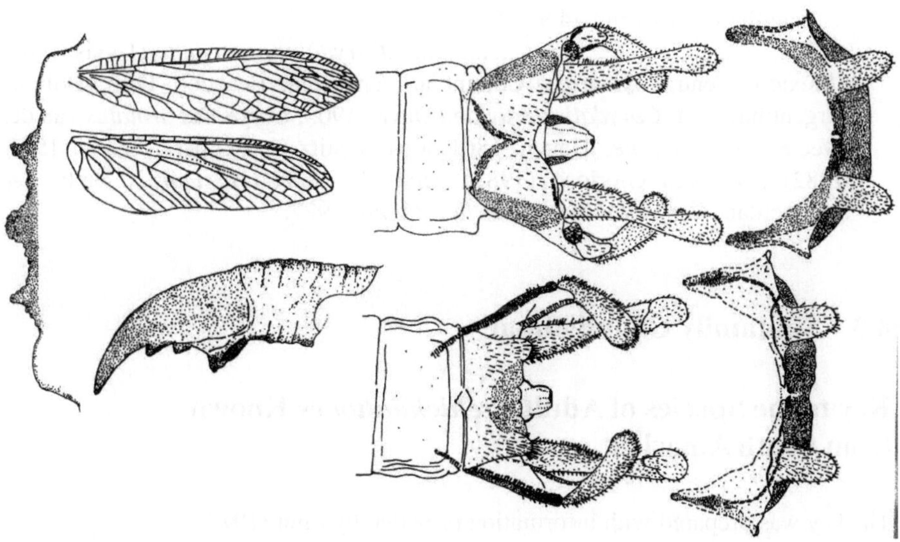

Fig. 4.44 *Corydalus peruvianus:* anterior margin of the clypeus (left), fore and hind wing (upper and middle left center), female mandible (lower left center), apex of the male abdomen in dorsal (upper right center) and ventral view (lower right center), and the tenth sternites of two male specimens (upper and lower right). Based on Contreras-Ramos (1998).

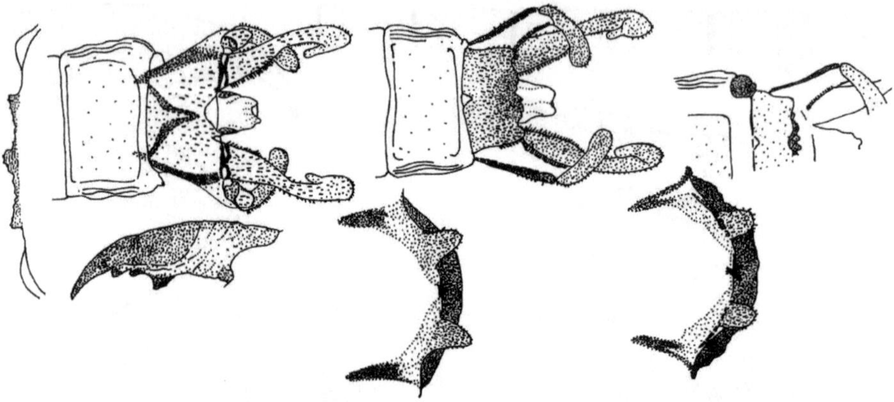

Fig. 4.45 *Corydalus armatus:* apex of the male abdomen in dorsal (upper left) and ventral view (lower left), male tenth sternites of a specimen from Venezuela (upper center) and one from Ecuador (lower center), female mandible (upper right), and the right side of the male genitalia in ventral view showing the pregenital sac (lower right). Based on Contreras-Ramos (1998).

– All segments of the antenna are yellowish or yellowish green with some infuscation on the distal third. The lobes of the tenth sternite are usually about twice as long as wide, with about half of the lobe extending beyond the posterior margin of the sternite. Pregenital sacs seem to be absent (Fig. 4.44). Length of fore-wing: 44.9–77.8 mm.
....*Corydalus peruvianus* (Davis, 1903) (Mexico, Central America, Colombia, Venezuela, Ecuador, Peru, Bolivia, Argentina). Syn: *Corydalis peruviana* Davis, 1903; *Corydalus armatus* van der Weele, 1910 pars, *nec* Hagen, 1861, also in Stitz (1914) pars, Penny (1977, 1982) pars, and Glorioso (1981) pars; *Corydalis crassicornis* auctt. *nec* McLachlan; *Corydalus primitivus fera* Navás, 1927.

4.3 Subfamily Chauliodinae

Key to the Species of Adult *Archichauliodes* Known from South America

The key was prepared with information provided by Flint (1973).

1. All muscle scars on the head are dark. The penis has a short, bilobed apex and lateral punctures (Fig. 4.2). Fore-wing length of male: 24–37 mm; female: 31–41 mm.
....*Archichauliodes chilensis* Kimmins, 1954 (Chile).

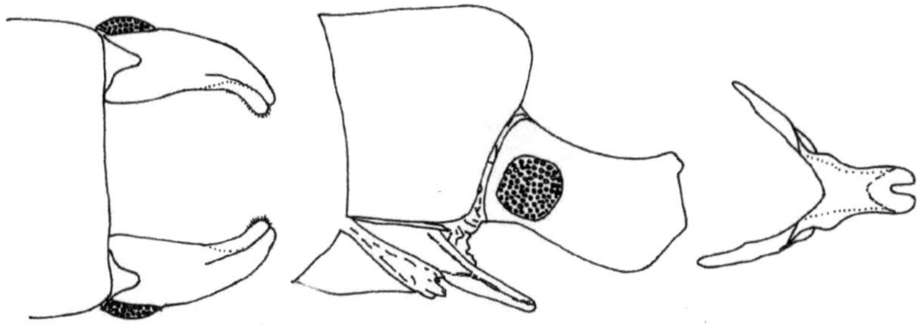

Fig. 4.46 *Archichauliodes pinares* male (left to right): apex of the abdomen in ventral and lateral view and the sclerotized structure of the penis in dorsal view. Based on Flint (1973).

– The posteromesal muscle scars on the head are pale, while the lateral muscle scars are dark. The penis has an elongated, bilobed apex and lacks lateral punctures (Fig. 4.46). Fore-wing length of male: c. 33 mm.
....*Archichauliodes pinares* Flint, 1973 (Chile).

Key to the Species of Adult *Protochauliodes* in South America

The key was prepared with information provided by Flint (1973).

1. The head is mainly yellowish with a fuscous marking only between the ocelli. The wings have fairly large dark spots. The penis is short and thick with a conical depression in its posterior surface (Fig. 4.47). Fore-wing length of male: 42–49 mm; female: 50–52 mm.
....*Protochauliodes bullocki* Flint, 1973 (Chile).

– Anterior to the antennae and sometimes also posterior to the ocelli, the head is fuscous. The wings are either uniformly fuscous or spotted. The penis is thin and elongate, and its posterior surface is unmodified (Fig. 4.48) 2

2. The part of the head posterior to the ocelli is uniformly yellow. The wings are infuscated. The anal plate of the male is somewhat enlarged and rounded at the apex (Fig. 4.48). Fore-wing length of male: 29–34 mm; female: 36–39 mm.
....*Protochauliodes humeralis* (Banks, 1908) (Chile). Syn: *Neohermes humeralis* Banks, 1908.

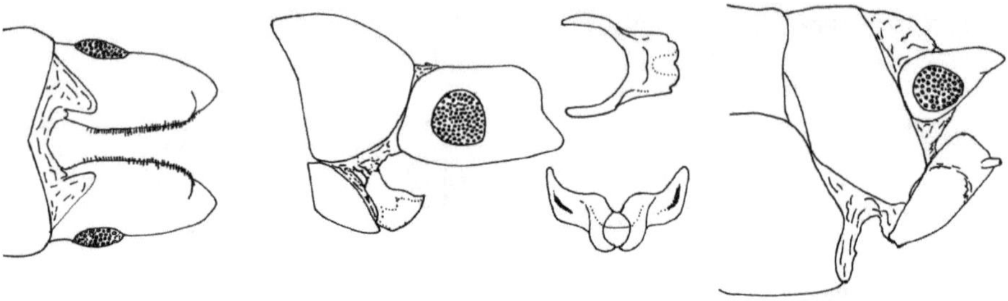

Fig. 4.47 *Protochauliodes bullocki* (left to right): male genitalia at the apex of the abdomen in dorsal and lateral view, penis in dorsal (above) and posterior view (below), and the female genitalia at the apex of the abdomen in lateral view. Based on Flint (1973).

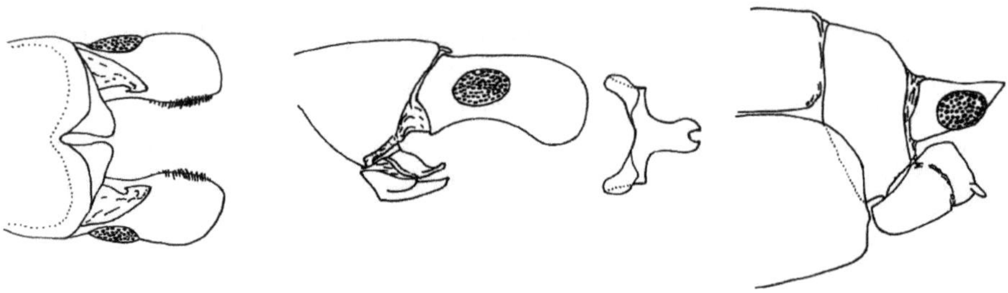

Fig. 4.48 *Protochauliodes humeralis* (left to right): male genitalia at the apex of the abdomen in dorsal and lateral view, penis in dorsal view, and the female genitalia at the apex of the abdomen in lateral view. Based on Flint (1973).

– The head is infuscate between the posterior muscle scars. The anal plate of the male has an apex with two shallowly divided lobes (Fig. 4.49).................... 3

3. The wings are infuscate or spotted, and if they are spotted, their humeral angles are yellowish. The penis lacks mid-dorsal lobes, or it has such lobes at about its mid-length (Fig. 4.5). Fore-wing length: 30–48 mm, combining both subspecies.
....*Protochauliodes cinerascens* (Blanchard, 1851) (Chile). Syn: *Chauliodes cinerascens* Blanchard 1851; *Chauliodes chilensis* Blanchard, 1851. Two subspecies have been described: *Protochauliodes cinerascens cinerascens* (Blanchard, 1851), which has spotted wings, and *Protochauliodes cinerascens fumipennis* Flint, 1973, characterized by infuscated wings.

– The wings are spotted, and the humeral angle is grayish. The penis has its mid-dorsal lobes in the apical half (Fig. 4.49). Fore-wing length of males and females: 41–47 mm, but Kimmins (1954) reported it to be 35–39 mm for males.
....*Protochauliodes reedi* Kimmins, 1954 (Chile). Syn: *Protochauliodes cinerascens reedi* Kimmins, 1954.

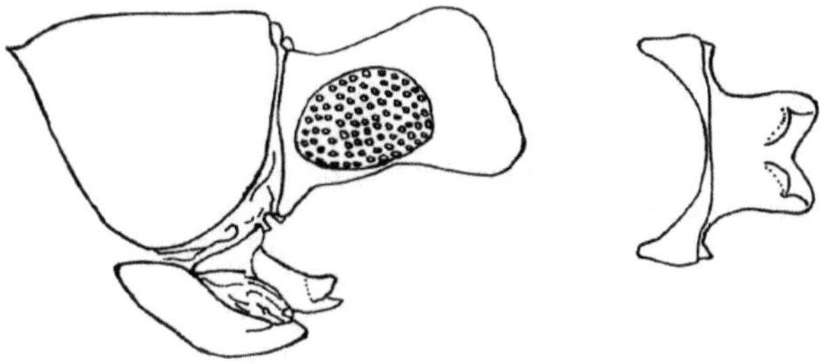

Fig. 4.49 *Protochauliodes reedi* male: genitalia in lateral view (left) and penis in dorsal view (right). Based on Flint (1973).

Literature

Azevêdo CAS de, Hamada N (2006) Description of the last instar larva of *Corydalus nubilus* Erichson, 1848 (Megaloptera: Corydalidae) and notes on its bionomics. Zootaxa 1177:57–68

Azevêdo CAS de, Hamada N (2007) Description of the larvae of *Corydalus batesii* MacLachlan and *C. ignotus* Contreras-Ramos (Megaloptera: Corydalidae). Zootaxa 1631:33–45

Banks N (1908) On the classification of the Corydalinae, with description of a new species. Proceedings of the Entomological Society of Washington 10:27–30

Banks N (1948) A new species of *Corydalus* (Neuroptera). Psyche 55:82–83

Blanchard CÉ (1851) Mirmeleonianos. In: Gay C (ed) Historia Fisica y Politica de Chile. Zoologia. E. Thunot, Paris, pp 119–129

Contreras-Ramos A (1993) Four unique dobsonflies (Megaloptera: Corydalidae, Corydalinae) from South America. Journal of the North American Benthological Soceity 10(1):130–131

Contreras-Ramos A (1995) New species of *Chloronia* from Ecuador and Guatamala, with a key to the species in the genus (Megaloptera: Corydalidae). Journal of the North American Benthological Soceity 14:108–114

Contreras-Ramos A (1998) Systematics of the dobsonfly genus *Corydalis* (Megaloptera: Corydalidae). Thomas Say Publications in Entomology (Monographs), Entomological Society of America, Lanham, Maryland. 360 pp.

Contreras-Ramos A (1999) List of species of Neotropical Megaloptera (Neuropterida). Proceedings of the Entomological Society of Washington 101(2):274–284

Contreras-Ramos A (2000) A new species of *Chloronia* Banks from southeastern Brazil, with a key to the species of Brazil (Neuropterida). Proceedings of the Entomological Society of Washington 102:919–923

Contreras-Ramos A (2002) Six new species of dobsonflies from Venezuela (Megaloptera: Corydalidae: Corydalinae). Aquatic Insects 24(1):55–75

Contreras-Ramos A (2004a) Is the family Corydalidae (Neuropterida, Megaloptera) a monophylum? Denisia 13:135–140

Contreras-Ramos A (2004b) Taxonomic and distributional notes on the dobsonflies of Ecuador (Megaloptera: Corydalidae). Dugesiana 11(1):1–11

Contreras-Ramos A (2006) A new species of *Chloronia* Banks (Megaloptera: Corydalidae) from Bolovia and Peru. Proceedings of the Entomological Society of Washington 108(4):808–813

Contreras-Ramos A (2011) Phylogenetic review of dobsonflies of the subfamily Corydalinae and the genus *Corydalus* (Megaloptera: Corydalidae). Zootaxa 2862:1–38

Contreras-Ramos A, von der Dunk K (2010) A new species of *Corydalus* Latreille from Venezuela (Megaloptera, Corydalidae). ZooKeys 67:11–19

Davis KC (1903) Sialididae (sic) of North and South America. Bulletin of the New York State Museum 68:442–486, 499+pl. 51–52

Erichson WF (1848) Insekten. In: M. R. Schomburgks Reisen in British-Guiana in den Jahren 1840–1884. Band 3. Weber, Leipzig, p 553–617

Flint OS Jr (1973) The Megaloptera of Chile (Neuroptera). Revista Chilena de Entomologia 7:31–45

Flint OS Jr (1977) Neuroptera. In: Hurlbert SH (ed) Biota Acuática de Sudamérica Austral. San Diego, San Diego State University, pp 187–188

Flint OS Jr (1983) *Nothochauliodes penai,* a new genus and species of Megaloptera from Chile (Neuroptera: Corydalidae). Entomological News 94:15–17

Flint OS Jr (1991) On the identity of *Chloronia bogatana* (sic) Weele (Neuropterida: Megaloptera: Corydalidae). Proceedings of the Entomological Society of Washington 93(2):489–494

Geijskes DC (1984) Notes on Megaloptera from the Guyanas, S. Am. In: Gepp J, Aspöck H, Hölzel H (eds) Progress in worlds Neuropterology. – Proceedings of the 1st international symposium on Neuropterology, Graz, p 79–84

Glorioso MJ (1981) Systematics of the dobsonfly subfamily Corydalinae (Megaloptera: Corydalidae). Systematic Entomology 6:253–290

Hagan HA (1861) Synopsis of the Neuroptera of North America, with a list of South American species. – Smithsonian miscellaneous collections 4(1):i–xx + 1–347

Kimmins DE (1954) A new genus and some new species of the Chauliodini (Megaloptera). Bull Br Mus (Nat Hist) Entomology 3:417–444

McLachlan R (1868 [1867]) New genera and species, &c., of neuropterous insects; and a revision of Mr. F. Walker's british museum catalogue of Neuroptera, part ii (1853), as far as the end of the genus *Myrmeleon*. Journal of the Linnean Society (Zoology) 9:230–281 + 1 pl

McLachlan R (1869) Considerations on the neuropterous genus *Chauliodes* and its allies; with notes and descriptions. Annals and Magazine of Natural History, Series 4(4):35–46

Navás L (1915) Neurópteros Sudamericanos. Segunda Serie. Brotéria, Série Zoológica 13(1):5–14

Navás L (1920) Insectos Sudamericanos (1ª Serie). Anales de la Sociedad Cientifica Argentina 90:33–43

Navás L (1925) Insectos exoticos nuevos o poco conocidos. 2ª Serie. Memorias de la Real Academia de Cíencias y artes de Barcelona, 3ª Serie 19:181–200

Navás L (1928a) Insectos del Museo de Estocolmo. – Revista de la Real Academia de Ciencias Exactas Fisicas y Naturales de Madrid 24:28–39

Navás L (1928b) Insectos del Museo de Hamburgo. Primera Serie. Boletin de la Sociedad Entomológica de España 11:59–67, 90–100, 121–138

Navás L (1934a) Insectos Suramericanos, octava serie. Revista de la Academia de Cíencias Exactas Físicas y Naturales de Madrid 31:9–28

Navás L (1934b) Insectos del Museo de Hamburgo. – Memorias de la Academia de Ciencias y Artes de Barcelona. Tercera Época – 23:499–508

Navás L (1935 [1934]) Insectos Suramericanos. Décima Serie. Revista de la Academia de Cíencias Exactas Físicas y Químicas y Naturales de Madrid 32:360–375

Navás L (1936 [1931]) Insectos del Brasil. 5ª Serie. Revista do Museu Paulista 20:721–734

New TR, Theischinger G (1993) Megaloptera (alderflies, dobsonflies). – Handbuch der Zoologie Band 4, Arthropoda: Insecta. Lieferung 33. - De Gruyter, Berlin. 97 pp

Penny ND (1978 [1977]) Lista de Megaloptera, Neuroptera e Raphidioptera do Mexico, America Central, ilhas Caraibas e America do Sul. Acta Amazonica, Supplement 7(4):1–61

Penny ND (1982) Neuroptera of the Amazon Basin. Part 7. - Corydalidae. Acta Amazonica 12:825–837

Penny ND (1999) *Madachauliodes ranomafana,* a new Madagascan species of Chauilodinae (Megaloptera: Corydalidae), with a key to the world genera of Chauliodinae. Entomological News 110(4):193–200

Penny ND, Flint OS Jr (1982) A revision of the genus *Chloronia* (Neuroptera: Corydalidae). Smithsonian Contributions to Zoology No. 348. 27 pp

Penny ND, Lee VF (1996) Types of Mecoptera, Megaloptera, and Neuroptera (Insecta) in the collection of the California Academy of Sciences. Proceedings of the California Academy of Sciences 49(4):127–149

Rambur JP (1842) Histoire naturelle des insectes: Névroptères. Librairie encyclopédique de Roret. – Fain et Thunot, Paris. 534 pp + 12 pl

Roback SS (1966) The catherwood foundation Peruvian-Amazon expedition. IX. - Megaloptera (Corydalidae, Neurominae). Monographs of the Academy of Natural Sciences of Philadelphia 14:217–223

Romero F (2003) Lista de Megaloptera de la Coleccion Instituto Fundacion Miguel Lillo (Tucuman, Argentina). Acta Zoológica Lilloana 47(1-2):77–83

Thouvenot M (2008) Mégloptères de Guyane. L'Entomologiste 64(2):65–68

Part III
Planipennia

Chapter 5
Coniopterygidae

Abstract An illustrated key is provided for identification of the genera of adult coniopterygids known to occur in South America. Illustrated keys are also provided to identify the species of each genus known to occur in South America. The countries or states in which specimens of each species have been found and invalid synonyms are reported.

Keywords Duskywing • Dusty lacewing • Coniopteryginae • Brucheiserinae • Aleuropteryginae

5.1 Key to the Subfamilies of Adult Coniopterygidae from South America

The English common names for most insects in this family are dusty lacewing and duskywing. Information for the key was provided by Meinander (1972) and Adams (1973a).

1. Adults are apparently flightless. A short distance from the base of the wing, vein R fuses with the costal margin. The fore and hind wings are brachypterous, heavily sclerotized along the entire margin, without trichosors, and with microtrichia covering the membranes. The fourth segment of each five-segmented tarsus is greatly expanded to form a pad on the ventral side. The thorax is sclerotized by a network of bars, which are separated by membranous areas, especially in the pleural region. There are patches of setae on the pronotum. Cerci are completely lacking. The apical segments of the five-segmented maxillary and three-segmented labial palps are enlarged and flattened (Fig. 5.1). The fore-wing is usually between 4 and 10 mm.

   .... Brucheiserinae..p. 127

   – Adults have fully developed wings and can fly. Vein R does not fuse with the costal margin just distal to the wing base. The fore and hind wings are not brachypterous and heavily sclerotized, or, if the fore-wing is so sclerotized, then the hind wing is reduced to a size and shape resembling the hind wings

C.W. Heckman, *Neuroptera (Including Megaloptera)*, Encyclopedia of South American Aquatic Insects, DOI 10.1007/978-3-319-35125-4_5

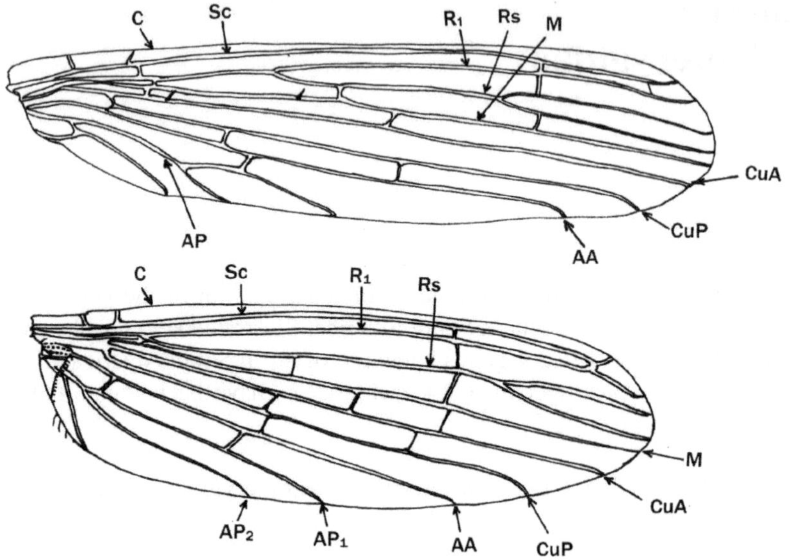

Fig. 5.1 Fore and hind wing of *Flintoconis gozmanyi*. Based on Sziráki (2007).

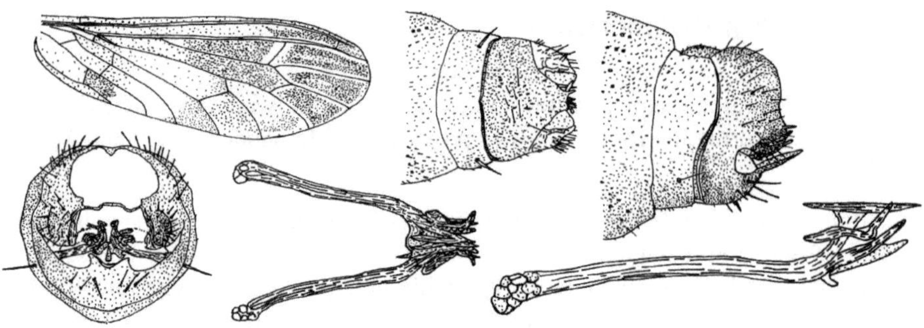

Fig. 5.2 *Semidalis maculosus* male (above, left to right): fore-wing, apex of the abdomen in ventral and lateral view, and (below, left to right), apex of the abdomen in posterior view and the internal parts of the genitalia in ventral and lateral view. Based on Sziráki (2009).

of dipterans. The thorax is usually completely sclerotized or has few membranous areas between the sclerites (Fig. 5.2). 2

2. (1) There are two r-m cross veins in the middle of the fore-wing. Vein R_s in the hind wing branches off vein R very close to the base of the wing. The galea consists of three segments. Plicaturae are present on the abdomen (Fig. 5.3). ....Aleuropteryginae, Enderlein, 1905..p. 130

– There is one r-m cross vein in the middle of the fore-wing. Vein R_s in the hind wing branches off vein R a considerable distance from the base of the wing. The galea consists of one segment. The abdomen lacks plicaturae (Fig. 5.2). ....Coniopteryginae, Burmeister, 1839..p. 140

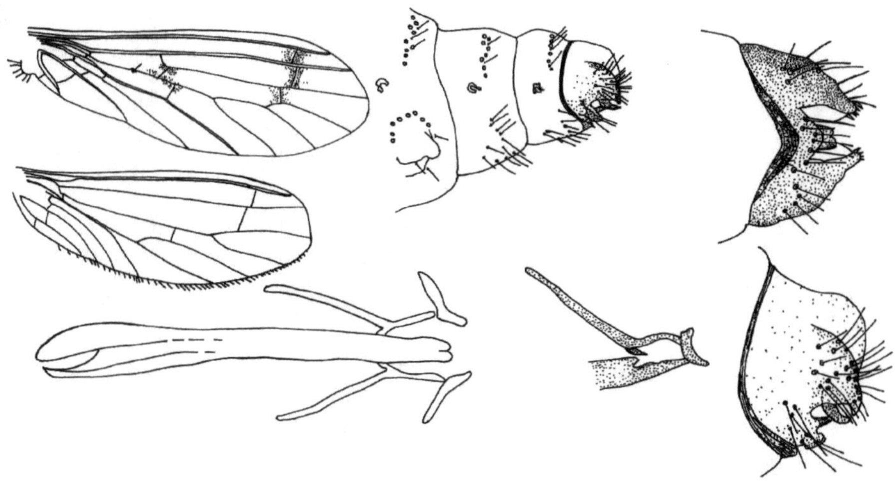

Fig. 5.3 *Neoconis tubifera* male: fore and hind wing (upper and middle left, respectively), apex of the abdomen in ventral (upper right) and lateral view (upper center) with an enlarged view of its apical segment in lateral view (lower right), internal genitalia in ventral (lower left) and lateral view with the inner parts not showing (lower right center). Based on Meinander (1980).

Key to the Subfamilies of Coniopterygidae Larvae from South America

Information for the key was provided by Meinander (1972). The characters in the key are those observed for specimens from other regions of the world.

1. Between the tarsal claws is a short, funnel-shaped empodium. The antenna and labial palp are approximately equal in length, or the antenna is as much as 20 % longer (Fig. 5.4). Total length: at least 3.2–4.2 mm, and possibly longer. .... Brucheiserinae..p. 127

   – The empodium between the tarsal claws of the known larvae is pad-like (Fig. 2.4). The total length of the fully developed larvae of the known species is shorter than about 3.2 mm. 2

2. The antenna and labial palp are approximately equal in length. The mandibles do not project anteriad from beneath the labrum (Fig. 5.5). ....Aleuropteryginae, Enderlein, 1905..p. 130

   – The antenna is approximately twice as long as the labial palp. The mandibles project anteriad from beneath the labrum (Fig. 5.4). ....Coniopteryginae, Burmeister, 1839..p. 140

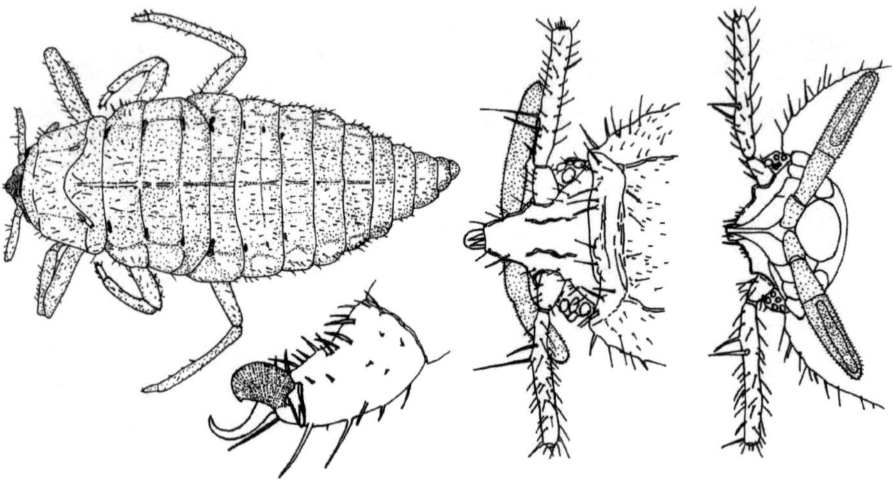

Fig. 5.4 *Brucheiser penai* larva: habitus in dorsal view (upper left), tarsal claw (lower left center), head and anterior part of the thorax in dorsal (right center) and ventral view (right). Based on Sziráki and Flint (2005).

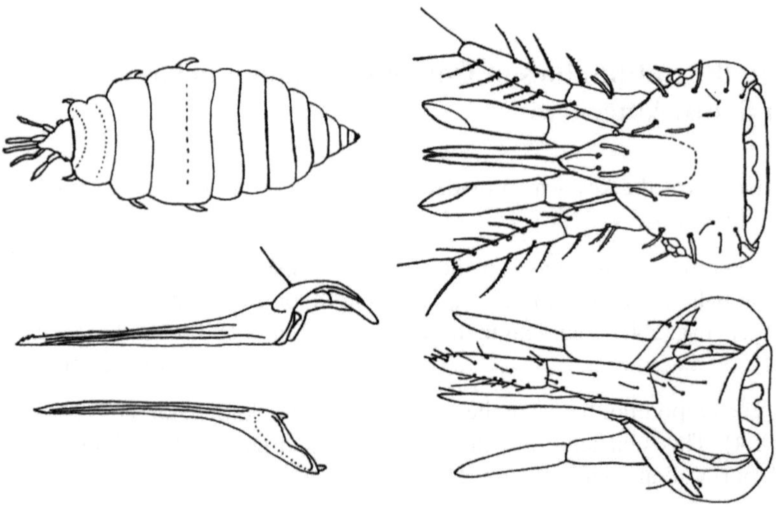

Fig. 5.5 Larva of *Aleuropteryx loewii* Klapalek, 1894, a European species: habitus in dorsal view (upper left), head in dorsal (upper right) and ventral view with one palp removed (lower right), right mandible in dorsal view (middle left), and right maxilla in ventral view (lower left). Based on Meinander (1972).

5.2 Subfamily Brucheiserinae

Key to the Genera of Adult Brucheiserinae Known from South America

Information for the key was provided by Sziráki (2007, 2009). There is not enough information to provide a key to the larvae because descriptions of the larvae of *Flintoconis* are still unavailable.

1. The wings appear reduced in size and extend only slightly beyond the apex of the abdomen. The male genitalia has a U-shaped sclerotization of the penis and appears relatively simple (Fig. 5.6).
   ....*Brucheiser* Navás, 1927..p. 128
   – The wings are relatively narrow and greatly elongated; the fore-wings are about twice as long as the body (Figs. 5.1 and 5.7). Male genitalia are notably complex.
   ....*Flintoconis* Sziráki, 2007..p. 128

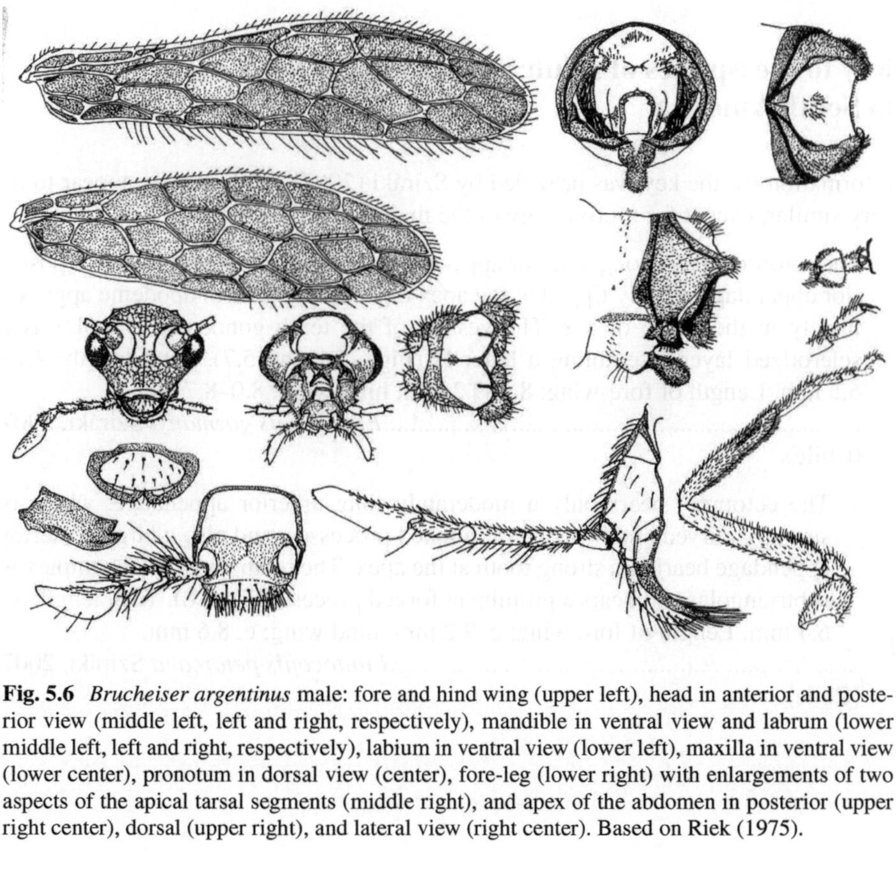

Fig. 5.6 *Brucheiser argentinus* male: fore and hind wing (upper left), head in anterior and posterior view (middle left, left and right, respectively), mandible in ventral view and labrum (lower middle left, left and right, respectively), labium in ventral view (lower left), maxilla in ventral view (lower center), pronotum in dorsal view (center), fore-leg (lower right) with enlargements of two aspects of the apical tarsal segments (middle right), and apex of the abdomen in posterior (upper right center), dorsal (upper right), and lateral view (right center). Based on Riek (1975).

Key to the Species of Adult Specimens of *Brucheiser* in South America

Information for the key was provided by Navás (1927a); Riek (1975), and Sziráki (2007).

1. The length of the fore-wing is less than 4 times its maximum width. The width of the labrum is more than twice its length, and its anterior surface is covered with fine setae (Fig. 2.7). Length of female: c. 4.2 mm. Length of fore-wing: c. 4.5 mm; hind wing: c. 4.2 mm. The male has not been described.
 .... *Brucheiser penai* Riek, 1975 (Chile).
 – The length of the fore-wing is more than 4 times its maximum width. The width of the labrum is less than twice its length, and its anterior surface bears well separated, coarse setae (Fig. 5.6). Length of male: c. 4.4 mm. Fore-wing length: c. 4.6 mm. The female has not been described.
 .... *Brucheiser argentinus* Navás, 1927 (Argentina).

Key to the Species of Adult Specimens of *Flintoconis* in South America

Information for the key was provided by Sziráki (2007). Both species appear to be very similar, except for morphology of the male genitalia.

1. The ectoproct has a long anterior appendage with pointed processes and a posterior appendage acutely tipped at the apex and bearing a basal apodeme approximately in the shape of a Y. The vestige of the tenth gonocoxal complex is a sclerotized layer resembling a letter H (Figs. 5.1 and 5.7). Total length: 4.1–5.5 mm. Length of fore-wing: 8.5–9.2 mm; hind wing: 8.0–8.7 mm.
 .... *Flintoconis gozmanyi* Sziráki, 2007 (Chile).
 – The ectoproct bears only a moderately long anterior appendage, which is strongly curved and bears short, hooked processes, and also a large posterior appendage bearing a strong tooth at the apex. The tenth gonocoxal complex is subtriangular and bears a prominent forked process (Fig. 5.8). Total length: c. 6.4 mm. Length of fore-wing: c. 9.2 mm; hind wing: c. 8.6 mm.
 .... *Flintoconis petorcana* Sziráki, 2007 (Chile).

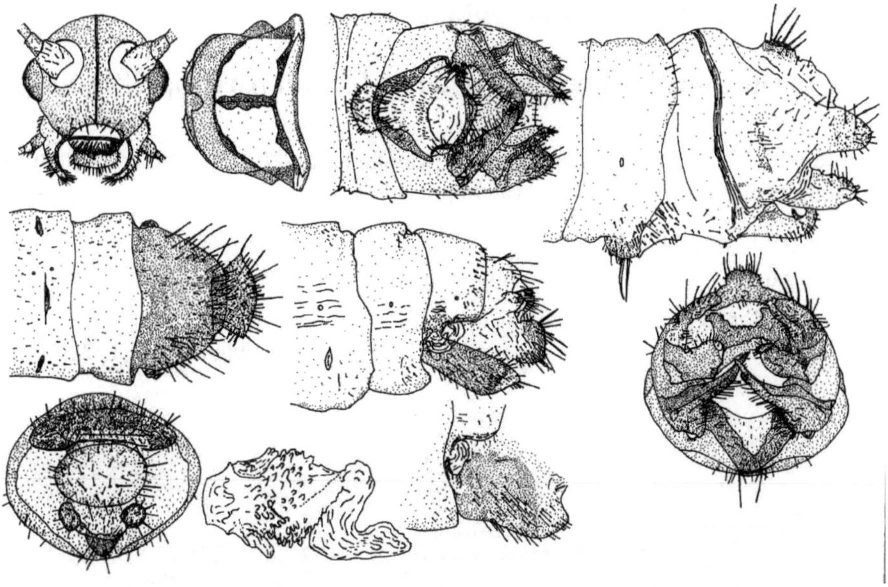

Fig. 5.7 *Flintoconis gozmanyi* (upper row, left to right): head of a female in anterior view, prothorax of a female in dorsal view, apex of the male abdomen in ventral and lateral view; (middle row, left and center respectively): apex of the female abdomen in ventral and lateral view; apex of male abdomen in posterior view (lower right); apex of female abdomen in posterior view (lower left); female spermatheca with tubules removed (lower left center); expanded area between the eight and ninth sternites of the female abdomen (lower center). Based on Sziráki (2007).

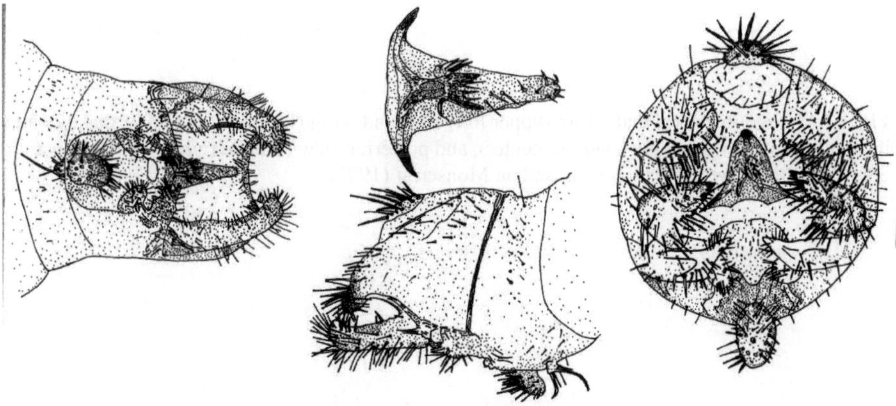

Fig. 5.8 *Flintoconis petorcana* male: apex of the abdomen in ventral (left), lateral (lower center), and posterior view (right), and the tenth gonocoxal complex in ventral view (upper center). Based on Sziráki (2007).

5.3 Subfamily Aleuropteryginae

Key to the South American Genera of Adult Aleuropteryginae

Information for the key was provided by Meinander and Penny (1982). There is not enough information about the larvae to provide a key.

1. The penis is tubular (Fig. 5.9).
   ....*Neoconis* Enderlein, 1929..p. 131

   – The dorsal part of the penis is open (Fig. 5.10).
   ....*Pampoconis* Meinander, 1972..p. 136

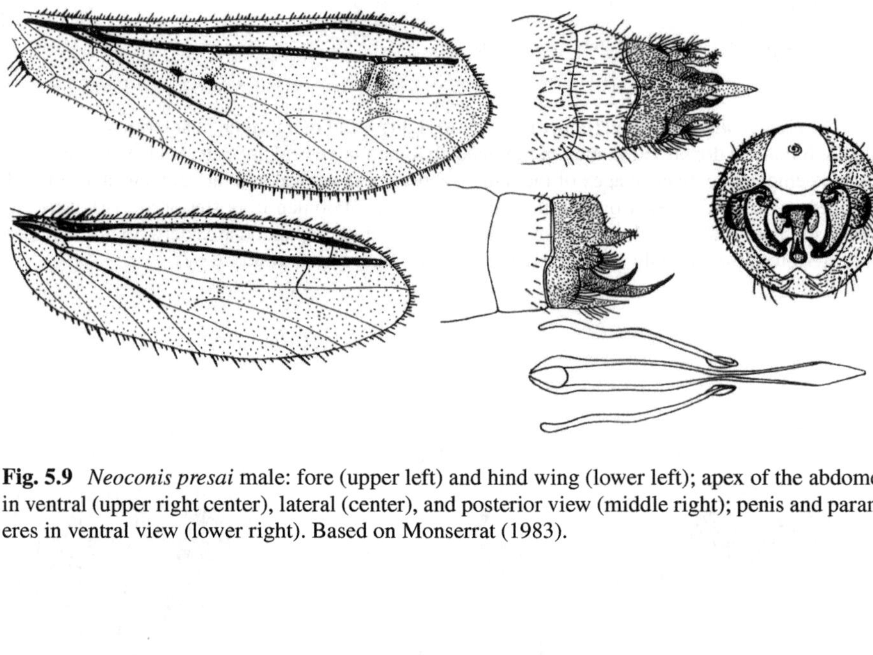

Fig. 5.9 *Neoconis presai* male: fore (upper left) and hind wing (lower left); apex of the abdomen in ventral (upper right center), lateral (center), and posterior view (middle right); penis and parameres in ventral view (lower right). Based on Monserrat (1983).

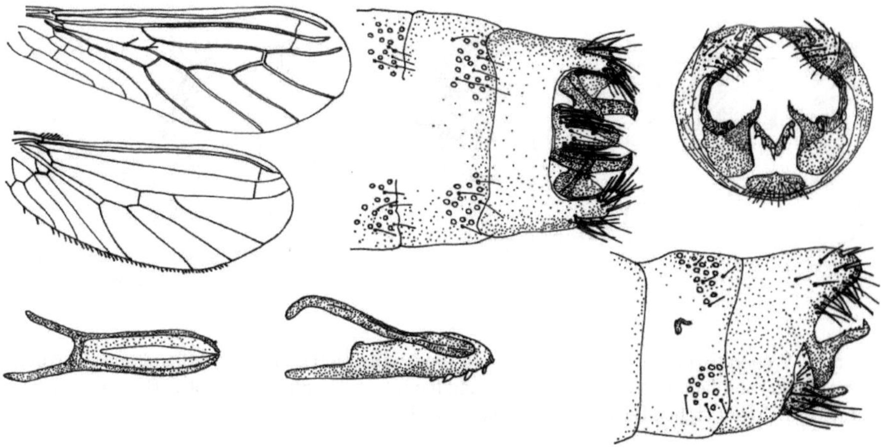

Fig. 5.10 *Pampoconis dentifera* male: fore and hind wing (upper and middle left), apex of the abdomen in ventral (upper center) and posterior view (upper right), and (below, left to right): penis in dorsal view, penis and paramere in lateral view, and the apex of the abdomen in lateral view. Based on Meinander (1973).

Key to the South American Species of Adult Male *Neoconis*

Information for the key was provided by Meinander (1972, 1980, 1983, 1990); Meinander and Penny (1982); and Monserrat (1981, 1983). The descriptions are cursory. A revision of this genus with descriptions of both sexes of each species is needed.

1. The length of the fore-wing is greater than 3.3 mm, and the hind wing is longer than about 3.0 mm. Only the male of one of the two species fitting this description has been described, and only the female of the other species is known. Until more about the natural intraspecific variability of these two species is known, identification based on wing length should be considered tentative, and the original descriptions should be consulted to confirm identifications 2
   – The fore-wing is shorter than 3.3 mm, and the hind wing does not exceed about 2.8 mm 3
2. (1) The fore-wing membrane is tinged with gray and has distinct grayish brown markings in most cells (Fig. 5.11). The antenna consists of about 30 segments. Fore-wing length of male: c. 3.6 mm; hind wing length: c. 3.1 mm. The female has not been described.
   .... *Neoconis unicornis* Meinander, 1990 (Colombia).

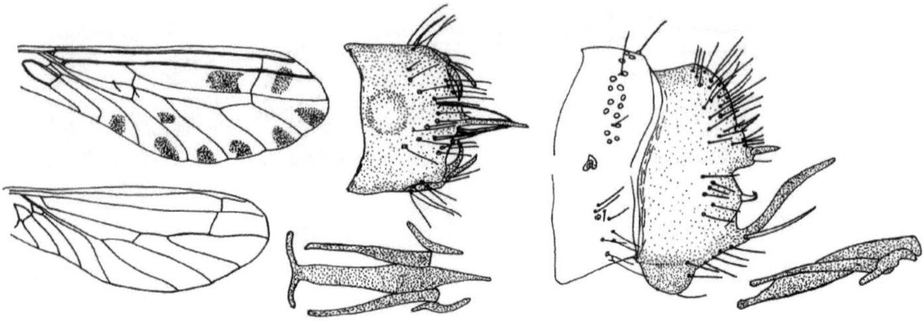

Fig. 5.11 *Neoconis unicornis* male: fore and hind wing (left), apex of the abdomen in ventral (upper center) and lateral view (upper right), and the internal genitalia in ventral (lower center) and lateral view (lower right). Based on Meinander (1990).

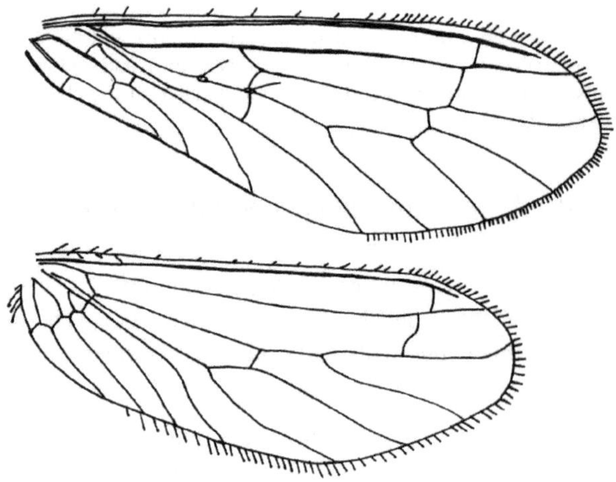

Fig. 5.12 Fore and hind wing of *Neoconis pistrix*. Based on Meinander (1972).

– Both wings have almost completely hyaline membranes (Fig. 5.12). The antenna consists of about 24 segments. Fore-wing length of female: c. 3.5 mm; hind wing length: c. 3.1 mm. The male has not been described. .... *Neoconis pistrix* (Enderlein, 1906) (Peru). Syn: *Helicoconis pistrix* Enderlein, 1906.

3. (1) A considerable area of the fore-wing membrane is darkened, but the middle parts of most cells are lighter (Fig. 5.13) 4

– Dark markings on the wing membranes, if present, are limited to narrow margins along a few cross veins in the fore-wing (Fig. 5.14). 5

4. (3) The wing membranes of the male are dark gray, darker toward the posterolateral margins of the wings, and lighter along most of the wing veins and in the middle of the cells (Fig. 5.13). The antenna is uniformly dark and consists of

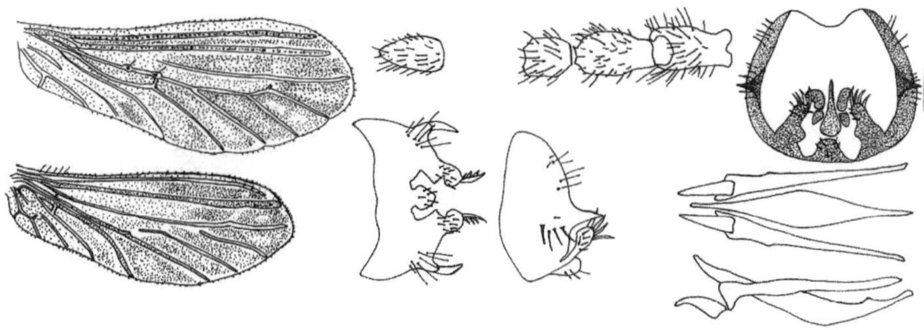

Fig. 5.13 *Neoconis gelesae* male: fore and hind wing (left), ventral views of the apical (upper left center) and basal segments of the antenna (upper right center), apex of the genitalia in posterior view (upper right), the ninth sternite in ventral (lower left center) and lateral view (lower right center), and the penis in ventral (middle right) and lateral view (lower right). Based on Monserrat (1981).

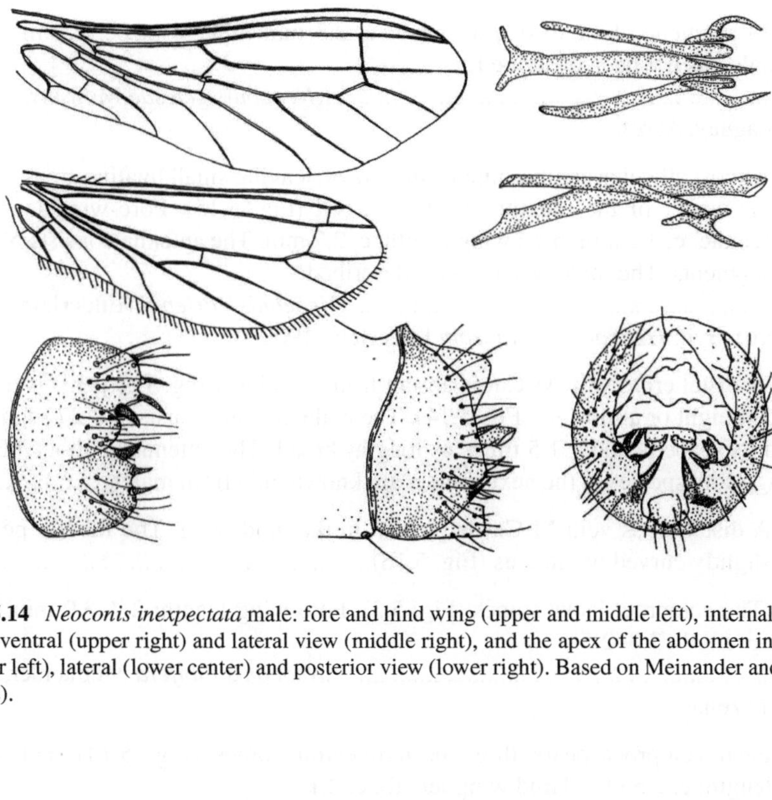

Fig. 5.14 *Neoconis inexpectata* male: fore and hind wing (upper and middle left), internal genitalia in ventral (upper right) and lateral view (middle right), and the apex of the abdomen in ventral (lower left), lateral (lower center) and posterior view (lower right). Based on Meinander and Penny (1982).

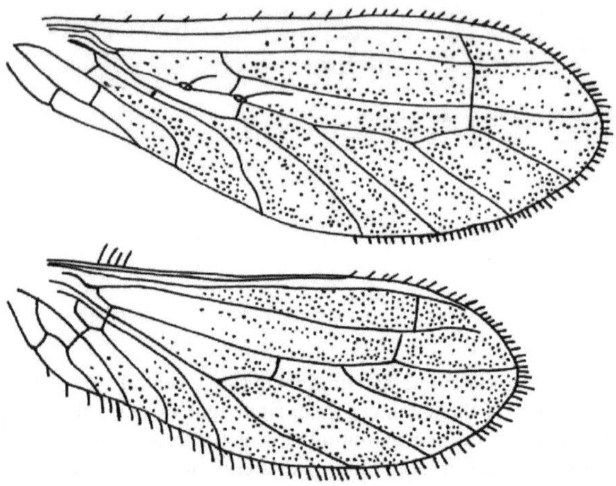

Fig. 5.15 Fore and hind wing of *Neoconis garleppi*. Based on Meinander (1972).

c. 23 segments. Length of fore-wing: c. 2.3 mm; hind wing: c. 2.0 mm. The female has not been described.
....*Neoconis gelesae* Monserrat, 1980 (Paraguay, Acre).

– Almost all veins are margined with brown, leaving small hyaline areas only in the center of the membrane of each cell (Fig. 5.15). Fore-wing length of female: c. 3.2 mm; hind wing length: c. 2.7 mm. The antenna consists of c. 27 segments. The male has not been described.
....*Neoconis garleppi* (Enderlein, 1906) (Peru). Syn: *Helicoconis garleppi* Enderlein, 1906.

5. (3) A distal cross vein M-Cu_1 is absent from the hind wing. The penis is tubular and straight or almost so (Fig. 5.14). The male antenna consists of 27 segments, and its scape is about 1.5 times as long as broad. The antenna is about 1.3 mm long. Both species in the next couplet are known only from males. 6

– A distal cross vein M-Cu_1 is present in the hind wing. The tubular penis is slightly curved or sinuous (Fig. 5.16)..7

6. (5) The ectoproct lacks spines (Fig. 5.3). Fore-wing length: 2.7–3.0 mm. Hind wing length: 2.3–2.5.
....*Neoconis tubifera* Meinander, 1980 (Amazonas).

– Each ectoproct bears three posteroventral spines (Fig. 5.14). Fore-wing length: c. 2.5 mm. Hind wing length: c. 2.1.
....*Neoconis inexpectata* Meinander, 1972 (United States, Amazonas).

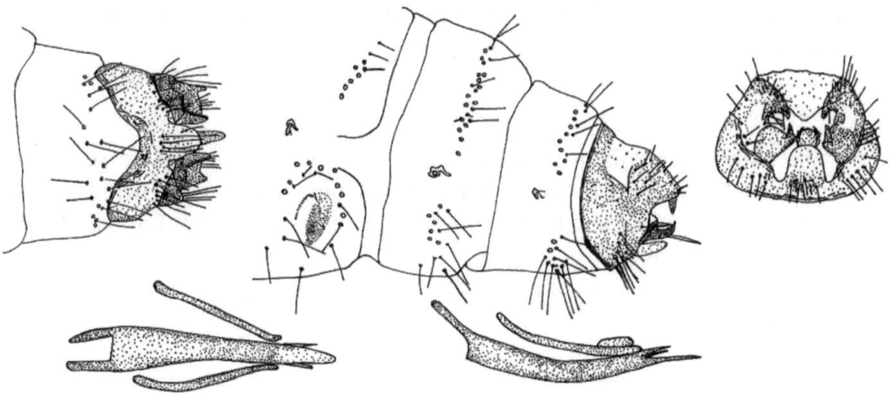

Fig. 5.16 *Neoconis amazonica* male (above, left to right): apex of the abdomen in ventral, lateral, and posterior view, and (below, left and right, respectively): the internal genitalia in ventral and lateral view. Based on Meinander (1983).

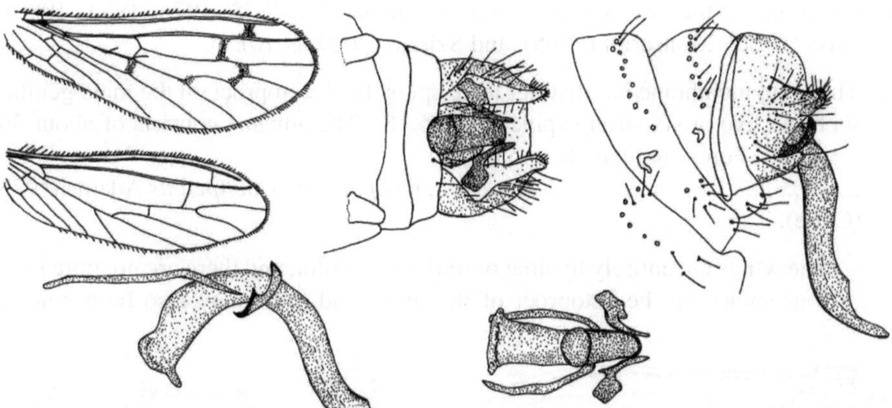

Fig. 5.17 *Neoconis brasiliensis* male: fore and hind wing (upper and middle left, respectively), apex of the abdomen in ventral (upper center) and lateral view (upper right), internal genitalia in ventral (lower right center) and lateral view (lower left). Based on Meinander (1980).

7. (5) The male antenna consists of about 29 segments, and its scape and pedicel are only slightly longer than broad. The penis is fairly straight and only slightly curved (Fig. 5.16). Fore-wing length: c. 2.8 mm; hind wing langth: c. 2.4 mm. ....*Neoconis amazonica* Meinander, 1983 (Amazonas, Pará).

   – The male antenna consists of about 24–26 segments, and its scape is longer than broad. The penis is tubular and sinuous (Fig. 5.17). 8

8. (7) The antenna of the male consists of 25 segments, and its scape is almost twice as long as broad. The penis appears tubular and sinuous in lateral view (Fig. 5.17).

The paramere lacks a sclerite. Fore-wing length: c. 3.1 mm. Hind wing length: 2.7 mm.
....*Neoconis brasiliensis* Meinander, 1980 (Amazonas).

– The male antenna consists of about 26 segments, including the scape and pedicel. The scape is fusiform and slightly longer than broad. There is a short sclerite in the paramere (Fig. 5.9). Fore-wing length: c. 3.1 mm. Hind wing length: c. 2.5 mm.
....*Neoconis presai* Monserrat, 1983 (Venezuela, Pará).

Key to the Adult Males of South American Species of *Pampoconis*

Information for the key was provided by Meinander (1972, 1973, 1974, 1980), Adams (1973a); Monserrat (2005), and Sziráki (2009, 2010).

1. The wing membrane has distinct large spots. Each ectoproct on the male genitalia bears five or six strong spines (Fig. 5.18). The antenna consists of about 36 segments. Fore-wing length: c. 5.6 mm.
....*Pampoconis punctipennis* Adams, 1973 (Chile).

– The wings are entirely hyaline or uniform in color, and there are no more than four spines on the ectoproct of the male, and there may also be a pair of

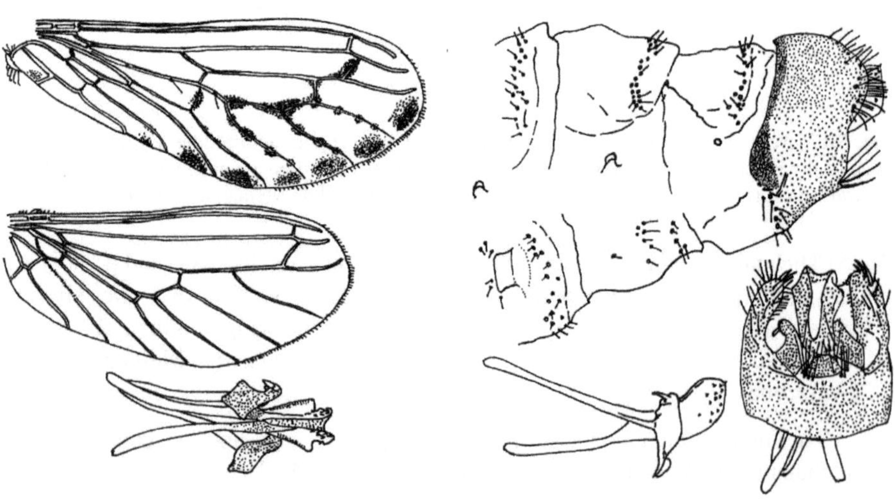

Fig. 5.18 *Pampoconis punctipennis* male: fore and hind wing (upper and middle left), apex of the abdomen in lateral (upper right) and ventral view (lower right), the penis and styli in ventral (lower left) and lateral view (lower center). Based on Adams (1973a).

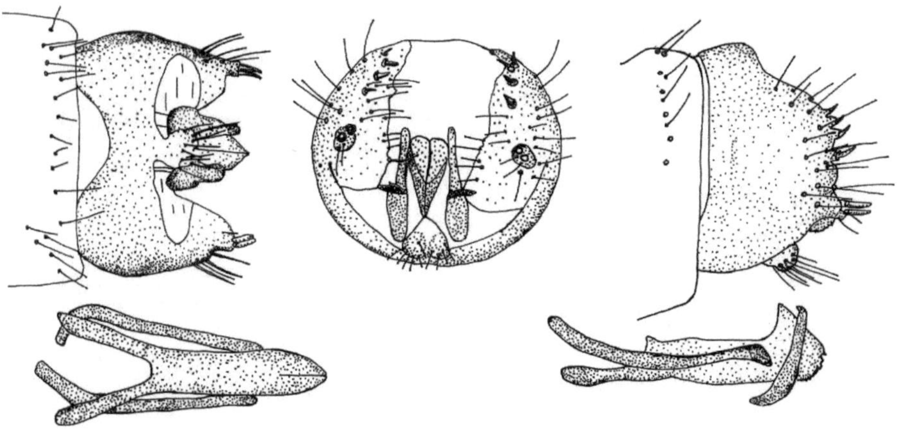

Fig. 5.19 *Pampoconis angustipennis* male (above, left to right): the apex of the abdomen in ventral, posterior, and lateral view, and (below, left and right, respectively): the internal genitalia in ventral and lateral view. Based on Meinander (1990).

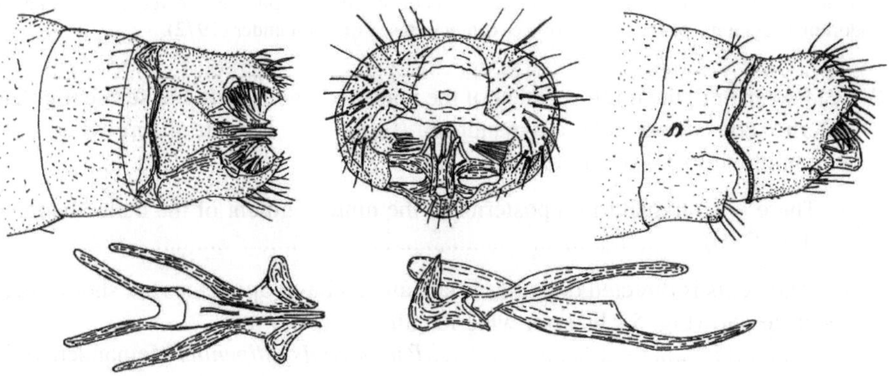

Fig. 5.20 *Pampoconis xerophila* male (above, left to right): apex of the male abdomen in ventral, posterior, and lateral view, and the internal male genitalia in ventral (lower left) and lateral view (lower right). Based on Sziráki (2009).

equally strong spines on a ventral process of the ectoproct (Fig. 5.19). The fore-wing is shorter than about 5 mm. 2

2. (1) There are four short, thick spines on the ectoproct and a pair of equally strong spines on a ventral process of the ectoproct (Fig. 5.19). A distal M-Cu cross vein is present in the hing wing. The wing is narrow, with a length of about 3.75 mm and a maximum width of c. 1.2 mm.
.... *Pampoconis angustipennis* Meinander, 1990 (Chile).

– There is no more than one spine present on the ectoproct (Fig. 5.20). 3

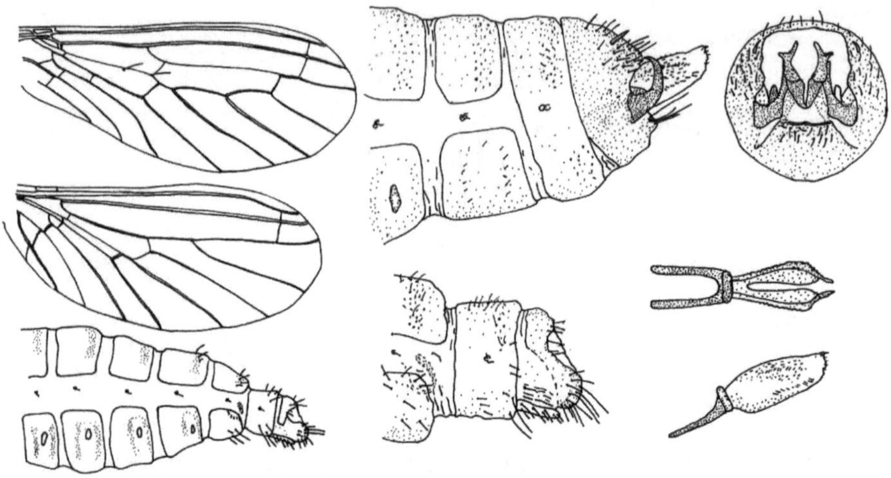

Fig. 5.21 *Pampoconis latipennis:* fore and hind wing (upper and middle left), apex of a male abdomen in lateral (upper center) and posterior view (upper right), penis in dorsal (middle right) and lateral view (lower right), posterior segments of the female abdomen (lower left), and apical abdominal segments of the female (lower center). Based on Meinander (1972).

3. (2) Posterior to the ninth segment of the male abdomen, there is a setose, ligulate sclerite, considered to be a hypandrium (Fig. 5.21). Length of fore-wing: 3.9–4.4; hind wing: 3.4–3.9 mm. 4

 – There is no hypandrium posterior to the ninth segment of the male abdomen (Fig. 5.20). 5

4. (3) The penis is directed dorsoposteriad, and its basal apodemes are shorter than its distal part (Fig. 5.21). Fore-wing length: c. 4.6–4.7 mm. .... *Pampoconis latipennis* Meinander, 1972 (Chile, Argentina).

 – The dorsoposterier part of the penis is broad and rounded, and its basal apodemes and posterior part are subequal in length (Fig. 5.22). The antenna consists of 26–27 segments. Fore-wing length: 4.3–4.7 mm. .... *Pampoconis uncinatus* Adams, 1973 (Chile). This species name was suggested to be a synonym of *P. latipennis* by Monserrat (2005a) but declared valid by Sziráki (2010).

5. (3) The penis is long and lacks dentiform structures. The ectoproct has an inner, setose swelling (Fig. 5.20). Fore-wing length: c. 3.2 mm; width: c. 1.1 mm. Length of hind wing: c. 2.7 mm; width of hind wing: c. 1.0 mm. .... *Pampoconis xerophila* Sziráki, 2009 (Argentina).

 – There are dentiform structures or serrations on the sclerotized parts of the male genitalia (Fig. 5.23). 6

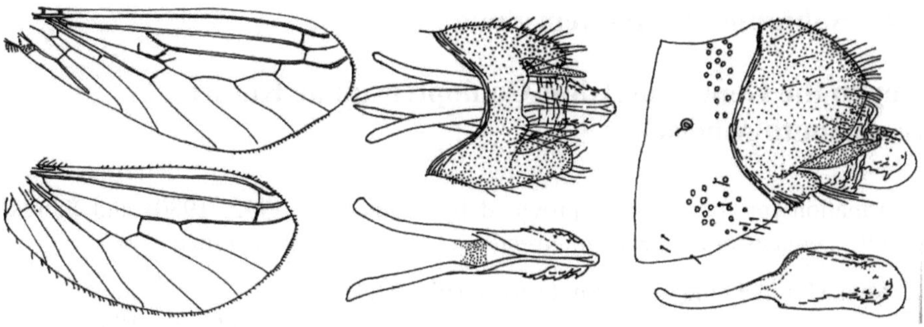

Fig. 5.22 *Pampoconis uncinatus* male: fore and hind wing (left), apex of the abdomen in ventral (upper center) and lateral view (upper right), and the penis in ventral (lower center) and lateral view (lower right). Based on Adams (1973a).

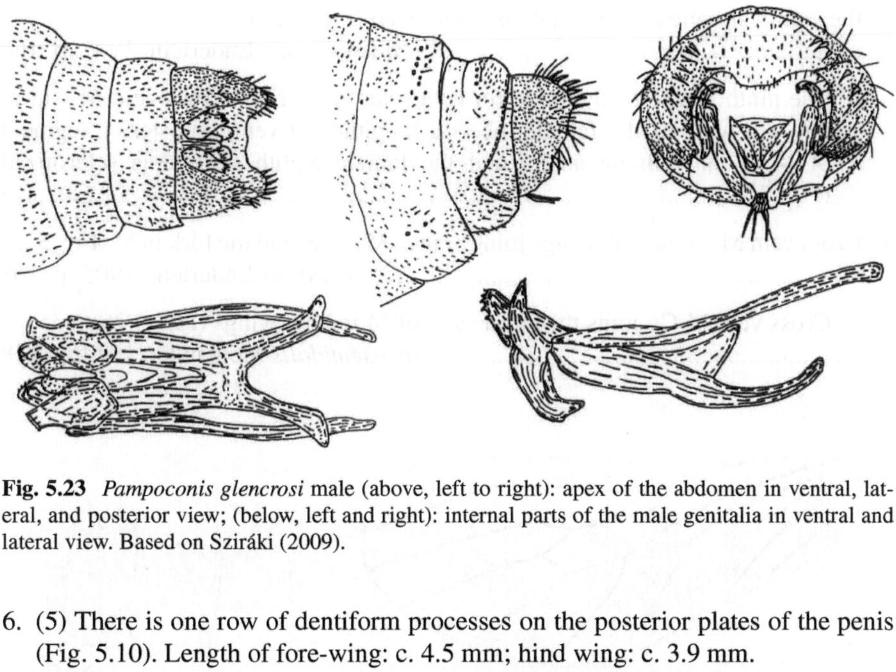

Fig. 5.23 *Pampoconis glencrosi* male (above, left to right): apex of the abdomen in ventral, lateral, and posterior view; (below, left and right): internal parts of the male genitalia in ventral and lateral view. Based on Sziráki (2009).

6. (5) There is one row of dentiform processes on the posterior plates of the penis (Fig. 5.10). Length of fore-wing: c. 4.5 mm; hind wing: c. 3.9 mm. .... *Pampoconis dentifera* Meinander, 1973 (Chile, Argentina).

– There are stout spines only on the dorsal part of the caudal edge of the ectoproct (Fig. 5.23). Length of fore-wing: c. 3.1 mm; width of fore-wing: c. 1.3 mm. Length of hind wing: c. 2.6 mm; width of hind wing: c. 1.2 mm. .... *Pampoconis glencrosi* Sziráki, 2009 (Argentina).

5.4 Subfamily Coniopteryginae

Key to the Genera of Adult Coniopteryginae Known from South America

Information for the key was provided by Meinander (1972, 1980) and Sziráki (2009). Females of the genus *Incasemidalis* cannot yet be identified.

1. Vein M in the hind wing is not forked (Fig. 5.24). ....*Coniopteryx* Curtis, 1834..p. 141
   – Vein M in the hind wing is forked (Fig. 5.25)...2
2. The ninth tergite of the male is obliterated. The cross vein M-Cu_1 in the fore-wing of the female meet a longitudinal vein, the stem of M, at a right angle, and the eighth sternite of the female is sclerotized (Fig. 5.25). ....*Parasemidalis* Enderlein, 1905..p. 166
   – The ninth tergite of the male forms a sclerotized ring. Cross vein M-Cu_1 in the fore-wing of the female meets a longitudinal vein, a posterior segment of M, at an oblique angle, and the female eighth sternite is sclerotized (Fig. 5.26)..3
3. Cross vein M-Cu in both wings joins branch M_{3+4} beyond the fork in M (Fig. 5.2). ....*Semidalis* Enderlein, 1905..p. 168
   – Cross vein M-Cu joins the basal stem of M in both wings (Fig. 5.26). ....*Incasemidalis* Meinander, 1972..p. 178

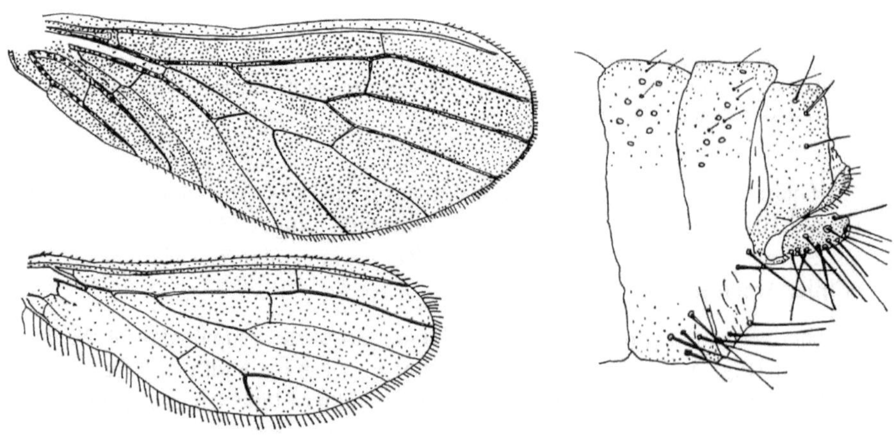

Fig. 5.24 *Coniopteryx phaeoptera* female: fore and hind wing (left) and the apex of the abdomen in lateral view (right). Based on Meinander (1972).

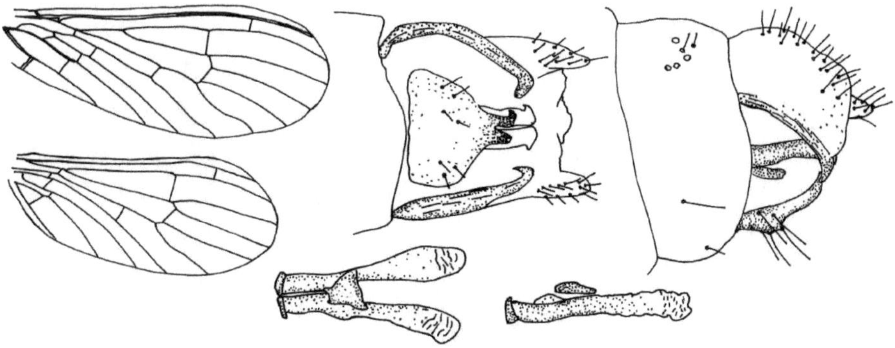

Fig. 5.25 *Parasemidalis subandina* male: fore and hind wing (left, above and below, respectively), apex of the abdomen in ventral (upper center) and lateral view (upper right), and internal genitalia in ventral (lower left center) and lateral view (lower right center). Based on Meinander (1990).

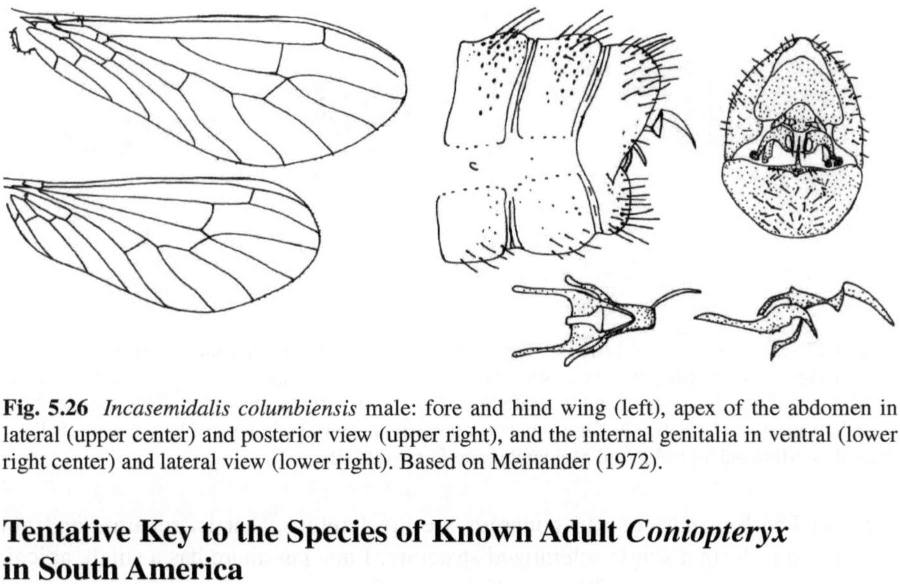

Fig. 5.26 *Incasemidalis columbiensis* male: fore and hind wing (left), apex of the abdomen in lateral (upper center) and posterior view (upper right), and the internal genitalia in ventral (lower right center) and lateral view (lower right). Based on Meinander (1972).

Tentative Key to the Species of Known Adult *Coniopteryx* in South America

Information for this key was provided by Navás (1928, 1930a); Meinander (1972, 1974, 1980, 1983, 1990); Meinander and Penny (1982), and Sziráki (2005, 2009). Females of many species have not yet been described, and males of two are unknown.

1. Males. The apex of the abdomen has a gonarcus and hypandrium, as illustrated in Fig. 5.27. 2
   – Females. The apex of the abdomen is simply rounded (Fig. 5.27). 36

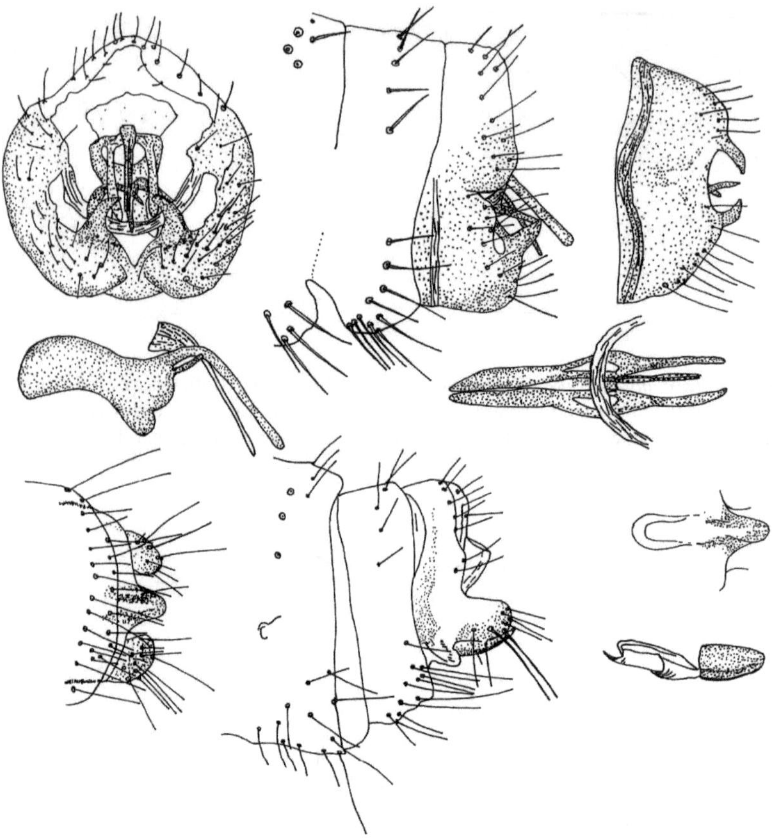

Fig. 5.27 *Coniopteryx ariasi* (above, left to right): apex of the male abdomen in posterior and lateral view; male hypandrium in ventral view; (middle row, left and right respectively): internal genitalia of the male in lateral and ventral view; (lower row, left to right): apex of the female abdomen in ventral and lateral view; female bursa copulatrix in ventral (above) and lateral view (below). Based on Meinander (1980) and Meinander and Penny (1982).

2. (1) The hypandrium and gonarcus are discleretous, that is, separate and not fused to form a single scleratized structure. Each paramere has a small, apical, dorsal process (Fig. 5.28). ....Subgenus *Coniopteryx* sensu stricto.............. 3

   – The sclerites of the gonarcus and hypandrium are synscleretous, that is, fused. There is no small, apical, dorsal process on each paramere, and both curve ventrad toward the apex (Fig. 5.27)... ....Subgenus *Scotoconiopteryx*.................................... 8

3. (2) In lateral view, the hypandrium tapers ventrad, and no processes can be seen. The posterior parts of the parameres are fused. The styli form a band ventral to the aedeagus. In ventral view, the median incision in the hypandrium appears very broad with an irregular but generally rectangular shape due to

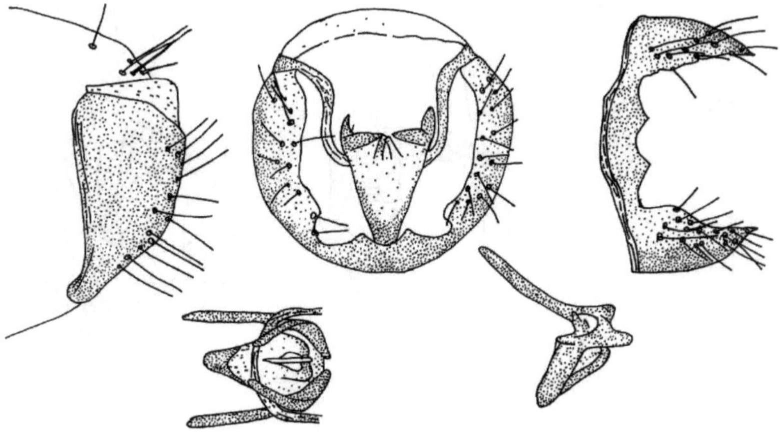

Fig. 5.28 *Coniopteryx gordica* male (above, left to right): apex of the abdomen in lateral and posterior view, hypandrium in ventral view, and (below, left and right): internal genitalia in ventral and lateral view. Based on Meinander (1983).

short terminal extensions along the posterior margin (Fig. 5.28). Length of fore-wing: c. 1.5 mm; hind wing: c. 1.3 mm. The female has not been described. .... *Coniopteryx gordica* Meinander, 1983 (Amazonas).

– In lateral view, the ventral part of the hypandrium does not narrow rapidly, and processes are sometimes apparent. The median incision in the hypandrium appears V-shaped or U-shaped in ventral view (Fig. 5.29). 4

4. (3) The angle between the two branches of the male stylus is greater than 90° in lateral view, and the anterior branch is approximately aligned along the longitudinal axis of the body. The middle apical incision of the hypandrium is broad and V-shaped. The posterior branch of the stylus is nearly straight (Fig. 5.29). The female has not been described, and the male holotype lacks most of its body. .... *Coniopteryx jorgei* Meinander, 1982 (Amazonas).

– The anterior branch of the male stylus is distinctly directed ventrad (Fig. 5.30). 5

5. (4) There is no inner transverse plate attached at the apical incision of the male hypandrium. There is a ventral break in the apodeme along the anterior margin of the hypandrium. The stylus is forked, and its outer branch is sigmoid with its apex pointed posteriad (Fig. 5.31). The antenna consists of about 25–30 segments. Length of fore-wing: 1.7–2.2 mm; hind wing: 1.4–1.8 mm. The female has not been described. .... *Coniopteryx callangana* Enderlein, 1906 (Mexico, Venezuela, Peru, Bolivia, Argentina, Amazonas, Pará, Rio de Janeiro, Santa Catarina).

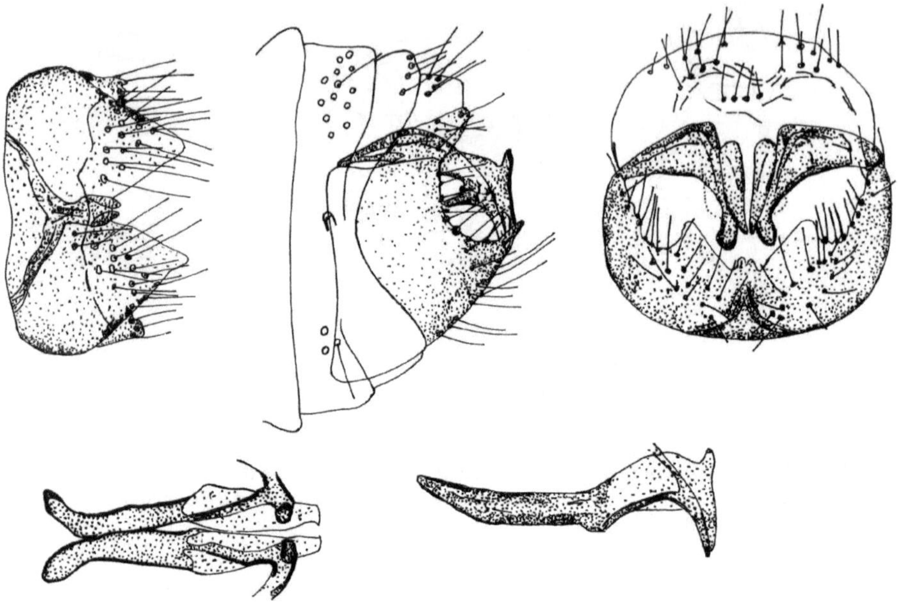

Fig. 5.29 *Coniopteryx jorgei* male (above, left to right): hypandrium in ventral view, apex of the abdomen in lateral and posterior view, and (below, left and right, respectively): the internal genitalia in ventral and lateral view. Based on Meinander and Penny (1982).

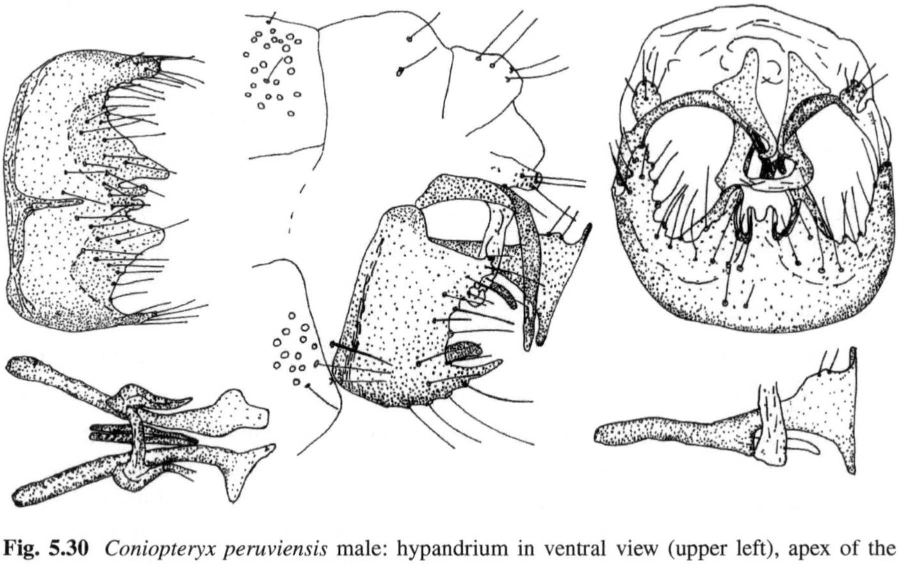

Fig. 5.30 *Coniopteryx peruviensis* male: hypandrium in ventral view (upper left), apex of the abdomen in lateral (center) and posterior view (upper right), and the internal genitalia in ventral (lower left) and lateral view (lower right). Based on Meinander (1990).

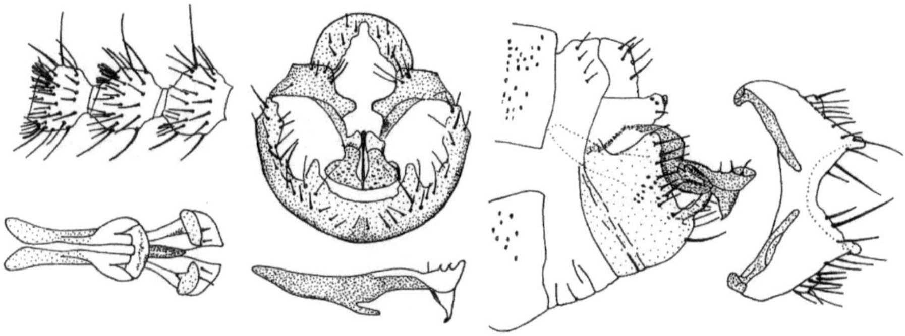

Fig. 5.31 *Coniopteryx callangana* male: eighth through tenth segments of the antenna (upper left), apex of the abdomen in posterior (upper left center) and lateral view (right center), hypandrium in ventral view (right), internal genitalia in ventral view (lower left), and a paramere in lateral view (lower left center). Based on Meinander (1972).

– There is an inner transverse plate, bifurcated distally, along the midline and attached at the U-shaped median apical incision of the male hypandrium (Fig. 5.30). 6

6. (5) In lateral view, the hypandrium appears slightly higher than long, and its apical incision is small and rounded (Fig. 5.30). The antenna consists of 29–30 segments. Length of fore-wing: 2.5–2.6 mm; hind wing: 2.0–2.1 mm. The female has not been described.
.... *Coniopteryx peruviensis* Meinander, 1990
(Peru).

– In lateral view, the hypandrium appears longer than high, and its apical incision is broadly rounded and deep (Fig. 5.32). If it does not appear longer than high, then a longitudinal branch of the apodeme runs ventrally from the anterior margin of the hypandrium to the ventralmost point of the median apical incision. 7

7. (6) The antenna consists of about 23–29 segments, and the basal segments of its flagellum are twice as wide as long (Fig. 5.32). Length of fore-wing: 2.0–3.3 mm; hind wing: 1.6–2.7 mm.
.... *Coniopteryx simplicior* Meinander, 1972
(North and Central America, Jamaica, Venezuela, Bolivia, Paraná).

– The antenna consists of about 25–26 segments, and its basal segments are longer than half their width (Fig. 5.33). Length of fore-wing: c. 2.5 mm; hind wing: c. 1.9–2.0 mm.
.... *Coniopteryx dorisae* Monserrat, 1983
(Venezuela, Colombia).

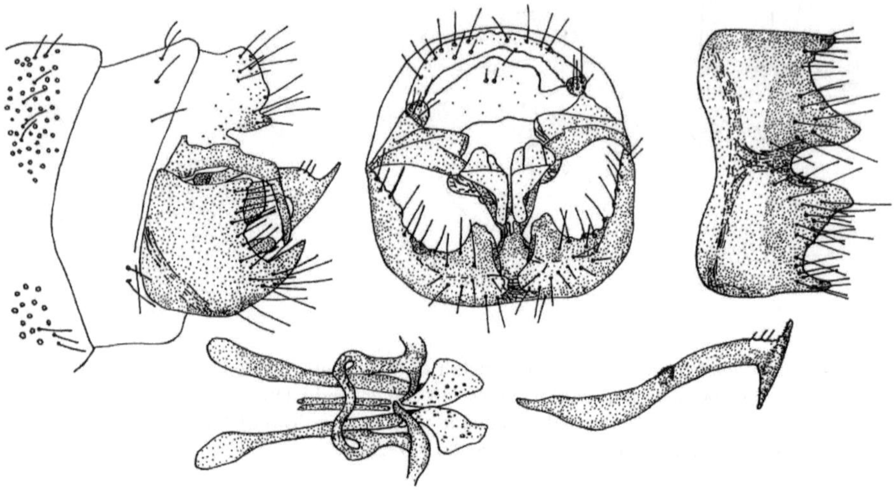

Fig. 5.32 *Coniopteryx simplicior* male (above, left to right): apex of the abdomen in lateral and posterior view, hypandrium in ventral view, and (below, left and right): internal genitalia in ventral and lateral view. Based on Meinander (1983).

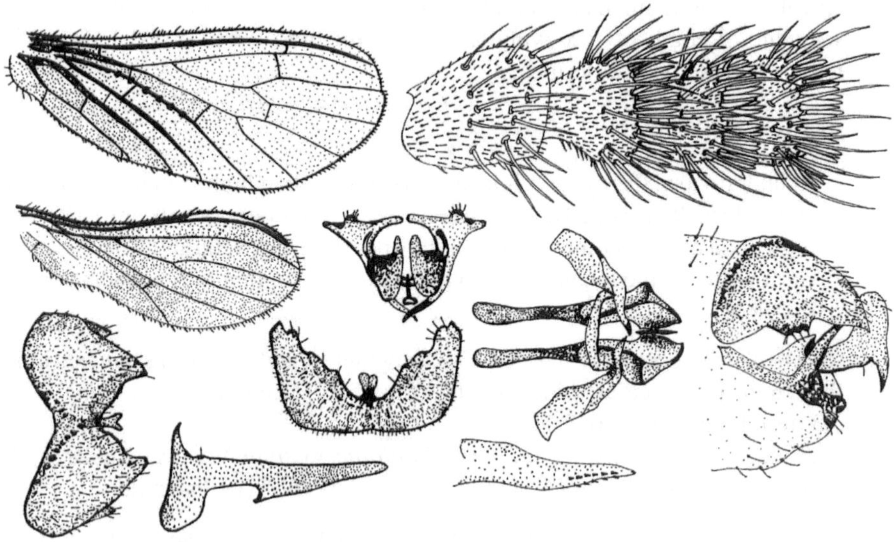

Fig. 5.33 *Coniopteryx dorisae* male: fore (upper left) and hind wing (middle left); basal segments of the antenna in dorsal view (upper right); apex of the abdomen in lateral view (lower right); gonarcus, style, and parameres in ventral (right center) and posterior view (center, right of hind wing); hypandrium in ventral (lower left) and posterior view (center, right of ventral view); paramere in lateral view (lower left center); apex of the posterior branch of the style (lower right center). Based on Monserrat (1983).

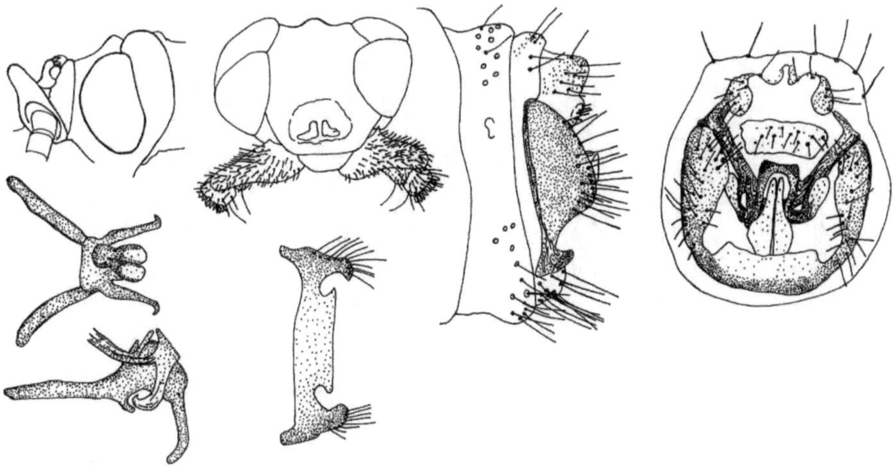

Fig. 5.34 *Coniopteryx paranana* male: head in lateral (upper left) and dorsal view (upper left center), the apex of the abdomen in lateral (right center) and posterior view (right), internal genitalia in ventral (middle left) and lateral view (lower left), and hypandrium in ventral view (lower left center). Based on Meinander (1990).

8. (2) There is an unsclerotized area on the frons bearing two tubercles. The antenna consists of about 29 segments. The very large scape has a dorsal finger-like projection directed laterad. The hypandrium is very short and lacks a lateral process. The styli are attached to the male gonarcus subapically and forked, and the inner branch of each is fused with the other to form an arch ventral to the penis. The male ejaculatory duct is not sclerotized (Fig. 5.34). Length of fore-wing: c. 2.2 mm; hind wing: c. 1.8 mm. The female has not been described. .... *Coniopteryx paranana* Meinander, 1990 (Argentina, Paraná). This species, formerly placed in the subgenera *Coniopteryx* and *Xeroconiopteryx*, has been assigned to *Scotoconiopteryx* by Sziráki (2009).

– Two tubercles and an unsclerotized area in the middle of the frons are absent, or the scape of the antenna is not notably enlarged. 9

9. (8) The styli are forked apically to form a V-shaped structure with two ventral apophyses at the base. The parameres are forked. The description is sketchy, so Figure 5.35 should be consulted as the primary reference. The dark brown antenna consists of about 28 segments. Length of fore-wing: c. 1.5 mm; hind wing: c. 1.2 mm. The female has not been described. .... *Coniopteryx furcata* Meinander, 1983 (Amazonas).

– The styli do not form a V-shaped structure with two ventral apophyses, or the parameres are not forked, or the fore-wing is about 2.0 mm long or longer (Fig. 5.36). 10

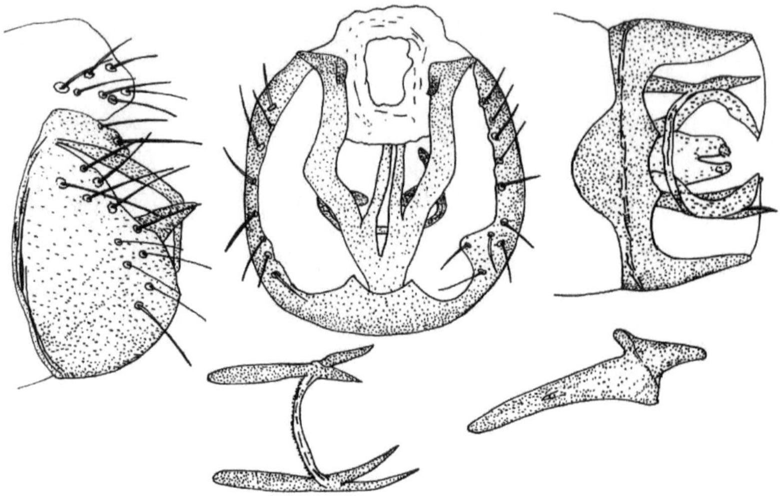

Fig. 5.35 *Coniopteryx furcata* male (above, left to right): apex of the abdomen in lateral and posterior view, hypandrium in ventral view, and (below, left and right): internal genitalia in ventral and lateral view. Based on Meinander (1983).

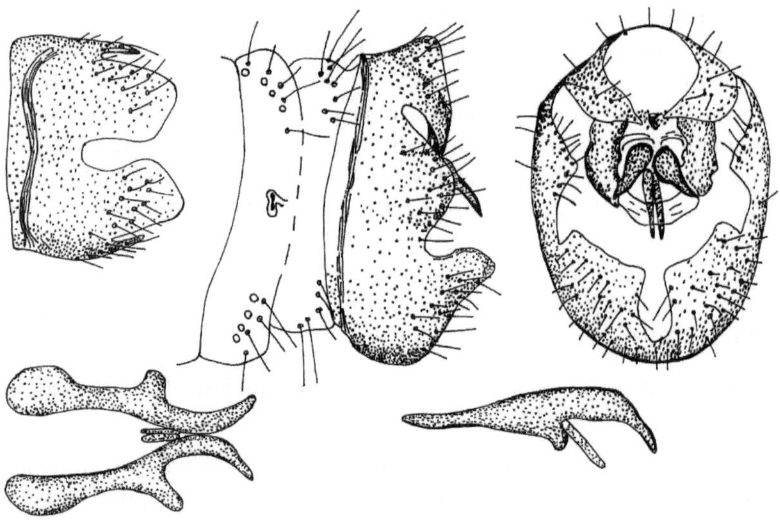

Fig. 5.36 *Coniopteryx paraensis* male (above, left to right): hypandrium in ventral view and apex of the abdomen in lateral and posterior view, and (below, left and right, respectively): internal genitalia in ventral and lateral view. Based on Meinander (1990).

10. (9) The gonarcus itself is discleritous, that is, the lateral plates are not fused, but the apices are connected by a band of sclerotized granules. The terminal processes of the hypandrium in lateral view are broadly rounded and distinctly evident. The sclerotized parts of the styli are not fused but are joined by an unsclerotized band. The parameres are stout, and toward their apices, they are curved ventrad and laterad (Fig. 5.36). Length of fore-wing: c. 2.1 mm; hind wing: c. 1.7 mm. The female has not been described. .... *Coniopteryx paraensis* Meinander, 1990 (Pará).

 – The gonarcus consists either of two separate lateral plates or fused at the apices of the plates (Fig. 5.37). 11

11. (10) The gonarcus itself is synscleritous, that is, with the lateral plates fused to a transverse apodeme (Fig. 5.37). 12

 – The gonarcus is discleritous with lateral plates separated from each other (Fig. 5.27). 20

12. (11) The lateral processes of the hypandrium are narrowly trianguler in lateral view, and between them in ventral view, a pair of long, thin, spine-like processes are evident (Fig. 5.37). The antenna consists of about 31 segments. The head, antennae, and plaps are pale brown. Length of fore-wing: c. 1.6 mm; hind wing: c. 1.4 mm. The female has not been described. .... *Coniopteryx rafaeli* Meinander, 1990 (Amazonas).

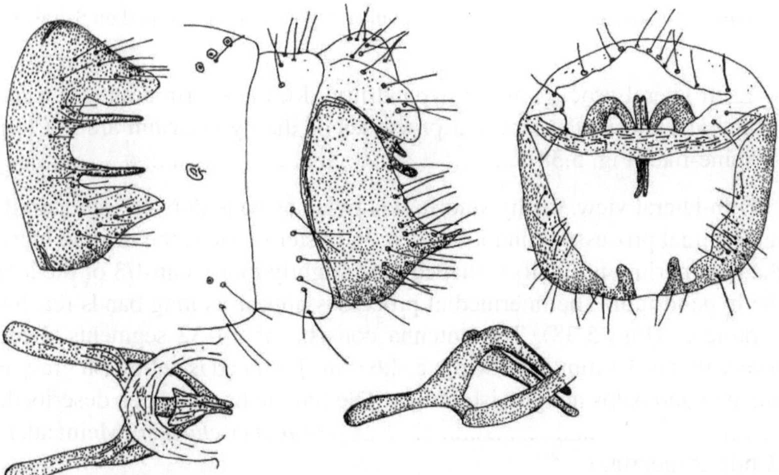

Fig. 5.37 *Coniopteryx rafaeli* male (above, left to right): hypandrium in ventral view and apex of the abdomen in lateral and posterior view, and (below, left and right, respectively): internal genitalia in ventral and lateral view. Based on Meinander (1990).

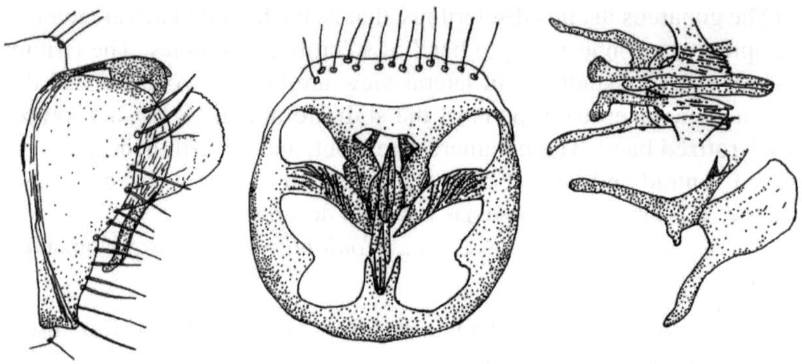

Fig. 5.38 *Coniopteryx chilensis* male: apex of the abdomen in lateral (left) and posterior view (center), and the internal genitalia in ventral (above right) and lateral view (lower right). Based on Meinander (1990).

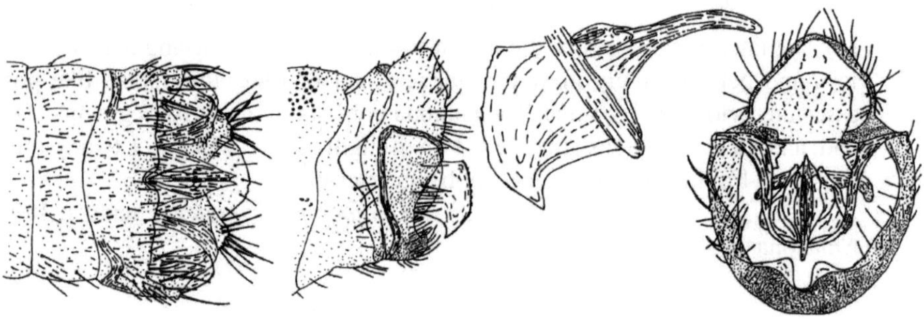

Fig. 5.39 *Coniopteryx calileguana* male (left to right): apex of the abdomen in ventral and lateral view, internal genitalia, and the apex of the abdomen in posterior view. Based on Sziráki (2009).

– Each lateral process of the hypandrium does not form a narrow triangle in lateral view, or the terminal processes of the hypandrium are not long and spine-like (Fig. 5.38). 13

13. (12) In lateral view, the hypandrium is twice as wide dorsally as ventrally, and its terminal processes cannot be seen. In posterior view, the median incision in the hypandrium is small, V-shaped, and slightly more than 1/3 of the length of the hypandrium. The intermedial processes appear as long bands reaching the parameres (Fig. 5.38). The antenna consists of 30–32 segments. Length of fore-wing: c. 3.1 mm; hind wing: c. 2.6 mm. The head is yellowish gray, and the antenna and palps are grayish brown. The female has not been described. .... *Coniopteryx chilensis* Meinander, 1990 (Chile, Argentina).

– In lateral view, the hypandrium appears as wide or wider ventrally than dorsally. In posterior view, the median incision is not approximately V-shaped (Fig. 5.39). 14

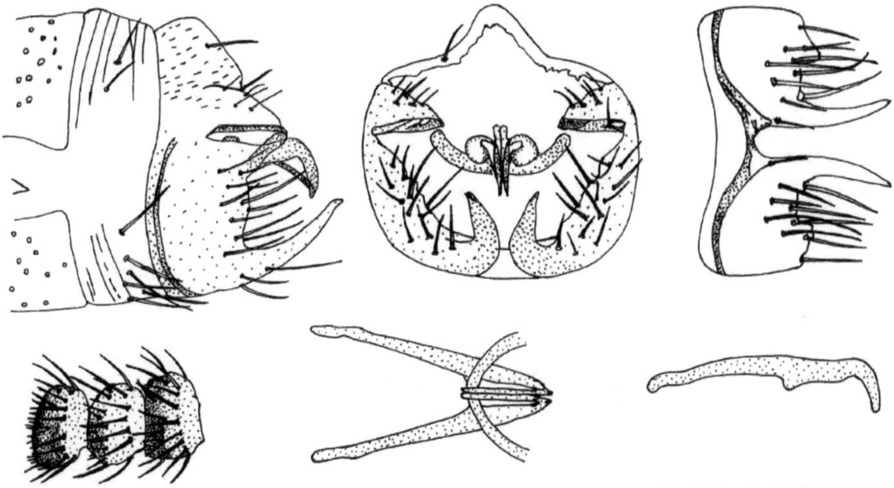

Fig. 5.40 *Coniopteryx isthmicola* male (above, left to right): apex of the abdomen in lateral and posterior view, hypandrium in ventral view, and (below, left to right): eighth through tenth segments of the antenna, internal genitalia in ventral view, and a paramere in lateral view. Based on Meinander (1972).

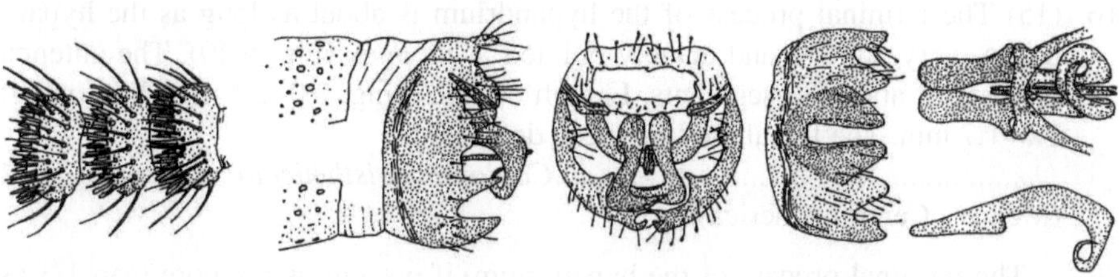

Fig. 5.41 *Coniopteryx fumicolor* male (left to right): eighth through tenth segments of the antenna, apex of the abdomen in lateral and posterior view, hypandrium in ventral view, and the internal genitalia in ventral (above) and lateral view (below). Based on Meinander (1972).

14. (13) The hypandrium of the male bears terminal and lateral processes, but sclerotized intermedial processes are absent. In posterior view, a deep and narrow incision is evident on the terminal process. In lateral view, the apical part of the paramere appears very wide (Fig. 5.39). Length of fore-wing: 2.5–2.7 mm; hind wing: 2.1–2.3 mm. The female has not been described. .... *Coniopteryx calileguana* Sziráki, 2009 (Argentina).

    – If two pairs of terminal process are present on the hypandrium, then one is terminal, and the other is lateral. Otherwise, only one process is obviously present (Fig. 5.40). 15

15. (14) Only one pair of apical process of the hypandrium is prominent in lateral view (Fig. 5.40). The fore-wing is less than about 2.5 mm long. 16

    – Two pairs of processes are prominent on the hypandrium in lateral view (Fig. 5.41). 18

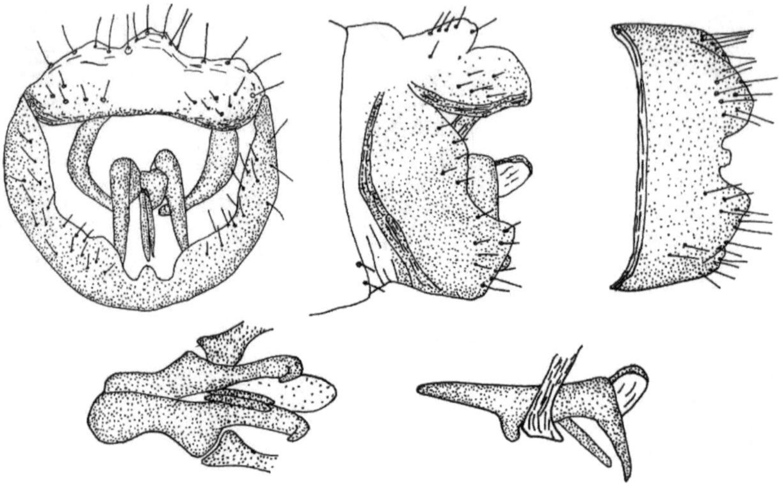

Fig. 5.42 *Coniopteryx indivisa* male (above, left to right): apex of the abdomen in posterior and lateral view, hypandrium in ventral view, and (below, left and right): internal genitalia in ventral and lateral view. Based on Meinander (1980).

16. (15) The terminal process of the hypandrium is about as long as the hypandrium, very narrow, and acutely pointed at its apex (Fig. 5.40). The antenna consists of about 32 segments. Length of fore-wing: 2.0–2.1 mm; hind wing: 1.6–1.7 mm. The female has not been described.
    .... *Coniopteryx isthmicola* Meinander, 1972
    (Mexico, Central America).

    – The terminal process of the hypandrium, if present, is no more than 1/3 as long as the hypandrium (Fig. 5.42). 17

17. (16) The hypandrium is incised along the midline to a depth equal to about 1/3 of its own maximum length, measured along its longitudinal axis. The ventral part of the apodeme is entire at the apex (Fig. 5.42). The antenna consists of about 30 segments. Fore-wing length: c. 1.8 mm; hind wing: c. 1.4 mm. The female has not been described.
    .... *Coniopteryx indivisa* Meinander, 1980
    (Amazonas, Pará).

    – The hypandrium is incised along the midline to a depth equal to about 2/3 of its own maximum length, measured along its longitudinal axis (Fig. 5.43). The antenna consists of about 28 segments. Length of fore-wing: 1.7–1.8 mm; hind wing: 1.2–1.4 mm. The female has not been described.
    .... *Coniopteryx silvicola* Meinander, 1982
    (Amazonas).

18. (15) The apical part of each paramere curves at least about 360° to form a spiral. It is strongly curved toward the apex but does not encircle the penis

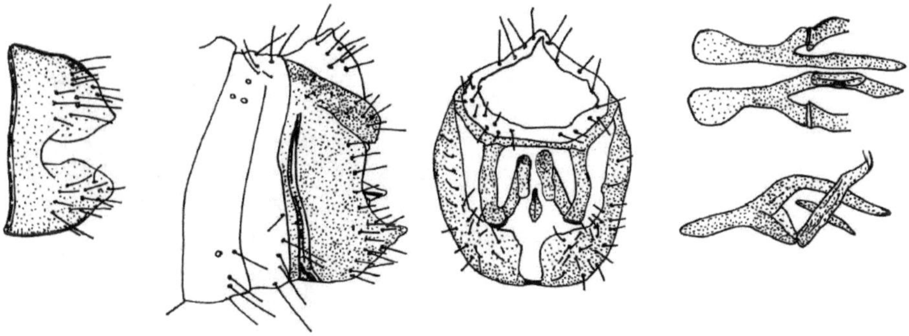

Fig. 5.43 *Coniopteryx silvicola* male (left to right): hypandrium in ventral view, apex of the abdomen in lateral and posterior view, and the internal genitalia in ventral (above) and lateral view (below). Based on Meinander and Penny (1982).

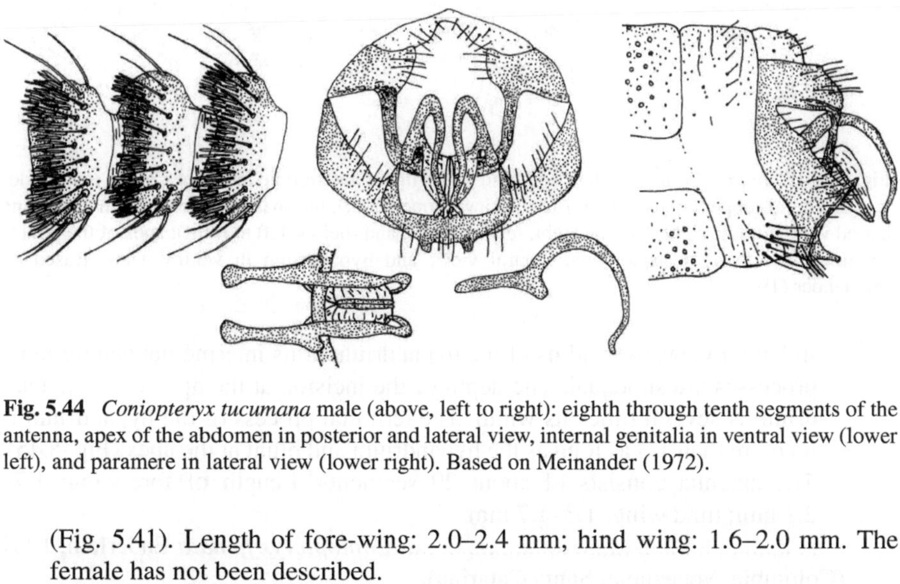

Fig. 5.44 *Coniopteryx tucumana* male (above, left to right): eighth through tenth segments of the antenna, apex of the abdomen in posterior and lateral view, internal genitalia in ventral view (lower left), and paramere in lateral view (lower right). Based on Meinander (1972).

(Fig. 5.41). Length of fore-wing: 2.0–2.4 mm; hind wing: 1.6–2.0 mm. The female has not been described.
.... *Coniopteryx fumicolor* Meinander, 1972 (Central America, Venezuela).

– The apical part of a paramere bends up to about 180° (Fig. 5.44). 19

19. (18) In lateral view, the hypandrium appears about twice as wide at its intermedial process as at its terminal process. The apex of the paramere curves to form a semicircle (Fig. 5.44). The median apical incision at the apex of the hypandrium is broader than long. The antenna consists of 31–36 segments. Length of fore-wing: 2.1–2.8 mm; hind wing: 1.8–2.4 mm.
.... *Coniopteryx tucumana* Navás, 1930 (Colombia, Venezuela, Argentina, Uruguay, Santa Catarina, Pará).

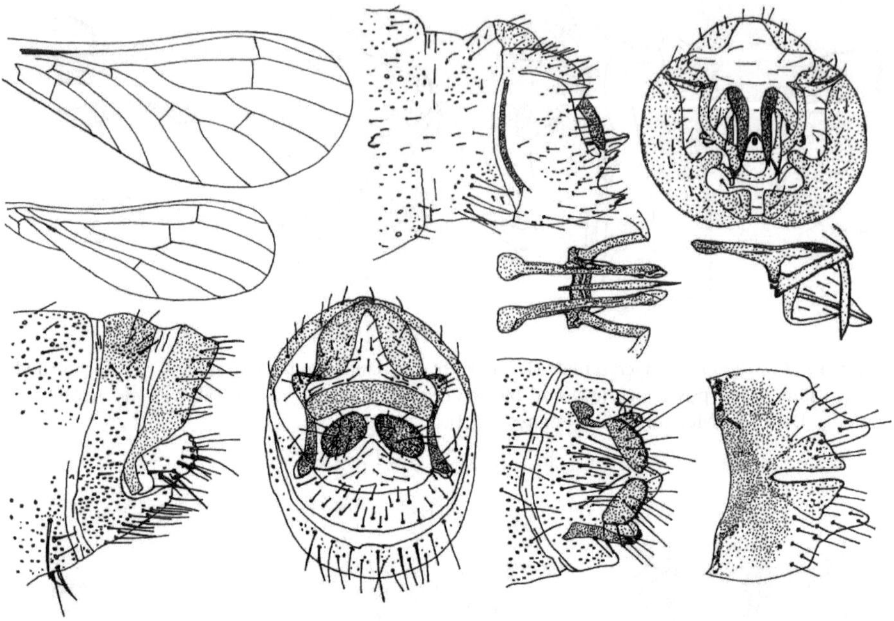

Fig. 5.45 *Coniopteryx fumata:* fore and hind wing (upper and middle left), apex of the male abdomen in lateral (upper center) and posterior view (upper right), internal male genitalia in dorsal and lateral view (middle right, left and right, respectively), and (below, left to right): apex of the female abdomen in lateral, posterior, and ventral view; and hypandrium in ventral view. Based on Meinander (1972).

– In lateral view, the widths of the hypandrium at its intermedial and terminal processes are subequal. The depth of the incision at the apex of the hypandrium is several times its width. The terminal process of the hypandrium is no more than 1/3 as long as the hypandrium and blunt at the apex (Fig. 5.45). The antenna consists of about 29 segments. Length of fore-wing: 1.7–2.1 mm; hind wing: 1.3–1.7 mm.
.... *Coniopteryx fumata* Enderlein, 1907 (Colombia, Venezuela, Santa Catarina).

20. (11) The terminal process of the hypandrium bears two or more finger-like apophyses, paired laterally on each side of the median apical incision. The gonarcus does not extend inward as far as the base of the styli. The paramere is not spirally coiled around the penis (Fig. 5.46). 21

– The terminal process of the hypandrium is simple and lacks digitate apophyses (Fig. 5.47). 23

21. (20) The hypandrium bears only two very long, finger-like apophyses, one on each side of the median apical incision (Fig. 5.46). Each apophysis is much longer than four times its own width. The antenna of the male consists of about 25 segments. The general coloration is grayish brown. The antenna is dark

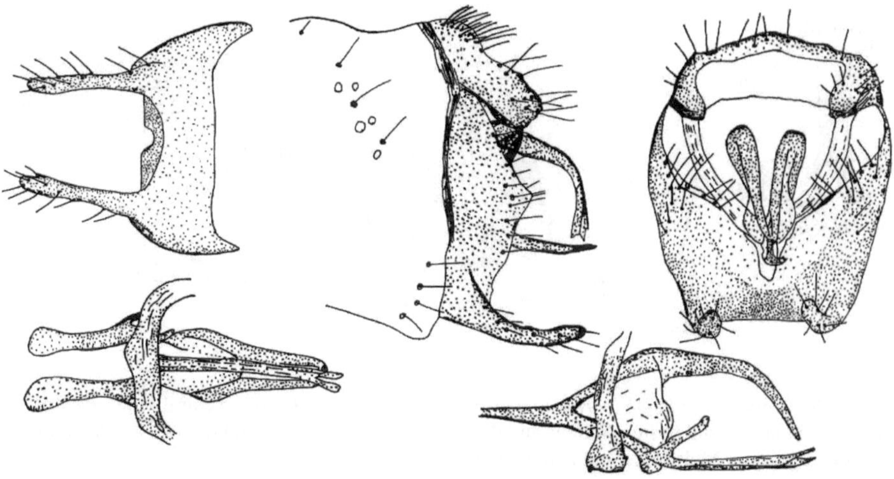

Fig. 5.46 *Coniopteryx bicornis* male (above, left to right): hypandrium and gonarcus in ventral view and apex of the abdomen in lateral and posterior view, and (below, left and right): internal genitalia in ventral and lateral view. Based on Meinander and Penny (1982).

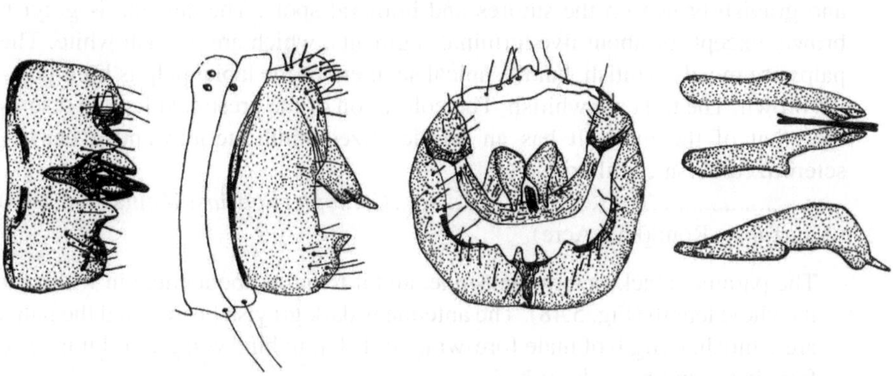

Fig. 5.47 *Coniopteryx rondoniensis* male (left to right): hypandrium in ventral view, apex of the abdomen in lateral and posterior view, and the internal genitalia in ventral (above) and lateral view (below). Based on Meinander and Penny (1982).

grayish brown, and the palps are lighter grayish brown. Length of male fore-wing: c. 1.7 mm; hind wing: c. 1.4 mm. The female has not been described. .... *Coniopteryx bicornis* Meinander, 1982 (Amazonas, Acre).

– There are four finger-like apophyses on the hypandrium, each relatively short, less than about four times its own width (Fig. 5.47). The antenna consists of about 31 or 32 segments. 22

22. (21) The paramere is very broad and has a dorsal, apical spine, which is long and narrow (Fig. 5.27). Length of male fore-wing: c. 1.8 mm; hind wing: c. 1.4 mm.

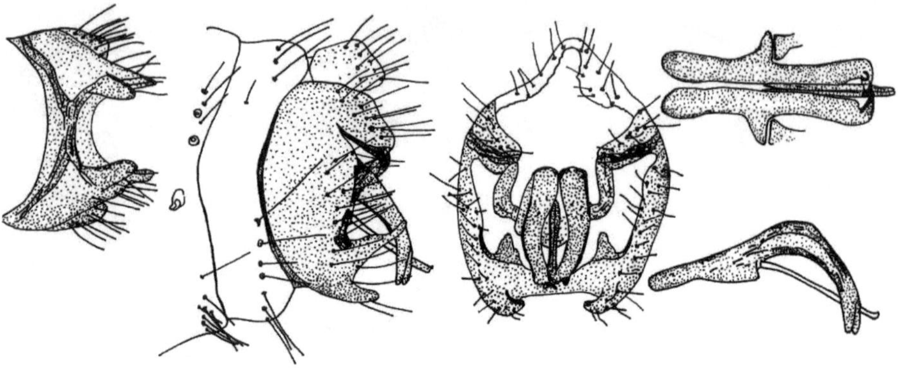

Fig. 5.48 *Coniopteryx quadricornis* male (left to right): hypandrium in ventral view, apex of the abdomen in lateral and posterior view, and the internal genitalia in ventral (above) and lateral view (below). Based on Meinander and Penny (1982).

Fore-wing length of females presumed to be conspecific with male type: c. 1.9 mm; hind wing: c. 1.5 mm. The color is light brown on the sclerotized parts and grayish brown on the sutures and humeral spots. The antenna is grayish brown, except for about five terminal segments, which are grayish white. The palps are mostly whitish, but the apical segment of the labial palp is light grayish brown. The tarsi are whitish. The coloration of the presumed female resembles that of the male. It has an unsclerotized ninth sternite and a heavily sclerotized bursa copulatrix.
.... *Coniopteryx ariasi* Meinander, 1980 (Amazonas, Rondônia, Acre).

– The paramere lacks a terminal spine, and it remains about equal in width for its whole length (Fig. 5.48). The antenna is dark grayish brown, and the palps are whitish. Length of male fore-wing: c. 1.8 mm; hind wing: c. 1.4 mm. The female has not been described.
.... *Coniopteryx quadricornis* Meinander, 1982 (Rondônia, Pará).

23. (20) Anterior to the median apical incision in the hypandrium, there is a strongly sclerotized plate (Fig. 5.47). 24

– A strongly sclerotized plate is not present anterior to the median apical incision of the hypandrium (Fig. 5.49). 25

24. (23) The paramere is slightly curved, broad, and blunt (Fig. 5.47). The antenna consists of about 34 segments and is light grayish brown. The head is light ochraceous brown, and the palps are light grayish. Length of male fore-wing: c. 1.7 mm; hind wing: c. 1.5 mm. The female has not been described.
.... *Coniopteryx rondoniensis* Meinander, 1982 (Rondônia).

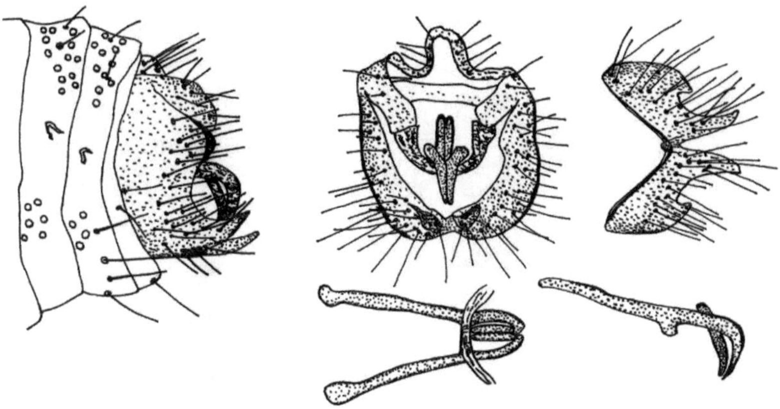

Fig. 5.49 *Coniopteryx biapicata* male (above, left to right): apex of the abdomen in lateral and posterior view, hypandrium in ventral view, and (below, left and right): internal genitalia in ventral and lateral view. Based on Meinander (1983).

– The paramere is rather slender and curves ventrad as it tapers toward the apex (Fig. 5.50). The antenna consists of about 35–36 segments and is dark brown. The palps are white. Length of male fore-wing: 1.8–1.9 mm; hind wing: 1.5–1.6 mm.
.... *Coniopteryx cucuminicola* Meinander, 1982 (Rondônia, Pará).

25. (23) The terminal process on the hypandrium is long, curves slightly outward and dorsad, and tapers to an acute point. The central incision on the hypandrium appears deep, narrow, and V-shaped in ventral view, and the lateral process appears broadly rounded. The parameres are bent ventrad toward the apices (Fig. 5.49). Length of fore-wing: c. 2.0 mm; hind wing: c. 1.7 mm. The head and antennae are ochraceous brown, and the eyes are black. The thorax is mainly light brown, and the legs are pale. The female has not been described.
.... *Coniopteryx biapicata* Meinander, 1983 (Uruguay).

– The terminal process on the hypandrium is not long, or it is not acutely pointed at the apex, or it does not curve slightly outward and dorsad. If in doubt, the median incision on the hypandrium is not narrow and V-shaped in ventral view (Fig. 5.51).. 26

26. (25) The terminal and intermedial processes of the hypandrium are both prominent in lateral view (Fig. 5.51). 27
Only one apical process of the hypandrium is prominent in lateral view (Fig. 5.52). 30

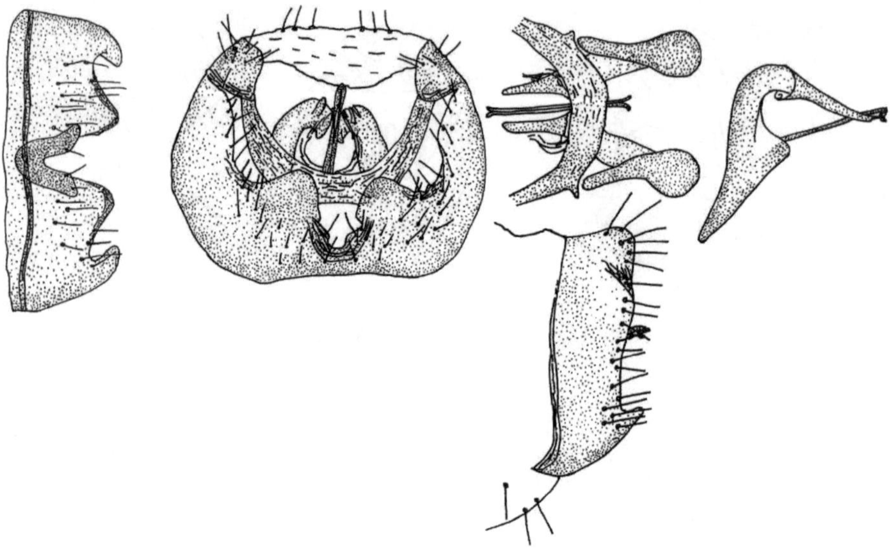

Fig. 5.50 *Coniopteryx cucuminicola* male (above, left to right): hypandrium in ventral view, apex of the abdomen in posterior view, internal genitalia in ventral and lateral view, and (below) apex of the abdomen in lateral view. Based on Meinander and Penny (1982).

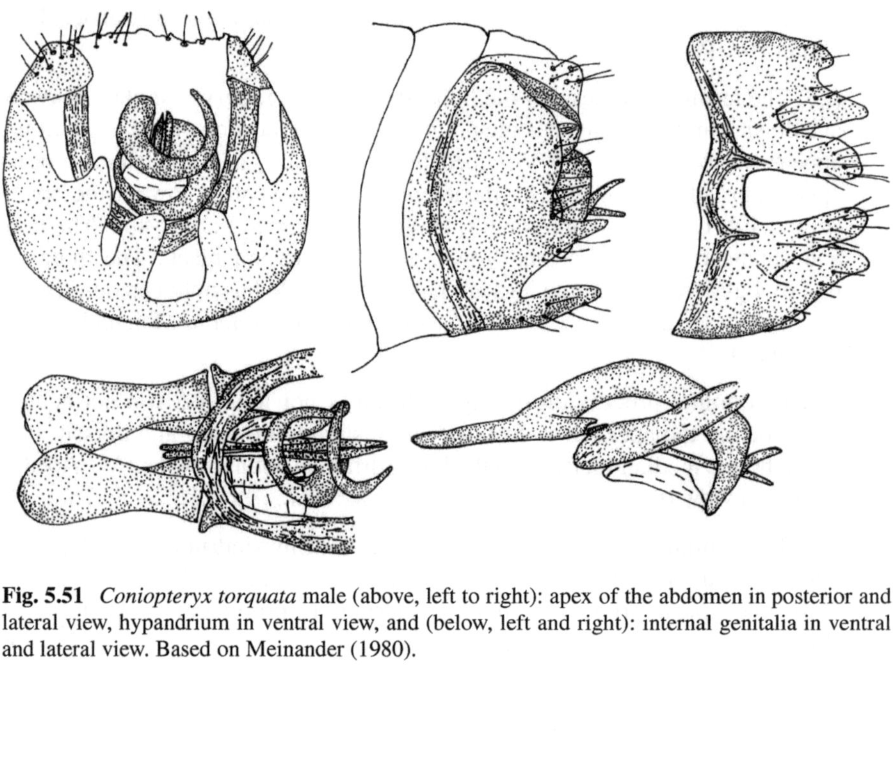

Fig. 5.51 *Coniopteryx torquata* male (above, left to right): apex of the abdomen in posterior and lateral view, hypandrium in ventral view, and (below, left and right): internal genitalia in ventral and lateral view. Based on Meinander (1980).

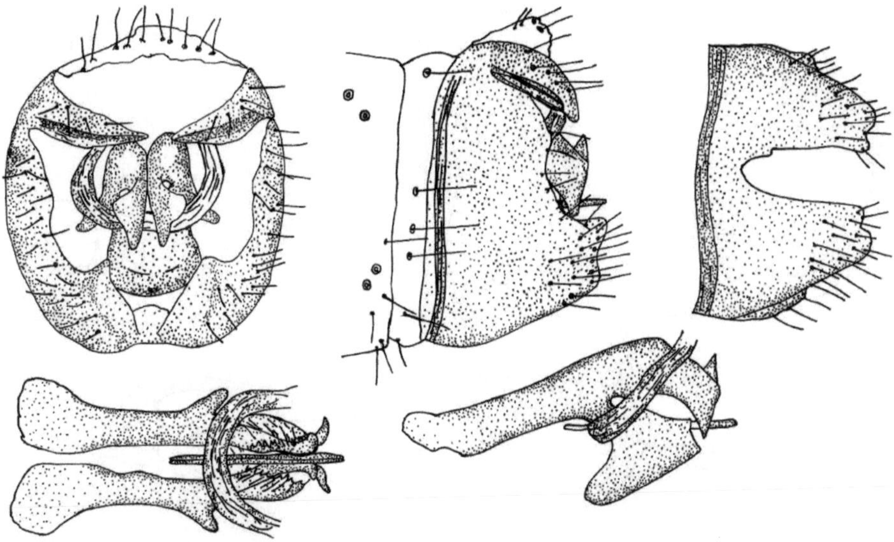

Fig. 5.52 *Coniopteryx pennyi* male (above, left to right): apex of the abdomen in posterior and lateral view, hypandrium in ventral view, and (below, left and right): internal genitalia in ventral and lateral view. Based on Meinander (1980).

27. (26) The apices of the parameres form about two complete spiral coils around the penis (Fig. 5.51). Length of fore-wing: c. 1.8 mm; hind wing: c. 1.4 mm. The female has not been described. .... *Coniopteryx torquata* Meinander, 1980 (Venezuela, Amazonas).

   – The apical part of each paramere does not encircle the penis and is usually bent only slightly ventrad (Fig. 5.53). 28

28. (27) In lateral view, the hypandrium appears about twice as wide at its intermedial process as at its terminal process. The paramere is simple and bears no processes; at its apex, it curves slightly so that it bends to a point 60° ventrad (Fig. 5.53). Length of fore-wing: c. 1.8 mm; hind wing: c. 1.4 mm. The female has not been described. .... *Coniopteryx amazonica* Meinander, 1980 (Amazonas).

   – In lateral view, the widths of the hypandrium at its intermedial and terminal processes are subequal (Fig. 5.54). 29

29. (28) The median apical incision in the hypandrium is much shallower than half the width of the hypandrium. The paramere broadens toward the apex to twice its width at the base (Fig. 5.54). Length of fore-wing: c. 1.9 mm; hind wing: c. 1.5 mm. The female has not been described. .... *Coniopteryx flinti* Meinander, 1975 (Mexico, Rondônia).

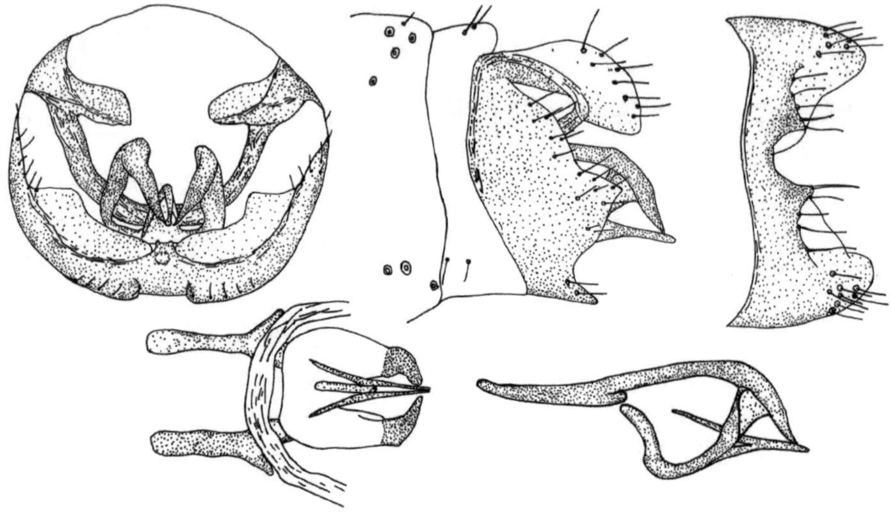

Fig. 5.53 *Coniopteryx amazonica* male (above, left to right): apex of the abdomen in posterior and lateral view, hypandrium in ventral view, and (below, left and right): internal genitalia in ventral and lateral view. Based on Meinander (1980).

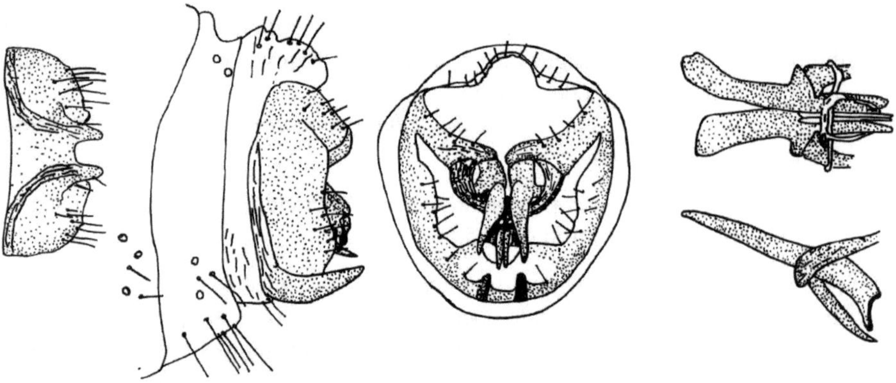

Fig. 5.54 *Coniopteryx flinti* male (left to right): hypandrium in ventral view, apex of the abdomen in lateral and posterior view, and the internal genitalia in ventral (above) and lateral view (below). Based on Meinander and Penny (1982).

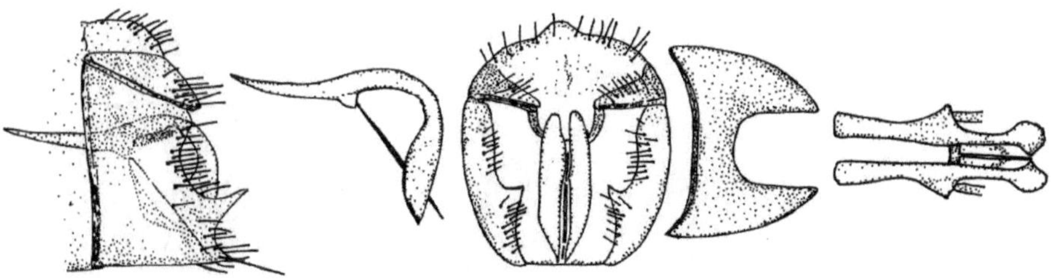

Fig. 5.55 *Coniopteryx cyphodera* male (left to right): apex of the abdomen in lateral view, partially cleared to show the position of the internal genitalia; internal genitalia in lateral view; apex of the abdomen in posterior view; hypandrium in ventral view; internal genitalia in ventral view. Based on Meinander (2002).

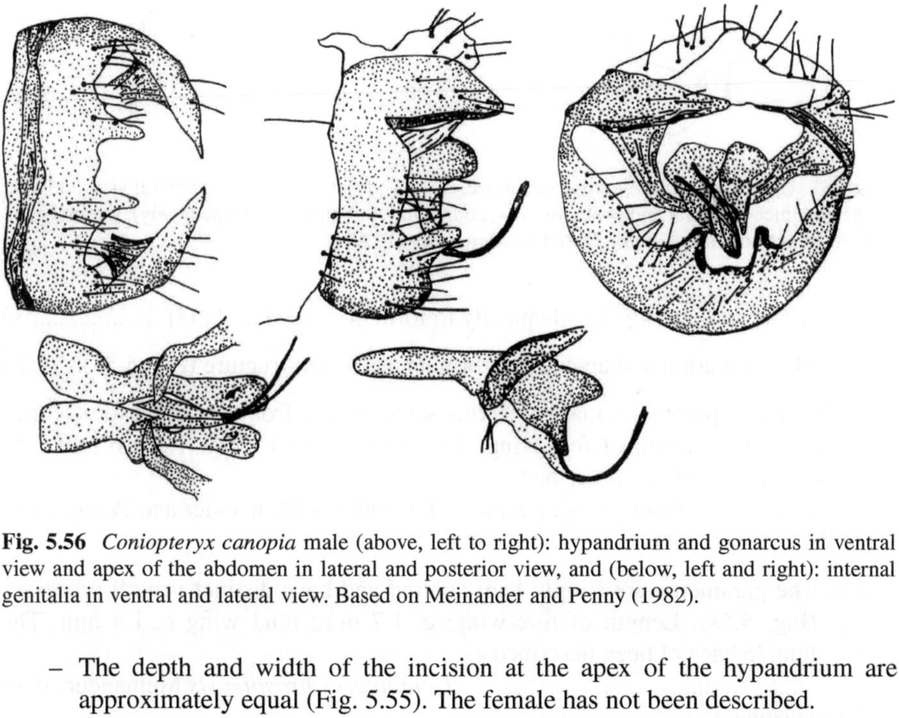

Fig. 5.56 *Coniopteryx canopia* male (above, left to right): hypandrium and gonarcus in ventral view and apex of the abdomen in lateral and posterior view, and (below, left and right): internal genitalia in ventral and lateral view. Based on Meinander and Penny (1982).

– The depth and width of the incision at the apex of the hypandrium are approximately equal (Fig. 5.55). The female has not been described. .... *Coniopteryx cyphodera* Johnson, 1978 (Panama).

30. (26) The apices of the parameres are connected to a large membranous, sclerotized structure ventral to the penis. The paramere bears a structure distally, which curves outward. The median incision at the apex of the hypandrium is about twice as deep as wide. (Fig. 5.52). Length of fore-wing: 1.8–1.9 mm; hind wing: 1.4–1.5 mm. The female has not been described. ....*Coniopteryx pennyi* Meinander, 1980 (Rondônia, Amazonas).

– There is no large membranous structure ventral to the penis to which the parameres are connected (Fig. 5.56). 31

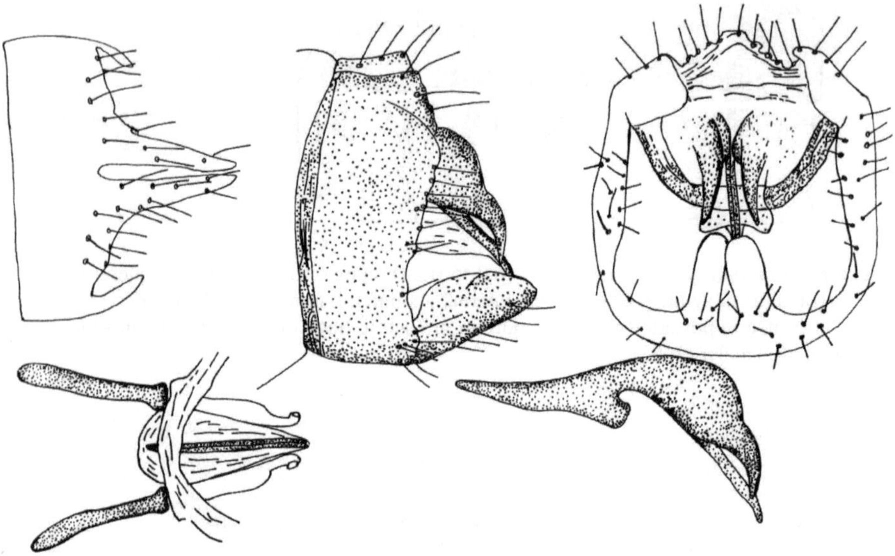

Fig. 5.57 *Coniopteryx bilinguata* male (above, left to right): hypandrium in ventral view and apex of the abdomen in lateral and posterior view, and (below, left and right, respectively): internal genitalia in ventral and lateral view. Based on Meinander (1990).

31. (30) The styli are not fused apically to form an arch (Fig. 5.57). 32

    – The styli are fused apically to form an arch-like structure (Fig. 5.58). 33

32. (31) Each paramere maintains the same width from its base to its apex (Fig. 5.56). Length of fore-wing: 1.8–2.0 mm; hind wing: 1.5–1.6 mm. The female has not been described.
    .... *Coniopteryx canopia* Meinander in Meinander and Penny, 1982 (Pará, Amazonas).

    – The parameres are simply bent ventrad and have distinct ventral processes (Fig. 5.58). Length of fore-wing: c. 1.7 mm; hind wing c. 1.4 mm. The female has not been described.
    .... *Coniopteryx brasiliensis* Meinander, 1983 (Amazonas).

33. (31) The terminal processes on the hypandrium consist of a pair of tongue-like processes as long as the ventral width of the hypandrium and bordering a deep and narrow median ventral incision (Fig. 5.57). There are 33 or 34 segments of the antenna. Length of fore-wing: 1.7–1.8 mm; hind wing: c. 1.4 mm. The female has not been described.
    .... *Coniopteryx bilinguata* Meinander, 1990 (Rondônia, Pará).

    – Terminal processes on the hypandrium are not well developed and border a broad median ventral incision (Fig. 5.59). 34

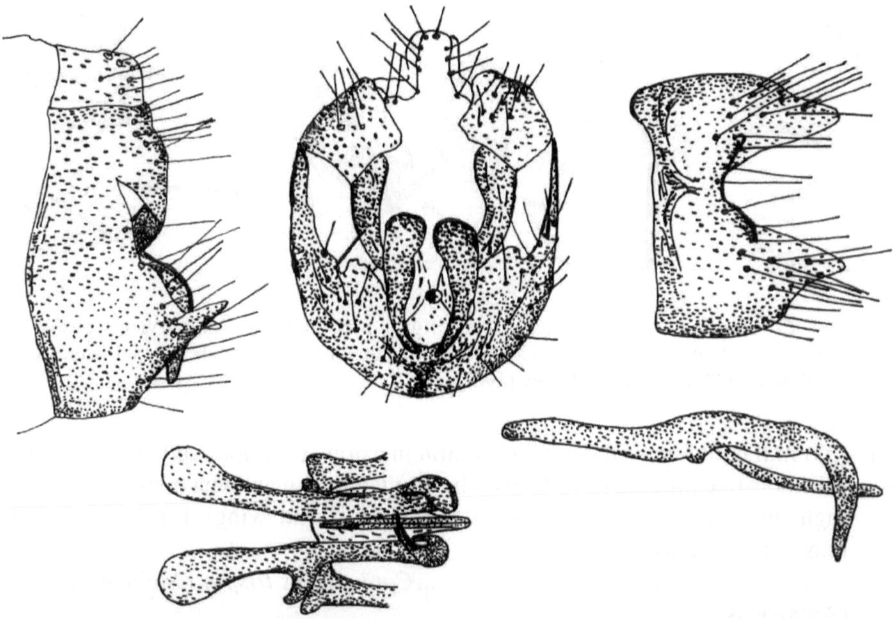

Fig. 5.58 *Coniopteryx brasiliensis* male (above, left to right): apex of the abdomen in lateral and posterior view, hypandrium in ventral view, and (below, left and right): internal genitalia in ventral and lateral view. Based on Meinander (1983).

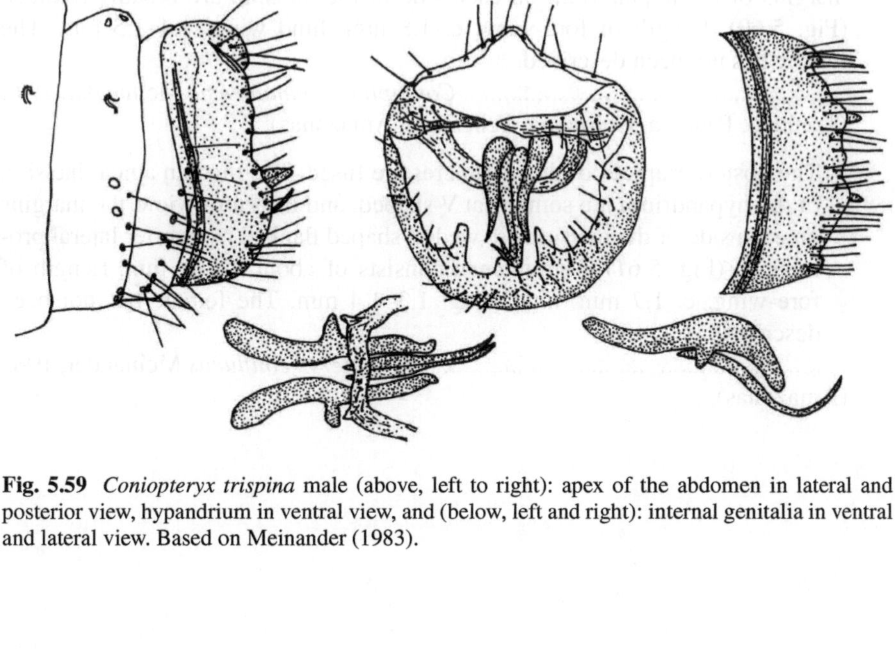

Fig. 5.59 *Coniopteryx trispina* male (above, left to right): apex of the abdomen in lateral and posterior view, hypandrium in ventral view, and (below, left and right): internal genitalia in ventral and lateral view. Based on Meinander (1983).

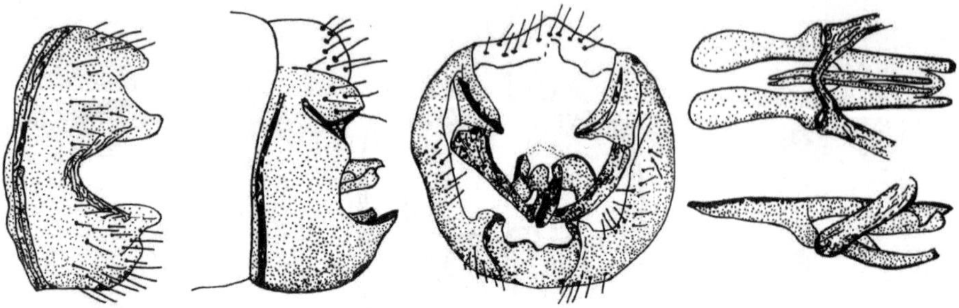

Fig. 5.60 *Coniopteryx panamensis* male (left to right): hypandrium in ventral view, apex of the abdomen in lateral and posterior view, and internal genitalia in ventral (above) and lateral view (below). Based on Meinander and Penny (1982).

34. (33) In the middle of the median ventral incision at the apex of the hypandrium is a knob-like process (Fig. 5.59). The grayish brown antenna consists of 20–29 sagments. Length of fore-wing: 1.6–1.7 mm; hind wing: 1.4–1.5 mm. The female has not been described.
 .... *Coniopteryx trispina* Meinander, 1983 (Amazonas).

 – The median ventral incision at the apex of the hypandrium is evenly curved without a median protuberance (Fig. 5.60). 35

35. (34) The parameres are joined by a membrane dorsal to the penis. The median apical incision of the hypandrium is broad and shallow, and in ventral view, the margins of the hypandrium on each side of the incision are broadly rounded (Fig. 5.60). Length of fore-wing: c. 1.8 mm; hind wing: 1.4–1.5 mm. The female has not been described.
 .... *Coniopteryx panamensis* Meinander, 1974 (Trinidad, Panama, Colombia, Venezuela, Amazonas).

 – The posterior apices of the parameres are fused. The median apical incision of the hypandrium are somewhat V-shaped, and in ventral view, the margins on each side of the incision are wedge-shaped flanked by shorter lateral protrusions (Fig. 5.61). The antenna consists of about 32–33 mm. Length of fore-wing: c. 1.7 mm; hind wing: 1.3–1.4 mm. The female has not been described.
 .... *Coniopteryx confluens* Meinander, 1983 (Amazonas).

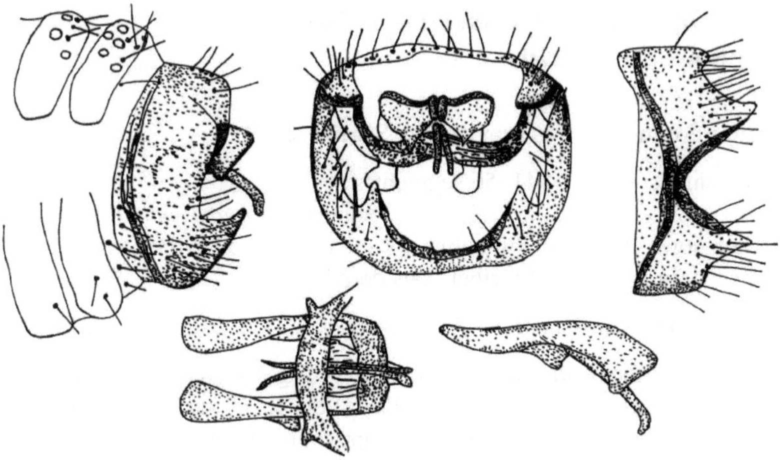

Fig. 5.61 *Coniopteryx confluens* male (above, left to right): apex of the abdomen in lateral and posterior view, hypandrium in ventral view, and (below, left and right): internal genitalia in ventral and lateral view. Based on Meinander (1983).

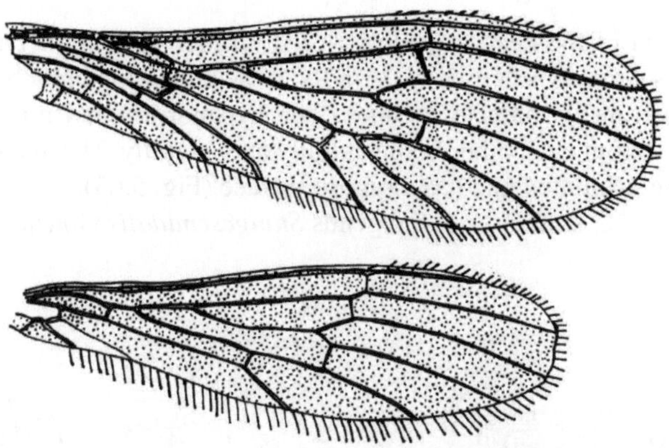

Fig. 5.62 Fore and hind wing of *Coniopteryx angustipennis*. Based on Meinander (1972).

36. (1) Length of fore-wing of female: 2.2–2.6 mm; hind wing: minimum, c. 1.8 mm. The color is dark gray, and the wings are narrow (Fig. 5.62). The male has not been described.
....*Coniopteryx angustipennis* Enderlein, 1906 (Paraguay, Argentina).

– Length of fore-wing of female: c. 3.2 mm; hind wing: c. 2.5 mm. The color is dark gray, and the wings widen toward the apex (Fig. 5.24). The antenna consists of about 39 segments.
....*Coniopteryx phaeoptera* (Enderlein, 1906) (Peru). Syn: *Parasemidalis phaeoptera* Enderlein, 1906; *Neosemidalis phaeoptera* (Enderlein, 1906) Enderlein, 1930.

Key to Subgenera of *Parasemidalis* Known Throughout the World

Information for the key was provided by Sziráki (2009).

1. The antenna consists of 41–56 segments. .... Subgenus *Canarisemidalis* Sziráki, 2009
   The known species in this subgenus are all confined to the Canary Islands and southwestern Africa. Not further considered.
   – The antenna consists of 27–34 segments. 2
2. The height of the head capsule is equal to or slightly greater than its length. The frons of the female is weakly sclerotized. Vein Rs in the hind wing branches at 1/3 of the distance from the base to the apex. The sclerotized parts of the hypandrium and ninth segment are fused. The anterior part of the paramere is slender and has a smooth surface. .... .Subgenus *Parasemidalis* Enderlein, 1905
   All known species in this subgenus inhabit North America or islands in the Western Palearctic Region. It is not further considered.
   – The height of the head capsule is 1.2–1.7 times its length. The frons of the female is strongly sclerotized. Vein Rs in the hind wing branches at 1/5 of the distance from its base to its apex. The hypandrium and ninth segment are separated from each other both ventrally and laterally. The anterior part of the paramere is inflated and has a rugose surface (Fig. 5.63). .... Subgenus *Stangesemidalis* González Olazo, 1985

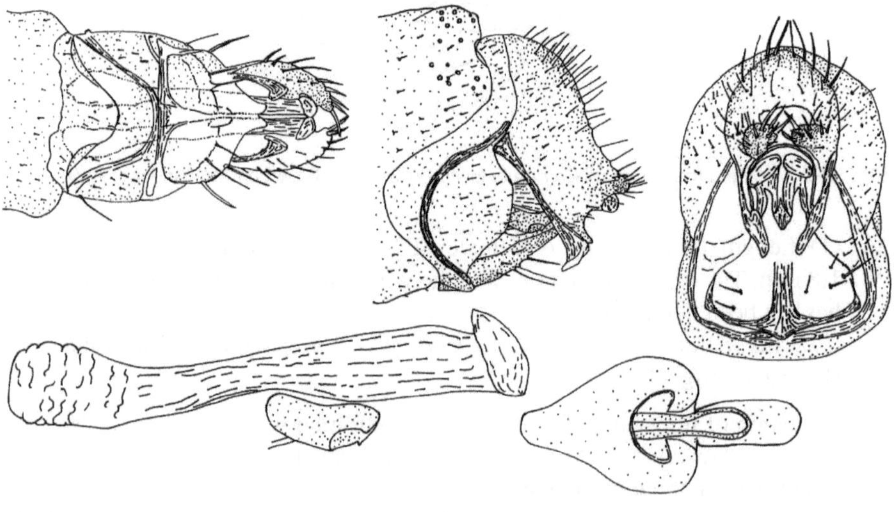

Fig. 5.63 *Parasemidalis enriquei* male (above, left to right): apex of the abdomen in ventral, lateral, and posterior view, and (below, left and right, respectively): a paramere and the penis in lateral view, and the penis in dorsal view. Based on Sziráki (2009).

Key to Known Species of *Parasemidalis (Stangesemidalis)* in South America

Information for the key was provided by Sziráki (2009), and Sziráki and Penny (2012).

1. The pedicel is much darker than the segments of the antennal flagellum. The caudal end of the penis narrows. On the caudal projection of the ectoproct, there are several subequal bristles. The paramere is truncated at its apical end, evident in lateral view. The hypandrium has dictinct proximal and lateral apodemes and wide median ones. The hypandrium is hooked at the apex and has a deep incision in the form of the letter V (Fig. 5.63). Total length: 1.5–2.3 mm. Fore-wing length: 1.7–2.9 mm, with the hind wing slightly shorter. Length of antenna: 1.2–1.8 mm
   .... *Parasemidalis enriquei* Sziráki, 2009 (Argentina).
   – The pedicel is the same shade and color as the segments of the antennal flagellum. Either the hypandrium has distinct proximal apodemes, wide median ones, but no lateral apodemes, or on the caudal projection of the ectoproct, if present, there is no more than one strong bristle. The hypandrium is not hooked at the apex and does not have a deep incision in the form of the letter V (Fig. 5.64). 2
2. (1) There is only one strong bristle on the caudal projection of the ectoproct. Median apodemes are present on the hypandrium. The lateral margins of the penis are generally convex (Fig. 5.64). The antenna is c. 1.7 mm long.
   .... *Parasemidalis principiae* Sziráki and Greve, 2001 (Chile, Argentina).

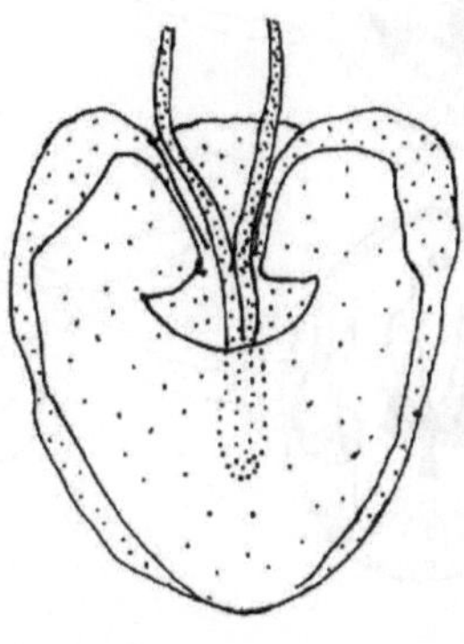

Fig. 5.64 The penis of *Parasemidalis principiae*. Based on Sziráki (2009).

- There are several strong bristles on the caudal projection of the ectoproct. Lateral apodemes are not present on the hypandrium (Fig. 5.25). The antenna is 1.2–1.4 mm long.

.... *Parasemidalis subandina* (González Olazo, 1984) (Chile, Argentina). Syn: *Stangesemidalis subandina* González Olazo, 1984.

Key to the Adults of South American Species of *Semidalis*

Information for the key was provided by Meinander (1972, 1974, 1980, 1983, 2002); Meinander in Meinander and Penny (1982); Monserrat (1981), and Sziráki (2005, 2009). Positive identifications are only possible with male specimens. *Semidalis nivosa* is only known from female specimens, so identification using this key can only be tentative until males have been described.

1. The wing membranes of the female are grayish brown with hyaline margins and streaks, which seem to vary from individual to individual. The antenna consists of about 33 segments. Fore-wing length: 2.8–4.0 mm; hind wing: 2.3–3.6 mm. The male has not been described.
   .... *Semidalis nivosa* Enderlein, 1906 (Peru). Syn: *Niphetia nivosa* Enderlein, 1930; *Semidalis pruinosa* Enderlein, 1906; *Semidalis scobis* Enderlein, 1906.

   - The wing membranes of the female are rather uniform in color, or the markings do not include hyaline wing margins and streaks (Fig. 5.65). 2

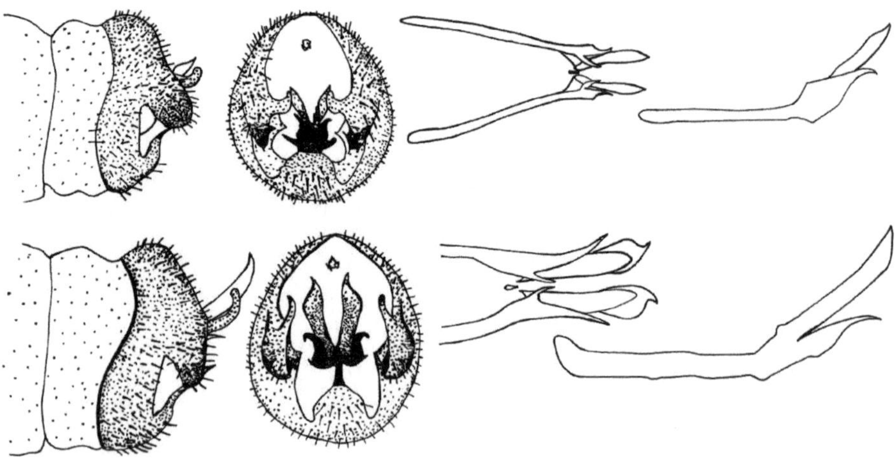

Fig. 5.65 *Semidalis intermedia* male: variability of the genitalia illustrated by comparing the same structures on two specimens, shown above and below each other (left to right): the apices of the abdomen in lateral and posterior view and the parameres in ventral and lateral view. Based on Monserrat (1983), who judged the two specimens depicted to be conspecific.

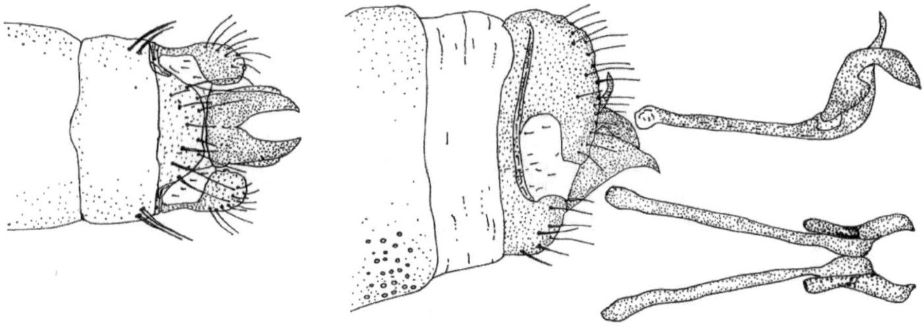

Fig. 5.66 *Semidalis kolbei* male: apex of the abdomen in ventral (left) and lateral view (center) and the parameres in ventral (lower right) and lateral view (upper right). Based on Meinander (1972).

2. (1) The paramere has a wide, divided distal part and lacks any other special modifications. Its shape is described as that of a smoking pipe (Fig. 5.65). 3

   – The distal part is not wide and divided, or there are notable processes or other modifications (Fig. 5.66). 4

3. (2) There is a process on the ectoproct, which curves dorsad and has subapical serrations. There are also serrations on the ventral side of the parameres, which diverge in ventral view (Fig. 5.65). The antenna of the male consists of 39 or 40 segments; that of the female has 34–36 segments. Fore-wing length: c. 3.4 mm: hind wing: c. 2.8 mm.
   .... *Semidalis intermedia* Monserrat, 1983
   (Venezuela, Colombia, Peru).

   – Serrations are not evident on processes of the ectoproct or on the ventral side of the parameres, but the paramere has an one apical and and one subapical dorsal tooth (Fig. 5.67). Fore-wing length: c. 3.0 mm; hind wing: c. 2.5 mm.
   .... *Semidalis ecuadoriana* Meinander, 1983
   (Ecuador). Meinander (2002) regards this nominal species to be a probable synonym of *Semidalis manausensis,* which is found in Couplet 13 of this key.

4. (2) The uncini are large and C-shaped. Toward their apices, the parameres curve dorsad, so that the vertical part is at least half as long as the horizontal part (Fig. 5.66). 5

   – Uncini, if present, are not large and C-shaped (Fig. 5.68). 6

5. (4) The dark brown antenna of the male consists of 42–47 segments, and that of the female, from 39 to 43 segments. The head and pronotum are dark brown with blackish brown shoulder spots on the pronotum. The hypandrium is small

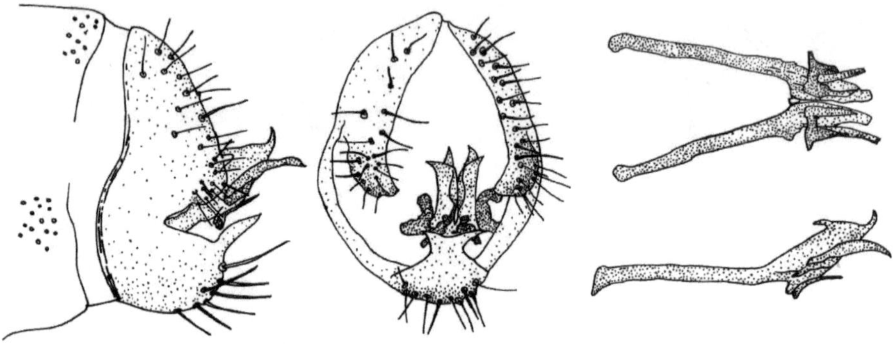

Fig. 5.67 *Semidalis ecuadoriana* male (left to right): apex of the abdomen in lateral and oblique posterior view, and the internal genitalia in ventral (above) and lateral view (below). Based on Meinander (1983).

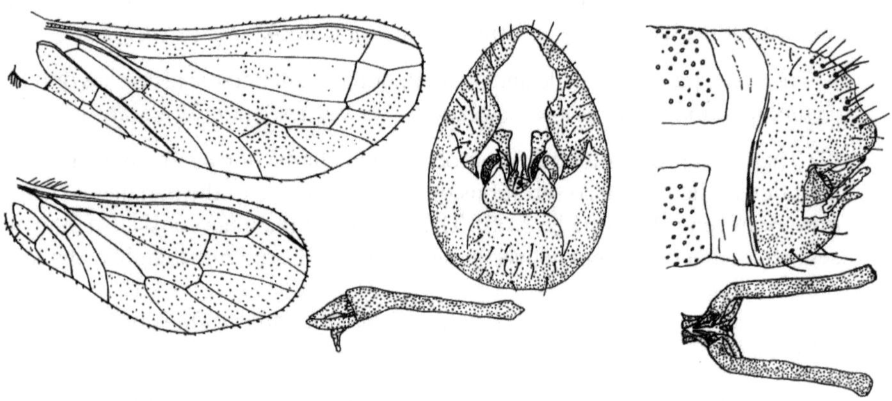

Fig. 5.68 *Semidalis boliviensis* male: fore and hind wing (left), apex of the abdomen in posterior (upper center) and lateral view (upper right), and parameres in lateral (lower center) and ventral view (lower right). Based on Meinander (1972).

and lacks spines (Fig. 5.66). Fore-wing length: 3.0–3.9 mm; hind wing length: 2.4–3.2 mm.
.... *Semidalis kolbei* Enderlein, 1906 (Chile, Argentina). Syn: *Niphetia kolbei* Enderlein, 1930. The plants on which this species has been found were reported by Rebolledo et al. (2009), which is of interest to researchers who evaluate insects in this family as potential agents of biological control of pests, which can be used to protect cultured plants.

– The antenna consists of about 32 segments; its scape and pedicel are whitish; the flagellum is blackish, except for the five apical segments, which are grayish. The head capsule is yellowish brown, and the clypeus is brown. The hypandrium is deeply bifurcate, and the uncinus is long and slender with 2 bends, each about 90° (Fig. 5.69). The sclerotized parts of the thorax are brown with dark brownish sutures. Fore-wing length: c. 2.0 mm; hind wing length: c. 1.5 mm.

..........................*Semidalis normani* Meinander in Meinander and Penny, 1982 (Argentina, Amazonas).

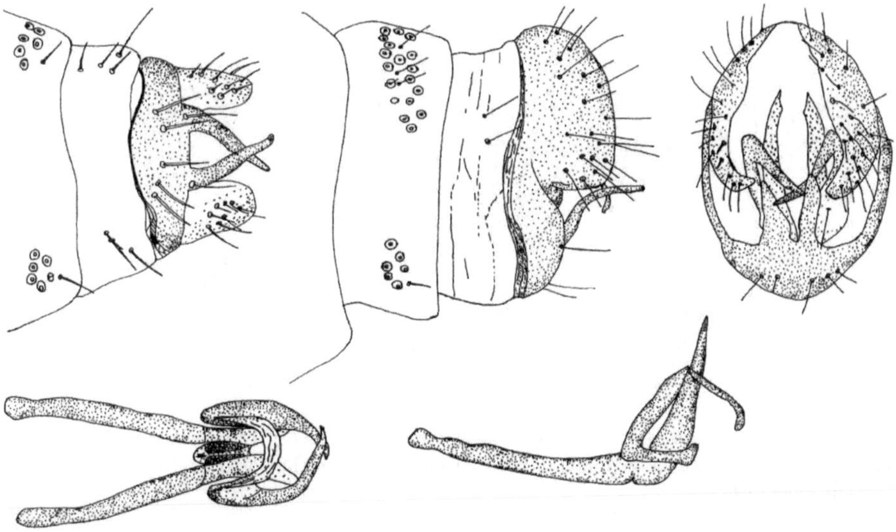

Fig. 5.69 *Semidalis normani* male (above, left to right): hypandrium in ventral view, apex of the abdomen in lateral and posterior view, and (below, left and right): the internal genitalia in ventral and lateral view. Based on Meinander and Penny (1982).

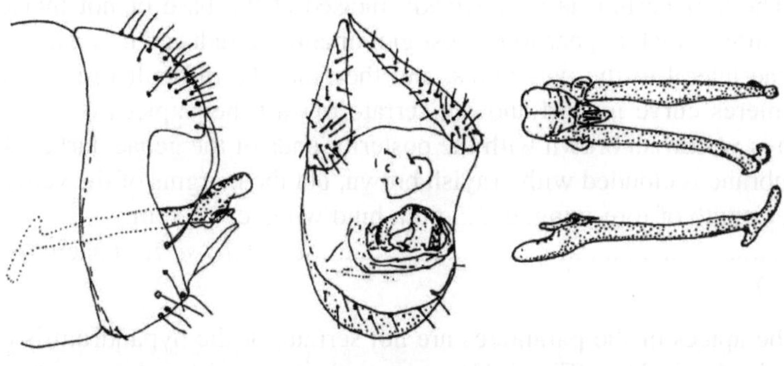

Fig. 5.70 *Semidalis peruviensis* male (left to right): apex of the abdomen in lateral and posterior view and the interial genitalia in ventral (above) and lateral view (below). Based on Meinander (1974).

6. (4) The apical part of each paramere runs caudad or curves dorsad, sometimes forming an obtuse angle; toward the apex, the paramere forks into narrow branches (Fig. 5.68). 7

   – The apical part of the paramere neither runs caudad nor dorsocaudad nor bends dorsad to form an obtuse angle (Fig. 5.70). 15

7. (6) Each antenna consists of more than 40 segments, and the uncini are fused to form a single structure. The hypandrium is evenly rounded. The parameres curve dorsad, but the vertical part is much less than half as long as the horizontal part (Fig. 5.68). Fore-wing length: 2.1–2.7 mm; hind wing: c. 1.5–2.1 mm. .... *Semidalis boliviensis* (Enderlein, 1906) (Mexico, Trinidad, Venezuela, Peru, Bolivia, Pará). Syn: *Alemella boliviensis* Enderlein, 1906.

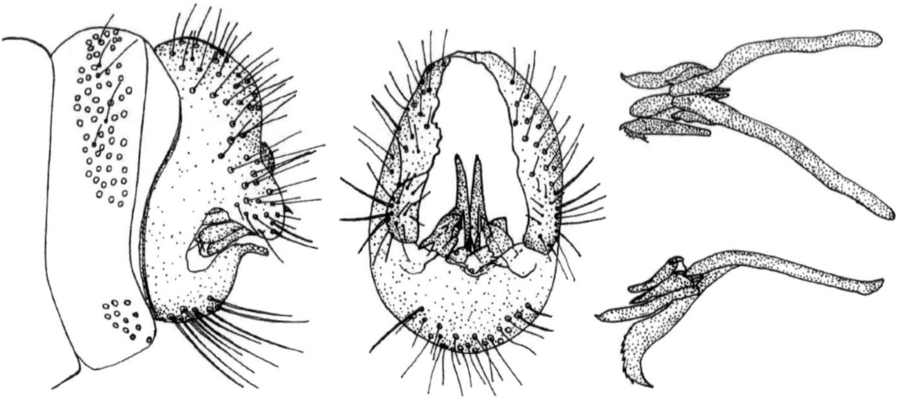

Fig. 5.71 *Semidalis serrata* male: apex of the abdomen in lateral (left) and posterior view (center) and the internal parts of the genitalia in ventral (upper right) and lateral view (lower right). Based on Meinander (1983).

– Each antenna consists of fewer than 40 segments, or the uncini are separate and distinct or completely absent (Fig. 5.71). 8

8. (7) The hypandrium is very weakly incised at the base or not incised at all. Structures which appear to be vestigial uncini are reduced to a pair of rod-like sclerae lateral to the parameres, and they may be difficult to recognize. The parameres curve inward and are serrate toward their apices (Fig. 5.71). The head is yellowish brown with the posterior ends of the genae darker. The wing membrane is clouded with grayish brown, but the margins of the veins are hyaline. Length of fore-wing: c. 2.5 mm; hind wing: c. 1.8 mm. .... *Semidalis serrata* Meinander, 1983 (Acre).

– The apices of the parameres are not serrate, or the hypandrum is distinctly incised at its base (Fig. 5.72). 9

9. (8) The ectoproct is short and broad with serrations along its dorsal margin and granulations. The hypandrium has a basal portion fused with the ninth sternite and a posterior portion shaped like an acute spine, slightly forked at the apex, which is located ventral to the uncini. The apices of the parameres converge (Fig. 5.72). The antenna consists of about 29 or 30 segments, including the scape and pedicel, and each flagellar segment bears a ring of setae. Length of fore-wing: c. 2.0 mm; hind wing: c. 1.6 mm. .... *Semidalis lolae* Monserrat, 1983 (Venezuela).

– The ectoproct is not short and broad with serrations along its dorsal margin (Fig. 5.73). 10

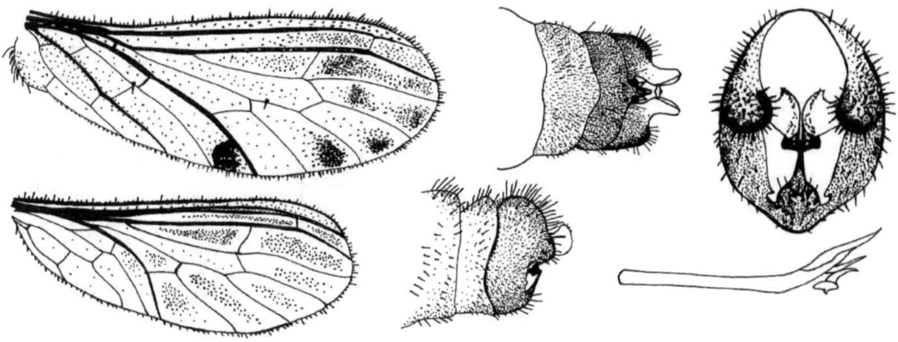

Fig. 5.72 *Semidalis lolae* male: fore (upper left) and hind wing (lower left); apex of the abdomen in ventral (upper center), lateral (lower center), and posterior view (upper right), and the parameres in lateral view (lower right). Based on Monserrat (1983).

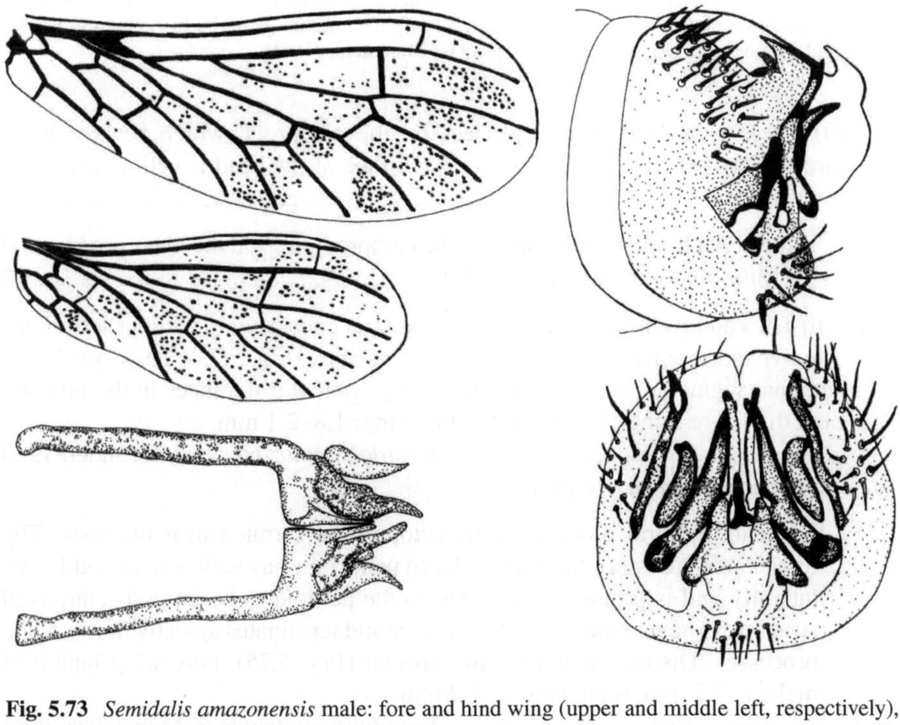

Fig. 5.73 *Semidalis amazonensis* male: fore and hind wing (upper and middle left, respectively), apex of the abdomen in lateral (upper right) and posterior view (lower right), and internal genitalia in ventral view (lower left). Based on Meinander (1980).

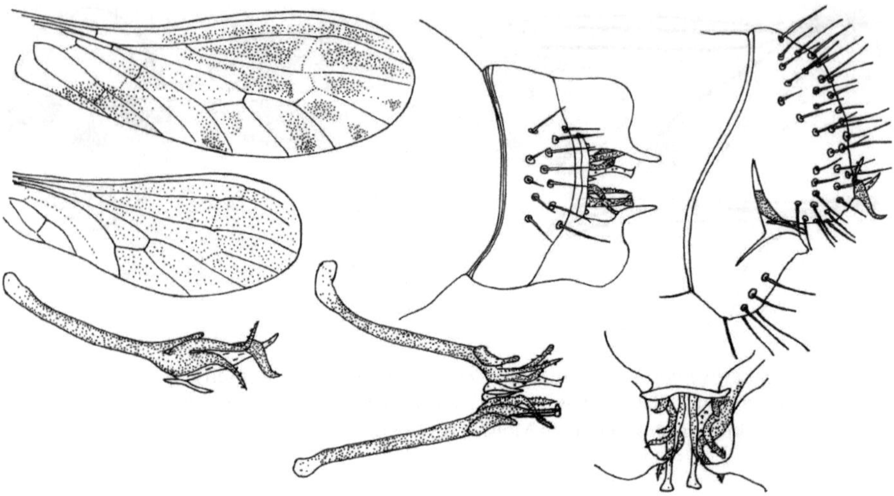

Fig. 5.74 *Semidalis manausensis* male (above, left to right): fore and hind wing, apex of the abdomen in ventral and lateral view, and (below, left to right): internal genitalia in lateral and ventral view, and the genitalia in posterior view. Based on Meinander (1980).

10. (9) There is a caudoventral process on the ectoproct, which bears a tooth directed dorsad (Fig. 5.73). The dorsal surface of the head is yellow with dark brown sutures. 11

   – If there is a tooth near the apex of the ectoproct, it is on the outer process and not on a caudoventral one (Fig. 5.74). 12

11. (10) The caudoventral process on the ectoproct ends acutely without serrations. The hypandrium has a deep bifurcation at its apex (Fig. 5.73). There are 27–36 antennal segments, with considerable geographical differences in the number. Length of fore-wing: 2.2–2.7 mm; hind wing: 1.8–2.1 mm. .... *Semidalis amazonensis* Meinander, 1980 (Rondônia, Amazonas, Acre).

   – The caudoventral process on the ectoproct is serrated near the apex. The hypanandrium is not bifurcate. The hypandrium has well sclerotized knobs laterally and is probably connected to the parameres by an odd sclerotized structure. The paramere is rather slender and terminates apically in two acute processes. Distinct uncini are not present (Fig. 5.75). Fore-wing length of male: c. 2.2 mm; hind wing: c. 1.7 mm. .... *Semidalis rondoniensis* Meinander, 1982 (Rondônia, Pará).

12. (10) The ectoproct lacks processes, but an inner apodeme may be present (Fig. 5.74). 13

   – The ectoproct has at least one process, although it may be small (Fig. 5.76). The eighth sternite does not bear a pair of long bristles. 14

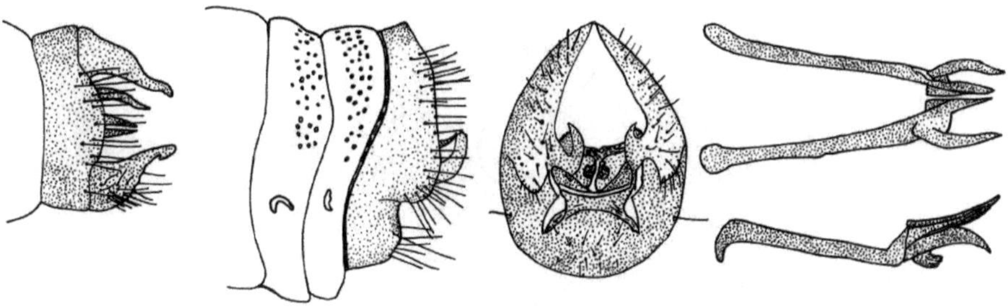

Fig. 5.75 *Semidalis rondoniensis* male (left to right): apex of the abdomen in ventral, lateral, and posterior view, and the internal genitalia in ventral (above) and lateral view (below). Based on Meinander and Penny (1982).

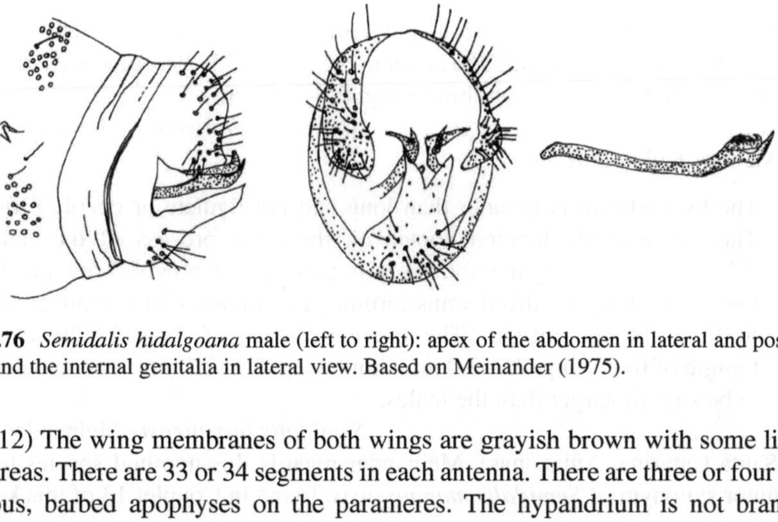

Fig. 5.76 *Semidalis hidalgoana* male (left to right): apex of the abdomen in lateral and posterior view and the internal genitalia in lateral view. Based on Meinander (1975).

13. (12) The wing membranes of both wings are grayish brown with some lighter areas. There are 33 or 34 segments in each antenna. There are three or four sinuous, barbed apophyses on the parameres. The hypandrium is not branched (Fig. 5.74). Length of fore-wing: c. 2.2 mm; hind wing: c. 1.8 mm. The female has not been described.
.... *Semidalis manausensis* Meinander, 1980 (Amazonas).
   – The wing membrane of the fore-wing has large fuscous areas, especially distally; these areas may contain lighter markings. The membrane of the hind wing is dark brown. Both wings have light brown veins. A broad inner apodeme is present on the ectoproct, which is short. In lateral view, the inner process of the hypandrium appears elongate and acute at the apex (Fig. 5.2). There is a pair of long bristles on the eighth abdominal sternite. The antenna consists of 33–37 segments. Length of fore-wing: 2.3–2.8 mm; hind wing: 1.8–2.3 mm.
.... *Semidalis maculosus* Sziráki, 2009 (Argentina).

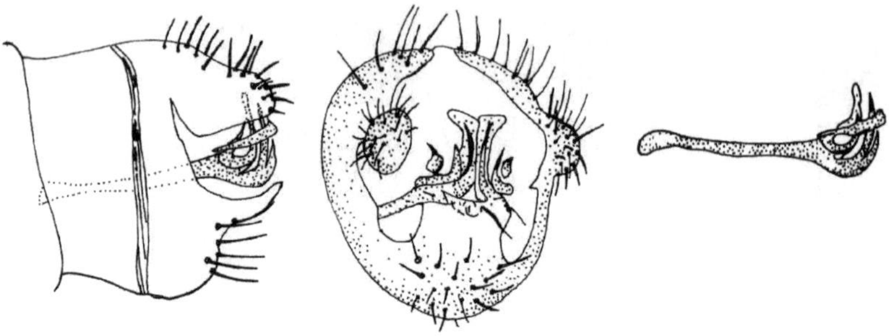

Fig. 5.77 *Semidalis brasiliensis* male (left to right): apex of the abdomen in lateral and posterior view and the interial genitalia in lateral view. Based on Meinander (1974).

14. (12) The hypandrium is bifurcate and ligulate (Fig. 5.76). The outer process on the ectoproct is very small. The antenna consists of 35–39 segments. Fore-wing length of male: 2.6–3.2 mm; hind wing: 2.0–2.5 mm.
.... *Semidalis hidalgoana* Meinander, 1975 (Mexico, Colombia).

 – The hypandrium is broader than long and not ligulate or deeply bifurcate. There is a tooth directed dorsad on the outer process of the ectoproct (Fig. 5.77). The wing membranes of both wings are grayish brown with hyaline areas along the distal veins forming a continuous band from cross vein $Sc\text{-}R_1$ to the apex of Cu_1. The antenna consists of about 27–29 segments. Length of fore-wing: 2.7–3.0 mm; hind wing: 2.1–2.5 mm. The females tend to be slightly larger than the males.
.... *Semidalis brasiliensis* Meinander, 1974 (Santa Catarina, Amazonas). Meinander regards this nominal species to be a junior synonym of *Semidalis manausensis,* found in Couplet 13 of this key.

15. (6) The hypandrium has distinctive lateral thorn-like structures (Fig. 2.9).
.... *Semidalis isabelae* Monserrat, 1981 (Argentina, Paraguay).

 – Lateral thorn-like processes are absent from the hypandrium. The ectoproct is short, and there are no distinct unci present (Fig. 5.70). 16

16. (15) On the vertex of the heads of both sexes, there are large, fingerlike projections pointing anteriad (Fig. 5.78). The scape of the antenna is 3½ times as long as its width; each antenna consists of about 33 segments. Fore-wing length: c. 2.7 mm.
.... *Semidalis absurdiceps* (Enderlein, 1908) (Colombia, Bolivia). Syn: *Niphas absurdiceps* Enderlein, 1908.

 – The head lacks large, fingerlike projections (Fig. 5.70). 17

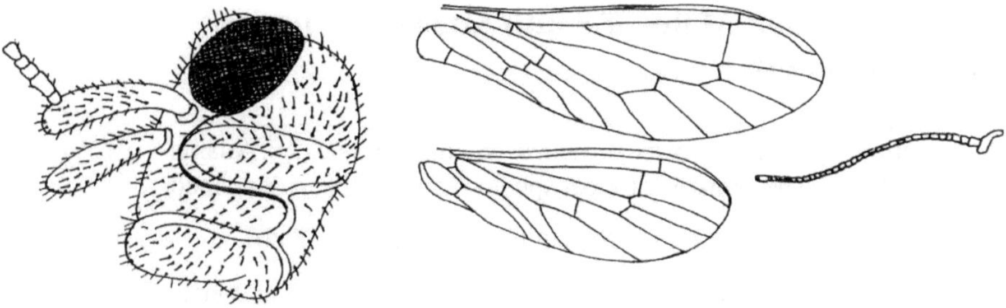

Fig. 5.78 *Semidalis absurdiceps* (left to right): head in oblique lateral view, fore and hind wing, and antenna. Based on Enderlein (1908).

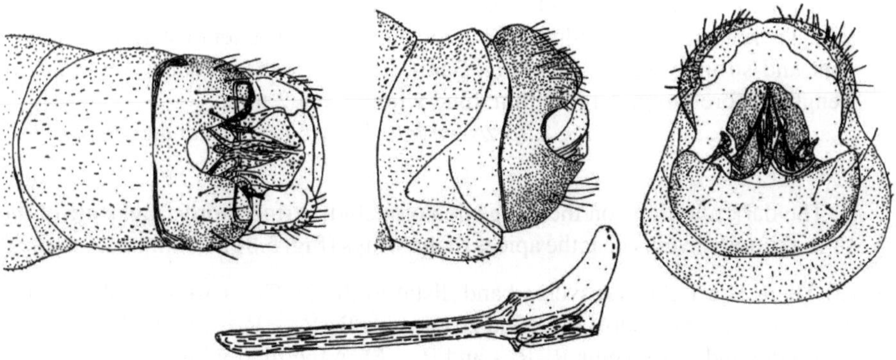

Fig. 5.79 *Semidalis jujuyana* male (above, left to right): apex of the abdomen in ventral, lateral, and posterior view, and a paramere in lateral view (below). Based on Sziráki (2009).

17. (16) There is a short, transverse rod at the base of each paramere (Fig. 5.70). The head is ochraceous. Each antenna consists of about 50 segments. The thorax has light brown spots on the shoulders, and the wing membranes are very light in color. Fore-wing length: c. 3.4 mm; hind wing: 3.0 mm. ....*Semidalis peruviensis* Meinander, 1974 (Peru).

   – There is no short, transverse rod at the base of the paramere, which has a forked apical process with a caudal tooth on the outer branch. An arch dorsal to the parameres connects the two sides of the hypandrium (Fig. 5.79). There are about 41 antennal segments. Total length: 1.6–2.0 mm. Fore-wing length: 2.5–2.8 mm; hind wing length: 1.8–2.0 mm. ....*Semidalis jujuyana* Sziráki, 2009 (Argentina).

Key to the Adult Males of South American Species of *Incasemidalis*

Information for the key was provided Meinander (1972); Adams (1973b), and Sziráki (2009). The females of most species have not been described.

1. The wings are marked with distinct brown spots (Fig. 5.80). 2

   – The wing membranes are relatively uniform in color and lack distinct markings, although areas of vaguely darker of lighter shading may be evident (Fig. 5.26). 4

2. (1) There are dark stripes at the end of some of the longitudinal veins on the wings with hyaline stripes between them. The ectoproct lacks a posterior projection. There is a ventral bridge joining the parameres. The tenth sternite is prominent and strongly sclerotized (Fig. 5.80). Length of the antenna: 1.9–2.1 mm. Length of fore-wing: 4.1–4.6 mm; hind wing: 3.5–4.2 mm. ....*Incasemidalis lineatellus* Sziráki and Penny, 2012 (Argentina).

   – The dark markings on the wings do not include stripes bordering some of the longitudinal veins near the apices of the wings (Fig. 5.81). 3

3. (2) There is a light, transverse band distal to the M-Cu cross vein of the fore-wing. Distinct dark spots surround cross veins R_1-R_{2+3}, R_{4+5}-M, and M-Cu in the fore-wing and cross veins R_1-R_{2+3} and R_{4+5}-M in the hind wing (Fig. 5.81). The

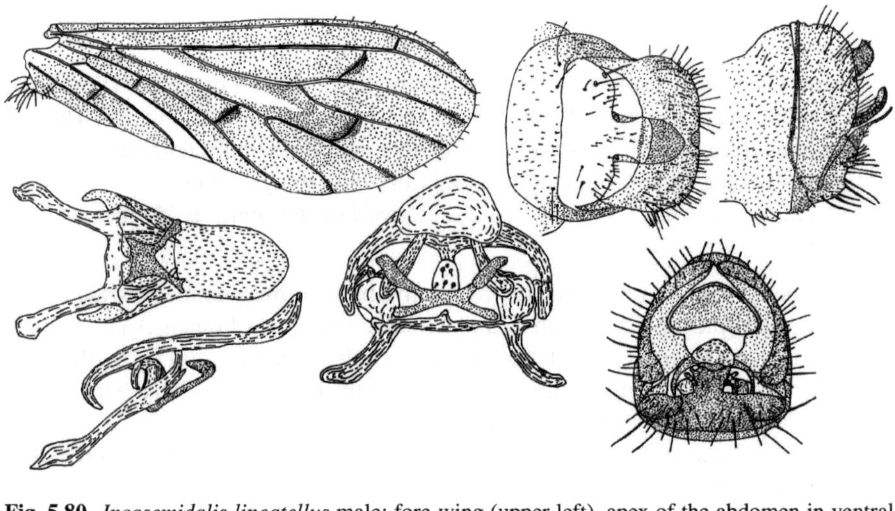

Fig. 5.80 *Incasemidalis lineatellus* male: fore-wing (upper left), apex of the abdomen in ventral (upper right center), lateral (upper right), and posterior view (lower right); and the internal genitalia in ventral (middle left), lateral (lower left), and posterior view (lower center). Based on Sziráki and Penny (2012).

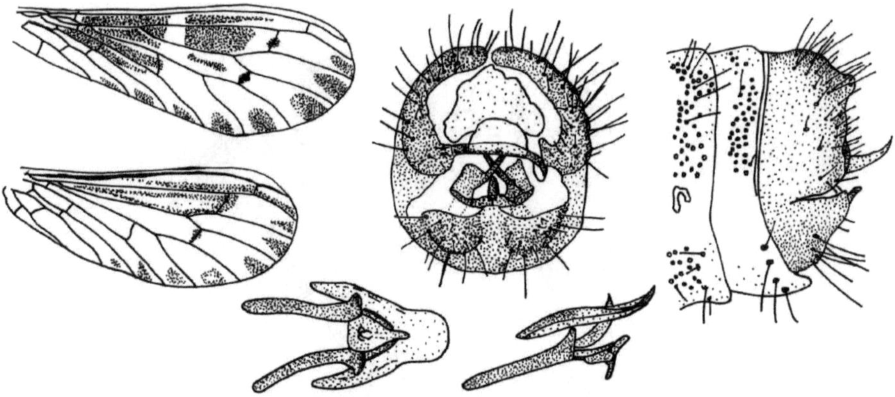

Fig. 5.81 *Incasemidalis chilensis* male: fore and hind wing (left), apex of the abdomen in posterior (upper center) and lateral view (right), and the internal genitalia in ventral (lower left center) and lateral view (lower right center). Based on Meinander (1990).

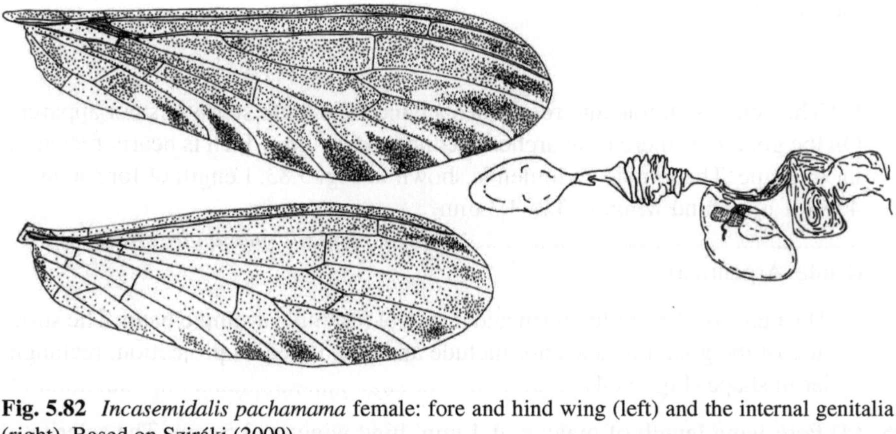

Fig. 5.82 *Incasemidalis pachamama* female: fore and hind wing (left) and the internal genitalia (right). Based on Sziráki (2009).

hypandrium has a plate-like posterior projection along the mid-line. There is a dorsal process on the penis bearing hair-like setae. The ectoproct has a posterior projection appearing in lateral view as a knob. Length of fore-wing: c. 3.5 mm; hind wing: c. 2.9 mm.

....*Incasemidalis chilensis* Meinander, 1990 (Chile).

– There is no light, transverse band distal to the M-Cu cross vein of the fore-wing, so the area in the middle of the anterior part of that wing is relatively uniform medium brown. Each of the cross veins has a narrow light margin on the membrane bordering it (Fig. 5.82). Length of fore-wing: c. 3.3 mm; hind wing: c. 2.9 mm.

....*Incasemidalis pachamama* Sziráki, 2009 (Chile).

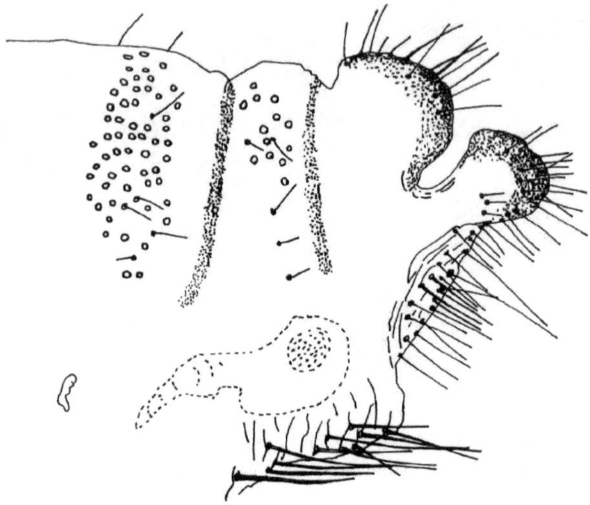

Fig. 5.83 Apex of the abdomen of a female *Incasemidalis meinanderi* in lateral view. Based on Meinander (1990).

4. (1) The penis is simple and resembles a band. The tenth sternite is not apparent. On the gonarcus, there is an arched median projection, which is nearly rectangular in shape. The female abdomen is shown in Fig. 5.83. Length of fore-wing: c. 4.0–4.1 mm; hind wing: c. 3.2–3.3 mm. .... *Incasemidalis meinanderi* Adams, 1973 (Chile, Argentina).

   – The penis is of complex form and does not resemble a simple band. The structure of the gonarcus does not include an arched median projection, rectangular in shape (Fig. 5.84). 5

5. (4) Fore-wing length of male: c. 4.4 mm; hind wing: c. 3.5 mm. The sclerite of the penis does not form a ring with some of the apothoses (Fig. 5.26). Only the male has been described from a specimen lacking a head. .... *Incasemidalis columbiensis* Meinander, 1972 (Colombia).

   – Fore-wing length of male: c. 3.6 mm; hind wing: c. 3.0 mm. The sclerite of the penis forms a ring with some of the apothoses (Fig. 5.84). .... *Incasemidalis peruviensis* Meinander, 1972 (Peru).

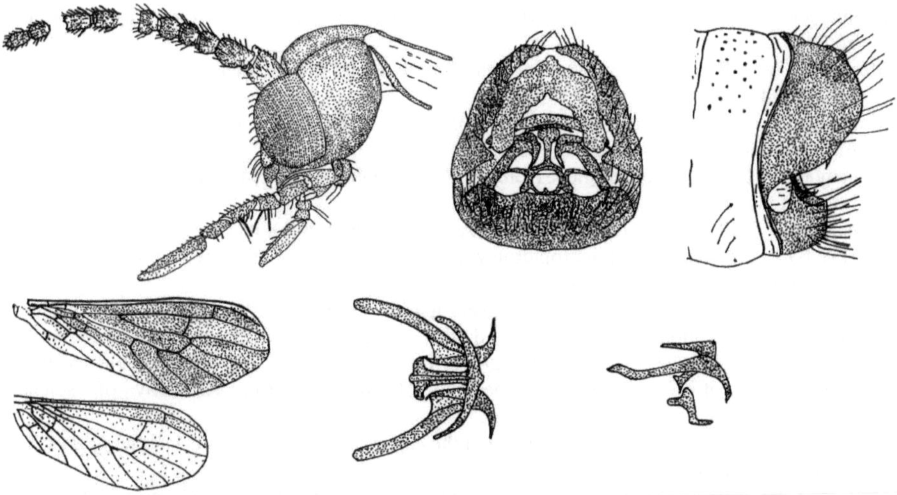

Fig. 5.84 *Incasemidalis peruviensis* male (above, left to right): the 37th and 38th segments at the apex of the antenna, the 17th and 18th segments, and the head and basal antennal segments, the apical segments of the abdomen in posterior and lateral view, and (below, left to right): fore and hind wing, and the internal genitalia in dorsal and lateral view. Based on Meinander (1972).

Literature

Adams PA (1973a) Synonymy in *Pampoconis* (Neuroptera: Coniopterygidae). Pan-Pacific Entomologist 49:324

Adams PA (1973b) Coniopterygidae from Chile, with three new species (Neuroptera). Pan-Pacific Entomologist 49:250–254

Burmeister H (1839) Handbuch der Entomologie. Neuroptera 2(2–3):397–1050. Theodor Friedrich Enslin, Berlin

Enderlein G (1906) Monographie der Coniopterygiden. Zoologische Jahrbücher, Abteilung für Systematik, Geographie und Biologie der Tiere 22:173–239+6 plates and 64 figures

Enderlein G (1907) Nachträge zur Monographie der Coniopterygiden. Stettiner Entomologische Zeitung 68:10–13

Enderlein G (1908) Familie Coniopterygidae. Genera Insectorum 67:1–18. M. P. Wytsman, Brussels

González Olazo EV (1984) Contribucion al conocimiento de *Semidalis absurdiceps* (Enderlein, 1908), (Neuroptera, Coniopterygidae). Acta Zoológica Lilloana 37(2):185–189

González Olazo EV (1985) *Stangesemidalis subandina*, nuevo genero y especie de Coniopterygidae de la Republica Argentina. (Neuroptera, Planipennia). Acta Zoológica Lilloana 38(1):59–63

Johnson V (1978) New Coniopterygidae from Central America. Annals of the Entomological Society of America 71:163–164

Meinander M (1972) A revision of the family Coniopterygidae (Planipennia). Acta Zoologica Fennica 136:1–357

Meinander M (1973) Notes on some types of Enderlein's species of Coniopterygidae, with description of a new species of *Pampoconis* (Neuroptera). Notulae Entomologicae 53:23–25

Meinander M (1974) Coniopterygidae from South and Central America (Neuroptera). Notulae Entomologicae 54:97–106

Meinander M (1975) Coniopterygidae from North America (Neuroptera). Notulae Entomologicae 55:28–32
Meinander M (1980) Coniopterygidae from Brazil (Neuroptera). Entomologica Scandinavica 11:129–144
Meinander M (1981) A review of the genus *Coniopteryx* (Neuroptera, Coniopterygidae). Annales Entomologici Fennici 47(4):97–110
Meinander M (1983) Coniopterygidae from South and Central America II (Neuroptera). Neuroptera International 2(4):179–198
Meinander M (1986) The Coniopterygidae of America (Neuroptera). In: Gepp J, Aspöck H, Hölzel H (eds) Recent research in neuropterology. Proceedings of the 2nd International Symposium on Neuropterology, Graz. pp 31–43
Meinander M (1990) The Coniopterygidae (Neuroptera, Planipennia). A check-list of the species of the world, descriptions of new species and other new data. Acta Zoologica Fennica 189:1–95
Meinander M (2002) Family Coniopterygidae. In: Penny ND (ed) A guide to the lacewings (Neuroptera) of Costa Rica. Proceedings of the California Academy of Sciences 53(12):227–236
Meinander M, Penny ND (1982) Neuroptera of the Amazon Basin. Part 5. – Coniopterygidae. Acta Amazonica 12(1):185–208
Monserrat VJ (1981 [1980]) Dos nuevos coniopterigidos de Paraguay (Insecta, Neuroptera, Coniopterygidae). Eos 55–56, 155–163
Monserrat VJ (1983) Contribución al conocimiento de los coniopterígidos de Venezuela (Neuroptera, Planipennia, Coniopterygidae). Boletín de Entomología Venezolana, N.S 2(19):137–156
Monserrat VJ (2005) New data on the dusty wings from Coquimbo, Patagonia and Tierra del Fuego (Neuroptera, Coniopterygidae). Entomofauna, Zeitschrift für Entomologie 26(25):421–436
Navás L (1927) Vienticinco formas neuvas de insectos. Boletin de la Sociedad Ibérica de Ciencias Naturales 26:48–75
Navás L (1928) Insectos de la Argentina. Cuarta Serie. Estudios 35:139–146
Navás L (1930) Insectos de la Argentina. Sexta Serie. Revista de la Sociedad Entomológica Argentina 3(13):125–132
Rebolledo R, Pinot P, Ojeda N, Aguilera A (2009) Distribución geográphica y plantas substratos de *Semidalis kolbei* Enderlein, 1906 (Neuroptera: Coniopterygidae) en la región de la Araucanía, Chile. IDESIA 27(1):41–46
Riek EF (1975) On the phylogenetic position of *Brucheiser argentinus* Navás 1927 and description of a second species from Chile (Insecta: Neuroptera). Studies on Neotropical Fauna 10(2):117–126
Sziráki G (2005) Species grouping of several genera and subgenera of Coniopterygidae (Neuroptera). Folia Entomologica Hungarica 66:101–136
Sziráki G (2007) Studies on Brucheiserinae (Neuroptera: Coniopterygidae), with description of the second genus of the subfamily. -. Acta Zoologica Academiae Scientiarum Hungaricae 53(Supplement 1):231–254
Sziráki G (2009) Data on Coniopterygidae of Argentina, with subgeneric division of *Parasemidalis* Enderlein, 1905. Folia Historico Naturalia Musei Matraensis 33:169–199
Sziráki G (2010) Taxonomic and nomenclatural questions of coniopterygid species (Neuroptera). Folia Historico Naturalia Musei Matraensis 34:141–150
Sziráki G, Flint OS, Jr (2005 [2007]) Larva of *Brucheiser penai* Riek, 1975 (Neuroptera Coniopterygidae). Annali del Museo Civico di Storia Naturale di Ferrara 8:45–48
Sziráki G, Greve L (2001) *Parasemidalis principiae* sp. n. – the first *Parasemidalis* species from South America (Neuroptera, Coniopterygidae). Journal of Neuropterology 3:5–8
Sziráki G, Penny ND (2012) Data on the dusty lacewing fauna of northwestern Argentina with description of a new species (Neuroptera: Coniopterygidae). Psyche 2012:1–4

Chapter 6
Ascalaphidae

Abstract An illustrated key is provided for identification of the genera of adult ascalaphids known to occur in South America. Illustrated keys are also provided to identify the species of each genus known to occur in South America. The countries or states in which specimens of each species have been found and invalid synonyms are reported.

Keywords Owlfly • Key to species • Albardinae • Haplogleniinae • Ascalaphinae

6.1 Key to the Genera of Known Adult Ascalaphidae from South America

The English common name for insects in this family is owlfly. The preferred habitats of the known species are not aquatic. The larvae form aggregations to repel potential predators (Henry 1972). Little is known about the development of the South American species, but some aggregations of Amazonian species have been described (Hogue and Penny 1988). Information for the key to the subfamilies and genera was provided by Navás (1909, 1913a, b, 1914; 1915, 1919, 1927a); Orfila (1949); Penny (1981a, b, 1983), and Machado and Rafael (2011).

1. The antennae are very short, typically much shorter than the head, and clubbed (Fig. 2.10). Much of the body is covered by dense, long setae. The length of the only known species is 35–45 mm, and its fore-wing length is 52–63 mm. Its abdominal tergites are deep red.
   .... Subfamily Albardinae
   The only known species in this subfamily, *Albardia furcata* van der Weele, 1903, inhabits Brazil and has been reported from Pará, Ceará, Minas Gerais, Espirito Santo, and Rio de Janeiro. Its larva is unknown, and no information about its habits is available. The species obviously differs from other known members of Neuroptera, and it was formerly placed in its own family: Stilbopterygidae.

   – The antennae are long, typically as long as or longer than the total length, and each bears a knob or club at the end. The anterior margins of the wings

C.W. Heckman, *Neuroptera (Including Megaloptera)*, Encyclopedia of South American Aquatic Insects, DOI 10.1007/978-3-319-35125-4_6

bear many cross veins. The large compound eyes meet along the midline on the dorsal part of the head (Fig. 6.1).. 2

2. (1) The eyes are not divided by a transverse suture. The ninth sternite of the male does not extend as far posteriad as the apices of the ectoprocts. The eighth tergite of the female is dorso-ventrally elongate, and it has a spiracle near the ventral margin (Fig. 6.1) .... Subfamily Haplogleniinae........................... 3

   – The eyes are divided into an upper and lower part by a transverse sulcum. The ninth sternite of the male does not extend as far posteriad as the apices of the ectoprocts. The ninth sternite of the male extends farther posteriad than the apices of the ectoprocts. The eighth tergite of the female is not dorsoventrally elongate, and it has a spiricle in the pleural membrane (Fig. 6.2) ....Subfamily Ascalaphinae....................... 6

3. (2) At the base of the anal margin of the fore-wing, there is a prominent axillary angle, apical to which the wing cells are shorter than the angle (Fig. 6.3). The wings are long and narrow.
   ....Tribe Verticillecerini..........*Verticillecerus* van der Weele, 1908

   – The only known South American representative of this tribe is *Verticillecerus gerstaeckeri* van der Weele, 1908, which has been found in Paraguay,

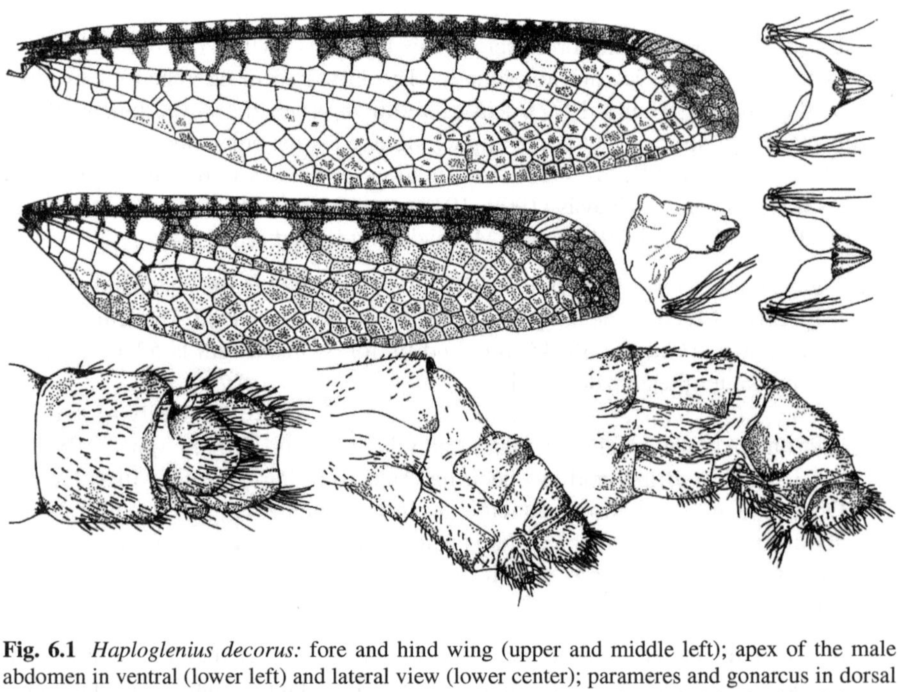

Fig. 6.1 *Haploglenius decorus:* fore and hind wing (upper and middle left); apex of the male abdomen in ventral (lower left) and lateral view (lower center); parameres and gonarcus in dorsal (upper right), ventral (middle right), and lateral view (right of center); apex of the female abdomen in lateral view (lower right). Based on Ábrahám (2013).

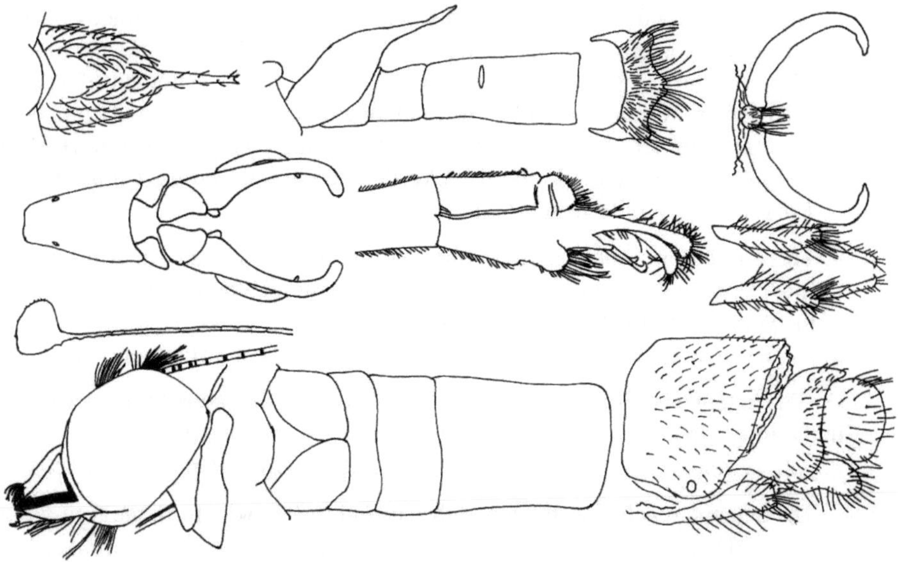

Fig. 6.2 *Fillus amazonicus* (upper row, left to right): dorsal projection from the first abdominal tergite of the male in dorsal view showing setae; apex of the thorax and first three abdominal segments of the male without their setae in lateral view; ninth abdominal segment of the male in ventral view, tarsal claws on the fore-leg; and (middle row, left to right): apex of the male abdomen without setae in posteroventral view and with setae in lateral view, and ventrovalvae and distivalvae in ventral view; and (lower row, left to right): and apical segments of the antenna with the apical bulb above the head in lateral view; apex of thorax and anterior abdominal segments of a female without setae in dorsal view; apex of the female abdomen in lateral view. Based on Machado and Rafael (2011).

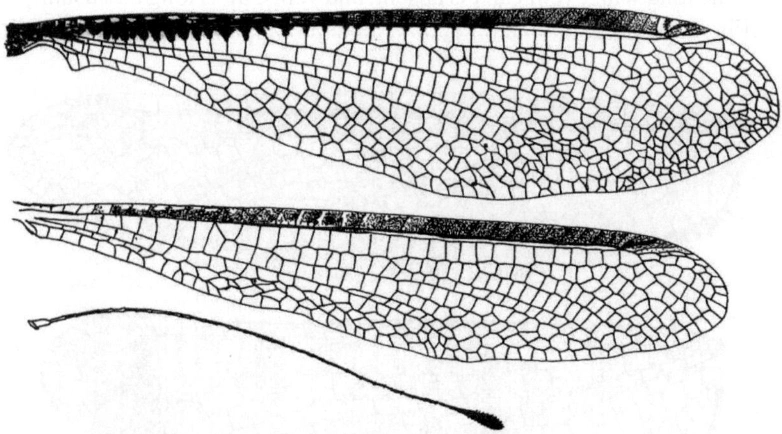

Fig. 6.3 *Verticillecerus gerstaeckeri:* fore and hind wing (above and middle) and the antenna (below). Based on Ábrahám (2013).

Argentina, Rondônia, Minas Gerais, and Rio de Janeiro. Specimens of this species are sometimes referred to in the literature as *Neohaploglenius rondonianus* Penny, 1981, which was reported to be a junior synonym of *V. gerstaeckeri* by Ábrahám (2013). The total length ranges from 26 to 31 mm, and the fore-wing length is 33 to 35 mm. The ground color is brown to reddish brown. Another genus in the tribe Verticillecerini, *Neohaploglenius* Penny, 1981, is known only from Mexico to Panama.

– There is no prominent axillary angle at the base of the anal margin of the fore-wing, and the wing cells apical to the base are equal to or longer than the veins along the basal curve (Fig. 6.1).
.... Tribe Haplogleniini 4

4. (3) The antenna is short and does not extend as far as the second fork in vein R_s in the fore-wing. The eyes are not divided into two sections (Fig. 6.4).
.... *Ascalobyas* Penny, 1981…p. 190

– The antenna reaches well beyond the second fork in vein R_s in the fore-wing (Fig. 6.1) 5

5. (4) Vein 2A in the hind wing is long (Fig. 6.1).
.... *Haploglenius* Burmeister, 1839..p. 191

– Vein 2A in the hind wing is very short or absent. The anal cells in the fore-wing proximal to the axillary angle are narrower than those distal to it, or the axillary angle is only slightly developed (Fig. 6.5).
.... *Amoea* Lefèbvre, 1842..p. 196

6. (2) In the hind wing, there is a well-developed fork in vein CuA, and vein CuP is almost straight (Fig. 6.3)..Tribe Suhpalacsini 7

– In the hind wing, vein CuA is absent, and vein CuP is long and usually curved (Fig. 6.6) .... Tribe Ululodini.....8

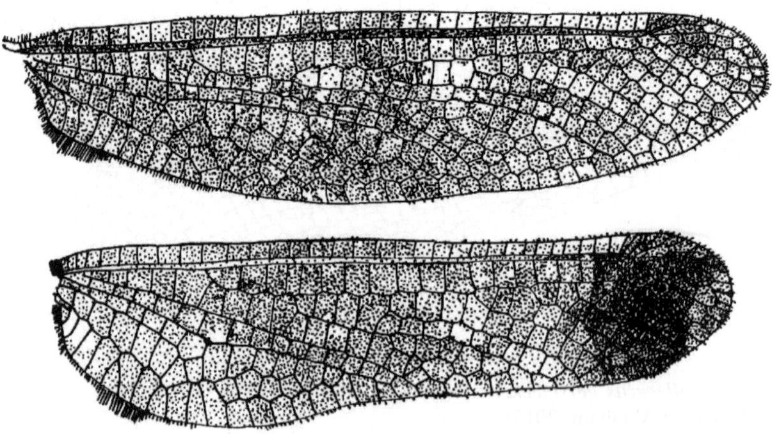

Fig. 6.4 Fore and hind wing of *Ascalobyas microcerus*. Based on Penny (2002).

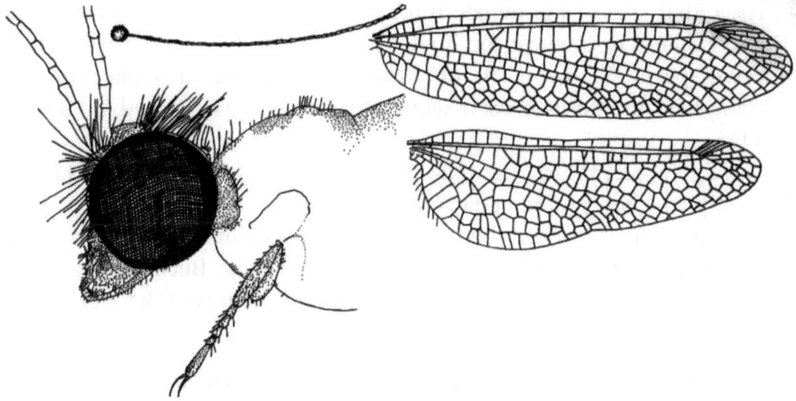

Fig. 6.5 *Amoea iniquus:* head with prothorax and bases of the antenna in oblique lateral view (lower left), antenna (upper left center), and fore and hind wing (right). Based on van der Weele (1908) and Penny (1981a, b).

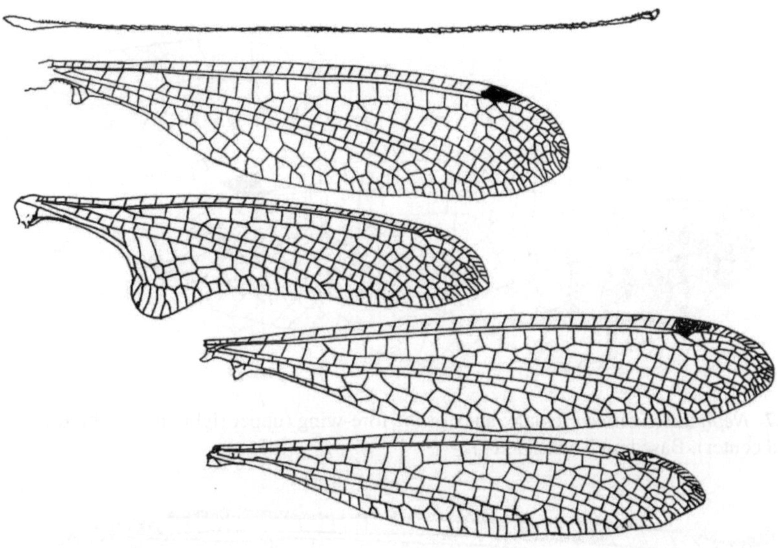

Fig. 6.6 *Ascalorphne macrocercus:* antenna of a male (above), fore and hind wing of a male (middle left to center) and a female (lower center to right). Based on van der Weele (1908).

7. (6) The radial area of the male wing has about 6 inner cross veins. The wing membrane is completely hyaline (Fig. 6.5). The legs are mainly black with fine white setae. There is a long dorsal process on the first abdominal segment, which forks near the apex to form two blunt knobs (Fig. 6.2). The two genera in this couplet cannot yet be satisfactorily distinguished from each other because of their sketchy descriptions and lack of specimens of both sexes. .... *Fillus* Navás, 1919..p. 199

– The radial area of the female fore-wing has about 5 inner, 16 intermediate, and 4 outer cross veins; the hind wing has about 2 inner cross veins in the radial space (Fig. 6.7). The only known species has a weakly tinted membrane near the base of each wing and yellow legs covered by fine yellow setae.

....*Nephelasca* Navás, 1914

The only species in this genus is *Nephelasca crocea* Navás, 1914, from Colombia, which was described from a female specimen. Its fore-wing is about 28 mm long, and its hind wing is about 23 mm long. Because the identifying characteristics of related genera depend upon structures of the male, the validity of this genus remains uncertain.

8. (6) The fore-wing has a very distinct axillary angle. The antenna is longer than the fore-wing (Fig. 6.6).

....*Ascalorphne* Banks, 1915..p. 201

– The base of the fore-wing is very evenly curved and lacks an axillary angle (Fig. 6.8) 9

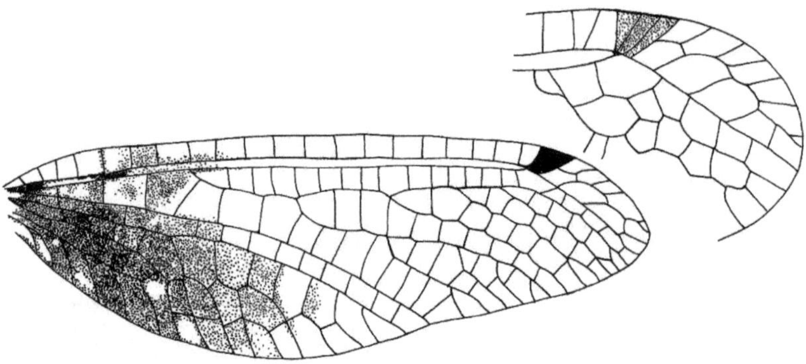

Fig. 6.7 *Nephelasca crocea* female: apex of the fore-wing (upper right) and the hind wing (lower left and center). Based on Navás (1914).

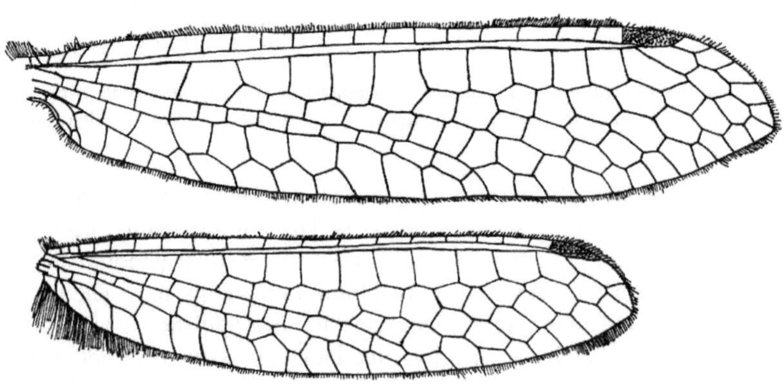

Fig. 6.8 Fore and hind wing of *Ameropterus breviantennus*. Based on Penny (2002).

9. (8) Vein CuP in the hind wing is straight (Fig. 6.8).
.... *Ameropterus* Esben-Petersen, 1922..p. 202

– Vein CuP in the hind wing is obviously sinuous (Fig. 6.9) 10

10. (9) The hind wing is approximately triangular with the width near the base much narrower than the width at mid-length (Fig. 6.9).
.... *Cordulecerus* Rambur, 1842..p. 211

– The hind wing is approximately as wide near the wing base as it is at mid-length (Fig. 6.10).
.... *Ululodes* Currie, 1899..p. 217

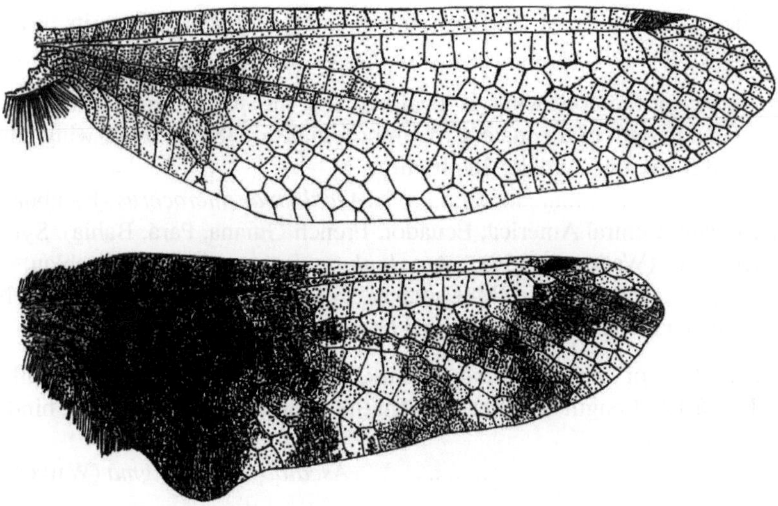

Fig. 6.9 Fore and hind wing of *Cordulecerus subiratus*. Based on Penny (2002).

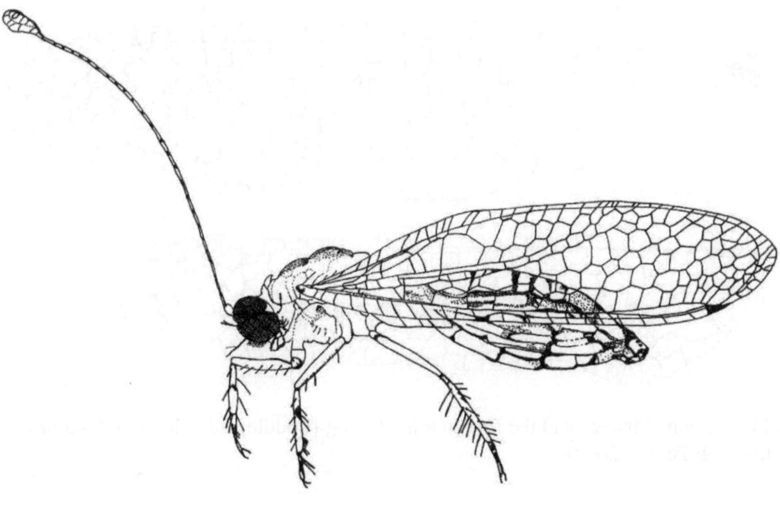

Fig. 6.10 Habitus of a female *Ululodes cajennensis* in lateral view. Based on Penny (1981a).

6.2 Subfamily Haplogleniinae

Key to the Species of Adult Ascalobyas in South America

Information for the key was provided by Navás (1928, 1930); Penny (1981a, b, 2002), and Thouvenot (2009).

1. The costal margin is clear and free from black pigment (Fig. 6.11). Length of the male: c. 35 mm. Fore-wing length of male: c. 34 mm.
   ....*Ascalobyas machadoi* Penny, 1981
   (Amazonas).
   – The costal margin of the fore-wing is darkly pigmented from its base to the apex (Fig. 6.12) 2
2. The apex of the fore-wing is hyaline, and the pterostigma is usually dark (Fig. 6.4). Length of the male: c. 39 mm; female: c. 38 mm. Fore-wing length of male: 33.5–40 mm; female: 38–46 mm.
   ....*Ascalobyas microcerus* (Rambur, 1942)
   (Caribbean, Central America, Ecuador, French Guiana, Pará, Bahia). Syn: *Byas microcerus* (Walker, 1853); *Ascalaphus leucostigma* McLachlan, 1858; *Haploglenius microcerus* (McLachlan, 1873); *Haploglenius camposi* Navás, 1928, male (Navás, 1930), described as *H. camposi*.
   – The apex of the fore-wing is dark, and the pterostigma is usually pale (Fig. 6.12). Length of male: 38. Length of male fore-wing: c. 33.5; hind wing: c. 30.5 mm.
   ....*Ascalobyas albistigma* (Walker, 1953)

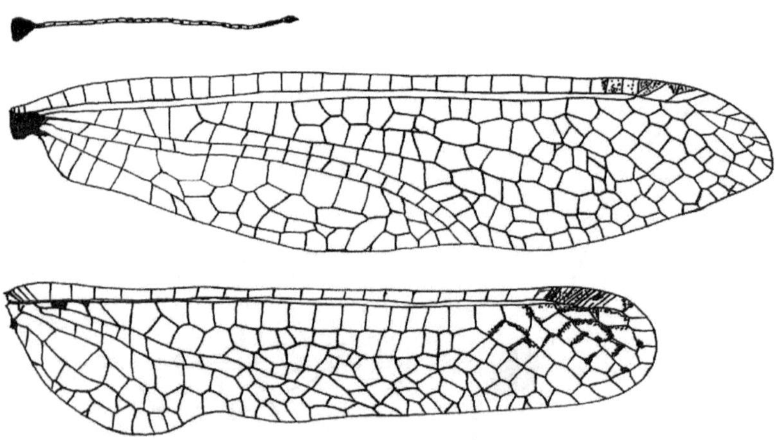

Fig. 6.11 Antenna (above) and the fore and hind wing (middle and below) of *Ascalobyas machadoi*. Based on Penny (2002).

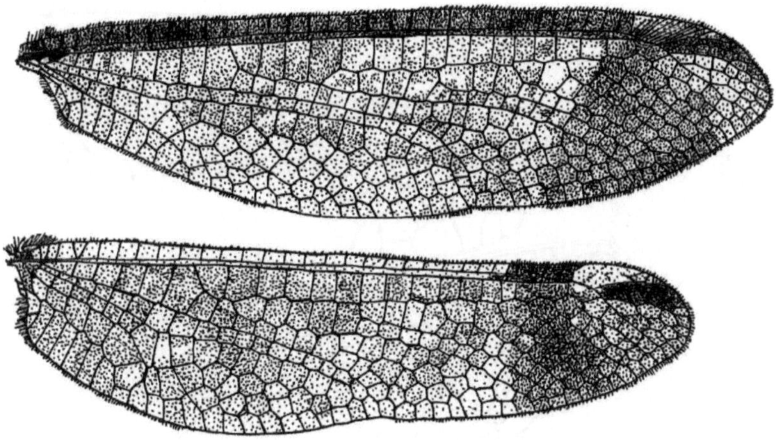

Fig. 6.12 Fore and hind wing of *Ascalobyas albistigma*. Based on Penny (2002).

(North and Central America, Colombia, Ecuador, French Guiana, Amazonas, Pará). Syn: *Ascalaphus albistigma* Walker, 1953; *Haploglenius albistigma* McLachlan, 1873; *Haploglenius terminalis* McLachlan, 1873; *Haploglenius hilaris* Gerstaecker, 1894; *Haploglenius fervidus* Gerstaecker 1894; *Byas albistigma* van der Weele, 1908; *Haploglenius camposi* Navás, 1928, female; probably also *Haploglenius dupuyi* Navás, 1923.

Key to the Species of Adult *Haploglenius* Species in South America

Information for the key was provided by Navás (1927a, 1932); (Penny 1981a, b); Monserrat (1985, 1986); Thouvenot (2009), and Ábrahám (2013).

1. There is a distinct deep, rounded indentation and a projection along the anal section of the fore-wing (Fig. 6.13).. 2

   – The anal margin of the fore-wing is nearly straight and lacks a projection (Fig. 6.14).. 4

2. (1) The projection at the base of the posterior margin of the fore-wing is not very large, and the costal area is tinged with brown. The apex of the hind wing is dark. There is only a slight darkening of the pterostigma, but the costal margin is very dark from the pterostigma to the apex. The infuscation in the costal area does not reach the radial cross veins (Fig. 6.13). Total length: 31–43 mm. Fore-wing length: 39–52 mm.

   .... *Haploglenius costatus* (Burmeister, 1839) (Central America, Colombia, Venezuela, Guyana, French Guiana, Ecuador, Peru, Bolivia, Argentina, Bahía, Pará, Amazonas, Tocantins, Roraima, Mato

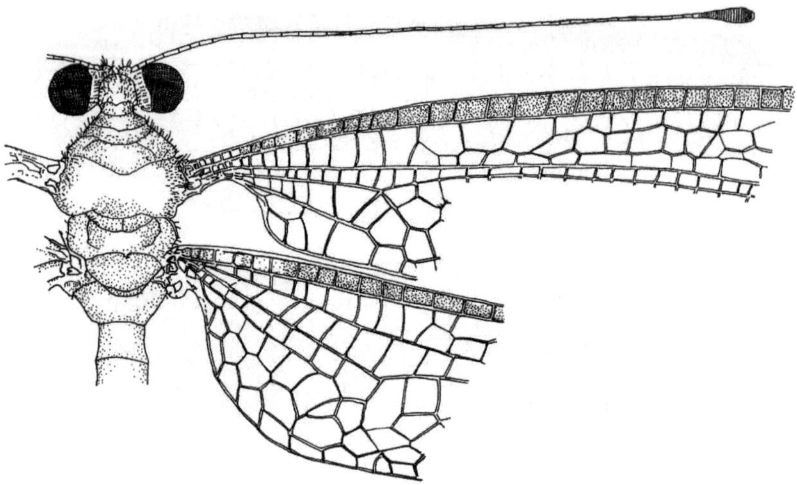

Fig. 6.13 Head, thorax, and proximal parts of the wing of *Haploglenius costatus* in dorsal view. Based on Penny (1981b).

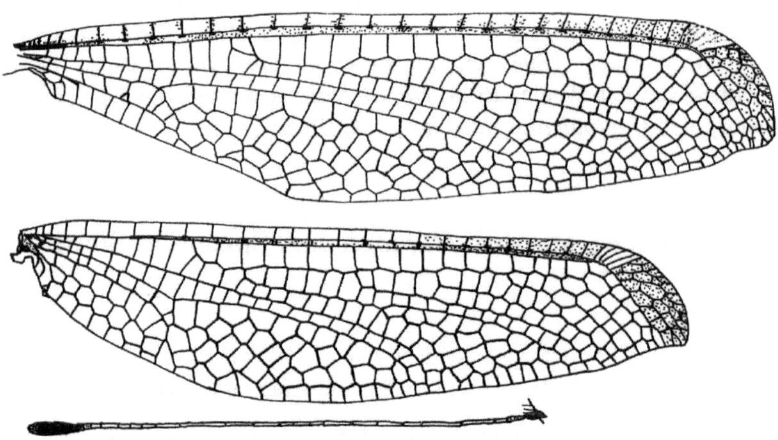

Fig. 6.14 The fore and hind wing (above and middle) and the antenna of *Haploglenius latoreticulatus* (below). Based on Ábrahám (2013), whose illustrations were of a specimen originally designated as a type of *Haploglenius neoguineensis,* now a junior synonym.

Grosso, Rio de Janeiro, Santa Catarina). Syn: *Ascalaphus costatus* Burmeister, 1839; *Ascalaphus costatus* Walker, 1853 partim; *Ascalaphus contrarius* Walker, 1853; *Ascalaphus imperator* Hagen, 1861 nomen nudum; *Ptynx costata* (Burmeister, 1839) Hagen, 1866; *Haploglenius pictus* Gerstaecker, 1884 *partim*; probably also *Haploglenius reticulatus* Navás, 1923.

– The projection at the base of the posterior margin of the fore-wing is prominent and well developed (Fig. 6.15) 3

3. (2) The costal areas of both wings are tinged with brown (Fig. 6.15).
....*Haploglenius flavicornis* McLachlan, 1873 (North and Central America, known as far south as Panama). Syn: *Haploglenius dentiger* Gerstaecker, 1894; *Neohaploglenius flavicornis* (McLachlan, 1873) Penny, 1982; *Haploglenius angulatus* Gerstaecker, 1884; *Neohaploglenius angulatus* (Gerstaecker, 1894), see Penny (1981b).

– The costal areas of both wings are hyaline, and the subcostal area has small brown spots at the bases of the cross veins (Fig. 6.16).
....*Haploglenius abdominevittatus* Ardila and Jones, 2012 (Colombia).

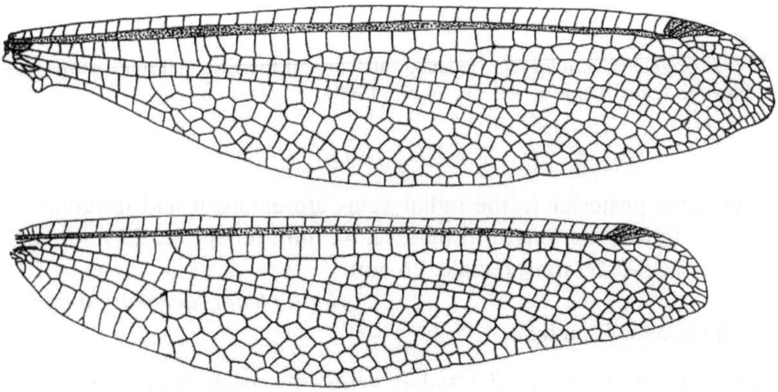

Fig. 6.15 Fore and hind wing of *Haploglenius flavicornis*. Based on Penny (2002).

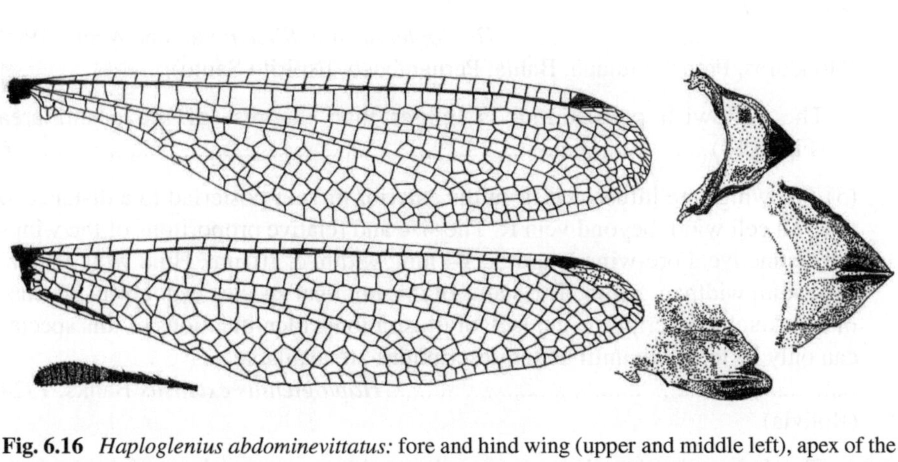

Fig. 6.16 *Haploglenius abdominevittatus:* fore and hind wing (upper and middle left), apex of the antenna (lower left), and male genitalia in the cleared apex of the abdomen in dorsal (upper right), ventral (middle right), and lateral view (lower right). Based on Ardila and Jones (2012).

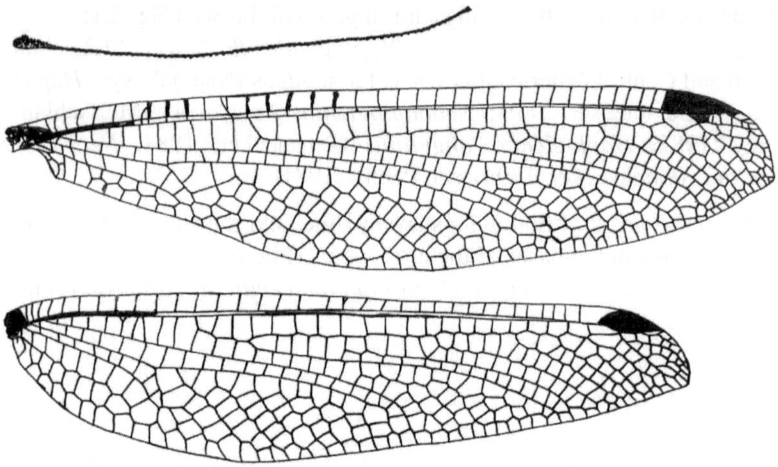

Fig. 6.17 *Haploglenius handlirschi:* antenna (above) and the fore and hind wing (middle and below), drawn to scale. Based on van der Weele (1908).

4. (1) The cells posterior to the radial veins are enlarged and irregular in shape (Fig. 6.1). Fore-wing length of male: c. 42 mm; female: c. 29 mm. Hind wing length of male: c. 39 mm; female: 26 mm.
 .... *Haploglenius decorus* Ábrahám, 2013 (French Guiana, Ecuador).

 – The cells posterior to the radial veins are rectangular and not enlarged (Fig. 6.14) 5

5. (4) The brown pterostigma is as dark or darker than the rest of the costal area (Fig. 6.17).
 .... *Haploglenius handlirschi* van der Weele, 1909 (Honduras, French Guiana, Bahía, Pernambuco, Espirito Santo).

 – The yellowish pterostigma is lighter than the rest of the costal area (Fig. 6.14) 6

6. (5) The wings are infuscated from the anterior margin posteriad to a distance ½ to a full cell width beyond vein R. The size and relative proportions of the wings are distinctive. Fore-wing length: c. 44 mm; width: c. 10 mm. Hind wing length: c. 41 mm; width: c. 8 mm. The status of this nominal species is doubtful because of the sketchy description and lack of illustrations. Identifications of this species can only be tentative until better descriptions are available.
 .... *Haploglenius extensus* Banks, 1924 (Bolivia).

 – The darkened stripe along the costal area, if present, reaches only as far posteriad as vein R (Fig. 6.14) 7

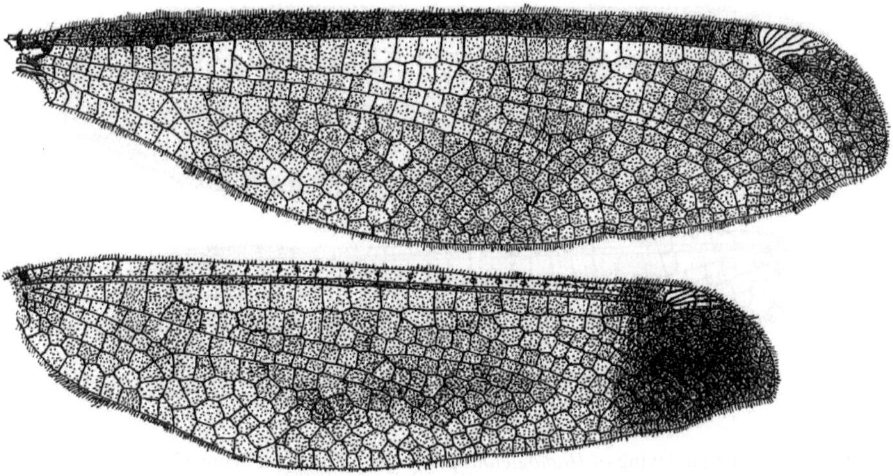

Fig. 6.18 Fore and hind wing of *Haploglenius luteus*. Based on Penny (2002).

7. (6) Fewer than 35 cross veins are in the costal area of the wing (Fig. 6.14). .... *Haploglenius latoreticulatus* van der Weele, 1909 (Central America, Ecuador, French Guiana; reports from New Guinea are apparently in error). Syn: *Haploglenius luteus* aberration *latoreticulatus* van der Weele, 1909; *Haploglenius neoguineensis* Navás, 1914, according to New (1986).
   – The wing has 35 or more cross veins in the costal area (Fig. 6.18) 8

8. (7) The costal area of the fore-wing is uniformly hyaline and lighter than that of the hind wing (Fig. 6.18). The pterostigma is yellow. Length of the male: 37–38 mm; female: c. 36 mm. Fore-wing length of male: 40–55 mm; female: c. 44 mm. .... *Haploglenius luteus* (Walker, 1853) (Mexico, Central America, Colombia, Venezuela, Ecuador, Bolivia, Peru, Guyana, French Guiana, Amazonas, Pará, Amapá). Syn: *Ascalaphus luteus* Walker, 1853; *Ascalaphus circumflexus* Walker, 1853; *Haploglenius costatus* McLachlan, 1873 partim; *Haploglenius pictus* Gerstaecker, 1884 *partim*; *Haploglenius eurypterus* Navás, 1920; *Haploglenius bolivianus* Navás, 1927. *H. luteus* may be conspecific with *H. costatus*, according to Ábrahám (2013).
   – The apex of the hind wing is hyaline. The area distal to the pterostigma is very dark. The infuscation on the costal margin often reaches the radial cross veins (Fig. 6.19). Length of the male: c. 34 mm. Fore-wing length of male: c. 33–40 mm. .... *Haploglenius peruvianus* van der Weele, 1908 (Central America, Peru, Bolivia, Argentina, Amazonas, Pará, Mato Grosso).

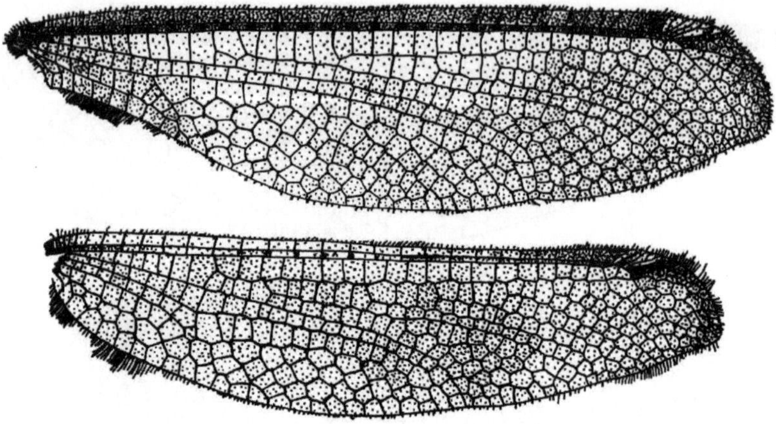

Fig. 6.19 Fore and hind wing of *Haploglenius peruvianus*. Based on Penny (2002).

Key to the Species of Adult *Amoea* in South America

Information for the key was provided by McLachlan (1873); van der Weele (1908); Navás (1911, 1913a, 1930, 1931). Thouvenot (2009). Penny (1981a, b; 2002) reported that there is little difference in the morphology and color pattern of the known species, and that some or all will prove to be conspecific after further study.

1. The thorax and anterior abdominal segments are snow white on the ventral surface. The wing membrane is darkened, and the pterostigma is dark grayish brown (Fig. 6.20). Total length: c. 32 mm. Fore-wing length: c. 30 mm; hind wing length: c. 25 mm. Length of antenna: c. 20 mm.
   .... *Amoea nivea* Navás, 1911
   (Paraguay).

   – The thorax and anterior abdominal segments are yellow or darkened on the ventral surface. The wing membranes are mainly hyaline, although the costal area, wing root, and pterostigma may be pigmented (Fig. 6.5)..................... 2

2. (1) The wing membrane is very slightly darkened near the apex of the wing, or the costal field is somewhat darkened and other wing markings might also be present (Fig. 6.21) 3

   – The wings are hyaline, and the pterostigma is yellow or yellowish brown (Fig. 6.5) 5

3. (2) The membranes are slightly tinted near the apex of the wing, and the pterostigma is yellow and supported by five cross veins, including ones that are forked. The base of the wing and all of the veins are dark (Fig. 6.21). The antenna is yellow with a narrow dark marking on the apical enlargement. Total length: c. 30.5 mm. Length of fore-wing: 28.5 mm; hind wing: c. 24.5 mm.
   .... *Amoea loretana* Navás, 1930
   (Argentina).

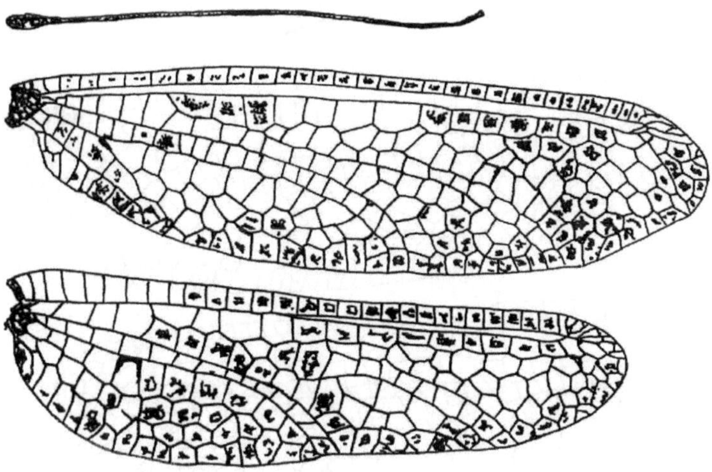

Fig. 6.20 *Amoea nivea:* antenna (above) and the fore (middle) and hind wing (below). Based on Navás (1911).

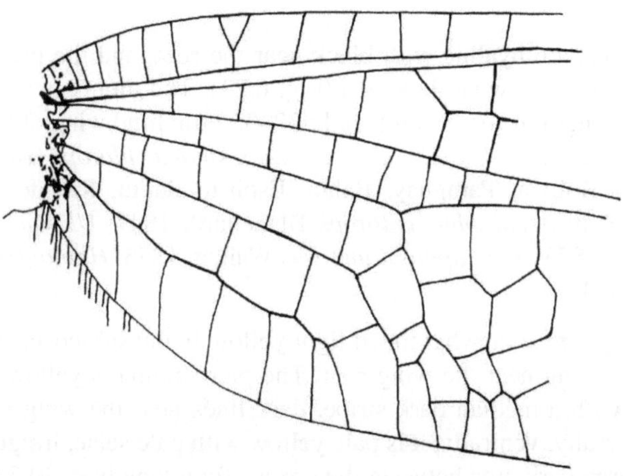

Fig. 6.21 Proximal part of the hind wing of a male *Amoea loretana*. Based on Navás (1930).

– The costal field of the wing is somewhat darkened, and other distinct wing markings may be present (Fig. 6.5). The apex of the wing may be hyaline with distinct markings but not faintly and indistinctly tinted. Whitish setae are present on the thorax 4

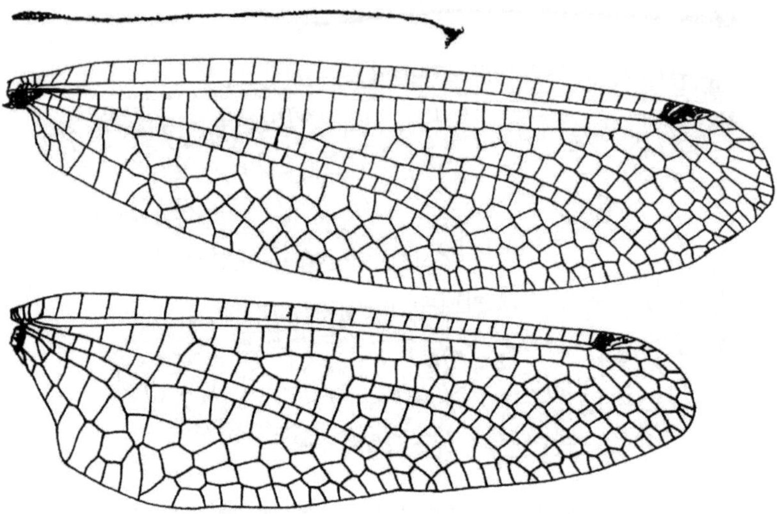

Fig. 6.22 *Amoea chlorops* male: antenna (above) and the fore and hind wing (middle and below). Based on van der Weele (1908).

4. (2) The wings are hyaline with black near the root, and the membranes in the costal area are somewhat darkened (Fig. 6.22). The thorax is dark with yellow markings. Length of fore-wing length: 32–35 mm; hind wing: 29–32 mm. .... *Amoea chlorops* (Blanchard, 1847) (Argentina, Bolivia, Paraguay, Bahia, Espirito Santo, Rio de Janeiro, Santa Catarina). Syn: *Ascalaphus chlorops* Blanchard, 1847; *Ululaus* (sic) *chlorops* McLachlan, 1873; *Ascalaphus damnosus* Walker, 1853; *Haploglenius damnosus* (Walker, 1853).

   – The wings are somewhat tinted light yellow in the sub-costal field but lack a black marking near the wing root. The pterostigma is yellow. The thorax is yellow with a median dark stripe, dark lines near the wing roots, and dark setae dorsally. Ventrally, it is pale yellow with pale setae, irregular dark lines, and a large dark area between the coxae. Total length: c. 30.5 mm: length of fore-wing: c. 30 mm; hind wing: c. 25.5 mm.

   .... *Amoea fulva* Navás, 1931 (Paraguay).

5. (2) The wings are hyaline, and the pterostigma is yellow (Fig. 6.23). The thorax is gray with yellowish markings and whitish setae. Fore-wing length: 35–49 mm; hind wing length: 32–37 mm. .... *Amoea immaculata* (Olivier, 1789) (French Guiana, Brazil). Syn: *Ascalaphus immaculatus* Olivier, 1789; *Haploglenius immaculatus* (Olivier, 1789); *Amoea subcostatus* Burmeister, 1839; *Haploglenius subcostatus* (Burmeister, 1839); *Ascalaphus subcostatus* (Burmeister, 1839); *Ascalaphus injurius* Walker, 1853; *Haploglenius injurius* (Walker, 1853).

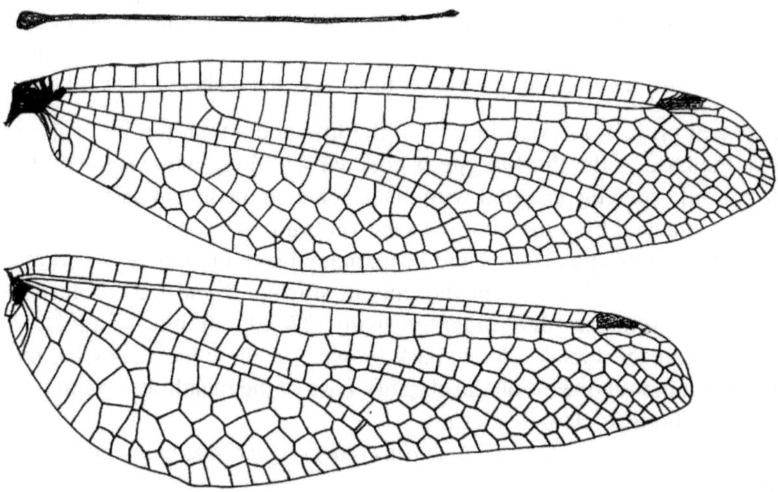

Fig. 6.23 *Amoea immaculata* male: antenna (above) and the fore and hind wing (middle and below). Based on van der Weele (1908).

– The wings are hyaline, and the pterostigma is yellowish brown (Fig. 6.5). The thorax is dark brown, sometimes with a transverse yellow stripe on the pronotum and yellow spots at the lateral margins of the scutum and scutellum, which sometimes are expanded to form one transverse or two longitudinal bands. Fore-wing length of male: 26–31 mm; female: 29–38 mm.
.... *Amoea iniquus* (Walker, 1853) (Venezuela, Peru, Amazonas, Pará). Syn: *Ascalaphus iniquus* Walker, 1853; *Haploglenius iniquus* (Walker, 1853) McLachlan, 1873; *Episperches iniquus* (Walker, 1853) van der Weele, 1908; *Ascalaphus impediens* Walker, 1853; *Haploglenius impediens* (Walker, 1853; McLachlan, 1873); *Episperches impediems* (Walker, 1853) van der Weele, 1908; *Ascalaphus arenosus* Walker, 1853; *Haploglenius arenosus* (Walker, 1853) McLachlan, 1873; *Episperches arenosus* (Walker, 1853) van der Weele, 1908; *Episperches arenosus* Gerstaecker, 1894; *Episperches irideus* Gerstaecker, 1894; *Episperches taeniatus* Gerstaecker, 1894.

6.3 Subfamily Ascalaphinae

Key to the Species of Adult *Fillus* in South America

Information for the key was provided by Machado and Rafael (2011). A third nominal species, presently assigned tentatively to *Fillus,* was described only from a female specimen from Paraguay, which may be the female of *Fillus brethesi,* still known only from the male (Penny, 1981b). The species was described as *Acmonotus paradoxus* van der Weele, 1908, but if it proves to be synonymous with *F. brethesi*, its valid name would be *Fillus paradoxus* (van der Weele, 1908). The species in this genus require further consideration after more specimens become available.

1. The dorsal projection extending from the first abdominal tergite of the male to above the base of the third tergite is covered with long white setae on the basal half but only sparcely covered with black setae on the apical half. The pulvinus, which is attached by a membrane to the ventral part of the ectoproct, is greatly elongated, and its apex curves and bears many black bristles. The parameres are long and thin but shorter than the pulvini (Figs. 1.5 and 6.2). Total length of male: 31–33 mm. Length of male fore-wing: 27–29 mm; hind wing: 22–24 mm. The predominant dorsal color is dark brown, but there are yellowish markings along the compound eyes, on the clypeus, base of the mandibles, basal part of the palps, and thoracic pleural sclerites.
   .... *Fillus amazonicus* Machado and Rafael, 2011 (Amazonas).
   – The dorsal projection extending from the first abdominal tergite of the male is densely covered by setae (Fig. 6.24). The pulvinus and parameres of the male are not much elongated. Length of fore-wing: c. 28.4 mm; hind wing: c. 23 mm.
   .... *Fillus brethesi* Navás, 1919 (Argentina).

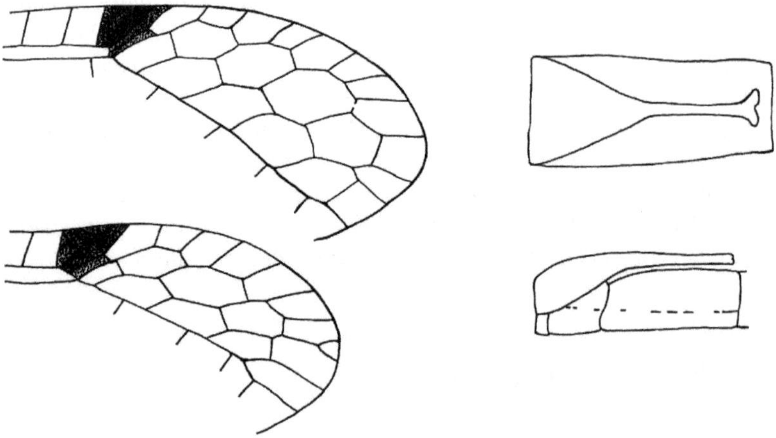

Fig. 6.24 *Fillus brethesi* female: apex of the fore-wing (upper left) and hind wing (lower left), and the apex of the thorax and basal segments of the abdomen in dorsal (upper right) and lateral view (lower right). Based on Navás (1919).

Key to the Species of Adult *Ascalorphne* Species in South America

Information for the key was provided by Navás (1911b, 1913a, b) and Penny (1981a, b, 2002).

1. The wings are hyaline with a blackish stigma and weakly spotted membranes in the subcostal field. The antenna has blackish rings (Fig. 6.25). Total length: c. 21–22 mm. Length of fore-wing: c. 25.5–26.5 mm; hind wing: c. 22.5–24 mm. Length of antenna: c: 28.5 mm.
   ....*Ascalorphne leisewitzi* (Navás, 1911) (Argentina). Syn: *Orphne leisewitzi* Navás, 1911.
   – The wings lack markings (Fig. 6.6)... 2
2. The antennae are black and notably longer than the wings. The stigma is intensly black, rhomboid, and contains six cross veins (Fig. 6.6).
   .... *Ascalorphne macrocercus* (Burmeister, 1839) (Bahía, Espirito Santo, Rio de Janeiro, Santa Catarina). Syn: *Ascalaphus macrocercus* Burmeister, 1839; *Suhpalacsa macroceroks* (sic) (Burmeister, 1839).
   – The antennae are only slightly longer than the wings, reddish, and with black rings. The stigma is dark, trapazoidal, and contains four to six cross veins. The fore-wing is 25.5–26.5 mm.
   .... *Ascalorphne impavida* (Walker, 1853) (Maranhão, Pará) Syn: *Ascalaphus impavidus* Walker, 1853; *Orphne impavida* (Walker, 1853) McLachlan, 1873; *Ascalorphne intempestivus* Walker, 1853.

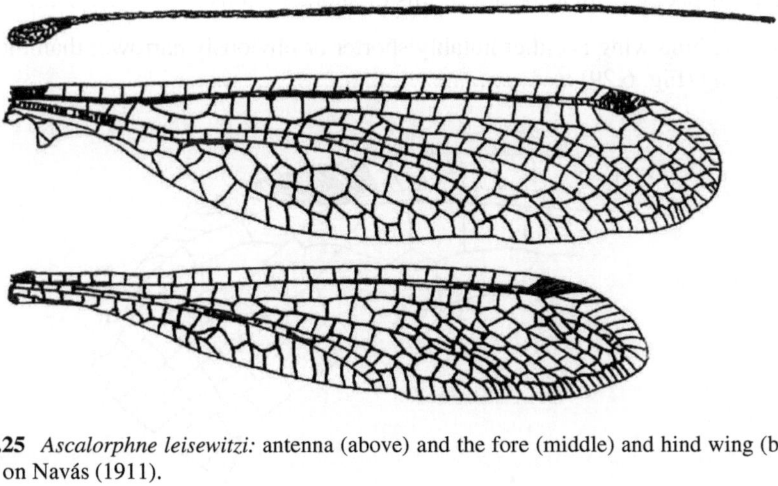

Fig. 6.25 *Ascalorphne leisewitzi:* antenna (above) and the fore (middle) and hind wing (below). Based on Navás (1911).

Key to the Species of Adult *Ameropterus* in South America

Information for the key was provided by Navás (1913a, 1919, 1932), Penny (1981a, 2002), Thouvenot (2008), and von Ellenrieder (2009).

1. The length if the pterostigma, which is darkened, along the anterior wing margin is more than twice its maximum width (Fig. 6.26)....................................2

 – The pterostigma is, at most, only slightly longer than its maximum width (Fig. 6.27)..3

2. (1) The legs are mainly testaceous with blackish markings. The tarsi are pale with black rings. The wings are acutely pointed at the apex (Fig. 6.26). Total length: c. 18.5 mm. Length of fore-wing: c. 22.5 mm; hind wing: c. 20 mm. Length of antenna: c. 22.5 mm.
 ....*Ameropterus longistigma* (Navás, 1913) (Argentina). Syn: *Colobopterus longistigma* Navás, 1913.

 – The legs are mainly yellowish with wide black rings and dark setae. The tarsi are testaceous with the apical segment black. The wings are subeliptically rounded at the apex (Fig. 6.28). Total length: c. 28 mm. Length of fore-wing: c. 27.5 mm; hind wing: c. 23 mm.
 ....*Ameropterus nigrostigma* Navás, 1932 (Peru).

3. (1) The fore-wing is only about 2 to 3 mm longer than the hind wing, and the fore-wing of the female is subequal in length to the antenna (Fig. 6.27). The species in the next couplet are poorly described, and one or both may prove to be conspecific with a species of this or another genus.....................................4

 – The hind wing is either notably shorter or obviously narrower than the fore-wing (Fig. 6.29)..5

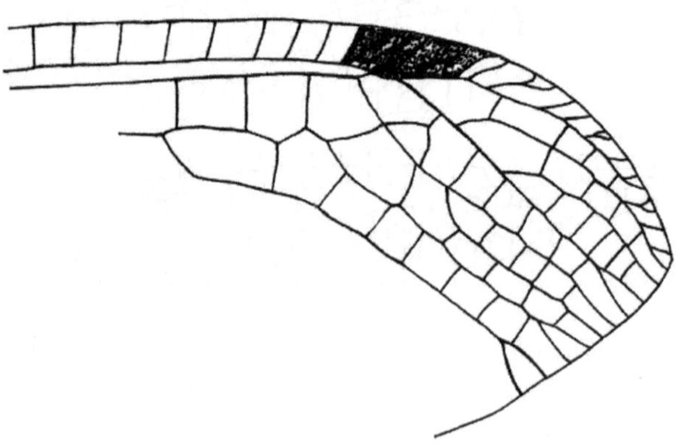

Fig. 6.26 *Ameropterus longistigma:* anterior part of the fore-wing apex. Based on Navás (1913a), who referred to the species as *Colobopterus longistigma*.

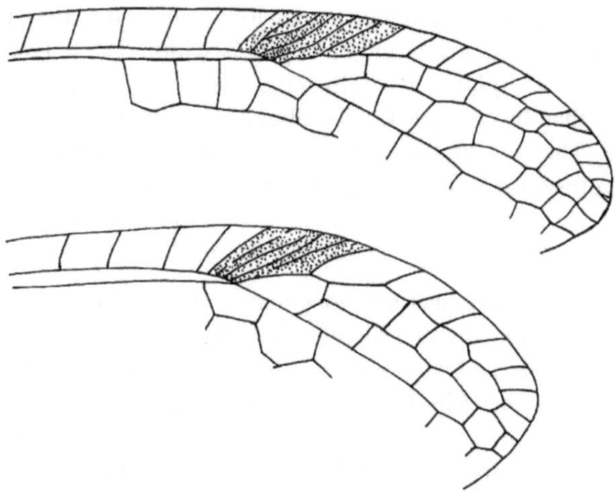

Fig. 6.27 *Ameropterus par:* anteroior part of the fore (above) and hind (below) wing, including the pterostigma. Based on Navás (1918), who referred to the species as *Colobopterus par*.

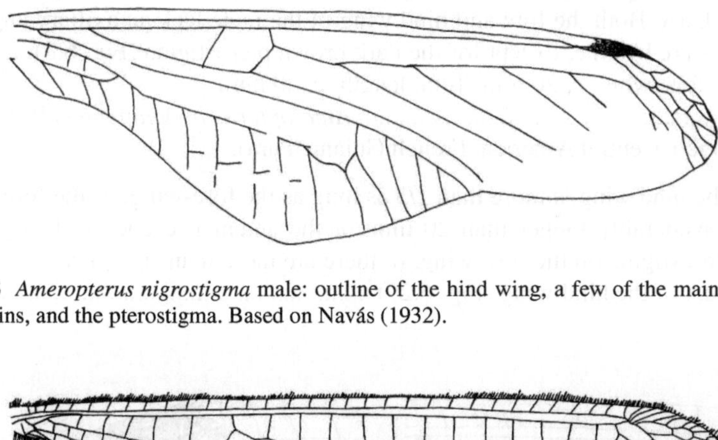

Fig. 6.28 *Ameropterus nigrostigma* male: outline of the hind wing, a few of the main veins, all apical veins, and the pterostigma. Based on Navás (1932).

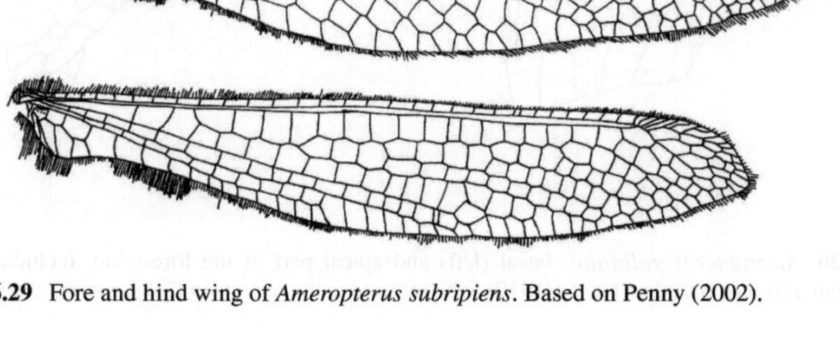

Fig. 6.29 Fore and hind wing of *Ameropterus subripiens*. Based on Penny (2002).

4. (3) The fore-wing is only about 2 mm longer than the hind wing, and both wings are relatively narrow and subequal is width. The antenna is c. 22 mm long, subequal or slightly shorter than the fore-wing. The pterostigma is somewhat longer than high, supported by about 5 strong cross veins, and yellowish (Fig. 6.27). The rest of the wings are hyaline with dark ferruginous veins. Length of fore-wing: c. 22 mm; hind wing: c. 20 mm.
....*Ameropterus par* (Navás, 1918)
(Argentina). Syn: *Colobopterus par* Navás, 1918.

   – The fore-wing is about 3 mm longer than the hind wing. The antenna and fore-wing are subequal in length. The pterostigma is blackish, somewhat longer than high, and supported by about 8 cross veins in the fore-wing and 6 in the hind wing (Fig. 6.30). Total length of the female: c. 18 mm. Fore-wing length of the female: c. 24.8 mm; hind wing: 21.8 mm. Length of the antenna: c. 24.8 mm. The male has not been described.
.... *Ameropterus gallardoi* (Navás, 1919)
(Argentina). Syn: *Colobopterus Gallardoi* Navás, 1919.

5. (3) The hind wing is only about 2/3 as long as the fore-wing, which is about 20 mm long. The antenna has a large terminal bulb, and it does not quite reach the pterostigma on the fore-wing. The hind wing has only six apical forks in vein CuA. Both the fore and hind wing of the male lack an axillary angle. The wings are hyaline, except for the dark brown pterostigma (Fig. 6.8). Length of male fore-wing: c. 20 mm. Total length: c. 20 mm.
....*Ameropterus breviantennus* Penny, 1981
(Mexico, Central America, French Guiana, Pará).

   – The hind wing is more than 2/3 as long as the fore-wing, or the fore-wing is considerably longer than 20 mm, or the antenna reaches well beyond the pterostigma on the fore-wing, or there are more than six apical forks in vein CuA of the hind wing (Fig. 6.29) 6

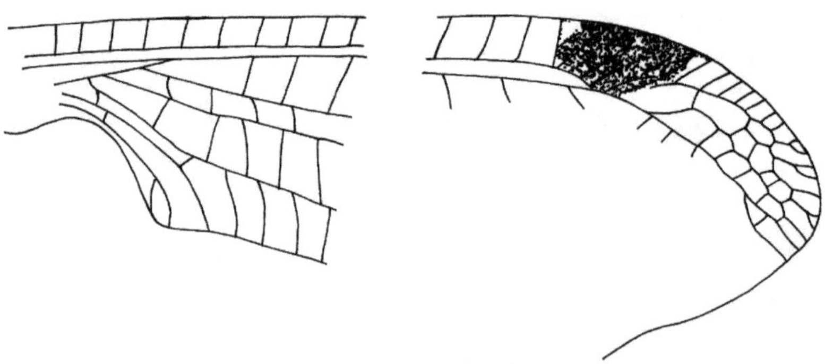

Fig. 6.30 *Ameropterus gallardoi:* basal (left) and apical part of the fore-wing, including the pterostigma (right). Based on Navás (1919), who referred to the species as *Colobopterus Gallardoi.*

6. (5) Both the fore and hind wings are narrowed forming a concave indentation along the posterior margin at or before the mid-length (Fig. 6.29) 7

   – Either one or both pairs of wings lack a concave indentation along the posterior margins (Fig. 6.31).. 8

7. (6) The wing membrane is dark and opaque. The indentation along the posterior margin of the female fore-wing is very slight. The hind wing has ten or more apical forks in vein CuA. The hind wing is almost as long as the fore-wing, but it is much narrower. The anal area of the hind wing has a distinct axillary angle, but it never forms a distinct lobe Vein Rs has five branches (Fig. 6.29). Total length: 20–21 mm. Length of fore-wing: 28–29 mm; hind wing: 26–27 mm. ....*Ameropterus subripiens* (Walker, 1853) (Central America, Venezuela, French Guiana). There are Central American species which are similar except for the shape of the hind wing, which is similar to that of the fore-wing, but they have not yet been reported from South America.

   – The wing membrane is hyaline or slightly clouded in the male. The concave depression along the posterior margin of the hind wing of the female is not very noticeable. Vein Rs has four branches (Fig. 6.32). Fore-wing length: c. 18 mm. The wings are hyaline, and the shape of the fore-wing is trianguloid. The pterostigma is almost transparent and crossed by 5 or 6 cross veins. Total length: 18–22 mm. Length of fore-wing: 18–19 mm; hind wing: 15–16 mm. ....*Ameropterus dissimilis* (McLachlan, 1873) (Peru, Amazonas). Syn: *Colobopterus dissimilis* McLachlan, 1873.

8. (6) The hind wings are obviously concave along their posterior margins (Fig. 6.31).. 9

   – The posterior margins of the hind wings are more or less straight or slightly sinuous (Fig. 6.33) 13

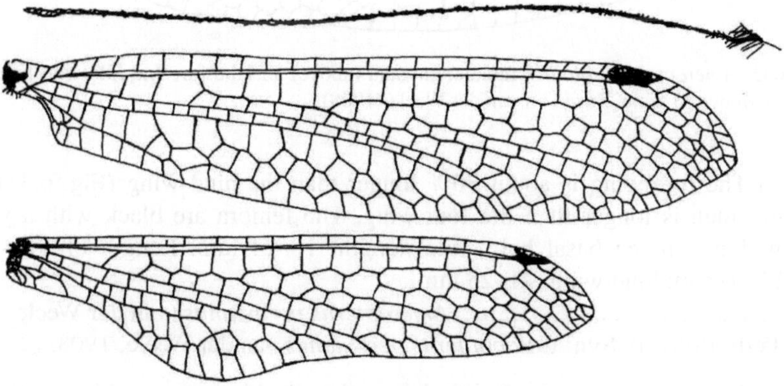

Fig. 6.31 *Ameropterus peruvianus* male: antenna (above) and the fore (middle) and hind wing (below), drawn to scale. Based on van der Weele (1908).

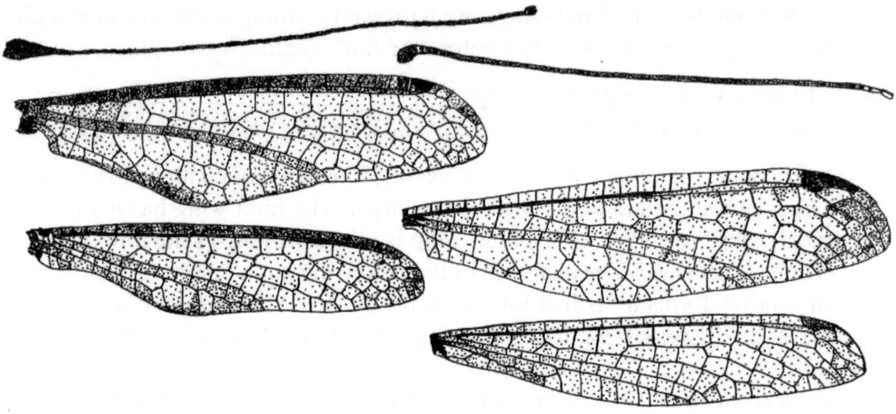

Fig. 6.32 *Ameropterus dissimilis:* antenna of a male and a female (above, left and right, respectively); and the fore (middle) and hind wing (below) of a male (left) and a female (right). Based on van der Weele (1908).

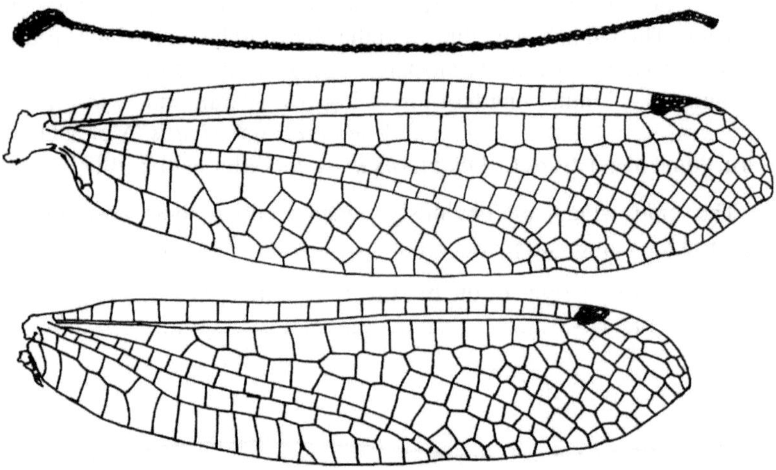

Fig. 6.33 *Ameropterus ululoides* female: antenna (above) and the fore (middle) and hind wing (below), drawn to scale. Based on van der Weele (1908).

9. (8) The fore-wing is about 30 % longer than the hind wing (Fig. 6.31). The abdomen is longer than the four-wing. The femora are black with a yellow marking on the basal half. Total length: 18–24 mm. Length of fore-wing: 27–30 mm; hind wing: 21–25 mm.
....*Ameropterus peruvianus* (van der Weele, 1908) (Peru, Bolivia). Syn: *Colobopterus peruvianus* van der Weele, 1908.
– The fore-wing is only slightly longer than the hind wing and longer than the abdomen (Fig. 6.34)... 10

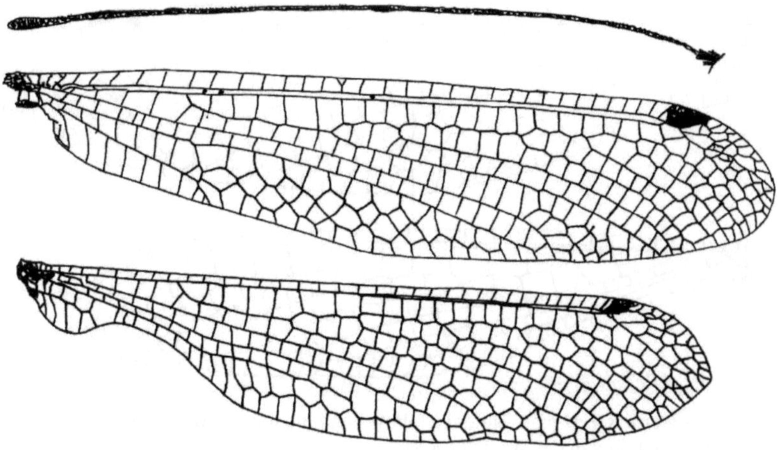

Fig. 6.34 *Ameropterus versicolor:* antenna of a male and a female (above, left and right, respectively); and the fore (middle) and hind wing (below) of a male (left) and a female (right). Based on van der Weele (1908).

10. (9) The depression along the posterior margin of the male hind wing is deep. The axillary angle of the fore-wing is very obtuse, rounded, and not prominent (Fig. 6.34). The antennae are yellow and finely ringed with black. The legs are yellow with black tarsi. Total length: 26–28 mm. Length of fore-wing: 28–31 mm; hind wing: 26.5–28 mm.
.... *Ameropterus versicolor* (Burmeister, 1839) (Argentina, Paraguay, Bahía, Espirito Santo, Minas Gerais, Rio de Janeiro, São Paulo, Santa Catarina, Rio Grande do Sul). Syn: *Ascalaphus versicolor* Burmeister, 1839; *Colobopterus versicolor* (Burmeister, 1839); *Ascalaphus appendifer* Lefebvre, 1842; *Colobopterus leptocerus* Rambur, 1842; *Ascalaphus leptocerus* (Rambur, 1842); *Ascalaphus nematocerus* Rambur, 1842.

– The concave depression along the posterior margin of the wing near the base of the wing is not deep (Fig. 6.35)... 11

11. (10) The antennae reach to the pterostigma and sometimes to the apex of the wing. Most of the antenna, including its entire apical knob, is white to dark brown in color with black setae, and the coxae are white to grayish brown. Both the fore and hind wing have no more than a slight trace of an axillary angle. The wings are entirely clear, except for the pterostigma, which ranges in color from yellowish brown to black and is crossed by four or five veins (Fig. 6.35). The legs are dark brown with fine black ringes on the tarsi. Total length: 20–23 mm. Length of fore-wing: 23–27 mm; hind wing: 21–24 mm.
.... *Ameropterus selysi* (van der Weele, 1908) (Venezuela, Pará, Amazonas) Syn: *Colobopterus selysi* van der Weele, 1908; *Ascalaphus sepultus* Walker, 1853; *Colobopterus sepultus* (Walker, 1853).

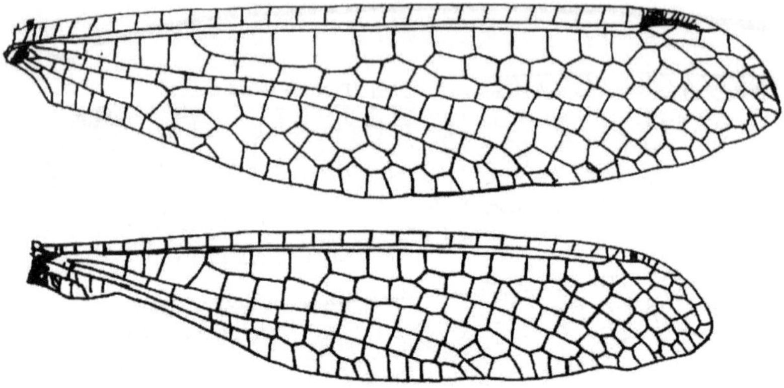

Fig. 6.35 *Ameropterus selysi* male: antenna (above) and the fore (middle) and hind wing (below), drawn to scale. Based on van der Weele (1908).

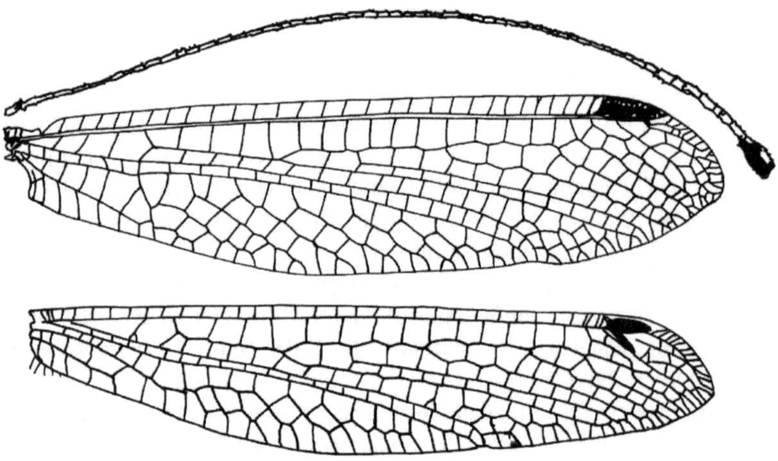

Fig. 6.36 *Ameropterus delicatulus* female: antenna (above); fore (middle) and hind wing (below), drawn to scale. Based on van der Weele (1908).

– The antenna is longer than the fore-wing (Fig. 6.36). The fore wing has an axillary angle of about 90°. The tarsi are blackish 12

12. (11) The hind wing is infuscate at the apex. The wings have membranes tinged with brown, darker apical to the black pterostigma (Fig. 6.36). The apex of the antenna is dark. Total length: 20–26 mm. Fore-wing length of male: 20–24 mm; female: 22–26 mm. Hind wing length: c. 23 mm. .... *Ameropterus delicatulus* (McLachlan, 1873) (Pará, Amazonas). Syn: *Colobopterus delicatulus* McLachlan, 1873.

– The membranes of the wings, including the apex of the hind wing. are hyaline, and the apices are narrowly rounded (Fig. 6.37). The pterostigma is dark brown. The axillary angle of the fore-wing is obtuse. The apex of the

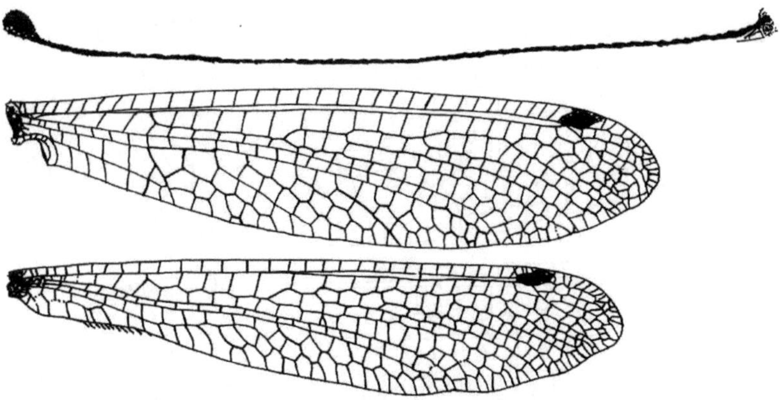

Fig. 6.37 *Ameropterus integer* female: antenna (above) and the fore (middle) and hind wing (below), drawn to scale. Based on van der Weele (1908).

antenna is yellowish with black rings. Total length: 20–26 mm. Length of fore-wing: c. 28 mm; hind wing: c. 27 mm.
....*Ameropterus integer* (McLachlan, 1873) (Paraguay, Brazil). Syn: *Colobopterus integer* McLachlan, 1873.

13. (8) The subcostal membrane is lightly tinged, and the rest of the wing is hyaline, except for the small pterostigma, which is dark brown and supported by three cross veins (Fig. 6.33). The antennae are dark and ringed with black. Total length: c. 28 mm. Length of fore-wing: c. 27 mm; hind wing: c. 24 mm.
....*Ameropterus ululoides* (van der Weele, 1908) (Brazil). Syn: *Colobopterus ululoides* van der Weele, 1908.

– The wing membranes, including those in the subcostal regions, are mainly hyaline, but there may be some opaque markings or tinges of color forming isolated markings (Fig. 6.38).. 14

14. (13) The entire wing membrane is hyaline, and the wing is rounded at the apex. Vein Rs has six branches. The pterostigma is yellow with three or four blackish cross veins supporting it (Fig. 6.38). Total length: c. 24 mm. Length of fore-wing: c. 30 mm; hind wing: c. 26.5 mm.
.... *Ameropterus scutellaris* (Gerstaecker, 1894) (Peru, Bolivia). Syn: *Colobopterus scutellaris* Gerstaecker, 1894.

– The wing is somewhat pointed at the apex. Vein Rs has three or four branches (Fig. 6.39) 15

15. (14) The pterostigma is white and supported by three or four yellowish cross veins (Fig. 6.39). There are sometimes opaque patches of color on parts of the wing membrane. The legs are reddish brown. Total length of female: c. 18 mm. Length of fore-wing: c. 22 mm; hind wing: c. 17 mm. A description of the male is not available.
....*Ameropterus muelleri* (van der Weele, 1908) (Bahía, Espirito Santo). Syn: *Colobopterus Mülleri* van der Weele, 1908.

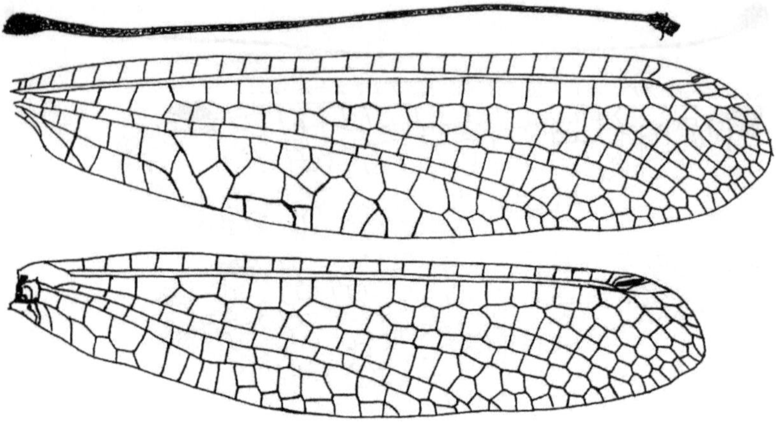

Fig. 6.38 *Ameropterus scutellaris* female: antenna (above) and the fore (middle) and hind wing (below), drawn to scale, with the apex of the antenna pointing left. Based on van der Weele (1908).

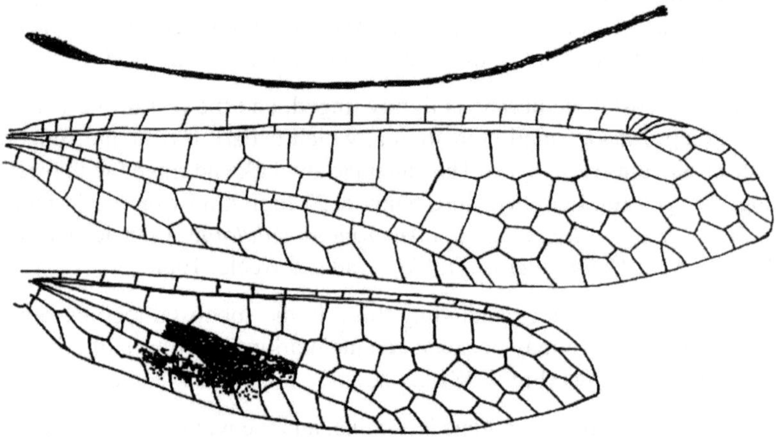

Fig. 6.39 *Ameropterus muelleri* female: antenna (above) and the fore (middle) and hind wing (below), drawn to scale. Based on van der Weele (1908).

– The pterostigma is yellow and supported by four blackish cross veins (Fig. 6.40). The legs are blackish. Total length of male: c. 21 mm. Length of fore-wing: c. 25 mm; hind wing: c. 21 mm. Length of antenna: c, 21 mm.
....*Ameropterus misionarius* (Williner, 1945) (Argentina). Syn: *Colobopterus misionarius* Williner, 1945.

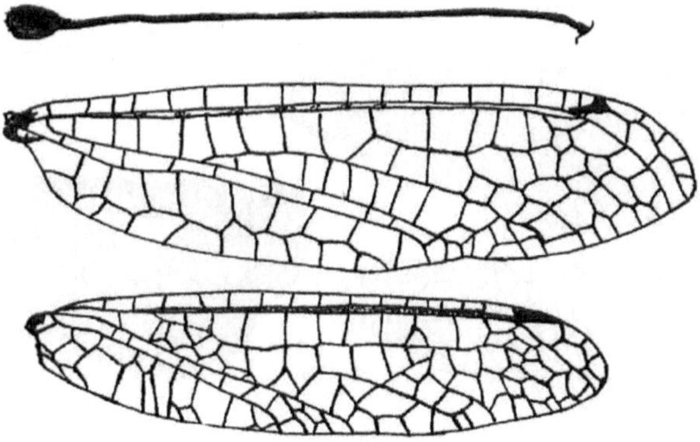

Fig. 6.40 *Ameropterus misionarius* female: antenna (above) and the fore (middle) and hind wing (*below*), drawn to scale. Based on Williner (1945).

6.4 Key to the Species of Adult *Cordulecerus* Species in South America

Information for the key was provided by van der Weele (1908); Navás (1913a, 1919); Penny (1981b, 2002), and Thouvenot (2008).

1. The basal part of the hind wing is notably wider than at mid-length, and the hind wing may have a roughly triangular shape (Fig. 6.41) 2

   – The fore and hind wings of both males and females are similar in shape: rather narrow and pointed at the apex. The hind wing is not notably wider in the basal half and does not become narrower at mid-length (Fig. 6.9) 7

2. (1) Both wings appear narrow and sharply pointed at the apex, with that of the male more triangular in form then that of the female. The hind wing of the male has some infuscation, and it lacks an obtuse angle excavated along the anal margin. There are dark markings across the basal and apical parts of the female hind wing (Fig. 6.41). Total length: c. 26 mm; length of fore-wing: 34–35 mm; hind wing: 31–32 mm.
   .... *Cordulecerus maclachlani* van der Weele, 1908 (French Guiana, Amazonas). Syn: *Cordulecerus surinamensis* Rambur, 1842, preoccupied.

   – The fore-wing is generally elongated and may be somewhat acutely pointed at the apex. The hind wing is narrowed proximal to the mid-length of the wing (Fig. 6.42) 3

3. (2) The head and thorax have a covering of long, dense setae. The hind wing of the female has a dark marking in the posterior corner (Fig. 6.42). The pterostigma and most of the legs are yellow 4

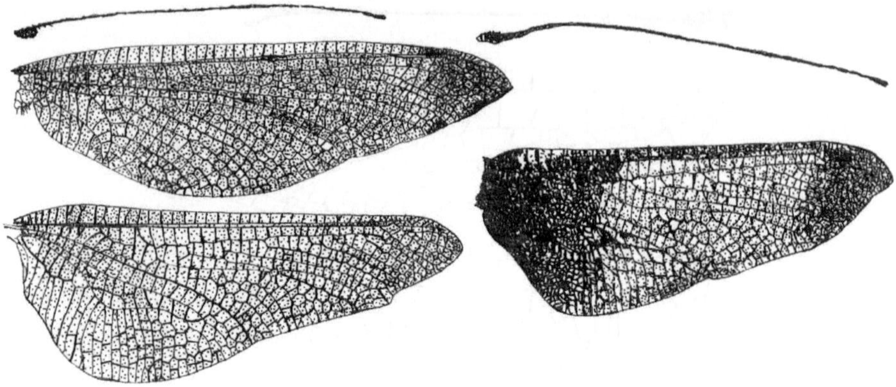

Fig. 6.41 *Cordulecerus maclachlani* (left side, top to bottom): antenna, fore, and hind wing of a male, and (right, top and bottom): antenna and hind wing of a female. Based on van der Weele (1908).

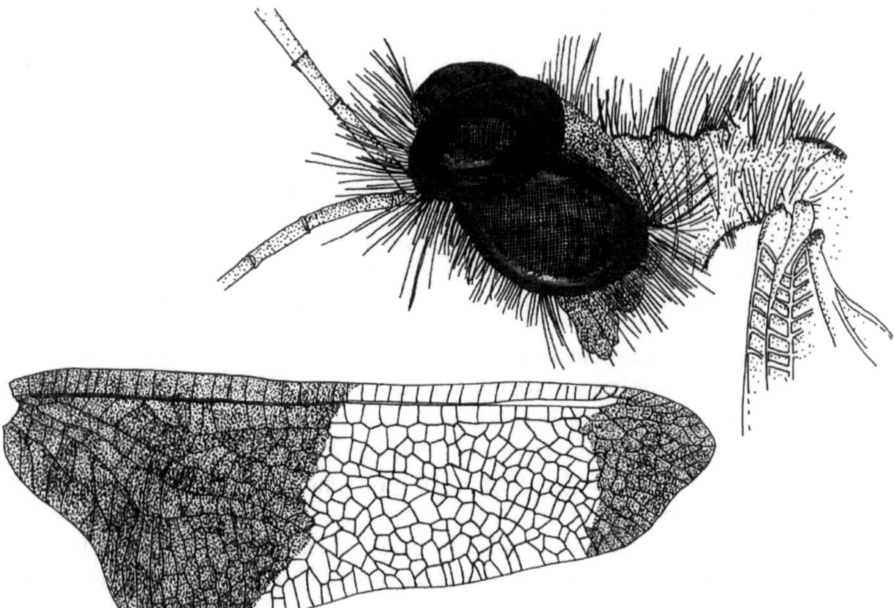

Fig. 6.42 *Cordulecerus elegans:* head and anterior part of the thorax in oblique dorsolateral view (*above*) and the hind wing of a female. Based on Penny (1981b, 2002).

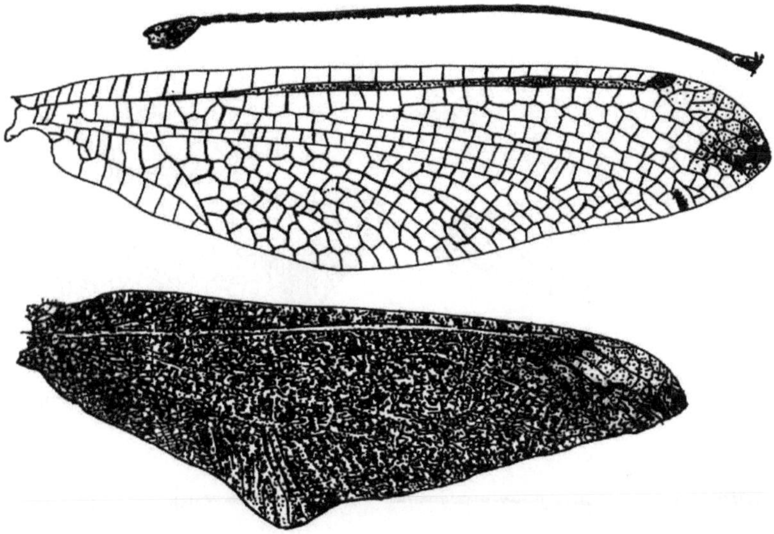

Fig. 6.43 *Cordulecerus unicus:* antenna (above) and the fore (middle) and hind wing (below) of a female. Based on van der Weele (1908).

– The head and thorax lack dense patches of long setae, or, if not, there is no dark marking at the posterior corner of the female hind wing, or the pterostigma and legs are not mainly yellow (Fig. 6.43) 5

4. (3) The wings are hyaline and rounded at the apices. The hind wing of the male is hyaline without dark markings and has an obtuse angle excavated along the anal margin. That of the female has a marginal spot at the base (Fig. 6.42). Total length: 30–33 mm; length of fore-wing: 35–39 mm; hind wing: 30–34 mm. .... *Cordulecerus elegans* van der Weele, 1908 (West Indies, Surinam, Guyana, Amazonas).

– The wings are acutely pointed at the apex, and their membranes are yellowish (Fig. 6.44). The femora of the fore and middle legs are darkened. Total length: 30–33 mm; length of fore-wing: 35–39 mm; hind wing: 34–36 mm. .... *Cordulecerus alopecinus* (Burmeister, 1839) (Argentina). Syn: *Ascalaphus alopecinus* Burmeister, 1839; *Ascalaphus vulpecula* Burmeister, 1839; *Cordulecerus vulpecula* (Burmeister, 1839); *Ascalaphus surinamensis* Guerin-Meneville, 1838, preoccupied; *Ascalaphus litigiosus* Walker, 1853; *Ascalaphus garrulus* Walker, 1853.

5. (3) Almost all of the hind wing is darkened. There is bulge at the posterior angle of the hind wing. There are blackish patches at the apex of the fore-wing (Fig. 6.43). Total length: 24–25 mm; length of fore-wing: 29–31 mm; hind wing: 24–26 mm. .... *Cordulecerus unicus* Walker, 1859 (Bahía, Espirito Santo).

– Most of the wing is colorless (Fig. 6.45) 6

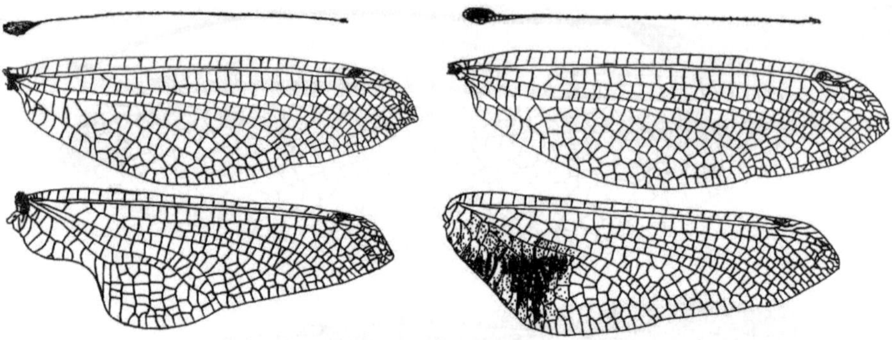

Fig. 6.44 *Cordulecerus alopecinus* male: antenna, fore, and hind wing of a male (left) and a female (right). Some females lack the dark marking on the hind wing. Based on van der Weele (1908).

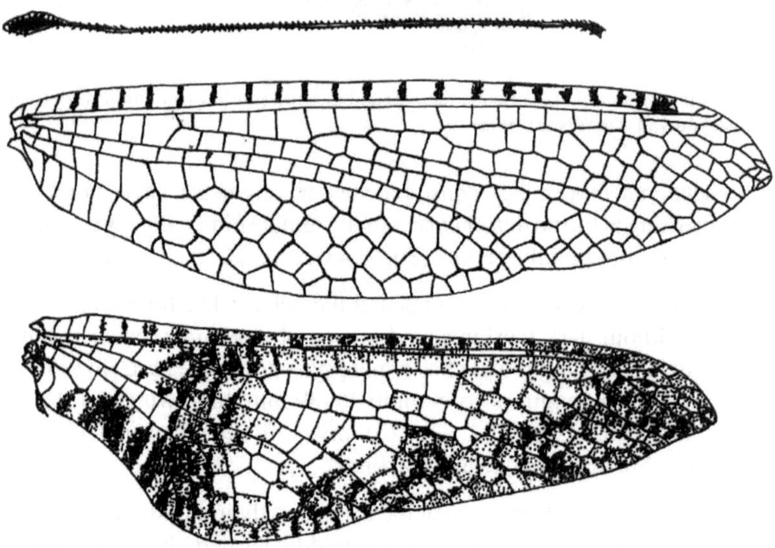

Fig. 6.45 *Cordulecerus dohrni* male: antenna (above) and the fore and hind wing (middle and below). Based on van der Weele (1908).

6. (5) The pterostigma is very small, pale brown, and supported by three cross veins. The costal cross veins are narrowly margined with brown (Fig. 6.45). Total length: c. 25 mm; length of fore-wing: 34 mm; hind wing: c. 31 mm. .... *Cordulecerus dohrni* van der Weele, 1908 (Ecuador).

– The fore and hind wings are of the male are hyaline but may become infuscated with brown as the adult becomes older. The base of the hind wing of the female is darkened, and there is a dark area along most of the posterior margin (Fig. 6.46). Total length: 25–33 mm; length of fore-wing: 33–35 mm; hind wing: 31–32 mm. .... *Cordulecerus inquinatus* Gerstaecker, 1888 (Mexico, Central America, Ecuador, Peru).

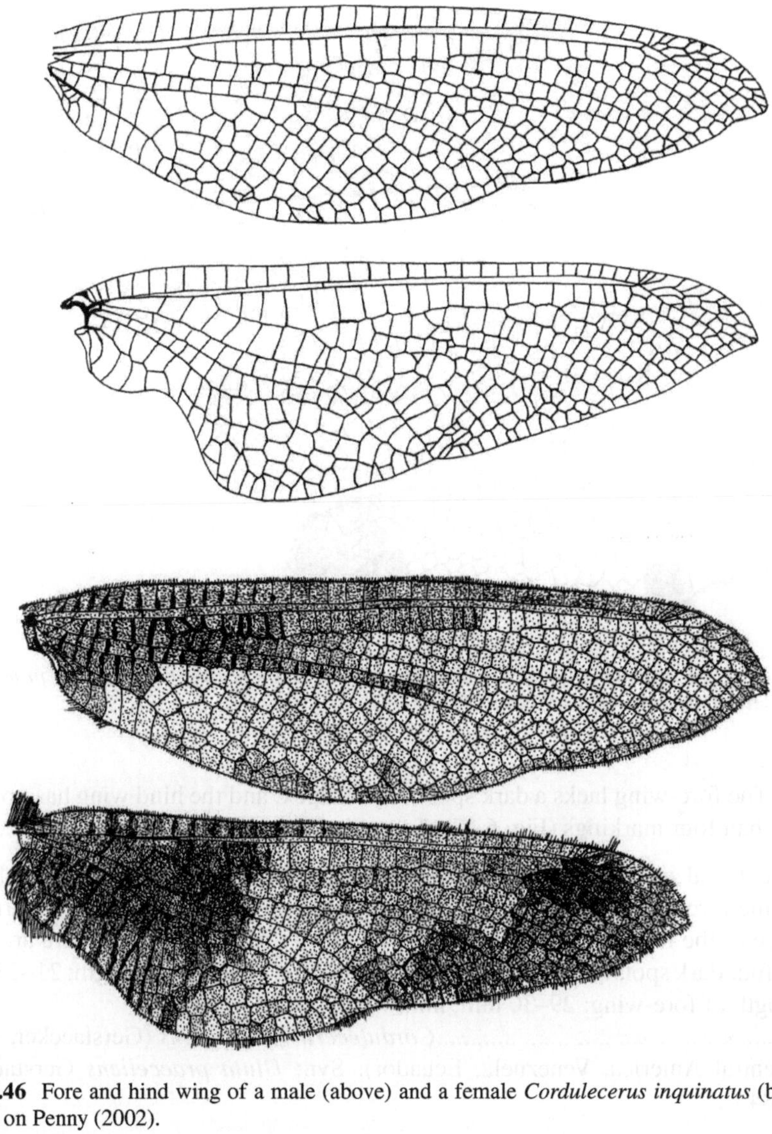

Fig. 6.46 Fore and hind wing of a male (above) and a female *Cordulecerus inquinatus* (below). Based on Penny (2002).

7. The fore-wing has a few small spots at the apex, and there is a series of four or more dark markings along the posterior margin of the hind wing, which may be separate or interconnected (Fig. 6.9). Total length: 20–24 mm; length of fore-wing: 20–28 mm; hind wing: 19–24 mm.
   .... *Cordulecerus subiratus* (Walker, 1853) (Central America, Argentina, Amazonas). Syn: *Ascalaphus subiratus* Walker, 1853.

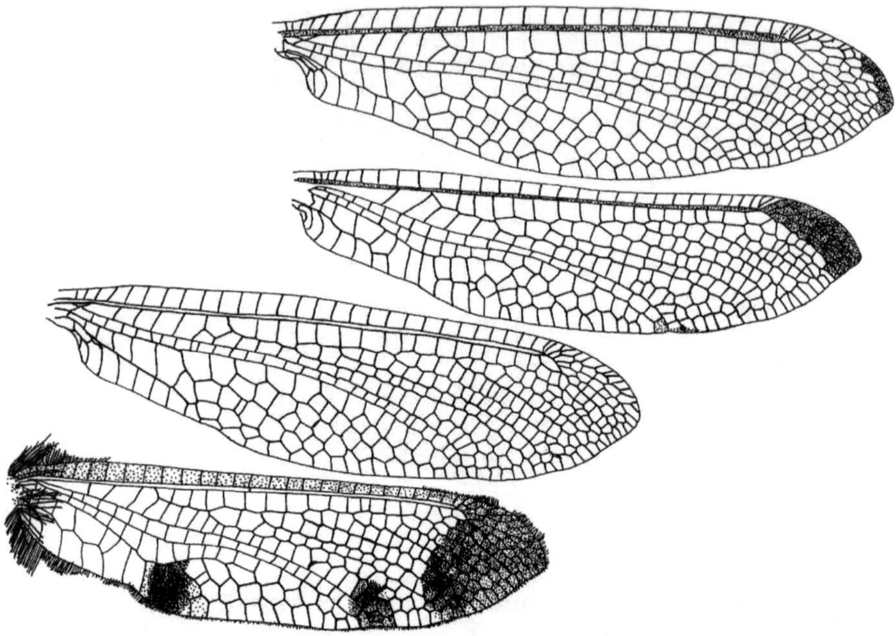

Fig. 6.47 Fore and hind wing of a male (upper right) and a female *Cordulecerus praecellens* (lower left). Based on van der Weele (1908).

– The fore-wing lacks a dark spot near the apex, and the hind wing has no more than four markings (Fig. 6.47). Fore-wing length is 25–30 mm 8

8. The dorsal surface of the abdomen is blackish. The brown tint on the apical parts of the fore and hind wings of the male is darker than that on the basal part. The base of the female hind wing is fairly uniformly translucent, and there are three of four dark spots along the posterior margin (Fig. 6.47). Total length: 21–23 mm; length of fore-wing: 29–30 mm; hind wing: 26–27 mm.
.... *Cordulecerus praecellens* (Gerstaecker, 1884) (Central America, Venezuela, Ecuador). Syn: *Ulula praecellens* Gerstaecker, 1884.

– The dorsal surface of the abdomen is orange. The hind wing of the female lacks dark spots along the margin (Fig. 6.48). Total length: c. 22 mm; length of fore-wing: c. 33 mm; hind wing: c. 29 mm.
.... *Cordulecerus surinamensis* (Fabricius, 1798) (Surinam, Brazil). Syn: *Ascalaphus surinamensis* Fabricius, 1798.

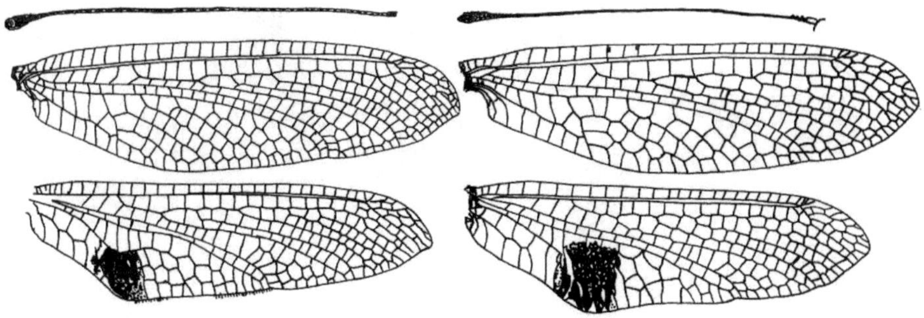

Fig. 6.48 *Cordulecerus surinamensis:* two females (left and right) with varying wing markings showing antennae (above), with apices pointing left, and fore (middle) and hind wing (below). Based on van der Weele (1908).

6.5 Key to the Species of Adult *Ululodes* Species in South America

Information for the key was provided by van der Weele (1908); Banks (1908), Navás (1911, 1913a, 1918, 1922); Penny (1981a, b, 2002), and Thouvenot (2008).

1. The anterior and posterior margins of middle third of both the fore and hind wing are subparallel. The apices of the wings are usually somewhat pointed, especially on the hind wing. The pterostigma is usually pale or yellow, but it apparently darkens with age to brown or black in the male. The pterostigmata on the hind wings of older males may be brown or blackish, while even older ones may have brown or blackish stigmata on all wings. The apex of the hind wing of the male is completely fuscous, while both wings of the female are almost hyaline (Fig. 6.10). The head is reddish brown with white pilosity. The antenna is reddish brown with a fuscous knob at the end. Fore-wing length: 24–25 mm. Hind wing length: 20–22 mm. Length of antenna: 19–22 mm. The species is abundant and noted for its variability.
   ....*Ululodes cajennensis* (Fabricius, 1787) (North and Central America, West Indies, French Guiana, Guyana, Surinam, Venezuela, Ecuador, Peru, Paraguay, Argentina, Maranhão, Pará, Amazonas, Rondônia, Mato Grosso). Syn: *Ascalaphus cajennensis* Fabricius, 1787; *Myrmeleon cajennensis* (Fabricius) Gmelin, 1793; *Ulula cajennensis* (Fabricius) McLachlan, 1873; *Ascalaphus hyalinus* Latreille, 1811; *Ascalaphus inhonestus* Walker, 1853; *Ascalaphus microcephalus* Walker, 1853; *Ascalaphus sublugens* Walker, 1858; *Ulula immersa* Gerstaecker, 1885.

   – The wings have a different shape, either being widened near the base and narrowing gradually toward the apex, widening at the middle, or otherwise enlarged along their length. If in doubt, the pilosity on the head is not white (Fig. 6.49)..2

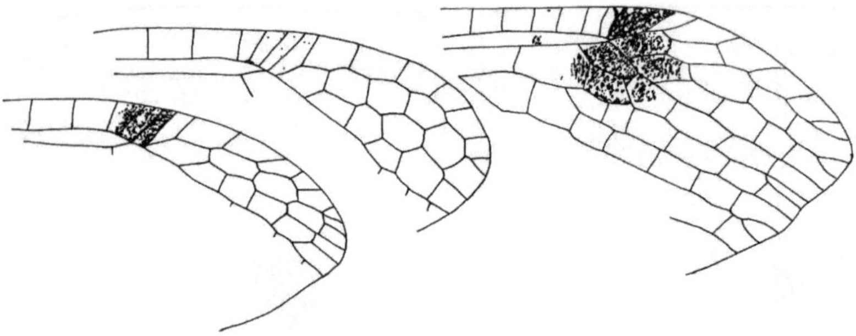

Fig. 6.49 *Ululodes brachycera* (left to right): pterostigmal area and apex of the hind and fore wing of a male and the hind wing of a female. Based on Navás (1918, 1920).

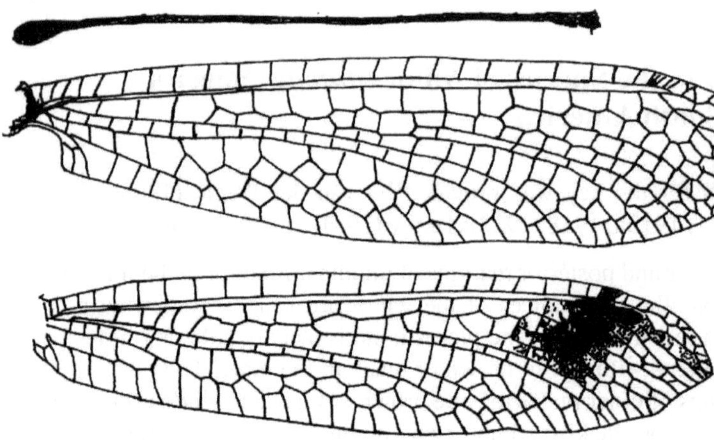

Fig. 6.50 *Ululodes pilosa* female: antenna (above) and the fore (middle) and hind wing (below). Based on van der Weele (1908).

2. (1) The pterostigma on the fore-wing is yellow and supported by five cross veins. That on the hind wing is dark and supported by three or four cross veins. The wings are elliptically rounded at the apex, and the membrane of the male is completely hyaline; on the hind wing of the female, there is a small, dark ferruginous spot on the membrane just posterior to the pterostigma (Fig. 6.49). Fore-wing length: c. 24.5 mm. Hind wing length: c. 21.5 mm. Antenna length: c. 17 mm. .... *Ululodes brachycera* Navás, 1918 (Argentina). Syn: *Ululodes brachicera* (sic) Navás, 1933

   – The pterostigmata on the fore and hind wings are similar in color. The antenna is usually longer than or subequal to the fore-wing, but if it is much shorter, then the color is predominantly yellowish (Fig. 6.50) 3

3. (2) The pterostigmata on both wings are yellowish red, small, rectangular, and supported by two or three cross veins. The thorax is gray and covered with a thick coat of white, hair-like setae. Long white setae are also conspicuous on the

abdomen. The wing veins are very dark (Fig. 6.50). Fore-wing length: 20–23 mm mm. Hind wing length: 21–23 mm. Antenna length: c. 21 mm. .... *Ululodes pilosa* van der Weele, 1908 (Bahía, Pernambuco).

– The pterostigma is not yellowish red (Fig. 6.51). The setae may not be conspicuously long and dense 4

4. (3) The pterostigma is dark brown or blackish (Fig. 6.51) 5

– The pterostigma is pale or yellowish and supported by fewer than seven cross veins (Fig. 6.52) 7

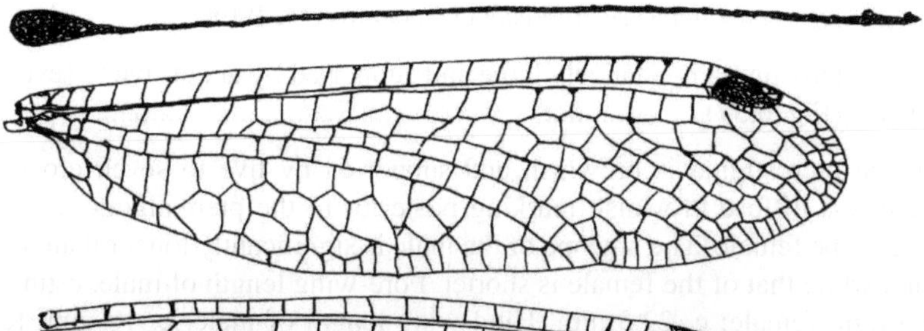

Fig. 6.51 *Ululodes roseni:* antenna (above) and the fore (middle) and hind wing (lower left), drawn to scale, without the veins, but with the position of vein Sc and the relative size of the pterostigma. Based on Navás (1911).

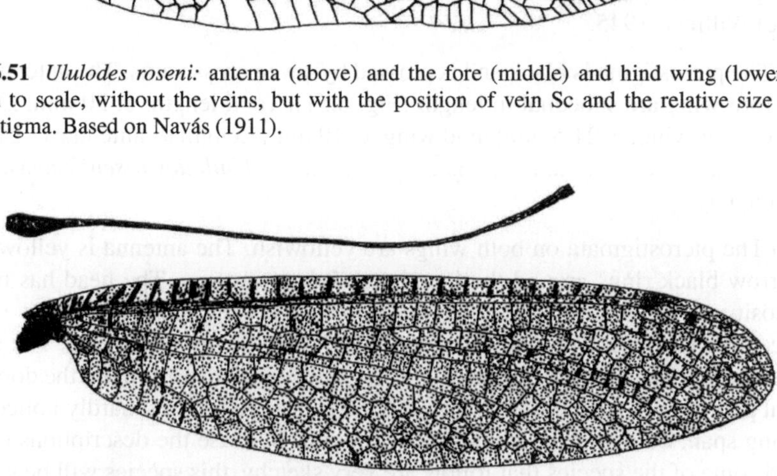

Fig. 6.52 *Ululodes vetula* female: antenna (above) and the fore (middle) and hind wing (below). Based on van der Weele (1908).

5. (4) The small pterostigma is black or dark brown, and there is a fuscous spot posterior to the pterostigma on the hind wing (Fig. 2.12). The pterostigma on the hind wing may have a weaker brown color, and those of a Caribbean population imay be light brown. The species is abundant and widespread, with some variations in color and pilosity among the populations. Fore-wing length: 21–28 mm. Hind wing length: 23–26 mm. Antenna length: 21–23 mm.
.... *Ululodes macleayanus* (Guilding, 1825) (North and Central America, West Indies, Colombia, Venezuela, Argentina, Amazonas, Espirito Santo, Rio de Janeiro, São Paulo, Rio Grande do Sul). Syn: *Ascalaphus macleayana* Guilding, 1825; *Ulula macleayana* (Guilding) McLachlan, 1873; *Ascalaphus limbata* Burmeister, 1839; *Ulula limbata* (Burmeister) Rambur, 1842; *Ululodes macleayana limbata* (Burmeister) van der Weele, 1908; *Ululodes macleayana venezolensis* (Burmeister) van der Weele, 1908.

– The pterostigma is relatively large and supported by about five to seven cross veins (Fig. 6.51).. 6

6. (5) The pterostigma is brownish and supported by five to seven cross veins. There is a diffuse brownish marking posterior to the pterostigma on the hind wing of the female. The antenna of the male is significantly longer than the fore-wing, while that of the female is shorter. Fore-wing length of male: estimated c. 19.5 mm; female: c. 22.5 mm. Hind wing length of male: c. 18 mm; female: estimated c. 19 mm. Length of antenna of male: c. 21.3 mm; female: c. 17.5 mm.
.... *Ululodes heterocera* Navás, 1916 (Argentina). Syn: *Vlulodes* (sic) *heterocera* Navás, 1916; *Ululodes heterocerca* (sic) Williner, 1945.

– The pterostigma is black and is crossed by seven cross veins. The antenna and fore-wing are subequal in length (Fig. 6.51). Total length: c. 20 mm. Length of fore-wing: c. 21.5 mm; hind wing: c. 19 mm; Length of antenna: c. 24 mm.
.... *Ululodes roseni* Navás, 1911 (Paraguay).

7. (4) The pterostigmata on both wings are yellowish. The antenna is yellow with narrow black rings around the junctions of the segments. The head has brown pilosity on the dorsal surface, vertex, and near the eyes and gray pilosity on the face. The thorax is dark brown or dull black with long grayish setae. The abdomen is yellowish brown dorsally with a curved, dark brown spot on the dorsolateral part of each segment. The pterostigma is pale yellow and hardly noticeable. Wing span: 60 mm. Fore-wing width: 7.5 mm. Because the descriptions of this and some of the species that follow are very sketchy, this species will be hard to identify with certainty. The sex of the type specimen was not specified.
.... *Ululodes flavistigma* Banks, 1907 (Ecuador).

– The antenna is usually uniform in color for at least part of its length (Fig. 6.52). The abdomen is not yellowish brown dorsally with a curved, dark brown spot on the dorsolateral part of each segment.. 8

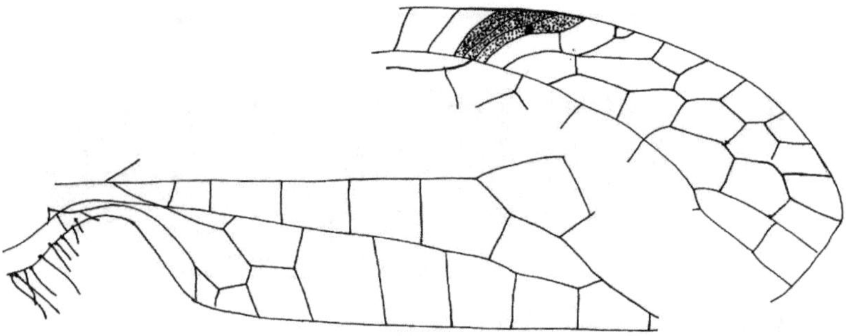

Fig. 6.53 *Ululodes apollinaris* male: axillary (lower left) and apical region of the fore-wing. Based on Navás (1927b).

8. (7) The antenna is about as long as or longer than the fore-wing. The pterostigma is light yellow and crossed by four simple cross veins (Fig. 6.53). Total length: c. 22.5. Length of fore-wing: c. 30 mm; hind wing: c. 26.4 mm; Length of antenna: c. 31.7 mm.
   .... *Ululodes apollinaris* Navás, 1927 (Colombia).
   – The antenna is shorter than the fore-wing (Fig. 6.52) 9
9. (8) The general color of the thorax and wings is yellowish (Fig. 6.52). The pterostigma is yellow, very small, and supported by three or four cross veins. Fore-wing length: 28–31 mm. Hind wing length: 24–28 mm. Length of antenna: 21–23 mm.
   .... *Ululodes vetula* (Rambur, 1842) (Paraguay, Argentina, Uruguay, Pará). Syn: *Ulula vetula* Rambur, 1842; *Ascalaphus vetula* (Rambur) Walker, 1953; *Ulula aurifera* McLachlan, 1873; *Ululodes subvertens* van der Weele *nec* Walker, 1908.
   – The thorax is dark brown, sometimes with ferruginous markings on the pronotum and mesonotum. The wings are broadly rounded at the apex and mainly hyaline, except for the hind wing of the female, which usually has dark markings toward the apex. The pterostigma is short, pale, and hardly noticeable. The axillary angle of the fore-wing is obtuse and not prominent (Fig. 6.54). Fore-wing length: c. 24 mm. Hind wing length: c. 21 mm. Length of antenna: c. 22 mm.
   .... *Ululodes subvertens* (Walker, 1853) (Colombia, Venezuela). Syn: *Ascalaphus subvertens* Walker, 1853.

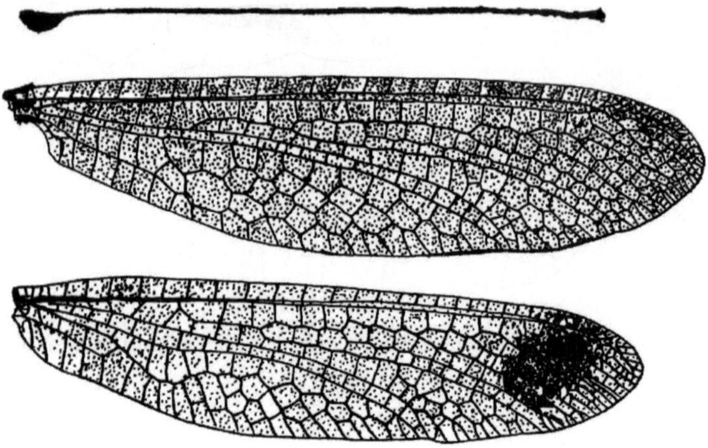

Fig. 6.54 *Ululodes subvertens:* antenna (above); fore (middle) and hind wing (below). Based on van der Weele (1908).

Literature

Ábrahám L (2013) Ascalaphid studies, IX. The genus *Haploglenius* from South America (Neuroptera: Ascalaphidae). Natura Somogyiensis 23:173–188

Ardila-Camacho A, Jones JR (2012) A new species of *Haploglenius* Burmeister, 1839 (Neuroptera: Ascalaphidae) from the Colombian Orinoquía. Zootaxa 3268:40–46

Banks N (1907) New tropical American Neuroptera. Proceedings of the Entomological Society of Washington 9:30–34

Banks N (1908) New tropical American Neuroptera. Proceedings of the Entomological Society of Washington 9(1–4):30–33

Banks N (1915 [1914]) New neuropteroid insects, native and exotic. Proceedings of the Academy of Natural Sciences of Philadelphia 66:608–632+pl. 27

Banks N (1924) Descriptions of new neuropteroid insects. Bulletin of the Museum of Comparative Zoology 65(12):421–455, +pl. 1–4

Gerstaecker A (1884) Vier Decaden von Neuroptera aus der Familie Megaloptera Burm. Mittheilungen aus dem Naturwissenschaftlichen Verein für Neu-Vorpommern und Rügen in Greifswald 16:1–49

Henry CS (1972) Eggs and rapagula of *Ululodes* and *Ascaloptynx* (Neuroptera: Ascalaphidae): a comparative study. Psyche 79(1–2):1–22

Hogue CL, Penny ND (1988) Aggregations of Amazonian owlflies (Neuroptera: Ascalaphidae: *Cordulecerus*). Acta Amazonica 18(1–2):359–362

Machado RJP, Rafael JA (2011) A new species of *Fillus* Navás, 1919 (Neuroptera: Ascalaphidae) from the Brazilian Amazon Basin. Zootaxa 2907:22–28

McLachlan R (1873) An attempt towards a systematic classification of the family Ascalaphidae. Journal of the Linnean Society 11:219–276

Monserrat VJ (1985) Lista de los tipos de Mecoptera y Neuroptera (Insecta) de la colección L. Navàs (sic), depositados en el Museo de Zoología de Barcelona. Miscellánea Zoológica 9:233–243

Monserrat VJ (1986) Longinos Navás, his neuropterological work and collection. In: Gepp J, Aspöck H, Hölzel H (eds) Recent research in neuropterology. Proceedings of the 2nd international symposium on neuropterology, Graz, pp 173–176

Navás L (1909) Notas neuropterológicas. X Sobre Ascaláfidos. 1. Monografia de los Ascaláfidos por el Dr. H. W. van der Weele. Bruselas, 1908 (Nota bibliográfica). Butlletí de la Institució Catalana d'Historia Natural 9:52–57

Navás L (1911) Notas sobre Neurópteros del Museo de Munich. I. Mitteilungen der Münchener Entomologischen Gesellschaft 2:22–28

Navás L (1912) Ascaláfidos sudamericanos. Brotéria Série Zoológica 10:203–231

Navás L (1913a) Sinopsis de los Ascaláfidos (Ins. Neur.). Arxivs de l'Institut de Ciencias, Barcelona 1(3):45–143, (reprint 1–98) + 2 pl

Navás L (1913b) Ascaláfidos sudamericanos. Revista Chilena de Historia Natural 17:41–74

Navás L (1914) Neurópteros Sudamericanos. Primera Serie. Brotéria, Série Zoológica 12(1):45–56, 215–234

Navás L (1915) Neurópteros nuevos o poco conocidos (Quinta Serie). Memorias de la Real Academia de Ciencias y Artes de Barcelona Tercera Época 11:455–480, + 1 pl

Navás L (1918) Algunos insectos de la República Argentina. Revista de la Real Academia de Ciencias Exactas Fisicas y Naturales de Madrid 16:491–504

Navás L (1919 [1918]) Algunos insectos Neurópteros de la República Argentina. Serie Tercera (III). Revista de la Real Academia de Ciencias Exactas Fisicas y Naturales de Madrid 17:287–305

Navás L (1922) Insectos nuevos o poco conocidos. Memorias de la Real Academia de Ciencias y Artes de Barcelona. Tercera Época 17(15):383–400

Navás L (1923a [1919]) Estudos sobre neuròpters (Insectes). Arxius de l'Institut de Ciències (Barcelona) 7(5):179–203

Navás L (1923b [1922]) Algunos insectos del Museo de París. Revista de la Academia de Ciencias de Zaragoza 7:15–51

Navás L (1927a) Insecta Nova. XII Series. Neuroptera. Memorie della'Pontifica Accademia Romana dei Nuovi Lincei, Roma (2)10:1–10

Navás L (1927b) Vienticinco formas neuvas de insectos. Boletin de la Sociedad Ibérica de Ciencias Naturales 26:48–75

Navás L (1930) Algunos insectos de Guayaquil (Ecuador). Revista Chilena de Historia Natural 34:18–19

Navás L (1931) Decadas de insectos nuevos. Decada 1.ª. Revista de la Real Academia de Ciencias Exactas, Fisicas y Naturals de Madrid 26:60–69

Navás L (1932) Insectos Suramericanos. Quinta Serie. Revista de la Real Academia de Ciencias Exactas, Fisicas y Naturals de Madrid 29(14):53–66

New TR (1986) A review of the biology of Neuroptera Planipennia. Neuroptera International, Supplemental Series 1:1–57

Orfila RN (1949) Notas criticas sobre Ascalaphidae. Anales de la Sociedad Cientifica Argentina 148:187–194

Penny ND (1981a) Review of the generic level classification of the New World Ascalaphidae (Neuroptera). Acta Amazonica 11(2):391–406

Penny ND (1981b) Neuroptera of the Amazon Basin, Part 3. Ascalaphidae. Acta Amazonica 11(3):605–651

Penny ND (1983) Neuroptera of the Amazon Basin. Part 9. – Albardinae. Acta Amazonica 13(3–4):697–699

Penny ND (2002) Family Ascalaphidae. In: Penny ND (ed) A guide to the lacewings (Neuroptera) of Costa Rica. Proceedings of the California Academy of Sciences 53(12):176–186

Rambur JP, (1842) Histoire naturelle des insectes: Névroptères. Librairie encyclopédique de Roret. Fain et Thunot, Paris. 534 pp + 12 pl

Thouvenot M (2008) Névroptères de Guyane (deuxième note). L'Entomologiste 64(5):289–291

Thouvenot M (2009) Névroptères de Guyane (troisième note). L'Entomologiste 65(2):65–66

van der Weele HW (1908) Ascalaphiden. In: Collections Zoologiques du Baron Edm. de Selys Longchamps, vol 8. HAYEZ, Brussels, p 326

von Ellenrieder N (2009) Type specimens of Insecta housed at the Museo de Ciencias Naturales de Salta, Argentina. Revista de la Sociedad Entomologica Argentina 68(3–4):253–262

Walker F (1853) List of the specimens of neuropterous insects in the collection of the British Museum. Part II (Sialidae – Nemopterides). Trustees, London, pp 193–476

Williner GJ (1945) Ascaláfidos argentinos (Neurópteros). Revista de la Sociedad Entomologica Argentina 12(5):425–437

Chapter 7
Myrmeleontidae

Abstract An illustrated key is provided for identification of the genera of adult ascalaphids known to occur in South America. Notes are also provided in a key to those larvae that have been collected and described. Illustrated keys are also provided to identify the species of each genus known to occur in South America. The countries or states in which specimens of each species have been found and invalid synonyms are reported.

Keywords Antlion • Key to species • Palparinae • Myrmeleontinae • Larvae

7.1 Key to the South American Subfamilies, Tribes, and Genera of Adults

The general common name for insects in this family is antlion. Information for the key was provided by Hagen (1866); Banks (1906); Navás (1911, 1912, 1913, 1915a, b, 1918a, 1927, 1928a, b, 1930a, 1932a); Porter (1916); Esben Petersen (1920); Stange and Miller (1985); Stange (1984, 1994, 2004, 2010), and Miller (2008).

1. The fore and middle femora lack hair-like sensory setae. The tarsal claws are longer than three times the width of a tarsal segment. In the hind wing, vein CuA curves anteriad to join with the posterior branch of vein MP_2, which fails to run discretely to the hind margin of the wing. In all known South American species, the hypostigmatic cell is at least five times as long as its bredth (Fig. 7.1). The palps are long and thread-like, and they have a marking that curves around the apex and almost reaches the opposite side. ....Subfamily Palparinae.................................2

   – The fore and middle femora each have a hair-like sensory seta. In the hind wing, vein CuA runs to the hind margin of the wing separately from MP_2, to which it is joined by a cross vein. The palps are not elongated or thread-like, and the marking on the palps does not curve around the apex toward the opposite side. In all known South American species, the hypostigmatic cell is at least five times as long as its width (Fig. 7.2). .... Subfamily Myrmeleontinae...............................3

C.W. Heckman, *Neuroptera (Including Megaloptera)*, Encyclopedia of South American Aquatic Insects, DOI 10.1007/978-3-319-35125-4_7

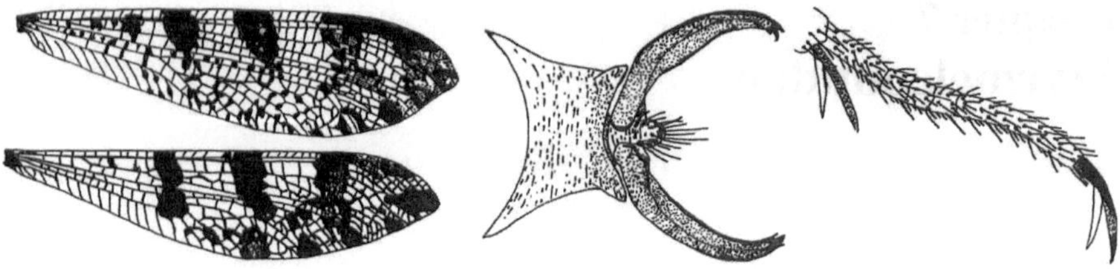

Fig. 7.1 *Dimares elegans* (left to right): fore and hind wing of a female which, unlike those of the male, have dark markings; the male genitalia; the hind tarsus. Based on Stange (1989).

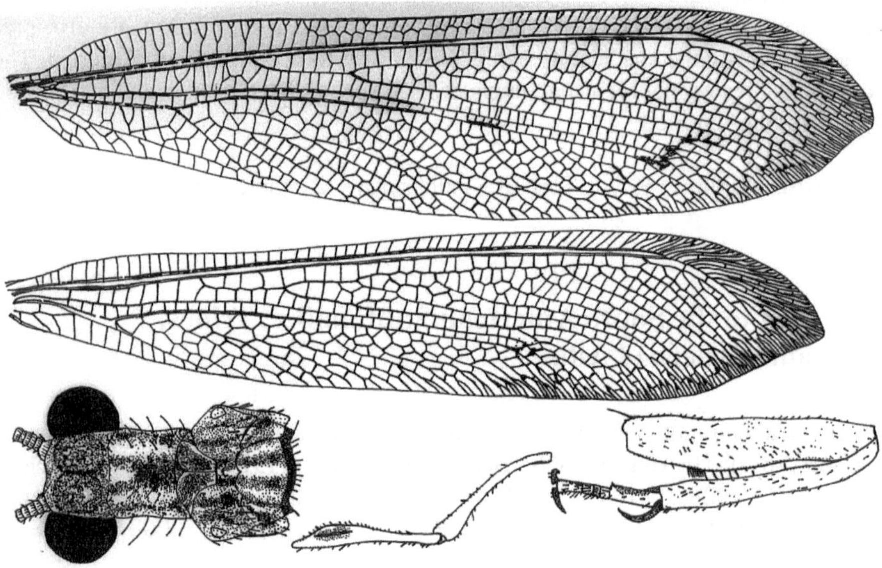

Fig. 7.2 *Vella fallax:* fore and hind wing (above), head and thorax in dorsal view (lower left), labial palp in lateral view (lower center), and hind leg in lateral view (lower right). Based on Stange (2002).

2. (1) The pretarsal claws on the hind leg are shorter than the basal tarsal segment. There is no hook on the paramere of the male, and there are at least 12 teeth along the mesal margin. The male has a well-developed pulula axillaris. The wings of both males and females have pigmented bands or spots (Fig. 7.3). .... *Millerleon* Stange, 1989..p. 250

   – The pretarsal claws on the hind leg are clearly longer than the basal tarsal segment. There is a long hook on the paramere of the male, which has an apical tooth, and there are at least no teeth along the mesal margin. The male lacks a pulula axillaris. The wings of males are nearly unmarked, while those of the females usually have any brown spots or bands (Fig. 7.1). .... *Dimares* Hagen, 1866

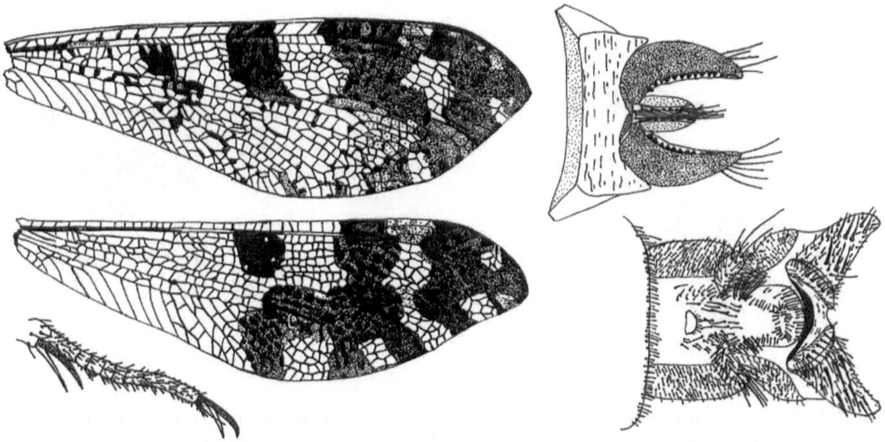

Fig. 7.3 *Millerleon subdolus:* fore and hind wing (upper and middle left), hind tarsus (lower left), male genitalia (upper right), and an exploded ventral view of the apex of the abdomen of a female (lower right). Based on Stange (1989).

At the present time, only one species is assigned to this genus: *Dimares elegans* (Perty, 1833), known from Argentina, Pará, Amazonas, and Mato Grosso. The species also appears in the literature under the following synonyms: *Myrmeleon elegans* Perty, 1833; *Myrmeleon myrmerophagus* Perty, a manuscript name according the Navás, 1912; *Myrmeleon conicollis* Walker, 1860; *Myrmeleon albidilinea* Walker, 1860; *Palpares elegans* (Perty, 1833) Hagen, 1860; *Myrmeleon congruus* Hagen, 1861; *Dimares elegans lepida* Navás, 1912; *Dimares lepidus* Navás, 1915; *Dimares erythrostigma* Navás, 1914; *Dimares hageni* Banks, 1920; *Dimares albidilinea* (Walker, 1860) Navás, 1922.

3. (1) Vein CuP in the fore-wing originates far distad from the basal cross vein, or it is vestigial. Vein 2A in the fore-wing continues in a relatively even curve all the way to the hind margin (Fig. 2.37). 4

   – Vein CuP in the fore-wing originates at or near basal cross vein, that is, no greater distance distad than the diameter of the cubital vein (Fig. 7.2). 18

4. (3) The ectoproct of the male has a postventral lobe, which is longer than its own diameter, or there is a row of clavate setae on the fore-femur, or the middle femur has hair-like sensory setae, which are longer than the femur. The posterior gonapophysis of the female genitalia is subcylindrical, or it is greatly reduced (Fig. 7.4). .... Tribe Brachynemurini................................5

   – The ectoproct of the male lacks a postventral lobe, or it is shorter than the maximum diameter of the lobe. There is never a row of clavate setae on the fore-femur. The hair-like sensory setae on the middle femur is much shorter than the femur in all known South American species. The posterior

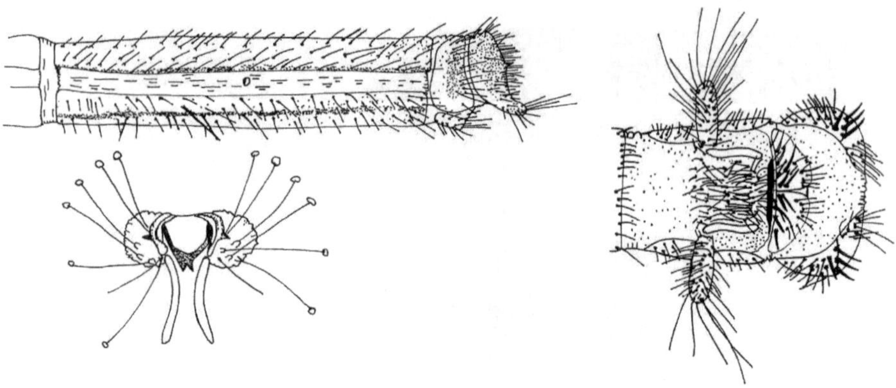

Fig. 7.4 *Venezueleon guaricus:* apex of the male abdomen in lateral view (upper left), male genitalia (lower left), and the apex of the female abdomen in ventral view (right). Based on Stange (1994).

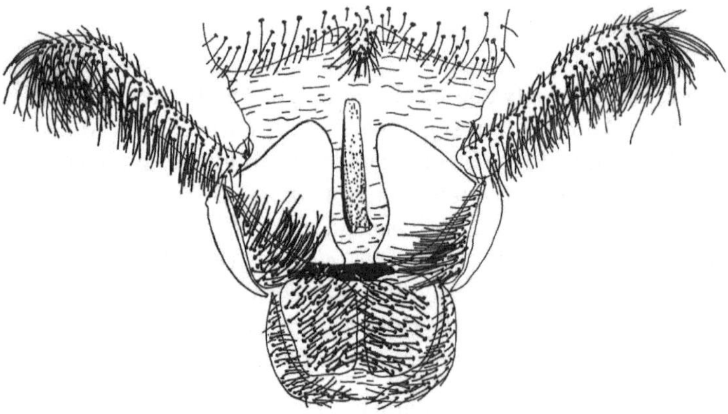

Fig. 7.5 The apex of the abdomen of a female *Neulatus porteri,* with the anterior end facing upwards. Based on Stange (1994).

gonapophysis of the female genitalia is swollen and thumb-like, or it curves strongly (Fig. 7.5).
.... Tribe Lemolemini.. 12

5. (4) The apical tarsal segment of the fore-leg is flattened, greatly enlarged, and bears enlarged setae, which come into contact with the tarsal claws. The 7th through 24th segments of the male antenna are covered by many calli. Along the pilula axillaris of the hind wing and the metanotum of the male is a distinctive series of peg-like setae. Vein CuA in the hind wing curves sharply toward the hind margin at or proximal to the medial fork, or there are only one or two cross veins between CuA and the posterior fork of vein MP_2 (Fig. 2.37). Length

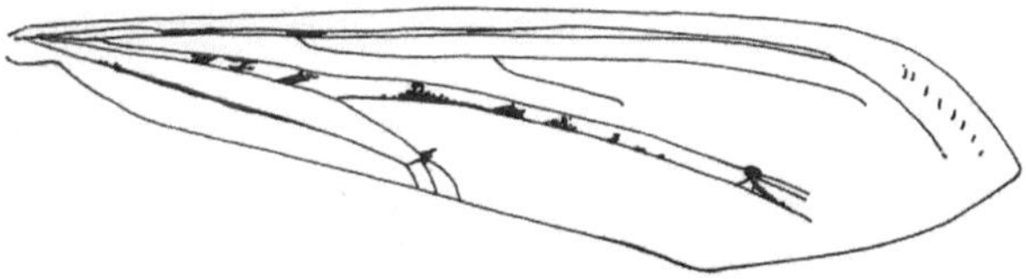

Fig. 7.6 *Peruveleon dolichogaster* male: fore-wing showing the shape and location of the dark markings but not most of the veins. Based on Navás (1915a).

of fore-wing of the only known species: c. 17 mm; greatest width: c. 4 mm. Length of hind wing: c. 15 mm; greatest width: c. 3.5 mm.
....*Stangeleon* Miller, 2008
The only known species in this genus is *Stangeleon longipalpus* Miller, 2008, from Venezuela.

– The apical tarsal segment of the fore-leg is not flattened, greatly enlarged, or covered with enlarged setae. There are no calli on the antenna. There are no peg-like setae along the pilula axillaris of the hind wing (Fig. 7.4). The arrangement of the wing veins is variable 6

6. (5) Vein CuA in the hind wing curves sharply toward the hind margin at or proximal to the medial fork, or there are only one or two cross veins between CuA and the posterior fork of vein MP_2. The posterior area of the hind wing is less than twice as wide as the same area in the fore-wing (Fig. 7.6). 7

– Vein CuA in the hind wing runs close and parallel to the posterior branch of vein MP_2 for a considerable distance, so that there are at least three cross-veins between CuA and the posterior fork of vein MP_2. The posterior area of the hind wing is at least twice as wide proximal to the fore in the medial vein as the same area in the fore-wing (Fig. 7.4). 11

7. (6) There are clavate setae on the fore-femur (Fig. 7.6). The female has a membranous pregenital plate 8

– There are no clavate setae on the fore-femur. If a rim surrounds each compound eye, it supports no setae (Fig. 7.7). 9

8. (7) The wings are longer than the abdomen. There is always a rim surrounding each compound eye supporting a few short setae that extend over the eye. The male ectoproct is much longer than the eighth abdominal segment, and its post-ventral lobe is longer than wide. The lateral gonapophysis of the female bears silky hairs.
....*Abatoleon* Banks, 1924..p. 251

– The wings are shorter than the abdomen and subacute or subfalcate at the apex (Fig. 7.6). Some species lack a rim surrounding each compound eye supporting a few short setae. The male ectoproct is less than half as long as

Fig. 7.7 *Ensorra verticalis*: apex of a male abdomen in lateral view (left) and apex of a female abdomen in ventral view (right). Based on Stange (1994).

the eighth abdominal segment. The lateral gonapophysis of the female bears weak digging setae.
....*Peruveleon* Miller and Stange, 2011..p. 251

9. (7) The hair-like sensory setae on the fore-femur are ¾ as long as the fore-femur or longer, or the claws on the fore-tarsi are longer than the basal tarsal segment on the hind leg. The male ectoproct has an elongated ventral lobe, and the eleventh sternite is especially long (Fig. 7.7). Wing span: c. 38–40 mm.
....*Ensorra* Navás, 1915
The only species in this genus is *Ensorra verticalis* (Banks, 1910) from Argentina and Bolivia. Banks (1910) originally named it *Austroleon verticalis,* and Penny (1978) called it *Brachynemurus verticalis* Banks, 1910. Stange (1994) regards Navás's genus, *Ensorra,* as valid. A junior synonym of this species is *Ensorra modestus* Navás, 1915.

– The hair-like sensory setae on the fore-femur, if present at all, are not longer than half as long as the fore-femur (Fig. 7.8). 10

10. (9) There is a postventral lobe on the ectoproct, which is at least eight times as long as its middle diameter. The fore-femur lacks a row of long white, bristle-like setae. The antenna apex is clubbed and flattened (Fig. 7.8).
....*Argentoleon* Stange, 1994..p. 252

– The ectoproct of the male lacks a postventral lobe, or, if a short one is present, it is less than twice as long as wide. The fore-femur bears a row of long, white setae (Fig. 7.9).
....*Austroleon* Banks, 1910..p. 253

11. (6) The fore-femur lacks hair-like, sensory setae (Fig. 7.4).
....*Venezueleon* Stange, 1994

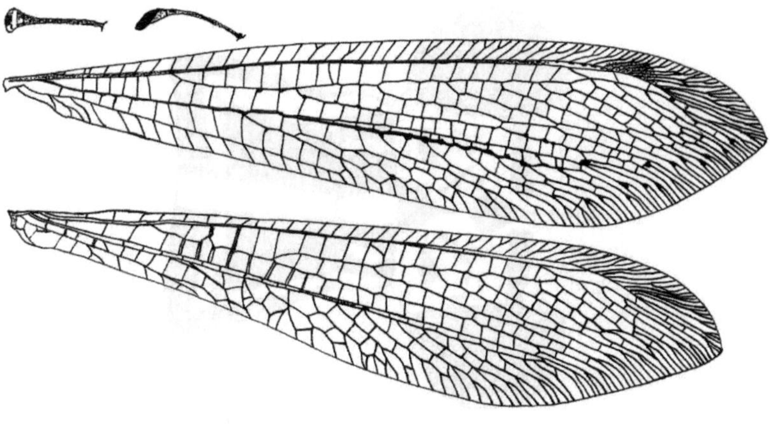

Fig. 7.8 *Argentoleon longitudinalis*: an antenna in dorsal and lateral view (above) and the fore (middle) and hind wing (below), based on Stange (2010).

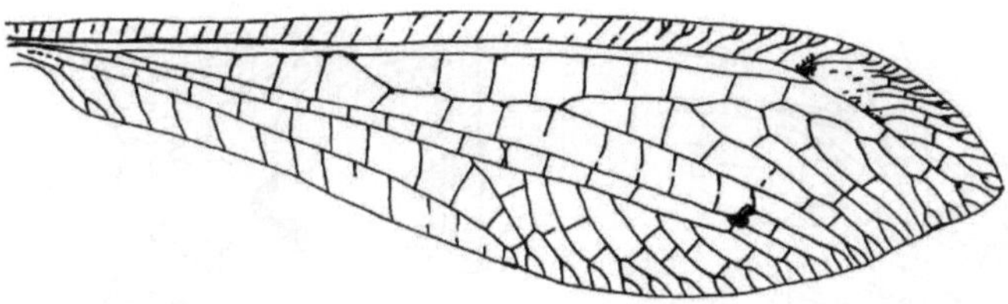

Fig. 7.9 Fore-wing of a female *Austroleon immitus*. Based on Navás (1914a), who called the species *Correa expansus*.

Only one species in this genus, *Venezueleon guaricus* Stange, 1994, has been described. It is known only from Venezuela. Stange (2004) indicated that congeneric species exist, but they have not yet been described.

– The fore-femur bears hair-like, sensory setae. The tarsal claws on the fore-legs are as long as the apical tarsal segment. There are numerous setae on the frons. Vein CuP + 1A on the fore-wing runs parallel to the posterior branch of CuA for a distance roughly equal to that between CuA and the posterior branch of MP_2 on the hind wing, or the hair-like sensory setae on the fore-femur is as long as the tarsus. The head and pronotum are strongly patterned (Fig. 7.10).
.... *Ameromyia* Banks, 1913..p. 254

12. (4) The pretarsal claws are shorter than the basal tarsal segment of the fore-leg. The hair-like, sensory setae on the fore-femur are about ¼ the length of the fore-femur. The antennal fossae are separated from each other by more than the width of the pedicel. The posterior gonopophysis of the female is not curved and is longer than the height of the ectoproct (Fig. 7.5). 13

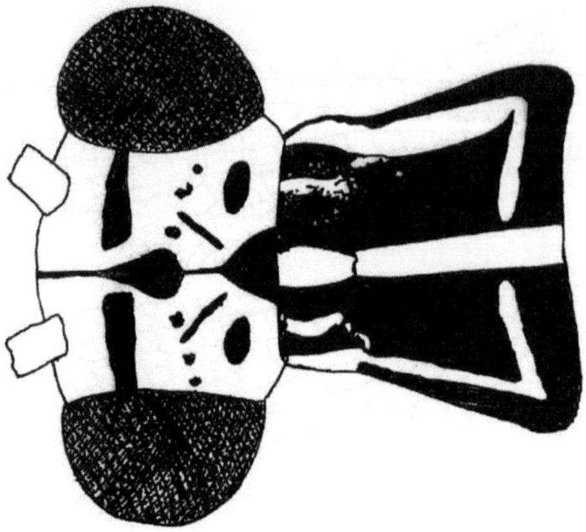

Fig. 7.10 Head without the antennae and the pronotum of *Ameromyia muralli* in dorsal view. Based on Navás (1932a).

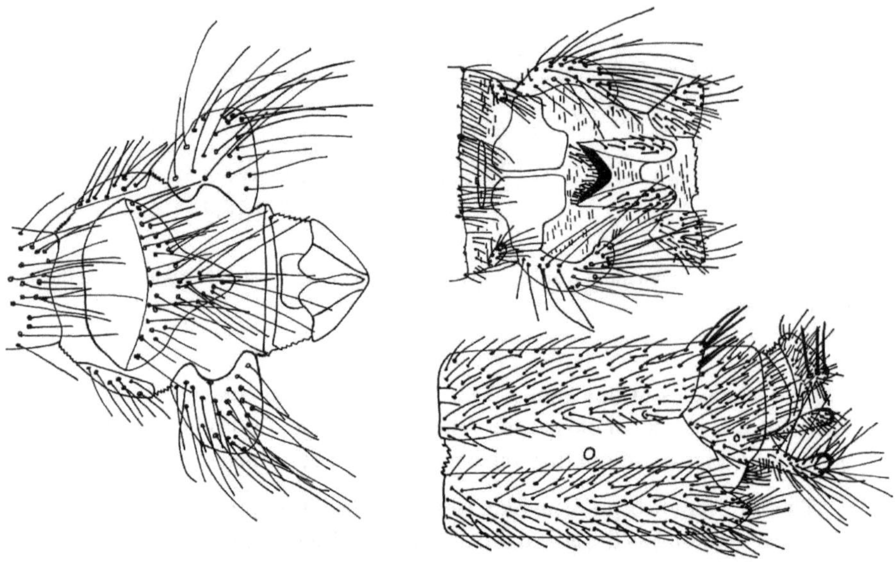

Fig. 7.11 *Ecualeon ovispargus:* apex of the female abdomen in ventral view (left) and the apex on the male abdomen in ventral (upper right) and lateral view (lower right). Based on Stange (1994).

– The pretarsal claws are longer than the basal tarsal segment of the fore-leg. The hair-like, sensory setae on the fore-femur are at least half as long as the fore-femur. The antennal fossae are separated from each other by no more than the width of the pedicel. The posterior gonopophysis of the female is strongly curved toward the mid-line and is shorter than the height of the ectoproct (Fig. 7.11). 14

13. (12) The costal area of the fore-wing has a single line of cells. The fore and hind wings are subequal in length. When the wings are at rest, the hind wing extends farther from the thorax than the fore-wing. The ectoproct of the male lacks a

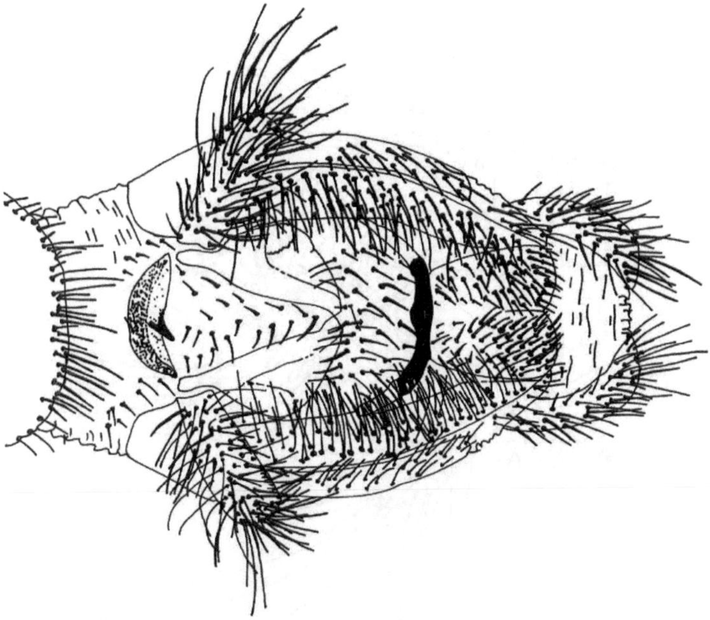

Fig. 7.12 Apex of a female *Jaffuelia chilensis* abdomen in ventral view. Based on Stange (1994).

ventral process. The gonapophysis of the female is swollen toward its apex, and it lacks scraping setae (Fig. 7.5).
.... *Neulatus* Navás, 1912
The only known species in this genus is *Neulatus porteri* (Brethes, 1908), from Chile. Its synonyms include *Glyptobasis porteri* Brethes, 1908; *Lincoya dealbatus* Navás, 1911 (preoccupied); *Nilcoya dealbatus* Navás, 1913.

- The costal area of the fore-wing has a double line of cells in places. The fore-wing is longer than the hind wing. When the wings are at rest, their apices extend almost an equal distance from the body. The ectoproct of the male has a large ventral process. The gonapophysis of the female is equally swollen throughout its entire length, and it bears scraping setae at the apex (Fig. 7.12).
.... *Jaffuelia* Navás, 1918..p. 258

14. (12) The ectoproct of the male has a vestige of a postventral lobe. The female posterior gonapophysis is longer than the height of the ectoproct and strongly curved mesad (Fig. 7.11). 15

- The ectoproct of the male lacks the slightest vestige of a postventral lobe. The female posterior gonapophysis is shorter than the height of the ectoproct and not strongly curved mesad (Fig. 7.13). 16

15. (14) The posterior area of the fore-wing is widest near the cubital fork. The male ectoproct has an emargination in the middle ventrally. The seventh

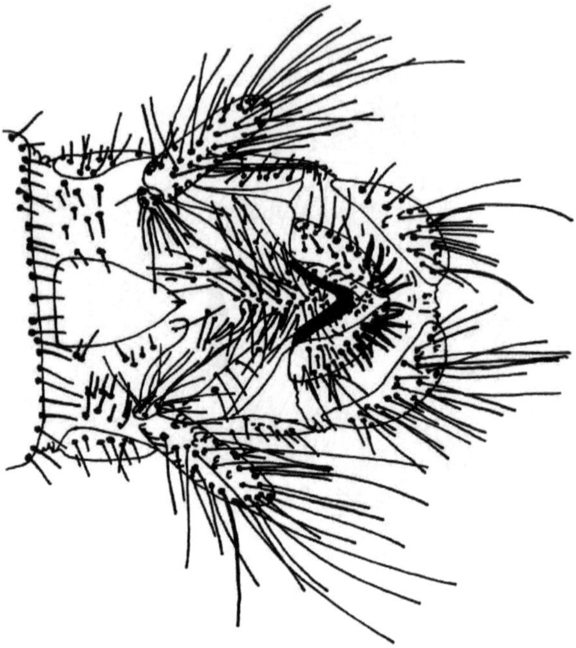

Fig. 7.13 Apex of a female *Elicura litigator* abdomen in ventral view. Based on Stange (1994).

abdominal tergite of the female has a prominent line of subapical setae (Fig. 7.11).
.... *Ecualeon* Stange, 1994
The only known species in this genus is *Ecualeon ovispargus* Stange, 1994, from Ecuador.

– The posterior area of the fore-wing is widest near the apex of vein 2A. The male ectoproct has no middle emargination. The seventh abdominal tergite of the female lacks a prominent line of subapical setae (Fig. 7.14).
.... *Galapagoleon* Stange, 1994
The only known species in this genus is *Galapagoleon darwini* (Stange, 1969), from the Galapagos Islands. It was known by its synonym: *Brachynemurus darwini* Stange, 1969.

16. (14) The fore-femur has long, white, bristle-like setae along its outer surface. The costal area of the fore-wing is wide with an area anterior to the cubital fork nearly three times as wide as the subcostal area and filled by as many as three lines of cells. The thin sensory setae on the middle femur are about half the length of that femur. The anterior gonapophysis on the female genitalia is plate-like. The pregenital plate is considerably longer than wide (Fig. 7.13).
.... *Elicura* Navás, 1918..p. 259

– The fore-femur does not have long, bristle-like setae on its outer surface, or it has one or two such setae just proximal to the apex. The coastal area

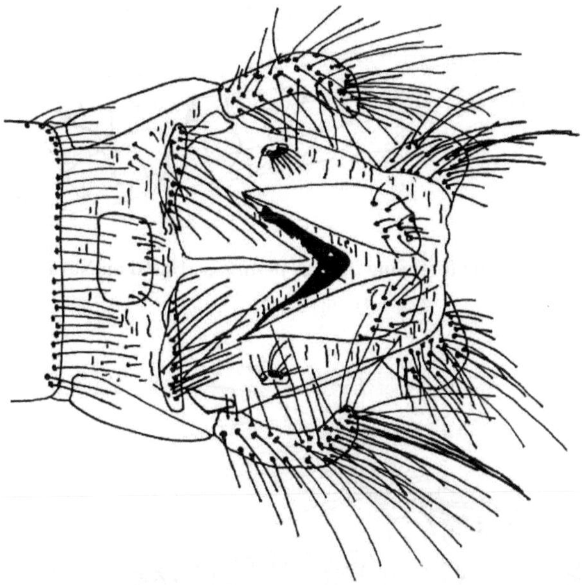

Fig. 7.14 Apex of the female abdomen of *Galapagoleon darwini* in ventral view. Based on Stange (1994).

of the fore-wing is less than twice as wide as the subcostal area and filled by no more than the length of that femur. The anterior gonapophysis on the female genitalia is lobe-like. The pregenital plate is wider than long (Fig. 7.15). 17

17. (16) The basal tarsal segment on the hind leg less than three times longer than wide. The costal area of the fore-wing has few interconnected cross veins in the basal half (Fig. 7.15), or it has none at all. The pregenital plate is large, and it lacks a median tooth.
.... *Sical* Navás, 1928..p. 260

– The basal tarsal segment on the hind leg at least four times longer than wide. Most of the cross veins in the costal area of the fore-wing are interconnected. The pregenital plate is small and has a median tooth (Fig. 7.16).
.... *Lemolemus* Navás, 1911

The only described species in this genus is *Lemolemus modestus* (Blanchard, 1851), from Chile. Its synonyms are *Myrmeleon modestum* Blanchard, 1851; *Puren modestus* (Blanchard, 1851) Navás, 1936; *Lemolemus nectator* Navás, 1911; *Lemolemus nectador* Navás, 1911; *Lemolemus necator* (sic) Navás, 1911; *Elicura ferus* Navás, 1918; *Puren bellator* Navás, 1911. The length of the fore-wing is c. 21.5 mm for the male and 24 mm for the female; The hind wing length of the male is c. 23.8 mm; that of the female is c. 23.0 mm. The coloration is mainly blackish and yellow. According to Stange (1994), several other species of *Lemolemus* are present in Chile but have not yet been named or described.

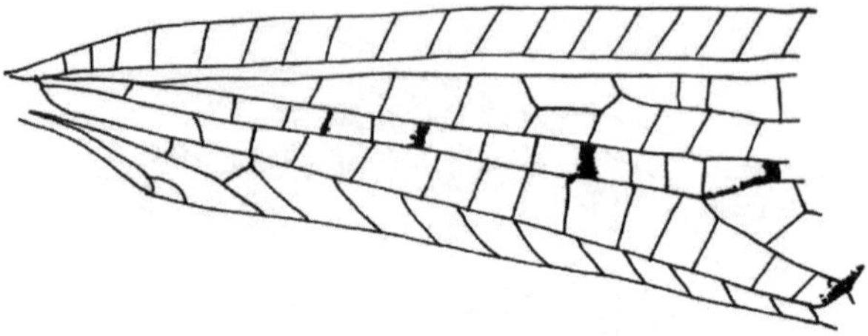

Fig. 7.15 Basal section of the fore-wing of *Sical peralinus*. Based on Navás (1928a).

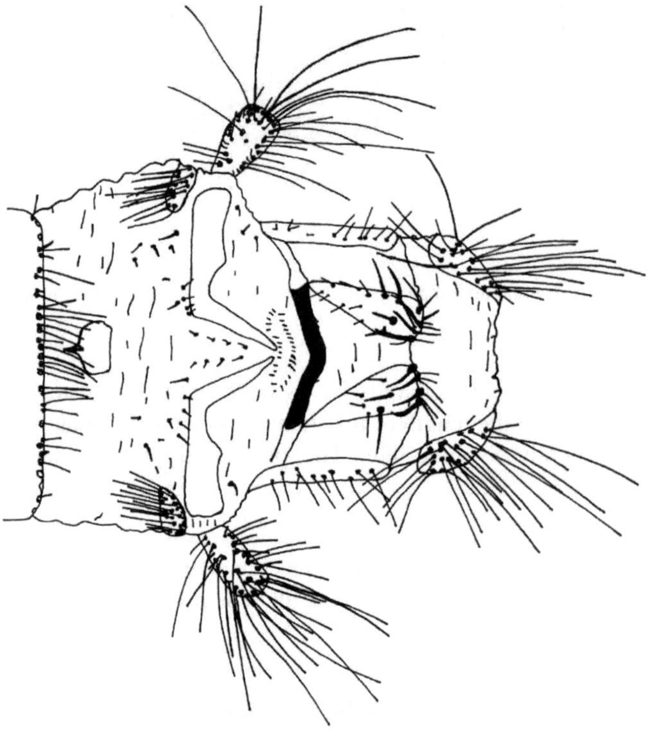

Fig. 7.16 Apex of the abdomen of a female *Lemolemus modestus* in ventral view. Based on Stange (1994), who referred to it by its synonym, *Lemolemus nectator*.

18. (3) In the hind wing of South American species, vein CuA joins the posterior branch of MP_2 slightly distal to its fork. The tarsal claws are strongly arched near their bases. Tibial spurs are present and strongly curved to an angle up to 90°. The hind femur bears elongate sensory hairs (Fig. 7.2). There are pencils of hair-like setae on the abdomen of the male. ....Tribe Acanthaclisini.........*Vella* Navás, 1913..p. 261

   – In the hind wing, vein CuA is joined to the posterior branch of MP_2 by a cross vein, or it does not reach the fork in MP_2 at all (Fig. 7.17). The hind

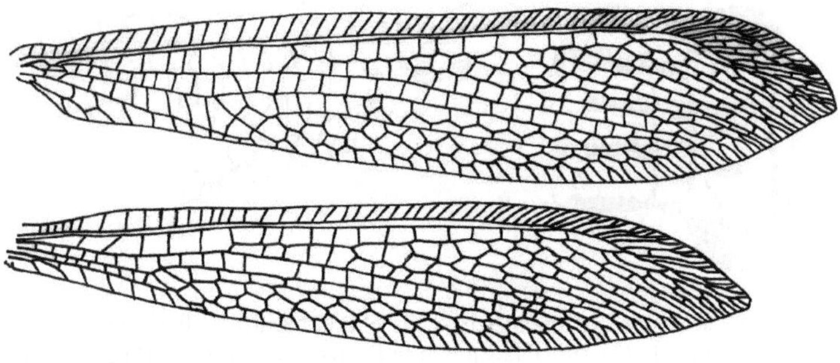

Fig. 7.17 The fore and hind wing of *Brasileon amazonica*. Based on Miller and Stange (1989a), who referred to the species as *Dimarella amazonica*.

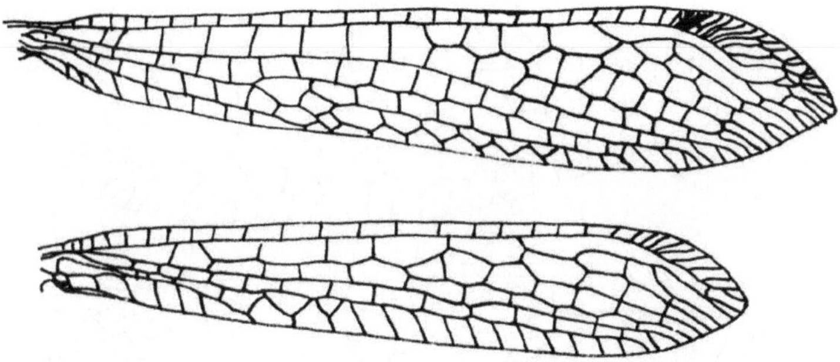

Fig. 7.18 The fore and hind wing of *Myrmeleon timidus*. Based on Stange (2002).

femora of all known South American species lack elongate sensory hairs. Pencils of hair-like setae are usually absent from the male abdomen 19

19. (18) The hind wing has one or rarely two or three presectoral cross veins (Fig. 7.17). A pilula axillaris is never present beside the tarsal claw of the male ....Tribe Nemoleontini..20

– The hind wing has four or more presectoral cross veins (Fig. 7.18). On the hind leg, the basal tarsal segment is shorter than the apical one. ....Tribe Myrmeleontini..31

20. (19) The fossa in the antenna is separated from the ocular rim by a distance greater than the maximum width of the pedicel. The middle leg is shorther than the fore-leg, which is shorter than the hind leg. The ectoproct of the male has a postventral lobe (Fig. 7.17). 21

– The fossa in the antenna is separated from the ocular rim by a distance shorter than the maximum width of the pedicel (Fig. 7.19). The middle leg is as long or longer than the fore-leg, which is shorter than the hind leg. The ectoproct of the male lacks a postventral lobe 22

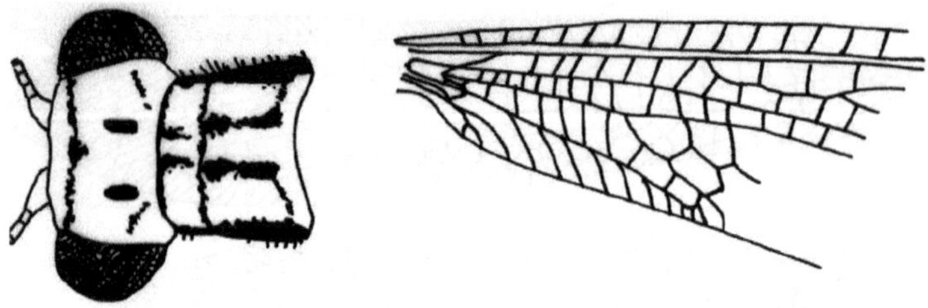

Fig. 7.19 *Navasoleon leptocerus:* head with pedicel and basal flagellar segments of the antenna and the pronotum in dorsal view (left); basal part of the fore-wing (right). Based on Navás (1915a), who named the species *Gymnocnemia leptocera.*

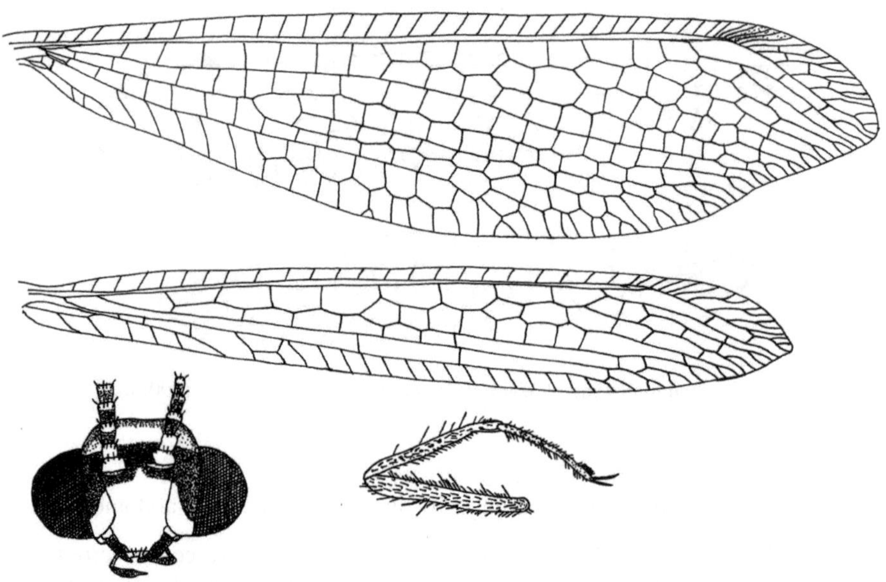

Fig. 7.20 *Dimarella angusta:* fore and hind wing (above and middle, respectively), head in anterior view (lower left), and the fore-leg in lateral view (lower right). Based on Stange (2002).

21. (20) There are no tibial spurs on the middle or hind legs. Vein CuP + 1A meet the hind margin of the fore-wing at an oblique angle well proximal to the origin of the radial sector. The basal tarsal segment of the hind leg is longer than the apical tarsal segment (Fig. 7.17).
.... *Brasileon* Miller and Stange, 1989..p. 262

 – All tibiae bear spurs. Vein CuP + 1A of the fore-wing runs parallel to vein CuA and the posterior wing margin for a long distance. The basal tarsal segment of the hind leg is shorter than the apical tarsal segment (Fig. 7.20).
.... *Dimarella* Banks, 1913..p. 263

22. (20) Tibial spurs on the hind legs, if present, are never much longer than the width of the first segment of the hind tarsus (Fig. 7.19). .... *Navasoleon* Banks, 1943..p. 268

   – The hind tibiae have well-developed spurs, at least twice as long as the width of the first segment of the hind tarsus (Fig. 7.21) 23

23. (22) The claws on the fore-tarsus can close against a brush of setae on the apical tarsal segment (Fig. 7.21)... 24

   – The claws on the fore-tarsus cannot close against an opposing brush of setae. All setae taper to an unmodified apex (Fig. 7.22). 26

24. (23) The setae in the brush on the fore-tarsus against which the claws can close are bent almost 90° near the apex (Fig. 7.21). .... *Elachyleon* Esben-Petersen, 1927..p. 269

   – The setae in the brush on the fore-tarsus against which the claws close are not bent nearly 90° but rather taper toward the apex (Fig. 7.23). 25

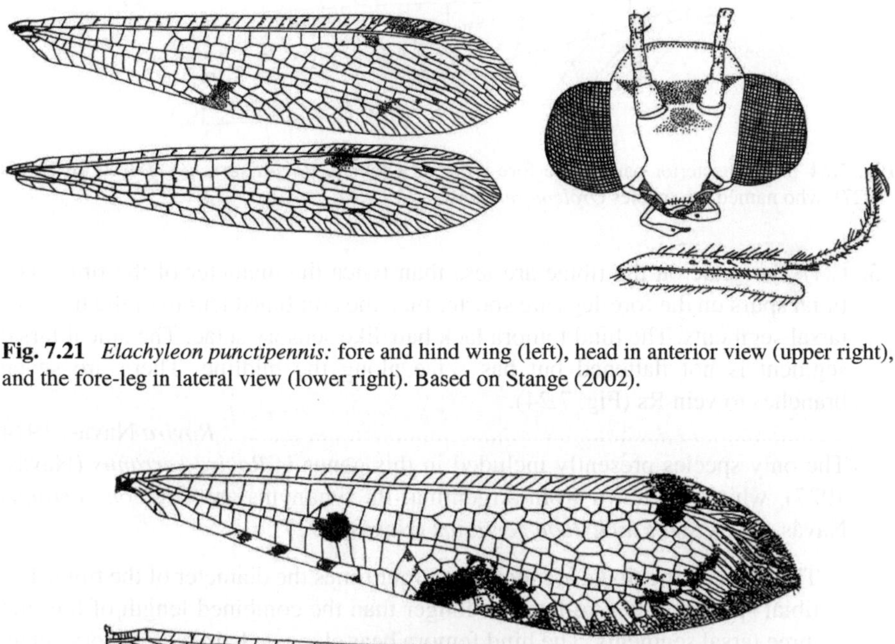

Fig. 7.21 *Elachyleon punctipennis:* fore and hind wing (left), head in anterior view (upper right), and the fore-leg in lateral view (lower right). Based on Stange (2002).

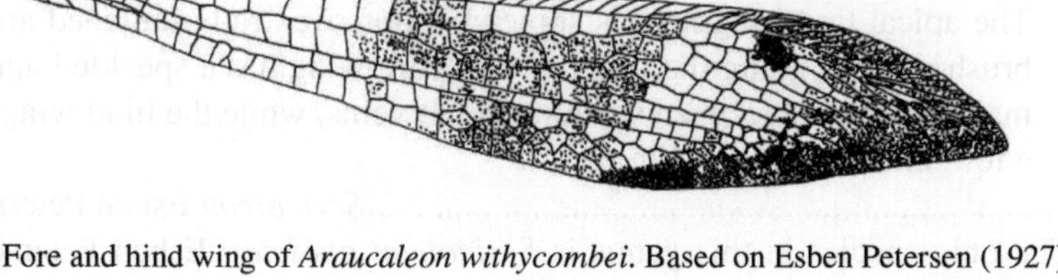

Fig. 7.22 Fore and hind wing of *Araucaleon withycombei*. Based on Esben Petersen (1927).

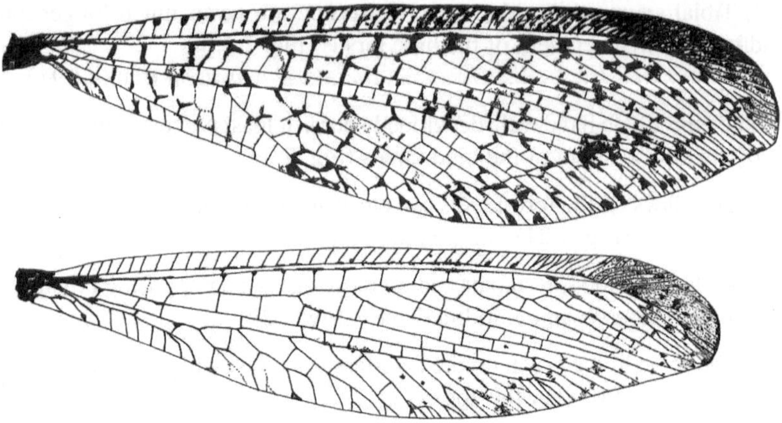

Fig. 7.23 Fore and hind wing of *Sericoleon paessleri*. Based on Esben Petersen (1933).

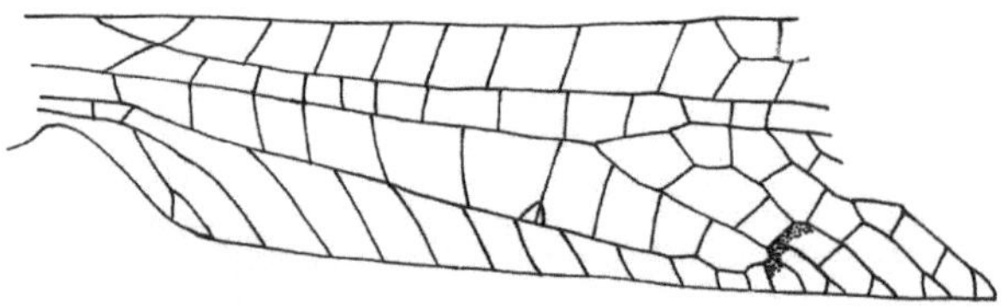

Fig. 7.24 Basal posterior part of the fore-wing of a female *Rovira serranus*. Based on Navás (1927), who named the species *Oroleon serranus*.

25. (24) The setae on the tibiae are less than twice the diameter of the tibia. The tibial spurs on the fore-legs are shorter than the combined length of the first two tarsal segments. The hind femora lack hair-like sensory setae. The apical tarsal segment is not flattened but has setae along the midline. There are seven branches to vein Rs (Fig. 7.24).
.... *Rovira* Navás, 1914
The only species presently included in this genus is *Rovira serranus* (Navás, 1927), which is known from Argentina. Its synonyms are *Oroleon serranus* Navás, 1927, and *Elachyleon serranus* (Navás, 1927).

– The setae on the tibiae are more than four times the diameter of the tibia. The tibial spurs on the fore-legs are longer than the combined length of the first three tarsal segments. The hind femora bear elongate hair-like sensory setae. The apical tarsal segment is somewhat dorsoventrally flattened and lacks a brush of setae along the midline. The fore-wing has a speckled appearance, mainly due to blackened sections of the veins, while the hind wing has only a few small markings (Fig. 7.23).
.... *Sericoleon* Esben Petersen, 1933
The only species in this genus is *Sericoleon paessleri* Esben Petersen, 1933, known only from Chile. The fore-wing is 28–33 mm long, while the hind wing varies from about 26–31 mm.

26. (23) The first segment of the hind tarsus is longer than the apical segment. The hind wing is falcate, and its costal area is twice as wide near the junction of veins Sc and R as it is near its mid-length (Fig. 7.22). 27

    – The proximal segment of the hind tarsus is shorter than the apical segment (Fig. 7.25). 28

27. (25) The fore-wing is only about 10 % shorter than the hind wing. The origin of vein Rs in the fore-wing is well proximal to the fork in vein CuA. Vein CuA in the hind wing ends less than half way to the fork in MP_2 (Fig. 7.22). ....*Araucaleon* Banks, 1938..p. 269

    – The fore-wing is less than ¾ as long as the hind wing, but it is about 20 % wider. ....*Ripalda* Navás, 1915

    The only South American species in this genus is *Ripalda insignis* (Rambur, 1842), originally described under the name, *Myrmeleon insignis* Rambur, 1842. Its fore-wing is about 27 mm, and its hind wing is about 40.5 mm, according to Navás (1915a). Its description was brief, and the type specimen is badly damaged. Its systematic position is doubtful. The species inhabits Brazil.

28. (26) If sensory, hair-like setae are present on the fore-femur, they are less than ¼ the length of the femur, which is elongated, cylindrical, and twice as long as the width of the head (Fig. 7.25). 29

    – Sensory, hair-like setae, at least 1/3 the length of the femur, are always present on the fore-femur, which is short, often swollen, and no longer than the width of the head (Fig. 7.26). 30

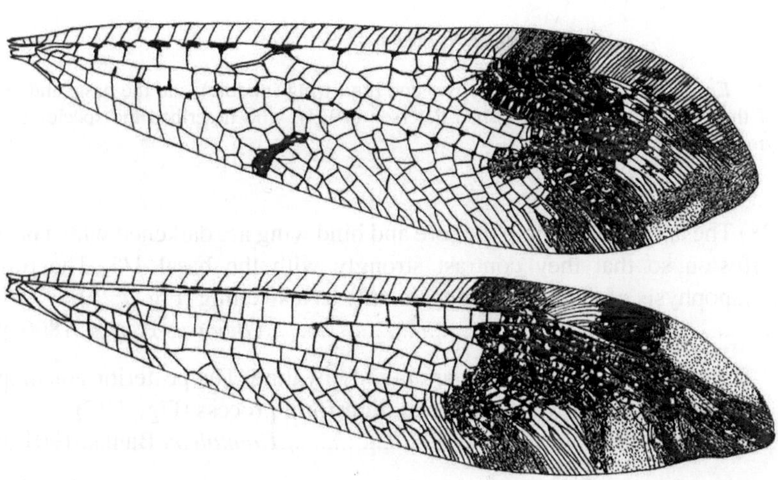

Fig. 7.25 Fore and hind wings of *Glenurus peculiaris*. Based on Stange (2010), who described a specimen originally called *Glenurus brasiliensis*.

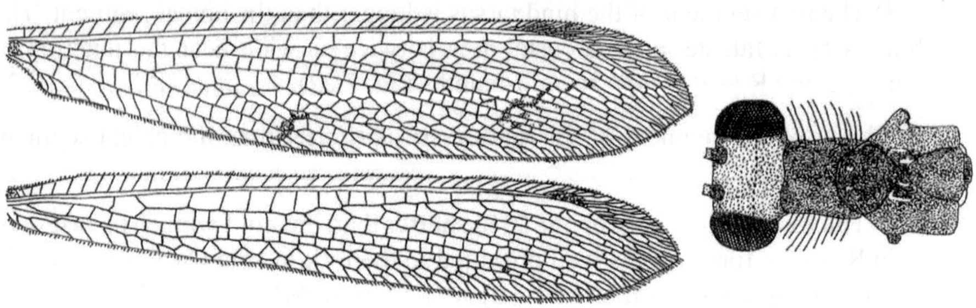

Fig. 7.26 *Purenleon parallelus:* fore and hind wing (left) and the head and thorax in dorsal view (right). Based on Stange (2002).

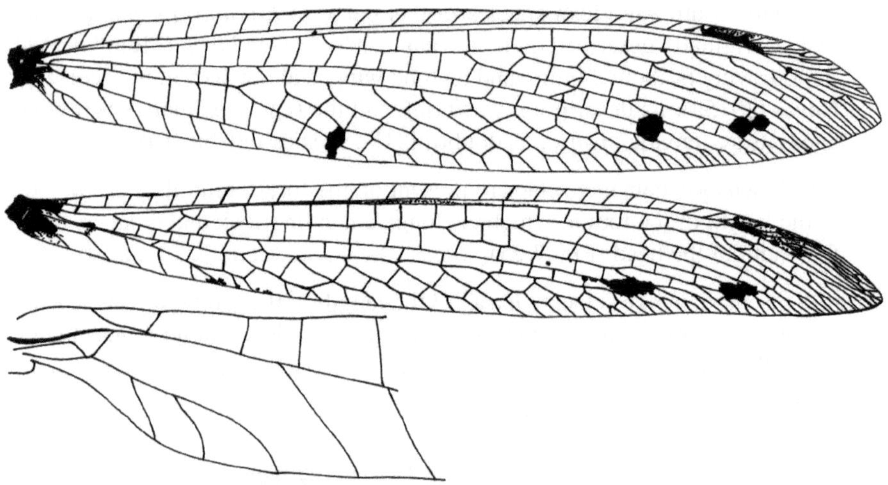

Fig. 7.27 *Eremoleon pulchra:* fore (above) and hind wing (middle), and the basal and anal sections of the fore-wing. Based on Esben-Petersen (1933), who described the species under the synonym: *Joergenia pulchra.*

29. (28) The apical 1/5 of both the fore and hind wing are darkened with a brownish suffusion so that they contrast strongly with the basal 1/5. The posterior gonapophysis of the female appears only as a swelling (Fig. 7.25). .... *Glenurus* Hagen, 1866..p. 270
   - The apical 1/5 of the wings are mainly hyaline. The posterior gonapophysis of the female is a well-developed digitiform process (Fig. 7.27). .... *Eremoleon* Banks, 1901..p. 271
30. (28) The first tarsal segment of the hind leg is no longer than twice its width and shorter than the pretarsal claws. The middle tibiae are not more inflated than the fore-tibiae. The eighth abdominal tergite of the female has a subapical row of

stout setae. The hair-like sensory setae on the middle femora are always longer than a tarsus.

....*Euptilon* Westwood, 1837

The species in this genus are all found in North America, but there is a questionable report of one of the species, *Euptilon ornatus* (Drury, 1773), being found in Colombia and Brazil. Synonyms for this species include *Hemerobius ornatus* Drury, 1773; *Myrmeleon ingeniosus* Walker, 1853; *Psammoleon guttipes* Banks, 1906; *Chauliodes ornatus* Hagen, 1866; *Formicaleo ingeniosus* (Walker, 1853) Navás, 1917.

– The length of the first tarsal segment of the hind leg is at least 2½ times its width, or, if not, the middle tibia is inflated to a thickness almost twice that of the fore-tibia. The eighth abdominal tergite of the female does not have a subapical row of stout setae (Fig. 7.26).

....*Purenleon* Stange, 2002…p. 274

31. (19) The ventral surface of the apical segments of the tarsus bears setae less than ¼ as long as the diameter of the segment. The distance between the antennal bases is at least as great as the maximum diameter of the scape. There is no postventral lobe on the male ectoproct (Fig. 7.18).

....*Myrmeleon* Linnaeus, 1767..p. 279

– The ventral surface of the apical segments of the tarsus bears setae more than half as long as the diameter of the segment. The distance between the antennal bases is less than 2.5 times the maximum diameter of the scape. There is a short postventral lobe on the male ectoproct.

....*Porrerus* Navás, 1913

The only known South American species assigned to this genus is *Porrerus famelicus* Navás, 1913, was reported to date only from Argentina and Paraguay. Its total length is c. 44 mm. The length of its fore-wing is c. 34.5 mm, and its hind wing is c. 36 mm long. Its head is yellow with dark markings, and its thorax and abdomen are mainly dark. Several South American species encountered in earlier literature which were assigned to this genus have since been transferred to other genera.

Key to South American Subfamilies and Genera of Known Myrmeleontidae Larvae

Information for the key was provided by Banks (1910, 1913a); Stange and Miller (1985, 2008) and Stange (2004, 2010). The characteristics shown in the key are applicable only to known South American species, and, needless to say, it is not known whether the features in the key are valid for larvae that have not yet been described. Where more than one genus is found under a description in the key, reliable ways to distinguish between them have still not been reported. The larvae in this family require considerable study because they are the life stage most frequently encountered in the field.

1. The median lobe of the ninth tergite bears a sclerotized median process, or the ninth sternite bears one or more pairs of double-bladed setae highly modified for digging. The eighth sternite bears a pair of submedian teeth near the posterior margin. The mesothoracic spiracle is never borne on a tubercle. ....Subfamily Palparinae.................................2

   – There is no median lobe bearing a sclerotized process on the ninth tergite; the ninth sternite lacks setae conspicuously modified for digging (Fig. 7.28), or, if such setae are present, the mandible has two teeth bearing vestigial apical setae. .... Subfamily Myrmeleontinae...............................3

2. (1) The mandible bears two teeth. The eighth sternite does not bear a well developed submedian tooth. The ninth sternite lacks greatly modified digging setae. .... *Millerleon* Stange, 1989..p. 250

   – The mandible bears three blunt teeth. All segments of the relatively short antenna are usually wider than long. The eighth sternite bears a well developed submedian tooth. The ninth sternite bears two pairs of fossoria, which

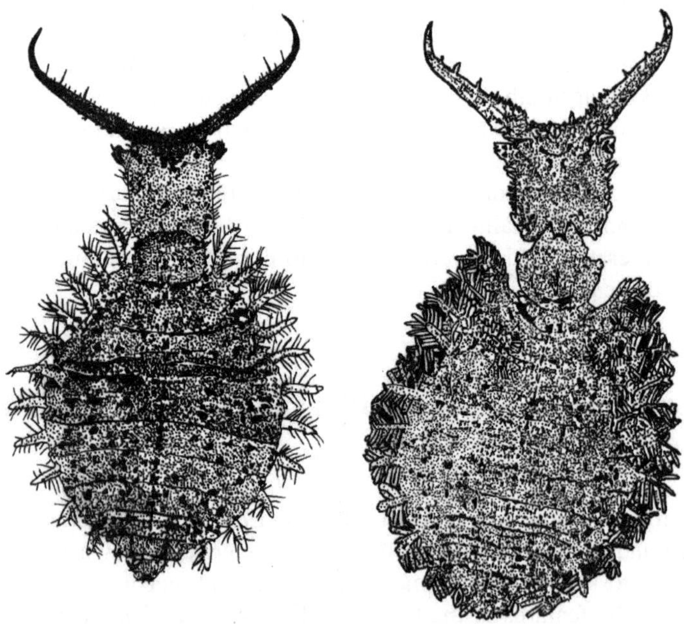

Fig. 7.28 *Neulatus porteri* larvae: earlier (left) and later stage instar (right). Based on Stange and Miller (1985).

are greatly modified digging setae. The larvae typically move slowly backward and foreward.

.... *Dimares* Hagen, 1866

At the present time, only one species is assigned to this genus: *Dimares elegans* (Perty, 1833), known from Pará, Amazonas, and Mato Grosso. The species also appears in the literature under the following synonyms: *Myrmeleon elegans* Perty, 1833; *Myrmeleon myrmerophagus* Perty, only a manuscript name according the Navás, 1912; *Myrmeleon conicollis* Walker, 1860; *Myrmeleon albidilinea* Walker, 1860; *Palpares elegans* (Perty, 1833) Hagen, 1860; *Myrmeleon congruus* Hagen, 1861; *Dimares elegans lepida* Navás, 1912; *Dimares erythrostigma* Navás, 1914; *Dimares hageni* Banks, 1920.

3. (1) The labial palp is longer than the width of a mandible at its base, or, if not, then the mesothoracic spiracle is borne on a tubercle (Fig. 7.28). Dolichasters are sometimes present on the head. The mandibles are straight and separated from each other by more than the width of one mandible. In the middle of the mesoscutum, there is no tuft of setae. The ninth abdominal sternite is shorter than its width. Members of all genera move both backwards and forewards 4

   – The labial palp is shorter than the width of a mandible at its base. The mesothoracic spiracle is never borne on a tubercle. The longest setae on the exterior margins of the mandibles are less than half the greatest width of the mandible. The subapical margin of the eighth sternite lacks teeth. The larvae do not build pitfall traps. Some species move only backwards 22

4. (3) Incurved setae or squat dolichasters are present on the sagital part of the first four abdominal tergites. The abdomen may bear lateral scoli (Fig. 7.28).

   .... Tribe Lemolemini 5

   – The sagital parts of the first four abdominal tergites do not bear specialized setae (Fig. 2.29). The abdomen never has lateral scoli. 9

5. (4) The first seven abdominal segments have well developed scoli, longer than their own width. The abdomen lacks a median row of tufted black setae, but it has a rosette of squat dolichasters (Fig. 7.28). 6

   – The first eight abdominal segments either lack scoli or bear some that are shorter then their width. The abdomen has a median row of tufted black setae and lacks a rosette of squat dolichasters (Fig. 1.24). 7

6. (5) The length of the abdominal scoli is at least three times their width, and they taper to the apex. Dolichasters are lacking. The head is longer than wide (Fig. 7.28). The larvae inhabit plants of the genus *Puya*.

.... *Neulatus* Navás, 1912

The only known species in this genus is *Neulatus porteri* (Brethes, 1908), from Chile. Its synonyms include *Glyptobasis porteri* Brethes, 1908; *Lincoya dealbatus* Navás, 1911 (preoccupied); *Nilcoya dealbatus* Navás, 1913.

– The length of the abdominal scoli is shorter than twice their width, and they are broadly rounded at the apex; dolichasters are present. The head is wider than long (Fig. 7.29). They are found on boulders.
.... *Jaffuelia* Navás, 1918..p. 258

7. (5) The first eight abdominal segments bear weak scoli. The head capsule and scoli bear many white, silky, hair-like setae (Fig. 1.24).
.... *Sical* Navás, 1928..p. 260

– The first eight abdominal segments lack scoli. The head capsule does not bear white, hair-like setae 8

8. (7) The mandible is pale with many dark brown spots. The anterior scolus of the mesothorax is twice as long as wide.
.... *Elicura* Navás, 1918..p. 259

– The mandible is mainly pale with most dark brown markings confined to the area near the base. The length of the anterior scolus is 1.5 times its width.
.... *Lemolemus* Navás, 1911

The only described species in this genus is *Lemolemus modestus* (Blanchard, 1851), from Chile. Its synonyms are *Myrmeleon modestum* Blanchard, 1851; *Lemolemus nectator* Navás, 1911; *Lemolemus nectador* Navás, 1911; *Elicura ferus* Navás, 1930; *Puren bellator* Navás, 1911, *Puren modestus* (Blanchard, 1851), Navás, 1911. The length of its fore-wing was reported to be c. 21.5–24 mm, and its hind wing is c. 23–23.8 mm (Navás, 1936b). According to Stange (1994), several other species of *Lemolemus* are present in Chile but have not yet been named or described.

9. (4) The distal tooth on the mandible is equal to or longer than the middle tooth, and the teeth are parallel to each other (Fig. 7.30).
.... Tribe Nemoleontini 10

– The distal tooth on the mandible is shorter than the middle tooth and often joins the mandible at an angle different from those of the other teeth (Fig. 2.29).
.... Tribe Brachynemurini 15

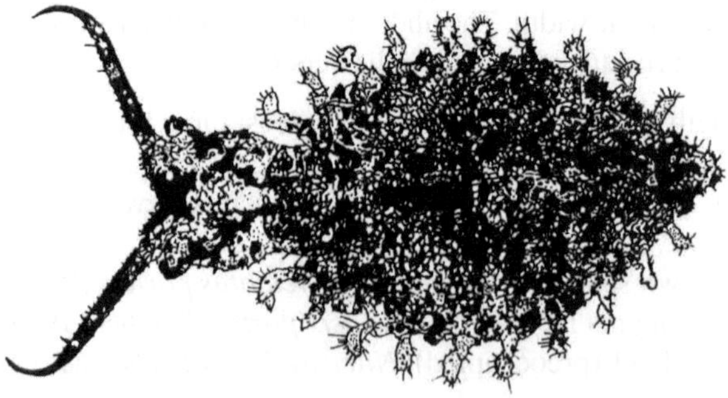

Fig. 7.29 Habitus of a *Jaffuelia chilensis* larva in dorsal view. Based on Stange (1994).

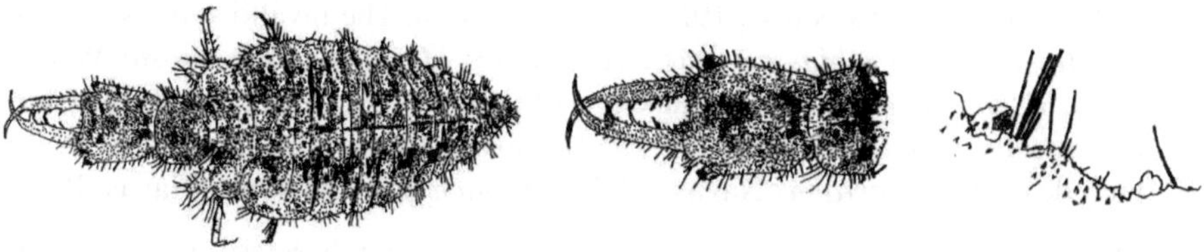

Fig. 7.30 *Purenleon debilis* larva (left to right): dorsal views of the habitus and the head and abdominal spiracles. Based on Miller and Stange (2014).

10. (9) The abdomen bears scoli 11
    – The abdomen lacks scoli 12
11. The mandible bears two teeth. The labial palp consists of two segments. The larvae inhabit rotting logs.
    .... *Glenurus* Hagen, 1866..p. 270
    – The mandible bears three teeth. The labial palp consists of three segments. The larvae live in sand.
    .... *Navasoleon* Banks, 1943..p. 268
12. (10) The middle tooth on the mandible is closer to the distal tooth than to the basal tooth. The head lacks dolichasters on its dorsal surface.
    .... *Sericoleon* Esben Petersen, 1933
    – The basal and distal teeth are about equidistant from the middle tooth. The head usually bears dolichasters on its dorsal surface 13
13. (12) The spiracles on the abdominal segments are borne on tubercles (Fig. 7.30). The larvae of the following two genera cannot yet be distinguisned.
    .... *Purenleon* Stange, 2002 pars..p. 274
    .... *Dimarella* Banks, 1913 pars..p. 263
    – The spiracles on the abdominal segments are not borne on tubercles 14
14. The basal tooth on the mandible is closer to the distal tooth than it is to the base of the mandible.
    .... *Araucaleon* Banks, 1938..p. 269
    – The basal tooth on the mandible is closer to the base of the mandible than to the distal tooth. The four genera here cannot yet be distinguished from each other from the larvae.
    .... *Dimarella* Banks, 1913 pars..p. 263
    .... *Eremoleon* Banks, 1901..p. 271
    .... *Euptilon* Westwood, 1837
    .... *Rovira* Navás, 1914

    Only one South American species is presently included in each of the last two genera. They are *Euptilon ornatus* (Drury, 1773) from Colombia and Brazil, and *Rovira serranus* Navás, 1927, from Argentina. The invalid synonyms of *E. ornatus* are *Hemerobius ornatus* Drury, 1773; *Myrmeleon ingeniosus* Walker, 1853; *Psammoleon guttipes* Banks, 1906; *Chauliodes ornatus* Hagen, 1866; *Formicaleo ingeniosus* (Walker, 1853) Navás, 1917a. Those of *R. serranus* include *Oroleon serranus* Navás, 1927 and *Elachyleon serranus* (Navás, 1927).

15. (9) The abdominal spiracles are enlarged, located on tubercles, or both........ 16
   – Abdominal spiracles are neither enlarged nor on tubercles........................ 18

16. (15) In dorsal view, the head capsule appears about twice as wide as long. In ventral view, it appears 1 1/3 times as wide as long. The abdominal spiracles are elevated and enlarged but not set on tubercles. The mesothoracic spriracle is set on a tubercle, about as long as width. The labial palp is shorter than the basal width of the mandible. The head bears many dolichasters on the dorsal surface and some simple setae ventrally. The middle tooth on the mandible is closer to the distal tooth than to the apical tooth (Fig.2.29). The eighth sternite of the abdomen bears many strong digging setae and submedian teeth that are about as long as wide. Length of the only known species: c. 7.7 mm. These larvae live just below the surface of the soil, waiting to catch passing prey.
   .... *Stangeleon* Miller, 2008
   The only known species in this genus is *Stangeleon longipalpus* Miller, 2008, from Venezuela.
   – In dorsal view, the head is much less than twice as wide as long. The length and width of the submedian teeth on abdominal segment VIII are unequal..17

17. The base and the basal tooth of the mandible are farther apart than the basal and distal teeth. Secondary tubular structures are sometimes associated with the abdominal spiracles. Submedian teeth on the eighth sternite are wider at their bases than their length. The larvae live in open sand.
   .... *Peruveleon* Miller and Stange, 2011 pars..p. 251
   – The distance between the base and the basal tooth of the mandible is about equal to the distance between the basal and distal teeth. Secondary tubular structures are never associated with abdominal spiracles. Submedian teeth on the eighth sternite are longer than the width at their bases. Larvae of these two genera with these characteristics cannot yet be distinguished.
   .... *Argentoleon* Stange, 1994 pars..p. 252
   .... and *Austroleon* Banks, 1909 pars..p.253

18. (15) There are dolichasters on the ventral surface of the head capsule 19
   – Dolichasters are absent from the ventral surface of the head capsule 20

19. (18) The length and width of the head capsule, measured in ventral view, are subequal. The abdominal segments lack thread-like setae.
   .... *Venezueleon* Stange, 1994

Only one species in this genus, *Venezueleon guaricus* Stange, 1994, has been described. It is known only from Venezuela. Stange (2004) indicated that congeneric species exist, but they have not yet been described.

– The head capsule, measured in ventral view, is longer than wide. The abdominal segments have ordinary setae and numerous long, thread-like setae. .... *Ameromyia* Banks, 1913 pars..p. 254

20. (18) The tubercle bearing the mesothoracic spiracle is longer than its diameter at the base. .... *Austroleon* Banks, 1910 pars..p. 253

– The tubercle bearing the mesothoracic spiracle is shorter than its diameter at the base 21

21. (20) The submedian teeth of the seventh abdominal sternite are shorter than the width at their bases. In dorsal view, the head length along the dorsal midline is greater than the width across the ocular tubercles. The distance between the base of the mandible and its first tooth is greater than the distance between the first and the distal tooth. The larvae live in open sand. .... *Peruveleon* Miller and Stange, 2011 pars..p. 251

– The submedian teeth of the seventh abdominal sternite are longer than their basal width. In dorsal view, the head length along the dorsal midline is less than its width across the ocular tubercles. The distance between the mandible base and its first tooth is less than that between the proximal and distal teeth. .... *Ameromyia* Banks, 1913 pars..p. 254

22. (3) The longest setae on the exterior margin of the mandible are less than half the maximal width of the mandible. The eighth sternite lacks teeth on the subapical margin. The larvae do not dig pits to trap prey. .... Tribe Acanthaclisini........ *Vella* Navás, 1913..p. 261

– There are setae on the exterior margin of the mandible, which are as long or longer than the width of the mandible. A pair of inconspicuous submedian teeth is located on the posterior margin of the eighth sternite. The larvae built pitfall traps. .... Tribe Myrmeleontini......... *Myrmeleon* Linnaeus, 1767..p. 279

This tribe also includes *Porrerus* Navás, 1913, in South America, but its larva has not been described.

7.2 Subfamily Palparinae

All known South American species in this subfamily are classified in the tribe Dimarini (Stange, 2004).

Key to the South American Species of Adults in the Genus *Millerleon*

Information for the key was provided by Stange (1989).

1. The length of the hind wing is less than 3.5 times its greatest width, which is at about the midlength of the wing (Fig. 7.3). Solid dark brown bands cover most the the apical 1/5 of the wing, and most of the hypostigmal cell is completely suffused with dark brown. There are numerous pore plates in most of the abdominal tergites and sternites of the male, usually associated with the bases of setae. These pore plates make the abdomen appear scaly and become more numerous toward the apex of the abdomen. Wing span: 52–76 mm.
   .... *Millerleon subdolus* (Walker, 1853)
   (Ecuador, Peru). Syn: *Myrmeleon subdolus* Walker, 1853; *Dimares formosus* Banks, 1907; *Dimares venustus* Banks, 1907; *Dimares nummatus* Navás, 1912; *Dimares decorus* Navás, 1913; *Dimares formosus punana* Navás, 1933.

   – The length of the hind wing is more than 4.0 times its greatest width, which is well distal to the midlength of the wing (Fig. 7.31). Pale brown with many areas suffused with dark brown covers most the the apical 1/5 of the wing, but most of the hypostigmal cell is not suffused with dark brown. There are no tufted pore plates distal to the third abdominal tergite, and the distal segments of the abdomen appear smooth.. 2

2. The pilula axillaris at the base of the fore-wing of the male has a terminal knob, which bears a mat of setae and is about three times wider than the base of the pedicel (Fig. 7.31). The abdomen appears scaly due to the presence of tufted pore plates. The fore-femur is dark brown. The wing span is about 64 mm.
   .... *Millerleon pretiosus* (Banks, 1908)
   (Peru, Chile). Syn: *Dimares pretiosus* Banks, 1908;

   – The pilula axillaris of the male has a terminal knob about twice as wide as the base of the pedicel. The fore-femur is light brown. The abdomen does not appear scaly. Wing span: c. 55 mm. Width of fore-wing: c. 8 mm.
   .... *Millerleon bellulus* (Banks, 1908)
   (Ecuador, Peru). Syn: *Dimares bellulus* Banks, 1908; *Dimares amoenus* Navás, 1913.

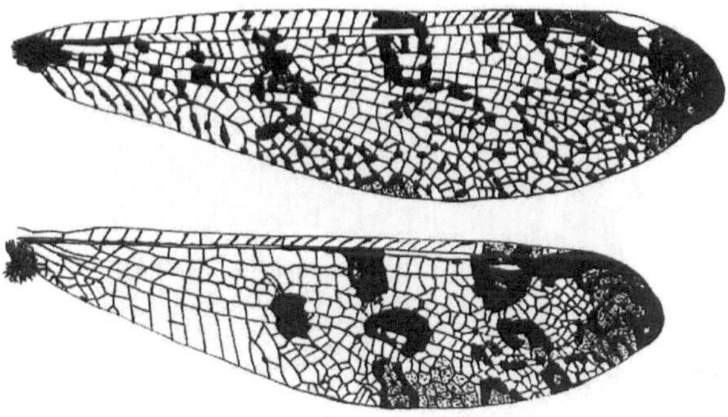

Fig. 7.31 Fore and hind wing of *Millerleon pretiosus*. Based on Stange (1989).

7.3 Subfamily Myrmeleontinae

Key to the South American Species of Adults in the Genus *Abatoleon*

Information for the key was provided by Banks (1910, 1924); Miller and Stange (1989, 2011), and Stange (2004). Since Stange (1994) reported on the distribution of the genus, species have been transferred to other genera.

1. Tibial spurs are absent. Expanse of wings: c. 28 mm.
 .... *Abatoleon deprivatus* Banks, 1924 (Argentina). Syn: *Brachynemurus deprivatus* (Banks, 1924).

 – Tibial spurs are present. Expanse of wings: 30–32 mm.
 .... *Abatoleon frontalis* (Banks, 1910) (Argentina). Syn: *Austroleon frontalis* Banks, 1910; *Brachynemurus frontalis* (Banks, 1910); *Austroleon helenicus* Navás, 1926.

Key to the South American Species of Adults in the Genus *Peruveleon*

Information for the key was provided by Banks (1907, 1909); Navás (1915a, 1928b, 1932b), and Miller and Stange (2011). Unfortunately, positive identification of the species is difficult due to omissions of details from the original descriptions.

1. The abdomen is yellowish brown without notable markings. The dorsal pattern on the head and pronotum is shown in Figure 7.32. The wings are hyaline and subfalcate at the apex. Wing span: c. 34 mm.
 .... *Peruveleon camposi* (Banks, 1907) (Ecuador, Peru). Syn: *Brachynemurus camposi* Banks, 1907; *Abatoleon camposi* (Banks, 1907) auctt.; *Austroleon heleninus* Navás, 1932; *Austroleon striolatus* Navás, 1913.

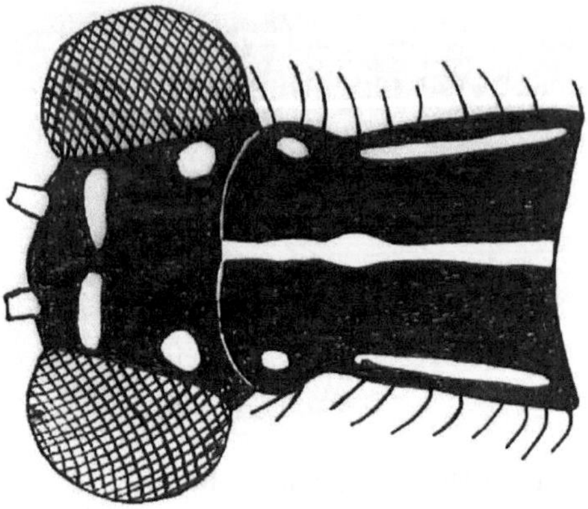

Fig. 7.32 Head and pronotum of *Peruveleon camposi* in dorsal view. Based on Navás (1932b), who described the species as *Austroleon heleninus*.

– The abdomen is whitish with a dark longitudinal stripe. The wings are subacute at the apex and hyaline, except for the sites of six short, dark intercubital markings (Fig. 7.6). Total length of male: c. 40 mm, of which 35 mm is accounted for by the abdomen. Length of male fore-wing: c. 20 mm; hind wing: c. 17.9 mm.

....*Peruveleon dolichogaster* (Navás, 1915) (Argentina, Uruguay). Syn: *Austroleon dolichogaster* Navás, 1915; *Brachynemurus dolichogaster* (Navás, 1915); *Austroleon elongatus* Navás, 1919 (preoccupied); *Austroleon elongatus* Navás, 1919; *Austroleon bridarollii* Navás, 1928; *Abatoleon dolichogaster* (Navás, 1915) auctt.; *Brachynemurus dispar* nec Banks, 1909 pars?

Key to the South American Species of Adults in the Genus *Argentoleon*

Information for the key was provided by Banks (1909); Navás (1914b, 1919) and Stange (2004, 2010).

1. There is a prominent dark brown stripe in the mediocubital area of the fore-wing, which extends well distal to the cubital fork (Fig. 7.8).

....*Argentoleon longitudinalis* (Navás, 1914) (Paraguay, Uruguay, Argentina). Syn: *Moza longitudinalis* Navás, 1914; *Brachynemurus longitudinalis* (Navás, 1914).

– There is no dark brown stripe in the mediocubital area of the fore-wing. Length of fore-wing: 19 mm; hind wing: 17.5 mm.
....*Argentoleon irrigatus* (Gerstaecker, 1894) (Venezuela, Bolivia, Paraguay, Uruguay, Argentina, Mato Grosso, São Paulo). Syn: *Formicaleo irrigatus* Gerstaecker, 1894; *Moza irrigatus* (Gerstaecker, 1894); *Brachynemurus irrigatus* (Gerstaecker, 1894); *Brachynemurus meridionalis* Banks, 1909; *Brachynemurus argentinus* Banks, 1910; *Austroleon argentinus* (Banks, 1910) Navás, 1917b; *Austroleon stictogaster* Navás, 1913; *Clathroneuria amazonica* Navás, 1914; *Austroleon clavatus* Navás, 1915; *Austroleon alienus* Navás, 1915; *Austroleon antennatus* Navás, 1919; *Austroleon sectorius* Navás, 1920; *Austroleon latipennis* Banks, 1923; *Austroleon graphonotus* Navás, 1934.

Key to the South American Species of Adults in the Genus *Austroleon*

Information for the key was provided by Navás (1914b, 1915b, 1917b, 1923, 1926a, 1932b); Stange (2010).

1. The sensory hair on the middle femora is longer than the tarsus and subequal in length to the seta on the fore-femur. For most of its length, the posterior area of the hind wing is much wider than the lengths of the setae fringing the posterior vein. There is a distinctive pattern of yellow and dark brown on the dorsal surface of the thorax (Fig. 7.33). Wing span: 38–40 mm.
....*Austroleon dispar* Banks, 1909 (Argentina). Syn: *Austroleon dorsalis* Banks, 1910; *Moza atomica* Navás, 1914; *Austroleon Lehmani* Navás, 1916; *Austroleon ternarius* Navás, 1919; *Clotus atomicus* Navás, 1922; *Austroleon francoi* Navás, 1926; *Brachynemurus dispar* (Banks, 1909).

– The sensory hair on the middle femur is much shorter than the tarsus and the seta on the fore-femur. For most of its length, the posterior area of the hind

Fig. 7.33 The dorsal surface of the thorax of a male *Austroleon dispar*. Based on Navás (1919), who described the species as *Austroleon ternarius* Navás (1919).

wing is much narrower than the lengths of the setae fringing the posterior vein (Fig. 7.9).

.... *Austroleon immitus* (Walker, 1853) (Venezuela, Paraguay, Bolivia, Argentina, São Paulo). Syn: *Myrmeleon immitus* Walker, 1853; *Moza immitus* (Walker, 1853); *Carreo immitus* (Walker, 1853); *Brachynemurus immitus* (Walker, 1853); *Macronemurus immitus* (Walker, 1853); *Austroleon compar* Banks, 1909; *Moza nubilis* Navás, 1912; *Carrea expansus* Navás, 1914; *Correa expansus* (Navás, 1914); *Austroleon stictogaster stigmatus* Navás, 1929; *Formicaleo ephemerinus* Gerstaecker, 1894; *Distoleon ephemerinus* (Gerstaecker, 1894); *Guipa columbiana* Navás, 1927.

Key to the South American Species of Adults in the Genus *Ameromyia*

Information for the key was provided by (Gerstaecker, 1894); Banks (1909, 1913a, 1943b); Navás (1914a, 1915b, 1917b, 1923a, 1926a, 1932b), and Stange (2010).

1. The wings are mainly hyaline and not infuscated, but the membrane may have a slight grayish clouding and the fore-wing may have short dark stripes or spots, or it may have the space between the subcostal and radial veins lightly tinted with a yellow ferrugineous color (Fig. 7.34). 2

   – Each wing is suffused uniformly with color, or it has a pattern of alternating light and dark areas. The head lacks whitish setae, or the fore-wing is not wide with a concave outer margin toward its apex (Fig. 7.10). 6

2. (1) The membranes of both wings are hyaline, except for the space between the subcostal and radial veins, which is lightly tinted with a yellowish ferrugineous color. Almost all of the wing veins are dark (Fig. 7.34). The head is mainly dark with blackish markings. Total length of the female: c. 36 mm. Length of fore-wing: 35 mm; hind wing 33.5 mm.

   .... *Ameromyia pleuralis* Navás, 1926 (Minas Gerais).

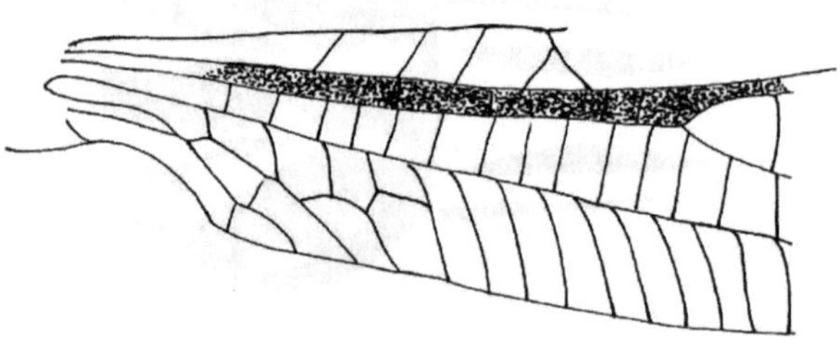

Fig. 7.34 Basal part of the fore-wing of *Ameromyia pleuralis*. Based on Navás (1926b).

– In addition to an occasional testaceous suffusion in the basal subcostal and redial areas, there are other obvious blackish spots or dark markings on the wing, or the veins are almost all pale. Figure 1.14 shows the head............. 3

3. (2) The pterostigma on the fore-wing is at least partially blackish; that on the hind wing may be dark brown. There are either about seven blackish spots along the posterior margin of vein R of the fore-wing from vein Rs to the pterostigma, or there are dark brownish markings along both cubital veins on the fore-wings 4

– The pterostigmata on both wings are pale 5

4. (3) The areas of the pterostigmata are a rust color with a blackish marking across it. Veins Sc and R of the fore-wing are a rusty yellow color with some darker markings, and the rest of the veins are dark brown. There are brownish markings along both cubital veins, which break up into spotted markings toward the apical parts of the veins. The wings are lancet-shaped with pointed apices. Total length of the male: c. 41 mm; female: 35–37 mm. Length of fore-wing of male: c. 29 mm; female: c. 35 mm. Length of hind wing of male: c. 28 mm; female: c. 33 mm. ....*Ameromyia tendinosa* (Gerstaecker, 1894) (São Paulo). Syn: *Myrmeleon tendinosus* Gerstaecker, 1894.

– The pterostigma on the fore-wing is black, and that of the hind wing is dark brown. The wings have testaceous suffusions in the basal subcostal and radial areas and about seven blackish spots along the posterior margin of vein R in the fore-wing from vein Rs and the pterostigma. The hind wings also has several dark spots along vein R and a dark streak at the apex of the wing. There is a W-shaped marking on the head ventral to the antenna insertions (Fig. 1.14). The wing span is from 60–70 mm. .... *Ameromyia nigriventris* (Walker, 1860) (Colombia, Venezuela, Argentina, Rio Grande do Sul, Amazonas?). Syn: *Myrmeleon nigriventris* Walker, 1860.

5. (3) The fore-wing is marked with short dark stripes, has dark veins, and is wide at the apex. The outer part of the fore-wing is widened, and its margin is strongly concave. The hind wing is very pale with vein R darkened. The head is mainly brownish yellow, but it bears fine whitish setae. Length of fore-wing: c. 29–30 mm; hind wing: c. 27.5 mm. ....*Ameromyia strigosa* (Banks, 1909) (Paraguay, Argentina, Chile). Syn: *Brachynemurus strigosus* Banks, 1909; *Ameromyia strigosa* Navás, 1922; *Ameromyia fidelis* Navás, 1915; *Moza strigosa* (Banks, 1909) Navás, 1917b.

– The fore-wing is subacute at the apex and not marked with dark stripes. The veins on both wings are almost all pale or ferruginous. The head is mainly yellowish with dark markings. The abdomen is yellow with yellow setae and narrow black longitudinal stripes. Length of fore-wing: c. 25 mm; hind wing: c. 23.5 mm. ....*Ameromyia pentheri* Navás, 1914 (Rio de Janeiro). Syn: *Ameromyia penthevi* (sic) Zoological Record, 1914.

6. (1) The bases of the antennae are separated by a distance longer than the diameter of one antenna. In the hind wing, there are three or four cross veins proximal to the origin of vein Rs in both wings. The first branch of Rs and the cubital fork are located at subequal distances from the wing base. The tibial spurs are equal in length to the first two tarsal segments, each of which is subequal to the third and fourth combined. There is a line running through the apices of both wings.. 7

   – The bases of the antennae are separated by a distance shorter than the diameter of one antenna, or the first branch of the radial sector in the fore-wing is located at a distance from the wing base equal to or less than the distance of the wing base to the apex of the anal vein. The tibial spurs are sometimes only as long as the first tibial segment (Fig. 7.10). 8

7. (6) The posterior margin of the fore-wing is not at all infuscate. The fore-wing is shorter than 30 mm. The tibiae are pale.
   ....*Ameromyia pubiventris* (Walker, 1860) (Venezuela, Surinam, Amazon Region of Brazil). Syn: *Myrmeleon pubiventris* Walker, 1860; *Amazoleon pubiventris* (Walker, 1860) Banks, 1913.

   – The posterior margin of the fore-wing is at least partially infuscate. The tibiae are dark. Fore-wing length: c. 33 mm; width: c. 6.8 mm. Length of abdomen: c. 35 mm.
   ....*Ameromyia modesta* (Banks, 1943) (Venezuela). Syn: *Amazoleon modestum* Banks, 1943.

8. (6) Vein Rs in the fore-wing is dark and contrasts strongly with the pale brown vein Sc. The width of the posterior area of the hind wing is subequal to that of the costal area at the level of the radial sector. The cells between vein CuA and the hind margin of the hind wing are much longer than wide, measured with the length in the direction of the long axis of the wing. Total length: c. 42–45 mm. Length of fore-wing: 35–38 mm; hind wing: 33–35 mm.
   ....*Ameromyia protensus* (Gerstaecker, 1894) (Paraguay, Uruguay, Argentina, São Paulo). Syn: *Myrmeleon protensus* Gerstaecker, 1894; *Nemotolus protensus* (Gerstaecker, 1894); *Foya trapezia* Navás, 1914; *Ameromyia alterna* Navás, 1919; *Ameromyia De Carloi* Navás, 1923.

   – Parallel segments of veins Sc and Rs are generally the same color, and/or the width of the posterior area of the hind wing is greater than that of the costal area at the level of the radial sector. The cells between vein CuA and the hind margin of the hind wing are shorter than wide, or the length and width are subequal (Fig. 7.10). 9

9. (8) The length and width of the prothorax are subequal. The abdomen is dark and covered by short gray setae. The radial and subcostal veins in the fore-wing remain nearly the same color for their entire lengths proximal to the radial sector. The mediocubital area of the fore-wing have cells that are almost

completely and uniformly suffused with color. Fore-wing length of female: c. 31 mm; fore-wing width: c. 6.5 mm.
.... *Ameromyia dimidiata* Navás, 1915 (Uruguay, Argentina, Paraguay). Syn: *Ameromyia baronei* Navás, 1921.

– The width of the posterior part of the prothorax is slightly greater than the length in dorsal view (Fig. 7.10). The greater wing length may be a better character for distinguishing the species 10

10. (9) The abdomen is dark with short gray setae. The pterostigma is yellowish. The fore-wing has a dark proximal margin of the pterostigma and other dark markings, while the hind wing is considerably paler. The head is mainly yellow with a dark triangle on the vertex bordered by narrow transverse markings between the compound eyes and two dark, oval spots on the occiput. The pronotum is mainly dark on the dorsal surface with a wide, pale median stripe and narrow yellow dorsolateral and lateral markings (Fig. 7.10). Total length: c. 23 mm. Fore-wing length of female: c. 25.5 mm; hind wing length: c. 24.4 mm. Only the female has been described.
.... *Ameromyia muralli* Navás, 1932 (Mato Grosso do Sul).

– The abdomen is dark with blackish markings and covered by yellowish or blackish setae. The pterostigma is white or pinkish and bordered by a darker color. The mid-dorsal stripe on the pronotum is narrow and pale, blackish, or dark brown bordered with darker or lighter brown areas (Fig. 7.35) 11

11. (10) The pterostigma is white with a dark border. The pronotum of the female is dark with large blackish areas on both sides of the narrow, pale median stripe. There is a small, pale ellipse near the anterolateral margin of each blackish dorsal area, and the wide, pale, lateral margins contain an elongated, dark ellipse on each side (Fig. 7.35).
.... *Ameromyia stevensi* Navás, 1914 (Argentina, Amazon Region of Brazil). The two species in this couplet require better descriptions to confirm their validity.

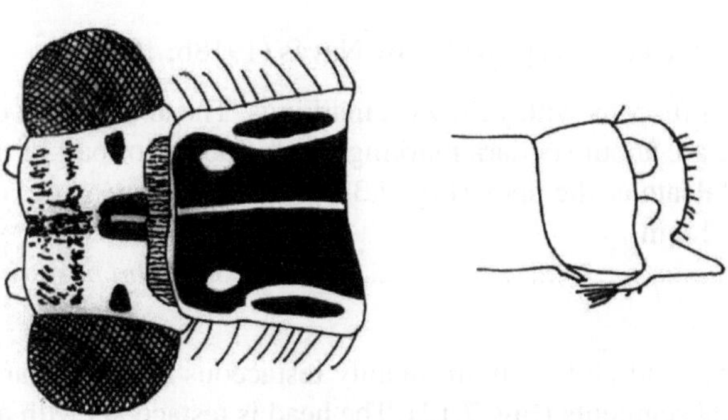

Fig. 7.35 *Ameromyia stevensi* female: head and pronotum in dorsal view (left) and the apex of the abdomen in lateral view (right). Based on Navás (1914a).

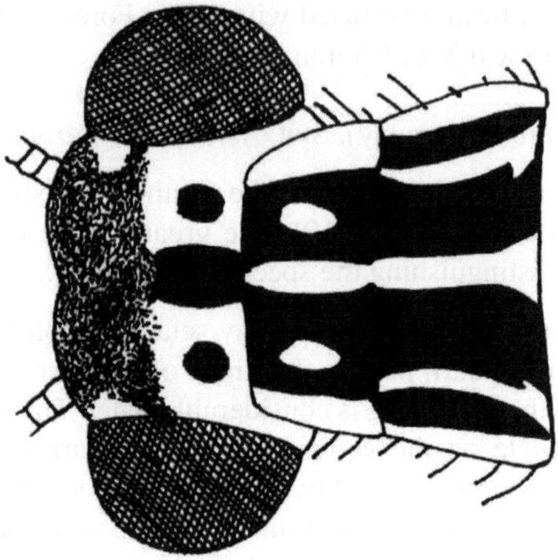

Fig. 7.36 Head and pronotum of a male *Ameromyia hirsuta* in dorsal view. Based on Navás (1917b), under its junior synonym, *Moza longiventris*.

– The pterostigma is light pinkish with a dark interior border. The pronotum is dark with a blackish median stripe and lateral markings. The radial and subcostal veins in the fore-wing have alternating lengths of light and dark brown. The mediocubital area of the fore-wing have cells that are each only about half covered with dark brown with spots (Fig. 7.36). Length of fore-wing: 32–34 mm; hind wing: 32–34 mm.

....*Ameromyia hirsuta* Navás, 1914 (Argentina, Uruguay, Rio Grande do Sul). Syn: *Ameromyia hirenta* Navás, 1914 (misspelling in the Zoological Record of 1914); *Moza longiventris* Navás, 1917; *Ameromyia stevensi striolata* Navás, 1922.

Key to the South American Species of Adults in the Genus *Jaffuelia*

Information for the key was provided by Navás (1918b, 1934a).

1. The thorax is fuscous with yellowish markings. The abdomen is dark with white setae. There are about six dark markings on the posterior part of the wing, which is alightly falcate at the apex (Fig. 7.37). Length of fore-wing: c. 23 mm; hind wing: c. 21.2 mm.

....*Jaffuelia porterina* Navás, 1934 (Chile).

– The thorax and abdomen are mainly testaceous red with dark apices of the abdominal segments (Fig. 7.12). The head is testaceous with an elevated vertex. Length of fore-wing: c. 27.5 mm; hind wing: c. 24.5 mm.

....*Jaffuelia chilensis* Navás, 1918 (Chile).

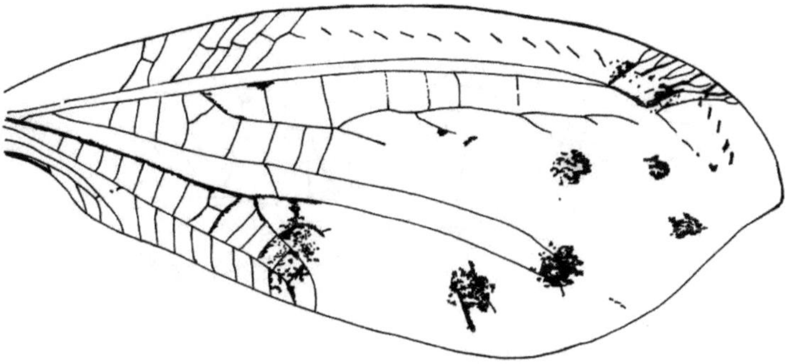

Fig. 7.37 A schematic illustration of the fore-wing of *Jaffuelia porterina,* not showing all veins in the anterior and apical parts but showing the location of the dark markings. Based on Navás (1934a).

Key to the South American Species of Adults in the Genus *Elicura*

Information for the key was provided by Navás (1911, 1918a, 1919, 1923a, 1930b, 1933b).

1. The fore-wing is c. 33 mm, and the hind wing, c. 34.4 mm. The dorsal surface of the thorax is yellowish red with dark markings; the ventral surface is yellowish green with large dark markings and both blackish and white pilosity. The pronotum has two dark longitudinal lines, dark pilosity, and is marked laterally with white (Fig. 7.13).

   .... *Elicura litigator* (Navás, 1911) (Chile). Syn: *Calinemurus litigator* Navás, 1911; *Elicura marganus* Navás, 1918; *Brachynemura marganus* (Navás, 1918); *Elicura ferus* Navás, 1930; *Brachynemura ferus* (Navas, 1930).

   – The fore-wing is usually between 20 and 25.5 mm long, and the hind wing is about 19–28 mm. The thorax is mainly dark with darker or yellowish markings 2

2. The hind wing is longer than the fore-wing. The fore-wing is 22–25.5 mm, and the hind wing, 23.5–28 mm. Most cells in the costal space are divided by one or more cross veins (Fig. 7.38). The thorax is dark with yellowish markings and a dark pilosity.

   .... *Elicura iniqua* (Navás, 1919) (Argentina, Uruguay). Syn: *Plater iniquus* Navás, 1919; *Lemolemus iniquus* (Navás, 1919); *Briderollus solers* Navás, 1933; *Lemolenus solers* (Navás, 1933).

   – The fore and hind wings are subequal in length. The thorax is mainly dark with two irregular dark brown stripes on the dorsal surface of the pronotum. Fore-wing: c. 21.8 mm. Hind wing: c. 21.8 mm.

   .... *Elicura justus* (Navás, 1923) (Argentina). Syn: *Plater justus* Navás, 1923.

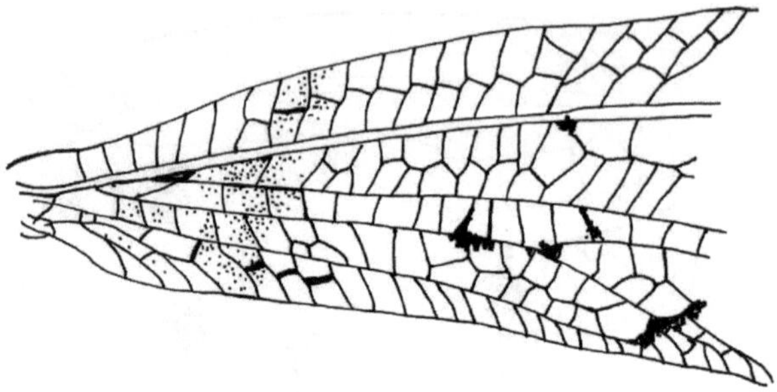

Fig. 7.38 Basal part of the fore-wing of *Elicura iniqua*. Based on Navás (1933), who called the species *Briderollus solers*.

Key to the South American Species of Adults in the Genus *Sical*

Information for the key was provided by Navás (1928a, 1934a, b).

1. The hind wing is longer than the fore-wing. The posterior margin of the fore-wing is slightly concave (Fig. 7.15). The palps are yellow, except for the apical segment of the labial palp, which is fusiform and black. The thorax is dark with whitish pilosity and short, narrow, dark yellow longitudinal stripes. Length of fore-wing: c. 22.5; hind wing: c. 24.5 mm.
   .... *Sical peralinus* Navás, 1928
   (Chile).

   – The fore-wing is longer than the hind wing 2

2. Length of fore-wing: c. 16.5; hind wing: c. 15 mm (Fig. 7.39). The palps are yellow, except for the apex of the apical segment, which is dark. The thorax is yellow with dark, longitudinal stripes on the dorsal surface.
   .... *Sical barrosi* (Navás, 1934)
   (Chile). Syn: *Neteta barrosi* Navás, 1934.

   – Length of fore-wing: c. 23 mm; hind wing: 21.5 mm. The palps are yellow with somewhat darker rings. The pronotum has a transverse sulcus in the anterior half. The dorsal surface of the pronotum is dark with a narrow light stripe and two wider yellowish stripes lateral to it. The mesonotum is dark reddish with a pale median stripe and a pair of reddish markings, one on each side of the median stripe (Fig. 7.40). The metanotum is dark reddish.
   .... *Sical stuardinus* (Navás, 1934)
   (Chile). Syn: *Chiloleon stuardinus* Navás, 1934.

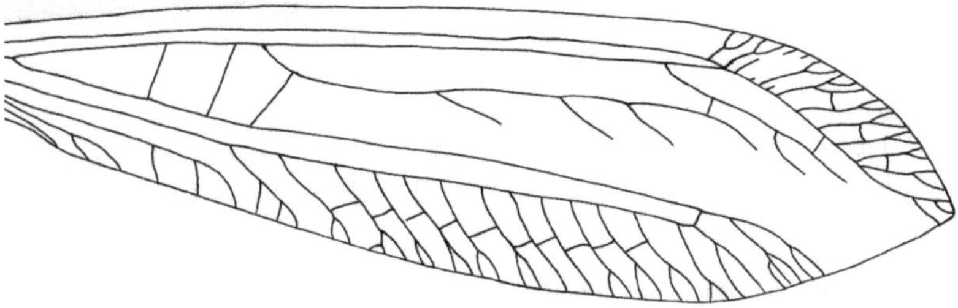

Fig. 7.39 Sketch of the hind wing of *Sical barrosi*, showing the cross veins only near the apex and along the posterior border of the wing. Based on Navas (1934b), who referred to the species as *Neteta barrosi*.

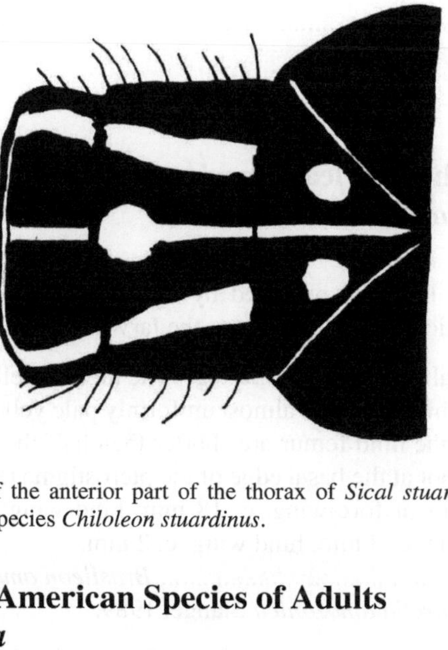

Fig. 7.40 Dorsal view of the anterior part of the thorax of *Sical stuardinus*. Based on Navás (1934b), who named the species *Chiloleon stuardinus*.

Key to the South American Species of Adults in the Genus *Vella*

Information for the key was provided by Navás (1914a, 1917a) and Stange (2002). The genus is in need of taxonomic revision.

1. Total length: 45–60 mm. The wing span reaches 160 mm. The biareolate veins of the costal area extend well into the basal third of the wing, which is more reticulated, producing sections that have three series of veins in places rather than two (Fig. 7.2).
   .... *Vella fallax* (Rambur, 1842)
   (North and Central America, Peru, Argentina). Syn: *Vella puerpera* Navás, 1914.

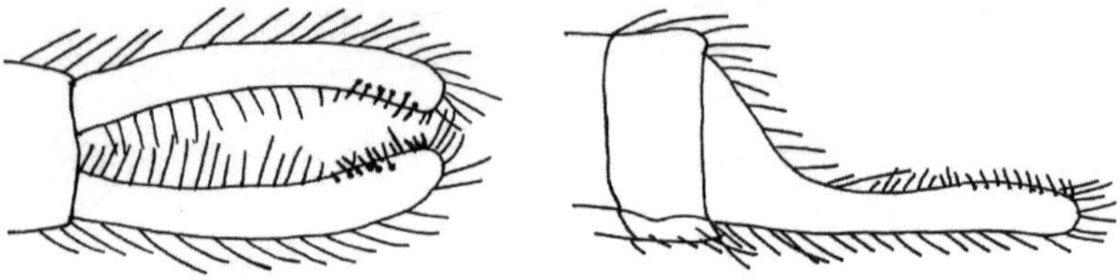

Fig. 7.41 Apex of the abdomen of a male *Vella flaccida* in dorsal (left) and lateral view (right). Based on Navás (1917a).

– The size is considerably smaller, with the male reaching about 35 mm in length. Length of male fore-wing: c. 43 mm; hind wing: c. 41.5 mm. The apical third of the male cerci bears black, spine-like setae along the internal margins (Fig. 7.41).
.... *Vella flaccida* Navás, 1917 (Colombia).

Key to the South American Species of Adults in the Genus *Brasileon*

Information for the key was provided by Miller and Stange (1989) and Stange (2004). No information is available about the larvae.

1. There are no tibial spurs on the fore-legs. The metascutellum is strongly inflated on the third. The hind femur is almost uniformly pale yellow on its basal 3/4. The longest setae on the hind femur are shorter than half the diameter of the femur. There is a dark spot at the basal edge of the pterostigma (Fig. 7.17). Total length: c. 16 mm. Length of fore-wing: c. 18 mm; hind wing: c. 14 mm. Maximum width of fore-wing: c. 3 mm; hind wing: c. 2 mm.
.... *Brasileon amazonica* (Stange, 1989) (Pará). Syn: *Dimarella amazonica* Stange, 1989.

– There are tibial spurs on the fore-legs. The anterior 1/3 of the metascutellum is weakly inflated. There are many black spots on the basal ¾ of the hind femur; the spots are at the bases of setae equal or greater in length to the diameter of the femur. Total length: c. 14 mm. Length of fore-wing: c. 17 mm; hind wing: c. 15 mm. Maximum width of fore-wing: c. 3 mm; hind wing: c. 2.5 mm.
.... *Brasileon pennyi* (Miller, 1989) (Chile, Pará). Syn: *Dimarella pennyi* Miller, 1989.

Key to the South American Species of Adults in the Genus *Dimarella*

Information for the key was provided by Banks (1913); Navás (1915a, b, 1918b, c, 1932b, 1936a), Miller and Stange (1989), and Stange (2004, 2010). The name *Dimarella tarsalis* (Guilding, 1933), originally *Formicaleo tarsalis* Guilding, 1833, is treated as a *nomen dubium* (Miller and Stange, 1989). Reports from Guyana and Jamaica are unconfirmed.

1. The hind wing is widest just proximal to the stigma and then narrows abruptly toward the apex, or, if in doubt, the claws on the fore-tarsus cannot close against the ventral surface of the apical tarsal segment (Fig. 7.42).
   .... Subgenus *Pachyleon* Stange, 1970.............................2
   – The hind wings are lanceolate. The claws on the fore-tarsus can close against the ventral surface of the apical tarsal segment (Fig. 7.43).
   .... Subgenus *Dimarella* Banks, 1913..............................5
2. (1) The fore-femur is greatly inflated; its greatest width is much more than the distance between the bases of the antennae. Its pubescence is especially dense

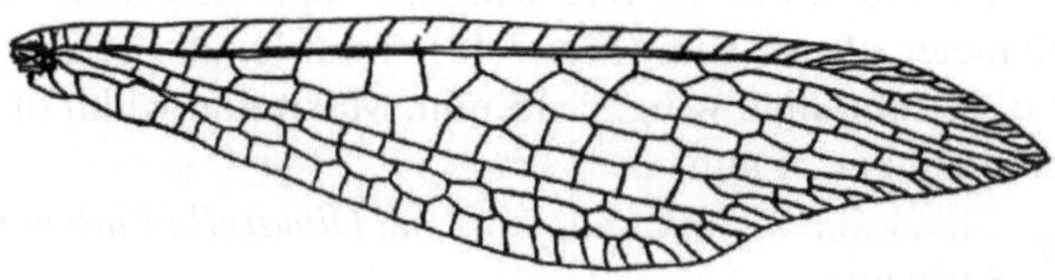

Fig. 7.42 The hind wing of *Dimarella blohmi*. Based on Miller and Stange (1989).

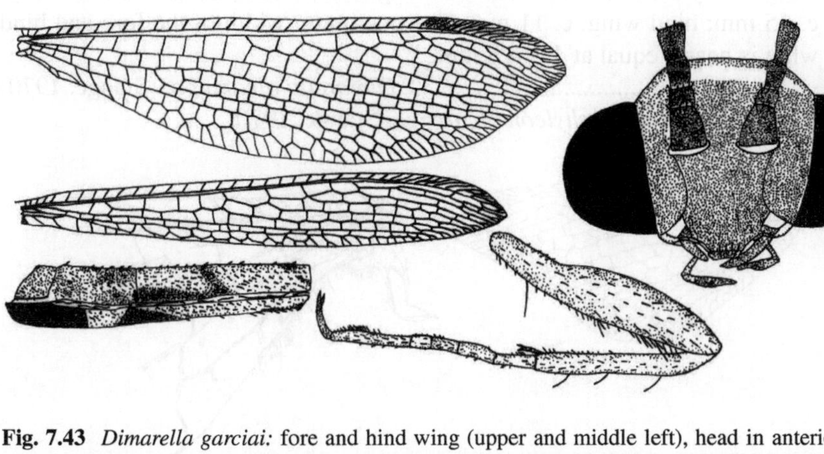

Fig. 7.43 *Dimarella garciai:* fore and hind wing (upper and middle left), head in anterior view (upper right), fore-leg in lateral view (lower center), and the three anterior abdominal segments of a male in lateral view (lower left). Based on Miller and Stange (1989) and Stange (2002).

on the surfaces that meet when the femur closes against the tibia. There is a sensory hair on the fore-femur, which is more than half as long as the femur. ....*Dimarella cautus* (Walker, 1853) (Peru, French Guiana, Brazil). Syn: *Myrmeleon cautus* Walker, 1853; *Formicaleo cautus* (Walker, 1853) Hagen, 1866; *Psammoleon cautus* (Walker, 1853) Banks, 1943; *Feinerus nebulosus* Navás, 1923.

– The fore-femur is not greatly enlarged, and its greatest width is much less than the distance between the bases of the antenna 3

3. (2) The fore-wing is c. 27.5 mm long, and the hind wing is c. 27 mm long. The pterostigma on the fore wing is darkened along its proximal margin, and there are two dark spots on the proximal part of that wing. ....*Dimarella bipunctatus* (Navás, 1915) (French Guiana, Ceará). Syn: *Formicaleo bipunctatus* Navás, 1915. Banks (1943) included this name as a junior synonym of *Psammoleon cautus* (Walker, 1853), a synonym of *Psammoleon cautus* Walker, 1853, but it was reinstated as a valid species by Stange (2002).

– The fore-wing is usually shorter with 20 mm; the hind wing is at least 1 mm shorter than the fore-wing (Fig. 7.42). 4

4. (3) The hind wing is much narrower than the fore-wing. Vein 2A in the fore-wing. Veins 2A and 3A in the fore-wing are separated from each other by at least the diameter of one vein (Fig. 7.42). Total length: c. 20 mm. Length of fore-wing: c. 17 mm; hind wing: c. 15 mm. Maximum width of the fore-wing: c. 5 mm; hind wing: c. 4 mm. .... *Dimarella blohmi* (Stange, 1989) (Venezuela, Surinam).

– The fore and hind wings are equal in width. Veins 2A and 3A in the fore-wing nearly touch (Fig. 7.44). Total length: c. 19 mm. Length of fore-wing: c. 15 mm; hind wing: c. 11 mm. The maximum width of the fore and hind wing is nearly equal at 4.7–4.8 mm. .... *Dimarella alvarengai* (Stange, 1970) (Mato Grosso). Syn: *Pachyleon alvarengai* Stange, 1970.

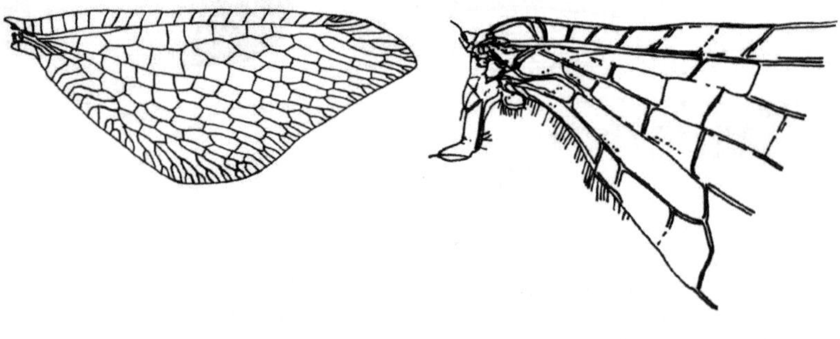

Fig. 7.44 *Dimarella alvarengai:* the hind wing (left) and the base of the fore-wing (right). Based on Miller and Stange (1989).

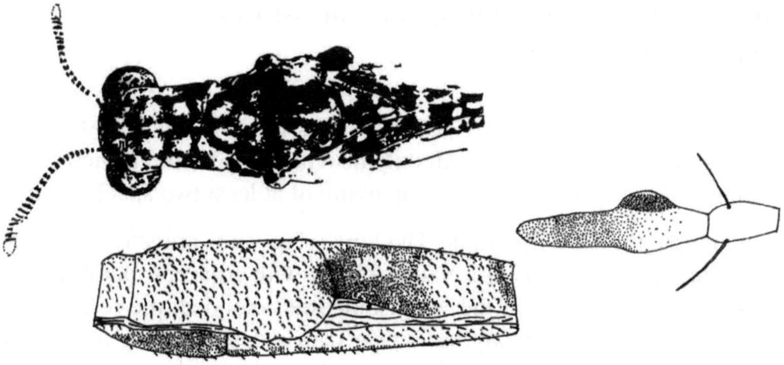

Fig. 7.47 *Dimarella effera:* head, thorax, and anterior part of the abdomen in dorsal view (upper left); three anterior abdominal segments in lateral view (lower left); and the apical segment of the palp of a larva (right). Based on Miller and Stange (1989).

11. (9) The basal tarsal segment on the hind leg is five times longer than its diameter at mid-length and much longer than a tibial spur. The second abdominal tergite of the male is almost as long as the third and much wider (Fig. 7.47). Total length: c. 21 mm. Fore-wing length: c. 20 mm. ....*Dimarella effera* (Walker, 1853) (Venezuela, French Guiana, Mato Grosso, Roraima, Pará). Syn: *Myrmeleon efferus* Walker, 1853; *Creagris effera* (Walker, 1853) Hagen, 1860; *Dimarella effusa* Banks, 1943. Banks (1943) listed this name as a junior synonym of *Dimarella tarsalis* (Guilding, 1833), herein treated as a *nomen dubium,* but it was later reinstated as a distinct species (Miller and Stange, 1989).

– The basal tarsal segment on the hind leg is four times longer than its diameter at mid-length and only slightly longer than a tibial spur. The second abdominal tergite of the male is half as long as the third and not much wider (Fig. 7.43). The hind femur has a complete fuscous, subbasal band. The stigmal area of the fore-wing is almost completely suffused with dark brown. The postventral lobe of the male ectoproct is longer than the longest seta on the apical swelling. Total length: 20–24 mm. Fore-wing length: 19–23 mm. ....*Dimarella garciai* (Navás, 1932) (Central America, Ecuador). Syn: *Nobra garciai* Navás, 1932. This species is similar to *Dimarella totoneca* (Miller, 1989), from Mexico, which differs by having its fuscous, subbasal band on the hind femur interrupted, a stigmal area only partially suffused with dark brown, and a postventral lobe on the male ectoproct shorter than the longest seta on the apical swelling.

Key to the South American Species of Adults in the Genus *Navasoleon*

Information for the key was provided by Banks (1920), Navás (1915b, 1922b). From the descriptions, the species can only be tentatively distinguished, and a revision of the genus is urgently needed. Synonymy of at least two species is likely.

1. Fore-wing length: c. 30 mm long. The long wing veins are alternately dark and pale, while the cross veins are mainly black, with adjacent membranes on the posterior and apical parts of the wing.sometimes clouded; the rest of the membrane is hyaline. The pterostigma is pale with a dark marking at its base. The head and thorax are pale dorsally with a black stripe between the antennae expanding into a large black spot dorsal to their bases, two dark markings posterior to it on the vertex, dark lateral margins on the pronotum, and two dark stripes parallel to them, leaving the middle of the pronotum broadly pale. The pronotum is longer than wide. The thorax has dark posterior markings. The abdomen is pale, but somewhat darker on the posterior half of each segment.
   .... *Navasoleon bolivianus* (Banks, 1920)
   (Bolivia). Syn: *Gymnocnemia boliviana* Banks, 1920.
   - The fore-wing is not longer than about 26 mm. The wing veins are mainly dark (Fig. 7.19). 2
2. The abdomen is light yellow with yellow setae and vaguely suffused with a ferrugenous coloration toward the apical margins of the segments. The wing veins are almost entirely dark, and the membranes are hyaline and iridescent. The pterostigma is pale. The head and pronotum are yellow with dark setae on the pronotum (Fig. 7.19). Length of fore-wing: c. 21 mm; hind wing: c. 19.5 mm.
   .... *Navasoleon leptocerus* (Navás, 1915)
   (Argentina). Syn: *Gymnocnemia leptocera* Navás, 1915.
   - The abdomen is predominantly yellow, darkened on the posterior parts of the segments ventrally and dark in the middle of the posterior part of the anterior segments dorsally. The three posteriormost segments of the abdomen are almost entirely yellow. The wing veins are almost all dark, and the membranes are hyaline. The head is yellow, and the pronotum is almost entirely dark dorsally and yellow ventrally. Length of fore-wing: c. 23 mm; hind wing: c. 22 mm.
   .... *Navasoleon bosqi* (Navás, 1922)
   (Argentina). Syn: *Gymnocnemia Bosqi* Navás, 1922.

Key to the South American Species of Adults in the Genus *Elachyleon*

Information for the key was provided by Esben Petersen (1927, 1933), and Banks (1943).

1. The prothorax is reddish or yellowish brown without markings. The mesothorax and metathorax are yellowish brown dorsally with three black streaks, one in the middle and broader ones lateral to it. The abdomen is dark brown, sometimes with dorsal markings on some segments. The pterostigma on the fore wing is yellowish bordered by a blackish marking. In addition, the fore-wing has about three other prominent dark marking, and the hind wing has two (Fig. 7.21). Length of fore-wing: 21–23 mm; hind wing: 20–23 mm.
   .... *Elachyleon punctipennis* Esben Petersen, 1927 (Mexico, Central America, Trinidad, Surinam, Argentina). Syn: *Elachyleon punctipennis pulchellus* Esben Petersen, 1933.
   - The prothorax is dark brown with a middle gray streak and two parallel ones that end at the groove. The rest of the thorax is mainly black. The abdomen is black dorsally, sometimes with pale posterior borders to the tergites. Length of fore-wing: c. 26 mm; hind wing: c. 25 mm.
   ....*Elachyleon gentilis* (Banks, 1943) (Guyana). Syn: *Sosa gentilis* Banks, 1943.

Key to the South American Species of Adults in the Genus *Araucaleon*

Information was taken from Esben-Petersen (1927) and Banks (1938).

1. Distal to the midlength of the fore-wing, there is an obvious bulge along the posterior margin. The apex of the fore-wing is broadly rounded, and its posterior margin has a distinctive rounded projection and a group of tiny veins near its midlength (Fig. 7.48). The wings are hyaline with dark brown spots. The veins are whitish in the hyaline areas. Length of fore-wing: c. 40 mm; hind wing: c. 44 mm. Width of fore-wing: c. 11 mm; hind wing: c. 8 mm.
   ....*Araucaleon inca* Banks, 1938 (Bolivia).

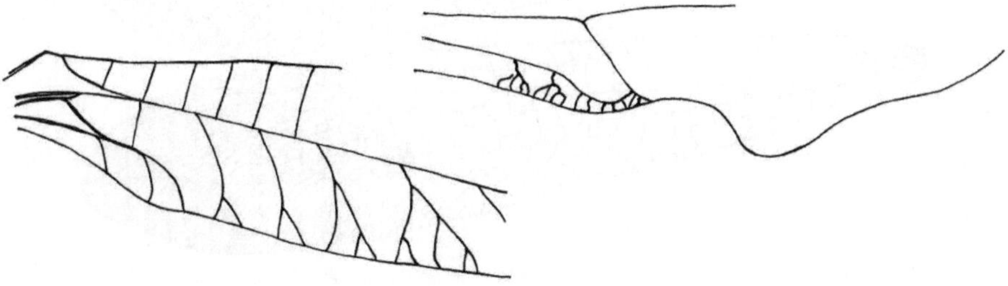

Fig. 7.48 *Araucaleon inca:* posterior margin of the fore-wing near its base (left) and near its midlength (right). Based on Banks (1938).

– The fore-wing is somewhat falcate toward the apex and lacks an obvious bulge along the posterior margin (Fig. 7.22).
....*Araucaleon withycombei* (Esben-Petersen, 1927) (Trinidad, Venezuela). Syn: *Glenurus withycombei* Esben-Petersen, 1927.

Key to the South American Species of Adults in the Genus *Glenurus*

Information for the key was provided by Navás (1918b, c, 1920); Banks (1923) and Stange (2000).

1. The fore-wing lacks a large dark marking at the apex (Fig. 7.49).
.... *Glenurus heteropteryx* Gerstaecker, 1885 (Central America, West Indies, Venezuela, Ecuador). Syn: *Glenurus discors* Navás, 1920.

   – Both wings have large dark markings at the apex (Fig. 7.25). 2

2. The space proximal to the large apical marking on the fore-wing is marked with many dark spots, with at least some of them associated by darkened sections of wing veins (Fig. 7.50). 3

   – The space proximal to the large apical marking on the fore-wing is mostly hyaline (Fig. 7.25). 4

3. The wing span of the type specimen was reported to be 72 mm. The color is deep black with a pale median line on the pronotum and yellow legs, becoming darker only on the apical tarsal segments.
.... *Glenurus croesus* Banks, 1923 (Bolivia, Argentina). It is not unlikely that this nominal species is a junior synonym of *G. penningtoni*.

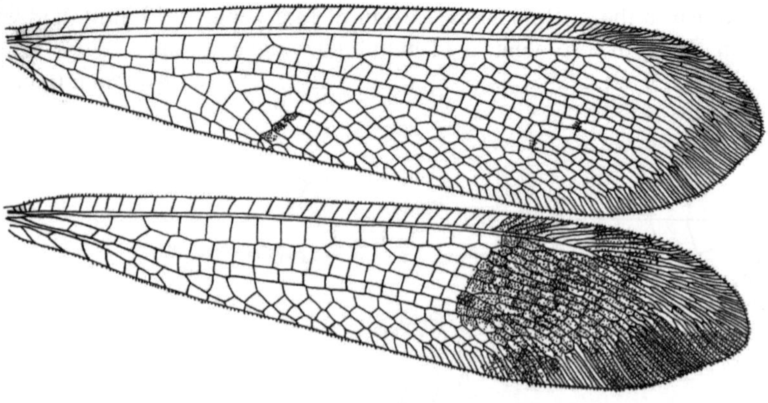

Fig. 7.49 Fore and hind wing of *Glenurus heteropteryx*. Based on Stange (2002).

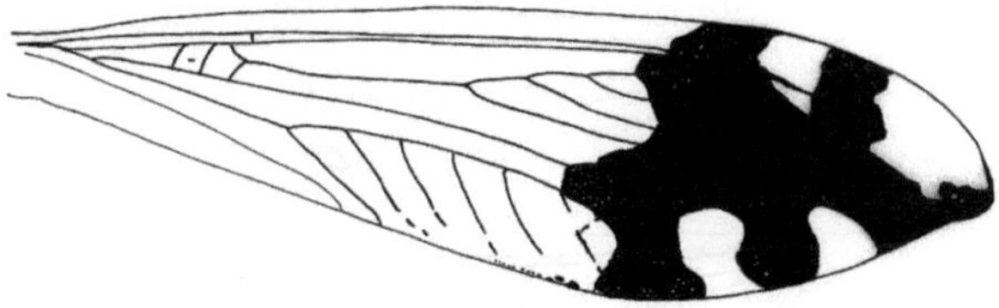

Fig. 7.50 The extent of the dark marking on the hind wing of *Glenurus penningtoni;* most cross veins are omitted from the simplified illustration. Based on Navás (1918b).

– The wing span is greater than c. 80 mm. The dark marking on the hind wing covers most of the apical 1/3 of the wing and encloses about four large paler areas, all contiguous with the wing margin (Fig. 7.50). The markings on the body are variable. Total length: c. Length of male fore-wing: c. 39.5 mm; hind wing: c. 38.4 mm.
....*Glenurus penningtoni* (Navás, 1918)
(Bolivia, Argentina). Syn: *Ledoscius Penningtoni* Navás, 1918.

4. There are two pale spots on the large dark marking at the apex of each fore-wing (Fig. 7.25).
.... *Glenurus peculiaris* (Walker, 1860)
(Guyana, Surinam, Argentina, São Paulo). Syn: *Myrmeleon peculiaris* Walker, 1860; *Glenurus brasiliensis* Navás, 1920.

– There is only one pale spot on the large dark marking at the apex of each fore-wing. The wing span of the type specimen was reported to be 84 mm.
....*Glenurus incalis* Banks, 1923
(Peru).

Key to the South American Species of Adults in the Genus *Eremoleon*

Information for the key was provided by Rambur (1842), Gerstaecker (1894), Navás (Navás 1913a, 1916, 1919); Banks (1910, 1923), Esben-Petersen (1933), and Stange (2004). A revision of this genus is necessary because few substantial differences can be found in the descriptions of most species.

1. The hind wings are clearly longer than the fore-wings....................................... 2

– The fore and hind wings are subequal in length, or the fore-wing is longer (Fig. 7.51). 3

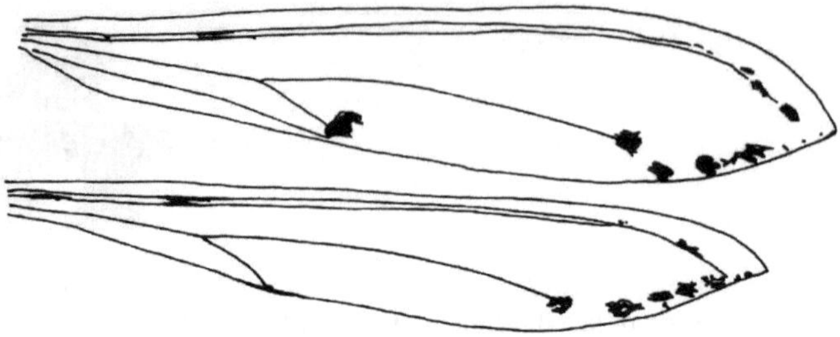

Fig. 7.51 The fore and hind wing of *Eremoleon punctipennis* sketched without most of the veins to show the shape and locations of the dark spots. Based on Navás (1919).

2. (1) The costal space of the fore-wing is greatly expanded and has many forked cross-veins and black flecks, like the rest of the wing surface. Total length: c. 32 mm. Length of fore-wing: c. 42 mm; hind wing: c. 47 mm. ....*Eremoleon impluviatus* (Gerstaecker, 1894) (Bolivia, Argentina). Syn: *Glenurus impluviatus* Gerstaecker, 1894.

   – The wings are very slender, and the costal space of the fore-wing is narrow near the base with almost all cross veins proximal to the pterostigma unforked; it is not marked with many black flecks. Total length: c. 31 mm. Length of fore-wing: c. 34 mm; hind wing: c. 37 mm. ....*Eremoleon anomalus* (Rambur, 1842) (Colombia, Venezuela, Argentina, Brazil). Syn: *Myrmeleon anomalus* Rambur, 1842; *Glenurus mollis* Gerstaecker, 1888; *Glenurus psilocerus* Gerstaecker, 1894; *Glenurus anomalus* (Rambur, 1842) Hagen, 1866; *Glenopsis anomalus* (Rambur, 1842) Banks, 1913.

3. (1) The wing span is greater than 70 mm...4

   – The wing span is less than 70 mm. Each wing may reach a length seldom, if ever, exceding 32 mm...5

4. (3) The fore and hind wings are subequal in length, and the fore-wing may be only very slightly longer in some specimens. The wings are hyaline with a pair of small brown clouds near vein Cu on the fore-wings. The fore-wing veins are also white with dark markings or spots along some of the cross veins. The hind wing has somewhat darker veins. All pterostigmata are white. The general coloration is yellow with blackish transvese bands both dorsal and ventral to the antenna insertions on the head; there are two dark spots near the center of the vertex. The antennae are dark with pale bands, becoming darker near the apex. The pronotum has dark streaks on each side and two dark median lines. The femora and tibiae are banded. The wing span is about 83 mm. ....*Eremoleon petersensi* (Banks, 1923) (Peru). Syn: *Glenopsis petersensi* Banks, 1923.

   – The fore-wing is clearly longer and wider than the hind wing, and both end in acute apices. The wings are hyaline with four spots along the posterior margin in the apical half and other small dark markings, including a black spot at the

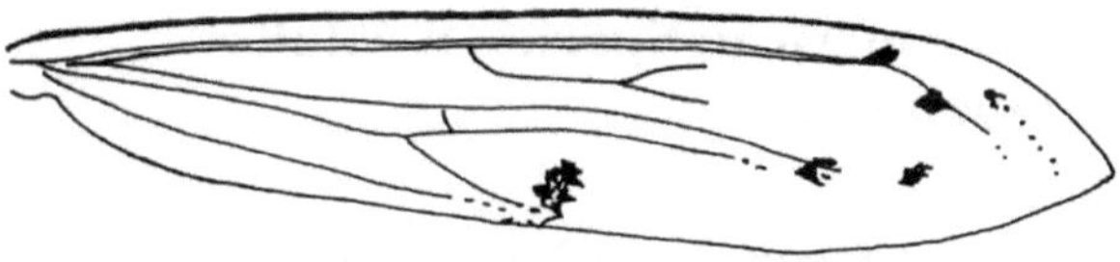

Fig. 7.52 The shape of the fore-wing of *Eremoleon capitatus* and locations of its dark markings, without most wing veins being shown. Based on Navás (1914a), who called the species *Sosa conspicuus*.

base of the pterostigma (Fig. 7.51). Spots are rarely missing, presumably on teneral specimens. The head is pale with black markings. The coloration of the thorax and abdomen is predoinantly blackish with a narrow yellowish midline stripe and two wider lateral stripes on the pronotum. There are yellow spots on the dorsal surfaces of the rest of the thorax, including one large spot at the apex of the scutellum. The abdomen also has pale markings on the middle and posterior parts of some segments. The apical abdominal segment has more prominent pale markings. The legs are pale with black dots on the external surfaces, especially on the fore and middle legs. The first, second, and fifth tarsal segments are pale; the third and fourth are blackish. The wing span is about 78 mm.

....*Eremoleon punctipennis* (Banks, 1910) (Colombia, Venezuela, Argentina). Syn: *Psammoleon punctipennis* Banks, 1910; *Incamoleon punctipennis* (Banks, 1910) Banks, 1913; *Formicaleo tetrastictus* Navás, 1913; *Formicaleo stictopterus* Navás, 1916; *Formicaleo punctipennis* (Banks, 1910) Navás, 1916.

5. (4) The head and thorax are brownish with yellow palps, clypeus, labrum, and rings around the compound eyes. The pronotum is brown with a rather narrow mid-dorsal stripe. The ventral part of the prothorax is yellow. Just proximal to the apex of the fore-wing, the posterior margin is slightly concave (Fig. 7.52). Total length: 22–24 mm. Lengh of fore-wing: 29–30.5 mm; hind wing: 27.5–29 mm; antenna: c. 7.5 mm.

....*Eremoleon capitatus* (Navás, 1913) (Rio de Janeiro). Syn: *Formicaleo capitatus* Navás, 1913; *Sosa conspicuus* Navás, 1914.

– The head and pronotum are mainly lemon yellow with a brown transverse band above the insertions of the antennae. The scape has a black marking on the ventral side, and the pedicel is entirely black. The basal 2/3 of the antenna is lemon yellow, and the apical 1/3 is yellow with broad black rings. The pronotum is yellow with brown lateral margins and indistinct brown spots near the posterior margins; the brown lateral stripes continue along the dorsolateral surfaces of the mesonotum and metanotum. The posterior margin of the fore-wing is entirely convex on the apical half (Fig. 7.27). Lengh of fore-wing: c. 27 mm; hind wing: c. 27 mm. Only the female has been described.

....*Eremoleon pulchra* (Esben-Petersen, 1933) (Paraguay). Syn: *Joergenia pulchra* Esben-Petersen, 1933.

Key to the South American Species of Adults in the Genus *Purenleon*

Information for the key was provided by Navás (1917a); Banks (1920, 1935, 1941); Stange (2002, 2004); Miller and Stange (2014).

1. The fore and hind wings are subequal in length or the hind wing is slightly longer: both are c. 46 mm long (Fig. 7.53). The costal space is rather narrow, and the basal cross veins are unforked, but some of those just proximal to the pterostigma of the fore-wing are forked. The thorax is mainly black with some brown dorsal spots, including four on the pronotum. The head is dark yellow with pale and darker markings.
 ....*Purenleon nubipennis* (Navás, 1917) (Colombia). Syn: *Formicaleo nubipennis* Navás, 1917a; *Psammoleon nubipennis* (Navás, 1917) Stange, 1970. This species was omitted from the latest revision by Miller and Stange (2014), apparently due to lack of suitable type specimens to examine. Because the initial description is sketchy, identifications are tentative.
   – The hind wing is as long or longer than the fore-wing, which is usually shorter than 35 mm (Fig. 7.54). 2

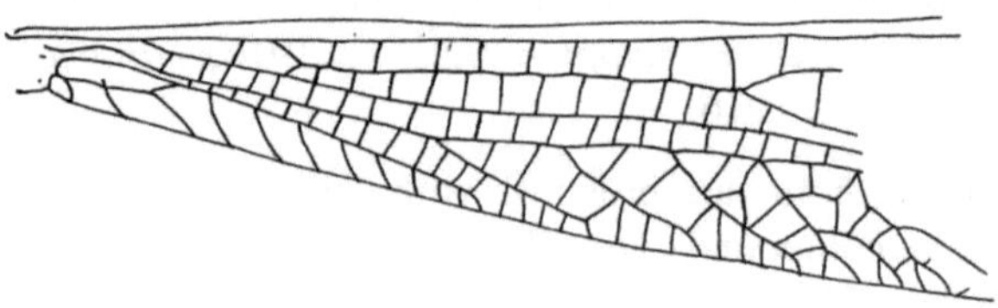

Fig. 7.53 Basal section of the hind wing of *Purenleon nubipennis,* showing the veins in the posterior part of the wing. Based on Navás (1917a).

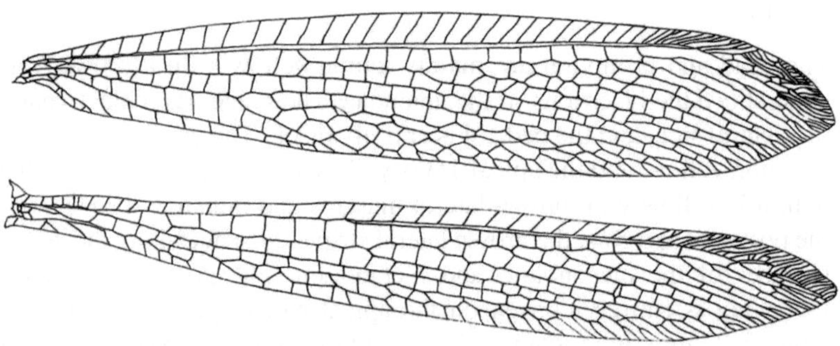

Fig. 7.54 Fore and hind wing of *Purenleon fernandezi*. Based on Miller and Stange (2014).

2. (1) In the fore-wing, vein CuP + 1A runs parallel to the posterior branch of CuA for a long distance beyond the origin of Rs. The veins in the costal space are parallel and not forked or connected by cross veins. The sensory seta on the middle femur is about half as long as the sensory seta on the fore-femur. The hind wing is slightly longer than the fore-wing (Fig. 7.26). Length of fore-wing: 22–33 mm; hind wing: 23–34 mm.
.... *Purenleon parallelus* (Banks, 1935) (Mexico, Central America, Colombia). Syn: *Psammoleon parallelus* Banks, 1935; *Psammoleon posticatus* Banks, 1941. The larvae have been found in sand dunes along the seacoast.

 – In the fore-wing, vein CuP + 1A runs obliquely to the posterior margin of the wing along the posterior branch of CuA at a point near the origin of Rs. The sensory setae on the fore and middle femora are about equal in length (Fig. 7.54). 3

3. (2) The tibia of the middle leg is considerably inflated, making it much wider than the fore-tibia. The length of the first tarsal segment of the hind leg is twice its greatest width. The eighth abdominal sternite of the female is produced laterally to form a broad lobe, which may be elongated or resemble an additional gonapophysis (Fig. 7.54). 4

 – The tibia of the middle leg is slightly wider than the fore-tibia, and the length of the first tarsal segment of the hind tibia is about 2.5 times its greatest width. The eighth abdominal segment of the female is not produced laterally (Fig. 7.55). 5

4. (3) The pronotum bears several long whitish setae along its lateral margin, as long or longer than those of the fore-coxae (Fig. 7.56). Total body length: c. 24 mm. Length of both the fore and hind wings: c. 27 mm.
.... *Purenleon tibialis* Miller and Stange, 2014 (Colombia).

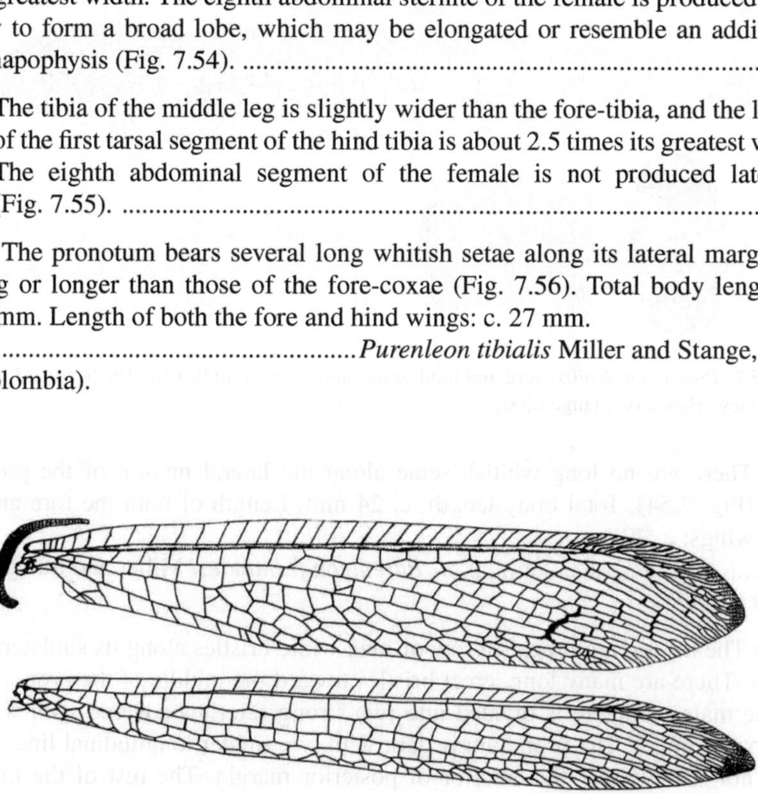

Fig. 7.55 *Purenleon clavatus:* antenna (upper left to left) and the fore (above right) and hind wing (below). Based on Miller and Stange (2014).

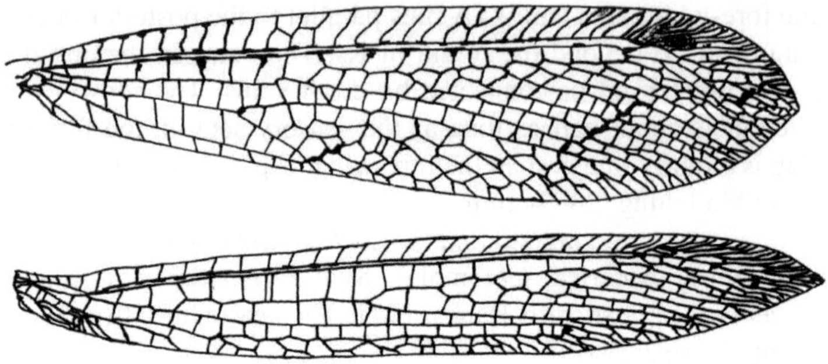

Fig. 7.56 Fore and hind wing of *Purenleon tibialis*. Based on Miller and Stange (2014).

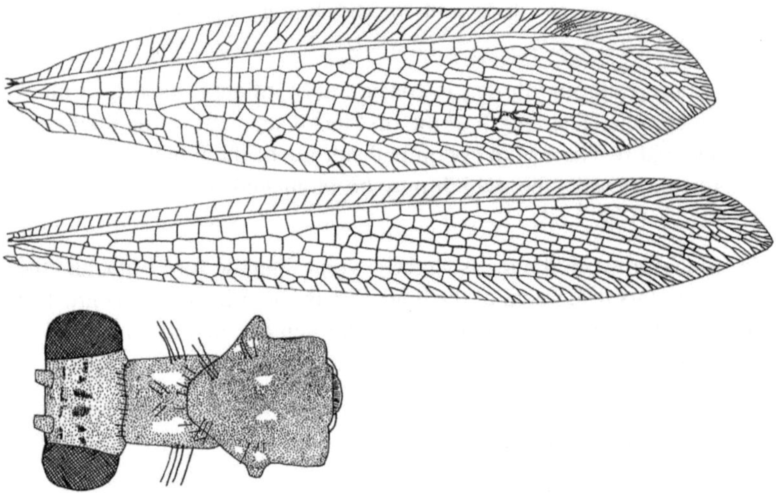

Fig. 7.57 *Purenleon debilis:* fore and hind wing (above and middle) and the head and thorax in dorsal view. Based on Stange (2002).

– There are no long whitish setae along the lateral margin of the pronotum (Fig. 7.54). Total body length: c. 24 mm. Length of both the fore and hind wings: c. 29 mm.
.... *Purenleon fernandezi* Miller and Stange, 2014 (Venezuela).

5. (3) The mesoscutum bears a row of long white bristles along its sublateral margin. There are many long, erect bristles toward the middle of the pronotal disc. The male paramere is divided into two strong sclerites covered with wart-like protuberances. The pronotum is dark with two lighter longitudinal lines, which to not reach either the anterior of posterior margin. The rest of the thorax is mainly dark with about six small, lighter spots on the mesonotum (Fig. 7.57). Total length: 20–28 mm. Length of fore-wing: 26–31 mm; hind wing: 27–32 mm. .... *Purenleon debilis* (Gerstaecker, 1894) (Central America, Colombia). Syn: *Formicaleo debilis* Gerstaecker, 1894; *Psammoleon debilis* (Gerstaecker, 1894) Banks, 1943.

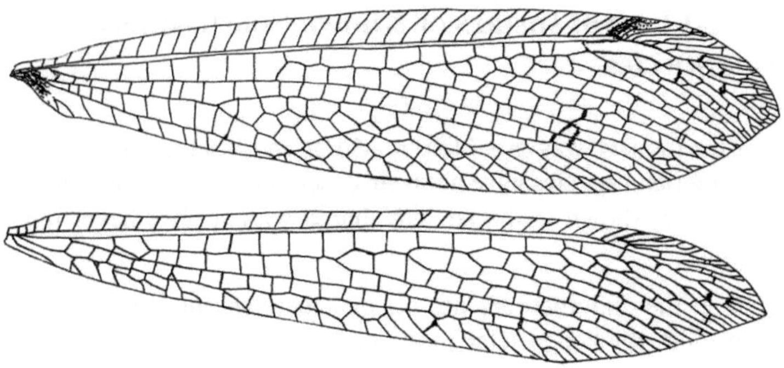

Fig. 7.58 Fore and hind wing of *Purenleon andinus*. Based on Miller and Stange (2014).

– The mesoscutum lacks white bristles along the sublateral margin of the mesoscutum. The pronotal disc lacks long, erect bristles (Fig. 7.55). The male paramere lacks warty protuberances 6

6. (5) There are several white bristles along the lateral margin of the pronotum, at least as long as those on the fore-coxae (Fig. 7.58). The abdominal sternites of the male lack scale-like sculpturing. The sensory hair-like setae on the fore and middle femora are subequal in length. Total length: c. 22 mm. Length of forewing: c. 21 mm; hind wing: c. 22 mm.
.... *Purenleon andinus* Miller and Stange, 2014 (Venezuela).

– The lateral margin of the pronotum lacks long white bristles, although there may be a white, hair-like seta on the pronotum. The fore-coxa bears hair-like sensory setae longer than the diameter of the coxa. There are no long setae on the exterior surface of the fore-femur. The spurs on the hind tibia extend beyond the end of the second tarsal segment, and there are elongate black bristles longer than the spurs. Total length: 18–21 mm. Length of fore and hind wing: 20–23 mm, usually with hind wing slightly shorter. The thorax is dark with five dark brown longitudinal lines (Fig. 7.55). The rest of the thorax is mainly dark.
.... *Purenleon clavatus* (Navás, 1914) (Trinidad, Venezuela, Ceara, Goias, Mato Grosso, Roraima). Syn: *Diazus clavatus* Navás, 1914; *Psammoleon parvulus* Banks, 1920; *Formicaleo serrei* Navás, 1920; *Formicaleo Chaperi* Navás, 1922; *Psammoleon clavatus* (Navás, 1914) Stange, 1970; *Psammoleon serrei* (Navás, 1920) Stange, 1970.

Key to Third Larval Instars of Known South American *Purenleon* Species

Information for the key was provided by Miller and Stange (2014). The larvae of most South American species have not been described, making identifications tentative. All characters shown in the key are shared with species never reported from South America.

1. Spiracles on the abdomen are not set on tubercles but rather on flat surfaces. There are short, thickened setae in the middle of the ventral surface of the head capsule (Fig. 7.59). The larvae have been found in sand dunes along the seacoast.
 .... *Purenleon parallelus* (Banks, 1935) (Mexico, Central America, Colombia). Syn: *Psammoleon parallelus* Banks, 1935; *Psammoleon posticatus* Banks, 1941.

 – The spiracles are located on tubercles bearing many short dolichasters (Fig. 7.30). 2

2. The length and width of the mesothoracic tubercles are subequal, and the tubercles are smaller than the largest scolus on the mesothorax (Fig. 7.30). The dorsal surface of the abdomen bears setae resembling sausages, which are approximately two to three times longer than wide.
 .... *Purenleon debilis* (Gerstaecker, 1894) (Central America, Colombia). Syn: *Formicaleo debilis* Gerstaecker, 1894; *Psammoleon debilis* (Gerstaecker, 1894) Banks, 1943.

 – The mesothoracic tubercles bearing spiracles are no less than twice as long as wide. The setae on the dorsal surface of the abdomen are spherical (Fig. 7.60).
 .... *Purenleon clavatus* (Navás, 1914) (Trinidad, Venezuela, Ceara, Goias, Mato Grosso, Roraima). Syn: *Diazus clavatus* Navás, 1914; *Psammoleon parvulus* Banks, 1920; *Formicaleo serrei* Navás, 1920; *Formicaleo Chaperi* Navás, 1922; *Psammoleon clavatus* (Navás, 1914) Stange, 1970; *Psammoleon serrei* (Navás, 1920) Stange, 1970.

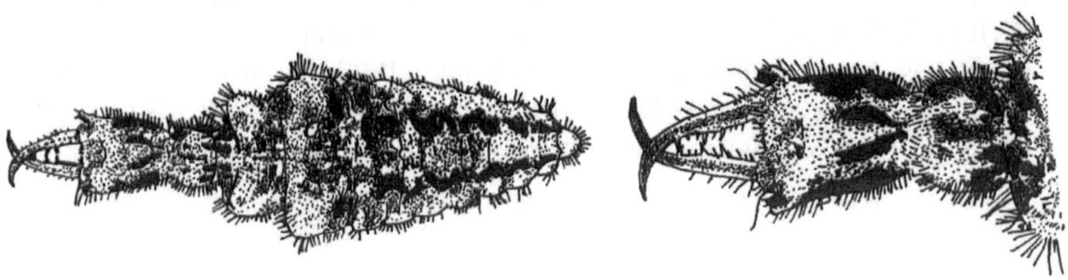

Fig. 7.59 *Purenleon parallelus* larva in dorsal view (left); dorsal (upper right) and ventral view of the head (lower right). Based on Miller and Stange (2014).

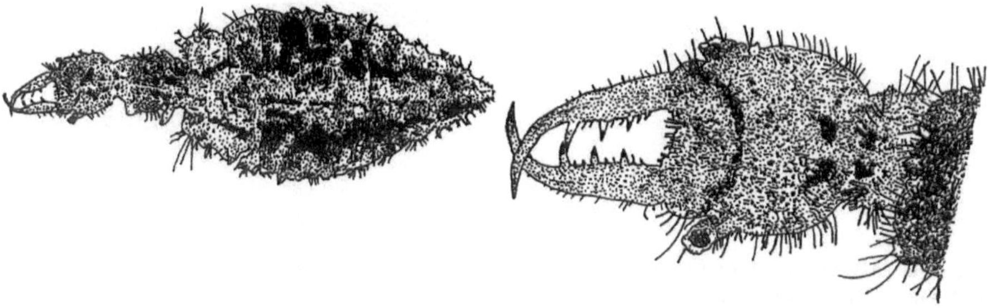

Fig. 7.60 *Purenleon clavatus* larva in dorsal view (left); dorsal (upper right) and ventral view of the head (lower right). Based on Miller and Stange (2014).

Key to the South American Species of Adults in the Genus *Myrmeleon*

Information for the key was provided by Walker (1853), Banks (1910); Navás (1913, 1914a, 1933a), and Stange (2002, 2010). Most species originally assigned to this genus have been transferred to other genera. Those remaining are not well defined because only old, sketchy descriptions are available.

1. The total body length ranges from 23 mm to 27 mm. The face has a wide, black band from just ventral to the antenna to a point ventral to the tentorial pits, or it is yellow adjacent to the mouth, or the entire face is dark (Fig. 7.18). 2
   – The total length exceeds 27 mm, or the face is pale and lacks a wide black band. The sclerotized plate on the ninth tergite of the male is not black and bilobed (Fig. 7.61). 6
2. (1) The color is predominantly ferrugineous. 3
   – The coloration is dark or light but not predominantly ferrugineous (Fig. 7.18). 5
3. (2) The color of the head, thorax, and first abdominal tergite is ferrugineous, lighter on the head and darker on the mesonotum and metanotum. The labrum and palps are yellow. The wings are hyaline with light ferrugineous veins and a whitish pterostigma. The pronotum has a distinct pattern, wth a broad, dark midline stript, which narrows considerably toward the anterior margin, and narrow laterodorsal stripes which bend toward the midline at their anterior and posterior ends. Total length: c. 24 mm. Length of fore-wing: c. 27 mm; hind wing: c. 24.5 mm.
   .... *Myrmeleon homsi* Navás, 1913 (Argentina).
   – The head is predominantly dark brown or black. Most of the surfaces of the thorax and abdomen are ferrugineous 4

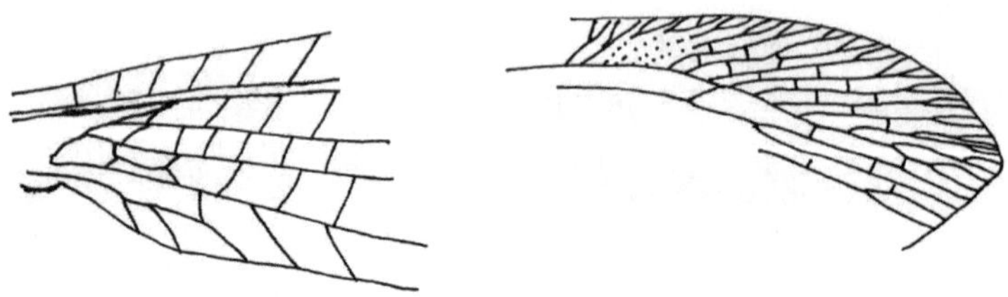

Fig. 7.61 *Myrmeleon brasiliensis:* base (left) and apex (right) of the fore-wing of a male. Based on Navás (1914a), who called the species *Moreyus brasiliensis*.

4. (3) The thorax is ferrugineous with short black setae; the abdomen is cylindrical and ferrugineous with short dark setae. The head is black; the clypeus, labrum, and palps testaceous, and compound eyes and antennae dark. Total length: c. 29.5 mm. Fore-wing length: c. 28.5 mm; hind wing: c. 28.5 mm. ....*Myrmeleon fischeri* (Navás, 1933) (Brazil). Syn: *Neohornius fischeri* Navás, 1933.
   – The dorsal and ventral surfaces of the thorax and the abdominal tergites are ferrugineous with darker and lighter markings. The head is dark with a dark labrum and pale testaceous palps. The abdomen is dark ferrugineous with a paler ferrugineous midline stripe. Total length: c. 24 mm. Length of fore-wing: c. 26 mm; hind wing: c. 24 mm.

   ....*Myrmeleon ambiens* Navás, 1913 (Brazil). Syn: *Myrmeleon tenuis* Perty, in manuscript. Illustrations of the two species in this couplet were not provided, and their descriptions are insufficient to distinguish or identify them with certainty. Specimens displaying their features and coloration should be preserved to provide new, detailed discriptions and perhaps reassignment to another genus.

5. (2) The face has a wide, black band from just ventral to the antenna to a point ventral to the tentorial pits. The male has a black, sclerotized plate on the ventral side of the ninth tergite, which forms two lobes. The cross veins in the costal space of the fore-wing are not forked or branched until just proximal to the pterostigma, which is inconspicuous (Fig. 7.18). ....*Myrmeleon timidus* Gerstaecker, 1888 (Mexico, Costa Rica, Panama, Peru, Galapagos Islands and probably continental Ecuador, Chile). Syn: *Grocus genini* Navás, 1925; *Grocus camposi* Navás, 1935; *Morter aricensis* Navás, 1933; *Morter peruanus* Navás, 1933; *Myrmeleon perpilosus* Banks, 1924; *Myrmeleon genini* (Navás, 1925); *Myrmeleon camposi* (Navás, 1935).
   – The general coloration is blackish; there is a yellow mark at the base of each antenna and yellow on the face adjacent to the mouth. A few cross veins in the costal space proximal to the pterostigma are crossed by transverse veinlets or

forked. The pterostigma is white and very conspicuous. The wings are iridescent; their veins are black and bear small setae.
....*Myrmeleon dolosus* Walker, 1853 (Venezuela, Guyana, Surinam, Pará). Syn: *Hagenomyia dolosus* (Walker, 1853) Banks, 1943.

6. (1) The wing span is greater than 60 mm. The wings are hyaline, and the pterostigma is whitish. The cross veins in the costal area of the fore-wing almost as far as the pterostigma are not forked or branched, and only those near the pterostigma may be forked (Fig. 7.61). The wing veins are almost entirely black with few pale sections. The posterior margin of the fore-wing is slightly concave. The palps are yellow with black rings, and the apical segment of the labial palps is greatly inflated and dark. The antenna is dark with yellow rings. Total length: c. 28 mm. Length of fore-wing: c. 32 mm; hind wing: c. 30 mm.
....*Myrmeleon brasiliensis* (Navás, 1914) (Brazil). Syn: *Moreyus brasiliensis* Navás, 1914; *Myrmeleon texanum* (Banks) sensu Esben Petersen, 1920.

 – The wing span is usually less than 55 mm. If not, some C-Sc cross veins are interconnected by cross veins.. 7

7. (6) The costal area of the fore-wing is greatly expanded at the stigma, so that the area is twice as wide as it is at the fork in vein Rs. Several cross veins proximal to the pterostigma are interconnected. The ectoproct of the male is not produced ventrally, and the ventral process is not bifurcated.
....*Myrmeleon obscurus* Navás, 1912 (Argentina). Syn: *Balaga obscurus* Navás, 1912.

 – The costal area of the fore-wing is not greatly expanded at the stigma, so that the area is not more than 1.5 times as wide as it is at the fork in Rs. Cross veins proximal to the stigma are usually not interconnected, and if any are, they do not number more than about two. The ectoproct of the male is produced ventrally, and the ventral process is bifurcated. Wing span: 50–52 mm.
....*Myrmeleon argentinus* Banks, 1910 (Bolivia, Paraguay, Uruguay, Argentina, Rio Grande do Sul). Syn: *Myrmeleon argentina* Banks, 1910 (error in Latin); *Morter argentinus* (Banks, 1910) Navás, 1914.

Literature

Banks N (1906) Three new species of Neuroptera. Psyche 13:98–100

Banks N (1907) New tropical American Neuroptera. Proceedings of the Entomological Society of Washington 9(1–4):30–34

Banks N (1909) New genera and species of tropical Myrmeleonidae. Journal of the New York Entomological Society 17(1):1–4

Banks N (1910) New South American neuropteroid insects. Proceedings of the Entomological Society of Washington 12:146–160

Banks N (1913) The Stanford Expedition to Brazil, 1911. Neuropteroid insects from Brazil. Psyche 20:83–89, +pl. 4

Banks N (1920) New neuropteroid insects. Bulletin of the Museum of Comparative Zoology 64(3):297–362, +pl. 1–7

Banks N (1923 [1922]) South American *Glenurus* and some other Myrmeleonidae. The Canadian Entomologist 54(3):58–60

Banks N (1924) Descriptions of new neuropteroid insects. Bulletin of the Museum of Comparative Zoology 65(12):421–455, +pl. 1–4

Banks N (1935) A few new North American Neuroptera. Psyche 42:53–57

Banks N (1938) Two new genera of Myrmeleonidae. Proceedings of the Entomological Society of Washington 40(5):125–129

Banks N (1941) Some new and interesting Neuroptera in the American Museum of Natural History. American Museum Novitates 1143:1–5

Banks N (1943) Neuroptera of northern South America. Part II Myrmeleonidae. Boletin de Entomologia Venezolana 2:161–173

Blanchard CÉ (1851) Mirmeleonianos. In: Gay C (ed) Historia Fisica y Politica de Chile. Zoologia. – E. Thunot, Paris, pp 119–129

Esben-Petersen P (1920) Revision of some type-specimens of Myrmeleonidae described by Navas and placed in the Vienna Museum. Bulletin & Annales de la Société royale d'entomologique de Belgique 60:190–196

Esben-Petersen P (1927) New and little-known species of Neuroptera in British collections. Part III. Annals and Magazine of Natural History, Series 9, 20(118):343–350+pl. X–XI

Esben-Petersen P (1933) New and little-known Neuroptera. Meddelelser fra Dansk Naturhistorisk Forening København 94:109–123

Gerstaecker A (1888 [1887]) Weitere Beiträger zur Artenkenntniss der Neuroptera Megaloptera. Mittheilungen aus dem Naturwissenschaftlichen Verein für Neu-Vorpommern und Rügen in Greifswald 19:39–130

Gerstaecker A (1894) Über neue und weniger gekannte Neuropteren aus der Familie Megaloptera Burm. Mittheilungen aus dem Naturwissenschaftlichen Verein für Neu-Vorpommern und Rügen in Greifswald 25:93–173

Guilding L (1833) The generic characters of *Formicaleo*; with the description of two new species. Transactions of the Linnaean Society of London 16:47–51

Hagan HA (1861) Synopsis of the Neuroptera of North America, with a list of South American species. Smithsonian Miscellaneous Collections 4(1):i–xx + 1–347

Hagan HA (1866) Hemerobidarum synopsis synonymica. Stettiner Entomologische Zeitung 27:369–462

Miller RB (2008) A new genus and species of Brachynemurini from Venezuela (Neuroptera: Myrmeleontidae). Insecta Mundi 0059:1–5

Miller RB, Stange LA (1989) Revision of the genus *Dimarella* Banks (Neuroptera: Myrmeleontidae). Insecta Mundi 3(1):11–40

Miller RB, Stange LA (2011) Antlions of Hispaniola (Neuroptera: Myrmeleontidae). Insecta Mundi 0182:1–28

Miller RB, Stange LA (2014) A revision of the genus *Purenleon* Stange (Neuroptera: Myrmeleontidae: Nemoleontini). Insecta Mundi 0384:1–67

Navás L (1911) Neurópteros chilenos (Series 2a.). Revista Chilena de Historia Natural 15:121–128

Navás L (1912) Notas sobre Myrmeleontidos. Brotéria Série Zoológica 10(29–75):85–97

Navás L (1913) Mirmeleónidos (Ins. Neur.) de Chile. Revista Chilena de Historia Natural 17:243–244

Navás L (1914a) Neurópteros Sudamericanos. Primera Serie. Brotéria, Série Zoológica 12(1):45–56, 215–234

Navás L (1914b) Neurópteros nuevos o poco conocidos. Tercera (III) Serie. Memorías de la Real Academia de Ciencias y Artes de Barcelona. Tercera Época 11:193–215

Navás L (1915a) Neurópteros Sudamericanos. Segunda Serie. Brotéria, Série Zoológica 13(1):5–14

Navás L (1915b) Neurópteros nuevos o poco conocidos (Quinta Serie). Memorias de la Real Academia de Ciencias y Artes de Barcelona. Tercera Época 11:455–480+1 pl
Navás L (1916) Neurópteros nuevos o poco conocidos (Séptima Serie). Memorias de la Real Academia de Ciencias y Artes de Barcelona. Tercera Época 12(13):219–243
Navás L (1917a) Neue Neuropteren. Dritte Serie. Entomologische Mitteilungen 6(7/9):274–282
Navás L (1917b) Algunos insectos Neurópteros de la Argentina. Serie I. Physis 3(13–15):186–196
Navás L (1918a) Algunos insectos de la República Argentina. Revista de la Real Academia de Ciencias Exactas Fisicas y Naturales de Madrid 16:491–504
Navás L (1918b) Insecta Nova. III Series. Neuroptera. Memorie della'Pontifica Accademia Romana dei Nuovi Lincei, Serie Seconda 4:1–11
Navás L (1918c [1917]) Insecta Nova. IV Series. Neuroptera. Memorie della'Pontifica Accademia Romana dei Nuovi Lincei, Serie Seconda 4:4–23, Fig. 10–19
Navás L (1919 [1918]) Algunos insectos Neurópteros de la República Argentina. Serie Tercera (III). Revista de la Real Academia de Ciencias Exactas Fisicas y Naturales de Madrid 17:287–305
Navás L (1920) Algunos insectos del Brasil 3ª Serie. Revista do Museu Paulista 12:413–472
Navás L (1922 [1921]) Insectos Sudamericanos Cuarta Serie (1). Revista de la Real Academia de Ciencias Exactas Fisicas y Naturales de Madrid 19:255–267
Navás L (1923) Algunos insectos del Brasil. Revista do Museu Paulista 13:767–774
Navás L (1924) Insectos de la América Central. Brotéria Série Zoológica 21:55–86
Navás L (1926a) Insecta Nova. XI Series. Neuroptera. Memorie della'Pontifica Accademia Romana dei Nuovi Lincei, Roma (2)9:101–110
Navás L (1926b) Algunos insectos del Brasil. (Serie 3). Brotéria (Série Zoológica) 23:5–15
Navás L (1927) Vienticinco formas neuvas de insectos. Boletin de la Sociedad Ibérica de Ciencias Naturales 26:48–75
Navás L (1928a) (1927): Insectos neotropicos, 3ª Serie. Revista Chilena de Historia Natural 31:316–328
Navás L (1928b) Insectos de la Argentina. Cuarta Serie. Estudios, Buenos Aires 35:139–146
Navás L (1930a) Insectos neotropicos, (Sexta Serie). Revista Chilena de Historia Natural 34:62–75, 299–307
Navás L (1930b [1929]) Algunos insectos de Chile. 4.ª Serie. Revista Chilena de Historia Natural 34:350–366
Navás L (1932a) Insectos Suramericanos. Quinta Serie. Revista de la Real Academia de Ciencias Exactas, Fisicas y Naturals de Madrid 29(14):53–66
Navás L (1932b) Alcuni insetti del Museo de Zoologia della R. Università de Torino. – Bollettino dei Musei de Zoologia e di Anatomia Comparata della R. Universita de Torino Serie 3, 42(26):1–38
Navás L (1933a) Insectos Suramericanos, Septima Serie. Revista de la Academia de Cíencias Exactas Físicas y Naturales de Madrid 31:303–314
Navás L (1933b [1932]) Insectos de la Argentina. Revista de la Academia de ciencias exactas, fisico-quimicas y naturals de Zaragosa 16:87–120
Navas L (1934a) Algunos insectos de Chile 5.a Serie. Revista Chilena de Historia Natural 38:9–12
Navás L (1934b) Insectos Suramericanos, Octava Serie. Revista de la Academia de Cíencias Exactas Físicas y Químicas y Naturales de Madrid 31:9–28
Navás L (1935 [1934]) Insectos Suramericanos. Décima Serie. Revista de la Academia de Cíencias Exactas Físicas y Químicas y Naturales de Madrid 32:360–375
Navás L (1936) Insectos neurópteros de Chile poco conocidos. Revista Chilena de Historia Natural 40:179–181
Penny ND (1978 [1977]) Lista de Megaloptera, Neuroptera e Raphidioptera do Mexico, America Central, ilhas Caraibas e America do Sul. Acta Amazonica Supplement 7(4):1–61
Perty M (1833) Delectus animalium articulatorum. Monachii, Arthor. pp. i-iii, 1–222
Porter CE (1916) Nueva localidad de *Glyptobasis porteri,* Brèthes. Revista Chilena de Historia Natural 20(4):78

Rambur JP (1842) Histoire naturelle des insectes: Névroptères. Librairie encyclopédique de Roret. Fain et Thunot, Paris. 534 pp + 12 pl

Stange LA (1969) Myrmeleontidae of the Galapagos Islands. Acta Zoológica Lilloana 25:187–198

Stange LA (1970) A generic revision and catalog of the Western Hemisphere Glenurini with the description of a new genus and species from Brazil. Los Angeles County Museum Contributions in Science 186:1–28

Stange LA (1984) Field studies of the insect order Neuroptera in South America. National Geographic Research Reports 17:51–76

Stange LA (1989) Review of the New World Dimarini with the description of a new genus from Peru (Neuroptera: Myrmeleontidae). Florida Entomologist 72(3):450–461

Stange LA (1994) Reclassification of the New World antlion genera formerly included in the tribe Brachynemurini (Neuroptera: Myrmeleontidae). Insecta Mundi 8:67–119

Stange LA (2000) Observations on the biology of the antlion genus *Glenurus* Hagen (Neuroptera: Myrmeleontidae). Insecta Mundi 14(4):228

Stange LA (2002) Family Myrmeleontidae. In: Penny ND (ed) A guide to the lacewings (Neuroptera) of Costa Rica. Proceedings of the California Academy of Sciences 52(12):275–289

Stange LA (2004) A systematic catalog, bibliography and classification of the world antlions. Memoirs of the American Entomological Institute 74:1–565

Stange LA (2010) Preliminary report on the Myrmeleontidae (Neuroptera) of Paraguay. Insecta Mundi 0114:1–14

Stange LA, Miller RB (1985) A generic review of the acanthaclisine antlions based on larvae (Neuroptera: Myrmeleontidae). Insecta Mundi 1(1):29–45

Stange LA, Miller RB (1990) Classification of the Myrmeleontidae based on larvae. In: Mansell MW, Aspöck H (eds) Advances in neuropterology. Proceedings of the third international symposium on neuropterology (Berg en Dal, Kruger National Park, RSA). Department of Agricultural Development, Pretoria. pp. 151–169

Thouvenot M (2008) Névroptères de Guyane (première note). L'Entomologiste 64(3):177–178

Walker F (1853) List of the specimens of neuropterous insects in the collection of the British museum (Sialidae – Nemopterides), 2:193–476

Walker F (1860) Characters of undescribed Neuroptera in the collection of W. W. Saunders, Esq., FRS, &c. Transactions of the Royal Entomological Society of London (N.S.) 5:176–199

Chapter 8
Crocidae

Abstract Illustrated keys are provided for identification of the genera of adult crocids and known larvae native to South America. Illustrated keys are also provided to identify the species of each genus known to occur in South America. The countries or states in which specimens of each species have been found and invalid synonyms are reported.

Keywords *Pastranaia* • *Moranida* • *Veurise* • *Amerocroce* • Key to species

8.1 Key to the Genera of Adults in the Family Crocidae in South America

Information for the key was provided by Navás (1910); Orfila (1955). Miller and Stange (1989); Monserrat (2008). Until recently, the species in the family Crocidae were included in Nemopteridae.

1. Vein 1A in the fore-wing reaches the wing's hind margin. The cubital vein forks near its base. The basal segment of the tarsus in the fore-leg is shorter than three times its width (Fig. 2.14).
   ....*Pastranaia* Orfila, 1955.
   The only known species in this family is *Pastranaia riojana* Orfila, 1955, found in Argentina.
   – Vein 1A in the fore-wing fuses with vein Cu near the base. The cubital vein forks distal to the origin of the radial sector (Fig. 8.1). The basal segment of the tarsus in the fore-leg is longer than six times its width. 2
2. Vein Rs in the fore-wing thickens progressively toward its fusion with the radial vein, and the cross veins joining the thickened part are sometimes thickened at their ends. The length of the rostrum is shorter than three times the interocular distance measured across the vertex. The setae near the middle of the hind wing of the male are much longer than those near the base and the apex (Fig. 8.1).
   ....*Moranida* Mansell, 1983..p. 289

C.W. Heckman, *Neuroptera (Including Megaloptera)*, Encyclopedia of South American Aquatic Insects, DOI 10.1007/978-3-319-35125-4_8

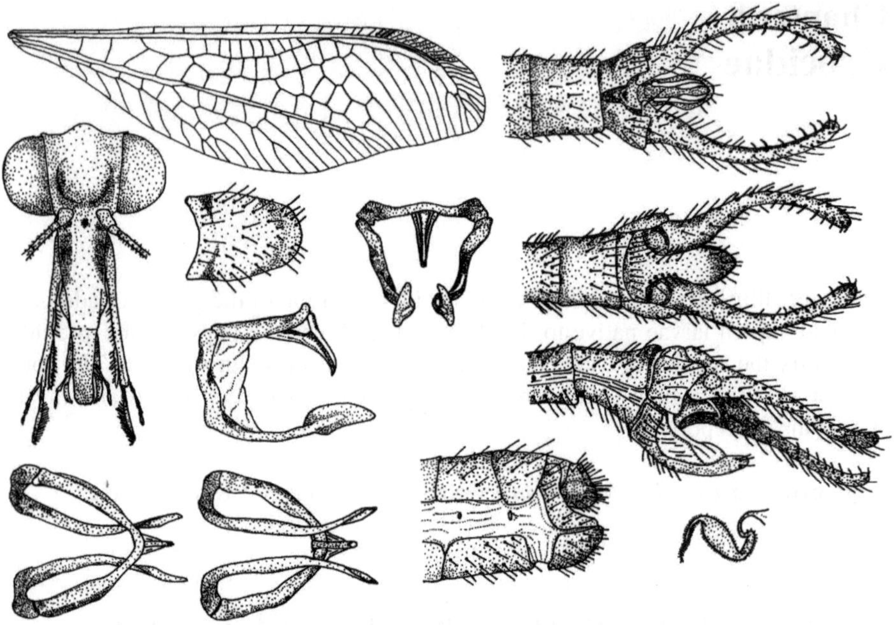

Fig. 8.1 *Moranida peruviensis* (upper row, left and right): fore-wing and apex of the male abdomen in dorsal view; (middle row, left to right): head in anterior view, pronotum in dorsal view (above), male gonarcus and parameres in lateral view (below); male gonarcus and parameres in posterior view, and apex of the male abdomen in ventral (above) and lateral view (below); (lower row, left to right): male gonarcus and parameres in dorsal and ventral view, apex of female abdomen in lateral view, and the female spermatheca. Based on Mansell (1983).

– Vein Rs in the fore-wing is not thickened and not joined by cross veins (Fig. 8.2). The rostrum is longer than three times the ocular distance measured across the vertex. The setae near the middle of the hind wing of the male are not much different than the setae on other parts of the wing. 3

3. The maxillary palp consists of five segments, and the seventh and eighth sternites of the female are fused. If this is not the case, then vein Rs in the fore-wing remains well separated from the radial vein, and vein CuA forks at about the same distance from the base as the origin of Rs. The length of the rostrum is about three times as long as the interocular distance (Fig. 8.3).
....*Veurise* Navás, 1927..p. 291

– The maxillary palp consists of four segments, and the seventh and eighth sternites of the female are separate. Vein Rs in the fore-wing curves anteriad to closely approach the radial vein, coming closer by a distance less than the width of the subcostal space. Vein CuA in the fore-wing forks well distal to the origin of Rs. The length of the rostrum is about four times as long as the interocular distance (Figs. 1.15 and 8.2).

....*Amerocroce* Mansell, 1983

The only known species in this genus is *Amerocroce boliviana* Mansell, 1983, known only from Bolivia.

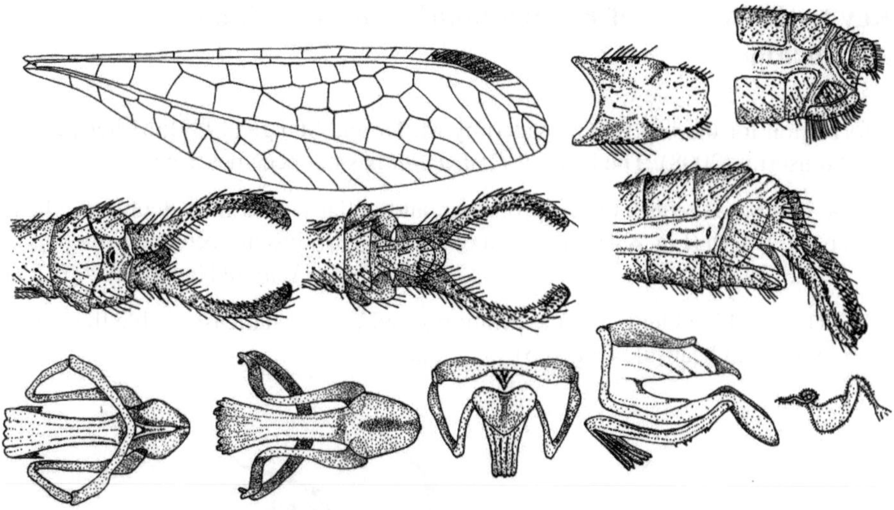

Fig. 8.2 *Amerocroce boliviana* (upper row, left to right): fore-wing, pronotum in dorsal view, apex of the female abdomen in lateral view; (middle row, left to right): apex of the male abdomen in dorsal, ventral, and lateral view; (lower row, left to right); internal male genitalia in dorsal, ventral, posterior, and lateral view, and the female spermatheca in lateral view. Based on Mansell (1983).

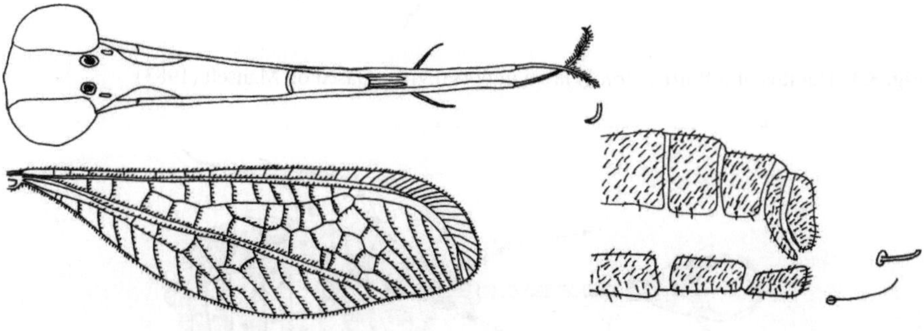

Fig. 8.3 *Veurise fritzi:* head in anterior view with the apical segment of the lacinia detached and enlarged (upper left), fore-wing (lower left), and apex of the female abdomen in lateral view with two different kinds of setae at the apex of the ninth sternite shown enlarged to the right (lower right). Based on Stange and Williner (1981).

Key to the Genera of Known South American Larvae of Crocidae

Information for the key was provided by Miller and Stange (1989); Miller (2008), and Monserrat (2008). The larva of *Pastranaia* has not been described.

1. The combined length of the prothorax and collar is less than double the head capsule length (Fig. 8.4). The larva of *Veurise fritzi* has not been described. .... *Veurise* Navás, 1927..p. 291

  – The combined length of the prothorax and collar is at least double the length of the head capsule (Fig. 8.5). 2

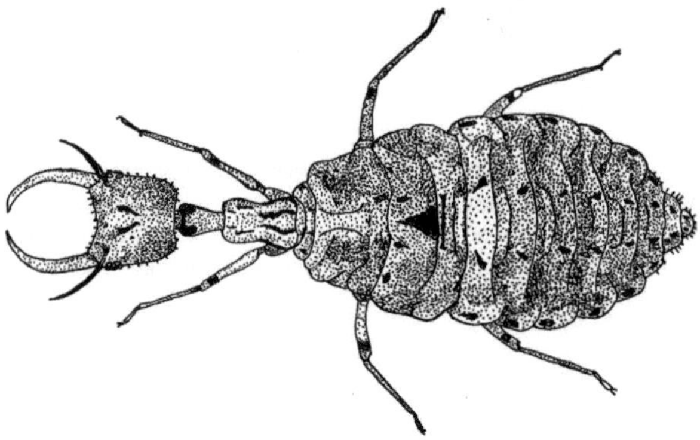

Fig. 8.4 Habitus of a *Veurise bruchi* larva in dorsal view. Based on Mansell (1983).

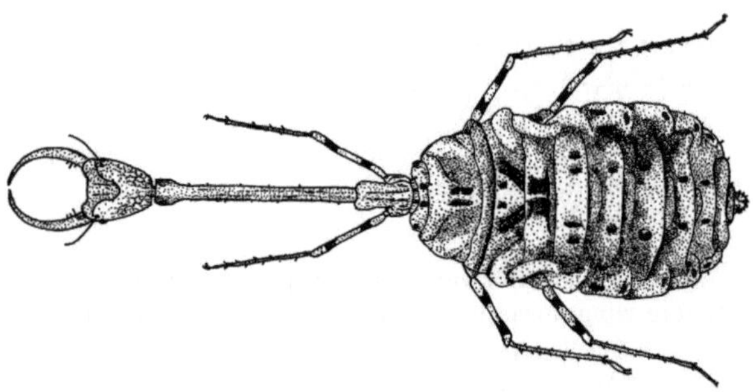

Fig. 8.5 Habitus of an *Amerocroce boliviana* larva in dorsal view. Based on Mansell (1983).

2. The head appears trapezoidal; it is as wide or wider than long, and it bears no noticeable papillae. There are at least a few dolichasters on the proximal part of the mesal margin of each mandible. The prothorax is about 3.8 times the length of the head (Fig. 8.5).
   ....*Amerocroce* Mansell, 1983
   The only known species in this genus is *Amerocroce boliviana* Mansell, 1983, known only from Bolivia.

   – The head is roughly triangular and longer than wide, or, if it is not, it is covered by many papillae, each bearing a minute dolichaster, and there are at least a few dolichasters on the proximal part of the mesal margin of each mandible. The prothorax is about 3.2 times the length of the head (Fig. 2.30).

   ....*Moranida* Mansell, 1983..p. 290

8.2 Key to the South American Species of Adults in the Genus *Moranida*

Information for the key was provided by Miller and Stange (1989).

1. The length of the rostrum equals 2.5 times the interocular distance measured across the vertex. The basal tarsal segment of the fore-leg is longer than the next four tarsal segments combined. There are no dark margins on the presectoral crossveins (Fig. 8.1). The abdomen of the male is banded and is more than half as long as the fore-wing.
   ....*Moranida peruviensis* Mansell, 1983
   (Peru).

   – The length of the rostrum equals 1.8 times the interocular distance measured across the vertex. The basal tarsal segment of the fore-leg is shorter than the next four tarsal segments combined. There are dark margins lining the presectoral crossveins (Fig. 8.6). The abdomen of the male is not banded and is about 1/3 as long as the fore-wing.

   ....*Moranida manselli* Miller and Stange, 1989
   (Venezuela).

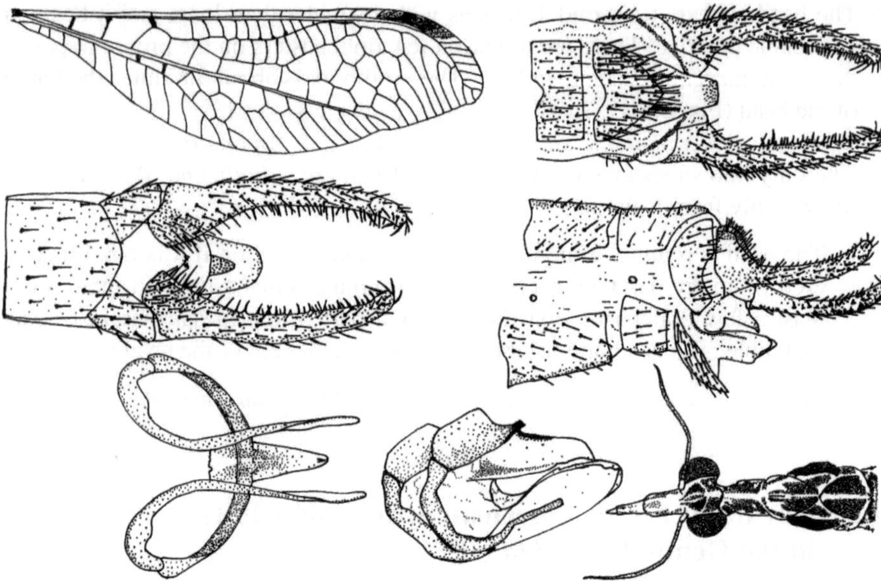

Fig. 8.6 *Moranida manselli* male: fore-wing (upper left); apex of the abdomen in dorsal (upper right), ventral (middle left), and lateral view (middle right); genitalia in ventral (lower left) and lateral view (lower center), and head and thorax in dorsal view (lower right). Based on Miller and Stange (1989).

Key to the South American Species of Larvae in the Genus *Moranida*

Information for the key was provided by Mansell (1983) and Miller and Stange (1989).

1. The head is longer than wide; the mandibles lack dolichasters (Fig. 8.7). ....*Moranida peruviensis* Mansell, 1983 (Peru).

   – The length of the head is equal to or less than the width, and the mandibles bear at least a few dolichasters on the proximal part of the mesal margin. The head capsule is covered with many papillae bearing minute dolichasters, especially abundant at the posterolateral margin. The prothorax is not longer than 2.5 times the length of the head (Fig. 2.30).

   ....*Moranida manselli* Miller and Stange, 1989 (Venezuela).

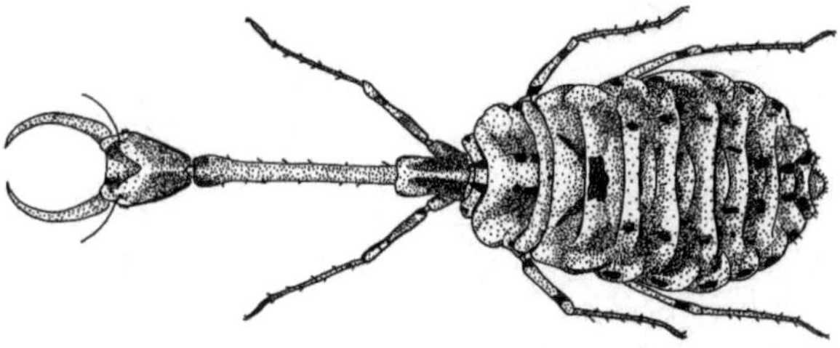

Fig. 8.7 Larva of *Moranida peruviensis* in dorsal view. Based on Mansell (1983).

8.3 Key to the South American Species of Adults in the Genus *Veurise*

Information for the key was provided by Miller and Stange (1989).

1. Vein Rs in the fore-wing curves anteriad to come closer to the radial vein at one point than the width of the subcostal space. Vein CuA in the fore-wing forks much closer to the apex of the wing than the location of the origin of the radial sector. The length of the rostrum is approximately four times the interocular space measured across the vertex (Fig. 8.3).
   .... *Veurise fritzi* Stange and Williner, 1983 (Bolivia).
   – Vein Rs in the fore-wing does not come close at all to the radial vein. Vein CuA in the fore-wing forks near the location of the origin of the radial sector. The length of the rostrum is approximately three times the interocular space measured across the vertex (Fig. 8.8).
   .... *Veurise bruchi* Navás, 1927 (Argentina).

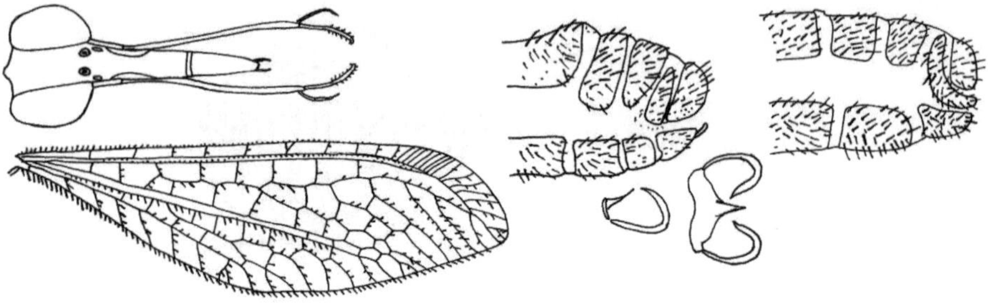

Fig. 8.8 *Veurise bruchi:* head in anterior view (upper left), fore-wing (lower left), abdomen of a male (upper center) and a female in lateral view (upper right), male genitalia in dorsal (lower right) and lateral view (lower right center). Based on Stange and Williner (1981).

Literature

Kirby WF (1900) Notes on the neuropterous family Nemopteridae. Annals and Magazine of Natural History: Zoology, Botany, and Geology, Seventh Series 6(31):456–464

Mansell MW (1983) New Crocinae (Neuroptera: Nemopteridae) from South America, with descriptions of larvae. Journal of the Entomological Society of Southern Africa 44:245–257

Mansell MW (1986) Biogeography and phylogeny of the Crocinae (Neuroptera: Nemopteridae). In: Gepp J, Aspöck H, Hölzel H (eds) Recent research in Neuropterology. Proceedings of the 2nd International Symposium on Neuropterology, Graz. pp 77–85

Miller RB, Stange LA (1989) A new species of *Moranida* Mansell from Venezuela (Neuroptera: Nemopteridae). Insecta Mundi 3(1):65–70

Monserrat VJ (2008) Nuevos datos sobre angunas especies de Nemopteridae y Crocidae (Insecta: Neuroptera). Heteropterus Revista de Entomologia 8(1):1–33

Navás L (1910) Monografia de los Nemoptéridos (Insectos Neurópteros). Memorias de la Real Academia de Ciencias y Artes de Barcelona. Tercera Época 8:341–403

Navás L (1927) Nemoptérido (Ins. Neur.) nuevo en América. Revista de la Sociedad Entomológica Argentina 1(3):3–4 + Table 1

Orfila RN (1955) Un nuevo Nemopteridae (Neuropt.) americano con una sinopsis de la familia. Revista de la Sociedad Entomológica Argentina 17:29–32

Stange LA, Williner GJ (1981) Una nueva especie de *Veurise* Navás de Bolivia con notas sobre Nemopteridae en América del Sur (Insecta: Neuroptera). Physis (Buenos Aires) Secc. C 39(97):35–39

Chapter 9
Nemopteridae

Abstract Both species of nemopterids known to inhabit South America are assigned to the genus *Stenorrhachus*. A simple key is provided to identify the adults of the two species, both of which have been found only in Chile. An invalid synonym is reported.

Keywords *Stenorrhachus* • Threadwing • Key to species • Chile • Brachyptery

The common name for insects in this family is threadwing. The only genus known to inhabit South America is *Stenorrhachus* McLachlan, represented by two known species in Chile. They have been found in South America only rarely. The family is better represented in other parts of the tropics, especially Africa. In older literature, the family Crocidae is usually included as a subfamily of Nemopteridae, making obsolete lists of members of this family in South America somewhat longer.

Key to the Species of Adult *Stenorrhachus* Known to Inhabit South America

Information for the key was provided by McLachlan (1885, 1886); Kirby (1900); Acker (1958); Oswald (1987), and Miller and Stange (2012).

1. The maxillary and labial palps are subequal in length. The basal segment of the labial palp is about 1.5 times as long as its diameter at the base. The setae on the second segment of the palp are much shorter than the segment. The length and maximal diameter of the basal postventral lobe of the male ectoproct is greater than its maximal diameter at the base, and the ninth sternite extends at least as far posteriad and the apex of this lobe. The paramere of the male is about eight times as long as its maximum width. The lateral margin of the gonarcus bends strongly posteriad at the location of a large, dorsolateral process (Fig. 2.15). Total length

C.W. Heckman, *Neuroptera (Including Megaloptera)*, Encyclopedia of South American Aquatic Insects, DOI 10.1007/978-3-319-35125-4_9

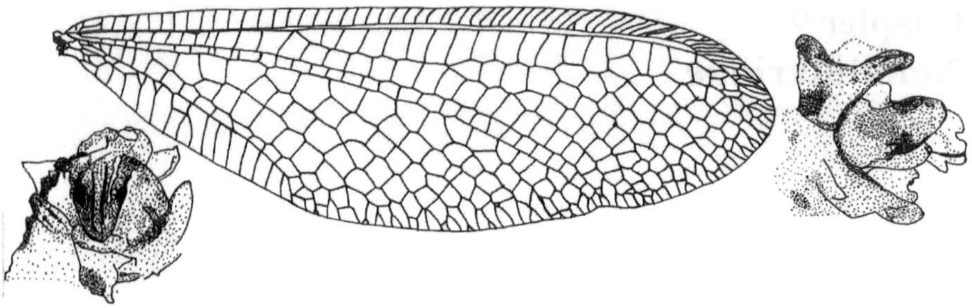

Fig. 9.1 *Stenorrhachus walkeri* male: fore-wing (upper left) and genitalia in dorsal (right) and posterior view (lower left). Based on Miller and Stange (2012).

of male: 7–12 mm. Length of the male fore-wing: 15–24 mm; hind filament: 42–60 mm. The females cannot fly and have only vestiges of wings. .... *Stenorrhachus chilensis* Miller and Stange, 2012 (Chile).

– The maxillary palps are shorter than the labial palps. The basal segment of the labial palp is twice as long as its diameter at the base. The second segment of the palp bears setae longer than the segment. The length of the basal postventral lobe of the male ectoproct is longer than its maximum diameter. The lateral margin of the gonarcus curves evenly and lacks a large, dorsolateral process (Fig. 9.1). Total length of male: 9–13 mm. Length of male fore-wing: 23–27 mm; hind filament: 52–65 mm. The wings of the females are vestigial, reaching a length of roughly 0.5 mm, and the filaments are slightly more than half the length of the male filament.

....*Stenorrhachus walkeri* (McLachlan, 1885) (Chile). Syn: *Stenotaenia walkeri* McLachlan, 1885.

Literature

Acker TS (1958) The comparative morphology of *Stenorrhachus walker* (McLachlan) and of *Nemopterella* sp. (Neuroptera: Nemopteridae). Microentomology 23:106–130 + Figs. 7–22

Kirby WF (1900) Notes on the neuropterous family Nemopteridae. Annals and Magazine of Natural History: Zoology, Botany, and Geology, Seventh Series 6(31):456–464

McLachlan R (1885) On the discovery of a species of the neuropterous family Nemopteridae in South America, with general considerations regarding the family. Transactions of the Royal Entomological Society of London 33:375–379

McLachlan R (1886) Note concerning certain Nemopteridae. Transactions of the Royal Entomological Society of London, Proceedings 34:lvii–lviii

Miller RB, Stange LA (2012) A new species of *Stenorrhachus* McLachlan from Chile (Neuroptera: Nemopteridae) with biological notes. Insecta Mundi 0226:1–8

Monserrat VJ (2008) Nuevos datos sobre angunas especies de Nemopteridae y Crocidae (Insecta: Neuroptera). Heteropterus Revista de Entomologia 8(1):1–33

Oswald JD (1987) *Hypochrysa* Hagen, 1866 (Chrysopidae) and *Stenorrhachus* Mel., 1886 (Nemopteridae) are valid names in the Neuroptera. Neuroptera International 4:225–229

Chapter 10
Mantispidae

Abstract Illustrated keys are provided for identification of the genera of adult mantispids native to South America. Illustrated keys are also provided to identify the species of each genus known to occur in South America. The countries or states in which specimens of each species have been found and invalid synonyms are reported.

Keywords Mantisfly • Key to species • Mantispinae • Symphrasinae • Drepanicinae

The species in this family are called mantisflies because of their resemblance to the unrelated preying mantises, which belong to the Order Orthoptera. The taxonomy of the species requires much revision, and many specimens still cannot be placed in genera because of poor initial descriptions and the lack of undamaged type specimens. Many revisions made in a doctoral dissertation by Hoffman cannot be considered valid because most of the dissertation was never published (Ohl 2007). Species originally placed in several genera, such as *Mantispilla,* are still recorded in lists of valid species, although there are none that were not found to be junior synonyms of species in other genera after their reexamination. Specimens of such species are best left unidentified, but they should be carefully preserved and stored until they can be provided to specialists who will require them to make the necessary taxonomic revisions. No nominal species of *Mantispilla* is presently confirmed for the South American fauna.

10.1 Key to the Subfamilies and Genera of Adult Mantispinae Known from South America

Information for the key was provided by Berg (1899); Navás (1909, 1932, 1934a, b); Banks (1912); Handschin (1960); Meinander (1980); Poivre (1982); Penny (1982, 1983); Penny and da Costa (1983); Hoffman (2002); Ohl (2004, 2007); Reynoso-Velasco and Contreras-Ramos (2008); Machado and Rafael (2007, 2010); Thouvenot (2010), and Ardila-Camacho and García (2015).

C.W. Heckman, *Neuroptera (Including Megaloptera)*, Encyclopedia of South American Aquatic Insects, DOI 10.1007/978-3-319-35125-4_10

1. There is one tarsal claw and no areolus on the fore-leg (Fig. 10.1). .... Subfamily Mantispinae....2

   – There are two tarsal claws on the fore-leg (Fig. 10.2). 9

2. (1) The segments of the antennal flagellum are at least three times as wide as long. At the maculae, the prothorax is two to three times as long as wide. In the hind wing, there is one cross vein between veins CuA and 1A. The anterior half of the fore-wing is amber to dark brown. The mesoscutal furrow is barely evident (Fig. 10.3). ....*Climaciella* Enderlein, 1910..p. 304

   – The width of the antennal flagellum segments at the midlength of the antenna is less than three times the length. At the maculae, the prothorax is more than four times as long as wide. In the hind wing, there are no cross veins between veins CuA and AA. The anterior half of the fore-wing is hyaline. There is a conspicuous mesoscutal furrow (Fig. 10.4). 3

3. (2) The pronotum is covered completely by setae (Fig. 10.4). 4

   – Setae cover only the anterior and posterior surfaces of the pronotum (Fig. 10.1). 6

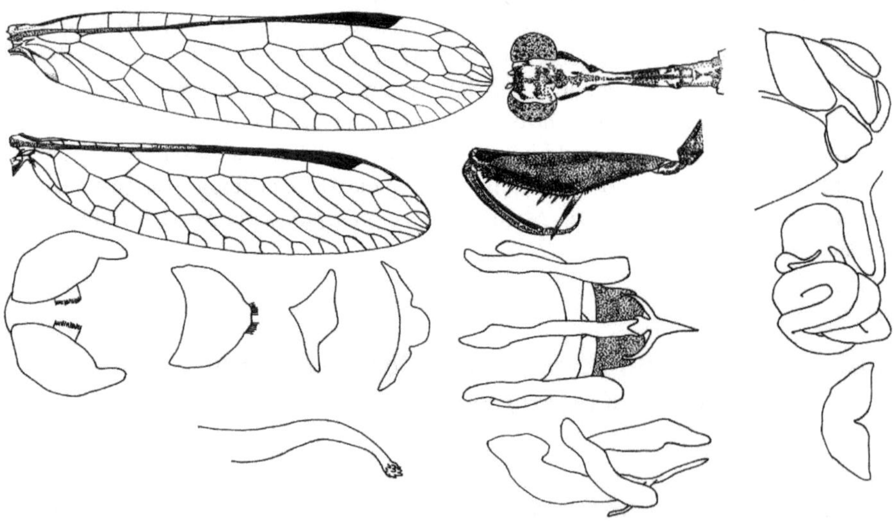

Fig. 10.1 *Buyda phthisica:* fore and hind wing (upper and middle left, respectively), head and pronotum (upper right center), outer surface of the fore-leg (below head and thorax), apex of the female abdomen in lateral view (upper right), and (lower middle row, left to right): male genitalia in dorsal and ventral view, male ectoproct in lateral view, gonarcus in dorsal view, male genitalia in ventral view, female spermatheca, and (lower row, left to right): male hypomere, male genitalia in lateral view, and the eighth sternite of the female in ventral view. Based on Machado and Rafael (2010).

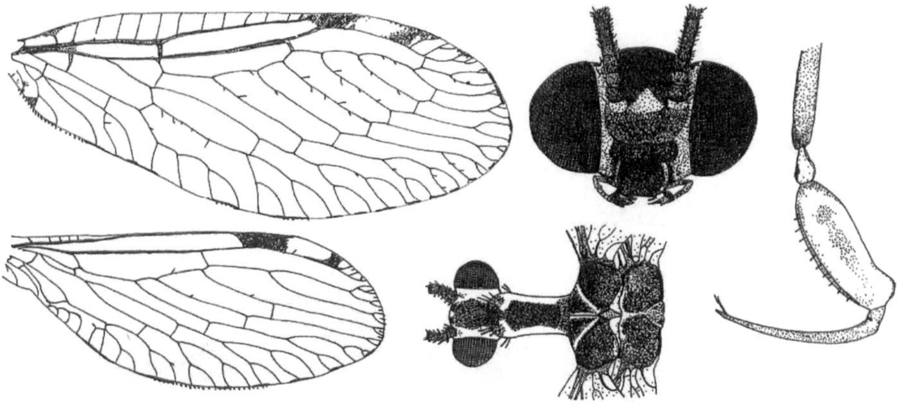

Fig. 10.2 *Anchieta remipes:* fore and hind wing (upper and lower left), head in anterior view (upper center), head and thorax in dorsal view (lower center), and the fore-leg in lateral view (right). Based on Hoffman (2002).

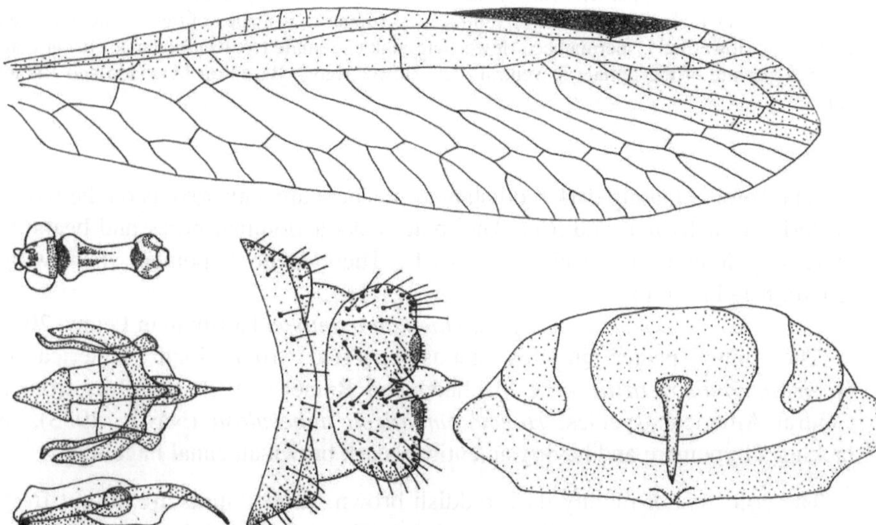

Fig. 10.3 *Climaciella amapaensis* male: fore-wing (above); head and pronotum in dorsal view (upper middle left); genitalia in ventral (lower middle left) and lateral view (lower left); apex of the abdomen in dorsal (lower center) and posterior view (lower right). Based on Penny (1983).

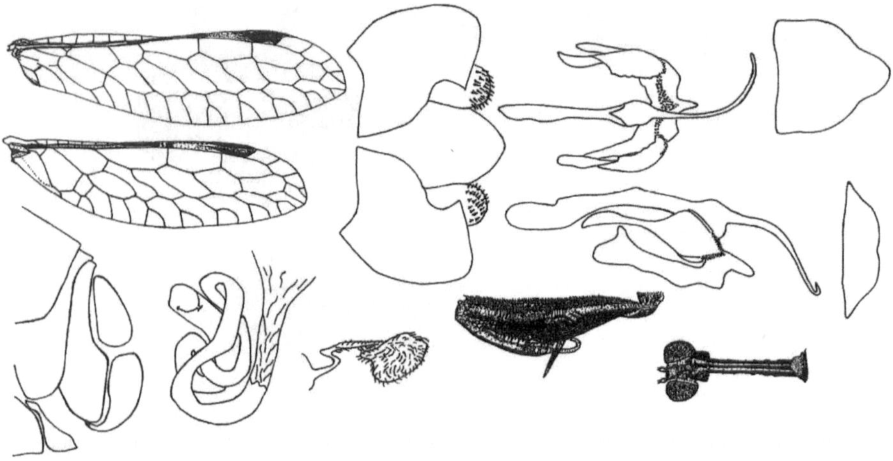

Fig. 10.4 *Haematomantispa amazonica:* fore and hind wing (upper and middle left, respectively), head and pronotum in dorsal view (lower right), fore-leg (lower right center), apex of the male abdomen in dorsal view (upper left center), male genitalia in ventral (upper right center) and lateral view (right center), male ninth sternite in ventral view (upper right), apex of the female abdomen in lateral view (lower left), spermatheca (lower left center), female fertilization canal (lower center), eighth sternite of the female in ventral view (lower right). Based on Machado and Rafael (2010).

4. (3) The color is mainly dark reddish brown. There are amber spots on the wings posterior to cells *imp* and *3 m*. The male lacks abdominal pores and bears an extremely long pseudopenis (Fig. 10.4). There is no hypomere. Fore-wing length: 8.1–11.3 mm.
   ....*Haematomantispa* Hoffman in Penny, 2002
   The only species in this genus reported from South America is *Haematomantispa amazonica* Machado and Rafael (2010). It differs from a Central American species, *Haematomantispa nubeculosa* (Navás, 1933), in lacking a hypomere and having an entirely dark brown antennal flagellum.

   – The color is not mainly dark reddish brown. Amber spots are absent from the wings posterior to cells *imp* and *3 m*. The male has abdominal pores and bears a rather short pseudopenis (Fig. 10.5) 5

5. (4) The body is mainly green. The setae on the pronotum arise from small protuberances (Fig. 10.5). The males have pores in the membranes between the third and fourth and fourth and fifth abdominal tergites.
   .... *Zeugomantispa* Hoffman in Penny, 2002. p. 307

   – The body is not mainly green. The setae on the pronotum arise from a smooth, flat surface (Fig. 10.6). The males have pores in the tergites themselves rather than in the intersegmental membranes.
   .... *Leptomantispa* Hoffman in Penny, 2002..p. 309

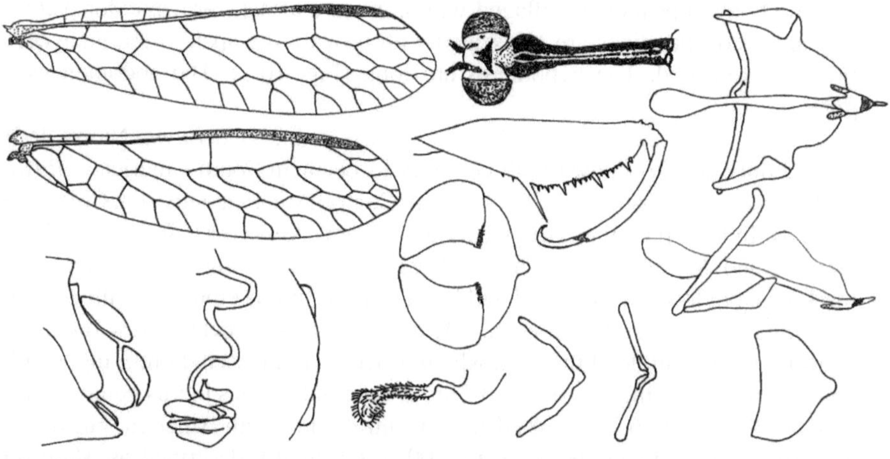

Fig. 10.5 *Zeugomantispa compellens:* fore and hind wing (upper and middle left), head and pronotum in dorsal view (upper right center), fore-leg (right center), male genitalia in ventral (upper right) and lateral view (middle right), apex of the male abdomen in dorsal view (center), and (below, left to right): apex of the female abdomen in lateral view, spermatheca, posterior border of the seventh and the eighth sternite of the female in ventral view, fertilization canal of the female, male gonarcus in dorsal and posterior view, and ninth sternite of a male in ventral view (lower right). Based on Machado and Rafael (2010).

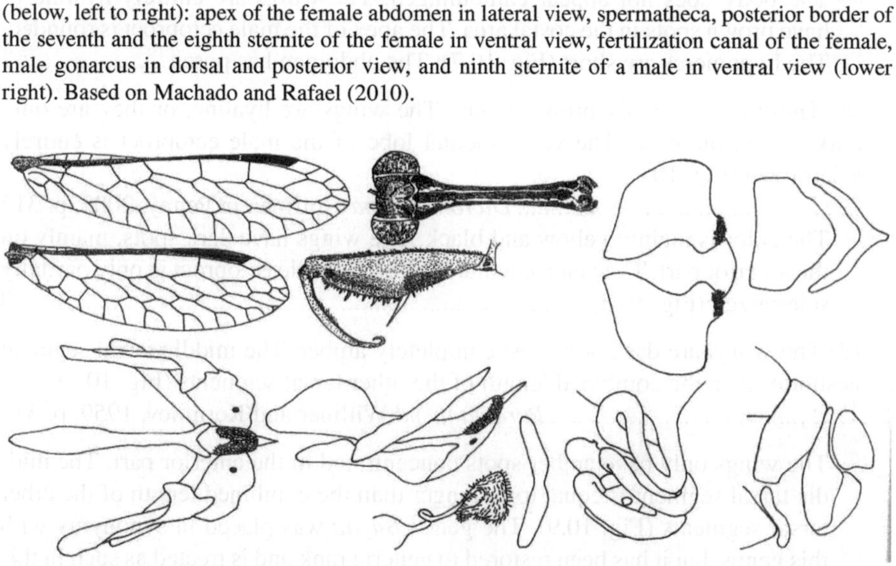

Fig. 10.6 *Leptomantispa nymphe:* fore and hind wing (upper and middle left), head and pronotum in dorsal view (top center), fore-leg in anterior view (above left center), apex of the male abdomen in dorsal view (upper right center), the ninth abdominal sternite of a male in ventral view and the gonarcus in dorsal view (upper right, left and right, respectively), male genitalia in ventral (lower left) and lateral view (below left center), female fertilization canal (lower left center), eighth female abdominal sternite in ventral view (lower center), spermatheca (lower right center), apex of the female abdomen in lateral view (lower right). Based on Machado and Rafael (2010).

6. (3) The body appears camouflaged with a green and dark brown color pattern. The apices of the wings are infuscated with brown. The apex of the male ectoproct is elongated. The hypomeres are long (Fig. 10.1). The abdomen lacks pores.
.... *Buyda* Navás, 1926

   The only species known with certainty to occur in South America is *Buyda phthisica* (Gerstaecker, 1885), known from Central America, Surinam, Colombia, Peru, Bolivia, Argentina, Uruguay, Bahia, Pará, Amazonas, Rondônia, Mato Grosso, and Minas Gerais. The prothorax has contrasting blackish and pale markings, obvious on the dorsal and lateral surfaces; it is elongate, with a length 4½ times its width at the prozona. The head has a prominent yellow rhomboid marking in the middle of the face, which is deeply impressed. Fore-wing length: 16–19 mm; hind wing length: 14–17 mm. A second species in this genus, *Buyda apicata* Navás, 1926, is reported from Central and South America, but it may actually be an African species (Ohl 2004). Originally described as *Mantispa phthisica* Gerstaecker, 1885, some authors have used the name *Entanoneura phthisica* (Gerstaecker, 1885) before *Buyda* was restored to generic rank.

   – The body does not appear camouflaged. The wings are entirely hyaline or have brown spots in the costal area. The apex of the male ectoproct is rounded. The hypomeres are short (Fig. 10.7). The abdomen has pores. 7

7. (6) The body is mainly brown or tan. The wings are hyaline, or they are only darkened at the base. The ventromedial lobe of the male ectoproct is entirely sclerotized (Fig. 10.7).
.... *Dicromantispa* Hoffman in Penny, 2002..p. 312

   – The color is mainly yellow and black. The wings have dark spots, mainly on the anterior part. The ventromedial lobe of the male ectoproct is only partially sclerotized (Fig. 10.8).. 8

8. (7) The wings are dark or almost completely amber. The middle tarsal segment is shorter than the combined length of the other tarsal segments (Fig. 10.8)
.... *Paramantispa* Williner and Kormilev, 1959..p. 316

   – The wings only have amber spots concentrated in the anterior part. The middle tarsal segment is equal to or longer than the combined length of the other tarsal segments (Fig. 10.9). The genus *Buyda* was placed in synonymy with this genus, but it has been restored to generic rank and is treated as such in this key.
.... *Entanoneura* Enderlein, 1910..p. 318

9. There is a dentiform process on the first tarsal segment of the fore-leg (Fig. 10.2). Length: 5–10 mm.
.... Subfamily Symphrasinae....10

   – There is no dentiform process on the first tarsal segment of the fore-leg (Fig. 10.10). Length: 17–25 mm.
.... Subfamily Drepanicinae....12

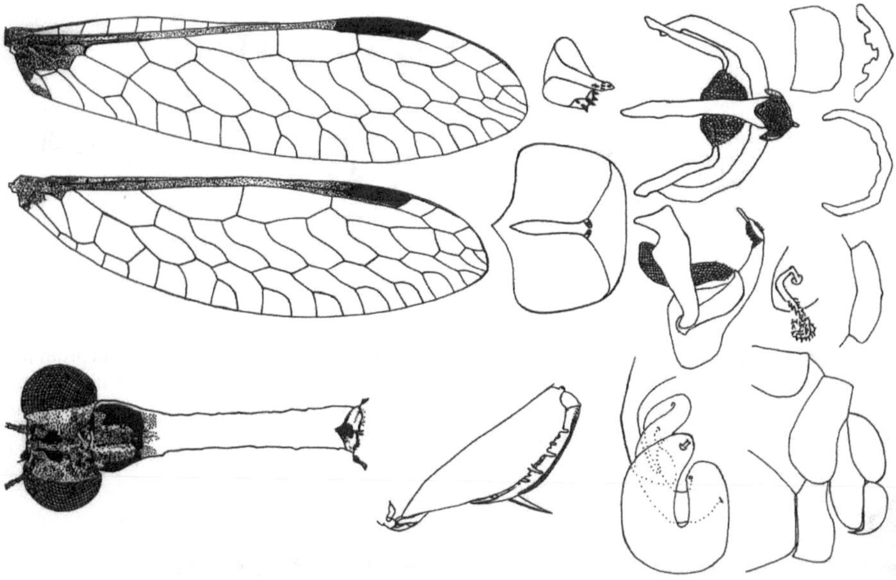

Fig. 10.7 *Dicromantispa leucophaea:* fore and hind wing (upper and middle left), head and prothorax in slightly oblique dorsal view (lower left), posterior view of the fore-leg (lower center), ventromedial lobe of the male in ventral view (upper center), male terminalia in dorsal view (center), male genitalia in ventral (upper right center) and lateral view (right center), male gonarcus in posterior (upper right) and dorsal view (upper middle right), outline of the male ninth sternite in ventral view (at top, between the male genitalia and gonarcus in posterior view), female terminalia in lateral view (lower right), female spermatheca (lower right center), posterior border of the seventh and the eighth sternite of the female in ventral view (middle right), and the fertilization canal of the female (middle row, between the male genitalia in lateral view and the female sternites). Based on Machado and Rafael (2010).

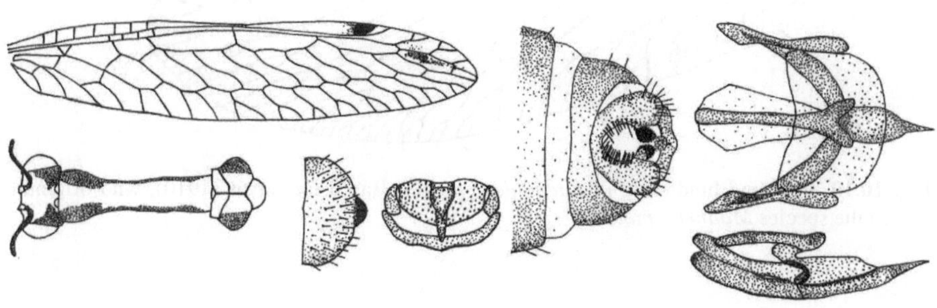

Fig. 10.8 *Paramantispa prolixa* male: fore-wing (upper left); head and pronotum in dorsal view (lower left); apex of the abdomen in dorsal view (right center); genitalia in ventral (upper right), lateral view (lower right), and posterior view (lower center) and the ninth sternite in ventral view (lower left center). Based on Penny and da Costa (1983).

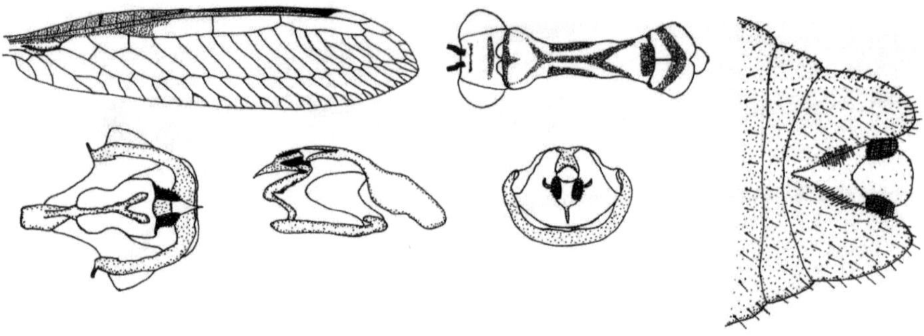

Fig. 10.9 *Entanoneura batesella:* fore-wing (upper left), dorsal surface of the head and prothorax (upper center), apex of the male abdomen in dorsal view (right), and the male genitalia in ventral (lower left), lateral (lower left center), and posterior view (lower right center). Based on Penny and da Costa (1983), who referred to the species by its synonym, *Entanoneura limbata.*

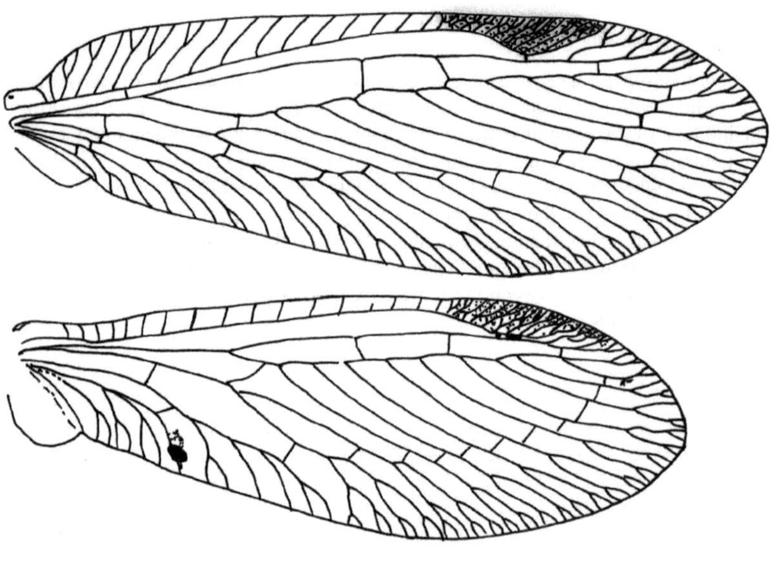

Fig. 10.10 Fore and hind wing of *Drepanicus moulti*. Based on Navás (1910), who originally named the species *Molinella moulti*.

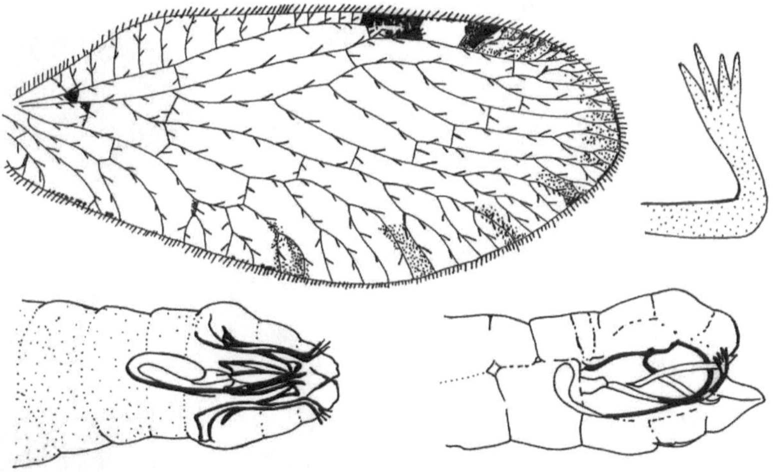

Fig. 10.11 *Trichoscelia anae* male: fore-wing (upper left), paramere (upper right), and apex of the male abdomen in ventral (lower left) and lateral view (lower right). Based on Penny (1983).

10. The femur of the fore-leg lacks a sub-basal spine (Fig. 10.11).
 .... *Trichoscelia* Westwood, 1852..p. 321

 – There is a sub-basal spine on the fore-femur (Fig. 10.2). 11

11. In the fore-wing, there are five radial veins branching from the second radial cell, which has straight sides. The first and second radial cells of the fore-wing frequently have setae on the membrane. The fore-femur may bear a row of more than ten long, dark, spine-like setae (Fig. 10.2).
 .... *Anchieta* Navás, 1909..p. 328

 – In the fore-wing, there are two radial veins branching from the second radial cell, which is slightly curved in the distal part of the wing (Fig. 10.12). The first and second radial cells of the fore-wing lack setae on the membrane. The hind tibia is never widened.
 .... *Plega* Navás, 1928..p. 331

12. The pterostigma of the fore-wing does not reach posteriad as far as vein R_1. There is only one Sc cross vein (Fig. 10.10).
 .... *Drepanicus* Navás, 1909..p. 334

 – The pterostigma of the fore-wing reaches vein R_1. There are two Sc cross veins (Fig. 10.13).
 .... *Gerstaeckerella* Enderlein, 1910..p. 336

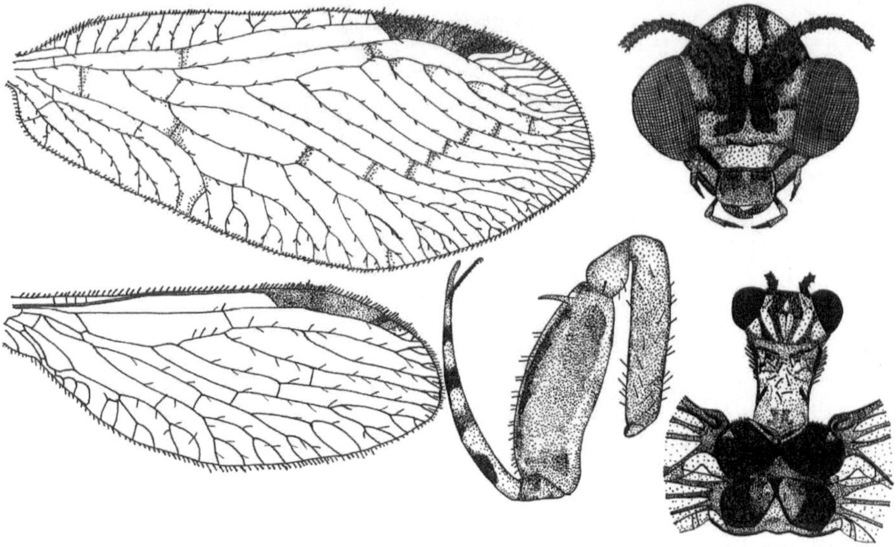

Fig. 10.12 *Plega hagenella:* fore and hind wing (left), head in anterior view without apical antennal segments (upper right), head and thorax in dorsal view showing only the basal segments of the antennae (lower right), and the fore-leg in lateral view (lower center). Based on Penny (2002).

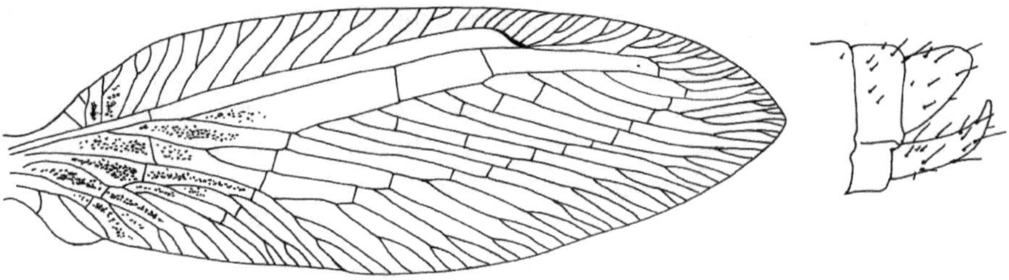

Fig. 10.13 *Gerstaeckerella implexa* male: fore-wing (left) and apex of the abdomen in lateral view (right). Based on Navás (1932).

10.2 Subfamily Mantispinae

Key to the Species of Adult *Climaciella* in South America

The key was prepared with information provided by Handschin (1960); Penny (1983); Penny and da Costa (1983); Hofmann in Penny (2002), and Ardila-Camacho and García (2015).

1. The color pattern includes a mixture of black, yellow, and reddish brown or orange (Fig. 10.14). 2
   – The color is mainly black or reddish brown with blackish and only small yellow markings, if any (Fig. 10.15). 3

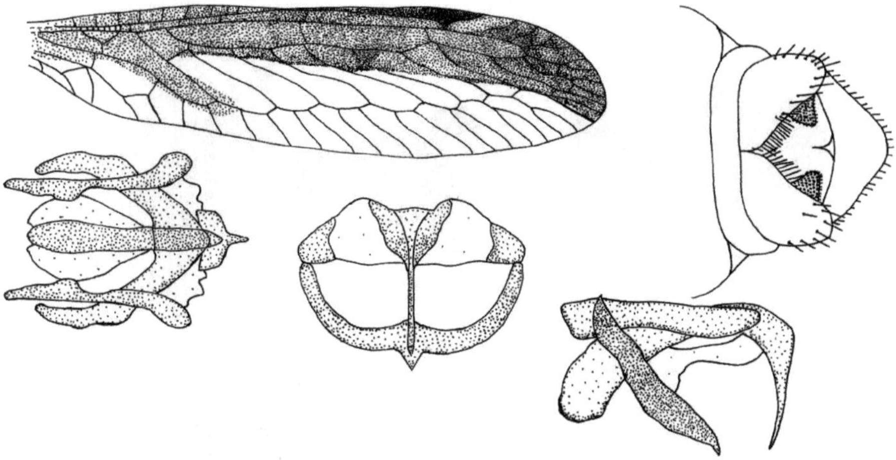

Fig. 10.14 *Climaciella duckei* male: fore-wing (upper left), apex of the abdomen in dorsal view (upper right), and (below, left to right): genitalia in ventral, posterior, and lateral view. Based on Penny (1983).

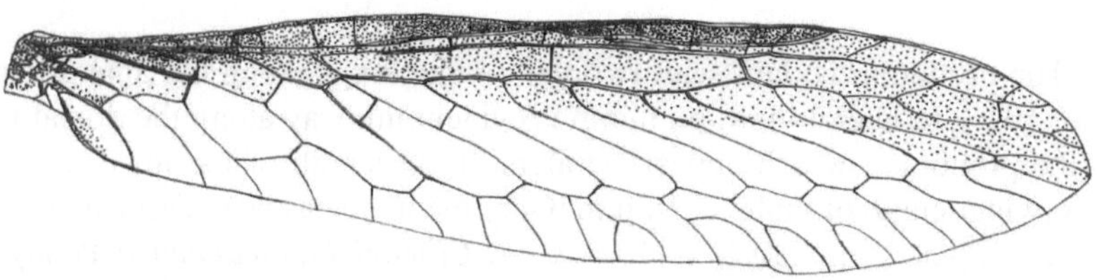

Fig. 10.15 *Climaciella porosa* male: fore-wing (upper right), apex of the abdomen in dorsal view (upper right), and (below, left to right): genitalia in ventral, posterior, and lateral view. Based on Ardila-Camacho and García (2015).

2. (1) The pronotum is black on the dorsum and ocher yellow laterally. The first six abdominal segments are brown with wide black posterior margins, and the seventh through tenth are entirely black. Length of fore-wing: c. 23 mm; hind wing: c. 21 mm. Only the male has been described.
.... *Climaciella personata* (Stitz, 1913) (Bolivia). Syn: *Euclimacia personata* Stitz, 1913. Handschin (1960) suggested that the type specimen of this species might actually be a fully pigmented specimen of *Climaciella semihyalina.*
– The body is yellowish (Fig. 10.14). Total length of male: c. 18 mm; female: c. 18.5 mm. Fore-wing length of male: c. 16 mm; female: c. 15 mm.
.... *Climaciella duckei* Navás, 1915 (Peru, Amazonas).

3. (1) The pronotum is mainly dark reddish brown marked with black and sometimes yellow (Fig. 10.15). 4
– The body is almost entirely black. In lateral view, the pronotum appears bent ventrad at mid-length (Fig. 10.16). 5

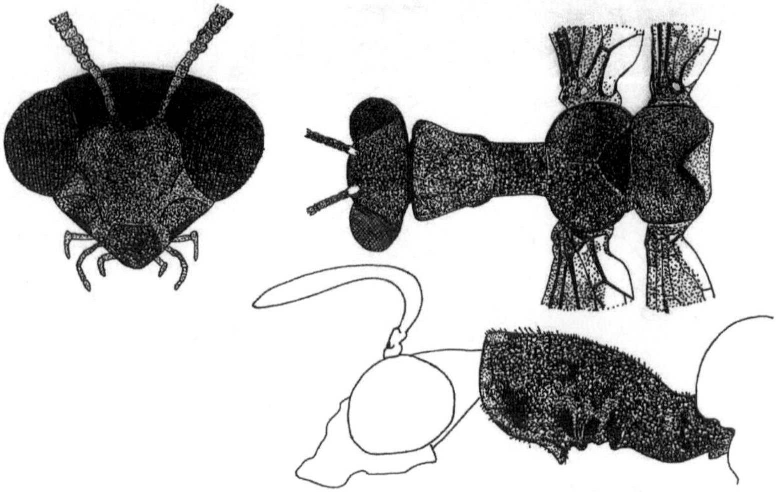

Fig. 10.16 *Climaciella obtusa* male: fore-wing (upper right), apex of the abdomen in dorsal view (upper right), and (below, left to right): genitalia in ventral, posterior, and lateral view. Based on Penny (1983).

4. (3) The pronotum is dark with two yellow markings along the anterior margin (Fig. 10.3). In lateral view, no hump is evident midway along the dorsal margin of the pronotum, which appears straight. Total length of the male: c. 15 mm. Fore-wing length of male: c. 14 mm. Only the male has been described. .... *Climaciella amapaensis* Penny, 1982 (Colombia, Amapá).
   - The color is mainly reddish brown. In lateral view, no hump is evident midway along the dorsal margin of the pronotum. The membrane is clouded light amber on the anterior 1/3 of the wing, and the rest is hyaline. (Fig. 10.15). ....*Climaciella porosa* Hoffman in Penny, 2002 (Costa Rica, Colombia).
5. (3) The head is black with dark reddish brown markings and yellow posterior margins around the eyes. The thorax is dark reddish brown and black, and the abdomen is black with yellow markings in the middle of the first two segments and usually along the posterior margins of the second tergite and sternite. The basal third of the fore-femur is dark reddish brown with a black spot in the middle. The anterior ¾ of each wing is dark brown, while the posterior ¼ is light amber or hyaline (Fig. 10.16). Fore-wing length: 15.8–24.2 mm. ....*Climaciella obtusa* Hoffman in Penny, 2002 (Ecuador, Colombia).

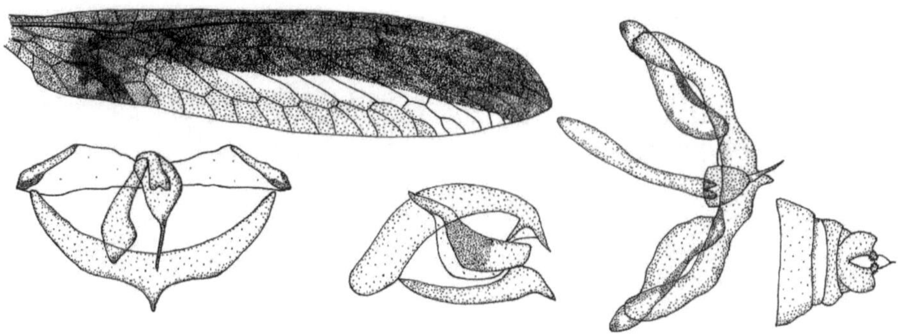

Fig. 10.17 *Climaciella semihyalina* male: fore-wing (upper left), apex of the abdomen in dorsal view (right), and the male genitalia in ventral (right center), posterior (lower left) and lateral view (lower center). Based on Penny (1983).

– The body is entirely black. The wing has a hyaline longitudinal band from the basal half to the margin (Fig. 10.17). Total length of male: c. 21 mm; female: 14–21 mm. Fore-wing length of male: c. 22 mm; female: 16–22 mm. ....*Climaciella semihyalina* (Le Peletier and Audinet-Serville, 1825) (Colombia, Surinam, French Guiana, Ecuador, Peru, Bolivia, Argentina, Amazonas, Rio de Janeiro, São Paulo, Paraná). Syn: *Mantispa semihyalina* Le Peletier and Audinet-Serville, 1825; *Mantispa chalybea* Erichson, 1839.

Key to the Species of Adult *Zeugomantispa* in South America

The key was prepared with information found in Navás (1930); Machado and Rafael (2007, 2010), and Ardila-Camacho and García (2015). The species are widespread and highly variable in size and appearance.

1. The pterostigma and most of the pronotum and wing veins are light green. There is a central invagination in the abdominal membranes, but the membranes are not divided into two distinct patches. Fore-wing length: 7.5–14.5 mm. .... *Zeugomantispa virescens* (Rambur, 1842) (North and Central America, Venezuela, Surinam, Peru, Paraguay, Uruguay, Argentina, Roraima, Rondônia, Amazonas, Pará, Bahia, Maranhão, Rio Grande do Norte, Mato Grosso, Goiás, Distrito Federal, Minas Gerais, Espirito Santo, São Paulo, Rio de Janeiro, Paraná, Santa Catarina, Rio Grande do Sul). Syn: *Mantispa viridula* Erichson, 1839; *Mantispa flaveola* Erichson, 1839; *Mantispilla virescens* Rambur, 1842; *Mantispa viridis* Walker, 1853; *Mantispilla punctata* var. *major* Stitz, 1913; *Mantispilla flavescens* Navás, 1914; *Mantispa pallescens* Navás, 1914; *Mantispilla trichostigma* Navás, 1921; *Mantispa punctata* Penny, 1977; *Mantispilla stigmata* Stitz, 1913; *Mantispilla viridis* Stitz, 1913;

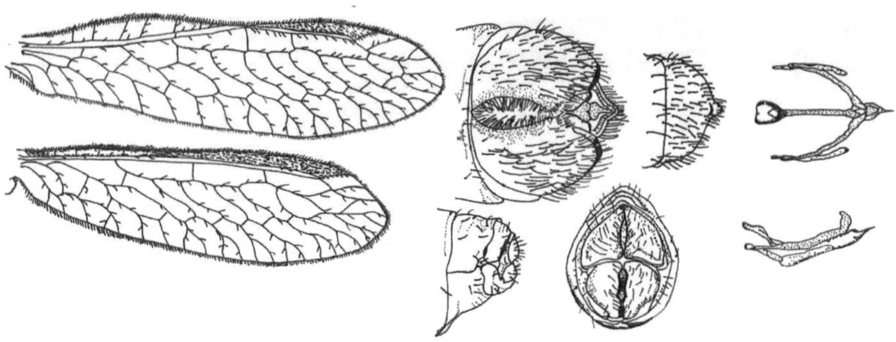

Fig. 10.18 *Zeugomantispa minuta:* fore and hind wing (left), apex of the male abdomen in dorsal view (upper left center), ninth sternite of the male in ventral view (upper right center), male genitalia in dorsal (upper right) and lateral view (lower right), and apex of the female abdomen in lateral (lower left center) and posterior view (lower right center).

Mantispilla rubricata Navás, 1924; *Mantispilla viridata* Navás, 1924; *Mantispilla flavicornis* Navás, 1930; *Mantispa stigmata* Penny, 1977; *Mantispa minuta* Penny, 1977 (*nec* Fabricius, 1775); *Zeugomantispa viridula* Hoffman in Penny, 2002 (*nec* Erichson, 1839).

– The pterostigma and most of the pronotum and wing veins are reddish brown. The abdominal membranes are completely divided into two distinct patches (Fig. 10.5). 2

2. (1) The membranes binding the third and fourth and fourth and fifth abdominal tergites are invaginated and bilobed; each contains two pores (Fig. 10.18).

.... *Zeugomantispa minuta* (Fabricius, 1775) (North and Central America, Colombia, Venezuela, Surinam, Peru, Paraguay, Uruguay, Argentina, Roraima, Rondônia, Amazonas, Pará, Bahia, Maranhão, Rio Grande do Norte, Mato Grosso, Goiás, Distrito Federal, Minas Gerais, Espirito Santo, São Paulo, Rio de Janeiro, Paraná, Santa Catarina, Rio Grande do Sul). Syn: *Mantis minuta* Fabricius, 1775; *Mantispa minuta* Erichson, 1839; *Mantispa flavomaculata* Latreille, 1805; *Mantis liliputana* Olivier, 1811; *Raphidia margaritacea* Fischer, 1834; *Mantispa margaritacea* Hagen, 1866; *Mantispa liliputana* Hagen, 1866; *Mantispilla punctata* Stitz, 1913; *Entanoneura margaritacea* Poivre, 1982.

– Each of the bilobed membranes binding the third and fourth and fourth and fifth abdominal tergites contain four to eight pores (Fig. 10.5). Fore-wing length: 6–13.5 mm.

.... *Zeugomantispa compellens* (Walker, 1860) (North and Central America, Colombia, Roraima, Rondônia, Amazonas, Pará, Bahia, Minas Gerais, Espirito Santo, Rio de Janeiro, Paraná). Syn: *Mantispa compellens* Walker, 1860; *Necyla uniformis* Navás, 1927; *Mantispa uniformis* Penny, 1982; *Mantispa parvula* Penny, 1982.

Key to the Species of Adult *Leptomantispa* in South America

The key was prepared with information found in Navás (1908); Penny (1983); Penny and da Costa (1983); Hoffman in Penny (2002), and Machado and Rafael (2007, 2010), and Ardila-Camacho and García (2015).

1. The wing has a red pterostigma and a hyaline space between veins Sc and R. The male hypomere is represented by four small sclerotized spots on the membrane of the pseudopenis (Fig. 10.19). Fore-wing length: 9–13.5 mm. ....*Leptomantispa axillaris* (Navás, 1908) (Rondônia, Amazonas, Pará, Tocantins, Maranhão, Rio Grande do Norte, Pernambuco, Espirito Santo, Goiás, Minas Gerais, São Paulo, Paraná). Syn: *Mantispa axillaris* Navás, 1908.
   – The wing has a light reddish brown or dark pterostigma and a light brown space between veins Sc and R. The male lacks a hypomere, or it is represented by two small sclerotized spots on the pseudopenis membrane (Fig. 10.20). 2

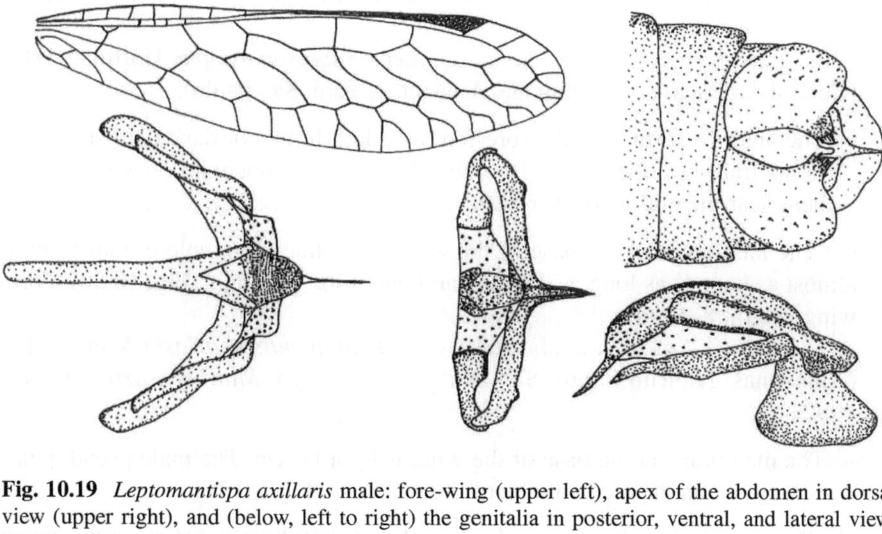

Fig. 10.19 *Leptomantispa axillaris* male: fore-wing (upper left), apex of the abdomen in dorsal view (upper right), and (below, left to right) the genitalia in posterior, ventral, and lateral view. Based on Penny and da Costa (1983).

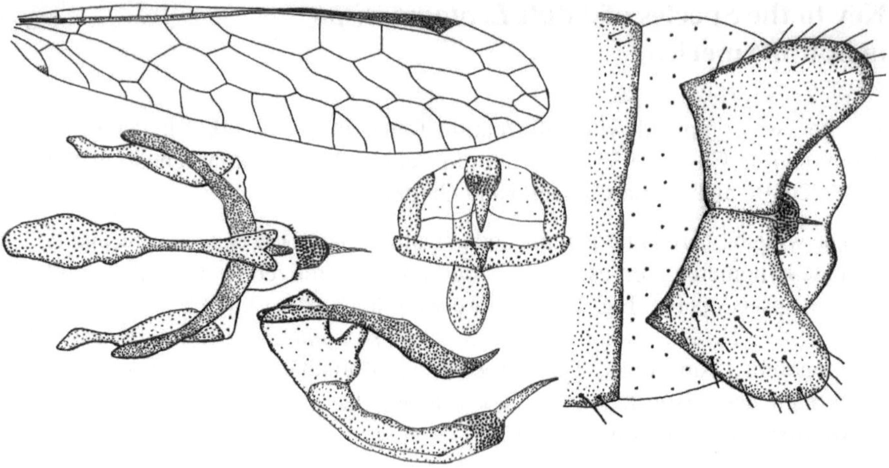

Fig. 10.20 *Leptomantispa ariasi* male: fore-wing upper left, apex of the male abdomen in dorsal view (right), and the genitalia in ventral (lower left), lateral (lower center), and posterior view (center). Based on Penny (1983).

2. (1) The anterior surface of the fore-femur is light brown or it has a light reddish brown spot near the basal spine. In ventral view, the external border of the male gonocoxite appears bent near the middle (Fig. 10.6). Fore-wing length: 8–11.5 mm.
   ....*Leptomantispa nymphe* Hoffman, 2002 (Central America, French Guiana, Amazonas, Pará, São Paulo).
   – The anterior surface of the fore-femur is dark brown or dark reddish brown. In ventral view, the external border of the male gonocoxite does not appear bent near the middle (Fig. 10.20). 3
3. (2) The membrane at the base of the wing is hyaline. The male pseudopenis is almost as long or as long as the median lobe of the gonarcus (Fig. 10.20). Fore-wing length: 8–11 mm.
   ....*Leptomantispa ariasi* (Penny, 1982) (Amazonas, Espirito Santo, São Paulo, Paraná) Syn: *Mantispa ariasi* Penny, 1982.
   – The membrane at the base of the wing is light brown. The male pseudopenis is longer than the median lobe of the gonarcus (Fig. 10.21). 4
4. (3) The male lacks a hypomere, has small scales at the apex of the gonocoxite, and lacks pores on the third abdominal tergite. The female ectoproct is as long as the gonocoxite, and the spermathecal duct has only a few bends. The pterostigma

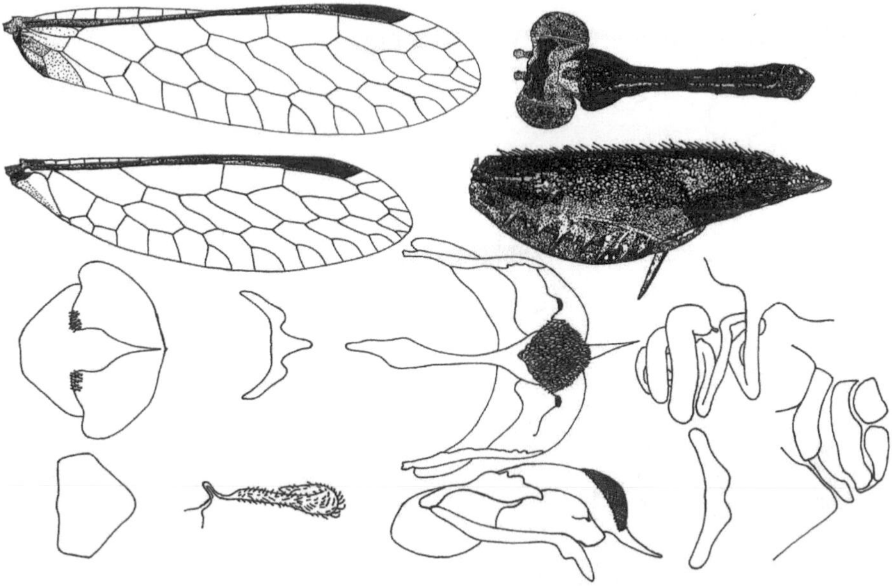

Fig. 10.21 *Leptomantispa chaos:* fore and hind wing (upper and upper middle left), head and pronotum in dorsal view (upper right), fore-leg in anterior view (upper middle right), male terminalia in dorsal view (lower middle left), male ninth sternite (lower left), male genitalia in ventral (center) and lateral view (lower center), male gonarcus in dorsal view (left of male genitalia), female genitalia in lateral view (lower right), female fertilization canal (lower left center), eighth sternite of a female (lower right center), and the spermatheca (upper middle right). Based on Machado and Rafael (2010).

is dark brown and very elongated. It joins with a light brown membrane in the space between Sc and R in the fore-wing and obscures the costal space in the hind wing (Fig. 10.22). Total length: 8.8–10.4 mm. Length of fore-wing: 6.0–9.0 mm.
.... *Leptomantispa catarinae* Machado and Rafael, 2007 (Amazonas).

– The male hypomere is represented by two small, sclerotized spot on the membrane of the pseudopenis. There are no small scales at the apex of the male gonocoxite, and there are pores on posterior margin of the third abdominal tergite. The female ectoproct is as shorter than the gonocoxite, and the spermathecal duct is coiled (Fig. 10.21). Fore-wing length: 7–10.5 mm.
.... *Leptomantispa chaos* Hoffman, 2002 (Central America, Venezuela, French Guiana, Pará, Amazonas).

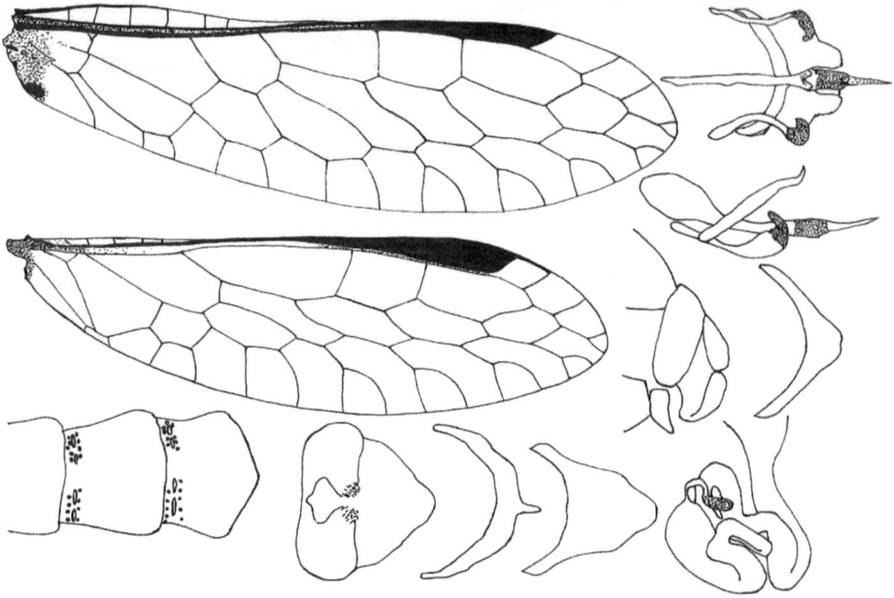

Fig. 10.22 *Leptomantispa catarinae:* fore and hind wing (upper and middle left), male genitalia in ventral (upper right) and lateral view (upper middle right), apex of the female abdomen in lateral view (right of hind wing), eighth sternite of a female in ventral view (lower middle right), and (lower row, left to right): fourth and fifth sternites and apex of the male abdomen in dorsal view, male gonarcus in dorsal view, ninth sternite of the male in ventral view, and spermotheca of a female. Based on Machado and Rafael (2007).

Key to the Species of Adult *Dicromantispa* in South America

The key was prepared with information found in Navás (1909); Carvalho and Corseuil (1995) and Machado and Rafael (2010).

1. At the base of the fore-wing, the wing membrane is hyaline (Fig. 10.23). 2

   – At the base of the fore-wing, the wing membrane is brownish (Fig. 10.24). 3

2. (1) The pterostigma is pale yellow, becoming brownish toward the apex of the wing (Fig. 10.23). The ventromedial lobe of the male is curved; the ectoproct is entirely dark brown. Fore-wing length: c. 7.9 mm.
   .... *Dicromantispa hyalina* Machado and Rafael, 2010 (Tocantins).

   – The pterostigma is reddish brown, becoming brownish toward the apex of the wing (Fig. 10.25). The ventromedial lobe of the male is straight; the ectoproct has a yellow external and brown internal border. Fore-wing length: 7–13 mm.
   .... *Dicromantispa gracilis* (Erichson, 1839)

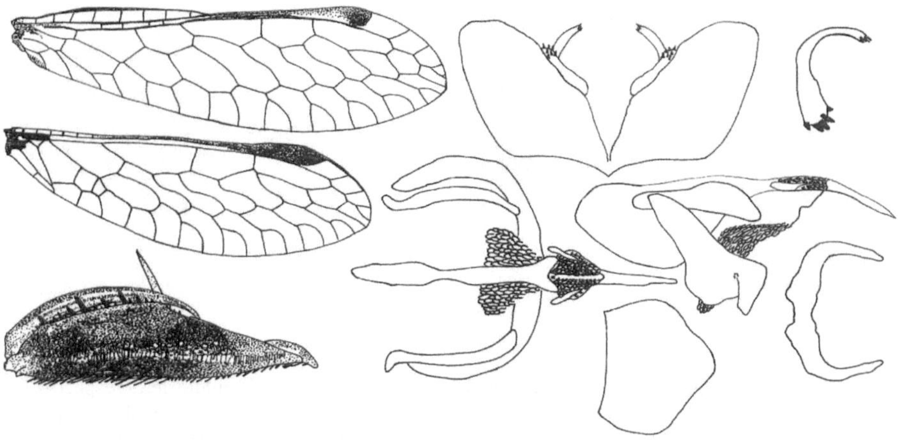

Fig. 10.23 *Dicromantispa hyalina* male: fore and hind wing (upper and middle left), fore-leg in posterior view (lower left), apex of the abdomen in dorsal view (upper right center), genitalia in ventral (lower center) and lateral view (middle right), ventromedial lobe (upper right), ninth sternite in ventral view (lower right center), and the gonarcus in posterior view (lower right). Based on Machado and Rafael (2010).

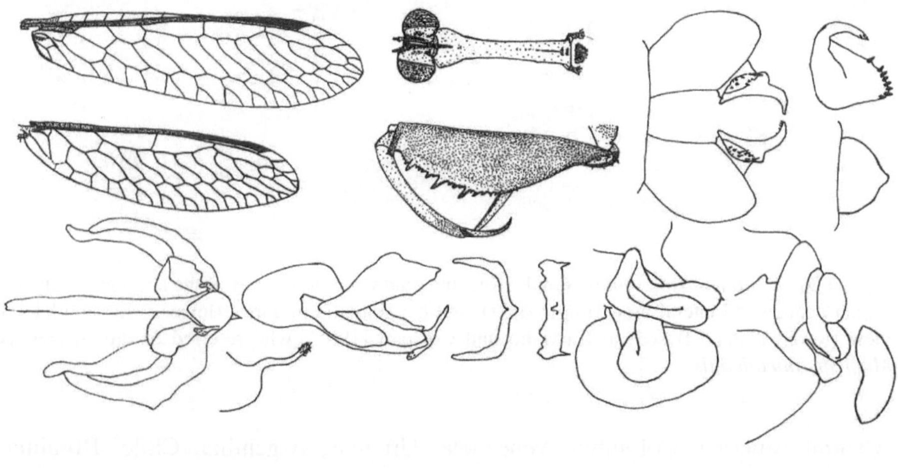

Fig. 10.24 *Dicromantispa moulti:* fore and hind wing (upper and middle left), head and pronotum in dorsal view (upper center), fore-leg in anterior view (center), apex of the male abdomen in dorsal view (upper right center), ventromedial lobe of a male (upper right), the ninth abdominal sternite of a male in ventral view (middle right), male genitalia in ventral (lower left) and lateral view (below left center), the gonarcus in dorsal and posterior view (lower center, left and right, respectively), female fertilization canal (lower left center), spermatheca and the apex of the female abdomen in lateral view (lower right center, left and right, respectively), and the eighth sternite of a female in ventral view. Based on Machado and Rafael (2010).

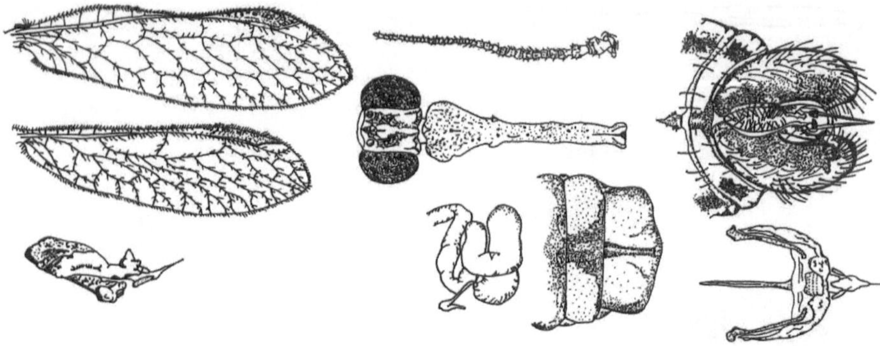

Fig. 10.25 *Dicromantispa gracilis:* fore and hind wing (upper and middle left, respectively), antenna (upper center), head and pronotum in dorsal view (center), apex of the male abdomen in dorsal view (upper right), eighth and ninth abdominal sternites of a male in ventral view (lower right center), male genitalia in dorsal (lower right) and lateral view (lower left), and the spermatheca in lateral view (lower center). Based on Carvalho and Corseuil (1995).

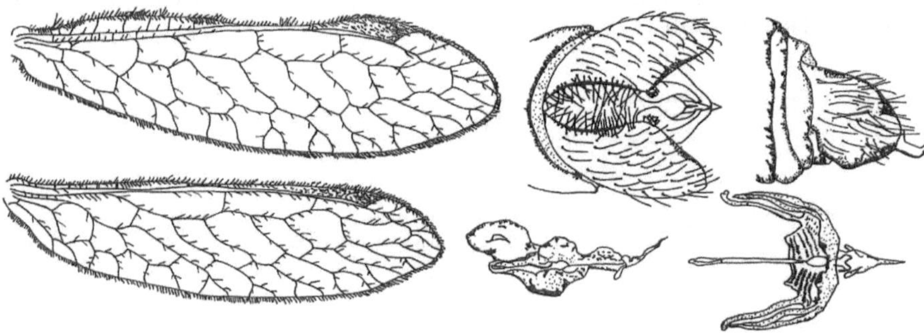

Fig. 10.26 *Dicromantispa synapsis* male: fore and hind wing (left), apex of the abdomen in dorsal (upper center) and lateral view (upper right), and the genitalia in dorsal (lower right) and lateral view (lower center). Based on Carvalho and Corseuil (1995), who referred to the species as *Mantispa lineaticollis*.

(Central America, Colombia, Venezuela, Uruguay, Argentina, Chile, Roraima, Amazonas, Bahia, Pernambuco, Tocantins, Mato Grosso, Distrito Federal, Minas Gerais, Espirito Santo, São Paulo, Rio de Janeiro, Paraná, Santa Catarina, Rio Grande do Sul). Syn: *Mantispa gracilis* Erichson, 1839; *Mantispa debilis* var. *nigricornis* Stitz, 1913; *Mantispa debilis* var. *rugicollis* Stitz, 1913; *Mantispa trilineata* Navás, 1914; *Mantispa Bruchi* Navás, 1915; *Mantispa calceata* Navás, 1917; *Mantispa mista* Navás, 1923; *Mantispilla mista* Navás, 1929; *Mantispilla gounellei* Navás, 1934; *Mantispa gounellei* Penny, 1977.

3. (1) The basal half or more of the space between veins C and Sc in the hind wing is brown (Fig. 10.24)..4
   – The entire space between veins C and Sc in the hind wing is hyaline (Fig. 10.26)..5

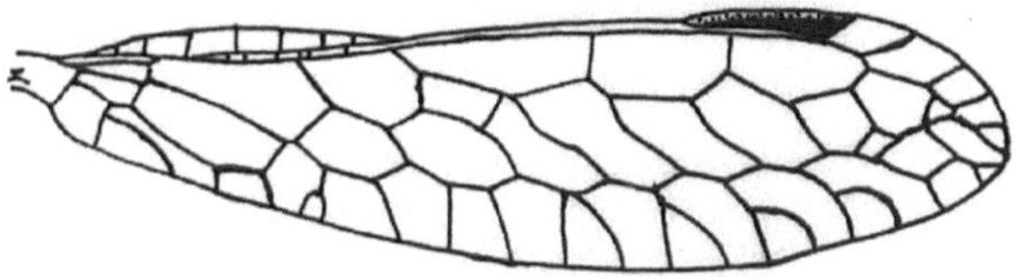

Fig. 10.27 Fore-wing of *Dicromantispa hoffmani*. Based on Ardila-Camacho and García (2015).

4. (3) The pronotum is light brown with a dark brown central stripe. The lateral lobes of the male gonarcus are not prominent in lateral view (Fig. 10.24). Fore-wing length: 13.5–20 mm.
   ....*Dicromantispa moulti* (Navás, 1909) (Colombia, French Guiana, Amazonas, Espirito Santo, São Paulo). Syn: *Mantispa moulti* Navás, 1909.

   – The pronotum is light yellow or white with a dark marking anterior to the maculae. The lateral lobes of the male gonarcus are plainly evident in lateral view (Fig. 10.7). Fore-wing length: 9.5–13 mm.
   ....*Dicromantispa leucophaea* Machado and Rafael, 2010 (Pará, Maranhão, Rondônia, Mato Grosso).

5. (3) There is a well-marked cross on the frons with bright red on the horizontal bars. The predominant color is dark brown (Fig. 10.27). Length of fore-wing: c. 12 mm; hind wing: c. 11 mm. The female has not been described.
   ....*Dicromantispa hoffmani* Ardila-Camacho, 2015 (Colombia).

   – There is no cross-shaped mark on the frons (Fig. 10.26)............................. 6

6. (5) The median longitudinal stripe on the head is interrupted at the frons. The head dorsal to the antennae has three confluent yellow markings, which form a broad, procurved band. The pronotum is dark brown anterior to the maculae. The dorsal membranes of the prothorax are yellow. The membrane of the male gonarcus has a medial group of spinules, which are wider than long and contact the apices of the gonocoxites. Anterior to the bursa of the female, there is a subrectangular plate covered with spinules (Fig. 10.26). Fore-wing length: 8.5–12 mm.
   ....*Dicromantispa synapsis* Hoffman in Penny, 2002 (Central America, Amazonas, Pará, São Paulo, Rio de Janeiro, Paraná, Santa Catarina, Rio Grande do Sul). Syn: *Mantispa lineaticollis* auctt. (*nec* Enderlein, 1910).

   – The median longitudinal stripe on the head is not interrupted at the frons. The head dorsal to the antennae has narrow yellow markings, which do not form a broad band. Anterior to the maculae, the pronotum is light brown with three dark brown spots. The dorsal membranes of the pterothorax are brown. The membrane of the male gonarcus has a medial group of spinules, which are longer than wide and do not contact the apices of the gonocoxites. Anterior to

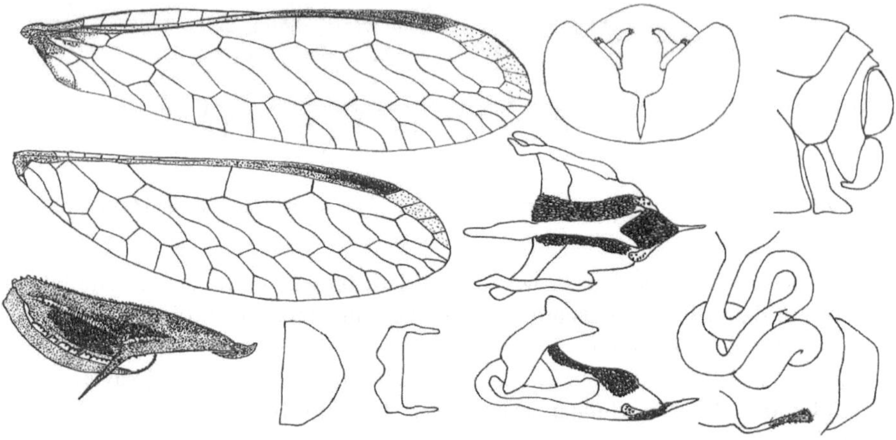

Fig. 10.28 *Dicromantispa debilis:* fore and hind wing (upper and middle left), fore-leg (lower left), apex of the male abdomen in dorsal view (upper right center), male genitalia in ventral (right center) and lateral view (lower right center), ninth sternite of a male in ventral view (lower left center), male gonarcus (lower venter), apex of the female abdomen in lateral view (upper right), spermatheca (right of center), eighth sternite of a female in ventral view (lower right), and female fertilization canal (below spermatheca). Based on Machado and Rafael (2010).

the bursa of the female, there is no subrectangular plate covered with spinules (Fig. 10.28). Fore-wing length of male: 8.5–12.5 mm; female: 11–13.5 mm. ....*Dicromantispa debilis* (Gerstaecker, 1888) (Central America, Colombia, Surinam, Bolivia, Amazonas, Pará, Bahia, Maranhão, Santa Catarina). Syn: *Mantispa debilis* Gerstaecker, 1888; *Mantispa debilis* var. *nuda* Stitz, 1913; *Mantispa (Mantispilla) lineaticollis* Enderlein, 1910.

Key to the Species of Adult *Paramantispa* in South America

The key was prepared with information found in Williner and Mariluis (1979) and Penny and da Costa (1983).

1. The compound eyes are small and clearly do not extend as far as the posterior margin of the head (Fig. 10.29)..2
   – The compound eyes occupy almost the entire lateral margins of the head, reaching approximately as far posteriad as the posterior margin of the head (Fig. 10.30) 3

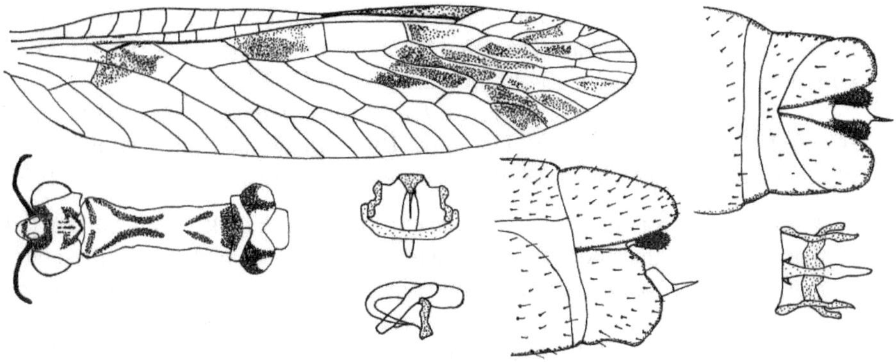

Fig. 10.29 *Paramantispa decorata* male: fore-wing (upper left), the head and prothorax in dorsal view (lower left), apex of the male abdomen in dorsal (upper right) and lateral view (lower left center), male genital structures in ventral (lower right), lateral (lower left center) and posterior view (left center). Based on Penny and da Costa (1983).

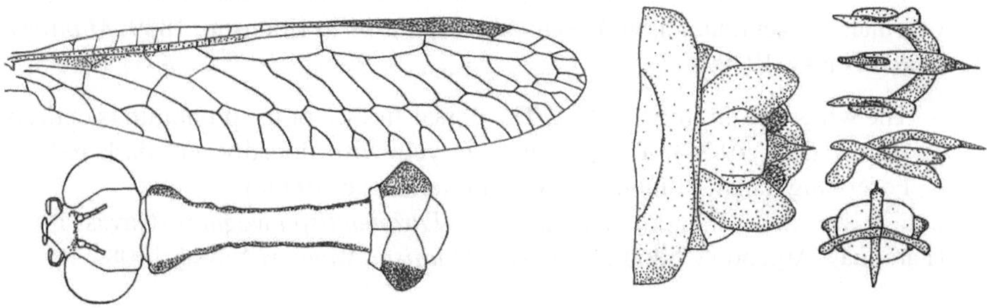

Fig. 10.30 *Paramantispa wagneri* male: fore-wing (upper left), the head and prothorax in dorsal view (lower left), apex of the male abdomen in dorsal view (right center), genital structures in ventral (upper right), lateral (middle right) and posterior view (left right). Based on Penny and da Costa (1983).

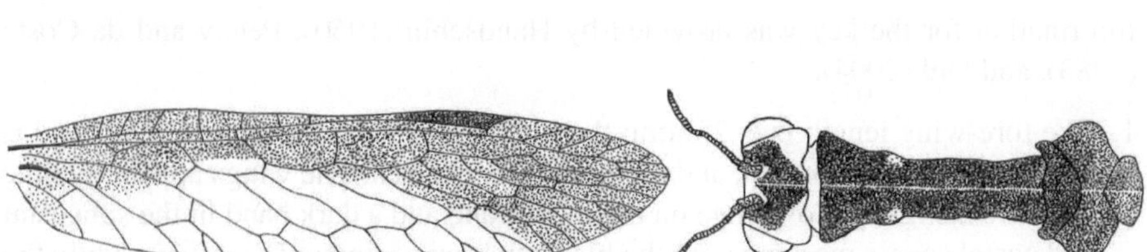

Fig. 10.31 *Paramantispa ambusta:* fore-wing (left), and the head and prothorax in dorsal view (right). Based on Penny and da Costa (1983).

2. (1) The head and pronotum are uniformly blackish. The fore-wing has a dark stripe running from the bases of veins Sc and R to the apex (Fig. 10.31). Fore-wing length of the female: 13–20 mm; hind wing: 11–14 mm.
   ....*Paramantispa ambusta* (Erichson, 1839) (Central America, Argentina, Uruguay, Rio Grande do Sul, Paraná). Syn: *Mantispa ambusta* Erichson, 1839; *Climaciella ambusta* (Erichson) Navás, 1932.
   – The head and pronotum are predominantly yellow with dark markings. The dark markings along the anterior part of the fore-wing are interrupted to form isolated markings (Fig. 10.29). Length of fore-wing: 12–20 mm; hind wing: 10–17 mm.
   ....*Paramantispa decorata* (Erichson, 1839) (Argentina, Uruguay, Rio Grande do Sul, Santa Catarina, Paraná, São Paulo, Minas Gerais). Syn: *Mantispa decorata* Erichson, 1839.
3. (1) The apex of the wing has a dark marking. The pronotum is yellow dorsally and dark reddish laterally (Fig. 10.8). The abdomen is dark red with a longitudinal series of yellow, diamond-shaped markings. Fore-wing length of male: c. 14 mm; female16–22 mm.
   ....*Paramantispa prolixa* (Erichson, 1839) (Surinam, Argentina, Paraná). Syn: *Mantispa prolixa* Erichson, 1839; *Mantispa mühni* Navás, 1930.
   – The wings are entirely hyaline. The dorsal surface of the pronotum is entirely yellow (Fig. 10.30). The abdomen is yellowish dorsally and dark reddish Fore-wing length of male: 17–18 mm; female: c. 19 mm.
   ....*Paramantispa wagneri* (Navás, 1908) (Paraguay, Argentina, São Paulo). Syn: *Mantispa Wagneri* Navás, 1908.

Key to the Species of Adult *Entanoneura* Known from South America

Information for the key was provided by Handschin (1960); Penny and da Costa (1983), and Ohl (2004).

1. The fore-wing length is c. 25 mm; the hind wing is c. 22 mm long. The head is uniformly blackish brown, and the prothorax is black. The wings appear hyaline but have a slight yellow tinge on the membrane and a dark band in the subcostal and costal spaces merging with the blackish brown stigma (Fig. 10.32). Only the female has been described.
   ....*Entanoneura brunneonigra* Handschin, 1960 (São Paulo). According to Penny and da Costa (1983), this name is considered a synonym of *Entanoneura januaria.* However, Ohl (2004) maintained it as a valid species and considered *E. januaria* to be a *nomen dubium.*
   – The fore-wing is no longer than c. 23 mm; the hind wing is not longer than c. 20 mm. In case of doubt, the head is black with yellow markings or predominantly yellow or pale (Fig. 10.33). 2

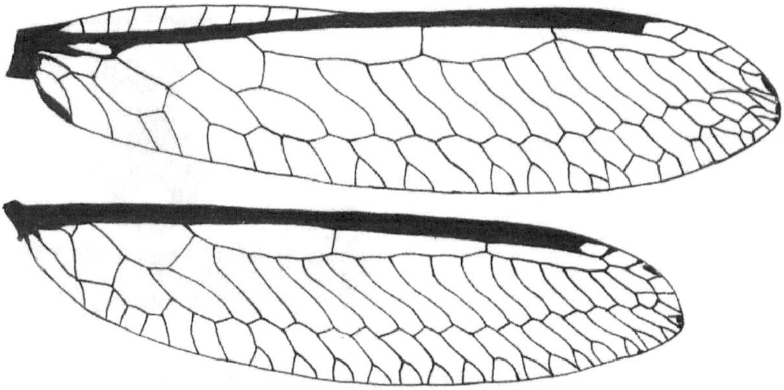

Fig. 10.32 Fore and hind wing of a female *Entanoneura brunneonigra*. Based on Handschin (1960).

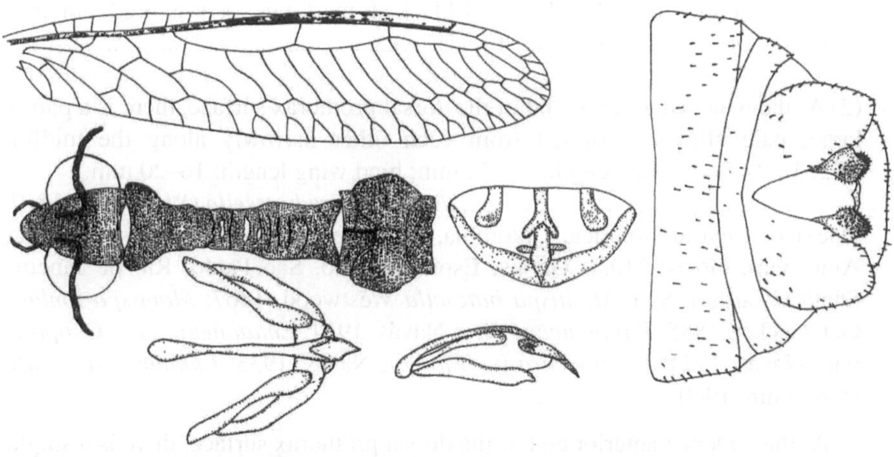

Fig. 10.33 *Entanoneura januaria* male: fore-wing (upper left), head and pronotum in dorsal view (middle left), apex of the abdomen in dorsal view (right), and the genitalia in ventral (lower left), lateral (lower center), and posterior view (center). Based on Penny and da Costa (1983).

2. (1) The head in anterior view is predominantly black or blackish brown with small yellow markings but without a complete yellow stripe between the compound eyes separating darker markings on the dorsal and ventral parts of the anterior surface of the head (Fig. 10.33). The prothorax is completely black on the dorsal surface and yellow ventrally. Fore-wing length: c. 20 mm; hind wing length: c. 18 mm.

   ....*Entanoneura januaria* (Navás, 1936)
   (Rio de Janeiro, São Paulo). Syn: *Entatoneura* (sic) *januaria* Navás, 1936. In his catalog of species, Ohl (2004) listed this species as a *nomen dubium.* Similar to this species but with a more brightly marked head is *Buyda phthisica,* formerly placed in *Entanoneura*. See key to the genera of Mantispidae, p. 295.

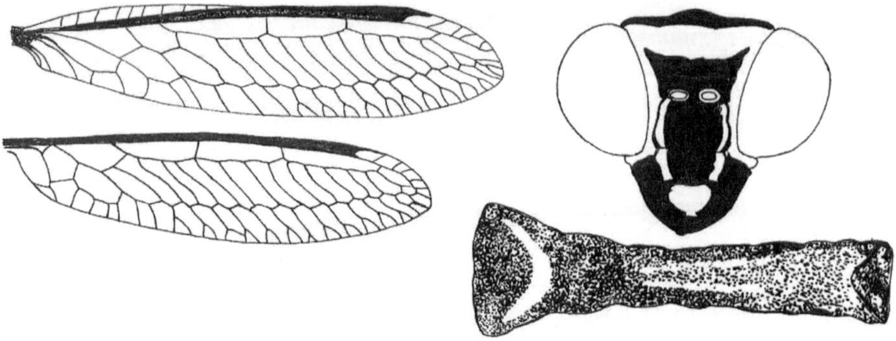

Fig. 10.34 *Entanoneura costalis:* fore and hind wing (left), head in anterior view (upper right), and the pronotum in dorsal view (lower right). Based on Handschin (1960).

– The head in anterior view has a yellow band across the middle from eye to eye and separating dorsal and ventral blackish markings, or it is predominantly yellow or yellow brown (Fig. 10.34).. 3

3. (2) At the widened anterior end of the dorsal prothorax surface, there is a pair of large, pale elipses separated from each other narrowly along the midline (Fig. 10.9). Fore-wing length: 17–23 mm; hind wing length: 16–20 mm. ....*Entanoneura batesella* (Westwood, 1867) (Mexico, Central America, Colombia, Venezuela, French Guiana, Argentina, Amazonas, Goiás, Minas Gerais, Espirito Santo, São Paulo, Rio de Janeiro, Santa Catarina). Syn: *Mantispa batesella* Westwood, 1867; *Mantispa limbata* Gerstaecker, 1885; *Entanoneura picta* Navás, 1914; *Entatoneura* (sic) *Chopardi* (sic) Navás, 1914; *Entatoneura* (sic) *jocosa* Navás, 1933; *Entanoneura similis* Handschin, 1960.

– At the widened anterior end of the dorsal prothorax surface, there is a single, pale, crescent-shaped marking with the anterior margin concave (Fig. 10.34). Fore-wing length: 19–22 mm; hind wing length: 18–21 mm. ....*Entanoneura costalis* (Erichson, 1839) (Argentina, Minas Gerais, Espirito Santo). Syn: *Mantispa costalis* Erichson, 1839; *Mantispa areolaris* Westwood, 1852; *Entanoneura areolaris* Enderlein, 1910.

10.3 Subfamily Symphrasinae

Key to the Species of Adult *Trichoscelia* Known from South America

Information for the key was provided by Gerstaecker (1888); Navás (1912a and b, 1936); Stitz (1913); Penny and Costa (1983); Ohl (2004), and Ardila-Camacho and García (2015).

1. The fore-wing has large, distinct markings or bands covering at least ¼ of the wings (Fig. 10.35) 2

   – The fore-wing lacks markings or has diffuse marks, mainly at the forks in the transverse veins (Fig. 10.11). 3

2. (1) The fore-wing has two markings and no bands (Fig. 10.35). Length: c. 8.5 mm. Fore-wing length of male: c. 8.0 mm; female: c. 8.1 mm.
   ....*Trichoscelia fenella* (Westwood, 1852) (Pará). Syn: *Mantispa (Trichoscelia) fenella* Westwood, 1867.

   – The fore-wings have two markings on the anterior part and a wide subapical stripe (Fig. 10.36). Length of male: c. 8.0 mm. Fore-wing length of male: 9.5–9.9 mm.
   ....*Trichoscelia latifascia* McLachlan, 1867 (Amazonas). Syn: *Anisoptera jocosa* Gerstaecker, 1888; *Trichoscelia jocosa* (Gerstaecker, 1888) Penny, 1977; *Symphrasis thaumasta* Navás, 1915; *Trichoscelia thaumasta* (Navás, 1915) Penny, 1977.

3. (1) The head of the female is pale yellow with two brown dorsal bands extending from the frons across the bases of the antennae to join and form a spot on the dorsal part of the head. The posterior part of the head is considerably inflated and lacks dark markings. The scape of the female antenna is yellow with brown margins, and the flagellum is brown with long setae on the apical half. The anterior

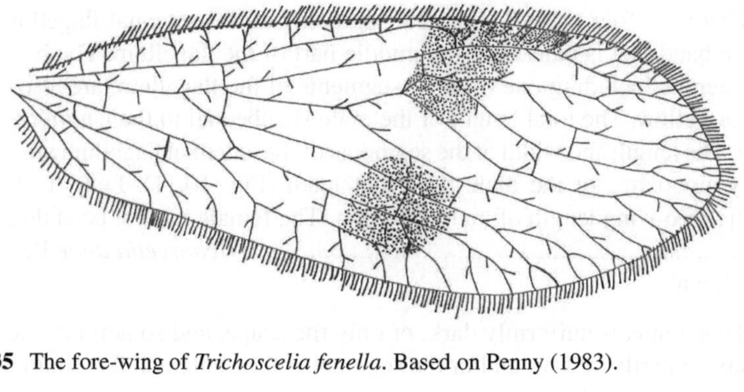

Fig. 10.35 The fore-wing of *Trichoscelia fenella*. Based on Penny (1983).

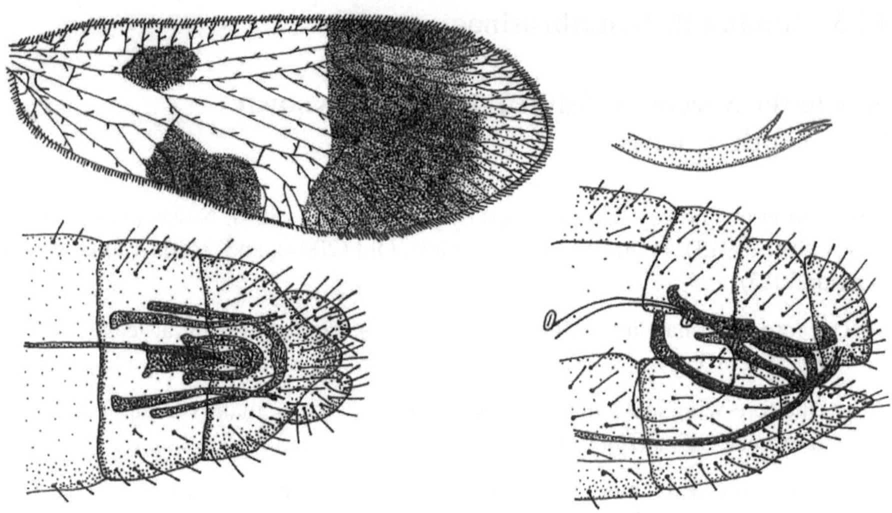

Fig. 10.36 *Trichoscelia latifascia* male: fore-wing (upper left), paramere (upper right), and apex of the abdomen cleared to show internal sclerotized structures in ventral (lower left) and lateral view (lower right).

part of the head has three dark markings and a brown band crossing anterior part of the frons, another on the clypeus; the labrum is entirely dark brown. The prothorax is light yellow with dark brown markings. The wings of the female are unmarked except for long darkened areas along the costal margins at the pterostigmas; the veins are darkened but have pale sections (Fig. 10.37). Only the female has been described.
....*Trichoscelia trifasciata* (Stitz, 1913) (Bolivia, Ecuador). Syn: *Symphrasis trifasciata* Stitz, 1913.

– The color pattern on the head is different, or the female has not been described. The ground color on the head and thorax is not pale yellow, or the dorsal part of the head is not notably inflated, or the dorsum of the pronotum is mostly dark (Fig. 10.37)..4

4. (3) There are four or five subapical segments of the antennal flagellum, which form a band that is paler than the middle part of the flagellum. The basal part of the antenna, including one or more segments of the flagellum, are also pale reddish or yellow. The total length of the male is subequal to the length of the fore-wing. The length and width of the segments of the antennal flagellum are subequal. The gonocoxites of the male bear four teeth (Fig. 10.11). Length of male: c. 8 mm. Fore-wing length of male: c. 8 mm. The female has not been described.
....*Trichoscelia anae* Penny, 1982 (Rondônia).

– The antenna is uniformly dark, or only the scape, and sometimes the pedicel, may be partly yellow, and/or the segments of the antennal flagellum are wider than long (Fig. 10.38)..5

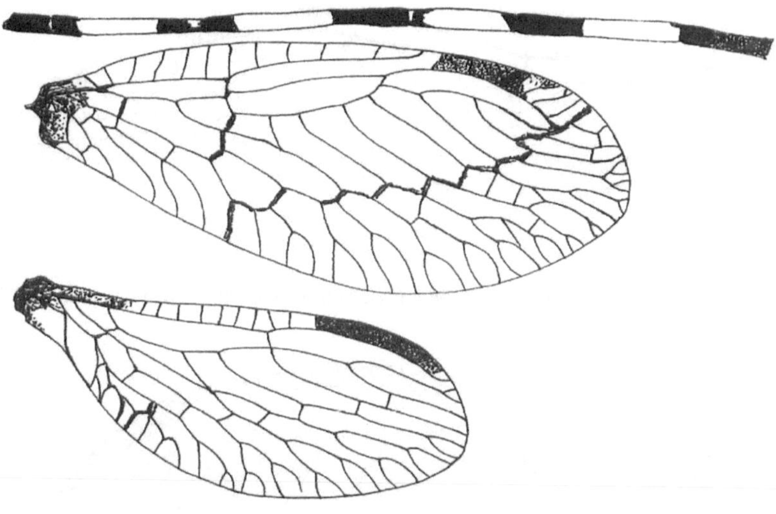

Fig. 10.37 *Trichoscelia trifasciata* female: section of the antenna (above) and the fore and hind wing (middle and below). Based on Stitz (1913).

Fig. 10.38 Fore-wing of *Trichoscelia gorgonensis*. Based on Ardila-Camacho and García (2015).

5. (4) The scape is yellow at the base and brownish toward the apex. The pedicel and flagellum are uniformly dark brown or black. The insect is marked with bright yellow and brown. The wings are hyaline with long, blackish pterostigmata, interrupted by bright orange markings. The male gonocoxite has three apical and two preapical spines (Fig. 10.38). Fore-wing length of male: 5.9–7.8 mm. ....*Trichoscelia gorgonensis* Ardila-Camacho, 2015 (Colombia).
   – The entire scape and sometimes the pedicel are yellow, while the flagellum is uniformly dark brown or black (Fig. 10.39)..6
6. (5) There are six spines at the apex of the male gonocoxite. Both the scape and pedicel are yellow, and the flagellum consists of about 35 segments, all dark

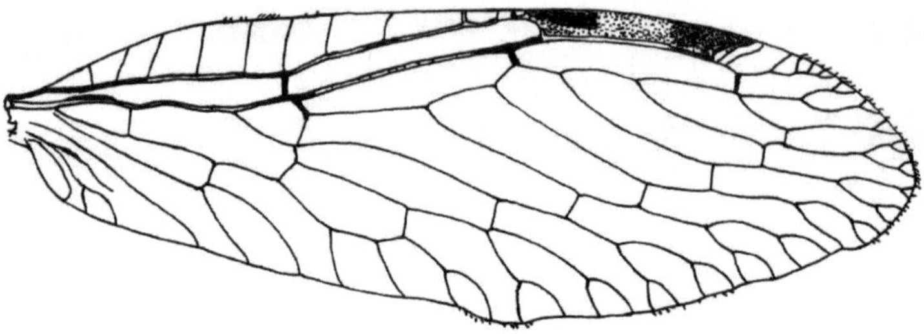

Fig. 10.39 Fore-wing of *Trichoscelia andina*. Based on Ardila-Camacho and García (2015).

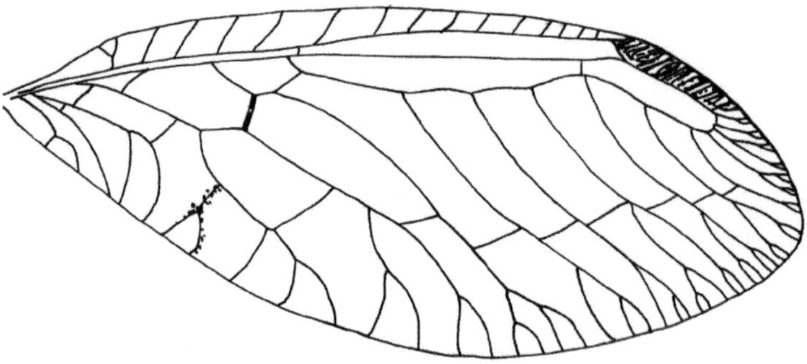

Fig. 10.40 Fore-wing of *Trichoscelia nassonovi*. Based on Navás (1912c), who referred to the species under its synonym, *Symphrasis nassonovi*.

brown and covered with dense dark brown setae. The gonocoxite has six spines, one apical and five pre-apical (Fig. 10.39). The head and thorax are mainly yellow. Both black and yellow wing veins are present. Length of male fore-wing: c. 9.7 mm; hind wing: c. 7.5 mm.
....*Trichoscelia andina* Ardila-Camacho, 2015 (Colombia).

– There are no more than five spines near the apex of the male gonocoxite. Usually, the entire antenna is uniform in color, and/or the middle and hind legs are predominantly yellow with a dark band on each tibia (Fig. 10.40) 7

7. (6) The head is yellow with a wide, transverse stripe anterior to the bases of the antennae. The anterior part of the prothorax is enlarged. The mesonotum and metanotum are dark with narrow yellow margins. The coxae of the fore-legs are darkened, and so is the lower apical spine on each femur. The middle and hind legs are yellow with a dark band on each tibia. There are about 15 cross veins in the space between vein C and Sc proximal to the pterostigma, which is dark with

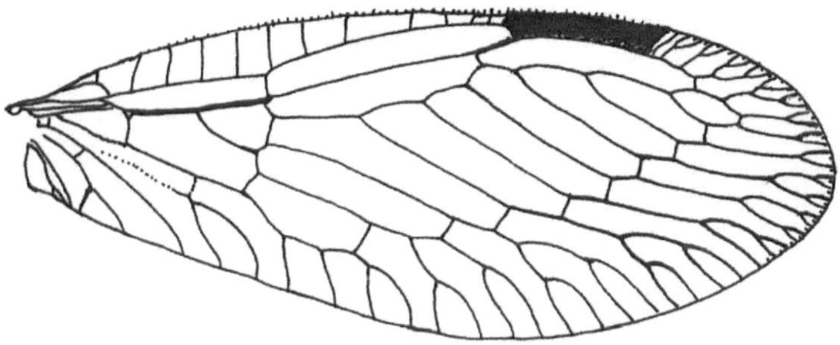

Fig. 10.41 Fore-wing of *Trichoscelia karijona*. Based on Ardila-Camacho and García (2015).

yellow markings in the center. Only the first of the cross veins is forked (Fig. 10.40).

....*Trichoscelia nassonovi* (Navás, 1912) (Peru). Syn: *Symphrasis nassonovi* Navás, 1912. The species requires a redescription of the male genitalia for positive identification.

– Either there are extensive orange markings on the coxae, femora, and dorsal surface of the thorax, or there are extensive yellow markings on all coxae, the pronotum, and the metanotum..8

8. (7) The wing veins and pterostigma are almost all dark brown or black, contrasting strongly with the immaculately hyaline membrane. Only short stretches of veins near the wing bases may be brownish yellow. Black or dark brown covered by blackish setae are predominant on the dorsal surfaces of the body (Fig. 10.41). Length of male fore-wing: 8 to.8.5 mm; hind wing: c. 6–6.5 mm.
   ....*Trichoscelia karijona* Ardila-Camacho, 2015 (Colombia).

– Many wing veins are light in color, and/or the dorsal surface is not predominently dark in color (Fig. 10.42)...9

9. (8) Two radial veinlets originate in the first radial cell of the fore-wing (Fig. 10.42). In the male, the gonocoxite has five apical teeth. The coxae and femora of the fore-legs, posterior third of the pronotum, the metanotum, and the metapleurae are all colored orange. Fore-wing length of male: c. 6.5 mm; female: 8.0–8.5 mm.
   ....*Trichoscelia egella* (Westwood, 1867) (Amazonas, Pará). Syn: *Mantispa (Trichoscelia) egella* Westwood, 1867; *Anchieta egella* (Westwood, 1867), *Anisoptera amoenula* Gerstaecker, 1888; *Trichoscelia amoenula* (Gerstaecker, 1888) in Penny (1977).

– Three radial veinlets originate in the first radial cell of the fore-wing. In the male, the gonocoxite bears two or three teeth (Fig. 10.43). The coxae and femora of the fore-legs, posterior third of the pronotum, the metanotum, and the metapleurae are dark.. 10

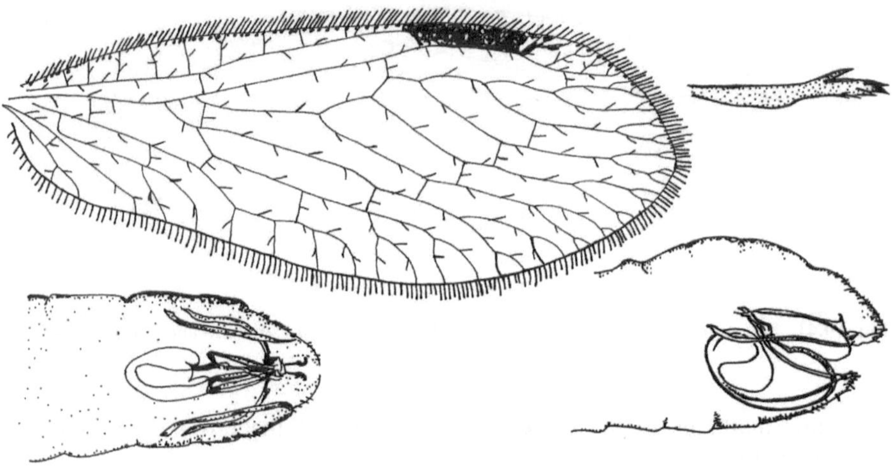

Fig. 10.42 *Trichoscelia egella* male: fore-wing (upper left), paramere (upper right), and the apex of the abdomen in ventral (lower left) and lateral view (lower right). Based on Penny (1983).

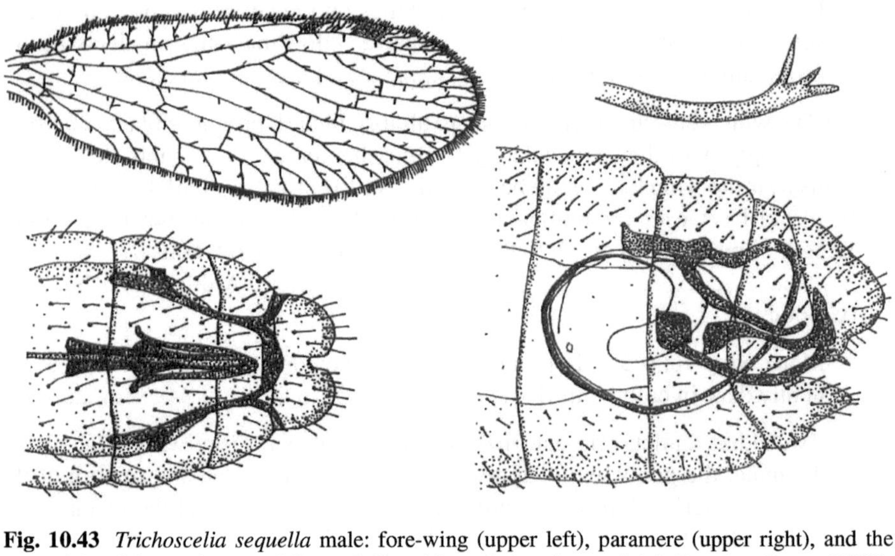

Fig. 10.43 *Trichoscelia sequella* male: fore-wing (upper left), paramere (upper right), and the apex of the abdomen in ventral (lower left) and lateral view (lower right). Based on Penny (1983).

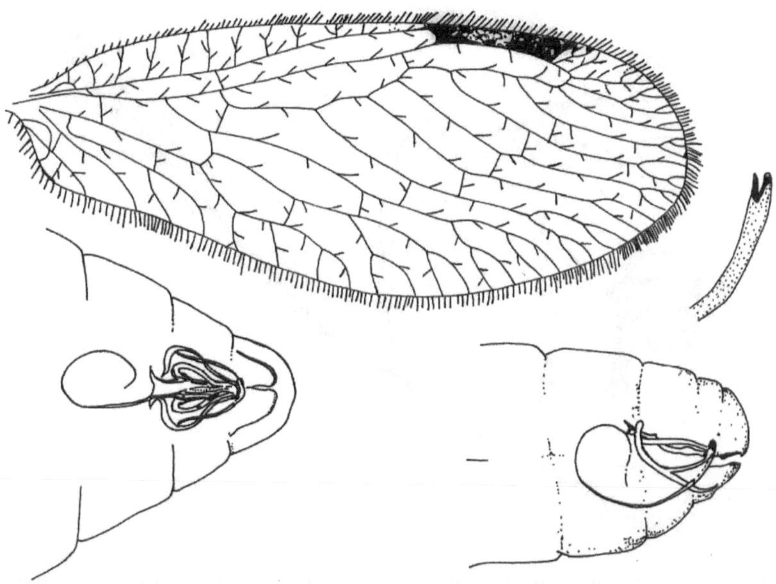

Fig. 10.44 *Trichoscelia iridella* male: fore-wing (upper left), apex of the abdomen in ventral after clearing in ventral (lower left) and lateral view (lower right), and the paramere (upper right). Based on Penny (1983).

10. (9) Only two radial veinlets originate in the second radial cell of the fore-wing. There are seven closed cells in the fore-wing posterior to the base of the first radial cell (Fig. 10.43). Length of male: c. 10 mm; female: c. 9 mm. Length of the fore-wing of both males and females: c. 10 mm. ....*Trichoscelia sequella* (Westwood, 1867) (Guyana, Amazonas). Syn: *Mantispa (Trichoscelia) sequella* Westwood, 1867.

   – Three radial veinlets originate in the second radial cell of the fore-wing. There are eight or nine closed cells in the fore-wing posterior to the base of the first radial cell (Fig. 10.44). 11

11. (10) The membrane of the fore-wing is completely transparent. In the male, the gonocoxite bears only the two apical teeth but no subapical tooth (Figs. 2.16, 7.44). Fore-wing length of male: 6.5–7.0 mm; female: 7–9 mm. ....*.Trichoscelia iridella* (Westwood, 1867) (West Indies, Ecuador, Amazonas). Syn: *Mantispa (Trichoscelia) iridella* Westwood, 1867; *Mantispa (Trichoscelia) basella* Westwood, 1867; *Trichoscelia basella* (Westwood, 1867) in Penny (1983).

   – The fore-wing membrane is clouded only along the transverse veins and between the branches of the forks along the fore-wing margin. In the male, the gonocoxite bears the two apical teeth and one subapical tooth. The spermatheca of the female has large and small ducts (Fig. 10.45).

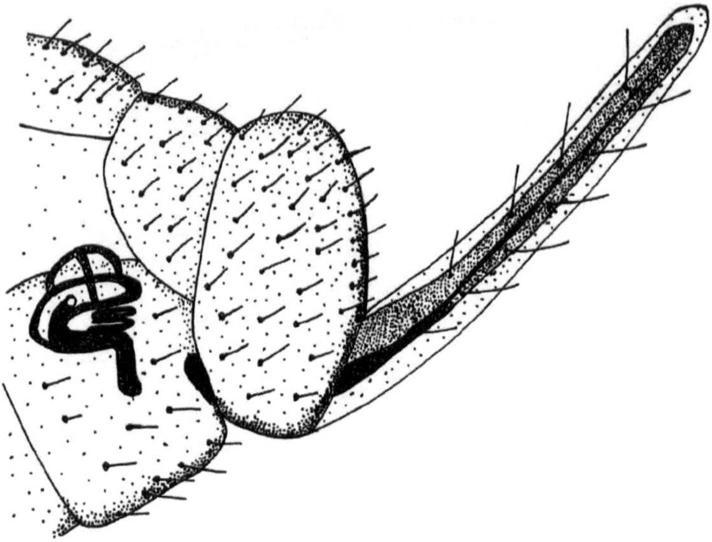

Fig. 10.45 Apex of the abdomen of a female *Trichoscelia varia* in lateral view. Based on Penny (1982, 1983).

Length of male: c. 5 mm; female: 5–7 mm. Fore-wing length of male: c. 7 mm; female: 8–9 mm.

....*Trichoscelia varia* (Walker, 1853) (Argentina, Uruguay, Surinam? Venezuela? Pará, Ceará, Mato Grosso, São Paulo, Santa Catarina, Rio Grande do Sul). Syn: *Rapphidia varia* Walker, 1853; *Symphrasis varia* (Walker, 1853) Enderlein, 1910; *Mantispa myrapetrella* Westwood, 1867; *Symphrasis myrapetrella* (Westwood, 1867) in Hagen (1877).

Key to the Species of Adult *Anchieta* Known from South America

Information for the key was provided by Navás (1909), Penny and da Costa (1983).

1. The fore-wing has dark markings along the radial and subcostal membrane (Fig. 10.46)..2
   – The fore-wing is hyaline or slightly tinted on the membranes, except sometimes on or adjacent to the pterostigma (Fig. 10.47) 3

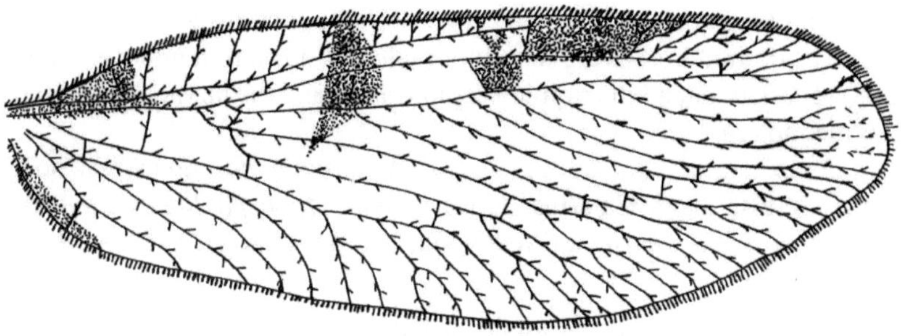

Fig. 10.46 The fore-wing of *Anchieta notha*. Based on Penny and da Costa (1983).

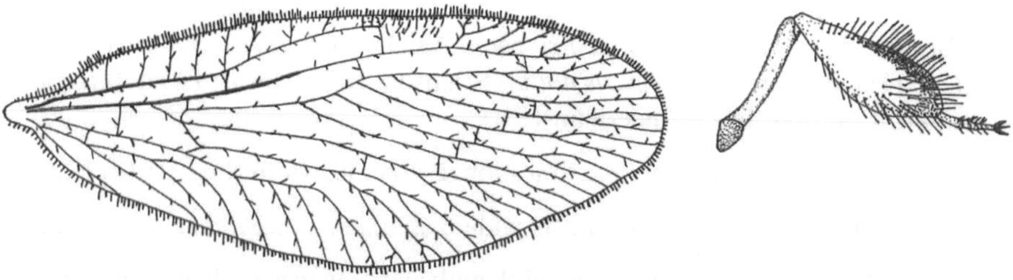

Fig. 10.47 The fore-wing (left) and hind leg (right) of *Anchieta bella*. Based on Penny and da Costa (1983).

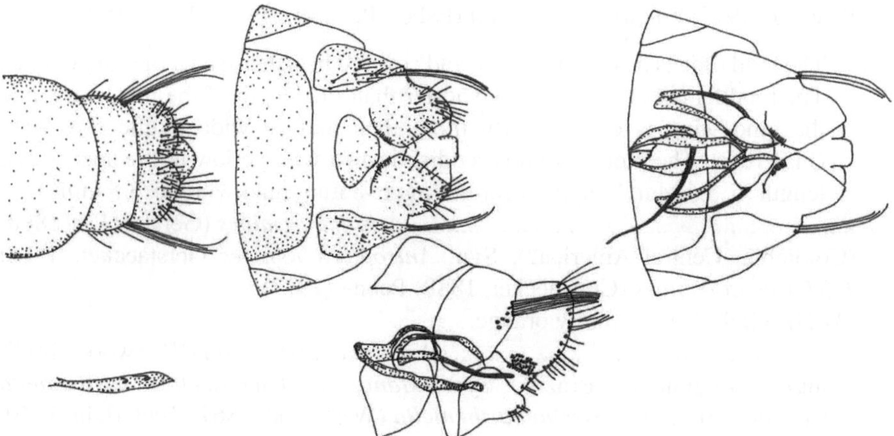

Fig. 10.48 *Anchieta fumosella* male (above, left to right): apex of the abdomen in dorsal and ventral view and apex in ventral view cleared to show the male genital structures, and (below, left and right) a paramere and the apex of the abdomen in lateral view cleared. Based on Penny (1983).

2. (1) The markings on the fore-wing form a continuous stripe (Figs. 10.2, 10.48). Length: c. 10 mm. Fore-wing length of male: 10–12 mm; female: c. 7 mm. Buys (2008) reported that adults mimic vespid wasps, and the larvae feed on the larvae of Hymenoptera.
....*Anchieta fumosella* (Westwood, 1867) (Minas Gerais, Santa Catarina, Goiás). Syn: *Mantispa fumosella* Westwood, 1867; *Trichoscelia fumosella* (Westwood, 1867) Enderlein, 1910; *Anchieta nobilis* Navás, 1909.

   – The dark areas on the fore-wing are discontinuous and form three distinct dark markings (Fig. 10.46). Fore-wing length: c. 11.9 mm. Only the female has been described.
....*Anchieta notha* (Erichson, 1839) (Brazil). Syn: *Mantispa notha* Erichson, 1839; *Trichoscelia notha* (Erichson, 1839) Hagen, 1861; *Symphrasis notha* (Erichson, 1839) Navás, 1909; *Anisoptera notha* (Erichson, 1839) Enderlein, 1910.

3. (1) The hind tibia is about three times longer than its width (Fig. 10.47) 4

   – The hind tibia is twice as long as wide 5

4. (3) The hind tibia is dark brown at midlength and orange at each end (Fig. 10.47). Length of the female: c. 7 mm. Fore-wing length of female: c. 8.5 mm.
....*Anchieta bella* (Westwood, 1867) (French Guiana, Amazonas, Pará). Syn: *Mantispa bella* Westwood, 1867; *Trichoscelia bella* (Westwood, 1867) Enderlein, 1910; *Anisoptera romani* Esben-Petersen, 1917; *Trichoscelia romani* (Esben-Petersen, 1917) Penny, 1977.

   – The hind tibia is brownish orange and covered by a dense coat of orange setae. The total length is significantly shorter than the length of the fore-wing, and the hind wing is only slightly more than half as wide as the fore-wing (Fig. 10.2). The pterostigma is reddish brown with yellow at the apex. Total length: c. 5.5 mm. Length of fore-wing: c. 8 mm; hind wing: c. 5.5 mm.
....*Anchieta remipes* (Gerstaecker, 1888) (Colombia, Central America?). Syn: *Anisoptera remipes* Gerstaecker, 1888; *Trichoscelia remipes* (Gerstaecker, 1888) Penny (1977).

4. (3) The hind tibia is entirely orange.
....*Anchieta partheniella* (Westwood, 1867) (Amazon Region of Brazil). Syn: *Mantispa (Trichoscelia) partheniella* Westwood, 1867; *Trichoscelia partheniella* (Westwood, 1867); Enderlein, 1910.

   – The swollen hind tibia is orange and black. Its fore-wing has a transparent membrane, with an orange band near the base and marking near the pterostigma (Fig. 10.49). Length of fore-wing: c. 9 mm; hind wing: c. 5.3 mm.
.... *Anchieta eurydella* (Westwood, 1867) (Colombia, Amazon Region of Brazil). Syn: *Mantispa (Trichoscelia) eurydella* Westwood, 1867; *Trichoscelia eurydella* (Westwood, 1867) Enderlein, 1910.

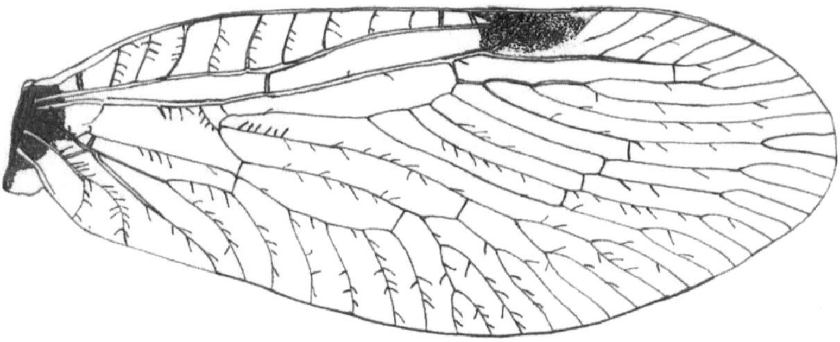

Fig. 10.49 Fore-wing of *Anchieta eurydella*. Based on Ardila Camacho and García (2015).

Key to the Species of Adult *Plega* Known from South America

Information for the key was provided by Penny (1982, 1983); Penny and da Costa (1983); Ohl (2004), and Ardila-Camacho and García (2015). *Plega* species mimic the adults of other insects in appearance, most notably members of Hymenoptera, on the larvae and pupae of which they feed.

1. The basal segments of the antennal flagellum are only half as long as wide. The filum of the penis is not coiled (Fig. 10.50). Length: 3.5–6.0 mm. Fore-wing length of male: 5.5–8.0 mm. The female has not been described. .... *Plega duckei* Penny, 1983 (Amazonas).
   - The length and width of the basal segments of the antennal flagellum are about equal (Fig. 10.12). 2
2. (1) Three or four of the subapical antennal segments are light yellow. The filum of the penis is simply recurved (Fig. 10.12). Fore-wing of male: c. 10.5 mm; female: c. 10.0 mm. ....*Plega hagenella* (Westwood, 1867) (Colombia, Venezuela, Amazonas). Syn: *Mantispa hagenella* Westwood, 1867; *Mantispa cognatella* Westwood, 1867.
   - All segments of the antenna are dark. The filum of the penis is either recurved or coiled (Fig. 10.51).. ..3
3. (2) There are no more than 40 segments of the antennal flagellum. The ten basal segments of the antennal flagellum are dark red (Fig. 10.51). 4
   - The antenna consists of more than 60 segments. Most of the fine setae on the mesonotum are dark brown or black (Fig. 10.52) 5

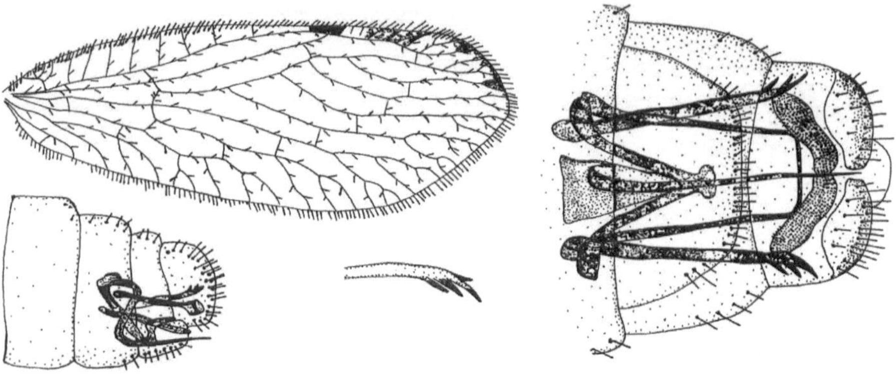

Fig. 10.50 *Plega duckei* male: fore-wing (upper left), apex of the abdomen cleared to reveal the genitalia in ventral (right) and lateral view (lower left), and the apex of the paramere (lower center). Based on Penny (1982, 1983).

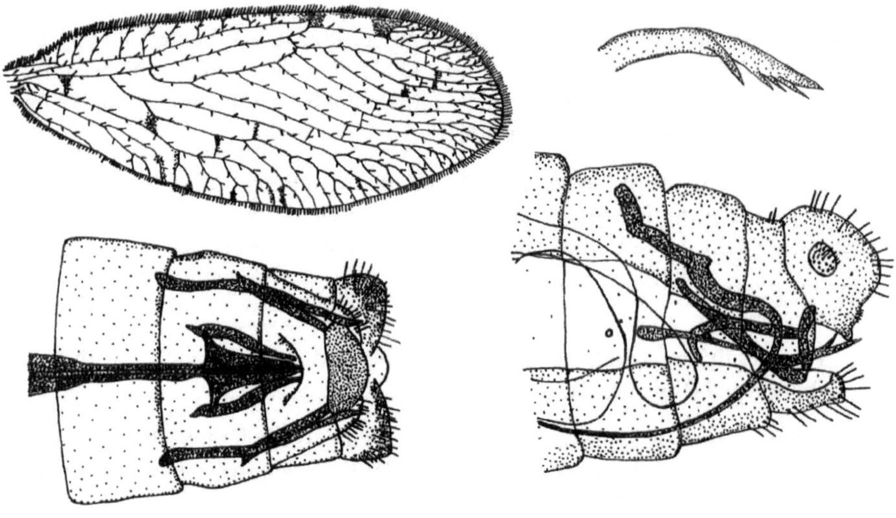

Fig. 10.51 *Plega beardi* male: fore-wing (upper left), paramere (upper right), and the apex of the abdomen in ventral (lower left) and lateral view (lower right). Based on Penny (1983).

4. (3) The antennal flagellum usually consists of 39 or 40 segments. Most of the fine setae on the mesonotum are black. The wing membrane lacks dark markings, although lightly clouded areas may be present. There are 7 or 8 strong setae on the anterior surface of the fore-femur (Fig. 10.51). The penis is coiled. Length of male: c. 9.0 mm. Fore-wing length of male: c. 11 mm. The female has not been described.
   ....*Plega beardi* Penny, 1982
   (Trinidad, Amazonas).

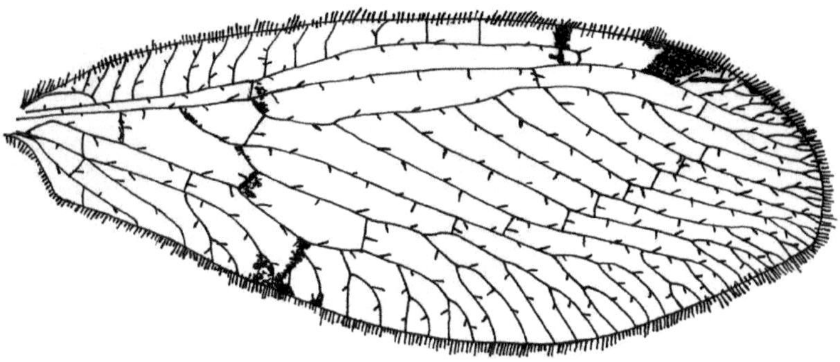

Fig. 10.52 The fore-wing of *Plega paraense*. Based on Penny (1983).

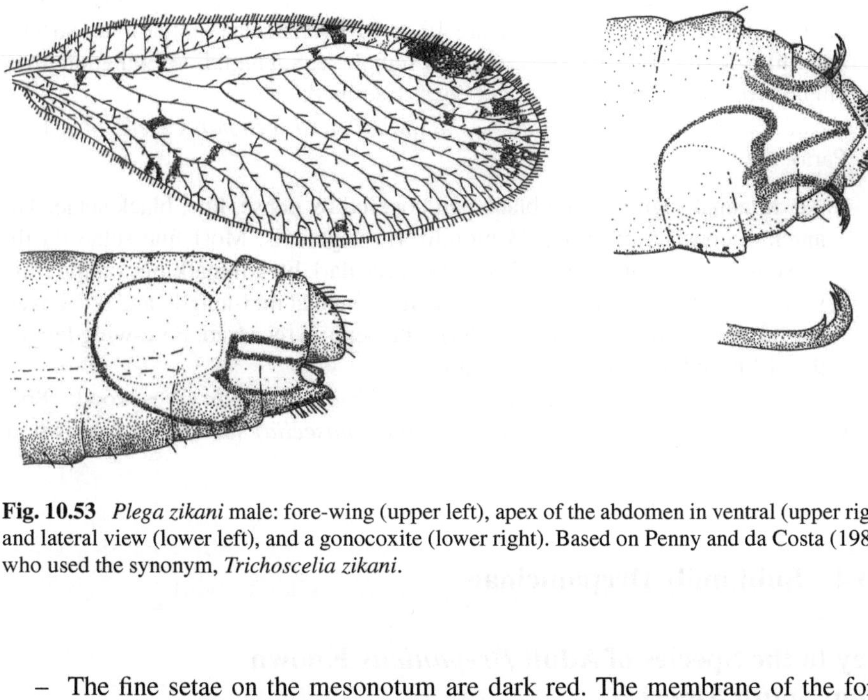

Fig. 10.53 *Plega zikani* male: fore-wing (upper left), apex of the abdomen in ventral (upper right) and lateral view (lower left), and a gonocoxite (lower right). Based on Penny and da Costa (1983), who used the synonym, *Trichoscelia zikani*.

– The fine setae on the mesonotum are dark red. The membrane of the fore-wing has small but distinct markings on the membrane in the apical and anal areas, especially between the long veins. The penis is recurved and not coiled (Fig. 10.53). The antennal flagellum consists of 33 or 34 segments. Length of male: 6–7 mm; female: c. 7 mm. Fore-wing length of male: 8–10 mm; female: 10–16 mm.

....*Plega zikani* Navás, 1936
(São Paulo, Rio de Janeiro, Minas Gerais). Syn: *Trichoscelia zikani* (Navás, 1936).

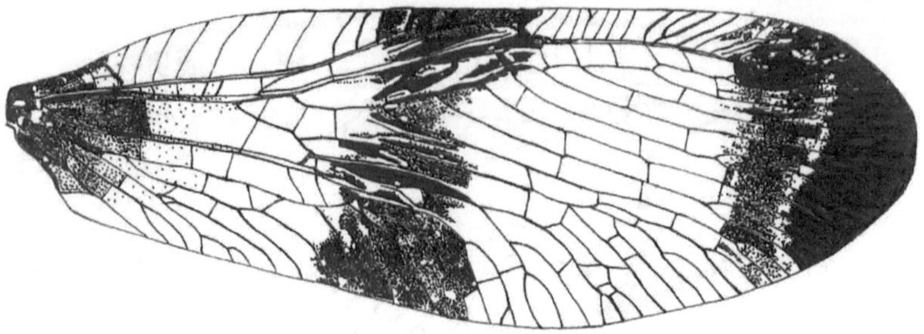

Fig. 10.54 The fore and hind wing of *Plega fasciatella*. Based on Ardila-Camacho and García (2015).

5. (3) Ten segments of the antennal flagellum are red at the base. Most fine setae on the mesonotum are yellow. There are 10 or 11 strong setae on the inner surface of the fore-femur (Fig. 10.52). Length of female: c. 8.5 mm. Fore-wing length of female: 10.5 mm. The male has not been described.
....*Plega paraense* Penny, 1983 (Pará).

– All antennal segments are black and covered by dense, fine, black setae. The antenna consists of about 73 moniliform segments. Most fine setae on the mesonotum are dark brown. There are three dark brown markings on the fore-wing: a small one at the base of the wing, a band at mid-length, and one covering the apex of the wing (Fig. 10.54). The penis is recurved. Fore-wing length: 9.9–13 mm; hind wing: 7.3–11 mm.
....*Plega fasciatella* (Westwood, 1867) (Colombia, Venezuela). Syn: *Mantispa (Trichoscelia) fasciatella* Westwood, 1867.

10.4 Subfamily Drepanicinae

Key to the Species of Adult *Drepanicus* Known from South America

Information for the key was provided by Brauer (1867a, b); Williner and Kormilev (1958), and Penny (1982, 1983). The genus is in need of revision. Williner and Kormilev (1958) reduced *Drepanicus chrysopinus* to a junior synonym of *Drepanicus gayi,* but subsequent authors maintained both names in lists of valid species. Judging by the wing lengths of the insects, it is more likely that *Drepanicus schajovskoyi* would be a synonym of *D. chrysopinus* if it is not a subspecies of *D. gayi.* The status of all three names remains in doubt.

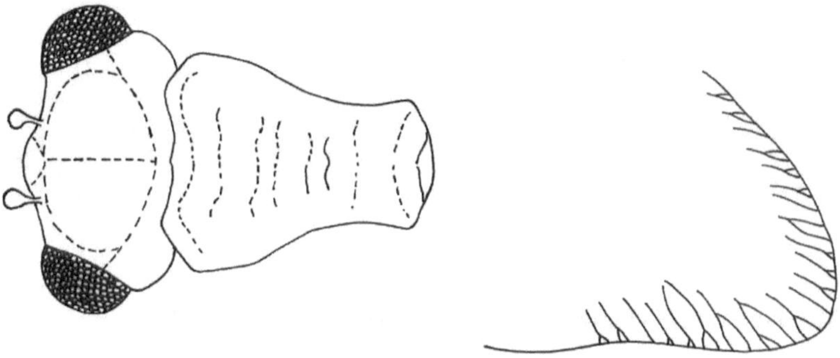

Fig. 10.55 *Drepanicus gayi:* head and pronotum in dorsal view (left) and the arrangement of the veins at the apex of the fore-wing (right). Based on Williner and Kormilev (1958).

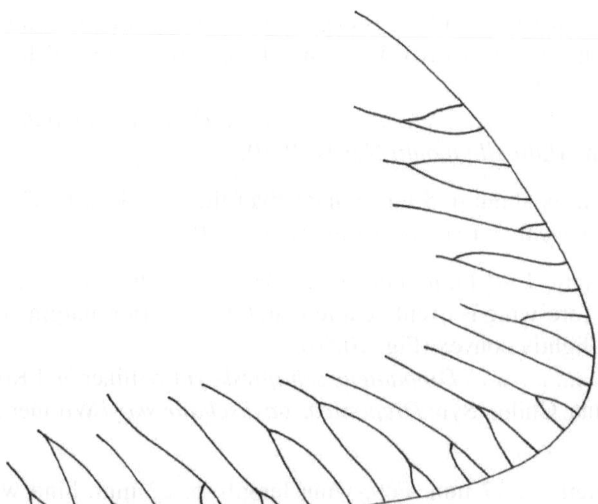

Fig. 10.56 Apex of the fore-wing of *Drepanicus schajovskoyi*. Based on Williner and Kormilev (1958), who called the species *Drepanicus gayi schajovskoyi.*

1. Total length: 20–24 mm. Fore-wing length: 60–70 mm; hind wing: 50–66 mm. The apex of the fore-wing is sharply bent, and its posterior margin near the apex is concave for a short distance (Fig. 10.55).
   ....*Drepanicus gayi* Blanchard, 1851
   (Argentina, Chile).
   – The total length is usually less than 20 mm. The length of the fore-wing is less than about 45 mm, and the hind wing is shorter than about 40 mm. The apex of the fore-wing is evenly curved, and the posterior margin near the apex is straight or slightly convex (Fig. 10.56) 2

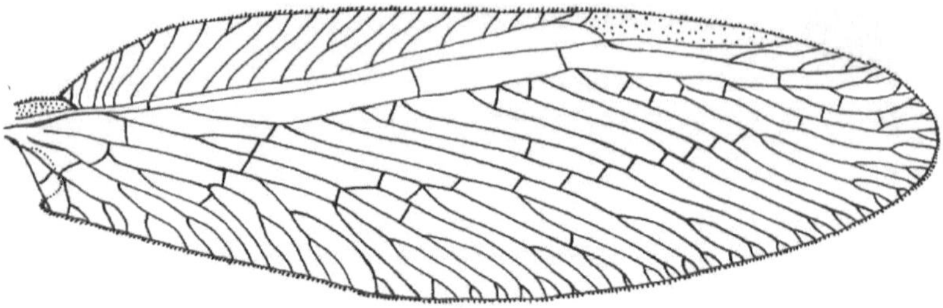

Fig. 10.57 Fore-wing of *Drepanicus chrysopinus*. Based on Penny (1982).

2. (1) The wings are only slightly longer than the total length. Length of fore-wing: c. 17 mm, hind wing: c. 15 mm. Total length: c. 14 mm. The wing membrane is uniformly hyaline, except for a yellowish posterior margin of the wing; the pterostigma, all veins, and the setae are greenish yellow. The apical part of the wing is elliptical (Fig. 10.10). The general coloration is greenish yellow without notable dark markings.
 ....*Drepanicus moulti* (Navás, 1910) (Chile). Syn: *Molinella moulti* Navás, 1910.

 – Each wing is about 4–5 mm longer than the total length. The apices of the wings are evenly curved or parabolic (Fig. 10.56). 3

3. (2) Total length: 13–14 mm. Fore-wing: 38–42 mm; hind wing: 34–39 mm. The apex of the fore-wing is evenly curved, and its posterior margin near the apex is straight or slightly convex (Fig. 10.56).
 ....*Drepanicus schajovskoyi* (Williner and Kormilev, 1958) (Argentina, Chile). Syn: *Drepanicus gayi schajovskoyi* Williner and Kormilev, 1958.

 – Total length: c. 17 mm. Fore-wing length: c. 22 mm; hind wing length: c. 21 mm. The apex of the wing is somewhat parabolic in shape (Fig. 10.57).
 ....*Drepanicus chrysopinus* Brauer, 1867 (Chile).

Key to the Species of Adult *Gerstaeckerella* Known from South America

Information for the key was provided by Banks (1913), Navás (1912a, 1930, 1932), and Penny and da Costa (1983).

1. The total length for the type specimens is between 10 and 11 mm. The fore-wing and hind wing are each shorter than c. 18 mm (Fig. 10.58) 2

 – The fore and hind wing are each longer than c. 18 mm 3

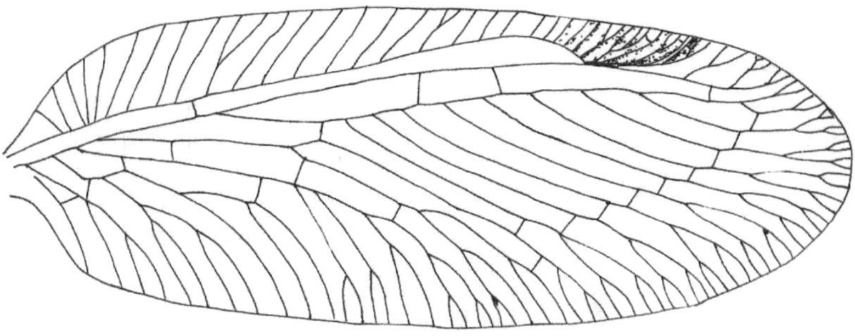

Fig. 10.58 Fore-wing of *Gerstaeckerella leonina.* Based on Navás (1930).

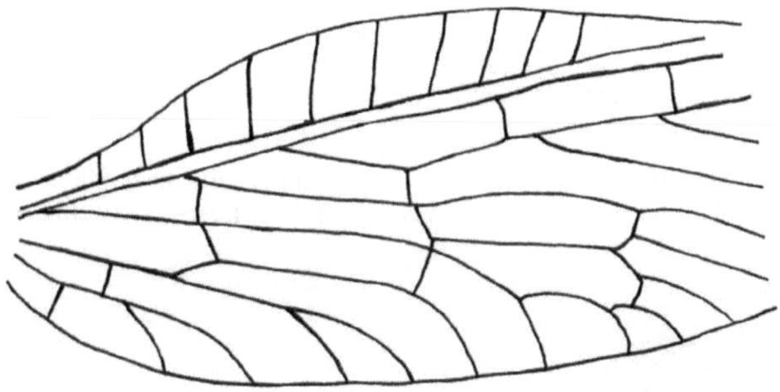

Fig. 10.59 The basal ¾ of the fore-wing of *Gerstaeckerella chilensis*. Based on Banks (1913), who referred to the species as *Mantispa chilensis*.

2. (1) The total length is about 10.2 mm. The length of the fore-wing: c. 16.5 mm; hind wing: c. 15.5 mm. The wing is broadly rounded, and the costal area is much wider near the base than near the pterostigma (Fig. 10.58). The wing membrane is hyaline, and the veins are yellowish. The color is almost entirely yellowish. ....*Gerstaeckerella leonina* Navás, 1930 (Chile).

   – The total length of the male is about 10.3 mm. Length of the fore-wing: c. 11 mm; hind wing: c. 9.4 mm. The fore-wing is hyaline with a dark pterostigma and a dark marking at about the mid-length of the posterior border of the wing. Between the pterostigma and the dark marking, there are six small dark clouds along the margin of the wing (Fig. 10.59). The color is mainly yellow with black markings, including a dense field of tiny black spots on the thorax.

   ....*Gerstaeckerella chilensis* (Hagen, 1859) (Chile). Syn: *Mantispa chilensis* Hagen, 1859; *Fusa Pirioni* (sic) Navás, 1925.

3. (1) The wings are hyaline with light yellow veins and yellowish membranes in the middle of cells near the base of the male fore-wing. The costal area is widest near the base and narrows to the margin of the pterostigma. Veins Sc curves

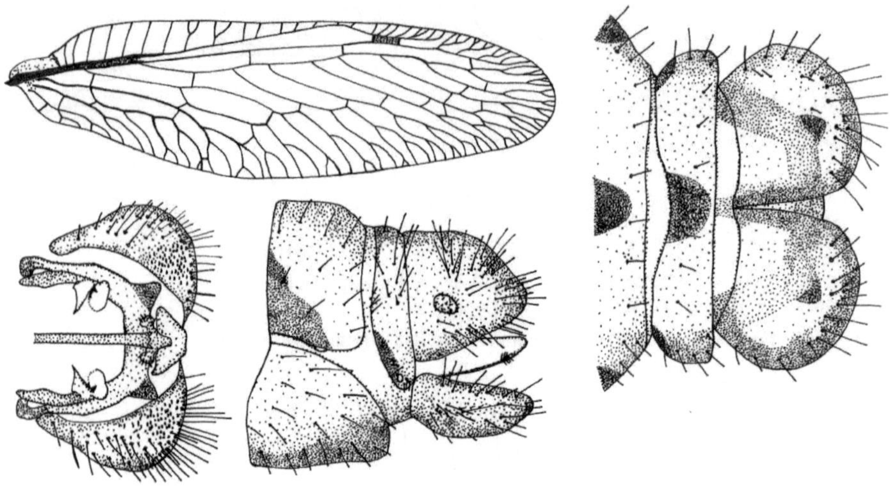

Fig. 10.60 *Gerstaeckerella irrorata* male: fore-wing (upper left), genitalia in ventral view (upper right), and apex of the abdomen in dorsal (lower left) and lateral view (lower right). Based on Penny and da Costa (1983).

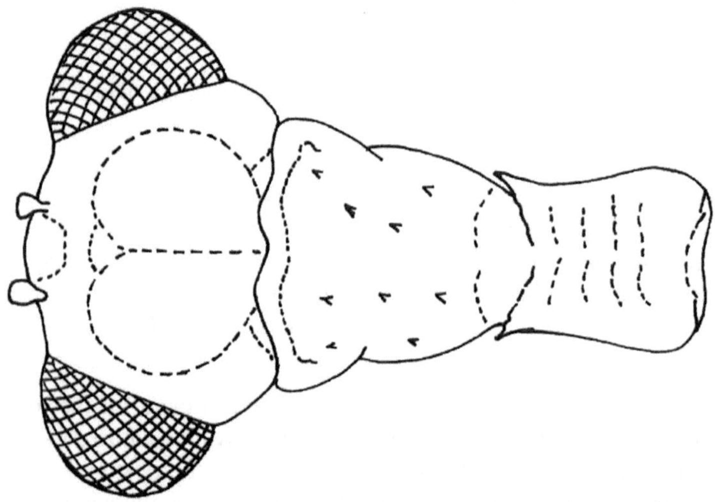

Fig. 10.61 Head and pronotum of *Gerstaeckerella gigantea* in dorsal view. Based on Williner and Kormilev (1958).

sharply at the pterostigma to join the radial vein (Fig. 10.13). The color is mainly yellow with the flagellum of the antenna somewhat darkened. Total length: 14–17.5 mm. Length of fore-wing: c. 21 mm; hind wing: c. 19 mm. The dimensions of the male and female are nearly equal.
....*Gerstaeckerella implexa* Navás, 1932 (Chile).

– The wings have dark markings, either along the wing margin or at the apex of vein M_1 and from the r-m cross vein to the apex of M_2 (Fig. 10.60).............. 4

4. (3) At the apex of vein M_1 in the fore-wing, there is a dark area. A straight dark stripe runs from cross vein r-m to the apex of M_2 (Fig. 1.4, 10.61). Total length of the female: c. 25 mm. Fore-wing length of the female: c. 35 mm. ....*Gerstaeckerella gigantea* Enderlein, 1910 (Colombia, Paraguay, Argentina, Santa Catarina, Paraná). Syn: *Gerstaeckerella Salonii* Navás, 1912; *Gerstaeckerella Anchietai* Navás, 1930.

– There are some dark markings along the wing margin but no large marking at the apex of vein MA and no straight marking across the cross veins from r-m to the apex of vein MP (Fig. 10.60). Total length of the male: c. 22 mm; female: c. 17 mm. Fore-wing length of the male: c. 25 mm; female: c. 22 mm. ....*Gerstaeckerella irrorata* (Erichson, 1839) (Cuba, Espirito Santo, Santa Catarina, Paraná, Distrito Federal). Syn: *Mantispa irrorata* Erichson, 1839; *Theristria irrorata* (Erichson, 1839) in Gerstaecker, 1885; *Raphidia riedeliana* Fischer, 1834.

Literature

Ardila-Camacho A, García A (2015) Mantidflies of Colombia (Neuroptera, Mantispidae). Zootaxa 3937(3):401–455

Banks N (1912) Exotic Neuroptera. Family Mantispidae. Transactions of the American Entomological Society 39:205–211

Banks N (1913) Synopses and descriptions of exotic Neuroptera. Transactions of the American Entomological Society 39:201–242, +pl. 23–26

Berg C (1899) Los mantíspidos Argentinos. Comunicaciones del Museo Nacional de Buenos Aires 1:139–145

Blanchard CÉ (1851) Mirmeleonianos. In: Gay C (ed) Historia Fisica y Politica de Chile. Zoologia. E. Thunot, Paris, pp 119–129

Brauer FM (1867a [1866]) Zusätze und Berichtigungen zu Hagen's Hemerobidarum Synopsis Synonymica und Beschreibung einer neuen Nymphiden-Gattung: *Myiodactylus osmyloides* aus Australien. Verhandlungen der Kaiserlich-Königlichen Zoologische-Botanischen Gesellschaft in Wien 16:983–992

Brauer FM (1867b) Beitrag zur Kenntniss der Mantispiden-Gattungen. Verhandlungen der Kaiserlich-Königlichen Zoologisch-Botanischen Gesellschaft in Wien 17:281–286

Buys SC (2008) Observations on the biology of *Anchieta fumosella* (Westwood 1867) (Neuroptera Mantispidae) from Brazil. Tropical Zoology 21:91–95

Carvalho AL, Corseuil E (1996 [1995]) Caracterização das espécies de *Mantispa* Illiger Ocorrentes no Rio Grande do Sul (Neuroptera, Mantispidae). Revista Brasileira de Zoologia 12(4):839–858

Enderlein G (1910) Klassifikation der Mantispiden nach dem Material des Stettiner Zoologischen Museums. Stettiner Entomologische Zeitung 71:341–379

Erichson WF (1839) Beiträge zu einer Monographie von *Mantispa*, mit einleitenden Betrachtungen über die Ordnungen der Orthopteren und Neuropteren. Zeitschrift für Entomologie 1:147–173

Esben-Petersen P (1917) Neue und wenig bekannte Mantispiden. Arkiv för Zoologi 11(10):1–15

Fabricius JC (1775) Systema entomologiae. Lipsiae, Flensburgi. 832 pp

Gerstaecker A (1888 [1887]) Weitere Beiträger zur Artenkenntniss der Neuroptera Megaloptera. Mittheilungen aus dem Naturwissenschaftlichen Verein für Neu-Vorpommern und Rügen in Greifswald 19:39–130

Hagan HA (1861) Synopsis of the Neuroptera of North America, with a list of South American species. Smithsonian Miscellaneous Collections 4(1):i-xx + 1–3410

Hagan HA (1866) Hemerobidarum synopsis synonymica. Stettiner Entomologische Zeitung 27:369–462

Hagan HA (1873) Die Larven von *Myrmeleon*. Stettiner Entomologische Zeitung 34:249–295

Handschin E (1960) Zur Revision süd-amerikanischer Mantispiden. Revue Suisse de Zoologie 67(32):523–558 + 2 pl

Hoffman KM (2002) Family Mantispidae. In: Penny ND (ed) A guide to the lacewings (Neuroptera) of Costa Rica. Proceedings of the California Academy of Sciences, 4th Series. 53(12):251–275

Hoffman KM, Brushwine JR (1992) Descriptions of the larvae and pupae of some North American Mantispidae (Neuroptera: Mantispidae) and development of a system of larval chaetotaxy for Neuroptera. Transactions of the American Entomological Society 118(2):159–196

Illiger JCW (1798) Versuche einer naturlichen Ordnungen und Gattungen der Insekten. In: Kugelann JO (ed) Verzeichnis der Käfer Preussans. J. J. Gerbauer, Halle, pp 489–501

Machado RJP, Rafael JA (2007) A new species of Mantispidae (Insecta: Neuroptera) from Central Amazonia, Brazil. Zootaxa 1530:37–40

Machado RJP, Rafael JA (2010) Taxonomy of the Brazilian species previously placed in *Mantispa* Illiger, 1798 (Neuroptera: Mantispidae), with the description of three new species. Zootaxa 2454:1–61

McLachlan R (1868 [1867]) New genera and species, &c., of neuropterous insects; and a revision of Mr. F. Walker's British Museum catalogue of Neuroptera, part ii (1853), as far as the end of the genus *Myrmeleon*. Journal of the Linnean Society (Zoology) 9:230–281 + 1 pl

Navás L (1908) Neurópteros nuevos. Memorias de la Real Academia de Ciencias y Artes de Barcelona. Tercera Época 6:401–423

Navás L (1909) Mantéspidos nuevos. Memorias de la Real Academia de Ciencias y Artes de Barcelona. Tercera Época 7:1–15

Navás L (1912a [1911]) Quelques Mantispides (Insectes Névroptères) du Musée Zoologique de l'Académie Impèriale des Sciences de St.-Pétersbourg. Annuaire du Musée Zoologique de l'Academie Impèriale des Sciences de St.-Pétersbourg 16:535–538

Navás L (1912b) Neurópteros nuevos o poco conodicos. Memorias de la Real Academia de Ciencias y Artes de Barcelona. Tercera Época 10(7):135–202

Navás L (1912c) Neurópteros nuevos de América. Brotéria, Série Zoológica 10:194–202

Navás L (1914) Neurópteros Sudamericanos. Primera Serie. Brotéria, Série Zoológica 12(1):45–56, 215–234

Navás L (1915) Neurópteros nuevos o poco conodicos. (Sexta Serie). Memorias de la Real Academia de Ciencias y Artes de Barcelona. Tercera Época 12(7):119–136

Navás L (1917) Neurópteros nuevos o poco conocidos. Novena Serie. Memorias de la Real Academia de Ciencias y Artes de Barcelona. Tercera Época 13(26):393–406

Navás L (1924) Insectos de la América Central. Brotéria Série Zoológica 21:55–86

Navás L (1925) Insectos Neotrópicos. 1.a Serie. Revista Chilena de Historia Natural 29:305–313

Navás L (1926) Insectos neotrópicos. 2.a Serie. Revista Chilena de Historia Natural 30:326–336

Navás L (1930) Insectos de la Argentina. Sexta Serie. Revista de la Sociedad Entomológica Argentina 3(13):125–132

Navás L (1932) Décadas de insectos nuevos. Década 21. Brotéria (Serie Trimestral, Ciências Naturais) 1:109–119

Navás L (1933) Insectos Neotropicos. 8ª Serie. Revista Chilena de Historia Natural 37:194–197

Navás L (1934a) Insectos Suramericanos, Octava Serie. Revista de la Academia de Cíencias Exactas Físicas y Naturales de Madrid 31:9–28

Navás L (1934b) Insectos Suramericanos, Novena Serie. Revista de la Real Academia de Cíencias Exactas Físicas y Naturales de Madrid 31:155–184
Navás L (1935 [1934]) Insectos Suramericanos. Décima Serie. Revista de la Academia de Cíencias Exactas Físicas y Químicas y Naturales de Madrid 32:360–375
Navás L (1936 [1931]) Insectos del Brasil. 5ª Serie. Revista do Museu Paulista 20:721–734
Ohl M (2004) Annotated catalog of the Mantispidae of the world (Neuroptera). Contrib Entomol Int 5:i–ii, + 131–262
Ohl M (2007 [2005]) Towards a global inventory of Mantispidae – the state-of-the-art in mantispid taxonomy. Museo Civico di Storia Naturale Ferrara – Annali 8:79–86
Penny ND (1982) Review of the generic level classification of New World Mantispidae (Neuroptera). Acta Amazonica 12(2):209–223
Penny ND (1983 [1982]) Neuroptera of the Amazon Basin. Part 6. – Mantispidae (1). Acta Amazonica 12(2):415–463
Penny ND, da Costa CA (1983) Mantispídeos do Brasil (Neuroptera: Mantispidae). Acta Amazonica 13(3–4):601–687
Poivre C (1982) Les Mantispides du Museum d'Histoire Naturelle de Genève. I. (Insecta, Planipennia). Revue Suisse de Zoologie 89(2):375–378, 667–672
Reynoso-Velasco D, Contreras-Ramos A (2008) Mantispidae (Neuroptera) of Mexico: distribution and key to genera. Annals of the Entomological Society of America 101(4):703–712
Stitz J (1913) Mantispiden der Sammlung des Berliner Museums. Mitteilungen aus dem zoologischen Museum in Berlin 7(1):1–49
Thouvenot M (2010) Mantispidae de Guyane (Neuroptera). L'Entomologiste 66(3):165–168
Walker F (1853) List of the specimens of neuropterous insects in the collection of the British Museum. Part II (Sialidae – Nemopterides). Trustees, London, pp 193–476
Walker F (1860) Characters of undescribed Neuroptera in the collection of W. W. Saunders, Esq., FRS, &c. Transactions of the Royal Entomological Society of London (NS) 5:176–199
Westwood JO (1852) On the genus *Mantispa,* with descriptions of various new species. Transactions of the Royal Entomological Society of London 1:252–270
Westwood JO (1867) Descriptions of new species of Mantispidae in the Oxford and British Museums. Transactions of the Royal Entomological Society of London 5(3):501–508
Williner GJ, Kormilev NA (1958) Notas sobre Mantispidae neotropicales, I (Neuroptera). Revista de la Sociedad Entomologica Argentina 21:1–18
Williner GJ, Mariluis JC (1979 [1978]) Revision del genero *Paramantispa* Williner y Kormilev, 1958 (Neuroptera, Mantispidae). Revista de la Sociedad Entomologica Argentina 37(1–4):39–46

Chapter 11
Dilaridae

Abstract All species known to inhabit South America are assigned to the genus *Nallachius*. An illustrated key is provided to identify the adults of the species. The countries or states in which each species has been found and lists of invalid synonyms are reported.

Keywords Pleasing lacewing • *Nallachius* • Key to species • Nallaiinae • Synonyms

11.1 Notes on *Nallachius*, the only known South American genus in the family

The English common name for species in this family is pleasing lacewing. According to the information on the taxonomy and discribution of the South American species (Adams 1970; Penny 1981; Oswald 1998; Bowles et al. 2015), two subfamilies are known, but only one of them, Nallaiinae, is known from the Western Hemisphere. It includes one genus, *Nallachius* Navás, 1909. Two other nominal genera, *Nulema* Navás, 1914, and *Neodilar* Carpenter, 1947, were reduced to junior synonyms by Adams (1970). In some earlier publications, the family name is incorrectly spelled Dilariidae.

Key to the Adult Males of the Known South American Species of *Nallachius*

Information for the key was provided by Navás (1909, 1911a, b, 1914, 1930); Carpenter (1947); Adams (1970); Penny (1981); Monserrat (2005); and Machado and Rafael (2010a, b). Because it is known only from Guatamala, *Nallachius championi* Navás, 1914, is not included. Should it occur in South America, it can be recognized by the 16 long processes on the antenna of the male. All known males of South American species, except for *Nallachius limai*, possess 14. The number in *N. limai* varies from 14 to 18.

C.W. Heckman, *Neuroptera (Including Megaloptera)*, Encyclopedia of South American Aquatic Insects, DOI 10.1007/978-3-319-35125-4_11

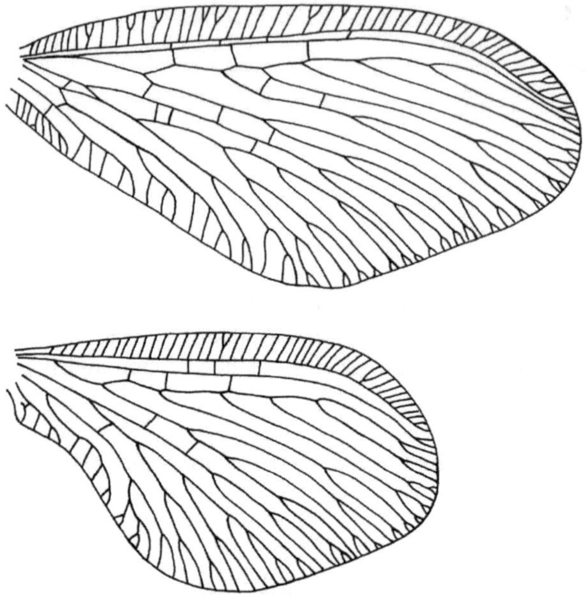

Fig. 11.1 Fore and hind wing of *Nallachius hermosus*. Based on Carpenter (1947).

1. Veins MP_2 and CuA in the fore-wing of the female are fused, and the costal veinlets are forked (Fig. 11.1). The labrum is black. The "wing spread" of the female is c. 25 mm. Only the female has been described. .... *Nallachius hermosus* (Banks, 1913) (Colombia). Syn: *Dilar hermosa* Banks, 1913; *Neodilar hermosa* (Banks, 1913).

   – Either veins MP_2 and CuA in the fore-wing are fused, or the costal veinlets are forked, but both do not occur simultaneously (Fig. 11.2)...................... 2

2. (1) Veins MP_2 and CuA in the fore-wing are fused. The veins in the costal area are not forked (Fig. 11.2). 3

   – Veins MP_2 and CuA in the fore-wing are not fused. A few of the small veins in the costal area may be forked (Fig. 11.3). The antenna bears no more than 14 long processes.. 7

3. (2) The apices of the wings are obviously asymmetrical, giving them angulate posterior margins (Fig. 11.2)... 4

   – The apical part of the wings is rounded and nearly symmetrical on both sides of its long axes (Fig. 11.4). 6

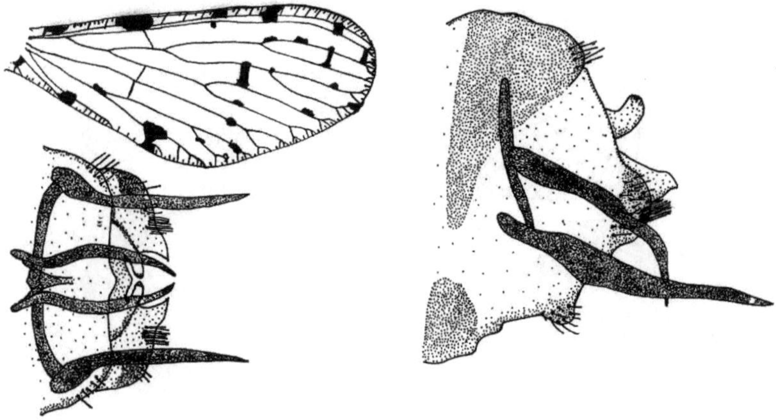

Fig. 11.2 *Nallachius adamsi* male: fore-wing with markings without showing hair-like setae (upper left); cleared apex of the abdomen in dorsal (lower left) and lateral view (right). Based on Penny (1981).

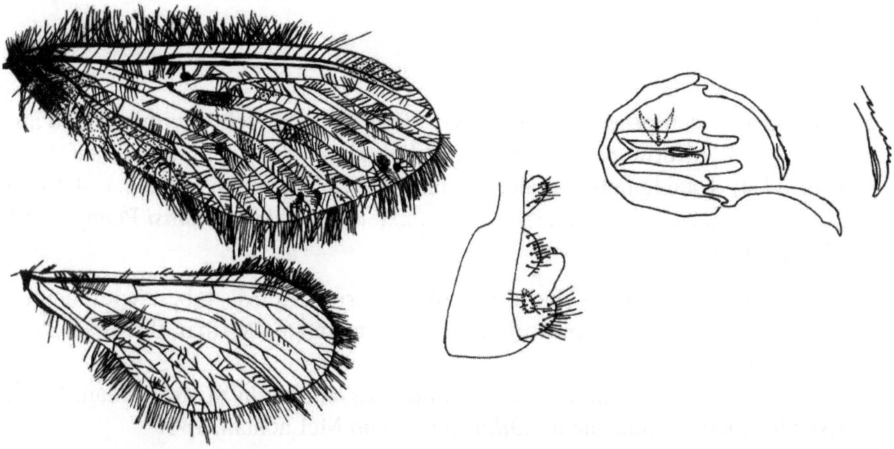

Fig. 11.3 *Nallachius phantomellus:* fore and hind wing of a female (left), apex of the male abdomen (center), internal genital structure (right center), and the apex of the gonocoxite (right). Based on Adams (1970).

4. (3) In the fore-wing, there are more than two radial crossveins. In both wings, there are several crossveins between the branches of Rs (Fig. 11.5). Fore wing length: c. 6.7 mm.
   ....*Nallachius loxanus* Navás, 1911 (Ecuador).
   – In the fore-wing, there are only two radial crossveins. In both wings, there are no crossveins between the branches of Rs (Fig. 11.2). 5

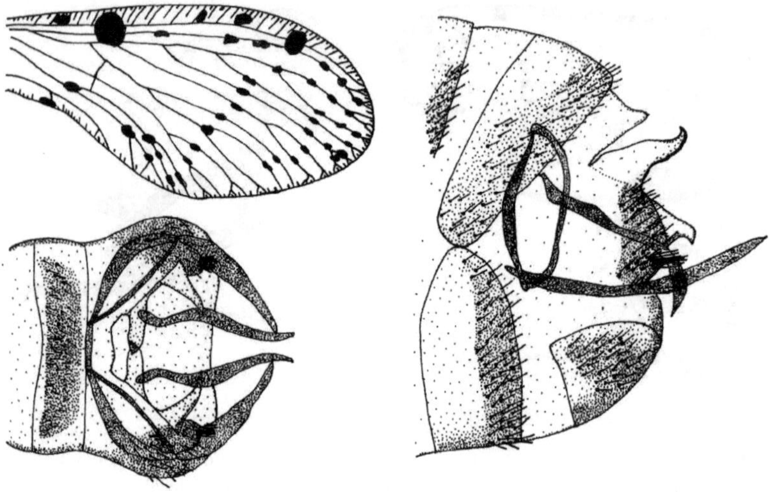

Fig. 11.4 *Nallachius maculatus* male: fore-wing with markings without showing hair-like setae (upper left); cleared apex of the abdomen is dorsal (lower left) and lateral view (right). Based on Penny (1981).

5. (4) The male ectoprocts have narrowly elongate dorsal lobes with rounded apices; the digitiform process is only slightly smaller than the dirsal lobes (Fig. 11.2). Total length of male: 2–3 mm. Fore-wing length of male: c. 4.1 mm. .... *Nallachius adamsi* Penny, 1981 (Amazonas).

   – The male ectoprocts have broadly tapering dorsal lobes with acutely pointed apices; the digitiform process is much smaller than the dirsal lobes (Fig. 11.6). .... *Nallachius americanus* (McLachlan, 1880). (North America, Venezuela). *Dilar americana* McLachlan, 1880

6. (3) There are four branches of Rs distal to the fused MA in the fore-wing (Fig. 11.4). Total length of male: c. 3.8 mm. Fore-wing length of male: c. 4.6 mm. .... *Nallachius maculatus* Penny, 1981 (Rondônia).

   – Vein Rs has three branches distal to a fused MA in the fore-wing (Fig. 11.7). .... *Nallachius bruchi* Navás, 1923 (Argentina, Paraguay). Syn: *Nulema pupillus* Navás, 1930; *Nallachius pupillus* (Navás, 1930) *auctt.* That *Nulema pupillus* is a junior synonym of *Nallachius bruchi* has been suggested by Monserrat (2005) and is presumed to be true. However, specimens of both nominal species are rare, and each type specimen may be unique after wartime losses of the collection holding a probable paratype. The type of *Nallachius bruchi* is a damaged female specimen from Argentina, while that of *Nulema pupillus* is a male from Paraguay.

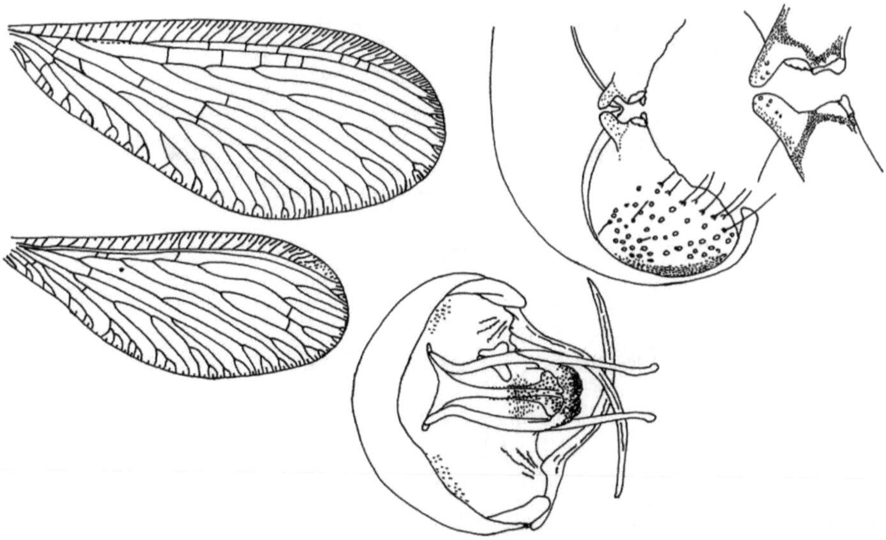

Fig. 11.5 *Nallachius loxanus* male: fore and hind wing not showing setae (left), middle and right side of the apex of the abdomen in dorsal view with an enlargement of the ectoprocts to the right (upper right), and the internal genitalia in dorsal view (lower center). Based on Adams (1970).

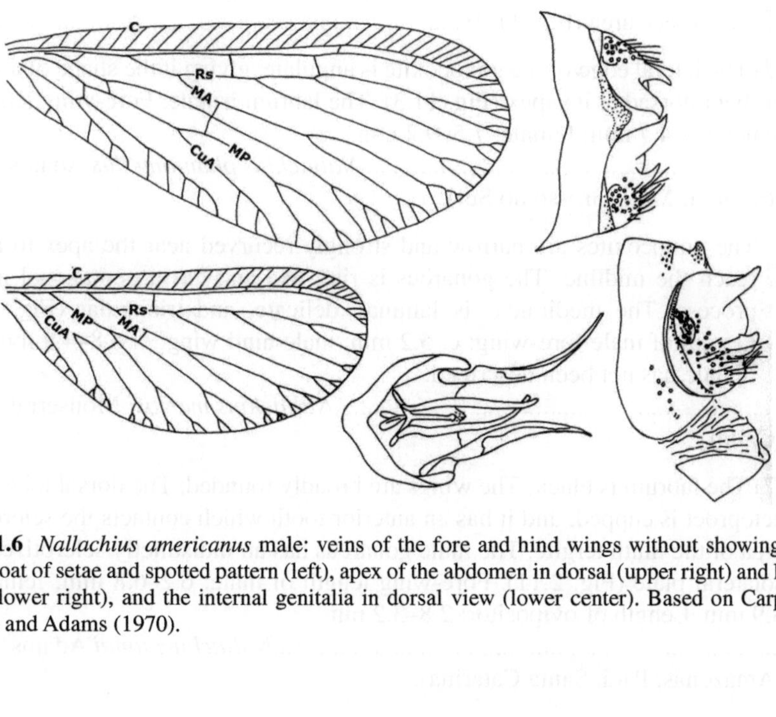

Fig. 11.6 *Nallachius americanus* male: veins of the fore and hind wings without showing their thick coat of setae and spotted pattern (left), apex of the abdomen in dorsal (upper right) and lateral view (lower right), and the internal genitalia in dorsal view (lower center). Based on Carpenter (1947) and Adams (1970).

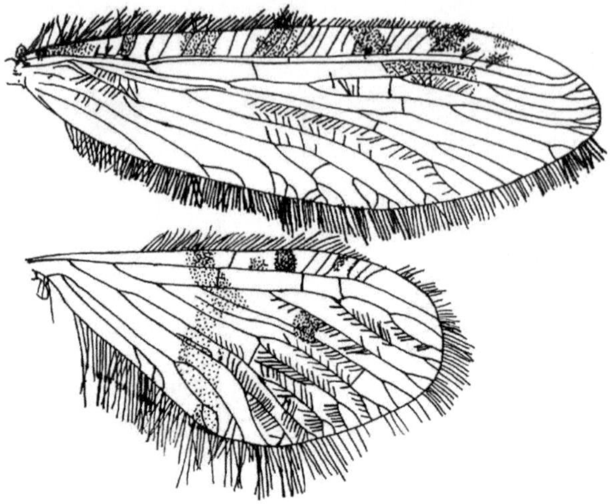

Fig. 11.7 Fore and hind wing of a female *Nallachius bruchi*. Based on Monserrat (2005).

7. (2) Both wings are uniformly smoky or pale with, at most, only a few small dark markings. The costal space is very narrow, especially on the hind wing (Fig. 11.8).. 8

   – The wings are darkened by many spots covering a considerable amount of the surface area (Fig. 11.9) 9

8. (7) The lateral edge of the gonocoxite is angulate, giving it the shape of a scimitar bent dorsad at its apex (Fig. 11.3). The labrum is pale. Fore-wing length of male: 4.6–4.7 mm; female: 7.5–9.2 mm.
   ....*Nallachius phantomellus* Adams, 1970
   (Ecuador, Mato Grosso do Sul).

   – The gonocoxites are narrow and strongly recurved near the apex to nearly reach the midline. The gonarcus is ring-like and has a pentagonal medial process. The mediuncus is laminar, delicate, and triangular (Fig. 11.8). Length of male fore-wing: c. 5.2 mm, male hind wing: c. 4.8–4.9 mm. The female has not been described.
   ....*Nallachius martosi* Monserrat, 2005
   (Chile).

9. (7) The labrum is black. The wings are broadly rounded. The dorsal lobe of the ectoproct is cupped, and it has an anterior tooth which contacts the sclerotized area of the ninth tergite. The male gonarcus has an indistinctly sclerotized dorsolateral plate (Fig. 2.11). Fore-wing length of male: 6.2–6.8 mm; female: c. 8.9 mm. Length of ovipositor: 2.8–3.2 mm.
   ....*Nallachius limai* Adams, 1970
   (Amazonas, Pará, Santa Catarina).

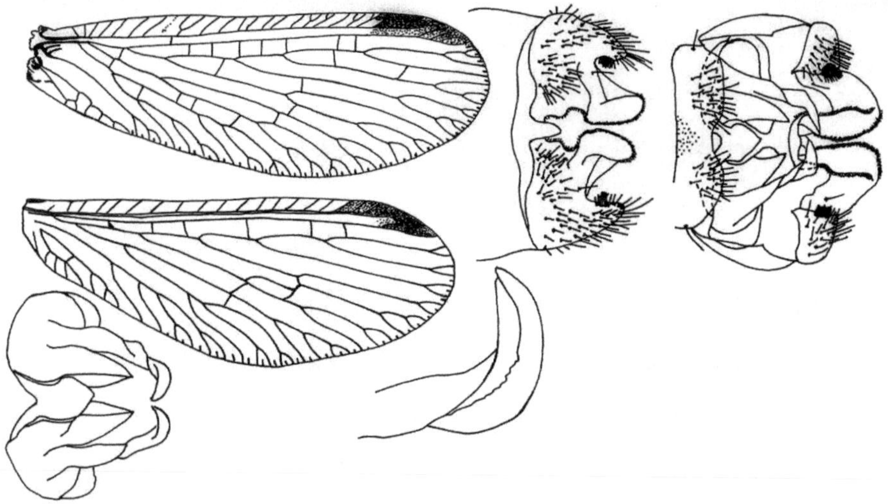

Fig. 11.8 Fore and hind wing of a male *Nallachius martosi* (upper and middle left); apex of the abdomen in dorsal view (upper right center) and posteroventral view cleared to show the internal genitalia (upper right), and the internal genitalia in ventral view (lower left) with the apical part of its gonocoxite enlarged (lower center). Based on Monserrat (2005).

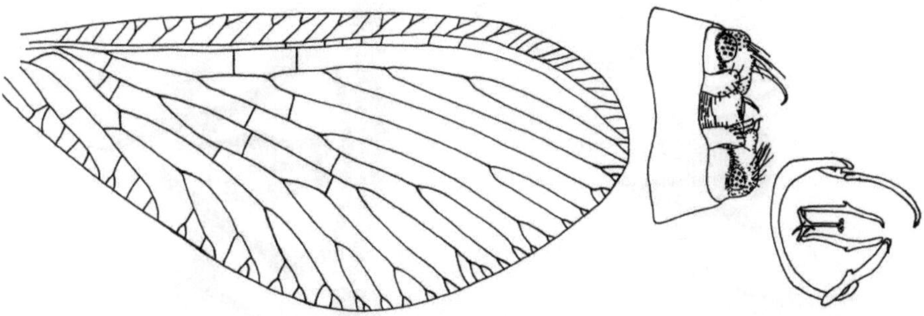

Fig. 11.9 *Nallachius prestoni* male (left to right): fore-wing without fine setae; apex of the abdomen and internal genital structures, both in dorsal view. Based on Adams (1970).

– The labrum is pale. The wings are somewhat elongate. The male gonarcus lacks a dorsolateral plate (Fig. 11.9). The dorsal lobe of the ectoproct is flat. A species known only from the southwestern United States and Cuba, *Nallachius pulchellus* (Banks, 1938), resembles the species in this group, but the male has an inflated dorsal lobe on the ectoproct 10

10. (9) There is a wide space between the dorsal lobes of the male ectoprocts across the dorsal midline (Fig. 11.9). 11

– The dorsal lobes of the male ectoprocts are very close at the dorsal midline (Fig. 11.10) 14

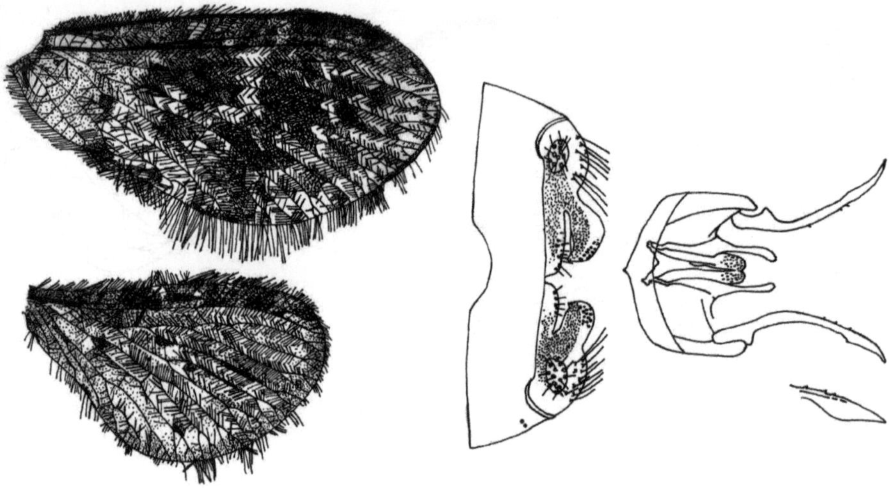

Fig. 11.10 *Nallachius dicolor* male (left to right): fore (above) and hind wing (below); apex of the abdomen and internal genital structures, both in dorsal view. Based on Adams (1970).

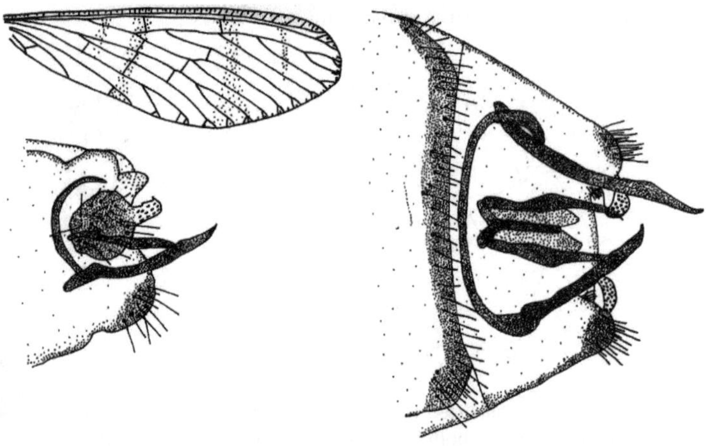

Fig. 11.11 *Nallachius infuscatus* male: fore-wing with infuscations but without showing hair-like setae (upper left); cleared apex of the abdomen in ventral (right) and lateral view (lower left). Based on Penny (1981).

11. (10) The costal area is narrow, has many forked costal veinlets, and is bordered by a straight subcosta (Fig. 11.9) 12

    – The subcosta curves posteriad at the stigma, making the costal area at the apex relatively wide. Few of the costal veinlets are forked (Fig. 11.11) 13

12. (11) The body is white but bears black tubercles. The fore-wing has narrow brown bands (Fig. 11.12). ....*Nallachius potiguar* Machado and Rafael, 2010 (Rio Grande do Norte).

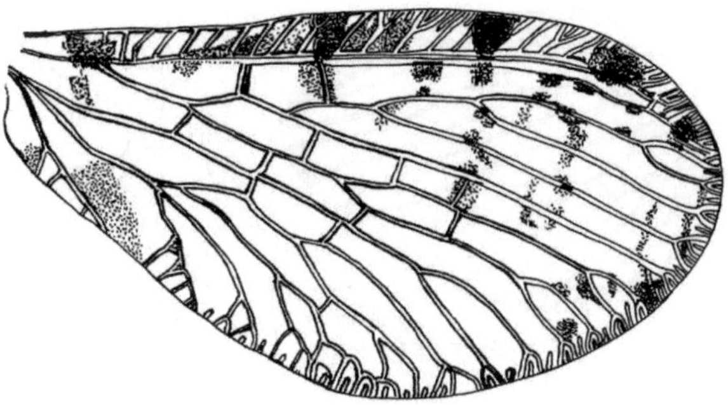

Fig. 11.12 Fore-wing of *Nallachius potiguar*. Based on Machado and Rafael (2010a).

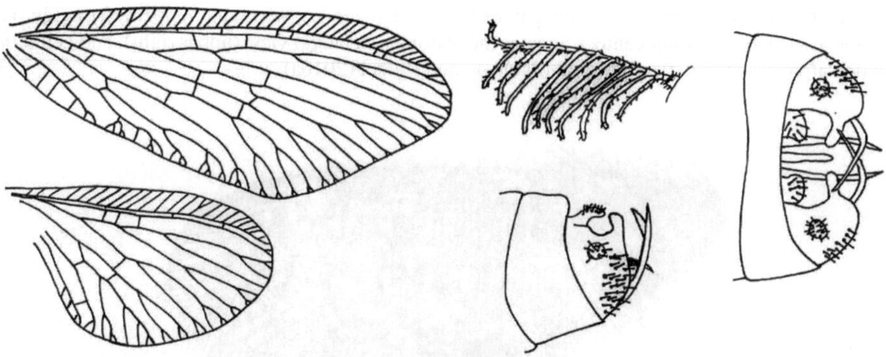

Fig. 11.13 *Nallachius reductus* male: fore and hind wing (left), antenna (upper center), and apex of the abdomen in dorsal (right) and lateral view (lower center). Based on Carpenter (1947).

– The body is all black. The fore-wing has broad brown bands (Fig. 11.9). ....*Nallachius prestoni* (McLachlan, 1880) (Rio de Janeiro). Syn: *Dilar prestoni* McLachlan, 1880.

13. (11) The gonocoxites are slender and do not expand apically. The finger-like process of the ectoproct lacks a median hook (Fig. 11.13). Fore-wing length of male: c. 5 mm. The female has not been described. ....*Nallachius reductus* Carpenter, 1947 (Paraguay).

– The gonocoxites expand apically to form a blade-like structure. The finger-like process of the ectoproct has a median hook (Fig. 11.11). Total length of male: 2.0–3.3 mm; female: 2.0–2.1 mm. Fore-wing length of male: c. 2.9–5.3 mm; female: 4.0–4.3 mm. ....*Nallachius infuscatus* Penny, 1981 (Amazonas, Rondônia).

14. (10) Vein Rs in the fore-wing has four branches. The apex of the median lobe of the male genitalia is forked (Fig. 11.14). ....*Nallachius furcatus* Machado and Rafael, 2010 (Paraíba).

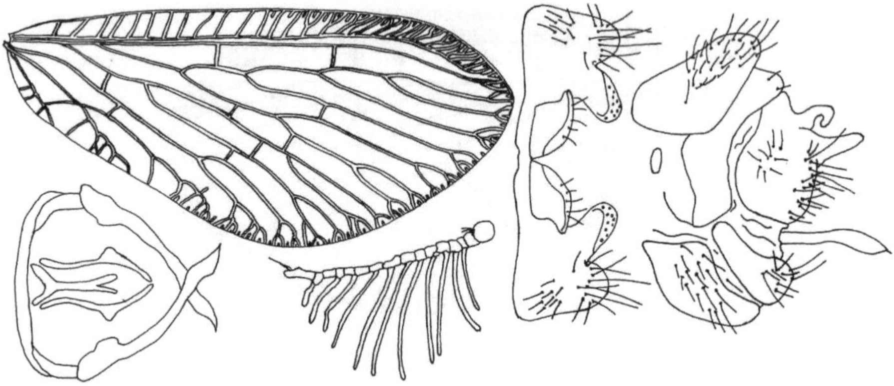

Fig. 11.14 *Nallachius furcatus* male: antenna (lower left center), fore-wing (upper left), ectoproct in dorsal view (upper right center), apex of the abdomen in lateral view (upper right), and genitalia in dorsal view (lower left). Based on Machado and Rafael (2010a).

Fig. 11.15 Fore and hind wing of a female *Nallachius ovalis*. Based on Adams (1970).

– Vein Rs in the fore-wing has five branches. The apex of the median lobe of the male genitalia is not forked, but its membrane is (Fig. 11.10) 15

15. (14) The costal area is relatively wide, and most of the costal veinlets are forked in the distal part of the wing (Fig. 11.15). Fore-wing length of male: c. 5.5 mm; female: 7.3–7.5 mm. Length of ovipositor: 2.8–3.2 mm. ....*Nallachius ovalis* Adams, 1970.

(Santa Catarina). Monserrat (2005)) examined the type specimens of the two nominal species in this couplet and concluded that they are very probably conspecific. However, he did not formally place them in synonymy.

– The costal area is relatively narrow, and most of the costal veinlets are not forked in the distal part of the wing (Fig. 11.10). Fore-wing length of male: 4.3–5.2 mm; female: 5.1–7.9 mm. Length of ovipositor: 1.6–2.2 mm.
....*Nallachius dicolor* Adams, 1970.
(Argentina, Goiás, Minas Gerais, Santa Catarina).

Literature

Adams PA (1970) A review of the New World Dilaridae. Postilla 148:1–30

Banks N (1913) Synopses and descriptions of exotic Neuroptera. Transactions of the American Entomological Society 39:201–242 + pl. 23–26

Banks N (1938) New West Indian neuropteroid insects. Revista de Entomologia, Rio de Janeiro 9:285–304

Bowles DE, Contreras-Ramos A, Sormiento-Cordero MA, Ferro ML (2015) New distributional records for pleasing lacewigs (Neuroptera: Dilaridae, *Nallachius* spp.) in the Americas. Insecta Mundi 0406:1–10

Carpenter FM (1947) Taxonomic notes on the Dilaridae (Neuroptera). Psyche 54:100–109

Machado RJP, Rafael JA (2010a) Two new species of Dilaridae (Insecta: Neuroptera) with additional notes on Brazilian species. Zootaxa 2421:61–68

Machado RJP, Rafael JA (2010b) Erratum: two new species of Dilaridae (Insecta: Neuroptera) with additional notes on Brazilian species. Zootaxa 2484:68

McLachlan R (1880) Occurrence of the neuropterous genus *Dilar* in South America. Entomologist's Monthly Magazine 17:39

Monserrat VJ (2005) Nuevos datos sobre angunas pequeñas familias de neurópteros (Insecta: Neuroptera: Nevrothidae, Osmylidae, Sisyridae, Dilaridae). Heteropterus Revista de Entomologia 5:1–26

Navás L (1909) Monografia de la familia de los Diláridos (Ins. Nevr.). Memorias de la Real Academia de Ciancias y Artes de Barcelona 7(3):619–671

Navás L (1911a) Dilariidae (Ins. Nevr.) neuveau. Annales de la Socièté Scientifique de Bruxelles 35(1):219–221

Navás L (1911b) Donnant suite aux conclusions favorables des rapports de M.M.F. Meunier et H. Lebrun, la section vote l'insertion aux Annales de qutre notes du R.B. Navas, S. J.: A) Dilaride (Ins. Névr.) nouveau; B) Deux Ėphémérides (Ins. Névr.) Nouveaux du Congo belge; C) Sur une nouvelle espèce de Némoptérides (Ins. névr.) du Conge Belge: D) Insecte Névroptère nouveau de l'Amérique méridionale. Annales de la Société scientifique de Bruxelles 3:1–10

Navás L (1914) Family Dilariidae. Neuroptera. Genera Insectorum 156:1–14

Navás L (1930) Insectos neotropicos, (Sexta Serie). Revista Chilena de Historia Natural 34:62–75, 299–307

Oswald JD (1998) Annotated catalogue of the Dilaridae (Insecta, Neuroptera) of the world. Tijdschrift voor Entomologie 141:115–128

Penny ND (1981) Neuroptera of the Amazon Basin, Part 2. Dilaridae. Acta Amazonica 11(2):383–390

Chapter 12
Hemerobiidae

Abstract An illustrated key is provided for identification of the genera of adult hemerobiids known to occur in South America. Illustrated keys are also provided to identify the species of each genus native to South America. The countries or states in which specimens of each species have been found and invalid synonyms are reported.

Keywords Brown lacewing • Key to species • Adelphohemerobiinae • Larvae

12.1 Key to the South American Genera of Adults in the Family Hemerobiidae

The English common name for the insects in this family is brown lacewing. Information for the key was provided by Navás (1912, 1915, 1921a, b, 1924a, b, 1927a, b, 1932a), Banks (1913), Nakahara (1965), Penny and Sturm (1984), Monserrat (1990a), Gonzalez Olazo (1992a), Oswald (1993a, b, 1994), and Garzón-Orduña et al. (2016). The key may be valid only for distinguishing known South American species from one another but not from species native to other continents.

1. The radius in the fore-wing has only one oblique branch. The female insemination canal is slit-like. The female has a distinct subanale. The stylus of the female ninth gonocoxite is elongate (Fig. 12.1). The fore-wing length of the female from the tegula to the apex is about 9.7 mm. Only the female has been described.

   .... *Adelphohemerobius* Oswald, 1994..p. 366

   This genus has been placed in its own subfamily: Adelphohemerobiinae.

   – The radius in the fore-wing has two or more oblique branches. The female insemination canal is pore-like. All females that have been described lack a subanale. The stylus of the female ninth gonocoxite is short or absent (Fig. 12.2). 2

C.W. Heckman, *Neuroptera (Including Megaloptera)*, Encyclopedia of South American Aquatic Insects, DOI 10.1007/978-3-319-35125-4_12

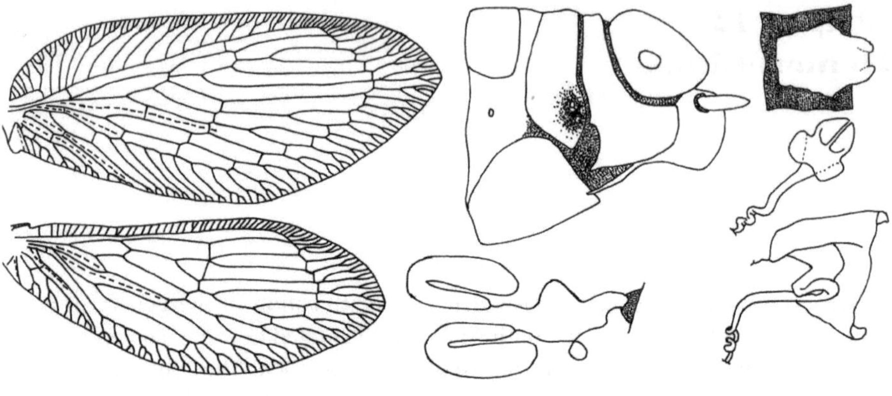

Fig. 12.1 *Adelphohemerobius enigmaramus* female: fore and hind wing (left), apex of the abdomen in lateral view (upper center), subgenitale in ventral view (upper right), bursal accessory glands and apex of the bursa (lower center), and the insemination, or fertilization canal in anterodorsal (middle right) and lateral view (lower right). Based on Oswald (1994).

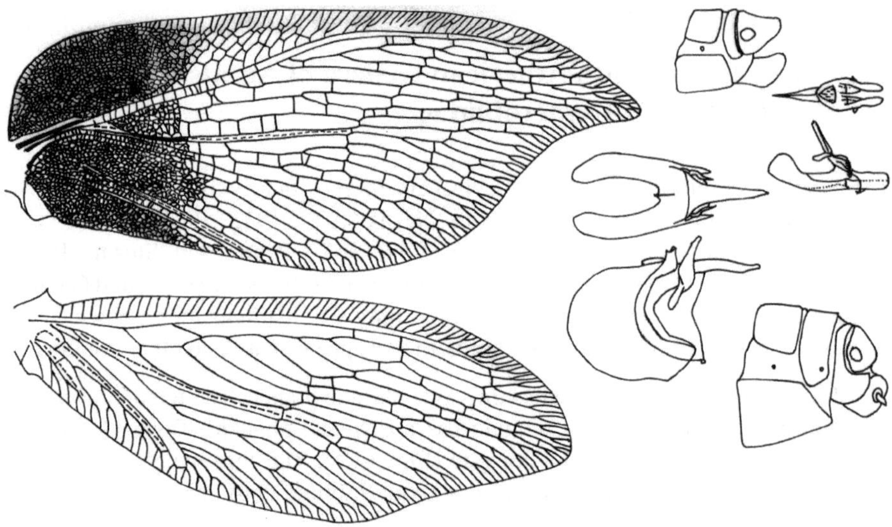

Fig. 12.2 *Gayomyia falcata*: fore and hind wing (left), apex of the male abdomen in lateral view (upper right center), parabaculum in dorsal (upper right) and lateral view (middle right), gonarcus in dorsal (right center) and lateral view (lower right center), apex of female abdomen in lateral view (lower right). Based on Oswald (1993a).

2. (1) Each fore-wing has at least ten sc-r cross veins proximal to the stigma, and the fore-wings are falcate and finely reticulate on the proximal portion. There is one row of cells in the subcostal space. The hind wing is not obviously smaller than the fore-wing (Fig. 12.2).

   ....*Gayomyia* Banks, 1915

   The only species included in this genus at the present time is *Gayomyia falcata* Banks, 1915. The name *Gayomyia cinerea* Krüger, 1922, is considered a *nomen*

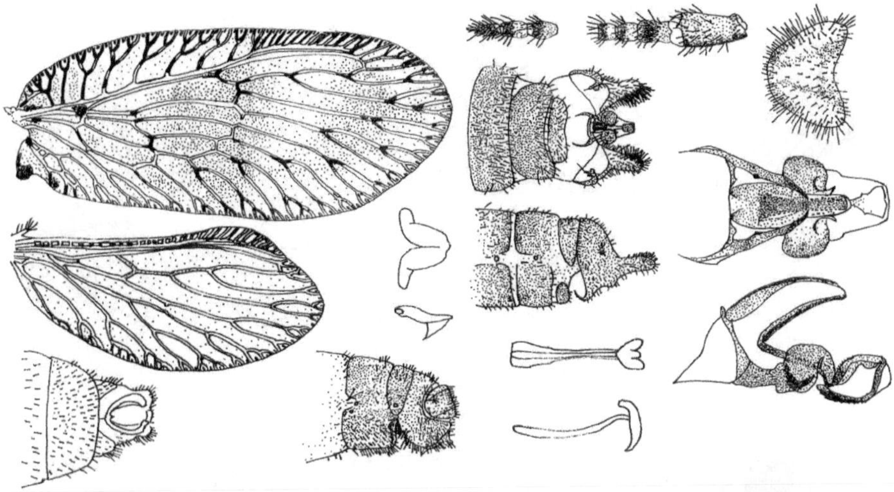

Fig. 12.3 *Notiobiella paddiae*: fore and hind wing (upper and middle left), basal (upper right center) and apical segments of the antenna (upper center), pronotum in dorsal view (upper right), apex of the male abdomen in ventral and lateral view (right center, above and below, respectively), internal male genitalia in dorsal (middle right) and lateral view (lower right), male hypandrium in ventral and lateral view (center, above and below, respectively), male paramere in ventral and lateral view (lower right center, above and below, respectively), and the apex of female abdomen in ventral (lower left) and lateral view (lower left center). Based on Monserrat (1984).

nudum (Monserrat, 1998). Faúndez (2005) detected a relationship between this insect and a plant in the family Saxifragaceae, *Ribes magellanicum* Poiret.

– The proximal portion of the fore-wing is not reticulate, and each fore-wing has no more than three sc-r cross veins proximal to the stigma (Fig. 12.3). 3

3. (2) The fore-wing of the male is much longer and larger than the hind wing. The costal space of the fore-wing is much wider near the base of the wing than just proximal to the pterostigma (Fig. 12.3). 4

– The fore-wing of the male is not much larger than the hind wing, and/or the costal space is not notably enlarged, and/or the space between subcostal and radial veins of the fore-wing is always at least double the size of the adjacent vein Sc (Fig. 12.4). 6

4. (3) The male has an eversible phallolingua on the gonarcus. The ectoprocts of the male are not large and flap-like, or their medial surfaces are not sclerotized. The space between veins Sc and R in the fore-wing of the male is no wider than the adjacent vein Sc for its entire length proximal to the stigma, except sometimes adjacent to a cross vein, where it is only slightly wider (Figs. 2.18 and 12.3). .... *Notiobiella* Banks, 1909..p. 367

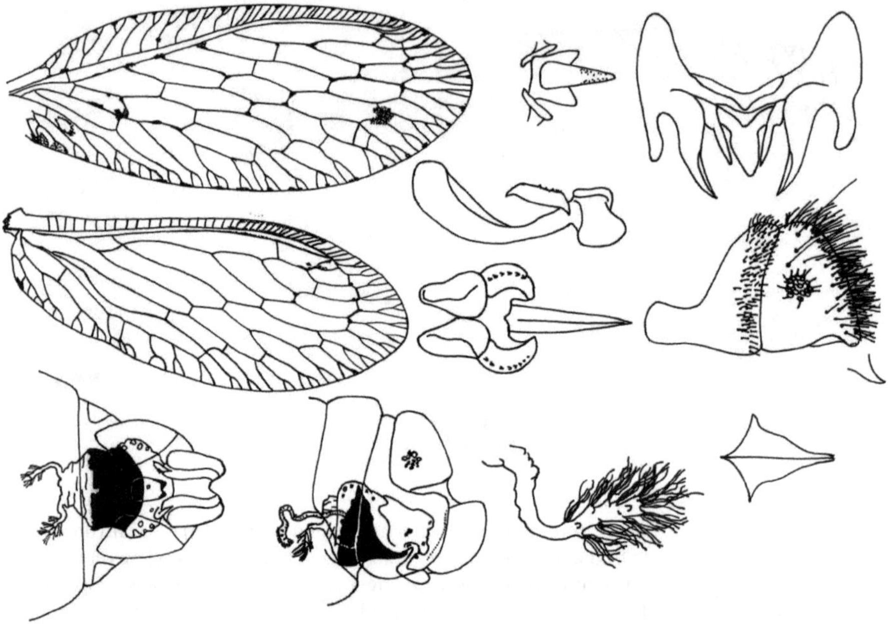

Fig. 12.4 *Nusalala erecta*: fore and hind wing (upper and middle left), internal genitalia of a male in ventral (upper center) and posterior view (upper right), paramere in dorsal and lateral view (center, below and above, respectively), hypandrium in ventral view (lower right), ninth tergite and ectoproct in lateral view (middle right) with the ventral apex enlarged shown to the lower right, apex of the female abdomen in ventral (lower left) and lateral view (lower left center), and the spermothecal gland of a female in dorsal view (lower center). Based on Monserrat (2000).

– The male lacks an eversible phallolingua on the gonarcus. The space between veins Sc and R is obviously wider than the adjacent veins. The costal vein of the male fore-wing is sharply curved near the wing base, making the costal space occupy at least one third of the wing width near the base. The costal space of the hind wing is narrow (Fig. 12.5). 5

5. (4) The costal area usually occupies more than half the width of the fore-wing and is crossed by more than 20 cross veins, usually branched more than once and sometimes connected by smaller cross veins. The ectoprocts of the male are large and flap-like, and their medial surfaces are well sclerotized. The gonarcus has a gonofenestral plate. The parabaculum is narrow and arches proximal to its apical lobes. The fore-wings of two species inhabiting Juan Fernandez Island are remarkable, being nearly circular with a coriaceous membrane and unusual arrangement of veins. The humeral area is extremely broad (Fig. 12.5). .... *Conchopterella* Handschin, 1955 pars..p. 369

– The costal area usually occupies somewhat more than 30 % of the proximal wing width and is often crossed by more than 14 cross veins, many of which branch more than once. The long veins are pale but may be irregularly

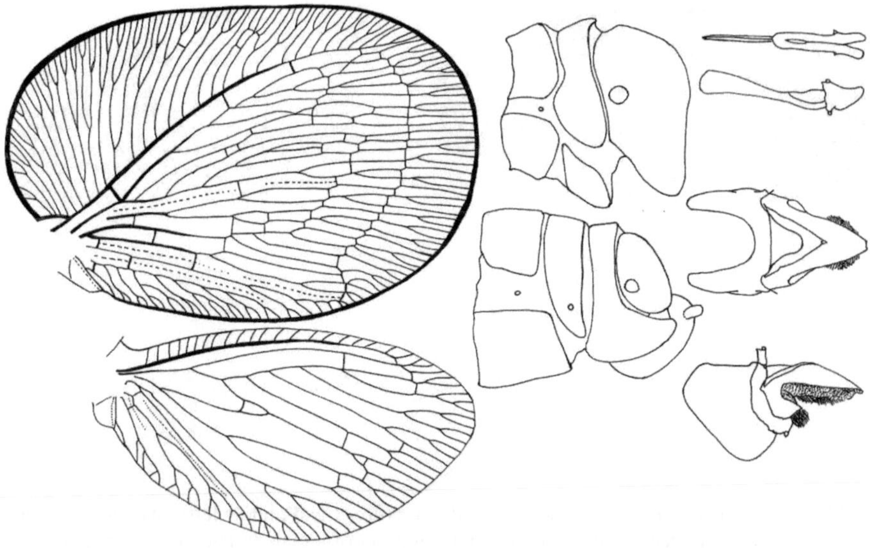

Fig. 12.5 *Conchopterella kuscheli*: fore (upper left) and hind wing (lower left), lateral view of the male (upper right center) and female (lower right center), parabaculum in dorsal (upper right) and lateral view (upper middle right), and the gonarcus in dorsal (lower middle left) and lateral view (lower right). Based on Oswald (1993a).

striated with blackish markings. The ectoproct of the male does not have a pecten (Fig. 12.6). Length of fore-wing: c. 8.4 mm; hind wing: c. 6.8 mm.
.... *Hemerobiella* Kimmins, 1940..p. 370

6. (3) The fore-wing has three series of gradate veinlets (Fig. 12.4).
....*Nusalala* Navás, 1913..p. 371
The description of this genus is similar to those of *Micromus* Rambur, 1842, and *Megalomina* Banks, 1909. At present, no species of either genus is known from South America. Species of *Micromus* are known with certainty from all continents except for Antarctica and South America. It would therefore not be surprising if one were discovered in South America, or if unconfirmed reports of the genus in South America were confirmed (Oswald 1993a).

– The fore-wing has one or two series of gradate veinlets (Fig. 12.7). 7

7. (6) Neither fore-wing has more than two radial sectors proximal to the stigma (Fig. 12.5). 8

– At least one of the fore-wings has four or more radial sectors, not including MA, proximal to the stigma (Fig. 12.8). 12

8. (7) In the fore-wing, the basal part of MA is present (Fig. 12.7).
.... *Hemerobius* Linnaeus, 1758..p. 376

– In the fore-wing, the basal part of MA is absent (Fig. 12.8). 9

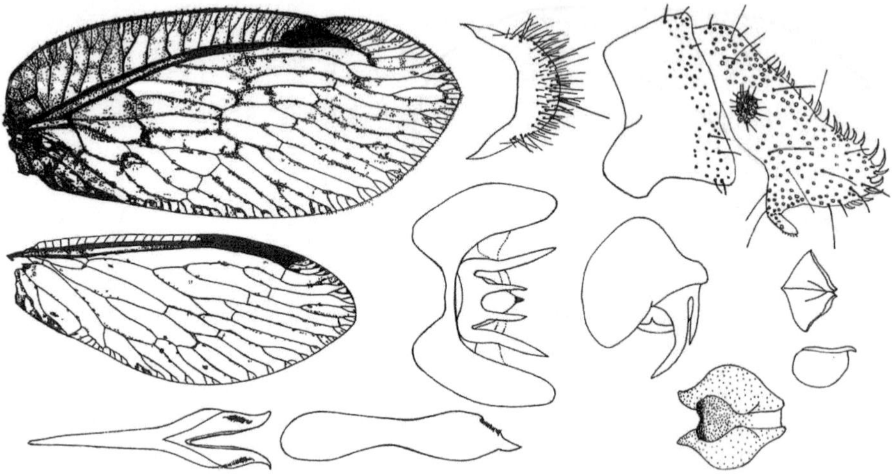

Fig. 12.6 *Hemerobiella oswaldi*: Fore and hind wing (upper and middle left), male ninth sternite in ventral view (upper center), ninth tergite and ectoproct of a male in lateral view (upper right), male internal genitalia in posterior (center) and lateral view (right center), male paramere in dorsal (lower left) and lateral view (lower left center), hypandrium in ventral (middle right) and lateral view (lower right), and the female subgenital plate in ventral view (lower right center). Based on Monserrat (1998).

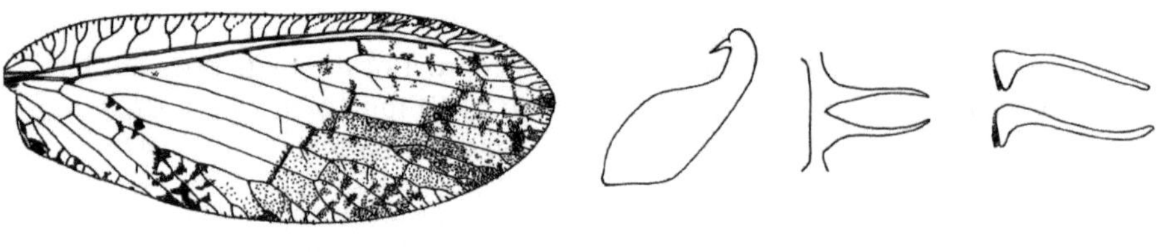

Fig. 12.7 *Hemerobius chilensis* male (left to right): fore-wing, anal plate, processes of the penis in dorsal view, and the parameres. Based on Nakahara (1965).

9. (8) The most proximal fork of the first oblique radial branch is distal to the most proximal fork of the second oblique radial branch. The subcostal space of the fore-wing is notably wider than the adjacent subcosta in the basal part of the wing (Fig. 12.9).
   ....*Biramus* Oswald, 1993..p.390
   - The most proximal fork of the first oblique radial branch is proximal to the most proximal fork of the second oblique radial branch (Fig. 12.10). 10

10. (9) The fourth series of cross veins in the fore-wing includes no more than two cross veins, and these cross veins are usually entirely absent. The head of the male has a bilobed frontal cavity (Fig. 12.10).
    ....*Neosympherobius* Kimmins, 1929
    The only known species in this genus is *Neosympherobius cinereus* Kimmins, 1929, from Argentina.

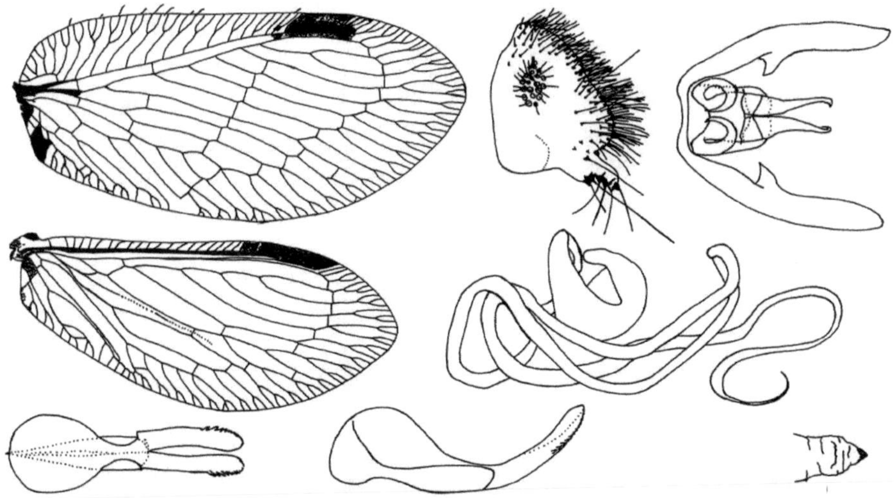

Fig. 12.8 *Megalomus marginatus*: fore and hind wing (upper and middle left), male ectoproct in lateral view (upper center), internal male genitalia in posterior view (upper right), male parameres in dorsal (lower left) and lateral view (lower center), female spermatheca (center to middle right), and subgenital plate of the female (lower right). Based on Monserrat (1997).

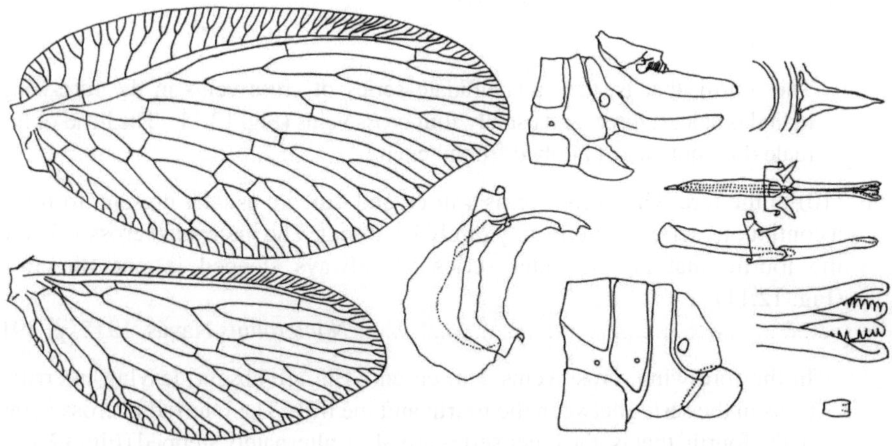

Fig. 12.9 *Biramus lunatus*: fore and hind wing (left), apex of the male abdomen in lateral view with its ectoproct in posterodorsal view above it (right center), gonarcus and mediuncus of a male specimen in lateral view (lower center), apex of a female abdomen in lateral view (lower right center), and (right, from top to bottom): male gonopons and mediuncus in dorsal view, male parabaculum in dorsal and lateral view, apex of the male parabaculum terminal lobes enlarged in dorsal view, and the female subgenitale in ventral view. Based on Oswald (1993b).

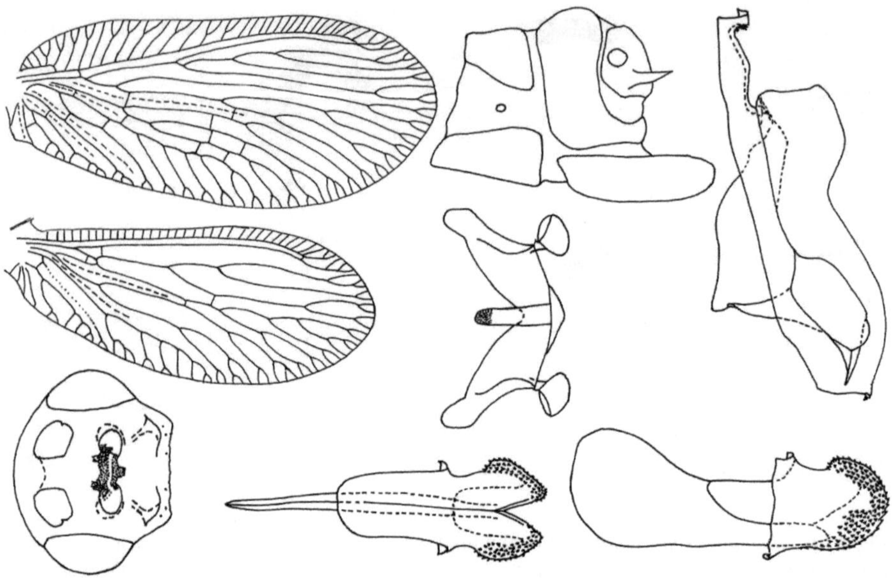

Fig. 12.10 *Neosympherobius cinereus* male: fore and hind wing (upper and middle left), head in anterior view (lower left), apex of the abdomen in dorsal lateral view (upper right center), gonarcus is dorsal (center) and lateral view (upper right), and parabaculum in dorsal (lower center) and lateral view (lower right). Based on Oswald (1993a).

– The fourth, that is, the outer gradate series of cross veins in the fore-wing includes at least three, and usually four cross veins (Fig. 12.11). The head of the male does not have a bilobed frontal cavity 11

11. (10) In the fore-wing, cross veins 4 m-cu and 4im are usually present, forming a continuous series, but one may rarely be absent. The interradial cross veins in the fourth, that is, the outer series are always stepped in one direction (Fig. 12.11).
.... *Nomerobius* Navás, 1915..p. 391

– In the fore-wing, cross veins 4 m-cu and 4im are absent, leaving interruptions in the series between the fourth and the fifth. The interradial cross veins in the fourth, that is, the outer series are often alternately stepped (Fig. 12.12).
.... *Sympherobius* Banks, 1904..p. 393

12. (8) The male seventh tergite bears a long, cylindrical process.
.... *Pirionius* Navás, 1915
No South American species are presently assigned to this genus. The species from South America appearing in older literature under this generic name are now included in *Megalomus*.

– The seventh tergite of the male does not bear a process (Fig. 12.8).
.... *Megalomus* Rambur, 1842..p.399

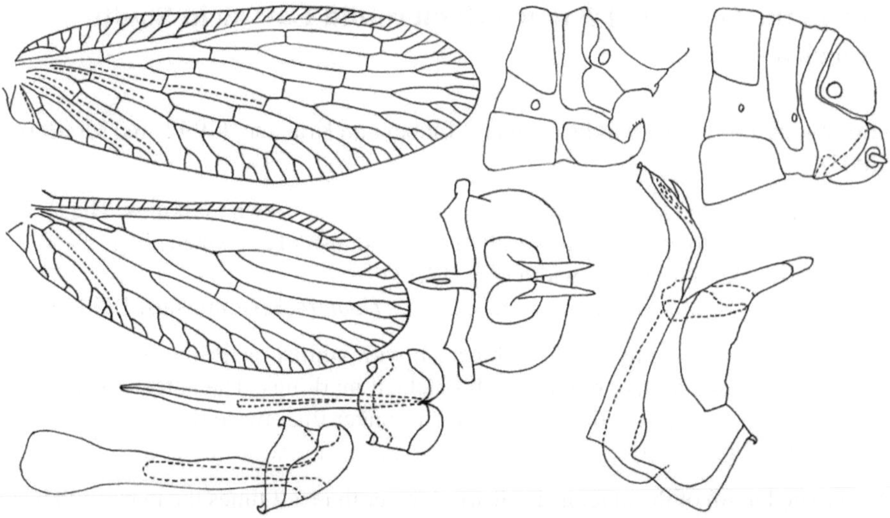

Fig. 12.11 *Nomerobius psychodoides*: fore and hind wings (upper and upper middle left), apex of the abdomen of a male (upper right center) and a female in lateral view (upper right), gonarcus in dorsal (center) and lateral view (lower right), and the parabaculum in dorsal (lower middle left) and lateral view (lower left). Based on Oswald (1993a).

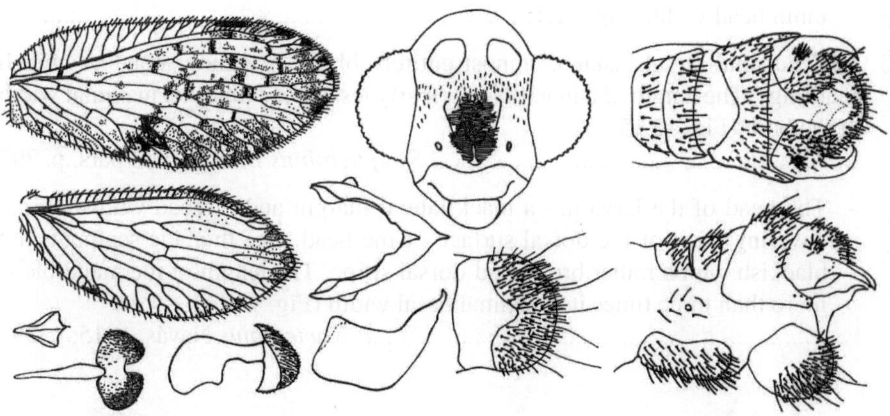

Fig. 12.12 *Sympherobius intermedius* male: fore and hind wing (upper and upper middle left), head in anterior view (upper center), apex of the abdomen in dorsal (upper right) and lateral view (lower right), ninth sternite in ventral view (lower right center), gonarcus-arcessus process in anterodorsal (center) and lateral view (lower center), paramere in ventral (lower left) and lateral view (lower left center), and the hypandrium in ventral view (lower middle left). Based on Monserrat (1998).

Key to the Genera of Known Hemerobiid Larvae in South America

Information for the key was provided by Tauber and Krakauer (1997) and Monserrat (2003).

1. The larvae of the known species are yellowish with brown markings. The antennae are much longer than the width of the head, usually between about 1.4 to 2.1 times the maximum head width (Fig. 12.13).
   .... *Hemerobius* Linnaeus, 1758..p. 376
   - The color is mainly blackish, blackish with whitish markings, or, in the case of first stage larvae, reddish with blackish markings. The antenna is roughly equal to the head width, or it is much greater than twice the maximum head width (Fig. 12.14) 2
2. (1) The length of the antenna is clearly greater than 2.3 times the maximum head width (Fig. 12.14).
   .... *Gayomyia* Banks, 1915
   The only species included in this genus at the present time is *Gayomyia falcata* Banks, 1915. The name *Gayomyia cinerea* Krüger, 1922, is considered a *nomen nudum* (Monserrat, 1898).
   - The length of the antenna is almost as long as or slightly longer than the maximum head width (Fig. 12.15) 3
3. (2) The final larval instar is almost entirely blackish without notable whitish markings. The length of a mandible is clearly less than 3 times its maximal width at the base (Fig. 12.15).
   .... *Sympherobius* Banks, 1904 pars..p. 393
   - The head of the larva has a black lateral margin and a broad whitish stripe covering most of the dorsal surface of the head. The thoracic segments are blackish with a rather broad mid-dorsal stripe. The length of the mandible is more than three times its maximal basal width (Fig. 12.16).
   .... *Nomerobius* Navás, 1915..p. 391

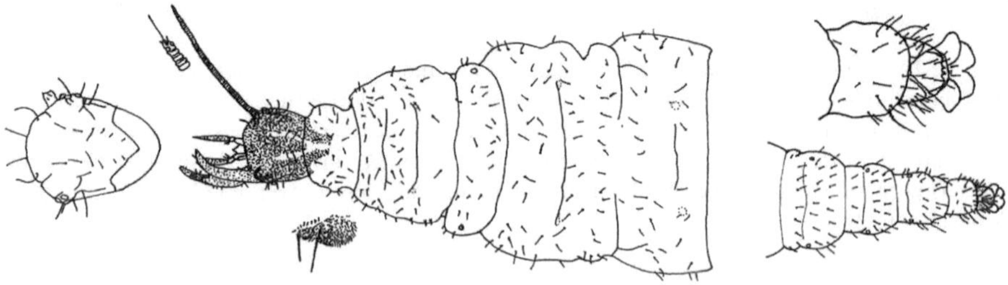

Fig. 12.13 *Hemerobius chilensis* last larval instar (left to right): head capsule in dorsal view, anterior segments of the larvae with an enlarged apex of the antenna and enlarged surface of the integument, and the apical segments (below) and an enlarged apex of the abdomen (above). Based on Monserrat (2003).

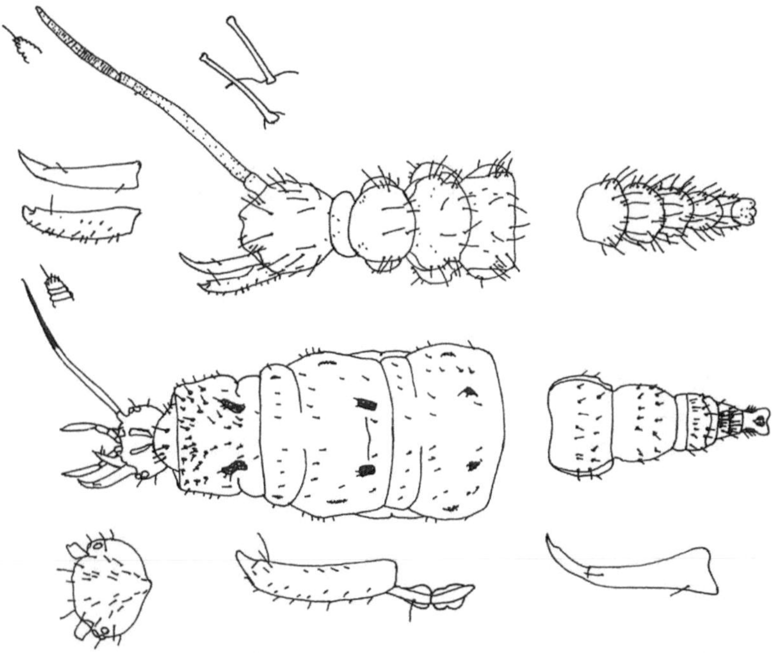

Fig. 12.14 *Gayomyia falcata* larvae: a first instar larva showing only the anterior and posterior segments and one antenna and a single set of mouthparts (upper left to upper right) with an enlargement of the apex of the antenna, two setae on the head enlarged (above the head), and a maxilla (below) and mandible in dorsal view (above) to the left of the head, a final instar larva showing only the anterior and posterior segments (below the middle, left to right) with an enlargement of the apex of the antenna, a dorsal view of the head capsule (lower left), and an enlarged maxilla (lower center) and mandible (lower right) in dorsal view. Based on Monserrat (2003).

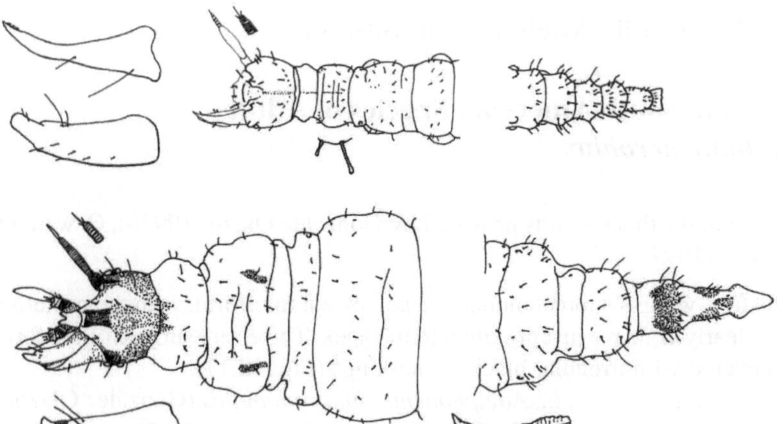

Fig. 12.15 *Sympherobius gayi* larvae: a first instar larva showing only the anterior and posterior segments and one antenna and a single set of mouthparts (upper left center to upper right) with an enlargement of the apex of the antenna, two setae on the thorax enlarged (below the thorax), and a maxilla (upper middle left) and mandible in dorsal view (upper left), a final instar larva showing only the anterior and posterior segments (lower middle, left to right) with an enlargement of the apex of the antenna, and an enlarged maxilla (lower left) and mandible (lower right) in dorsal view. Based on Monserrat (2003).

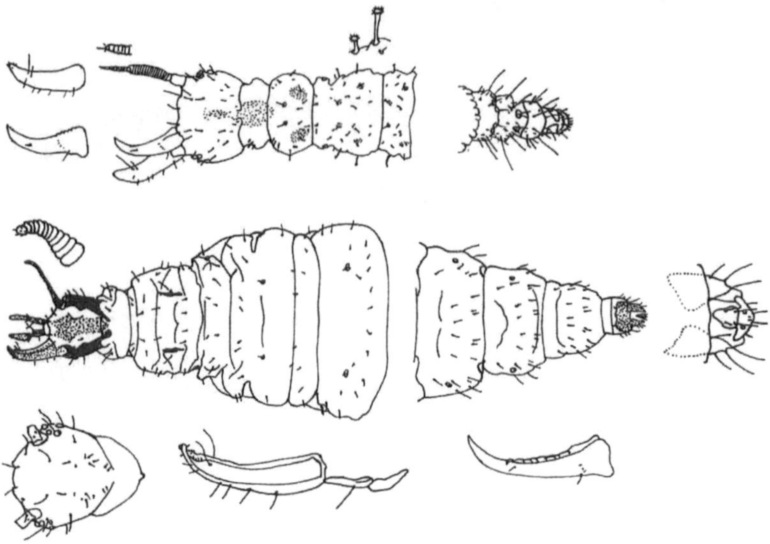

Fig. 12.16 *Nomerobius cuspidatus* larvae: a first instar larva showing only the anterior and posterior segments and one antenna and a single set of mouthparts (left center to upper right) with an enlargement of the apex of the antenna (above the antenna), two setae on the thorax enlarged (upper center), and a maxilla (upper left) and mandible in dorsal view (upper middle left), a final instar larva showing only the anterior and posterior segments (lower middle, left to right) with an enlargement of the apex of the antenna at its far right, a dorsal view of the head capsule (lower left), and an enlarged maxilla (lower center) and mandible (lower right) in dorsal view. Based on Monserrat (2003).

12.2 Subfamily Adelphohemerobiinae

Key to the South American Species of Adult *Adelphohemerobius*

Information for this key was provided by Gonzalez Olazo (1992b), Oswald (1994), Monserrat (1997).

1. The fore-wing is subrectantular and somewhat falciform, with the posterior margin clearly concave just proximal to the apex. The entire surface of the fore-wing is covered with irregular blackish markings (Fig. 12.17).
   ....*Adelphohemerobius anomalus* (Gonzãlez Olazo, 1992) (Argentina). Syn: *Pirionius anomalus* Gonzãlez Olazo, 1992.

   – The fore-wing is broadly elliptical with hyaline membranes and a few dark markings, including small spots along the wing margin (Fig. 12.1). Length of fore-wing: c. 9.7 mm.
   ....*Adelphohemerobius enigmaramus* Oswald, 1994 (Chile).

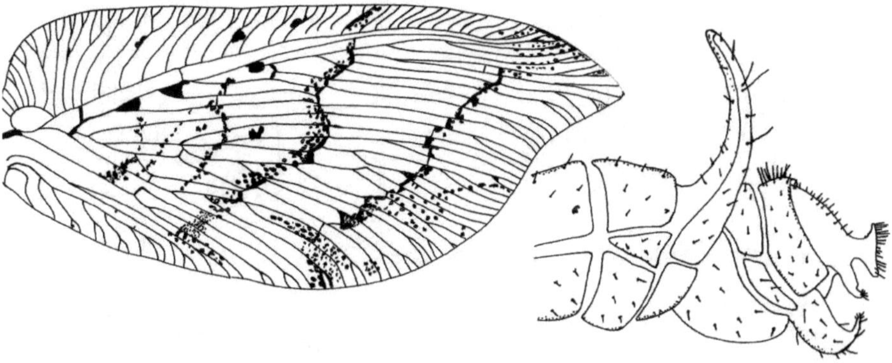

Fig. 12.17 *Adelphohemerobius anomalus* male: fore-wing (left) and apex of the abdomen in lateral view (right). Based on González-Olazo (1992b).

12.3 Subfamily Notiobiellinae

Key to the Species of Adult *Notiobiella* Known from South America

Information for the key was provided by Monserrat and Penny (1983), Penny and Monserrat (1985), Monserrat (1984), and Gonzalez Olazo (1992c).

1. There is a subapical black spot on both the fore and middle tibiae (Fig. 12.18). Length of fore-wing: c. 5.2 mm. .... *Notiobiella maculata* Monserrat and Penny, 1983 (Amazonas).

   – The color of the fore and middle tibiae is uniformly pale or dark, but without subapical spots on the fore and middle tibiae (Fig. 12.19)............................ 2

2. (1) The pterostigma of the fore-wing is pale. There is sometimes faint streaking in the costal area of the fore-wing. The male paramere is longer than twice its width (Fig. 12.20). .... *Notiobiella brasiliensis* Monserrat and Penny, 1985 (Colombia, Pará, Amazonas, Rondônia, Acre).

   – The pterostigma of the fore-wing is dark and sometimes streaked, or it is dark yellow with red at its proximal and distal end. The male paramere is distinctly shaped with three ridges on its posterior surface (Fig. 12.19........................ 3

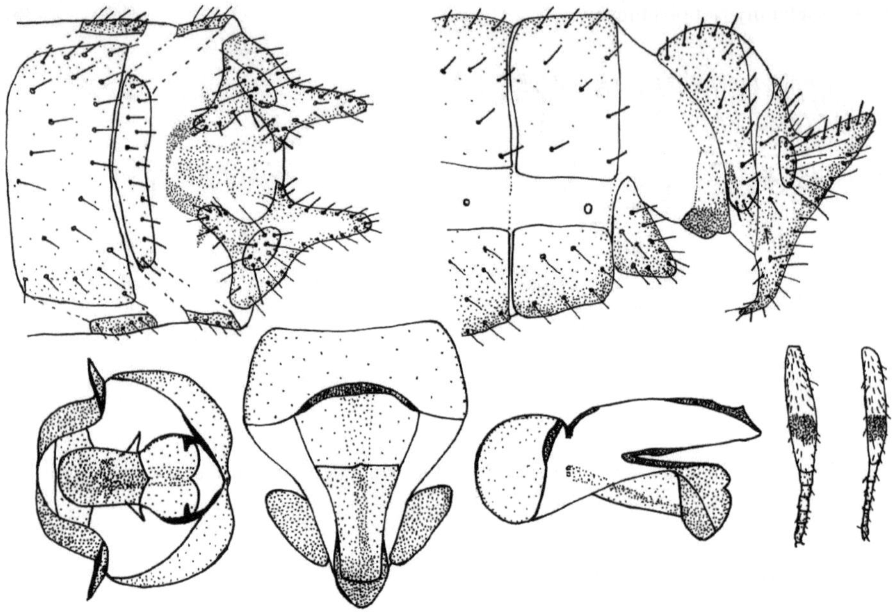

Fig. 12.18 *Notiobiella maculata* male (above, left and right): apex of the male abdomen in ventral and lateral view, and (below, left to right): male genitalia in ventral, posterior, and lateral view; the left fore and middle leg. Based on Monserrat and Penny (1983).

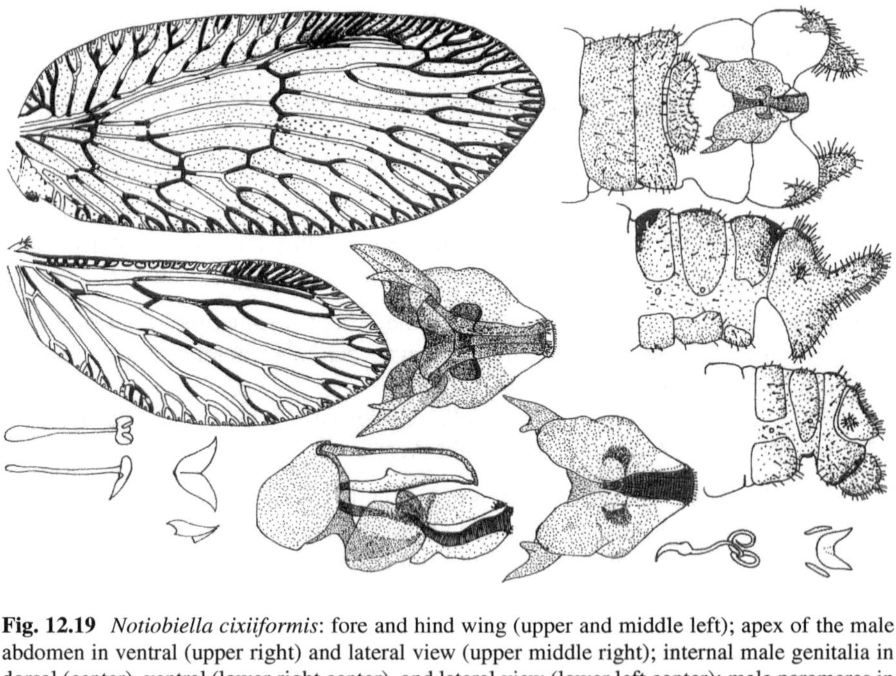

Fig. 12.19 *Notiobiella cixiiformis*: fore and hind wing (upper and middle left); apex of the male abdomen in ventral (upper right) and lateral view (upper middle right); internal male genitalia in dorsal (center), ventral (lower right center), and lateral view (lower left center); male parameres in dorsal and lateral view (lower left, above and below, respectively); male hypandrium in ventral and lateral view (right of parameres, above and below, respectively); apex of the female abdomen in lateral view (lower middle right); female spermatheca and subgenitale (lower right, left and right, respectively) (Based on Monserrat (1984), who referred to the species by its probable junior synonym, *Notiobiella rubrostigma*.

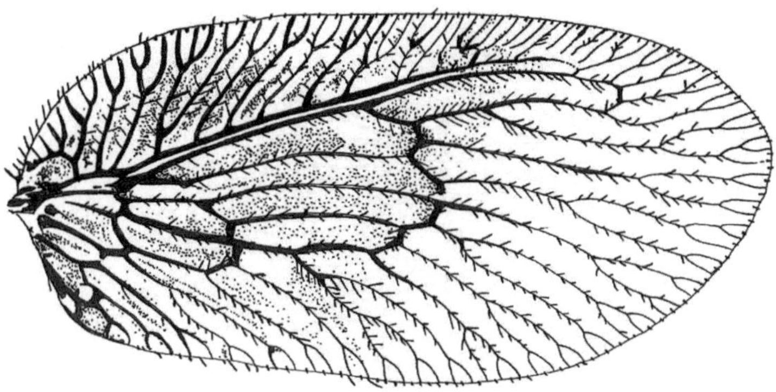

Fig. 12.20 Fore-wing of *Notiobiella brasiliensis*. Based on Penny and Monserrat (1985).

3. (2) The pterostigma is dark yellow in the middle and reddish at both the proximal and distal end. The fore-wing is hyaline with dark markings on the membrane only near the base of the wing and near bifurcations in the costal area (Fig. 12.19). Fore-wing length: c. 5.5 mm.
   ....*Notiobiella cixiiformis* (Gerstaecker, 1888) (Venezuela, Peru, Bolivia, Paraguay, Argentina, Amazonas, São Paulo). Syn: *Hemerobius cixiiformis* Gerstaecker, 1888; *Notiobiella callangana* Kimmins, 1940; probably *Notiobiella rubrostigma* Navás, 1914.

   – The pterostigma of the fore-wing is dark, sometimes streaked with red. Otherwise, there is no streaking in the costal area of the fore-wing (Figs. 2.18 and 12.3). The male paramere is oval. Fore-wing length: c. 6.1 mm.
   ....*Notiobiella paddiae* Monserrat, 1984 (Paraguay, Pará, Amazonas, Rondônia, Acre).

12.4 Subfamily Depanacrinae

Key to the South American Species of Adults in the Genus *Conchopterella*

Information for the key was provided by Handschin (1955).

1. The outer gradate series of veins runs parallel to the outer margin of the wing. The series is complete, and the veins appear as sharp, pale lines (Fig. 12.5). Length of fore-wing: 6 to 7 mm; hind wing: 4 to 5 mm.
   .... *Conchopterella kuscheli* Handschin, 1955 (Chile).

   – The outer gradate series is incomplete, and the veins appear as simple lines (Fig. 12.21). Length of fore-wing: c. 6.5 mm; hind wing: c. 5.5 mm.
   ....*Conchopterella maculata* Handschin, 1955 (Chile).

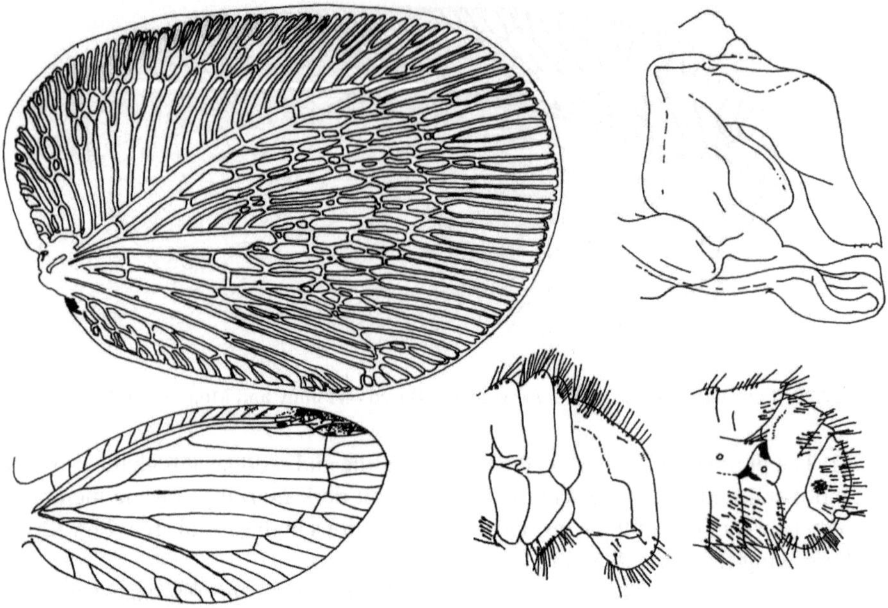

Fig. 12.21 *Conchopterella maculata*: fore and hind wing (left), apex of the male (lower center) and the female abdomen (lower right), and the male genitalia (upper right). Based on Handschin (1955).

12.5 Subfamilies Hemerobiinae and Microminae

Key to the South American Species in the Genus *Hemerobiella*

Information for the key was provided by Monserrat (1998), and Lara and de Freitas (2003).

1. The pterostigmata are dark and opaque. The male ectoproct bears coarse spines along its posterior margin, and it is divided at the apex into a broad posterodorsal process and a narrow, curved anteroventral process (Fig. 12.6). The lateral gonopophysis of the female is relatively short, and the subgenital plate is almost spherical. Length of fore-wing: c. 8.4 mm; hind wing: c. 6.8 mm. The color is generally black with an orange marking on the mesonotum. .... *Hemerobiella oswaldi* Monserrat, 1998 (Venezuela). The two species in this couplet were reported by Monserrat (1998) to be very similar in their features, making it necessary to use the genitalia as distinguishing characteristics.

   – The male ectoproct lacks coarse spines along its posterior margin, and its anteroventral process is short and triangular. The lateral gonopophysis of the female is relatively long, and the subgenital plate is approximately pentagonal (Fig. 12.22). Length of fore-wing: c. 6.1 mm; hind wing: c. 4.9 mm. ....*Hemerobiella sinuata* Kimmins, 1940 (Mexico, Ecuador).

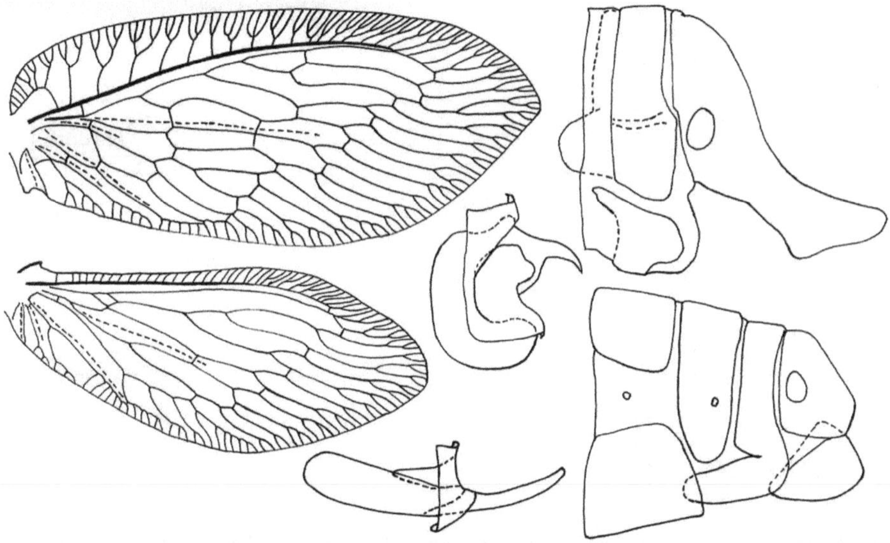

Fig. 12.22 *Hemerobiella sinuata*: fore and hind wing (left), apex of the male (upper right) and female abdomen in lateral view (lower right), gonarcus in lateral view (center), and parabaculum in lateral view (lower center). Based on Oswald (1993a).

Key to the South American Species of Adults in the Genus *Nusalala*

Information for the key was provided by Navás (1910a), Penny and Monserrat (1983, 2000), Penny and Sturm (1984), Souza (1999), and Lara and Freitas (2002).

1. The wings are brachypterous and shorter than the abdomen (Fig. 12.23). Only the male has been described.
   .... *Nusalala andina* Penny and Sturm, 1984 (Colombia). Syn: *Nusalala andinus* Penny and Sturm, 1984. A second brachypterous species, *Nusalala brachyptera* Oswald, 1996, is known only from Costa Rica.

   – The wings are fully developed and longer than the abdomen (Fig. 12.24). 2

2. (1) The fore-wings are strongly falcate, and there is only one blackish marking on the hind femur and one on the hind tibia (Fig. 12.24). Length of fore-wing: c. 9.1 mm; hind wing: c. 8.2 mm.
   .... *Nusalala falcata* Kimmins, 1940, *nec* Banks, 1910 (Paraná, Rio Grande do Sul).

   – The fore-wing is not strongly falcate, or, if it is, there are three blackish markings on the hind femur and three more on the hind tibia (Fig. 12.25). 3

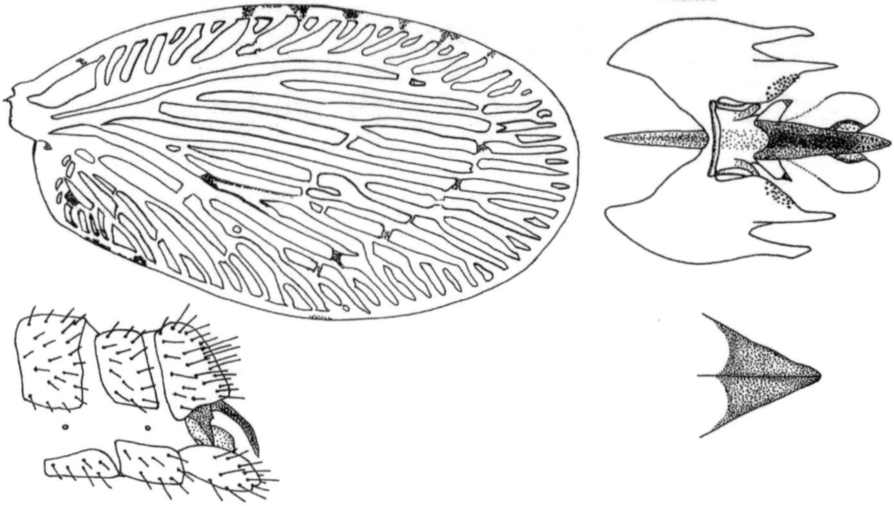

Fig. 12.23 *Nusalala andina* male: fore-wing (upper left), apex of the abdomen in lateral view (lower left), internal genitalia in dorsal view (upper right), and the hypandrium in dorsal view (lower rightr). Based on Penny and Sturm (1984) and Monserrat (2000).

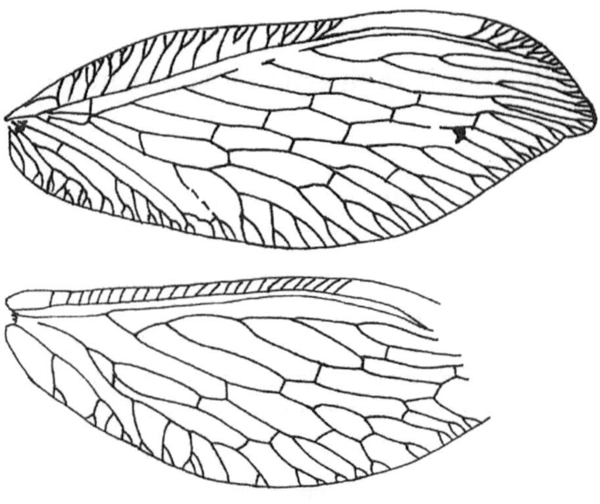

Fig. 12.24 Fore and hind wing of *Nusalala falcata*. Based on Monserrat (2000).

3. (2) There are three blackish markings on the hind femur and three more on the hind tibia. The fore-wing is strongly falcate, or it is broadly rounded (Fig. 12.25). Length of fore-wing: c. 12.2 to 11.0 mm; hind wing: c. 9.0 mm. .... *Nusalala dispar* (Banks, 1910) (Colombia). Syn: *Micromus dispar* Banks, 1910; *Micromus falcatus* Banks, 1910, nec Zimmermann, 1957.

   – The hind femur and tibia do not each have three blackish markings (Fig. 12.26)... 4

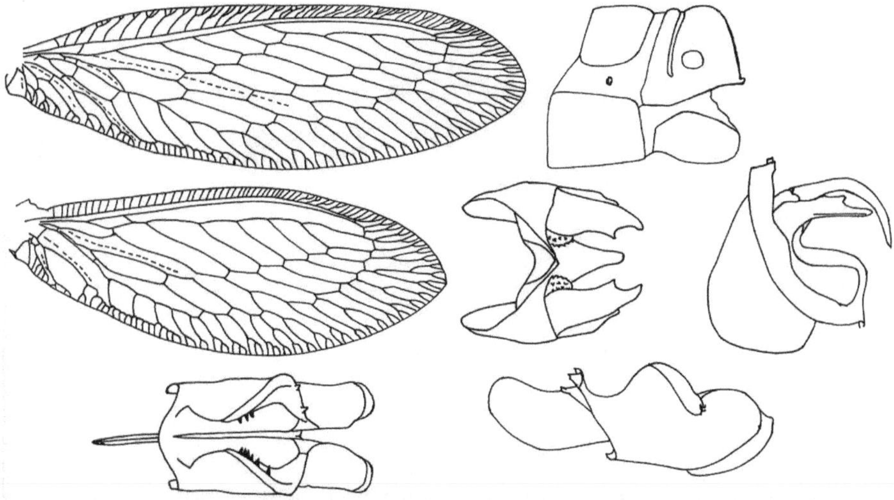

Fig. 12.25 *Nusalala dispar* male: fore and hind wing (upper and middle left, respectively), apex of the abdomen in lateral view (upper right center), gonarcus in dorsal (center) and lateral view (right), and the parabaculum in dorsal (lower left) and lateral view (lower right center). Based on Oswald (1993a).

Fig. 12.26 *Nusalala neotropica*: fore and hind wing (upper and middle left), internal genitalia of a male in ventral (upper center) and posterior view with the apices of its processes enlarged to its lower right (upper right), paramere in dorsal and lateral view (right of center, below and above, respectively), hypandrium in ventral view (left of center), ninth tergite and ectoproct in lateral view (middle right), apex of the female abdomen in ventral (lower left) and lateral view (lower center), and spermothecal gland of a female in dorsal view (lower right). Based on Monserrat (2000).

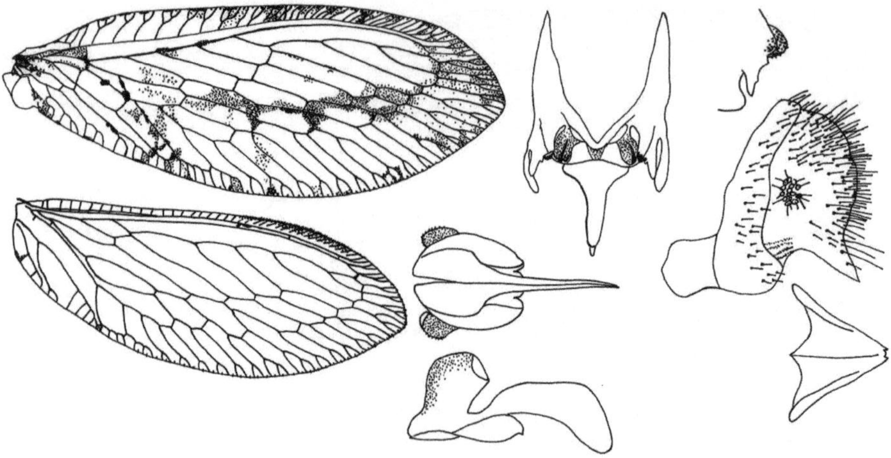

Fig. 12.27 *Nusalala colombiensis* male: fore and hind wing (left), internal genitalia in lateral (upper right) and posterior view (upper right center), paramere in dorsal (center) and lateral view (lower center), hypandrium in ventral view (lower right), ninth tergite and ectoproct in lateral view (middle right). Based on Monserrat (2000).

4. (3) The fore-wing has five or six radial sectors (Fig. 12.26). Length of fore-wing: c. 10.9 mm; hind wing: c. 9.7 mm.
 .... *Nusalala neotropica* (Esben-Petersen, 1914) (Venezuela, Colombia, Ecuador, Peru, Bolivia, Bahia, Santa Catarina). Syn: *Haarupiella neotropica* Esben-Petersen, 1914. Several species known only from Central America also have five or six radial sectors in the fore-wing.
 – The fore-wing has three or four radial sectors (Fig. 12.27). 5
5. (4) The radial sector farthest distal has three bifurcations proximal to the external gradate series. The fore-wings are lanceolate and pointed at the apex, and they have a blackish longitudinal stripe in the middle (Fig. 12.27). Length of fore-wing: c. 9.1 mm; hind wing: c. 8.5 mm.
 ....*Nusalala colombiensis* (Banks, 1910) (Colombia, Peru). Syn: *Boriomyia colombiensis* Banks, 1910; *Haarupiella colombiensis* (Banks, 1910) Esben-Petersen, 1914.
 – The radial sector farthest distal has two bifurcations proximal to the external gradate series. The fore-wings are ovoid and broadly rounded at the apex, and they have markings and clouds on the membrane that do not form a blackish longitudinal stripe in the middle (Fig. 12.28). 6
6. (5) On each fore-wing, there is a dark, circular marking located on the external gradate series between the first and second radial sector (Fig. 12.4). Length of fore-wing: c. 8.9 mm; hind wing: c. 8.1 mm.
 ....*Nusalala erecta* Navás, 1913

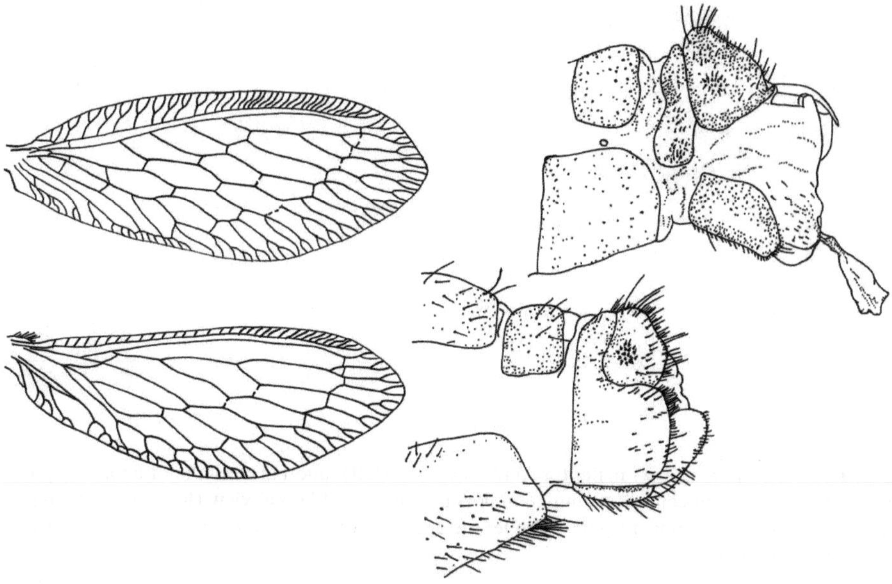

Fig. 12.28 *Nusalala tessellata*: fore (upper left) and hind wing (lower left), apex of the male (upper right) and female abdomen (lower center), both in lateral view. Based on Adams (1970).

(Ecuador, Peru, Paraguay, Argentina, Brazil). Syn: *Haarupiella pallida* Esben-Petersen, 1914; *Huarupiella pallida* Esben-Petersen, 1914 misspelling by Navás, 1920; *Nusalala pallida* (Esben-Petersen, 1914) auctt.; *Nusalala escomeli* Navás, 1922.

– Dark markings on the fore-wing do not include a dark, circular marking located on the external gradate series between the first and second radial sector (Fig. 12.28). 7

7. (6) The fore and middle legs have one blackish spot on the coxa and one on the trochanter, while the hind legs have one blackish marking on the femur and one on the tibia (Fig. 12.29). Length of fore-wing: c. 8.3 mm; hind wing: c. 7.3 mm. .... *Nusalala marini* Monserrat, 2000 (Peru).

– The fore and middle legs have no blackish spots on the coxa and trochanter, while the hind legs have no blackish markings on the femur and tibia (Fig. 12.28). Length of fore–wing: c. 7.9 mm; hind wing: c. 7.0 mm. .... *Nusalala tessellata* (Gerstaecker, 1888) (Central America, West Indies, Colombia, Venezuela, Ecuador, Peru, Bolivia, Paraguay, Argentina, Rondônia, Rio Grande do Sul, Santa Catarina, Paraná, São Paulo, Rio de Janeiro, Minas Gerais, Espirito Santo). Syn: *Micromus tessellatus* Gerstaecker, 1888; *Micromus reticulatus* Navás, 1910; *Neomicromus tessellatus* (Gerstaecker, 1888); *Meomicromus tessellatus* (Gerstaecker, 1888) Krüger, 1922

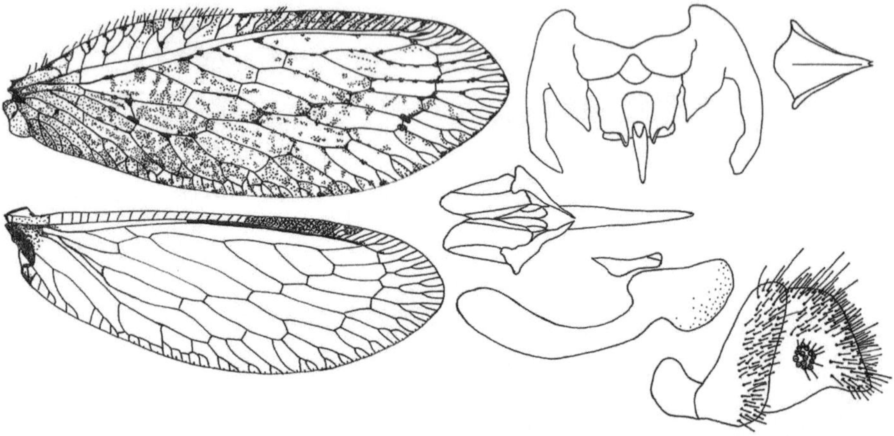

Fig. 12.29 *Nusalala marini* male: fore and hind wing (left), internal genitalia of a male in posterior view (upper right center), paramere in dorsal (center) and lateral view (lower center), hypandrium in ventral view (upper right), ninth tergite and ectoproct in lateral view (lower right). Based on Monserrat (2000).

(spelling error); *Haarupiella uruguaya* Navás, 1923; *Haarupiella gradata* Navás, 1925; *Nusalala gradata* (Navás, 1925); *Nusalala uruguaya* (Navás, 1923); Kimmins, 1936; *Micromus reticulatus* Navás, 1910; *Nusalala reticulata* (Navás, 1910); Penny and Monserrat, 1983; *Nusala rhegmatica* Navás, 1914; *Haarupiella rhegmatica* (Navás, 1914) Esben-Petersen, 1914. Three species known only from Central America or Cuba also fit this description but can be distinguished by differences in the wing markings and male genitalia.

Key to the South American Species of Adults in the Genus *Hemerobius*

Information for the key was provided by Navás (1917a, 1918a, b, 1921b, 1929a), González Olazo (1993), Monserrat (1990a, b, 1996, 1998), Monserrat et al. (2013), Lara and Freitas (2003), and Letardi et al. (2013). The description of *Hemerobius tibialis* Navás 1917, from Colombia, is not detailed enough to place this nominal species in the key or determine its synonymy with any other species.

1. The veins and membranes of the wings of the female are completely hyaline without a trace of dark markings or clouding. Crossvein CuA-CuP in the forewing is located at the level of the fork in vein M (Fig. 12.30). The head and thorax are dark and very shiny, almost black. Only the female has been described. Wing span: c. 16 mm.
   .... *Hemerobius albipennis* Banks, 1910 (Colombia).

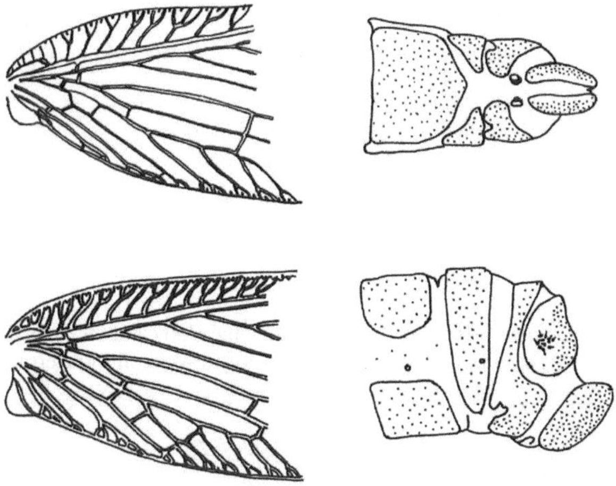

Fig. 12.30 *Hemerobius albipennis* female: bases of the fore and hind wing (left) and the apex of the abdomen in ventral (upper right) and lateral view (lower right). Based on Monserrat (1996).

– The veins and membranes of the wings are dark or have dark markings and clouding. Crossvein CuA-CuP in the fore-wing is located distal to the fork in vein M (Fig. 12.31). The head and thorax are yellow, yellow with dark markings, or dark 2

2. (1) The fore-wing has no more than 11 cross veins in the costal field between the recurrent vein and the pterostigma, and the second and third sections of the radius are slightly curved toward the base. In the hind wing, crossvein CuA has one or two forks proximal to the marginal fork. The pterostigma is very dark and prominent (Fig. 12.31). 3

– The fore-wing has more than 11 cross veins in the costal field between the recurrent vein and the pterostigma, and the second and third sections of the radius are straight near the base. In the hind wing, cross vein CuA usually has three or more forks proximal to the marginal fork. The pterostigma is usually inconspicuous but is rarely somewhat darker than the rest of the wing membrane (Fig. 12.32). 7

3. (2) There is a dark marking across the fore-wing membrane in the basal part of the costal field. The shape of the distal sternite of the female is semi-circular in outline and approximately twice as wide and its length along the ventral midline (Fig. 12.31). Length of fore-wing: c. 5.0 mm; hind wing: c. 4.2 mm. .... *Hemerobius pennii* Monserrat, 1996 (Central America, Ecuador).

– There is no dark marking across the membrane in the basal part of the costal field (Fig. 12.33). 4

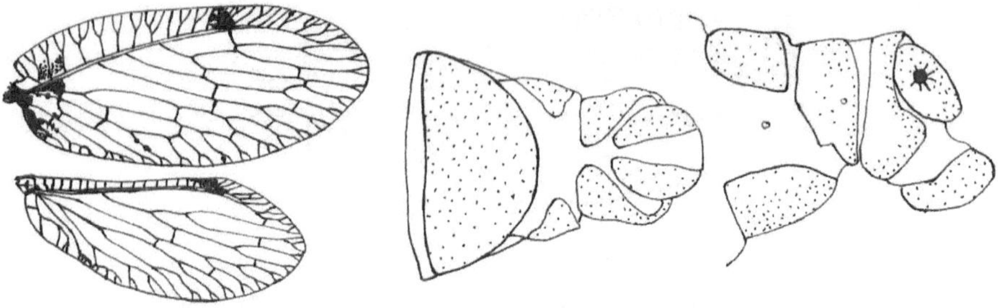

Fig. 12.31 *Hemerobius pennii* female: fore and hind wing (left) and the apex of the female abdomen in ventral (upper right) and lateral view (lower right). Based on Monserrat (1996).

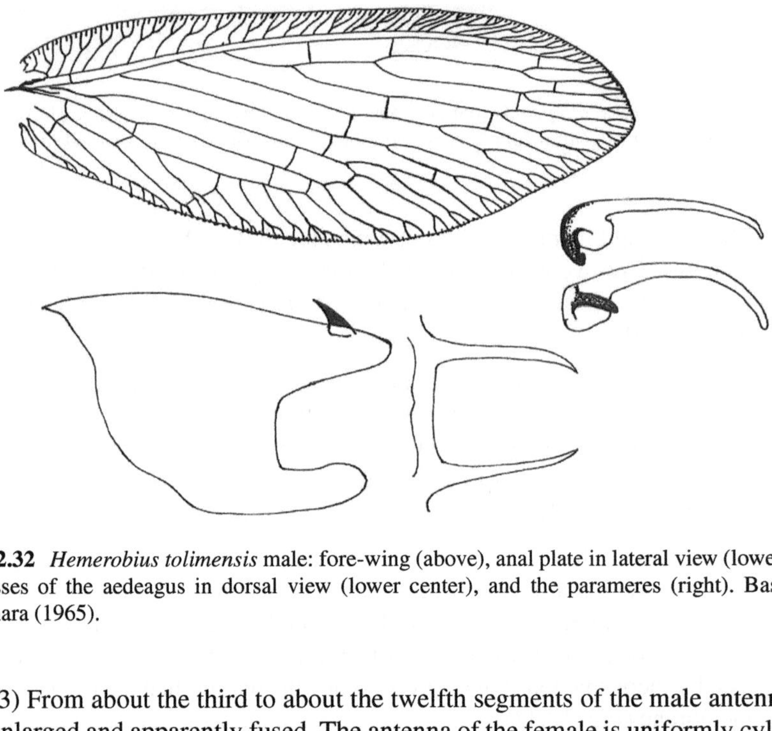

Fig. 12.32 *Hemerobius tolimensis* male: fore-wing (above), anal plate in lateral view (lower left), processes of the aedeagus in dorsal view (lower center), and the parameres (right). Based on Nakahara (1965).

4. (3) From about the third to about the twelfth segments of the male antenna are enlarged and apparently fused. The antenna of the female is uniformly cylindrical. The fore-wing is elliptical and its membrane is somewhat variegated; its pterostigma is not conspicuous, but that of the hind wind is more uniformly darkened (Fig. 12.33). Length of fore wing: c. 6.0 mm; hind wing: c. 5.0 mm. ....*Hemerobius pectinicornis* Monserrat, 1998 (Mexico, Central America, Ecuador).

   – The male antenna shows no enlargement near the base of the flagellum. The wings are at least weakly variegated (Fig. 12.34). 5

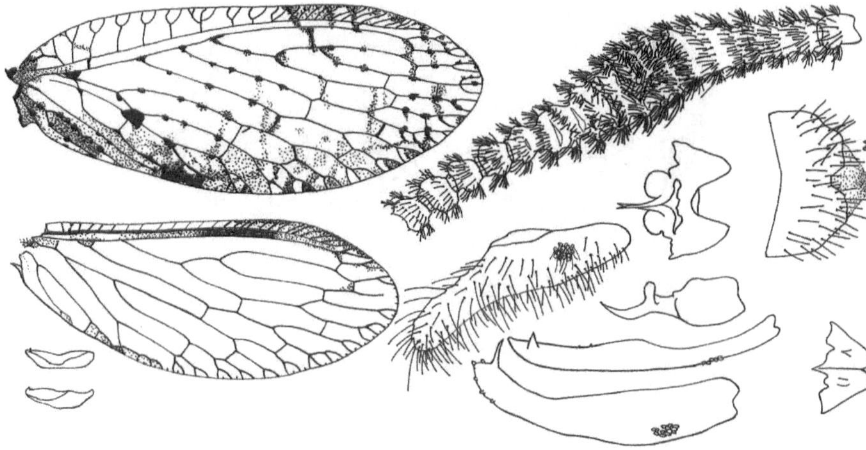

Fig. 12.33 *Hemerobius pectinicornis* male: fore (upper left) and hind wing (middle left), basal segments of the antenna (center to upper right), the gonarcus-entoprocessus complex in posterior and lateral view (right center, above and below, respectively), the ninth sternite in ventral view (middle right), the hypandrium in ventral view (lower right), the ectoproct in lateral view showing the setae (right center to lower center), and the shape of the ectoproct as seen in dorsal and dorsolateral view (lower right center, above and below, respectively), and a paramere in ventral and lateral view (lower left, above and below, respectively). Based on Monserrat (1998).

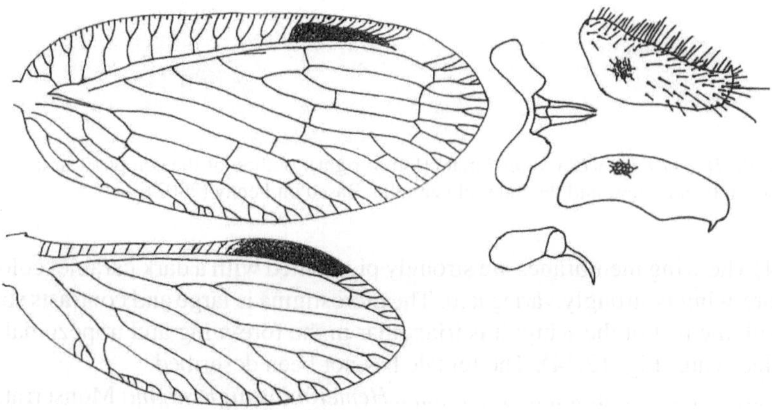

Fig. 12.34 *Hemerobius nigrostigma* male: fore and hind wing (left), the gonarcal and entoprocessus complex of a male in caudal (upper right center) and lateral view (lower right center), and male ectoproct with setae in lateral view (upper right) and without setae in dorsoposterior view (lower right). Based on Monserrat (1996).

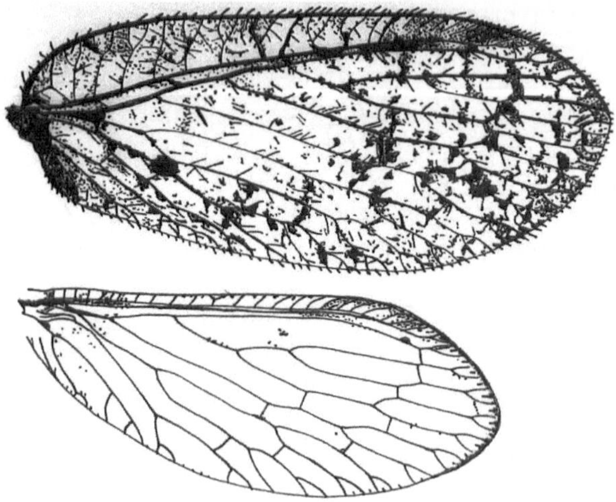

Fig. 12.35 Fore and hind wing of *Hemerobius edui*. Based on Monserrat (2002).

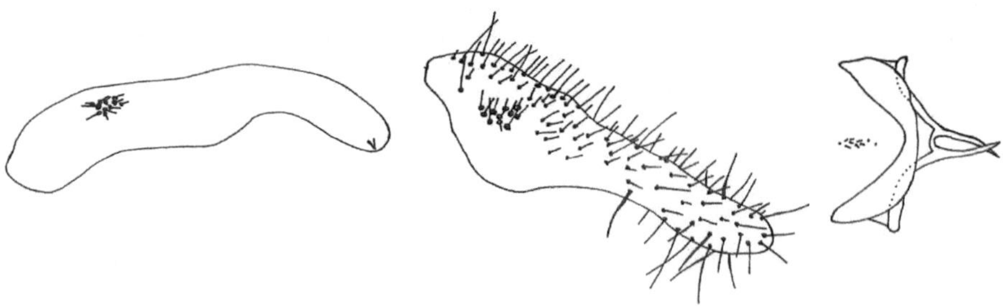

Fig. 12.36 *Hemerobius withycombei* male (left to right): outline of the ectoproct in dorsal view, ectoproct in lateral view, and the internal genitalia. Based on Penny (2002).

5. (4) The wing membranes are strongly pigmented with a dark caramel color. The fore-wing is strongly variegated. The pterostigma is large and contrasts strongly with the rest of the wing; it is triangular in the fore-wing and trapezoidal in the hind wing (Fig. 12.34). The female has not been described. .... *Hemerobius nigrostigma* Monserrat, 1990 (Venezuela).

   – The membranes of the wings have a weak, dark pigmentation. The fore-wing has a weak, scattered variegation. The pterostigma is small and appears circular (Fig. 12.35). 6

6. (5) The scape is yellow with a dark lateral stripe on its outer surface. The dorsal surface of the thorax is uniformly dark. The outline of the ectoproct appears somewhat sigmoid in dorsal view (Fig. 12.36). .... *Hemerobius withycombei* (Kimmins, 1928) (Mexico, Central America, Colombia). Syn: *Anotiobiella withycombei* Kimmins, 1928.

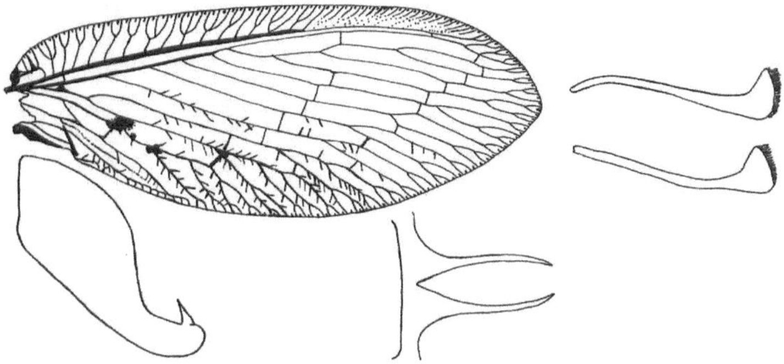

Fig. 12.37 *Hemerobius exceptatus* male: fore-wing (above), anal plate in lateral view (lower left), processes of the aedeagus in dorsal view (lower center), and the parameres (right). Based on Nakahara (1965).

– The scape is uniformly yellow without a dark stripe on its outer surface. The dorsal surface of the thorax is dark with a median yellow band (Fig. 12.35). ....*Hemerobius edui* Monserrat, 1991 (Mexico, Colombia, Peru).

7. (2) The fore-wing is elongate with a straight anterior margin; it is almost three times as long as its width. The cubital cell is always closed (Fig. 12.32). The membrane and veins of the fore-wing are strongly variegated. Wing span: c. 20 mm. ....*Hemerobius tolimensis* Banks, 1910 (Mexico, Central America, Colombia, Venezuela, Peru, Bolivia, Ecuador). Syn: *Hemerobius pinnatus* Navás, 1914 *non* 1918.

– The anterior margin of the fore-wing is convex (Fig. 12.37). 8

8. (7) The cubital cell in the fore-wing is closed (Fig. 12.38). .... *Hemerobius solidarius* Monserrat, 1996 (Colombia, Venezuela).

– The cubital cell in the fore-wing is open (Fig. 12.37). 9

9. (8) There is no yellow median band on the dark dorsal surface of the thorax (Fig. 12.39). ....*Hemerobius nigridorsus* Monserrat, 1996 (Central American, Venezuela).

– The dorsal surface of the thorax is dark with a median yellow band (Fig. 12.37). 10

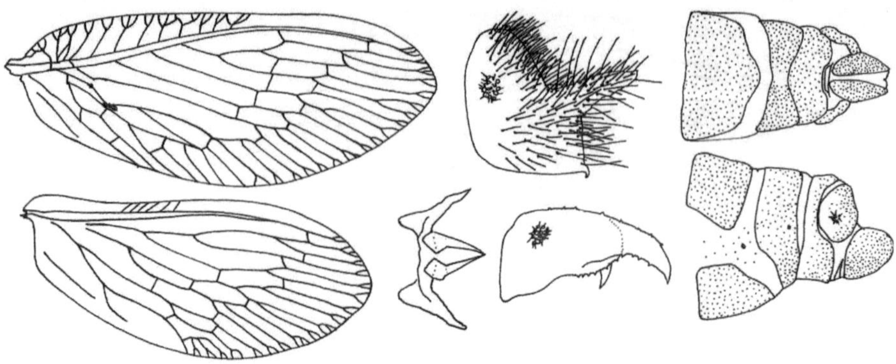

Fig. 12.38 *Hemerobius solidarius*: fore and hind wing (left), the gonarcal and entoprocessus complex of a male in caudal view (lower center), male ectoproct with setae in lateral view (upper right center) and without setae in dorsoposterior view (lower right center), and the apex of the female abdomen in ventral (upper right) and lateral view (lower right). Based on Monserrat (1996).

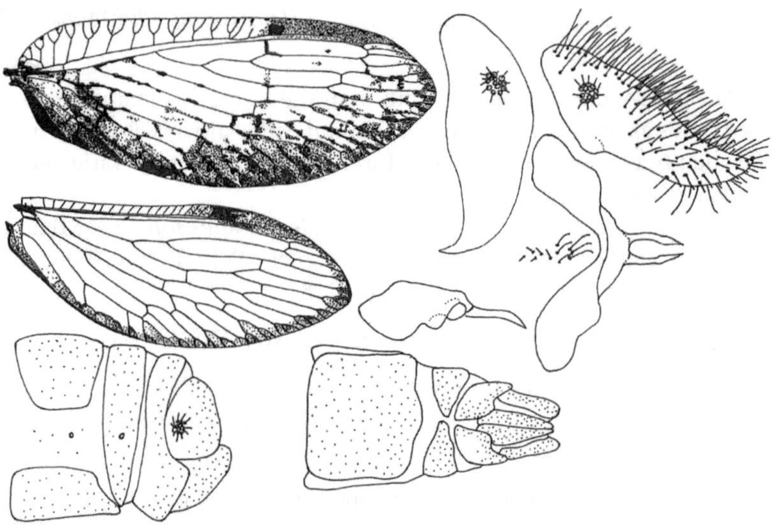

Fig. 12.39 *Hemerobius nigridorsus*: fore and hind wing (upper and middle left), the male ectoproct without setae in dorsal view (upper right center) and in lateral view shown with its setae (upper right), male gonarcus in posterior (middle right) and lateral view (center), and the apex of the female abdomen in ventral (lower cnter) and lateral view (lower left). Based on Monserrat (1996).

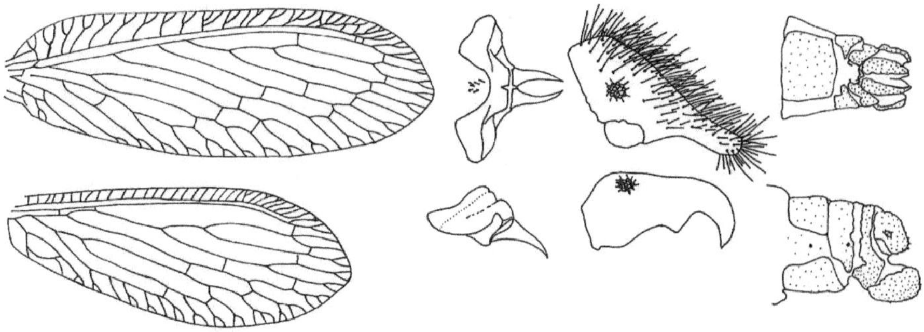

Fig. 12.40 *Hemerobius nekoi*: fore and hind wing (left), the gonarcal and entoprocessus complex of a male in caudal (upper center) and lateral view (lower center), male ectoproct with setae in lateral view (upper right center) and without setae in dorsoposterior view (lower right center), and the apex of the female abdomen in ventral (upper right) and lateral view (lower right). Based on Monserrat (1996).

10. (9) There are four or five sectors of the radial vein in the fore-wing, and the distalmost of these has four or five forks (Fig. 12.38). .... *Hemerobius exceptatus* Nakahara, 1965 (Colombia, Venezuela, Rio de Janeiro, Santa Catarina).

 – The fore-wing usually has two, three, or rarely four sectors of the radial vein, and the distalmost of these has two or three forks (Fig. 12.40). 11

11. (10) The wings are narrow and ovoid; the proximal third of the anterior margin of the fore-wing is strongly convex. The width of the space between CuA and the posterior margin of the hind wing is approximately equal to the width of the space between CuA and MP. The pterostigmata are pale and indistinct. In the fore-wing, the costal area occupies about 30 % of the wing width proximally and is crossed by about 11 to 14 cross veins, typically branched once, proximal to the stigmal area. The cross veins in the costal area of the hind wing are typically unbranched, and the proximal part of the costal area is not expanded. The fore and hind wings obviously differ in shape, and the hind wings may be longer (Fig. 12.41). The wings are hyaline with light yellow long veins and some light red gradate veins. Fore-wing length: c. 8.0 mm; width: c. 3.6 mm. Hind wing length: c. 8.3 mm; width: c. 1.7 mm. .... *Hemerobius domingensis* Banks, 1941 (Dominican Republic, Cuba, São Paulo, Paraná).

 – The wings are elongate and elliptical or fusiform. The basal third of the anterior margin of the fore-wing is only weakly convex (Fig. 12.40). 12

12. (11) The fore-wings are elongate and fusiform, with subparallel anterior and posterior margins. The space between vein CuA and the hind wing margin is just as narrow as the space between veins CuA and MP (Fig. 12.40). 13

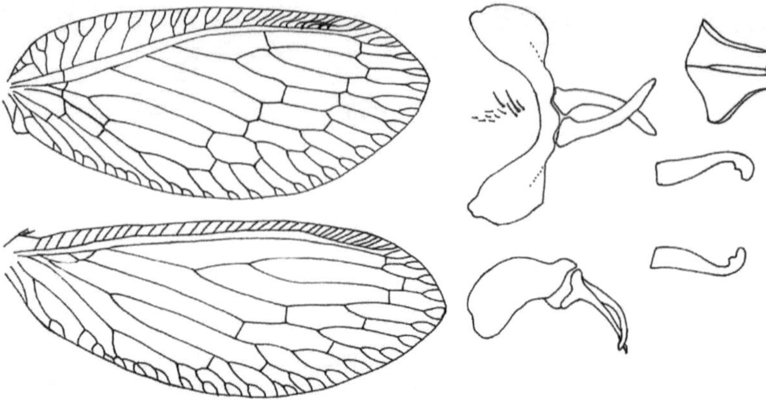

Fig. 12.41 *Hemerobius domingensis*: fore and hind wing (left), gonarcus in dorsal (upper center) and lateral view (lower left), internal hypandrium (upper right), and parameres (middle and lower right). Based on Lara and de Freitas (2003).

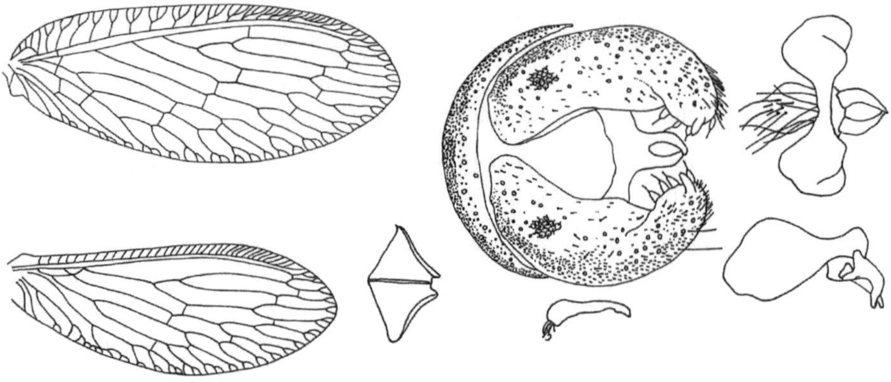

Fig. 12.42 *Hemerobius gaitoi* male: fore (upper left) and hind wing (lower left) showing the arrangement of the veins but omitting the color pattern, apex of the abdomen in dorsal view (upper center), gonarcus in dorsal (upper right) and lateral view (lower right), paramere in dorsal view (lower right center), and the hypandrium in dorsal view (lower center). Based on Lara and Freitas (2003).

– The fore-wings are elongate ovoid or triangular; their anterior and posterior margins are clearly divergent. In the hind wing, the space between CuA and MP is wider than the space between CuA and the posterior margin of the wing (Fig. 12.42). 15

13. (12) The head is entirely black, or, sometimes, it has narrow yellowish stripes along the dorsal midline and ventral surface of the scape. The fore-wing is slightly clouded near the apex and blackish on most of the veins, or, sometimes, paler lines on the veins. The cross veins CuA – CuP and CuA – MP are divergent (Fig. 12.40).
....*Hemerobius nekoi* Monserrat, 1996 (Chile).

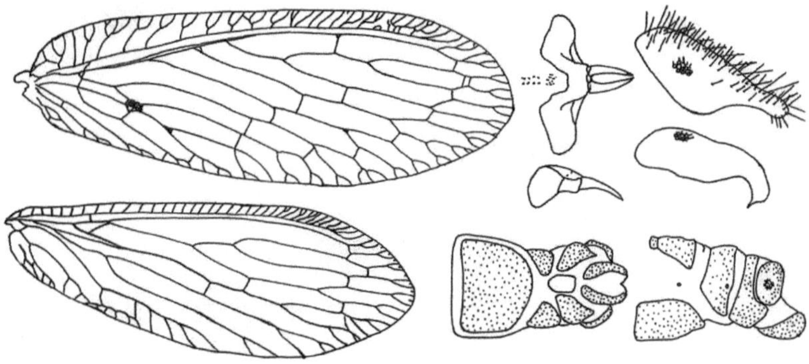

Fig. 12.43 *Hemerobius stenopterus*: fore and hind wing (left), the gonarcal and entoprocessus complex of a male in caudal (upper right center) and lateral view (right center), male ectoproct with setae in lateral view (upper right) and without setae in dorsoposterior view (middle right), and the apex of the female abdomen in ventral (lower right center) and lateral view (lower right). Based on Monserrat (1996).

– The head is yellow but may have some black spots on the vertex, and the scape is yellow on the ventral surface. The fore-wing is rounded at the apex. The veins are yellowish with abundant dark markings and lines. The cross veins CuA – CuP and CuA – MP are parallel (Fig. 12.42). 14

14. (13) The vertex lacks blackish markings, and the scape is entirely yellow (Fig. 12.42).
.... *Hemerobius stenopterus* Monserrat, 1996 (Chile, Argentina).

– The vertex has blackish spots on a yellowish background. The scape has a blackish stripe on its outer surface (Fig. 12.43).
.... *Hemerobius montsae* Monserrat, 1996 (Argentina, Bolivia, Peru).

15. (12) There are usually two cross veins proximal to the fork in M in the fore-wing, but the one closest to the base may be barely visible (Fig. 12.41). The vertex, frons, scape, and sometimes the dorsal surface of the thorax have dark blotches, stripes, or spots, which are usually conspicuous. 16

– There is usually only one cross vein proximal to the fork in M in the fore-wing. The vertex, frons, scape, and the dorsal surface of the thorax are usually uniformly yellow of dark, but there may be some faint dark spots on the vertex (Fig. 12.44). 17

16. (15) The fore-wings are elliptical and irregularly variegated, with blackish markings over the gradate veins and on vein CuA. The gradate cells are irregular and elongated. There are usually two cross veins proximal to the fork in vein M in the fore-wing, although the most proximal of these may be faint

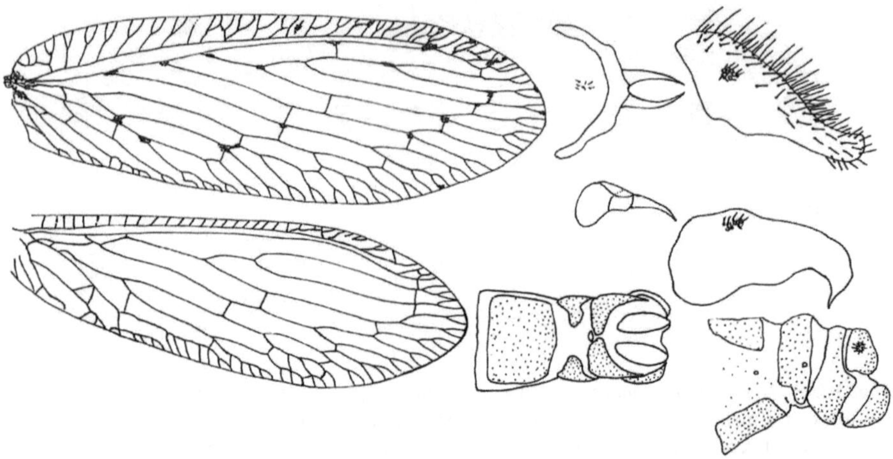

Fig. 12.44 *Hemerobius montsae*: fore and hind wing (left), the gonarcal and entoprocessus complex of a male in posterior (upper right center) and lateral view (right center), male ectoproct with setae in lateral view (upper right) and without setae in dorsoposterior view (middle right), and the apex of the female abdomen in ventral (lower right center) and lateral view (lower right). Based on Monserrat (1996).

(Fig. 12.42). The vertex, frons, scape, and sometimes the thoracic tergites have blackish markings, stripes, and tiny spots, usually very conspicuous.
.... *Hemerobius gaitoi* Monserrat, 1996
(Mexico, Central America, West Indies, Ecuador, Venezuela, São Paulo).

– The fore-wings are narrow, subtriangular, and uniformly blackish variegated (Fig. 12.46). The legs have striking black markings, especially on the tibiae.
.... *Hemerobius centralis* Navás, 1913
(Venezuela, Colombia, Ecuador, Peru, Bolivia).

17. (15) The fore-wings are wide, subtriangular in shape, and slightly bent at the apex. The pterostigma is not pigmented. The dark variegation on the wing is irregular over the gradate series (Fig. 12.45). Wing span: c. 16 mm.
....*Hemerobius bolivari* Banks, 1910
(Mexico, Central America, Colombia, Bolivia, Chile, Argentina, Uruguay, Minas Gerais, Rio de Janeiro, São Paulo, Paraná, Santa Catarina, and introduced to southern Africa, Australia, and Europe). Syn: *Hemerobius hageni* Navás, 1918; *Hemerobius hageni* var. *distincta* Navás, 1920; *Megalomus pallidus* Blanchard, 1851 non Uddman, 1790, nec Stephens, 1836; *Schneiderobius pallidus* (Blanchard, 1851) Kruger, 1922; *Hemerobius pinnatus* Navás, 1918 non Navás 1914; *Schneiderobius bolivianus* Kruger, 1922 nomen nudum; *Hemerobius skottsbergi* Navás, 1924; *Hemerobius sjostedti* Navás, 1921;

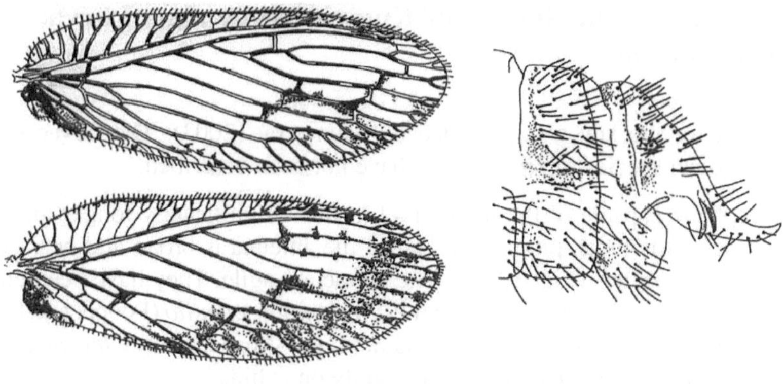

Fig. 12.45 *Hemerobius bolivari* male: fore and hind wing (left), and the apex of the male abdomen in lateral view (right). Based on Handschin (1955), who referred to the species by its synonym, *Hemerobius sjöstedti*.

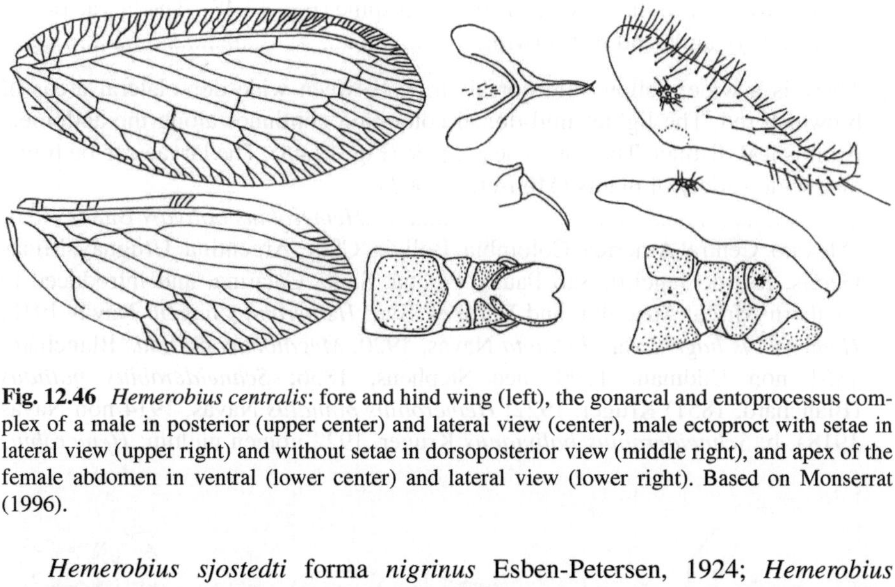

Fig. 12.46 *Hemerobius centralis*: fore and hind wing (left), the gonarcal and entoprocessus complex of a male in posterior (upper center) and lateral view (center), male ectoproct with setae in lateral view (upper right) and without setae in dorsoposterior view (middle right), and apex of the female abdomen in ventral (lower center) and lateral view (lower right). Based on Monserrat (1996).

Hemerobius sjostedti forma *nigrinus* Esben-Petersen, 1924; *Hemerobius fumosus* Esben-Petersen, 1924; *Hemerobius nigrinus* Esben-Petersen, 1924; *Hemerobius brethesi* Navás, 1929; *Hemerobius piunatulus* Navás, 1935; *Hemerobius pinnatulus* Navás, 1935; *Hemerobius blanchardi* Nakahara, 1965; *Hemerobius topali* Steinmann, 1965.

– The fore-wings are elongate and rounded at the apex. The pterostigma is well pigmented and dark. The variegated pattern is relatively uniform over the gradate series (Fig. 12.7).
....*Hemerobius chilensis* Nakahara, 1965 (Chile, Argentina, Peru).

Tentative key to the Species of Known *Hemerobius* Larvae in South America

Information for the key was provided by Monserrat (2003). The larvae of most South American species of *Hemerobius* have not been described.

1. The cephalic capsule of the final larval instar is significantly longer than wide. The base of the long spine at the end of the antenna is broad and bears two or three smaller spines (Fig. 12.47). The larvae are agile. They have been encountered on *Desfontainia spinosa* R. and P., *Nothofagus pumilio* (Poepp. and Endel.), *Nothofagus nitida* (Phil.), *Nothofagus dombeyi* (Mirbel), and *Chusquea culeou* E. Desk, where they apparently feed mainly on aphids.

   ....*Hemerobius stenopterus* Monserrat, 1996 (Chile, Argentina).

   – The length and width of the head capsule are subequal. The antenna narrows progressively to the base of the terminal spine (Fig. 12.48). The larvae do not seem very agile and fast-moving. 2

2. There is a wide, yellow mid-dorsal stripe between wide dorsolateral areas of brownish red. The lighter mid-dorsal coloration continues along the abdomen, but it is less distinct. The antennae are pale (Fig. 12.48). The larvae can be found on a wide variety of plants (Monserrat 2003).
   ....*Hemerobius bolivari* Banks, 1910 (Mexico, Central America, Colombia, Bolivia, Chile, Argentina, Uruguay, Minas Gerais, Rio de Janeiro, São Paulo, Paraná, Santa Catarina, and introduced to southern Africa, Australia, and Europe). Syn: *Hemerobius hageni* Navás 1918; *Hemerobius hageni* var. *distincta* Navás, 1920; *Megalomus pallidus* Blanchard, 1851 non Uddman, 1790, nec Stephens, 1836; *Schneiderobius pallidus* (Blanchard, 1851) Kruger, 1922; *Hemerobius pinnatus* Navás, 1914 non Navás 1918a, b; *Schneiderobius bolivianus* Kruger, 1922 nomen nudum; *Hemerobius*

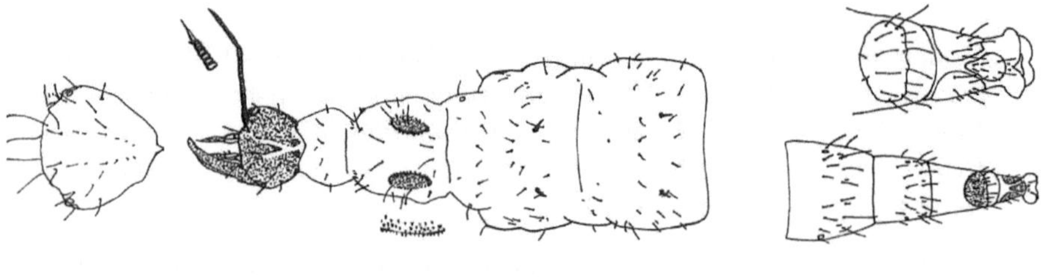

Fig. 12.47 *Hemerobius stenopterus* last larval instar (left to right): head capsule in dorsal view, anterior segments of the larvae with an enlarged apex of the antenna and enlarged surface of the integument, and the apical segments (below) and an enlarged apex of the abdomen (above). Based on Monserrat (2003).

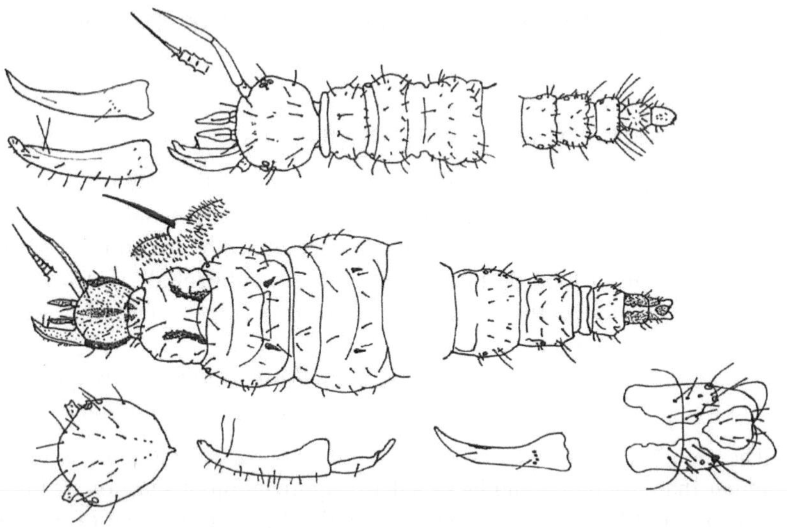

Fig. 12.48 *Hemerobius bolivari* larvae: a first instar larva showing only the anterior and posterior segments and one antenna and a single set of mouthparts (upper left center to upper right) with an enlargement of the apex of the antenna, and a maxilla (upper middle left) and mandible in dorsal view (upper left), a final instar larva showing only the anterior and posterior segments (lower middle, left to right) with an enlargement of the apex of the antenna and a magnified surface ot the thoracic integument above the thorax, a dorsal view of the head capsule (lower left), an enlarged maxilla (lower left center) and mandible (lower right center) in dorsal view, and an enlarged apex of the abdomen (lower right). Based on Monserrat (2003).

skottsbergi Navás, 1924; *Hemerobius sjostedti* Navás, 1921; *Hemerobius sjostedti* forma *nigrinus* Esben-Petersen, 1924; *Hemerobius fumosus* Esben-Petersen, 1924; *Hemerobius nigrinus* Esben-Petersen, 1924; *Hemerobius brethesi* Navás, 1929; *Hemerobius piunatulus* Navás, 1935; *Hemerobius pinnatulus* Navás, 1935; *Hemerobius blanchardi* Nakahara, 1965; *Hemerobius topali* Steinmann, 1965.

– The dark brownish red markings on the thorax and abdomen are almost randomly arranged on the dorsum, and a well-defined mid-dorsal stripe is not evident. The antennae are dark (Fig. 12.13). The larvae apparently show a strong preference for species of *Nothofagus* as a substrate. They were found on *Nothofagus pumilio* (Poepp. and Endel.) and *Nothofagus antarctica* (Forst.). ....*Hemerobius chilensis* Nakahara, 1965 (Chile, Argentina, Peru).

Key to the South American Species of Adults in the Genus *Biramus*

Information for the key was provided by Oswald (2004).

1. The elongate posteroventral lobe on the ectoproct of the male is roughly in the shape of an obtuse triangle and bears a dense oval aggregation of setae on its inner face near its mid-length. The female, recognized by the absence of an elongate posteroventral lobe on the ectoproct, has a subgenitale which is roughly twice as long as its width and rounded at the apex (Figs. 1.10 and 12.49). Forewing length: 8.0–8.5 mm.
   .... *Biramus aggregatus* Oswald, 2004
   (Costa Rica, Panama, and thought to range through northern South America).

   – The elongate posteroventral lobe on the ectoproct of the male is finger-like rather than triangular and lacks a dense aggregation of setae on its inner face near its mid-length. The female, recognized by the absence of an elongate posteroventral lobe on the ectoproct, has a subgenitale which is roughly 1.5 times as long as its width and truncate at the apex (Fig. 12.9).
   .... *Biramus lunatus* Oswald, 1993
   (Venezuela).

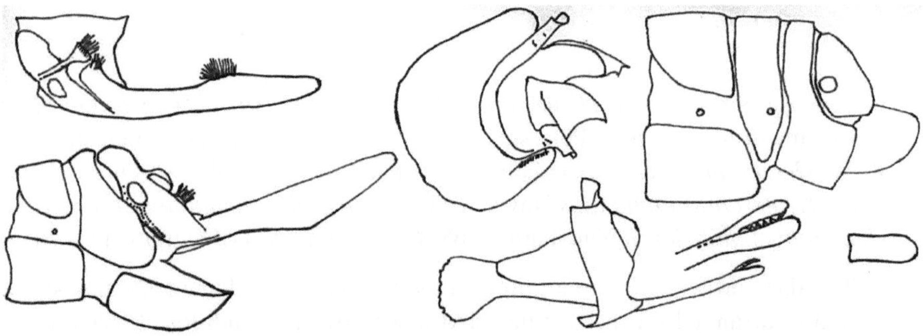

Fig. 12.49 *Biramus aggregatus*: apex of the male abdomen in lateral view (lower left), male ectoproct in dorsal view (upper left), parabaculum and supraparabacular process in lateral view with an enlargement of the apex of its terminal lobe in dorsolateral view to its upper right (lower center), gonarcus and mediuncus in lateral view (upper center), apex of the female abdomen in lateral view (upper right), and subgenitale in ventral view (lower right). Based on Oswald (1993b, 2004).

12.6 Subfamily Sympherobiinae

Key to the South American Species in the Genus *Nomerobius*

Information for the key was provided by Oswald (1990) and Gonzalez Olazo (1992d).

1. The ninth sternite of the male is not elongated upward at the apex; its áposterior margin has a denticulated rim (Fig. 12.50). Length of fore-wing: c. 3.6 mm; hind wing: c. 3.2 mm. The female has not been described. ....*Nomerobius golbachi* González Olazo, 1992 (Argentina).

   – The ninth sternite is not elongated upward toward its apex (Fig. 12.51). 2

2. (1) At the posterior margin of the ninth sternite of the male, there are spine-like setae. The spermatheca of the female is arranged as a series of arched loops, which are long, regularly arranged parallel to each other, and arched about 270° (Fig. 12.51). 3

   – At the posterior margin of the ninth sternite of the male, no spine-like setae are present. The spermatheca of the female has loops only weakly arched or not forming arches at all (Fig. 12.52). 4

3. (2) At the posterior margin of the ninth sternite, there are various kinds of dark, spine-like setae. The posterior margin of the seventh abdominal sternite of the female lacks a midventral cusp (Fig. 12.11). Fore-wing length: 3.9–5.3 mm. ....*Nomerobius psychodoides* (Blanchard in Gay, 1851)

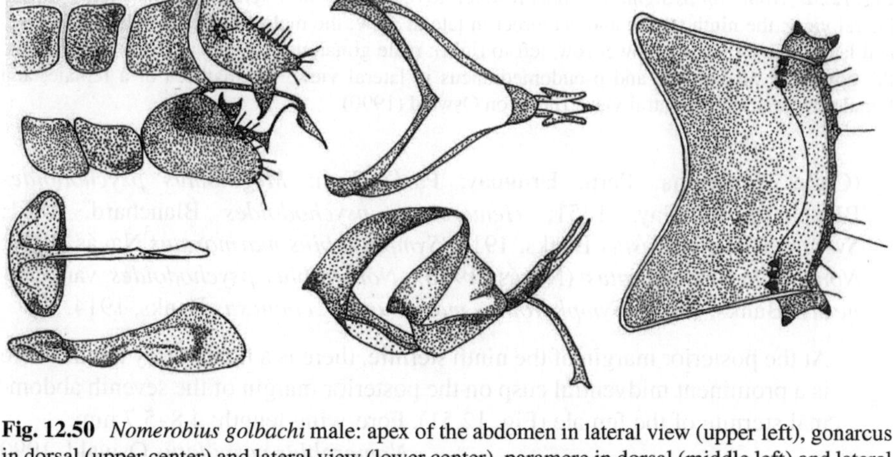

Fig. 12.50 *Nomerobius golbachi* male: apex of the abdomen in lateral view (upper left), gonarcus in dorsal (upper center) and lateral view (lower center), paramere in dorsal (middle left) and lateral view (lower left), and the ninth sternite (right). Based on González Olazo (1992d).

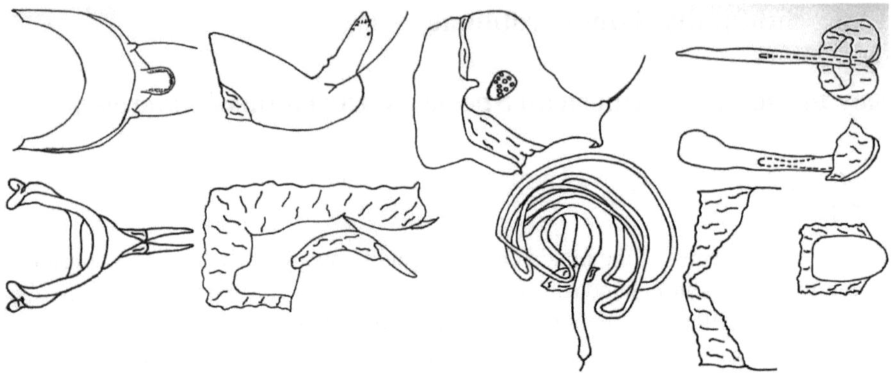

Fig. 12.51 *Nomerobius cuspidatus* (upper row, left to right): the ninth sternite of a male in dorsal and lateral view; the ninth tergite and ectoproct in lateral view; the male parameres in dorsal (above) and lateral view (below); (lower row, left to right): male gonarcus and mediuncus in dorsal view; the gonarcus, mediuncus, and pseudomediuncus in lateral view; spermatheca of a female; seventh sternite of a female in ventral view, and female subgenitale in ventral view. Based on Oswald (1990).

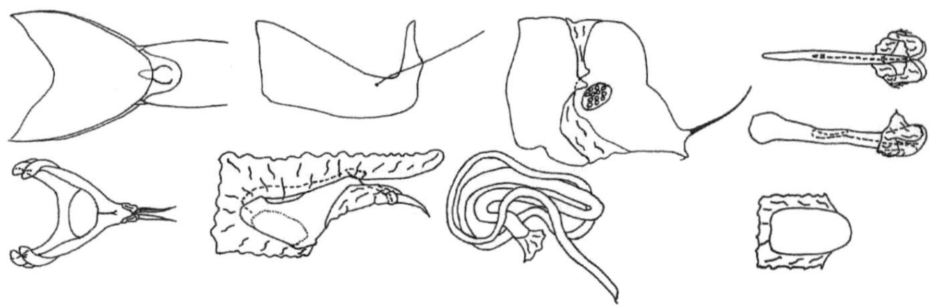

Fig. 12.52 *Nomerobius signatus* (upper row, left to right): the ninth sternite of a male in dorsal and lateral view; the ninth tergite and ectoproct in lateral view; the male parameres in dorsal (above) and lateral view (below); (lower row, left to right): male gonarcus and mediuncus in dorsal view; the gonarcus, mediuncus, and pseudomediuncus in lateral view; spermatheca of a female; and female subgenitale in ventral view. Based on Oswald (1990).

(Chile, Argentina, Peru, Uruguay, Pará). Syn: *Megalomus psychodoides* Blanchard in Gay, 1851; *Hemerobius psychodoides* Blanchard, 1851; *Sympherobius modestus* Banks, 1910; *Sympherobius marmoratus* Navás, 1910; *Nomerobius marmoratus* (Navás, 1910); *Nomerobius psychodoides* var. *connexus* (Banks, 1914); *Sympherobius modestus* var. *connexus* Banks, 1914.

– At the posterior margin of the ninth sternite, there is a field of tiny teeth. There is a prominent midventral cusp on the posterior margin of the seventh abdominal sternite of the female (Fig. 12.51). Fore-wing length: 4.8–5.7 mm.
....*Nomerobius cuspidatus* Oswald, 1990
(Argentina, Chile, Bolivia).

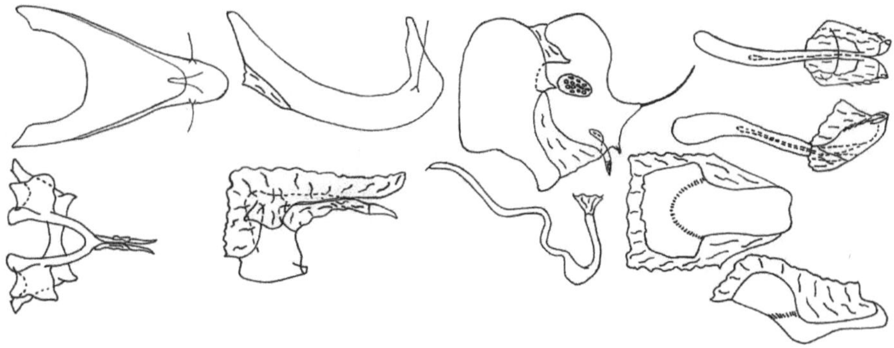

Fig. 12.53 *Nomerobius spinosus* (upper row, left to right): the ninth sternite of a male in dorsal and lateral view; the ninth tergite and ectoproct in lateral view; the male parameres in dorsal (above) and lateral view (below); (lower row, left to right): male gonarcus and mediuncus in dorsal view, the gonarcus, mediuncus, and pseudomediuncus in lateral view; the spermatheca of a female; female subgenitale in ventral and lateral view (below and right of ventral view). Based on Oswald (1990).

4. (2) The rim at the posterior margin of the ninth sternite bears no setae. The spermatheca of the female has weakly arched loops (Fig. 12.52). Fore-wing length: 4.9–6.6 mm.
   .... *Nomerobius signatus* (Hagen, 1861) (Chile, Argentina, Peru). Syn: *Hemerobius signatus* Hagen, 1861; *Nomerobius psychodoides* Nakahara, 1965 pars, *nec* Blanchard, 1861.
   – The rim at the posterior margin of the ninth sternite bears a row of fine setae on each side. The subgenitale of the female is large and shaped like a T; the spermatheca is convoluted but not arranged in a series of arched loops (Fig. 12.53). Fore-wing length: 4.8–6.6 mm.
   .... *Nomerobius spinosus* Oswald, 1990 (Chile, Argentina).

Key to the South American Species in the Genus *Sympherobius*

Information for the key was provided by Navás (1910b, 1917b, 1920, 1929b), Kimmins (1929), Penny and Monserrat (1985), Oswald (1988, 1990), Klimaszewski et al. (1987), Monserrat (1990b, 1998, 2002, 2008), Monserrat et al. (2013), Letardi et al. (2013). The key is based mainly on the male, and some female specimens cannot yet be identified. *Sympherobius blanchardi* (Navás, 1930), from Chile, originally named *Coloma blanchardi* Navás, 1930, was considered to be a likely synonym of *Sympherobius gayi*, but due to a lack of satisfactorily preserved type specimens, Monserrat (1998) preferred to treat it as a *nomen dubium.* Monserrat (1998) also

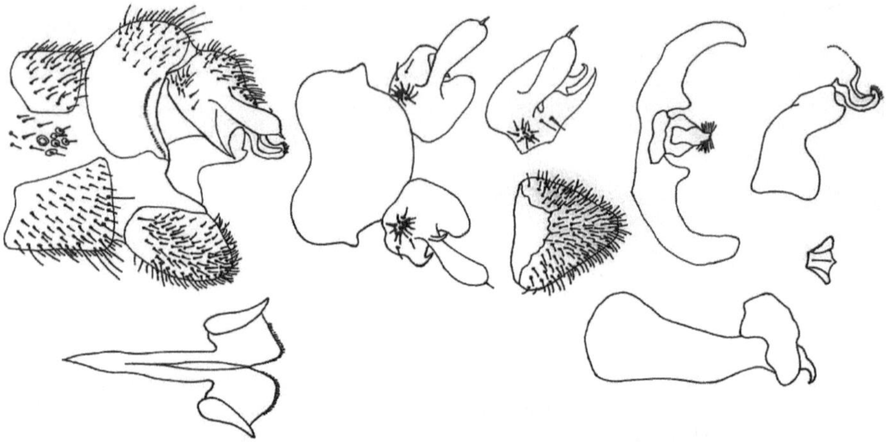

Fig. 12.54 *Sympherobius innoceus* male: apex of the abdomen in lateral view (upper left), ninth tergite and ectoprocts in dorsal (upper left center) and an ectoproct in dorsoposterior view (above right from dorsal view), ninth sternite in ventral view (below the dorsoposterior view of the ectoproct), internal genitalia in posterior (upper right center) and lateral view (upper right center), the hypandrium in ventral view (middle right), and the parameres in dorsal (lower left) and lateral view (lower right). Based on Monserrat (1998).

considered *Sympherobius humilis* Navás, 1914, from Paraguay and Argentina, to be a likely synonym of *S. gayi,* but the only known specimen was destroyed in the Hamburg Museum during World War II, so he also treated it as a *nomen dubium.* Synonyms found in the literature for *S. humilis* include *Coloma scriptus* Navás, 1917, and *Sympherobius scriptus* (Navás, 1917).

1. The dorsolateral process on the hypandrium of the male is fusiform (Fig. 12.54). The antennae are entirely dark. Fore-wing length of male: c. 3.7–3.8 mm; female; c. 3.3 mm. The genitalia of the female have not been described.
   .... *Sympherobius innoceus* Steinmann, 1965 (Argentina, Amazonas).
   - The dorsolateral process on the male hypandrium is not fusiform, or the male fore-wing is c. 5.0 mm long, or the antennae are not all dark (Fig. 12.55)..........2
2. (1) There is a roughly triangular area in the center of the frons, in which there are dense patches of hair-like setae. The fore-wing has an unusual arrangement of dark-margined cross veins. The hypandrium is somewhat triangular in ventral view but has a projection along the midline of the triangle's base (Fig. 12.12). Length of fore-wing: c. 5.4 mm.
   .... *Sympherobius intermedius* Monserrat, 1998 (Chile).
   - The frons does not have a triangular area of dense, hair-like setae, and/or the fore-wing length does not usually exceed 5.0 mm (Fig. 12.55). 3

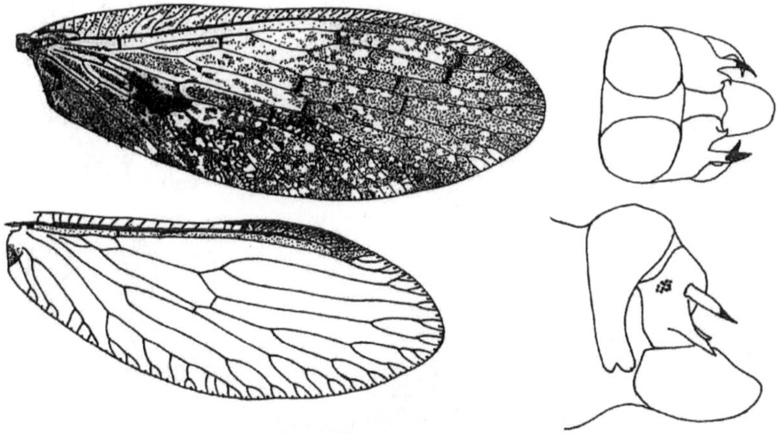

Fig. 12.55 *Sympherobius similis* male: fore and hind wing (left) and the apex of the abdomen in dorsal (upper right) and lateral view (lower right). Based on Monserrat (2008).

3. (2) The male ectoproct bears two dorsomedial processes, each bearing only one modified terminal seta. The paramere lacks anterior lobes, and the apical margins of the median lobes are not toothed. The fore-wing has a proximal but lacks a distal radial cross vein. The pterothorax lacks a pale chevron on the dorsal surface between the wings. The basal third of vein 1A of the fore-wing is uniformly darkened and sometimes has a brown margin; the anal veins are otherwise hyaline. The long radial veins are uniformly brown but lack wide margins on the membranes (Fig. 12.55). There are no dark spots at the bases of the setae on the wings. Fore-wing length: c. 3.8 mm. ....*Sympherobius similis* Carpenter, 1940 (North and Central America, Colombia, Peru).

– The proximal third of vein 1A is not darkened, or there are other dark markings on the other anal veins, which are not hyaline (Fig. 12.56), or the wings have a hyaline membrane with a yellowish tinge, except sometimes for dark marks at the confluences of a few wing veins. If in doubt, the fore-wing length is usually significantly shorter or longer than 3.8 mm 4

4. (3) The antenna is dark at the base and becomes lighter toward the apex. The fore-wings have brown spots, which are sometimes more numerous on the apical part, and clouded along some of the gradate cross veins (Fig. 12.56). 6

– The antennal flagellum is uniform in color or pale brown with a darker external band, and/or the wings lack distinct dark markings near veins 2A and 3A or along the cross veins near the middle of the wing, or the wing veins are all dark (Fig. 12.58). 8

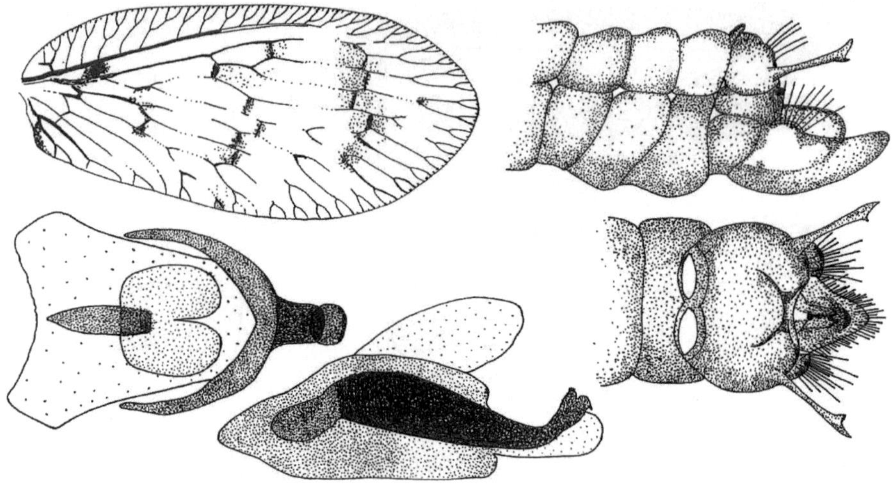

Fig. 12.56 *Sympherobius ariasi* male: fore-wing (upper left), apex of the abdomen in dorsal (lower right) and lateral view (upper right), the aedeagal complex with the parameres in dorsal view (lower left), and without the parameres in lateral view (lower center). Based on Penny and Monserrat (1985).

5. (4) The antennal flagellum is brown proximally and yellow distally. The wings lack distal radial cross veins and usually also proximal ones. The fore-wings have yellow veins with brown spots at the bases of the setae, dark brownish near veins 2A and 3A, and dark variable markings along some of the crossveins (Fig. 12.57). Fore-wing length: 3.5–3.7 mm.
   .... *Sympherobius barberi* (Banks, 1903)
   (North and Central America, introduced to Peru, Galapagos Islands, and Hawaii). Syn: *Hemerobius barberi* Banks, 1903.

   – The basal part of the antenna is dark brown, and the flagellum becomes lighter brown toward the apex. The membranes of the fore-wings are brownish hyaline with many dark spots on the apical half and darkened gradate cross veins. The hind wings are hyaline with pale long veins and colorless cross veins (Fig. 2.19). Length of fore-wing: c. 5 mm; hind wing: c. 4.5 mm.
   .... *Sympherobius gayi* Navás, 1910
   (Peru, Chile, Argentina, Uruguay, Paraguay, Bolivia, Rio de Janeiro, São Paulo, Rio Grande do Sul, and introduced to Portugal, Nigeria, and Easter Island). Syn: *Sympherobius maculipennis* Kimmins, 1929; Monserrat (1998) reported that *Coloma marmoratipennis* Blanchard, 1851, also known as *Sympherobius marmoratipennis* (Blanchard, 1851) in several publications, is very likely a synonym of *S. gayi,* as well. Both intentional and accidental introductions of this species occurred during attempts at pest control (Monserrat 1991).

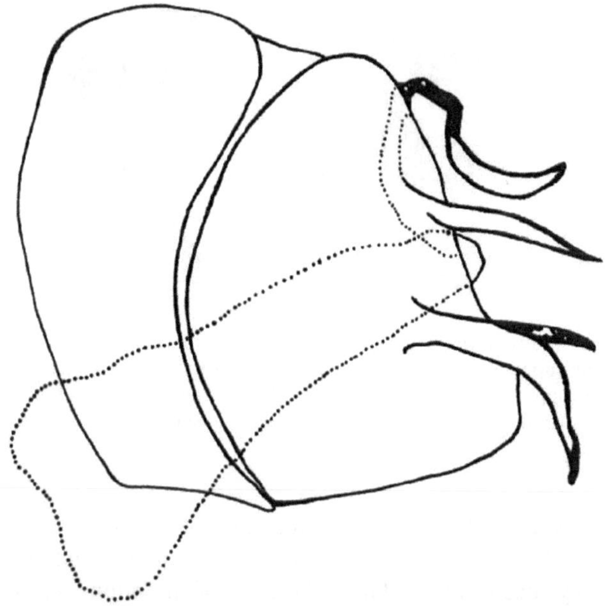

Fig. 12.57 Apex of the abdomen of a male *Sympherobius barberi* in lateral view. Based on MacLeod and Stange (1981).

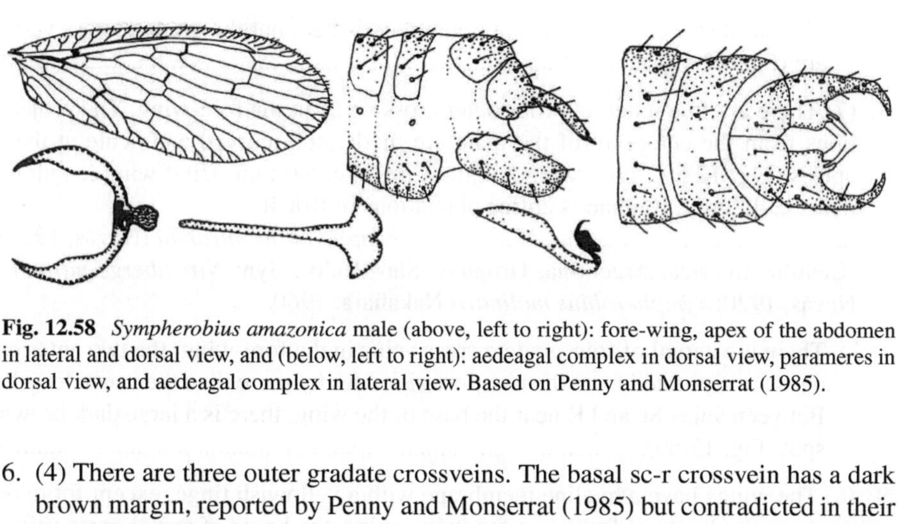

Fig. 12.58 *Sympherobius amazonica* male (above, left to right): fore-wing, apex of the abdomen in lateral and dorsal view, and (below, left to right): aedeagal complex in dorsal view, parameres in dorsal view, and aedeagal complex in lateral view. Based on Penny and Monserrat (1985).

6. (4) There are three outer gradate crossveins. The basal sc-r crossvein has a dark brown margin, reported by Penny and Monserrat (1985) but contradicted in their key. Projections from the ectoproct of the male curve inward toward the apex, are not apically swollen or bifurcate, and reach well beyond the apex of the ninth sternite (Fig. 12.58). Fore-wing length of male: c. 3.1 mm. .... *Sympherobius amazonica* Penny and Monserrat, 1985 (Amazonas).

   – There are four outer gradate cross veins, or the fore-wing of the male is from 4.0 to 5.5 mm long (Fig. 12.56). 7

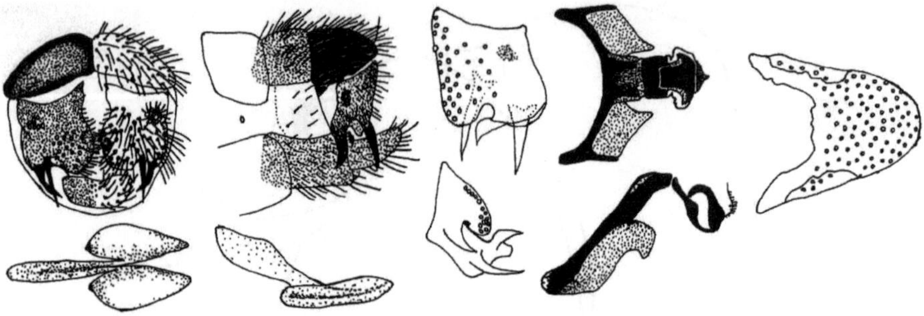

Fig. 12.59 *Sympherobius miranda* male: abdomen in posterior (upper left) and lateral view (upper left center), median (upper center) and ventrolateral view of an ectoproct (lower center), gonarcal complex in dorsal (upper right center) and lateral view (lower right center), ninth sternite (right), and paremeres in dorsal (lower left) and lateral view (lower left center). Based on Monserrat (1990b).

7. (6) There is a very dark margin along the basal sc-r cross vein. Projections from the ectoproct of the male are straight, apically swollen at the apex, and slightly bifurcate; they reach only as far as the apex of the ninth sternite (Fig. 12.56). Fore-wing length of male: 4.2 to 5.0 mm; female: 4.0–4.5 mm. .... *Sympherobius ariasi* Penny and Monserrat, 1985 (Amazonas).

   – The basal sc-r cross vein is not notably darkened, but the gradate cross veins are (Fig. 12.59). 8

8. (7) There are five inner and four outer cross veins in the fore-wing. The projections from the ectoproct of the male are moderately curved and acute at their apices (Fig. 12.59). Fore-wing length of male: c. 5.0 mm. Hind wing length of male: c. 4.2 mm. It inhabits coffee plantations in Brazil. ....*Sympherobius miranda* (Navás, 1920) (Central America, Argentina, Uruguay, São Paulo). Syn: *Niremberge miranda* Navás, 1920; *Sympherobius molinarii* Nakahara, 1960.

   – There are a total of four gradate cross veins in the fore-wing; the two anterior ones are separated from the two posterior ones by more than their own length. Between veins Sc and R near the base of the wing, there is a large dark brown spot (Fig. 12.60). 9

9. (8) The wings have a hyaline membrane with a yellowish tinge, except for dark marks at the bases of forks in a few wing veins, the bases of costal cross veins, and all cross veins of M and Cu. The distal margin of the pterostigma is also dark. There are dots between the endings of the veins around the entire margin of the wing. The antenna is pale, except for three or four segments about one

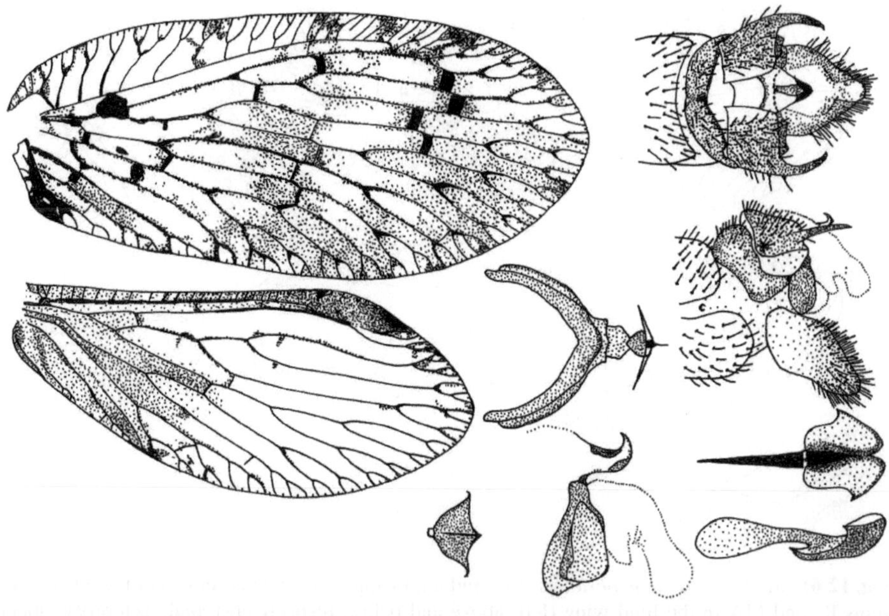

Fig. 12.60 *Sympherobius subcostalis* male: fore and hind wing (left); apex of the abdomen in dorsal (upper right) and lateral view (upper middle right); arcessus, gonarcus, and entoprocessus in posterior (center) and lateral view (lower right center); hypandrium (lower center). Based on Monserrat (1990b).

third of the way from the base to the apex. The palps are brown. The wing span is c. 11 mm.
....*Sympherobius intervenalis* Banks, 1915 (Colombia). Syn: *Nomerobius annulicornis* Navás, 1929. Because of the brevity of the description and a lack of figures, it cannot be ruled out that this species is identical with one of the other species in this genus.

– The wing membrane is hyaline and somewhat darker only along the costal margins. The outer gradate cross veins are darkened, but the other cross veins are not (Fig. 12.60). Length of fore-wing: c. 5.6 mm; hind wing: c. 4.0 mm.
.... *Sympherobius subcostalis* Monserrat, 1990 (Mexico, Argentina).

12.7 Subfamily Megalominae

Key to the South American Species of Adults in the Genus *Megalomus*

Information for the key was provided by Navás (1910a, 1926, 1927b, 1928, 1929c, 1932b, c, 1933a, b), Monserrat (1990a, b, 1997), and Gonzalez-Olazo (1992b). *Megalomus nebulosus* Navás, 1926, described from Peru, cannot be placed at this time, and it is considered a *nomen nudum*.

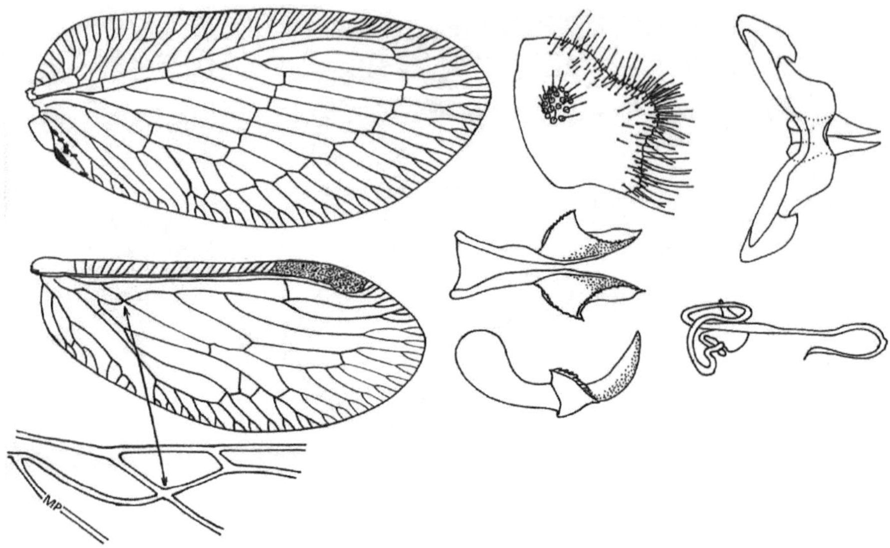

Fig. 12.61 *Megalomus sammnesianus*: fore and hind wing with an enlargement of the junction of veins Rs and MA in the hind wing (left, above and below, respectively), male ectoproct (upper right center), internal male genitalia in posterior view (upper right), male paramere in dorsal (right center) and lateral view (lower center) and the sclerotized duct of the spermatheca of a female in lateral view (lower right). Based on Monserrat (1997).

1. Vein MA in the hind wings appears to simply cross the first section of vein Rs. The inner and outer gradate veins are subparallel and do not converge (Fig. 12.61). 2

   – Vein MA in the hind wing does not appear to simply cross the first section of vein Rs at the junction with the second (Fig. 12.62). 3

2. (1) The gradate veins in the fore-wing are only slightly darkened, and the wings are only faintly variegated. The ectoproct of the male has a strongly convex margin. Denticles are present on the lateral surfaces of the parameres (Fig. 12.61).
   .... *Megalomus sammnesianus* González Olazo, 1987 (Peru).

   – The gradate veins in the fore-wing are heavily darkened, and the wings are conspicuously variegated. The ectoproct of the male has a slightly convex margin. Apical denticles are present on the parameres (Fig. 12.63).
   .... *Megalomus ricoi* Monserrat, 1997 (Santa Catarina).

3. (1) In the hind wing, vein MA fuses with the first section of vein Rs for a short distance before its first bifurcation (Fig. 12.64). 4

   – After the first bifurcation of vein MA in the hind wing, its anterior branch fuses with vein Rs (Fig. 12.62). 9

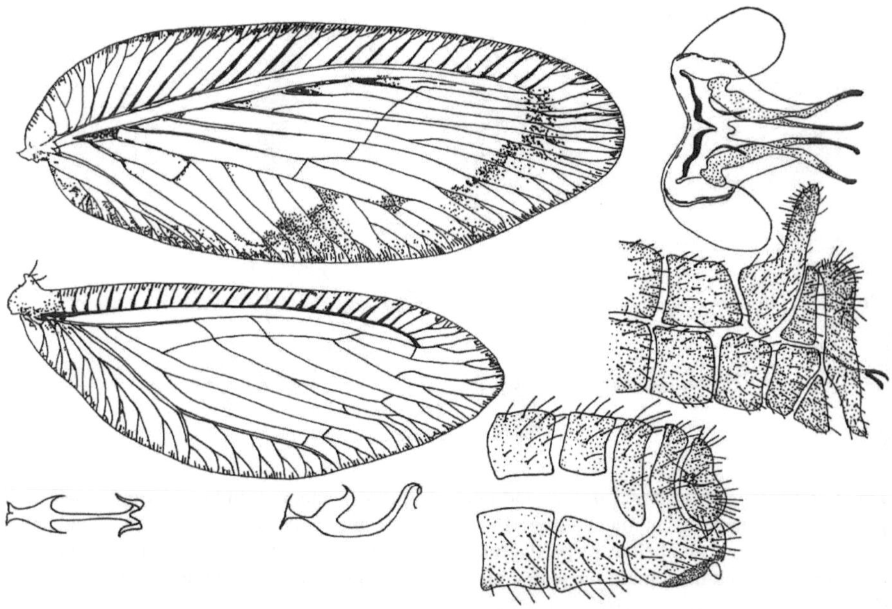

Fig. 12.62 *Megalomus flinti*: fore and hind wing (upper and middle left), male genitalia in dorso-posterior view (upper right), apex of the male (middle right) and female abdomen (lower right), and the male parameres in dorsal (lower left) and lateral view (lower left center). Based on Gonzalez-Olazo (1992b).

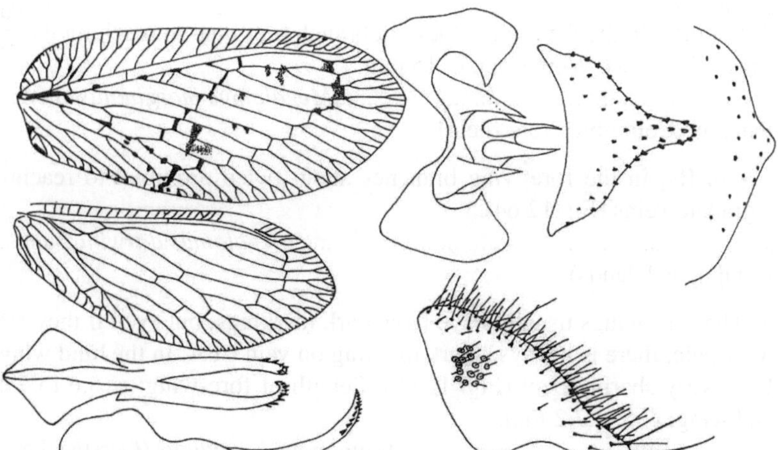

Fig. 12.63 *Megalomus ricoi* male: fore and hind wing (upper and middle left), internal male genitalia in posterior view (upper center), ventral views of the posterior parts of the eighth (upper right) and ninth sternites (upper right center), ectoproct in lateral view (lower right), and the parameres in dorsal (lower middle left) and lateral view (lower left). Based on Monserrat (1997).

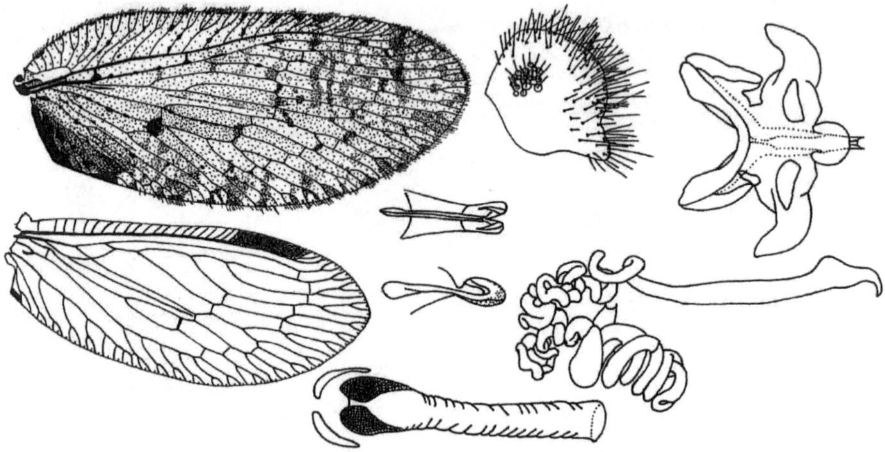

Fig. 12.64 *Megalomus darwini*: fore and hind wing (left), ectoproct in lateral view (upper right center), male internal genitalia in posterior view (upper right), male parameres in dorsal and lateral view (center, above and below, respectively), and female spermatheca in lateral view (right center to lower right), and the female subgenital plate (lower center). Based on Monserrat (1997).

4. (3) In the fore-wing, the first and second crossveins in the subcostal space are parallel (Fig. 12.64). 5
   – In the fore-wing, the first and second crossveins in the subcostal space are strongly oblique to each other (Fig. 12.65). 6
5. (4) Vein Rs_1 in the fore-wing does not branch proximal to reaching the gradate veins (Fig. 12.8). Wing span: c. 15 mm.
   ....*Megalomus marginatus* Banks, 1910
   (Panama, Colombia, Venezuela).
   – Vein Rs_1 in the fore-wing branches about twice proximal to reaching the gradate veins (Fig.12.64).
   .... *Megalomus darwini* Banks, 1924
   (Galapagos Islands).
6. (4) The fore-wings usually have large dark markings, but even if they are relatively pale, there is always a dark marking on vein CuA. In the hind wing, vein Rs_2 is very sharply bent (Fig. 12.65). Length of fore-wing: c. 5.6 to 7.4 mm; hind wing: c. 5 to 7.2 mm.
   ....*Megalomus impudicus* (Gerstaecker, 1888)
   (Peru, Bolivia, Argentina, Espirito Santo, Minas Gerais, Paraná, Santa Catarina). Syn: *Hemerobius impudicus* Gerstaecker, 1888; *Haarupiella impudica* (Gerstaecker, 1888) Esben-Petersen 1914; *Nusalala impudica* (Gerstaecker, 1888), assigned to the wrong genus by Monserrat 1990a, b; *Megalomus linguatus* Navás, 1910; *Megalomus infuscatus* Navás, 1914; *Megalomus*

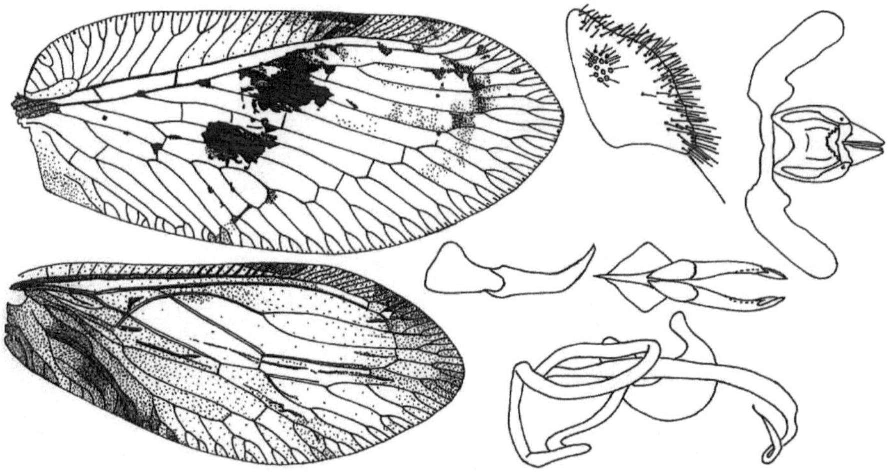

Fig. 12.65 *Megalomus impudicus*: fore and hind wing (left), male ectoproct in lateral view (upper right center), male internal genitalia in posterior view (upper right), male parameres in dorsal (middle right) and lateral view (center), and female spermatheca in lateral view (lower right). Based on Monserrat (1997).

intricatus Navás, 1914; *Megalomus sinuatus* Krüger 1922 (nomen nudum); *Megalomus decoratus* Krüger, 1922 (nomen nudum); *Megalomus acuminatus* Krüger, 1922 (nomen nudum); *Megalomus princeps* Navás, 1932; *Megalomus bridarollius* Navás 1933; *Megalomus bridarolli* Navás, 1932, in González Olazo, 1991; *Megalomus nanus* Navás, 1933.

– The fore-wings lack large dark markings, and there is never one on vein CuA. Vein Rs_2 in the hind wing is straight or very slightly bent (Fig. 12.66). 7

7. (6) The parameres of the male are not joined with each other, and each bears an external denticle at the apex (Fig. 12.66). Fore-wing length: c. 5 mm; hind wing: c. 4.4 mm, measurement by Navás (1924a, b).
.... *Megalomus minor* Banks, 1905 (North and Central America, West Indies, Colombia, Bolivia). Syn: *Megalomus lioni* Navás, 1927; *Megalomus exterior* Navás, 1929; *Megalomus cubanus* Banks, 1930; *Megalomus angustatus* Navás, 1932; *Megalomus parvulus* Kimmins, 1935; *Megalomus serrinus* Navás, 1936.

– The parameres of the male are joined to a caudal lamina, and neither has an apical denticle (Fig. 12.67). 8

8. (7) The ectoproct of the male is subtriangular and bears a distinct terminal process; its posterodorsal is very convex. The internal genital structures of the male appear short and robust. The external margins of male the parameres lack denticles (Fig. 12.67).
.... *Megalomus amnistiatus* Monserrat, 1997 (Colombia, Peru).

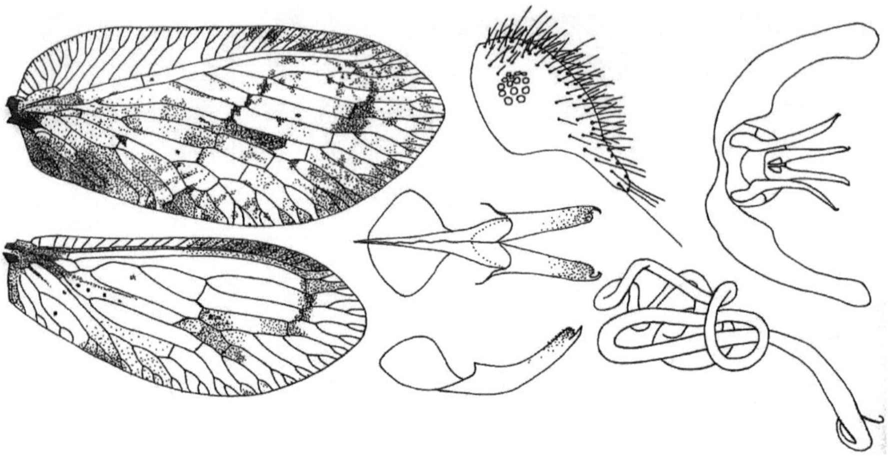

Fig. 12.66 *Megalomus minor*: fore and hind wing (left), male ectoproct in lateral view (upper center), male internal genitalia in posterior view (upper right), parameters in dorsal (center) and lateral view (lower center), and the sclerotized part of the female spermatheca in lateral view (lower right). Based on Monserrat (1997).

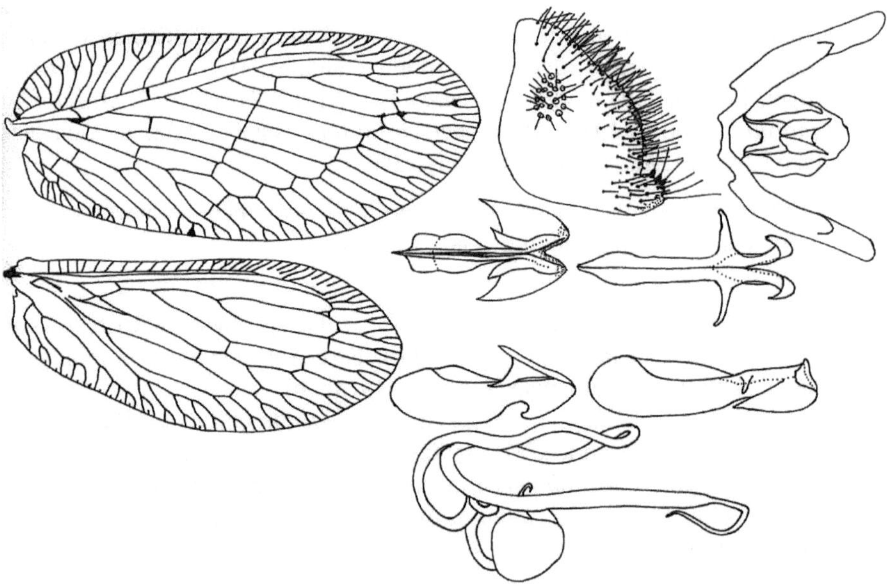

Fig. 12.67 *Megalomus amnistiatus*: fore and hind wing (left), epiproct (upper right center), male genitalia in posterior view (upper right), dorsal (above) and lateral (below) views of the parameres of a specimen from Colombia (center) and one from Peru (right center to right), the sclerotized duct of the spermatheca (lower center to right). Based on Monserrat (1997).

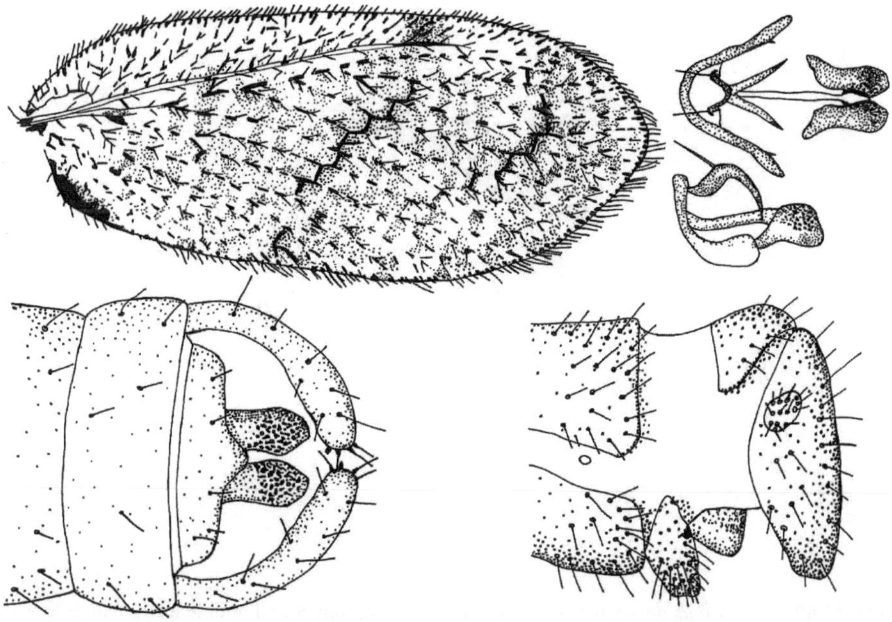

Fig. 12.68 *Megalomus rafaeli* male: fore-wing (upper left), apex of the abdomen in ventral (lower left) and lateral view (lower right), and the internal genitalia in dorsal (upper right) and lateral view (upper middle right). Based on Penny and Monserrat (1985).

– The ectoproct of the male is subeliptical, and bears no distinct terminal process; its posterodorsal is only slightly convex. The internal genital structures of the male appear thin and sinuous. The external margins of male the parameres bear denticles (Fig. 12.68).
....*Megalomus rafaeli* Penny and Monserrat, 1985 (Amazonas).

9. (3) Between veins R and Rs in the hind wing, there is only one crossvein (Fig. 12.62). Fore-wing length: c. 8.0 mm; hind wing: c. 7.0 mm.
....*Megalomus flinti* (Nakahara, 1965) (Chile, Argentina). Syn: *Spinomegalomus flinti* Nakahara, 1965; *Megalomus (Pirionius) flinti* (Nakahara, 1965). Two species known only from North and Central America are characterized by two crossveins between veins R and Rs in the hind wing.

– Between veins R and Rs in the hind wing, there are three crossveins (Fig. 12.69). 10

10. (9) The fore-wings are obviously falcate (Fig. 12.69). 11

– The fore-wings are not falcate (Fig. 12.70). 12

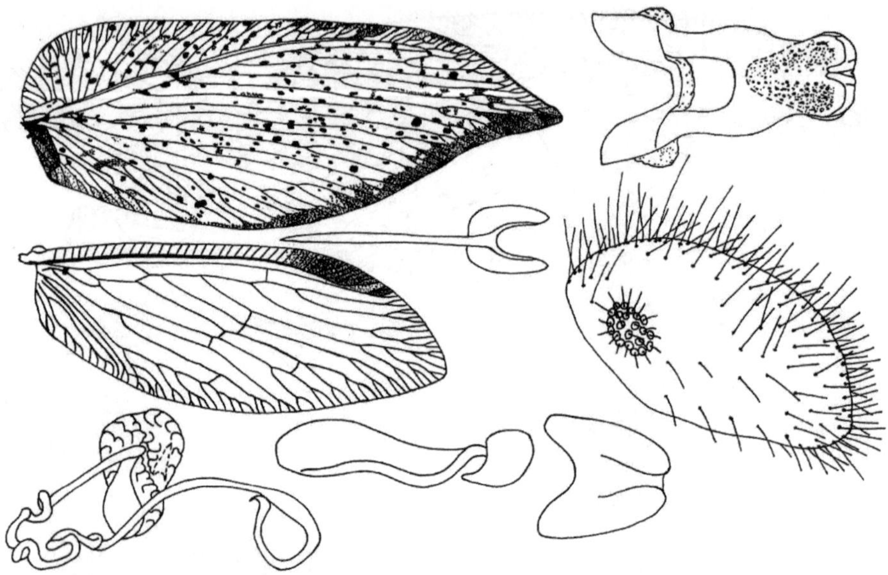

Fig. 12.69 *Megalomus democraticus*: fore and hind wing (upper and middle left), male ectoproct (middle right), male parameres in dorsal (center) and lateral view (lower center), male internal genitalia in posterior view (upper right), spermatheca of female in lateral view (lower left), and the female subgenitale (lower right center). Based on Monserrat (1997).

11. (10) The fore-wings are lanceolate but not very much narrowed toward the apex; their external margins are markedly sinuous. The external gradate veins are darkened. The tergites of the mesothorax and metathorax are blackish. The fore-femora are darkened distally on their exterior margins (Fig. 12.71). ....*Megalomus stangei* González Olazo, 1981 (Chile, Argentina). Syn: *Megalomus (Megalomus) stangei* González Olazo, 1981; *Conchopterella stangei* (González Olazo, 1981) González Olazo, 1991; *Falcomegalomus stangei* (González Olazo, 1981) González Olazo, 1992b.

– The fore-wings are lanceolate and narrow to an acuminate apex; their external margins are only slightly sinuous. The external gradate veins are barely darkened or completely hyaline. The mesothorax and metathorax are yellowish. The fore-femora are darkened on their dorsal surfaces (Fig. 12.69). Length of fore-wing: c. 11 mm; hind wing: c. 9 mm. ....*Megalomus democraticus* Monserrat, 1997 (Chile, Argentina).

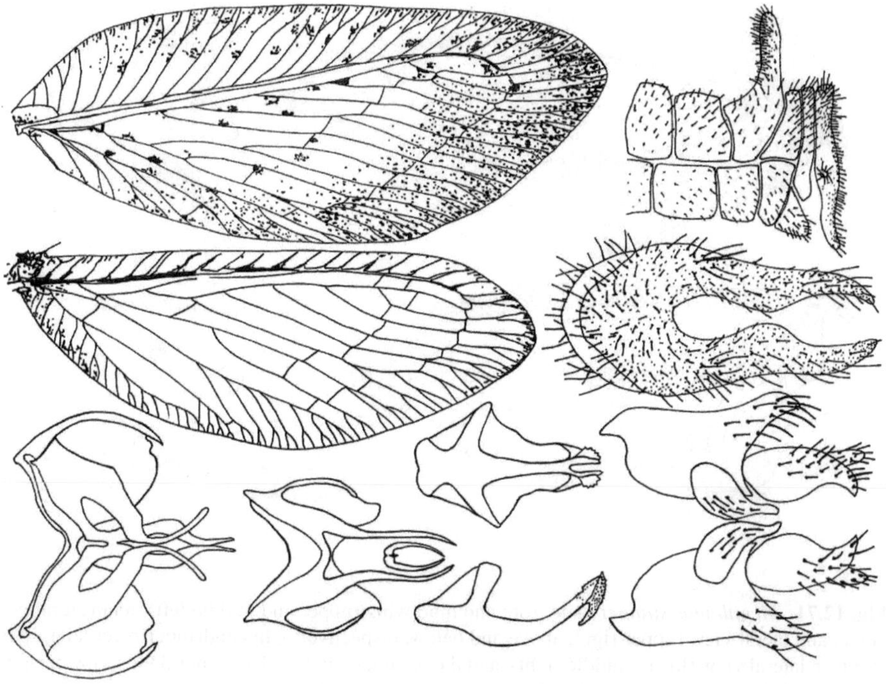

Fig. 12.70 *Megalomus nigratus*: fore and hind wing (upper and middle left); apex of the male abdomen in lateral view (upper right) and that of a specimen identified under the synonym, *Pirionius gentiliorum*, in posterior view (upper middle right); tenth sternite of a male in dorsal view (lower left) and that of a specimen identified under the synonym, *Pirionius gentiliorum* (lower right center); apex of the female abdomen of a specimen identified as *Pirionius gentiliorum* in ventral view (lower right); male parameres in dorsal and lateral view (lower right center, above and below, respectively). Based on Gonzalez Olazo (1992b) and Monserrat (1997).

12. (10) The wings are barely twice as long as wide. The basal part of the costal field of the fore-wing is strongly convex (Fig. 12.70).
.... *Megalomus nigratus* (Navás, 1928) (Chile, Argentina, Rio de Janeiro). Syn: *Pirionus nigratus* Navás, 1928; *Megalomus (Pirionius) nigratus* (Navás, 1928); *Pirionius nigratus* (Navás, 1928); *Pirionus gentiliorum* González Olazo, 1992 (spelling mistake in generic name); *Pirionius gentilliorum* González Olazo, 1992, (alternate spelling in the description).

– Each wing is considerably longer than twice its width. The basal part of the costal field of the fore-wing is weakly convex and very sinuous (Fig. 12.72).
.... *Megalomus australis* (González Olazo, 1992) (Argentina). Syn: *Navasius australis* González Olazo, 1992.

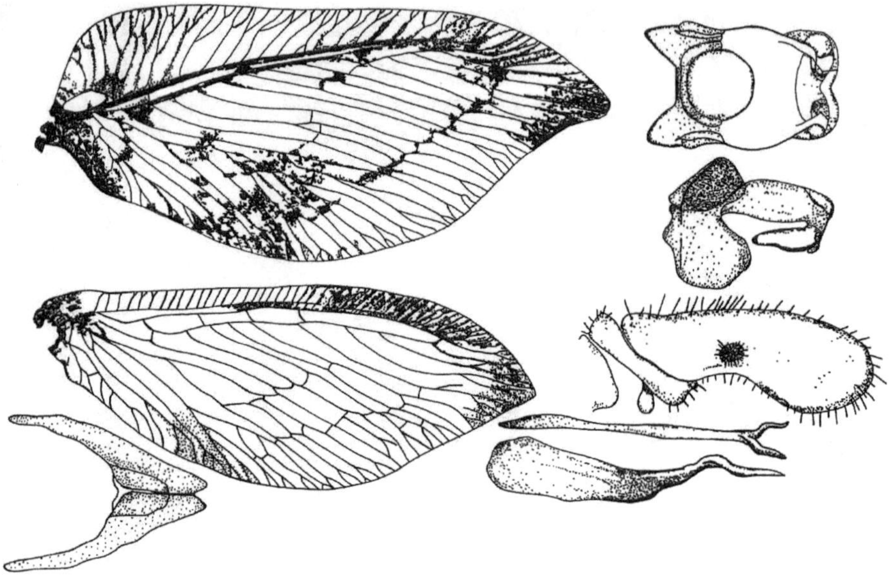

Fig. 12.71 *Megalomus stangei* male: fore and hind wing (upper and middle left), tenth sternite in dorsal and laeral view (upper right, above and below, respectively), hypandrium (lower left), ectoprocts in lateral view (lower middle right), and the paramere in dorsal and lateral view (lower right, above and below, respectively). Based on Gonzalez-Olazo (1992b).

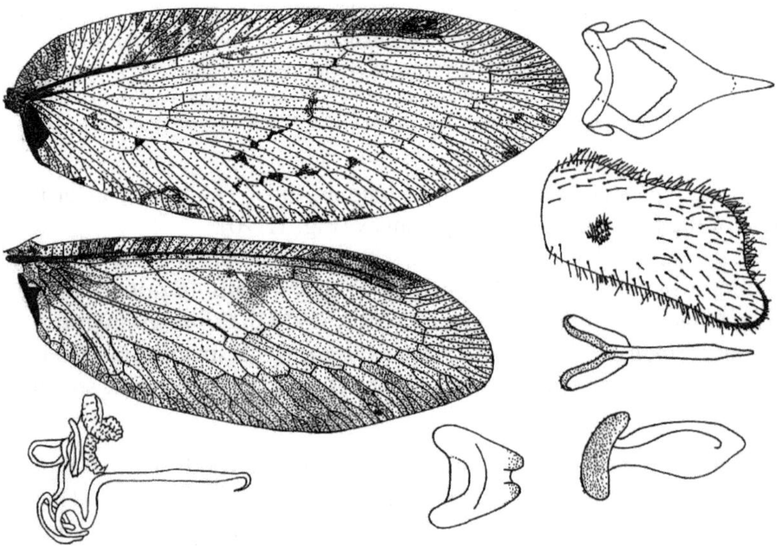

Fig. 12.72 *Megalomus australis*: fore and hind wing (upper and middle left), male ectoproct in lateral view (upper middle right), male genital structures in posterior view (upper right), male parameres in dorsal (lower middle right) and lateral view (lower right), sclerotized duct of the female spermatheca in lateral view (lower left), and female subgenital place in ventral view (lower right center). Based on Monserrat (1997).

Literature

Banks N (1904) Neuropteroid insects of New Mexico. Transactions of the American Entomological Society 30:97–110
Banks N (1909) Hemerobiidae from Queensland, Australia. Proceedings of the Entomological Society of Washington 11:76–82
Banks N (1910) New South American neuropteroid insects. Proceedings of the Entomological Society of Washington 12:146–160
Banks N (1913) Synopses and descriptions of exotic Neuroptera. Transactions of the American Entomological Society 39:201–242 + pl. 23–26
Banks N (1915 [1914]) New neuropteroid insects, native and exotic. Proceedings of the Academy of Natural Sciences of Philadelphia 66:608–632 + pl. 27
Banks N (1924) Neuroptera from the Williams Galapagos expedition. Zoologica, Scientific Contributions of the New York Zoological Society 5:177–180
Banks N (1930) Some new Neotropical neuropteroid insects. Psyche 37:183–191
Esben Petersen P (1914) Descriptions of a new genus and some new or interesting species of Planipennia. Notes from the Leyden Museum 36:263–270
Esben Petersen P (1924) More Neuroptera from Juan Fernandez and Easter Island. In: The Natural History of Juan Fernandez and Easter Island 3:309–313. Uppsala
Faúndez E (2005) Asociación críptica de *Gayomyia falcata* (Blanchard, 1851) (Neuroptera: Hemerobiidae) con la Zarzaparrilla *Ribes magellanicum* Poiret, 1812 (Saxifragaceae) en la región de Magallanes. Anales Instituto Patagonia (Chile) 33:63–64
Garzón-Orduña IJ, Menchaca-Armenta I, Contreras-Ramos A, Liu X, Winterton SL (2016) The phylogeny of brown lacewings (Neuroptera: Hemerobiidae) reveals multiple reductions in wing venation. BMC Evolutionary Biology 16:192–210
Gerstaecker A (1888 [1887]) Weitere Beiträger zur Artenkenntniss der Neuroptera Megaloptera. Mittheilungen aus dem Naturwissenschaftlichen Verein für Neu-Vorpommern und Rügen in Greifswald 19:39–130
González Olazo EV (1981) El genero *Megalomus* Rambur (Neurop.-Hemerobiidae) en Argentina y Chile. Acta Zoológica Lilloana 36:97–113
González Olazo EV (1992a) Sobre dos nuevos generos australes de Hemerobiidae (Neuroptera, Planipennia). Acta Zoológica Lilloana 42(1):149–154
González Olazo EV (1992b) Contribucion al conocimiento del genero *Pirionius* Navás, 1928 (Neuroptera, Planipennia). Acta Zoológica Lilloana 42(1):155–164
González Olazo EV (1992c) Notas sobre *Notiobiella rubrostigma* Navás, 1914 (Neuroptera, Hemerobiidae). Acta Zoológica Lilloana 42(1):177–178
González Olazo EV (1992d) Una nueva especie de *Nomerobius* (Neuroptera, Hemerobiide) de la Argentina. Acta Zoológica Lilloana 42(1):109–111
González Olazo EV (1994) Contribucion al conocimiento de las especies de *Nusalala* (Neuroptera, Planipennia). Acta Zoológica Lilloana 42(2):215–224
Handschin E (1955) Los insectos de la Islas Juan Fernandez. 15. Neuroptera. Revista Chilena de Entomologia 4:3–20
Kimmins DE (1928) New and little known Neuroptera of Central America. Eos 4(3–4):363–370
Kimmins DE (1929) Some new and little known Argentine Neuroptera. Revista de la Sociedad Entomológica Argentina 2:187–192
Kimmins DE (1935) Notes on the genera *Megalomus* Rambur and *Nesobiella* gen. nov. (Neuroptera) with descriptions of new species. Annals and Magazine of Natural History 10(16):602–619
Klimaszewski J, McE. Kevan DK, Peck SB (1987) A review of the Neuroptera of the Galápagos Islands with a new record for *Sympherobius barberi* (Banks) (Hemerobiidae). Canadian Journal of Zoology 65:3032–3040
Krüger L (1922) Beiträge zu einer Monographie der Neuropteren. Familie der Hemerobiiden. III. Literatur und Katalog. Stettiner Entomologische Zeitung 83:138–172

Lara RIR, de Freitas S (2002) Caracterização morphológica de adultos de *Nusalala tessellata* (Gerstaecker, 1888) (Neuroptera, Hemerobiidae). Revista Brasileira de Entomologia 46(4):521–528

Lara RIR, de Freitas S (2003) Caracterização morphológica de espécies de *Hemerobius* Linnaeus, 1758 (Neuroptera, Hemerobiidae) associadas a cultivos de café (*Coffea aribica* L.), milho (*Zea mays* L.) e erva-mate (*Ilex paraguariensis* St. Hill.). Revista Brasileira de Entomologia 47(3):427–434

Lara RIR, Perioto NW (2003a) Bioecologia de hemerobiídeos (Neuroptera, Hemerobiidae). Arquivo do Instituto de Biologia, São Paulo 70(4):517–523

Lara RIR, Perioto NW (2003b) Primeiro Registro de ocorrência dis gêneros *Megalomus*, *Nomerobius* e *Sympherobius* (Neuroptera, Hemerobiidae) para o estado do Paraná, Brasil. Arquivo do Instituto de Biologia, São Paulo 70(4):513–515

Lara RIR, Perioto NW (2003c) Primeiro Registro de ocorrência de *Sympherobius miranda* (Navás, 1920) (Neuroptera, Hemerobiidae) para o Brasil. Arquivo do Instituto de Biologia, São Paulo 70(4):511–512

Lara RIR, Perioto NW, de Freitas S (2010) Diversidade de hemerobiídios (Neuroptera) e suas associações com presas em cafeeiros. Pesquisa Agropecuária Brasileira, Brasilia 45(2):115–123

Letardi A, Almeida JM, Badano D, Silva RR, Machado E (2013) Contributing to a checklist of Neuropterida in Portugal: the Naturdata Project. Açoreana, Supplement 9:29–38

Linnaeus C (1758) Systema Naturae. 10(1), 824 pp, Holmiae

MacLeod EG, Stange LA (1981) The brown lacewings of Florida. Entomology Circular No. 227, Florida Department of Agriculture and Consumer Services, Division of Plant Industry, 4 pp

Monserrat VJ (1984) Sobre las especies Americanas del género *Notiobiella* Banks, 1909 I. (Neuroptera, Planipennia, Hemerobiidae). In: Gepp J, Aspöck H, Hölzel H (eds) Progress in World's Neuropterology. Proceedings of the 1st international symposium on neuropterology, Graz, pp 85–97

Monserrat VJ (1990a) A systematic checklist of the Hemerobiidae of the world (Insecta: Neuroptera). In: Mansell MW Aspöck H (eds) Advances in neuropterology. Proceedings of the 3rd international symposium on neuropterology, Pretoria, pp 215–262

Monserrat VJ (1990b) Systematic studies on Hemerobiidae (Insecta: Neuroptera). In: Mansell MW Aspöck H (eds) Advances in Neuropterology. Proceedings of the 3rd international symposium on neuropterology, Pretoria, pp 67–88

Monserrat VJ (1991) Nuevos datos sobre algunas especies del género *Hemerobius* L., 1758 (Insecta, Neuroptera: Hemerobiidae). Graellsia 46:61–70

Monserrat VJ (1996) Revision del genero *Hemerobius* de Latinoamérica (Neuroptera, Hemerobiidae). Fragmenta Entomologica, Roma 27:399–523

Monserrat VJ (1997) Revision del genero *Megalomus* de Latinoamérica (Neuroptera, Hemerobiidae). Fragmenta Entomologica, Roma 29(1):123–206

Monserrat VJ (1998) Nuevos datos sobre los hemeróbidos de América (Neuroptera: Hemerobiidae). Journal of Neuropterology 1:109–153

Monserrat VJ (2000) Revisión del género *Nusalala* (Neuroptera, Hemerobiidae). Fragmenta Entomologica, Roma 32(1):83–162

Monserrat VJ (2002) Family Hemerobiidae. In: Penny ND (ed) A guide to the lacewings (Neuroptera) of Costa Rica. Proceedings of the California Academy of Sciences 53(12):238–251

Monserrat VJ (2003) Contribución al conocimiento de los hemeróbidos de Patagonia y Tierra del Fuego (Insecta, Neuroptera, Hemerobiidae). Graellsia 59(1):37–56

Monserrat VJ (2004) Nuevos datos sobre angunas especies de hemeróbidos (Insecta: Neuroptera: Hemerobiidae). Heteropterus Revista de Entomología 4:1–26

Monserrat VJ (2008) Nuevos datos sobre algunas especies de Hemeróbidos (Insecta, Neuroptera, Hemerobiidae). Graellsia 64(2):233–253

Monserrat VJ, Deretsky Z (1999) New faunistical, systematic and taxonomic data on brown lacewings (Neuroptera: Hemerobiidae). Journal of Neuropterology 2:45–66

Monserrat VJ, Penny ND (1985 [1983]) Sobre las especies americanas del género *Notiobiella* Banks, 1909, II (Neuroptera, Planipennia, Hemerobiidae). Nouvelle Revue d'Entomologie 13:125–132

Monserrat VJ, Triviño V, Acevedo F, Garcia A (2013) Nuevos datos sobre algunos especies de Hemeróbidos de la Península Ibérica e Islas Canarias, incluyendo una nuena especie invasora de origen neotropical en Portugal (Insecta, Neuroptera, Hemerobiidae). Graellsia 69(2):157–168

Nakahara W (1965) Neotropical Hemerobiidae in the United States National Museum. Proceedings of the United States National Museum 117(3509):107–122 + 2 pl

Navás L (1910a) Hemeróbidos (Ins. Neur.) nuevos con clave de las tribus y géneros de la familia. Brotéria, Série Zoológica 9(2):69–90

Navás L (1910b) Neurópteros chilenos (1ª Serie). Revista Chilena de Historia Natural 14(4–6):235–241

Navás L (1912) Crisópidos y Hemeróbidos (Ins. Neur.) nuevos o críticos. Brotéria Série Zoológica 10:98–113

Navás L (1913) Nevroptères. In: Mission l'Equateur 10:69–77. Gauthier-Villars, Paris

Navás L (1914) Neurópteros Sudamericanos. Primera Serie. Brotéria, Série Zoológica 12(1):45–56, 215–234

Navás L (1915) Neurópteros nuevos o poco conodicos. (Sexta Serie). Memorias de la Real Academia de Ciencias y Artes de Barcelona. Tercera Época 12(7):119–136

Navás L (1917a) Neue Neuropteren. Dritte Serie. Entomologische Mitteilungen 6(7/9):274–282

Navás L (1917b) Neurópteros nuevos o poco conocidos. Novena Serie. Memorias de la Real Academia de Ciencias y Artes de Barcelona. Tercera Época 13(26):393–406

Navás L (1918a) Algunos insectos de la República Argentina. Revista de la Real Academia de Ciencias Exactas Fisicas y Naturales de Madrid 16:491–504

Navás L (1918b) Insectos Chilenos. Boletin Sociedad Aragonesa de Ciencias Naturales 17:212–230

Navás L (1920) Insectos Sudamericanos (1ª Serie), (2ª Serie), (3ª Serie). Anales de la Sociedad Cientifica Argentina 90:33–72

Navás L (1921a) Insectos Americanos nuevos o críticos. Brotéria, Série Zoológica 19:113–124

Navás L (1921b) Névroptéres des Islas Juan Fernandez et de L'Isle de Pâques. In: Skottsberg C (ed) The natural history of Juan Fernandez and Easter Island 3:125–128. Uppsala

Navás L (1922) Sur des Nevroptères nouveaux ou critiques. V. Annales de la Société Scientifique de Bruxelles 3:250–256

Navás L (1924a) Algunos insectos de Chile. II. Revista Chilena de Historia Natural 28:12–16

Navás L (1924b) Insectos de la América Central. Brotéria, Série Zoológica 21:55–86

Navás L (1925) Neuropteren aus Brasilien. Mitteilungen der Münchener Entomologischen Gesellschaft 15:64–68

Navás L (1926) Insectos exóticos Neurópteros y afines. Brotéria, Série Zoológica 23:79–93

Navás L (1927a) Insectos neotropicos, 3ª Serie. Revista Chilena de Historia Natural 31:316–328

Navás L (1927b) Insectos des Museo de Paris (4ª Serie). Brotéria, Série Zoológica 24:5–33

Navás L (1928) Insectos neotropicos, 4.ª Serie. Revista Chilena de Historia Natural 32:106–128

Navás L (1929a) Neurópteros críticos. Butlletí de la Instiució Catalana de Historia Natural 9:113–115

Navás L (1929b) Insectos Neurópteros del Museo de Hamburgo. Memorias de la Real Sociedad España de Historia Natrual 15:315–322

Navás L (1929c) Insecta nova. Series XIV. Memorie dell'Accademia Pontifica dei Nuovi Lincei, Rome (2) 12:24–32

Navás L (1929d) Insecta nova. Series XIII. Memorie dell'Accademia Pontifica dei Nuovi Lincei, Rome (2) 12:15–23

Navás L (1929e) Insectos exóticos neurópteros y afines del Museo Civico di Storia Naturale di Génova. Annali del Museo Civico di Storia Naturale di Génova 53:354–389

Navás L (1930 [1929]) Insectos neotropicos, 5.ª serie. Revista Chilena de Historia Natural 33:17–24

Navás L (1932a) Décadas de insectos nuevos. Década 21. Brotéria (Serie Trimestral, Ciências Naturais) 1:109–119
Navás L (1932b) Insectos de la Argentina y Chile. Revista de la Sociedad Entomológica Argentina 22:79–86
Navás L (1932c) Alcuni insetti del Museo de Zoologia della R. Università de Torino. Bollettino dei Musei de Zoologia e di Anatomia Comparata della R. Universita de Torino Serie 3, 42(26):1–38
Navás L (1933a [1932]) Insectos de la Argentina. Revista de la Academia de ciencias exactas, fisico-quimicas y naturals de Zaragosa 16:87–120
Navás L (1933b) Neurotteri e tricotteri del "Deutsches Entomologisches Institut" di Berlino – Dahlem. Bollettino della Società Entomologica Italiana 65(5):105–113
Navás L (1935 [1934]) Insectos Suramericanos. Décima Serie. Revista de la Academia de Cíencias Exactas Físicas y Químicas y Naturales de Madrid 32:360–375
Navás L (1936) Décadas de insectos nuevos. Decada 28. Brotéria (Serie Trimestral. Ciencias Naturais 5:161–170
Oliveira RC, Lara RIR, Fonseca AR, Perioto NW (2013) Hemerobiidae (Neuroptera) in the mid-western Region of Minas Gerais State, Brazil. Revista Colombiana de Entomologia 39(2):256–259
Oswald JD (1988) A revision of the genus *Sympherobius* Banks (Neuroptera: Hemerobiidae) of America North of Mexico with a synonymical list of the world species. Journal of the New York Entomological Society 96(4):390–451
Oswald JD (1990) Revision of the Neotropical brown lacewing genus *Nomerobius* (Neuroptera: Hemerobiidae). Annals of the Entomological Society of America 83:18–29
Oswald JD (1993a) Revision and cladistic analysis of the world genera of the family Hemerobiidae (Insecta: Neuroptera). Journal of the New York Entomological Society 101(2):143–299
Oswald JD (1993b) A new genus and species of brown lacewing from Venezuela (Neuroptera: Hemerobiidae), with comments on the evolution of the hemerobiid forewing radial vein. Systematic Entomology 18:363–370
Oswald JD (1994) A new phylogenetically basal subfamily of brown lacewings from Chile (Neuroptera: Hemerobiidae). Entomologica Scandinavica 25(3):295–302
Oswald JD (1996) A new brachypterous *Nusalala* species from Costa Rica, with comments on the evolution of flightlessness in brown lacewings (Neuroptera: Hemerobiidae). Systematic Entomology 2:343–352
Oswald JD (2004) Review of the brown lacewing genus *Biramus* (Neuroptera: Hemerobiinae), with the description of a new species from Costa Rica and Panama. Tijdschrift voor Entomologie 147:41–47
Penny ND (1978 [1977]) Lista de Megaloptera, Neuroptera e Raphidioptera do Mexico, America Central, ilhas Caraibas e America do Sul. Acta Amazonica, Supplement 7(4):1–61
Penny ND, Monserrat VJ (1985 [1983]) Neuroptera of the Amazon Basin. Part 10. – Hemerobiidae. Acta Amazonica 13(5–6):879–909
Penny ND, Sturm H (1984) A flightless brown lacewing from Colombia (Neuroptera, Hemerobiidae). Spixiana 7(1):19–22
Reguilón C (2002) Morfología de los estadios inmaduros de *Hemerobius bolivari* (Neuroptera: Hemerobiidae). Revista de la Sociedad Entomologica Argentina 61(1–2):63–68.9
Reguilón C, Nuñez Campero SR (2005) Morfología de los estadios larvales de *Sympherobius marmoratipennis* (Neuroptera: Hemerobiidae). Revista de la Sociedad Entomologica Argentina 64(3):81–85
Souza B (1999) Aspectos morfológicos de adultos de *Nusalala uruguaya* (Navás, 1923) (Neuroptera: Hemerobiidae). Ciência e Agrotecnologia, Lavras 23(2):252–257
Stange LA (1968 [1967]) Catálogo de Neuroptera de Argentina y Uruguay. Acta Zoológica Lilloana 22:5–86
Steinmann H (1965) The zoological results of Ty. Topal's collectings in South Argentina. 19. Neuroptera, Hemerobiidae. Folia Entomologica Hungarica N.S 18:567–574
Tauber CA, Krakauer AH (1997) Larval characteristics and generic placement of endemic Hawaiian hemerobiids (Neuroptera). Pacific Science 51:413–423

Chapter 13
Osmylidae

Abstract An illustrated key is provided for identification of the genera of adult osmylids known to occur in South America. Illustrated keys are also provided to identify the species of each genus native to South America. The countries or states in which specimens of each species have been found and invalid synonyms are reported.

Keywords Key to species • Protosmylinae • Kalosmylinae • Stenosmylinae

13.1 Key to the Subfamilies and Genera of Adults in the Family Osmylidae in South America

Information for the key was provided by Navás (1912), Krüger (1913a, b), Kimmins (1940), Adams (1969, 1977), Penny (1981), Oswald (1994), and Monserrat (2005). Most larvae of South American species have not been described. Most larvae on other continents live in or near freshwater.

1. There are relatively few longitudinal veins and cross veins. Series of cross veins are recognizable, and there are three veins in the first row. There are cross veins along the edges of the wings. A better description of the only known South American species is needed.
   ....Subfamily Protosmylinae...............*Paryphosmylus* Krüger, 1913
   The only species in this genus is *Paryphosmylus ornatus* Krüger, 1913, from Ecuador. The name of the genus was misspelled as *Parosmylus* in the 1913 issue of the Zoological Record.
   – There are a great many longitudinal veins in both wings, and only the outermost row of gradate cross veins are recognizable. There are no cross veins along the edges of the wing (Fig. 13.1). 2
2. (1) Vein M in the fore-wing forks near the base but apical from the origin of the first branch of vein Rs. There is a basal cross vein between M and Rs on the hind

C.W. Heckman, *Neuroptera (Including Megaloptera)*, Encyclopedia of South American Aquatic Insects, DOI 10.1007/978-3-319-35125-4_13

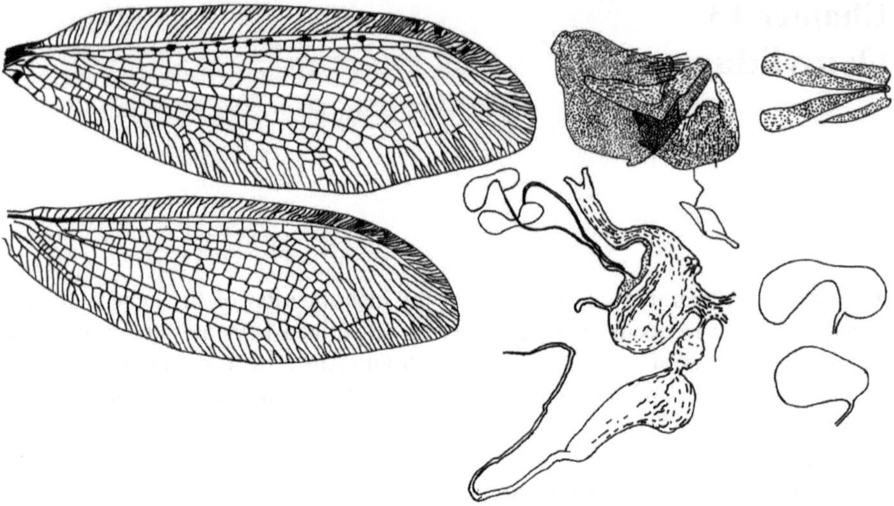

Fig. 13.1 *Kempynus crenatus:* fore and hind wing (left), male genitalia in lateral view (upper right center), impandrium internum (upper right), internal female reproductive system in lateral view (lower right center), and the spermathecae (middle and lower right). Based on Adams (1971).

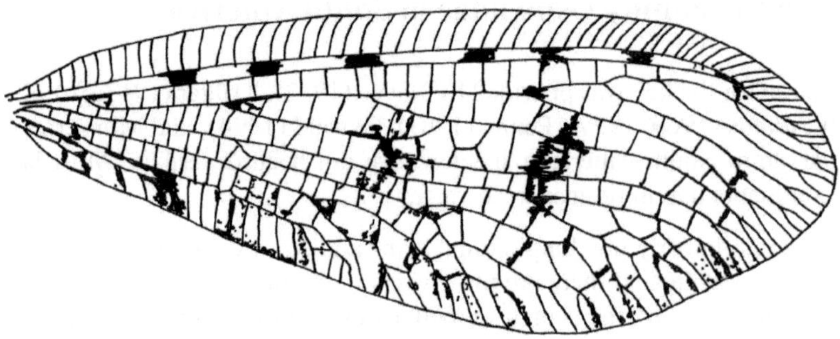

Fig. 13.2 The fore-wing of *Gumilla adspersus*. Based on Navás (1914).

wing, which is somewhat sinuous. There is a pair of eversible scent glands between the eighth and ninth tergites of the male (Fig. 13.1).
.....................Subfamily Kalosmylinae................*Kempynus* Navás, 1912..p. 416

– Vein M in the fore-wing forks far from the wing base. There is no sinuous cross vein or a feeble and incomplete one between the base of vein M and vein Rs on the hind wing. No eversible scent glands are apparent between the eighth and ninth tergites of the male (Fig. 13.2).
....Subfamily Stenosmylinae.................3

3. (2) Vein Rs of the male has about six branches. The antenna is extremely long and consists of filiform segments. Fore-wing length of the only known species:

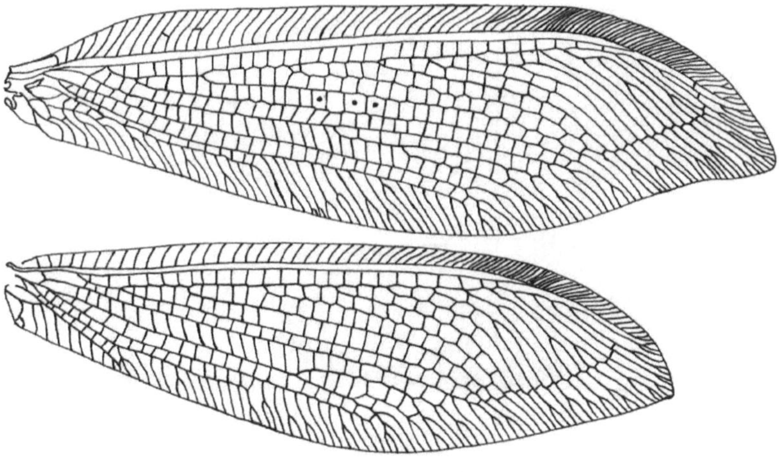

Fig. 13.3 The fore and hind wing of *Phymatosmylus caprorum*. Based on Adams (1969).

c. 20 mm. There are about six dark, rectangular spots in the space between veins Sc and R (Fig. 13.2).
....*Gumilla* Navás, 1912
The only species in this genus known with certainty to inhabit South America is *Gumilla adspersus* Navás, 1912, from Santa Catarina, which was assigned to this family based on a single male specimen (Navás 1912). Adams (1969a) did not discuss this genus in his review because it is too poorly known and differs markedly from other South American osmylids. A second species, *Gumilla longicornis* (Walker, 1853), was believed by Adams (1977) to have been found in Brazil, although the only specimen available for examination had a label showing that it was collected in Georgia. It is noted for having very long antennae and lacking ocelli, which resulted in its being treated as a member the Chrysopidae for many years. If this species was indeed found in Brazil and is conspecific with *Gumilla adspersus,* then its correct name would be *Gumilla longicornis* (Walker, 1853). Its junior synonyms would then be *Osmylus longicornis* Walker, 1853; *Meleoma longicornis* (Walker, 1853) Hagan, 1861; *Leucochrysa longicornis* (Walker, 1853) Banks, 1907; *Allochrysa longicornis* (Walker, 1853) Bickley and MacLeod, 1956; and *Gumilla adspersus* (Navás, 1912) *nec* Navás, auctt. There is still too little information to propose these changes, and additional specimens of the members of this genus are needed for study.

– Vein Rs of the male has about ten or more branches (Fig. 13.3). The antenna is not very long, and its segments are not filiform. 4

4. (3) The eighth and ninth abdominal tergites of the male are only partially fused (Fig. 1.11). The bases of veins Cu and A_1 are slightly thickened (Fig. 13.3). Fore-wing length of the only known species: 20–26 mm. Length of antenna:

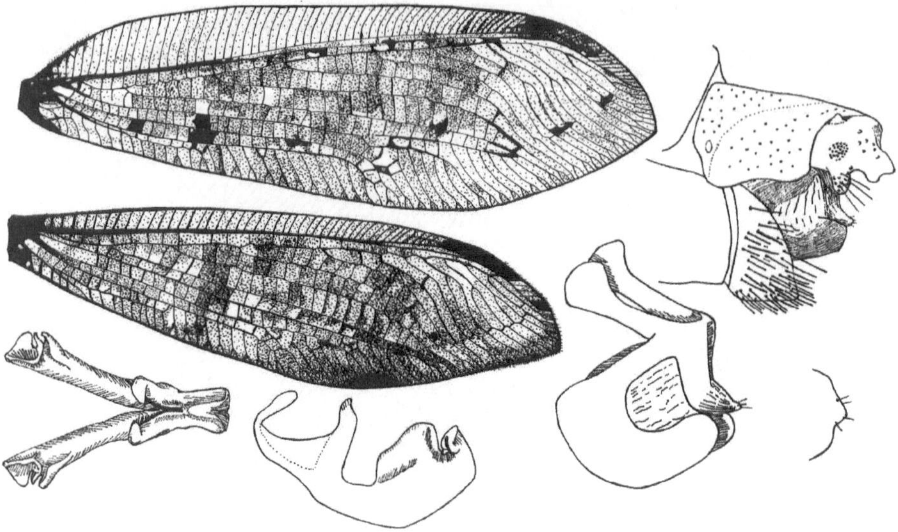

Fig. 13.4 *Isostenosmylus fasciatus* male: fore and hind wing (upper and middle left, respectively), apex of the abdomen in lateral view (upper right), and (below, left to right): parameres in dorsal and lateral view, ninth sternite in lateral view, and apex of the ninth sternite in dorsal view. Based on Kimmins (1940).

6.5–8.4 mm. The color of the body and wings is pale to yellowish with darker markings.
....*Phymatosmylus* Adams, 1969
The only known species in this genus is *Phymatosmylus caprorum* Adams, 1969, reported from Chile and Argentina.

– The eighth and ninth abdominal segments of the male are completely fused, and the suture at the fusion is sometimes indistinct (Fig. 13.4). The bases of veins Cu and A_1 are never thickened in either sex.
....*Isostenosmylus* Krüger, 1913..p. 418

13.2 Subfamily Kalosmylinae

Key to the Species of Adult *Kempynus* Known from South America

Information for the key was provided by Adams (1971) and Oswald (1994).

1. The fore-wing is broadly rounded at the apex, and its apical area has a pale lunate region, well delimited posteriorly from the rest of the wing (Fig. 13.1). The distal part of the wing lacks a fuscous, biangulate marking.
.... *Kempynus crenatus* Adams, 1971 (Chile, Argentina).

– The fore-wing is angulate at the apex, and the apical area lacks a pale lunate region well delimited from the rest of the wing (Fig. 13.5). 2

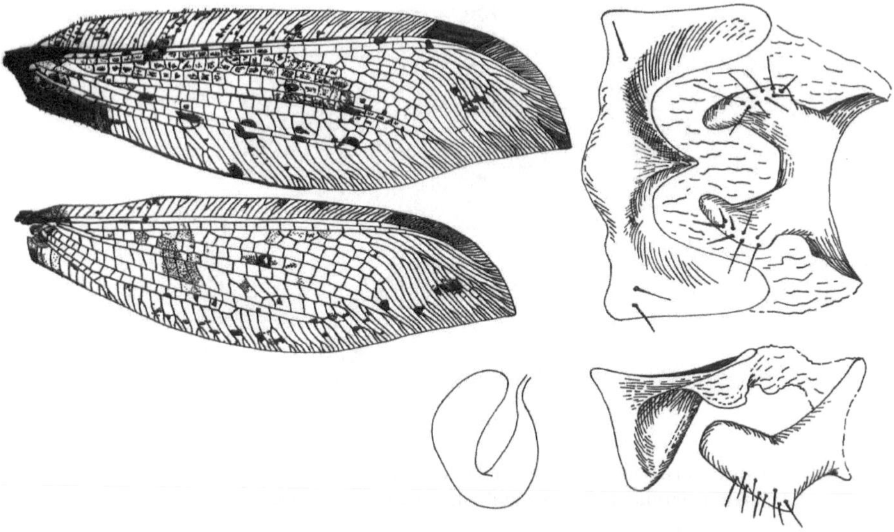

Fig. 13.5 *Kempynus falcatus* female: fore and hind wing (left), eighth sternite in ventral (upper right) and lateral view (lower right), and the apex of the accessory gland (lower center). Based on Kimmins (1940).

2. (1) The distal part of the fore-wing of the female lacks a prominent, fuscous, biangulate marking, although dark markings of other shapes may be present. The proximal third of the space between Sc and R contains several well-defined fuscous spots (Fig. 13.5). A description of the male is unavailable.
   .... *Kempynus falcatus* Navás, 1912 (Chile, Argentina). Syn: apparently *Kalosmylus falcatus* (Navás, 1912), assigned to a new genus by Krüger (1913b); *Stenosmylus* sp. MacLachlan, 1894.

   – The fore-wing has a prominent, fuscous, biangulate marking at its distal end. The proximal third of the space between Sc and R contains several well-defined fuscous spots (Fig. 13.6). 3

3. (2) There are two or three well-defined fuscous spots in the proximal third of the subcostal space, i.e., the space between veins Sc and R. The free distal lobes of the female subgenitale surround a narrow, proximally acute space, visible in ventral view (Fig. 13.6). Fore-wing length of female: c. 22 mm. The male has not been described.
   .... *Kempynus tjederi* Oswald, 1994 (Chile).

   – There are no well-defined fuscous spots in the proximal third of the subcostal space, although tiny brownish clouds may be present. The free distal lobes of the female subgenitale surround a broad, proximally rounded space, visible in ventral view (Fig. 13.7). Fore-wing length: 29–34 mm.
   .... *Kempynus digoniostigma* Oswald, 1994 (Chile, Argentina).

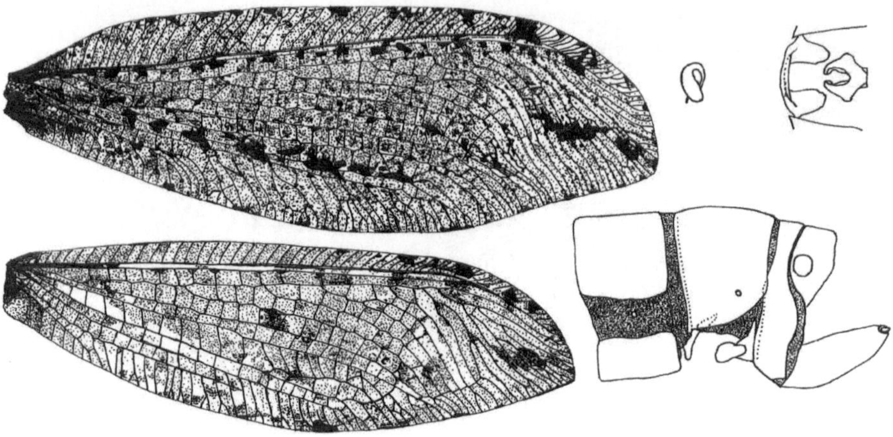

Fig. 13.6 *Kempynus tjederi* female: fore and hind wing (left), apex of the abdomen in lateral view (lower right), eighth sternite and subgenitale in ventral view (upper right), and the spermathecal bulb (upper right center). Based on Oswald (1994a).

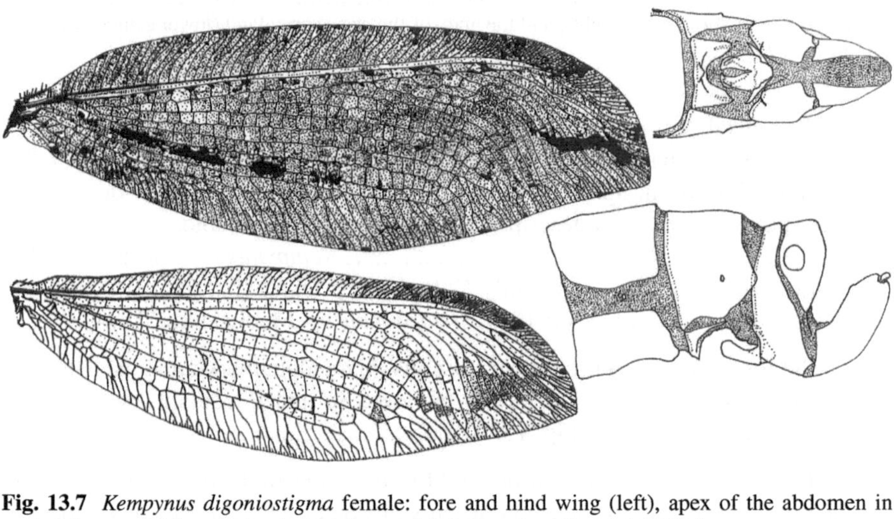

Fig. 13.7 *Kempynus digoniostigma* female: fore and hind wing (left), apex of the abdomen in ventral (upper right) and lateral view (lower right). Based on Oswald (1994a).

13.3 Subfamily Kalosmylinae

Key to the Species of Adult Male *Isostenosmylus* in South America

The key was prepared with information found in Navás (1928), Kimmins (1940), and Ardila-Camacho and Noriega (2014).

1. The basal third of the fore-wing has a brown band extending obliquely from vein R_1 to the posterior margin of the wing (Fig. 13.8). Length of fore-wing:

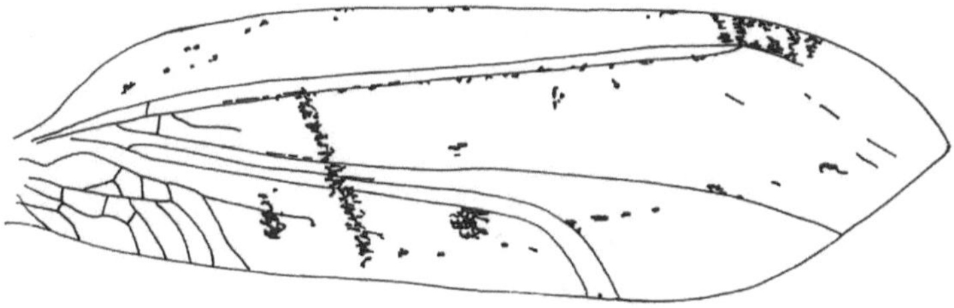

Fig. 13.8 Outline of the fore-wing of *Isostenosmylus morenoi*, showing the outline, some veins, and locations of *dark* markings. Based on Navás (1928).

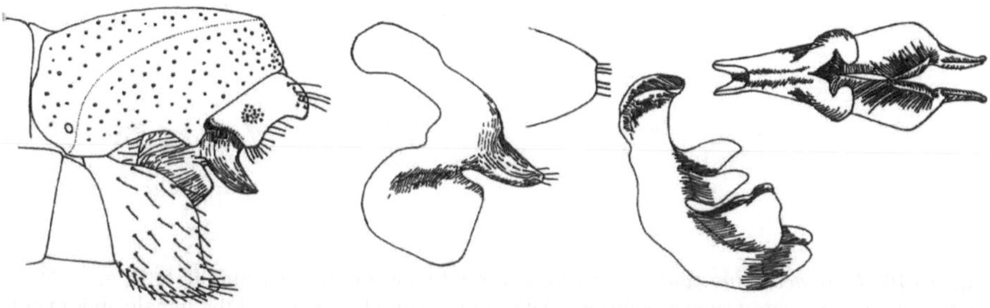

Fig. 13.9 *Isostenosmylus nigrifrons* male (left to right): apex of the abdomen in lateral view, tenth sternite in lateral view, apex of the tenth sternite in dorsal view, and the parameres in lateral and dorsal view. Based on Kimmins (1940).

c. 21.5 mm; hind wing: c. 19.0 mm. Maximum width of fore-wing: c: 7.0 mm; hind wing: c. 6.1 mm.
....*Isostenosmylus morenoi* (Navás, 1928) (Ecuador). Syn: *Oedosmylus morenoi* Navás, 1928.

– The fore-wing has other dark markings or none at all (Fig. 13.9). 2

2. (1) There are markings on the hind wing between veins C and R_1, and the posterior margin is infuscated with brown. The frons of the male is fulvous with a transverse black stripe between the compound eyes. The male parameres have fused bases and a V-shaped excision, best seen in dorsal view (Fig. 13.9). Fore-wing length: c. 22 mm; hind wing: c. 20 mm.
....*Isostenosmylus nigrifrons* Kimmins, 1940 (Equador).

– The hind wing membrane is uniformly hyaline (Fig. 13.10). 3

3. (2) The considerable area of the fore-wing is darkly marked (Fig. 13.10). 4

– A small area of the fore-wing is weakly marked (Fig. 13.11). 5

4. (3) Posterior to the second fork in vein Rs, there is an amber spot covering the area between vein MP and the posterior edge of the wing. In lateral view, the

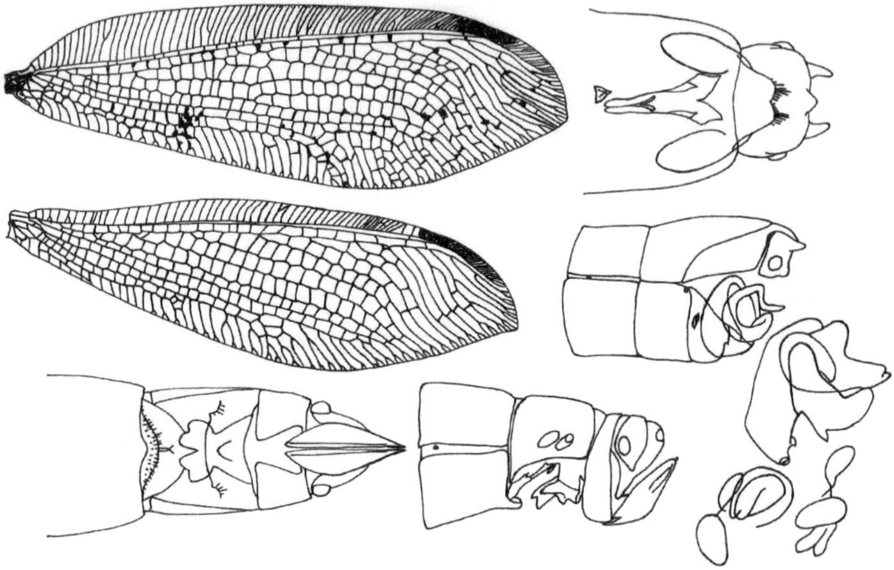

Fig. 13.10 *Isostenosmylus contrerasi:* fore and hind wing (above and middle left), apex of the male abdomen in ventral (upper right) and lateral view (middle right) and the female abdomen in ventral (lower left) and lateral view (lower center), male genitalia in lateral view (lower middle right), and the female spermathecae in dorsal view (lower right). Based on Ardila-Camacho and Noriega (2014).

male ectoproct can be seen to bear a prominent, acuminate posterodorsal process. Gonocoxite VIII of the female has bilobed lateral lobes (Fig. 13.10). Fore-wing length: 30–31 mm; hind wing: 28–29 mm. Length of the pterostigma in the fore-wing: c. 4 mm.
..........................*Isostenosmylus contrerasi* Ardila-Camacho and Noriega, 2014 (Colombia).

– Posterior to the first fork in vein Rs, there is an amber spot covering the area between veins MP and 1A. The ectoproct of the male lacks is obliquely truncate and lacks an upper branch near the apex. The lateral lobes on gonocoxite VIII of the female are not bilobed (Fig. 13.12). Fore-wing length: 21–25 mm.
....*Isostenosmylus pulverulentus* (Gerstaecker, 1894) (Rio de Janeiro). Syn: *Osmylus pulverulentus* Gerstaecker, 1894.

5. (3) Many cross veins in the costal space between veins C and Sc fork before reaching vein C (Fig. 13.11).
....*Isostenosmylus septemtrionalandinus* Ardila-Camacho and Noriega, 2014 (Colombia).

– The cross veins in the costal space are unforked, with no more than a few exceptions (Fig. 13.4). 6

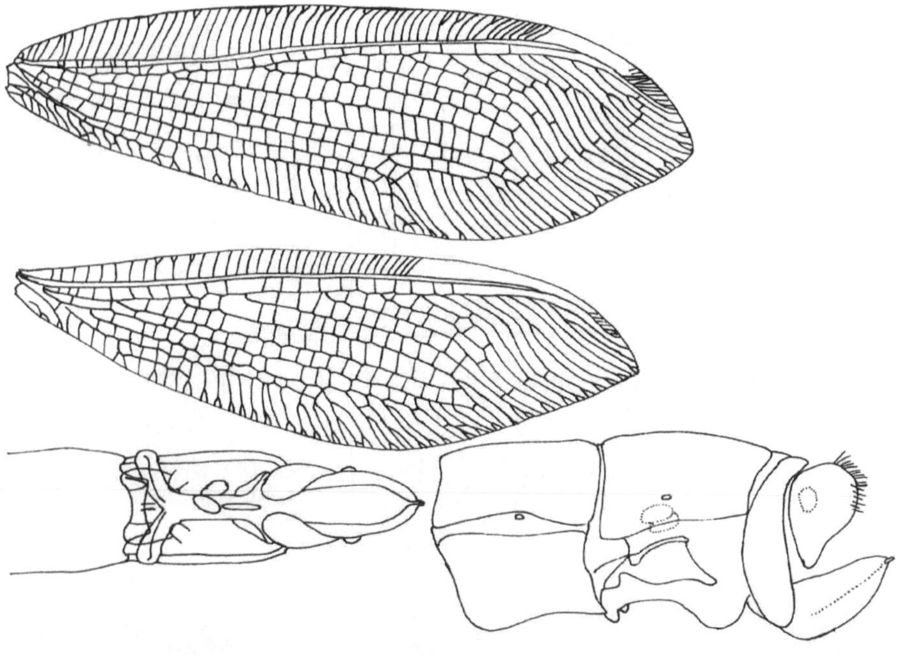

Fig. 13.11 *Isostenosmylus septemtrionalandinus* female: fore and hind wing (above and middle), apex of the abdomen in ventral (lower left) and lateral view (lower right). Based on Ardila-Camacho and Noriega (2014).

6. (5) There is a conspicuous dark spot on vein CuP posterior to the first fork in vein Rs. Both wings are solidly dark brown at their bases. The pterostigma is strongly sclerotized and fairly uniformly darkened. The membrane of the hind wing is fairly uniform grayish hyaline, except at the base; it is almost hyaline or slightly clouded just beyond the dark coloration at the base. In lateral view, a small, triangular anterodorsal process is evident on the ectoproct of the male, and a rounded, posteroventral lobe is produced into a broad process. The vertex, genae, and frons to below the antennae are fuscous, and the rest of the frons is fulvous (Fig. 2.17). Fore-wing length: c. 25 mm; hind wing: c. 23 mm.
   ....*Isostenosmylus fusciceps* Kimmins, 1940 (Peru).
   – Only small spots are present on vein CuP of the fore-wing. The pterostigma is only lightly sclerotized. In lateral view, the male ectoproct appears rounded and bears complex folded lobes near its apex; posteroventrally, it appears rounded and only slightly produced; posterodorsally, it bears a sharply pointed process (Fig. 13.4). Length of fore-wing length: c. 23 mm; hind wing: c. 21 mm.
   ....*Isostenosmylus fasciatus* Kimmins, 1940 (Peru, Bolivia).

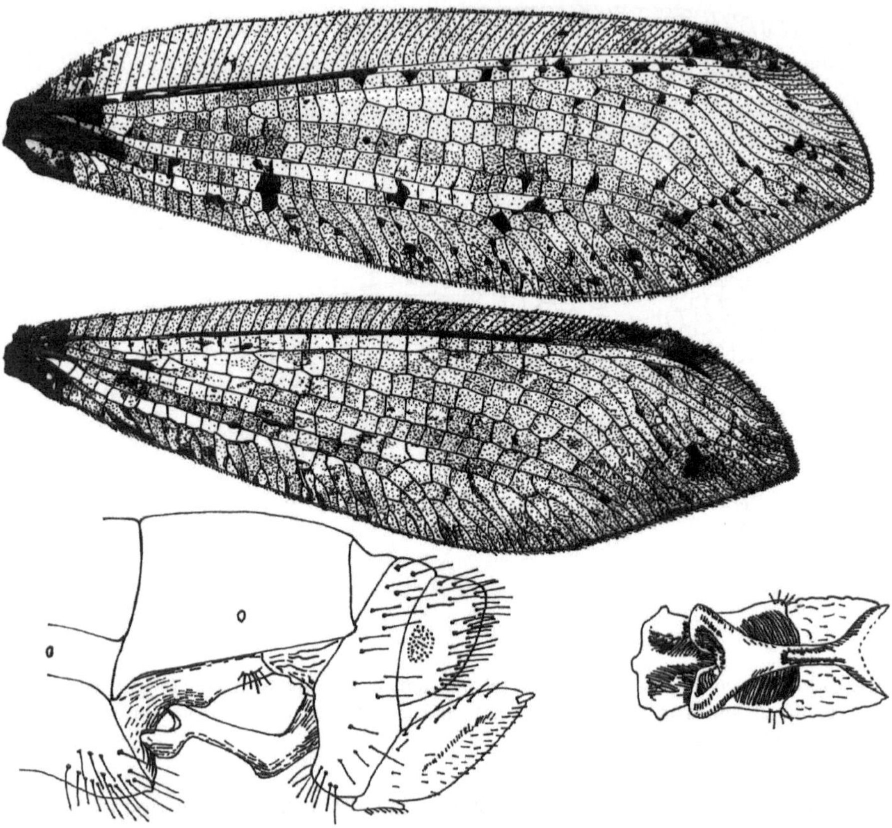

Fig. 13.12 *Isostenosmylus pulverulentus* female: fore and hind wing (above and middle), apex of the abdomen in lateral view (lower left), and the eighth sternite in ventral view (lower right). Based on Kimmins (1940).

Literature

Adams PA (1969) A new genus and species of Osmylidae [Neuroptera] from Chile and Argentina, with a discussion of planipennian genitalic homologies. Postilla 141:1–11

Adams PA (1971) Variation and geographic distribution in some Argentine and Chilean Osmylidae, with a new species of *Kempynus* (Neuroptera). Bulletin of the Southern California Academy of Sciences 70:45–49

Ardila-Camacho A, Noriega JI (2014) First record of Osmylidae (Neuroptera) from Colombia and description of two new species of *Isostenosmylus* Krüger, 1913. Zootaxa 3826(2):315–328

Gerstaecker A (1894) Über neue und weniger gekannte Neuropteren aus der Familie Megaloptera Burm. Mittheilungen aus dem Naturwissenschaftlichen Verein für Neu-Vorpommern und Rügen in Greifswald 25:93–173

Hagan HA (1861) Synopsis of the Neuroptera of North America, with a list of South American species. Smithsonian Miscellaneous Collections 4(1):i–xx + 1–347

Kimmins DE (1940) A revision of the osmylid subfamilies Stenosmylinae and Kalosmylinae (Neuroptera). Novitates Zoologicae (London) 42(1):165–201 + pl. I–VIII

Krüger L (1913a) Osmylidae. Beiträge zu einer Monographie der Neuropteren. Familie der Osmyliden. II. Charakteristik der Familie, Unterfamilien und Gattungen auf Grund des Geäders. Entomologische Zeitung herausgegeben von dem entomologischen Vereine zu Stettin 74(1):3–127

Krüger L (1913b) Osmylidae. Beiträge zu einer Monographie der Neuropteren. Familie der Osmyliden. III. Literatur und Katalog. – Entomologische Zeitung herausgegeben von dem entomologischen Vereine zu Stettin 74(2):193–224

Monserrat VJ (2005) Nuevos datos sobre angunas pequeñas familias de neurópteros (Insecta: Neuroptera: Nevrothidae, Osmylidae, Sisyridae, Dilaridae). Heteropterus Revista de Entomologia 5:1–26

Navás L 1912 (1911) Insectos Neurópteros nuevos ó poco conocidos. – Memorios de la Real Academia de Ciencias y Artes de Barcelona. Tercera Época 10:135–202

Navás L (1914) New Neuroptera from the United States. Bull Brooklyn Entomol Soc 9:13–20

Navás L (1928) Insectos del Museo de Hamburgo. Primera Serie. Boletin de la Sociedad Entomológica de España 11:59–67, 90–100, 121–138

Oswald JD (1994) Two new South American species of the genus *Kempynus* Navás (Neuroptera: Osmylidae: Kempyninae). Proceedings of the Entomological Society of Washington 96(2):367–372

Walker F (1853) List of the specimens of neuropterous insects in the collection of the British Museum. Part II (Sialidae – Nemopterides). Trustees, London, pp 193–476

Chapter 14
Polystoechotidae

Abstract An illustrated key is provided for identification of the genera of adult polystoechotids known to occur in South America. Illustrated keys are also provided to identify the species of each genus which includes more than one species native to South America. The countries or states in which specimens of each species have been found and invalid synonyms are reported.

Keywords Giant lacewing • *Fontecilla* • *Polystoechotes* • Key to species • Chile

Key to the Genera of Adults in the Family Polystoechotidae in South America

The common English name for insects in this family is giant lacewing. Information for the key was provided by Navás (1932); Oswald (1998), and Winterton and Makarkin (2010).

1. The apex of the fore-wing is falcate, and two nygmata are usually present. On the hind wing, there are dark markings, especially near the posterior margin, and sometimes one nygma is present. The median part of the mesothorax is yellowish.
 .... *Fontecilla* Navás, 1932
 There is only one known species in this genus: *Fontecilla graphicus* Navás, 1932, known only from Chile.

 – The apex of the fore-wing is not falcate, and no more than one nygma is present. The hind wing lacks prominent dark markings and a nygma (Fig. 14.1). The mesonotum has dark markings.
 .... *Polystoechotes* (Burmeister, 1839)

C.W. Heckman, *Neuroptera (Including Megaloptera)*, Encyclopedia of South American Aquatic Insects, DOI 10.1007/978-3-319-35125-4_14

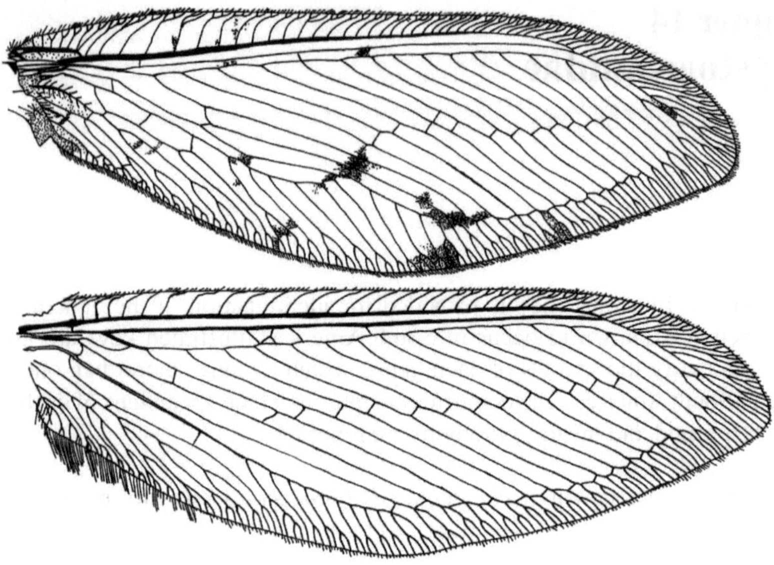

Fig. 14.1 Fore and hind wing of *Polystoechotes punctatus*. Based on Penny (2002).

Key to the Species of Adult *Polystoechotes* in South America

Information for the key was provided by Navás (1924) and Oswald (1998). A third known species in this genus is *Polystoechotes lineatus* Carpenter, 1940, known from California, but is is unlikely to occur in South America.

1. The anterior and posterior margins of the basal third of the fore-wing are approximately parallel, and its humeral region is broad. The subgenitale of the female consists of one median and two basolateral sclerites; the median one is deeply incised at the apex and has a small process in the middle of the apical incision (Fig. 2.20). Length of fore-wing: c. 19.5 mm; hind wing: c. 17.0 mm. The species is considered rare.

   ....*Polystoechotes gazullai* Navás, 1924 (Chile).

   – The posterior margin of the fore-wing is prominently rounded in the basal third and not nearly parallel with the anterior margin; the humeral region of that wing is hardly wider than the distal part of the subcostal area (Fig. 14.1). The subgenitale consists of one wide sclerite with a heavily sclerotized lateral lobe on each side.

   ....*Polystoechotes punctatus* (Fabricius, 1793) (North and Central America as far south as Panama).

Literature

Burmeister H (1839) Handbuch der Entomologie. Neuroptera 2(2–3):397–1050. Theodor Friedrich Enslin, Berlin

Carpenter FM (1940) A revision of the Nearctic Hemerobiidae, Berothidae, Sisyridae, Polystoechotidae and Dilaridae (Neuroptera). Proceedings of the American Academy of Arts and Sciences 74(7):193–280

Fabricius JC (1793) Enomologia Systematica 2. C. G. Proft, Hafniae. 599 pp

Navás L (1924 [1922]) Insectos de la Argentina y Chile. Estudios (Buenos Aires) 22:358–368

Navás L (1932 [1931]) Insectos de Papudo (Aconcagua) recogidos por don Arturo Fontecilla en Febrero de 1930. Revista Chilena de Historia Natural 35:71–73

Oswald JD (1998) Rediscovery of *Polystoechotes gazullai* (Neuroptera: Polystoechotidae). Proceedings of the Entomological Society of Washington 100(3):389–394

Penny ND (2002) Family Polystoechotidae. In: Penny ND (ed) A guide to the lacewings (Neuroptera) of Costa Rica. Proceedings of the California Academy of Sciences 53(12):290

Winterton S, Makarkin VN (2010) Phylogeny of moth lacewings and giant lacewings (Neuroptera: Ithonidae, Polystoechotidae) using DNA sequence data, morphology, and fossils. Annals of the Entomological Society of America 103(4):511–522

Chapter 15
Sisyridae

Abstract Keys are provided to distinguish the adults, pupae, and larvae of the two genera of sisyrids known to inhabit South America. Illustrated keys were prepared to identify the adults in each genus. The countries or states of Brazil in which each species has been found and invalid synonyms are reported.

Keywords Spongillafly • *Climacia* • *Sisyra* • Larvae • Pupae

The common name of the species in this family is spongillafly, named because the larvae of *Sisyra* species live inside freshwater sponges of the genus *Spongilla*. Sometimes the name is also applied to species of *Climacia*, although the habits of few South American species of *Climacia* are known. Pupae of South American *Climacia* species have been found among debris along the margins of water bodies. The spongillaflies are the most fully aquatic species of Planipennia. The larvae of most or possibly all of the species in South America appear to be aquatic or littoral (Cover and Resh 2008).

15.1 Key to the Genera of Adults in the Family Sisyridae in South America

Information for the key was ürpvoded by Banks (1913a), Chandler (1956), Parfin and Gurney (1956), and Penny (1981). The larvae and pupae of few South American sisyrid species have been described, so the key to the genera depends primarily of information about larvae from other continents. It cannot be ruled out that the morphology of some South American larvae in this family differ from those that have already been described. Undescribed larvae should be carefully preserved for further study.

1. Vein Rs in the fore-wing has one fork posterior to the pterostigma. Outer gradate cross veins are present. The fore-wing has brown markings. The apical segment of the palp is swollen, so that it is widest near the middle but only slightly wider than the other segments (Fig. 15.1). The known larvae are found along the shores

C.W. Heckman, *Neuroptera (Including Megaloptera)*, Encyclopedia of South American Aquatic Insects, DOI 10.1007/978-3-319-35125-4_15

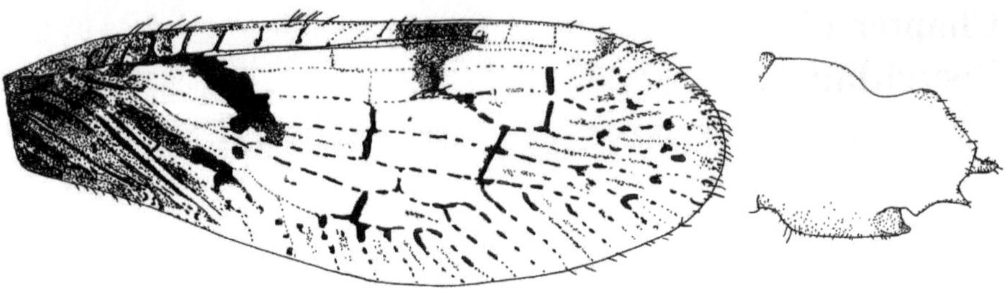

Fig. 15.1 *Climacia negrense* male: fore-wing (above) and the apex of the abdomen in lateral view (lower left). Based on Penny (1981).

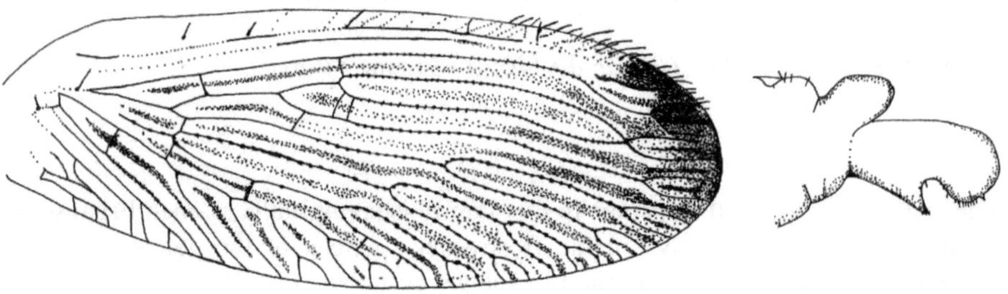

Fig. 15.2 *Sisyra ariasi* male: fore-wing (above) and the apex of the abdomen in lateral view (lower left). Based on Penny (1981).

of water bodies. Pupae are usually encountered slightly above the water level along the shores. However, the larval and pupal stages of species in this genus are known mainly from species inhabiting other continents.
....*Climacia* McLachlan, 1869..p. 433

– Vein Rs in the fore-wing has two main forks, both proximal to the pterostigma. Outer gradate cross veins are absent. The fore-wing is uniformly brown. The apical segment of the palp is widest at its base and triangular in shape (Figs. 2.21 and 15.2). The common name for species in this genus is *spongillafly*, named for a genus of freshwater sponges in which the larvae develop. Larval development of most South American species has not been studied. Larvae have been found inside of freshwater sponges, but whether they should be considered parasites or symbionts has not yet been determined with certainty. They have also been found in colonies of ectoprocts, but it is not yet known whether they are utilized by the larvae as food.
....*Sisyra* Burmeister, 1839..p. 442

Key to the Genera of Known Larvae in the Family Sisyridae in South America

Information for the key taken from Chandler (1956) and Penny (1981).

1. There are conspicuous tubercles on the dorsum, each bearing two or three setae (Fig. 15.3). The larvae are usually found at or near the edge of freshwater bodies or in wetlands. Little is known about the food sources or habits of the South American species.
....*Climacia* McLachlan, 1869..p. 433

– Setae on dorsum not inserted in conspicuous tubercles (Fig. 2.27). The larvae are aquatic and inhabit freshwater sponges or, possibly, ectoproct colonies. They apparently feed by sucking small organisms or cells of the sponges through their pipette-like mouthparts. Studies of the larval development of South American species have rarely been undertaken.

....*Sisyra* Burmeister, 1839..p. 442

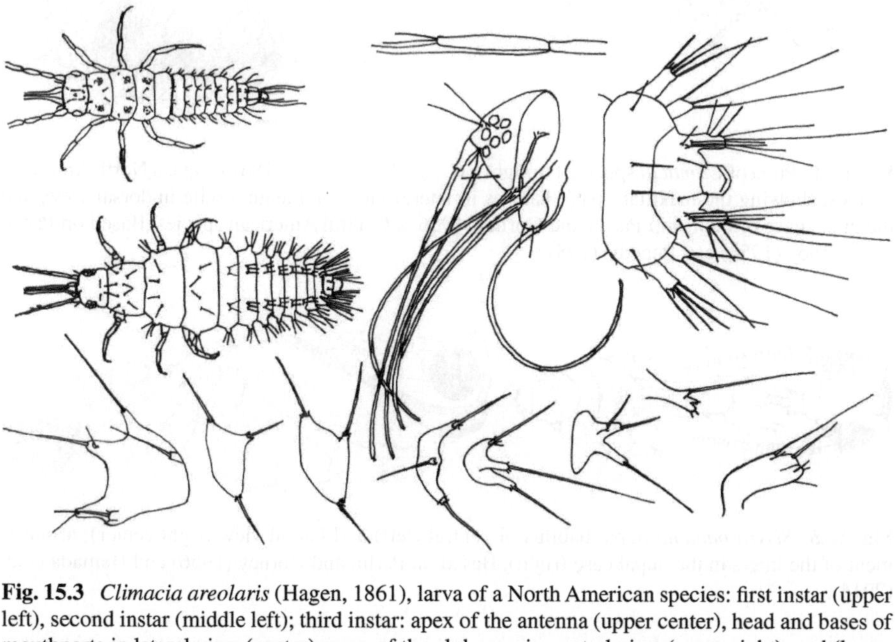

Fig. 15.3 *Climacia areolaris* (Hagen, 1861), larva of a North American species: first instar (upper left), second instar (middle left); third instar: apex of the antenna (upper center), head and bases of mouthparts in lateral view (center), apex of the abdomen in ventral view (upper right), and (lower row, left to right): dorsal pronotal, mesonotal, and metanotal plates; dorsal plates with tubercles and setae on the second and sixth abdominal segments; lateral plates on the mesonotum; lateral tubercles on the sixth abdominal segment; dorsal tubercles on the eighth abdominal segment. Based on Parfin and Gurney (1956).

Key to the Genera of Pupae in the Family Sisyridae in South America

Information for the key taken from Chandler (1956), Parfin and Gurney (1956), and Hamada et al. (2014). Because pupae of most South American species have not yet been described; the feature described in the key might not be valid for all.

1. The outer net is woven of hexagonal mesh-like bobbinet (Fig. 15.4). ....*Climacia* McLachlan, 1869..p. 433
   – The outer net is woven of crisscrossing threads (Fig. 15.5). ....*Sisyra* Burmeister, 1839..p. 442

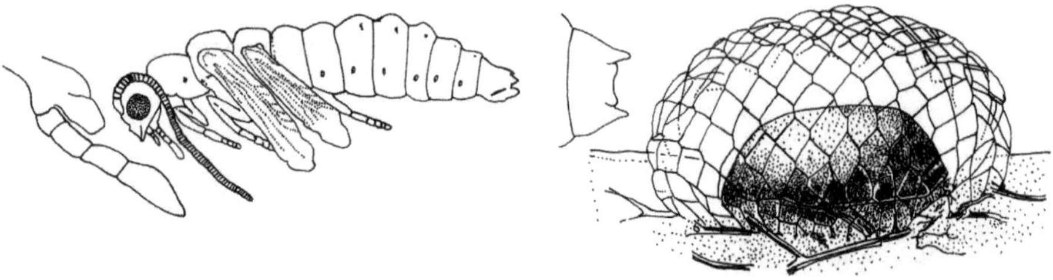

Fig. 15.4 Pupa of *Climacia* spp. (left to right): *C. areolaris* (Hagen, 1861) pupa, a North American species, showing the maxillary palp, habitus in lateral view, and tenth tergite in dorsal view, and the pupal case of *C. striata* Parfin and Gurney, 1956, a Central American species. Based on Parfin and Gurney (1956) and Pupedis (1981).

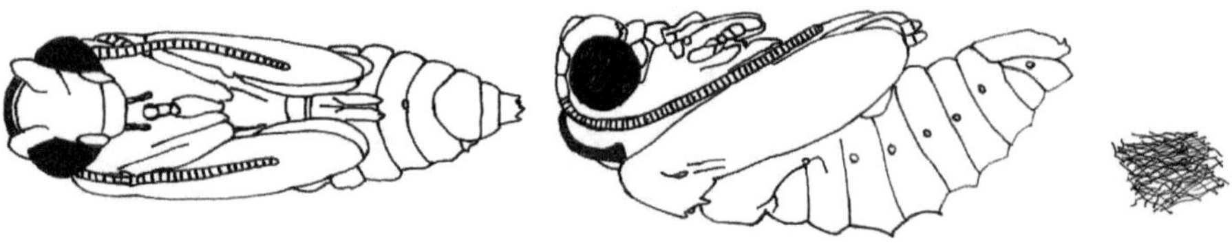

Fig. 15.5 *Sisyra panama* pupa: habitus in ventral (left) and lateral view (right center); arrangement of the fibers in the pupal case (right). Based on Parfin and Gurney (1956) and Hamada et al. (2014).

15.2 Key to the Species of Adult *Climacia* in South America

The key was prepared with information found in Banks (1913b), Parfin and Gurney (1956), Penny (1981), Penny and Rafael (1983), Flint (2006), and Bowles (2015). The characters important for identification are those of the female in most cases because males of some species have not been described.

1. The fore-wing lacks a conspicuous pale radiomedial streak extending from the basal pigmentation (Fig. 15.6) 2

   – The fore-wing has a conspicuous pale radiomedial streak extending from brownish black basal pigmentation (Fig. 15.1).. 6

2. (1) There is heavy brownish black streaking in the anal area of the fore-wing. The second m vein is directly posterior to the first r-m. Spotting at the setae along the long veins is not conspicuous (Fig. 15.6). Fore-wing length: c. 4.2 mm; hind wing: c. 3.8 mm. The male has not been described.
   ....*Climacia townesi* Parfin and Gurney, 1956 (Venezuela, Guyana, Peru, Amazonas).

   – There is no heavy brownish black streaking in the anal area of the fore-wing. Blackish spotting at the bases of the setae along the long veins is conspicuous (Fig. 15.7)... 3

3. (2) The ectoproct of the male is approximately trapezoidal; the angle formed by its dorsal and posterior margins is acute and pointed (Fig. 15.7). Fore-wing length: 4–4.5 mm.
   .... *Climacia punctulata* Flint, 2006 (Minas Gerais, Rondônia).

   – The ectoproct of the male is not approximately trapezoidal and lacks a pointed, acute angle formed by its dorsal and posterior margins (Fig. 15.8)............... 4

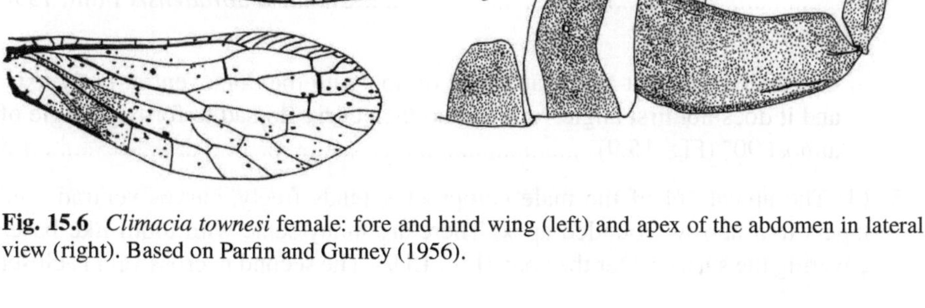

Fig. 15.6 *Climacia townesi* female: fore and hind wing (left) and apex of the abdomen in lateral view (right). Based on Parfin and Gurney (1956).

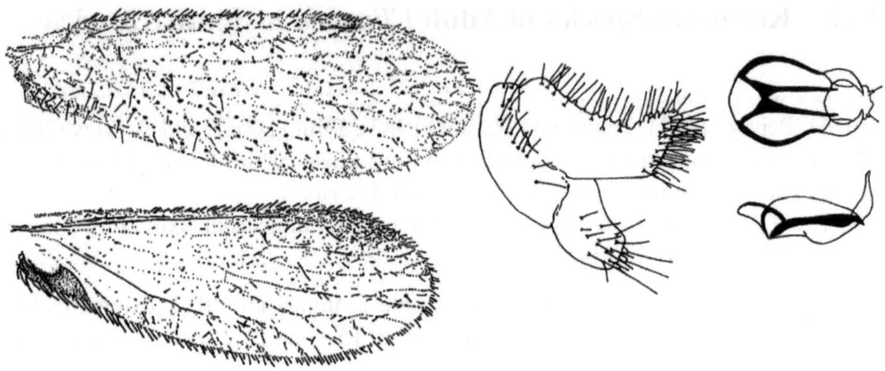

Fig. 15.7 *Climacia punctulata* male: fore and hind wing (left), genitalia in lateral view (right center), and the conarcal complex in dorsal (upper right) and lateral view (lower right). Based on Flint (2006).

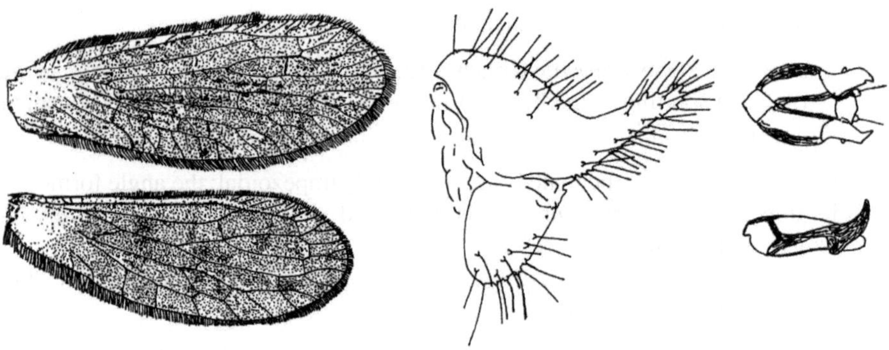

Fig. 15.8 *Climacia doradensis* male: fore and hind wing (left), genitalia in lateral view (center), and the gonarcal complex in dorsal (upper right) and lateral view (lower right). Based on Flint (1998).

4. (3) The basal half of the dorsal margin of the male ectoproct is angled ventrad, and the apical half is narrow and curves dorsad almost 90° near the apex (Fig. 15.8).
   .... *Climacia doradensis* Flint, 1998 (Venezuela).
   – The male ectoproct forms an angle of 45° with the dorsoventral body axis, and it does not first angle ventrad and then curve dorsad to form an angle of almost 90° (Fig. 15.9) 5
5. (4) The apical 1/4 of the male ectoproct extends freely, curves ventrad, and tapers to a narrow, rounded apex. There are many setae with tooth-like bases covering the surface near the apex (Fig. 15.9). The second m cross vein is closer

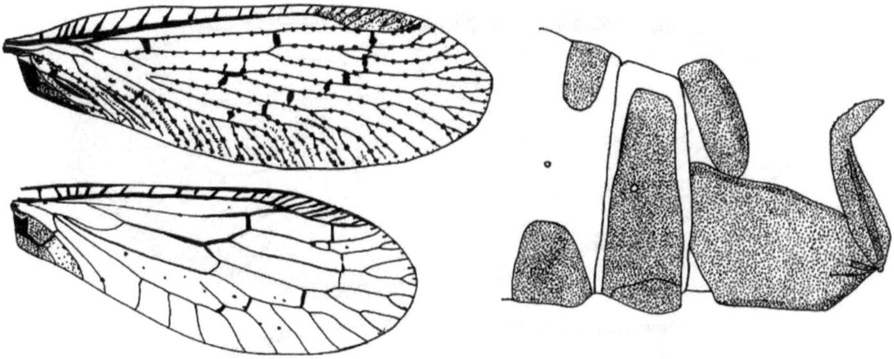

Fig. 15.9 *Climacia chilena* female: fore and hind wing (left) and apex of the abdomen in lateral view (right). Based on Parfin and Gurney (1956).

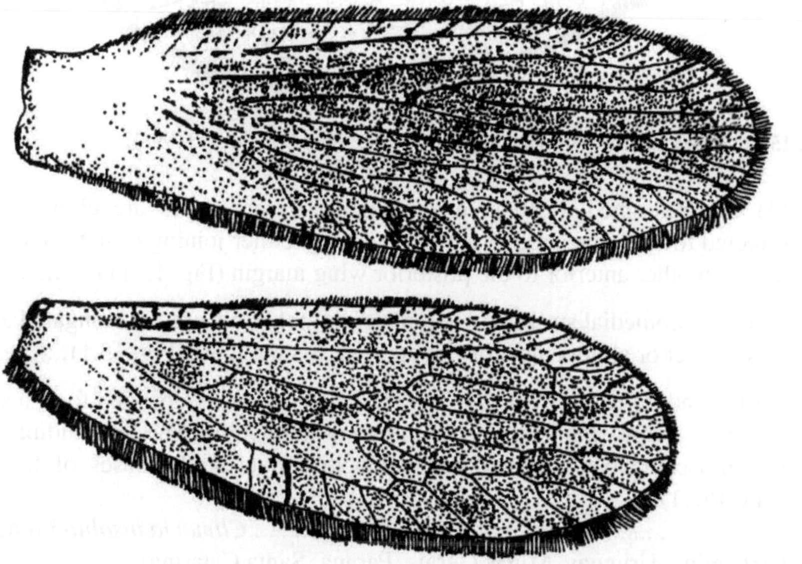

Fig. 15.10 Fore and hind wing of *Climacia lemniscata*. Based on Flint (1998).

to the wing base than the first r-m. There is a small radial streak on the fore-wing. Fore-wing length: c. 6 mm; hind wing: c. 5 mm.
....*Climacia chilena* Parfin and Gurney, 1956 (Chile, Argentina).

– The apical 2/5 of the male ectoproct extends freely and tapers to a narrow, rounded apex. There are many setae with tooth-like bases covering the surface near the apex. There is no small radial streak on the fore-wing (Fig. 15.10).
....*Climacia lemniscata* Flint, 1998 (Argentina).

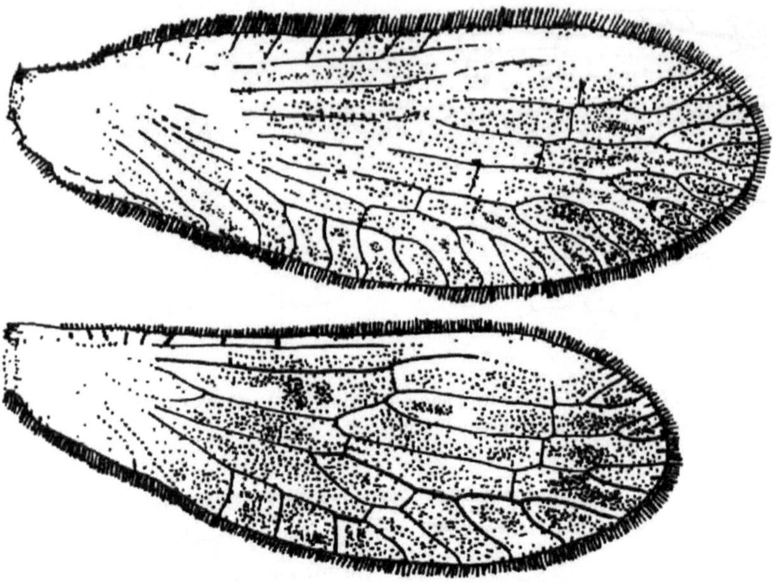

Fig. 15.11 Fore and hind wing of *Climacia insolita*. Based on Penny (1998).

6. (1) The radiomedial streak and antepterostigmal fascia are elongated and directed toward one another across the wing, either joining or not quite reaching each other anterior to the posterior wing margin (Fig. 15.11) 7

   – The radiomedial streak and antepterostigmal fascia are not elongated and do not meet or almost meet one another across the wing (Fig. 15.1) 8

7. (6) The radiomedial streak and antepterostigmal fascia meet and form a single, broad, U-shaped marking across the wing membrane. The longitudinal veins are infuscated but not marked by blackish dots at the bases of the setae (Fig. 15.11).

   .... *Climacia insolita* Flint, 1998 (Argentina, Uruguay, Minas Gerais, Paraná, Santa Catarina).

   – The radiomedial streak and antepterostigmal fascia do not quite meet, leaving a break at the base of the U. There are dark spots at the bases of setae along the longitudinal veins distal to the pterostigmal fascia. The male ectoproct has a broadly rounded apex, and the apicodorsal half of the male gonarcal complex has an odd, inflated apicodorsal half (Fig. 15.12).

   .... *Climacia versicolor* Flint, 1998 (Argentina, Uruguay).

8. (6) The radiomedial marking on the fore-wing is elongated transversely toward the posterior margin of the wing (Fig. 15.13) 9

   – The radiomedial marking on the fore-wing is elongated from the base toward the apex of the wing (Fig. 15.1) 13

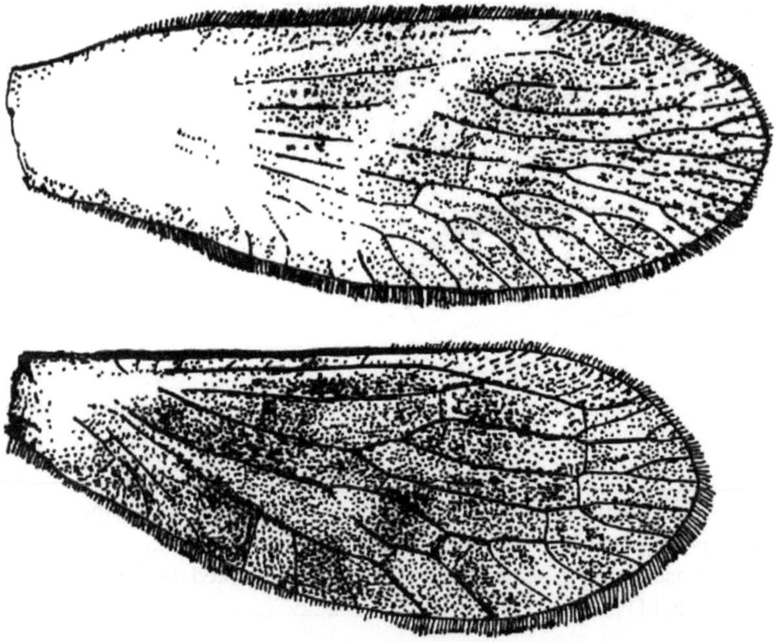

Fig. 15.12 Fore and hind wing of *Climacia versicolor*. Based on Flint (1998).

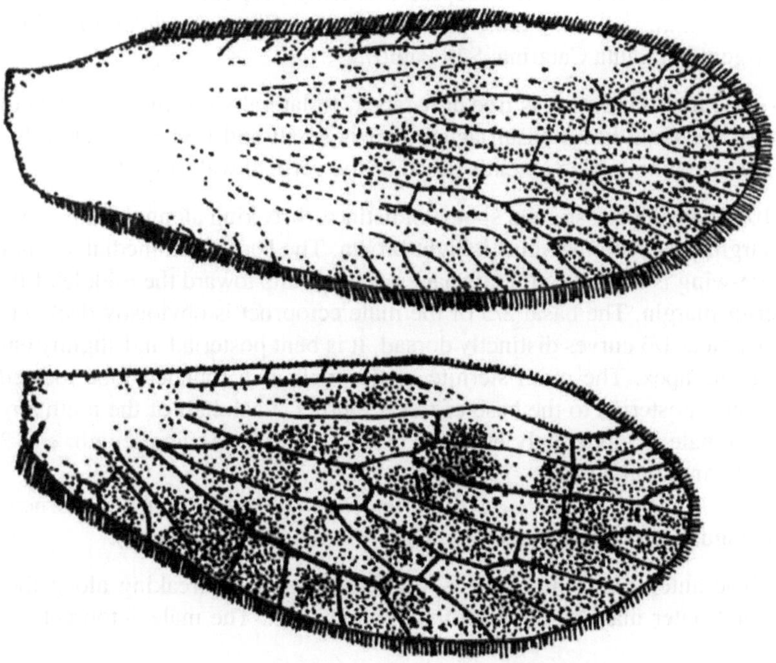

Fig. 15.13 Fore and hind wing of *Climacia triplehorni*. Based on Flint (1998).

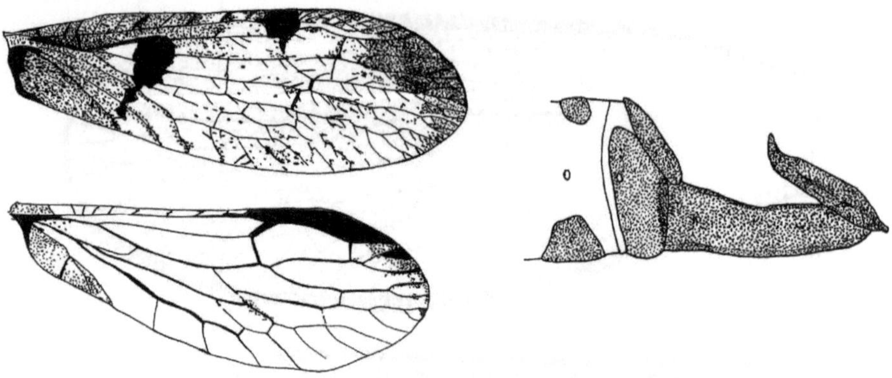

Fig. 15.14 *Climacia bimaculata* female: fore and hind wing (left) and apex of the abdomen in lateral view (right). Based on Parfin and Gurney (1956).

9. (8) There is distinct spotting at the bases of the setae along the longitudinal veins, at least near the margins of the wings (Fig. 15.13)............................... 10
   – The bases of the setae are not distinctly marked by many black spots (Fig. 15.14) 12

10. (9) The male ectoproct is simply rounded and not free at the apex. In lateral view, the gonarcal complex appears with a large basal lacuna and an angulate vertical wall with an external spine extending posteriad from it (Fig. 15.13). ....*Climacia triplehorni* Flint, 1998 (Argentina, Santa Catarina, São Paulo).

   – The male ectoproct is free at its apex. In lateral view, the gonarcal complex of the male has a basal lacuna that is small and truncated ventrally by an inner spine, which extends basad of the vertical wall (Fig. 15.15)............. 11

11. (10) The antepterostigmal spot is indistinct. Streaking along the inner and outer margins between the veins is light brown. The basal radiomedial streak on the fore-wing is angled so that its narrow end points toward the middle of the posterior margin. The basal 2/3 of the male ectoproct is obviously declivous, and the apical 1/3 curves distinctly dorsad. It is bent posteriad and slightly enlarged near its apex. The ninth sternite of the male is a rounded lobe located only slightly posteriad to the base of the ectoproct. Each half of the ninth tergite of the female is moderately elongate (Fig. 15.15). Fore-wing length: c. 4.3 mm; hind wing: c. 3.8 mm.
    ....*Climacia carpenteri* Parfin and Gurney, 1956 (Paraguay, Uruguay, Argentina, Minas Gerais, Santa Catarina).

   – The antepterostigmal spot is small but distinct. Streaking along the inner and outer margins between the veins is dark. The male ectoproct is points

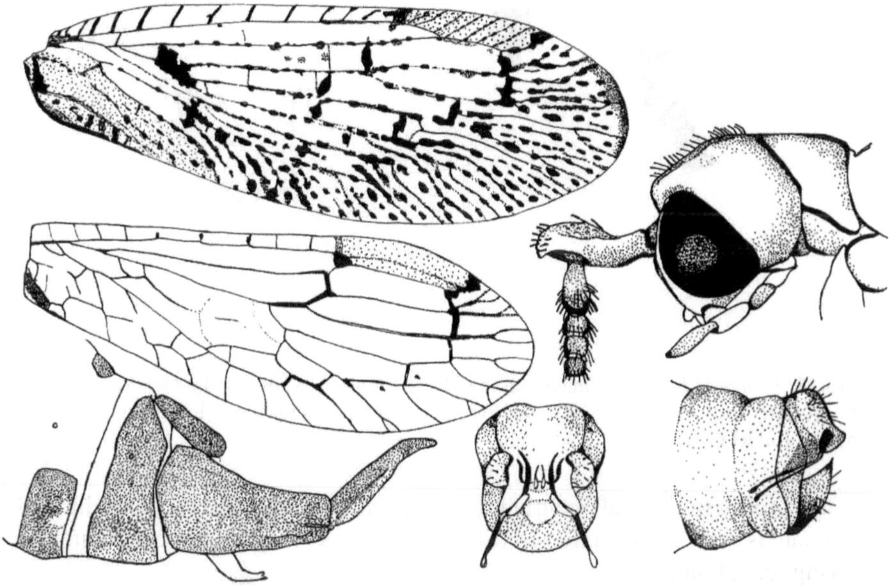

Fig. 15.15 *Climacia carpenteri:* fore and hind wing of a female (upper and middle left); head, collar, and basal antennal segments or a male in lateral view (upper right); apex of the male abdomen in dorsal (lower right center) and lateral view (lower right); apex of the female abdomen in lateral view (lower left). Based on Parfin and Gurney (1956) and Gonzalez-Olazo (1983).

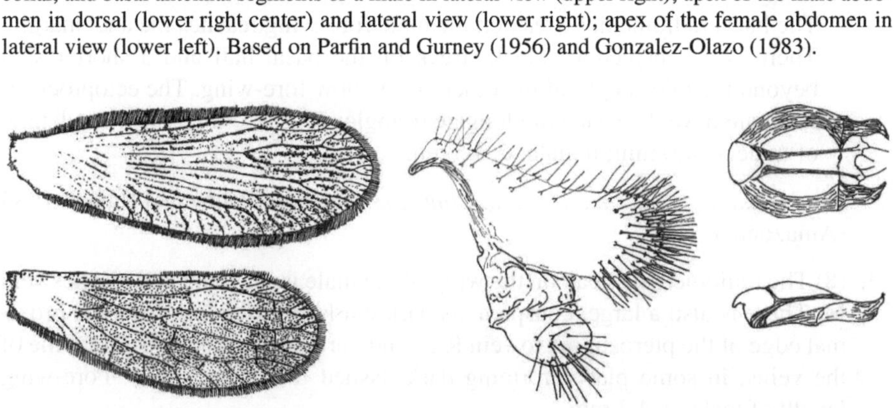

Fig. 15.16 *Climacia amalla* male: fore and hind wing (left), genitalia in lateral view (center) and the gonarcal complex in dorsal (upper right) and lateral view (lower right). Based on Flint (1998).

posteriad and curves only slightly dorsad. The ninth sternite of the male is a rounded lobe obviously displaced posteriad to the base of the ectoproct (Fig. 15.16).

....*Climacia amalla* Flint, 1998 (Venezuela, Peru).

12. (9) The basal radiomedial streak on the male fore-wing is angled so that its narrow posterior end points toward the basal part of the posterior margin, but it fails to reach the anal margin. The rest of the wing is mainly hyaline except for the antepterostigmal spot. The ectoprocts of the male extending posteriad

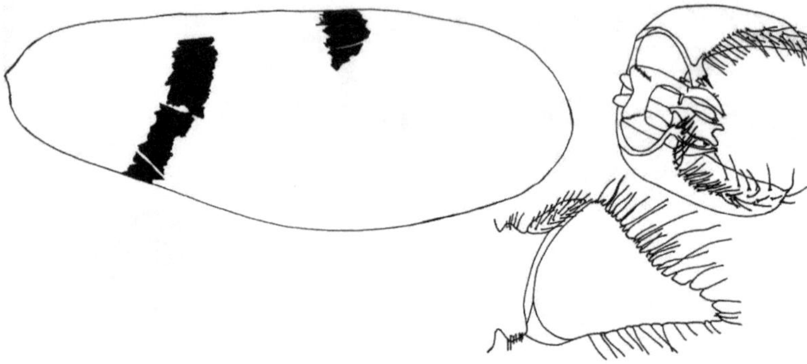

Fig. 15.17 *Climacia bifasciata* male: outline of the fore-wing showing the location of the dark markings (left) and the genitalia in dorsal (upper right) and lateral view (lower right). Based on Penny and Rafael (1983).

appear only to be small ventral projections. Each half of the ninth tergite of the female is very much elongated (Fig. 15.14). Fore-wing length: c. 3.9 mm; hind wing: c. 3.4 mm.
....*Climacia bimaculata* Banks, 1913
(Guyana, Surinam, Amazonas).

– The basal radiomedial streak on the male fore-wing reaches the anal margin. There is a conspicuous black streak on the basal half and a shorter spot beyond the mid-length of the otherwise yellow fore-wing. The ectoprocts of the male have the shape of elongate triangles (Fig. 15.17). Fore-wing length of male: c. 4.0 mm; female: c. 4.5 mm.

....*Climacia bifasciata* Penny and Rafael, 1983
(Amazonas).

13. (8) The radiomedial streak on the wing of the male is diagonal and reaches vein M. There is also a large, conspicuous dark marking extending from the proximal edge of the pterostigma to vein R_{2+3}, and dark dots and streaks line some of the veins, in some places forming dark dashed lines (Fig. 15.1). Fore-wing length of male: c. 4.3 mm.
....*Climacia negrense* Penny, 1981
(Guyana, Amazonas).

– The radiomedial streak of the female usually reaches only to the first r cross vein. If a spot is present on the proximal side of the pterostigma, it is faint and inconspicuous. Any dark shading on the wing veins is usually uniform and not broken up into dashes (Fig. 15.18) 14

14. (13) The radiomedial streak of the female is straight and short, so it barely reaches the first r cross vein. There is no other conspicuous marking on the fore-wing, except for the spots at the bases of the setae along the long veins

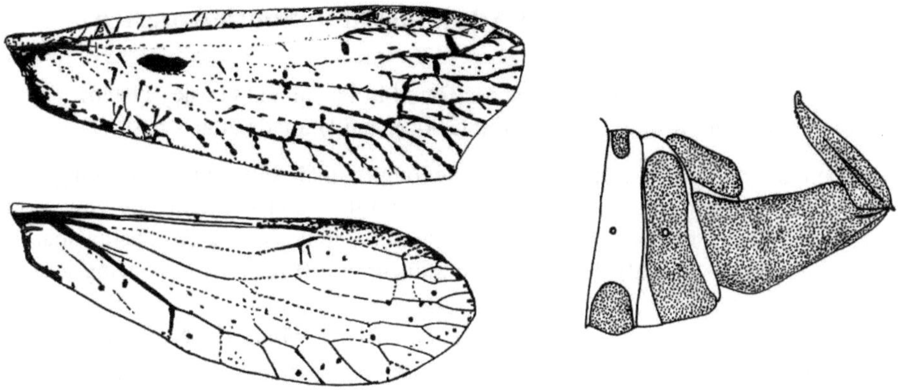

Fig. 15.18 *Climacia nota* female: fore-wing with the apex damaged and the hind wing (left) and apex of the abdomen in lateral view (right). Based on Parfin and Gurney (1956).

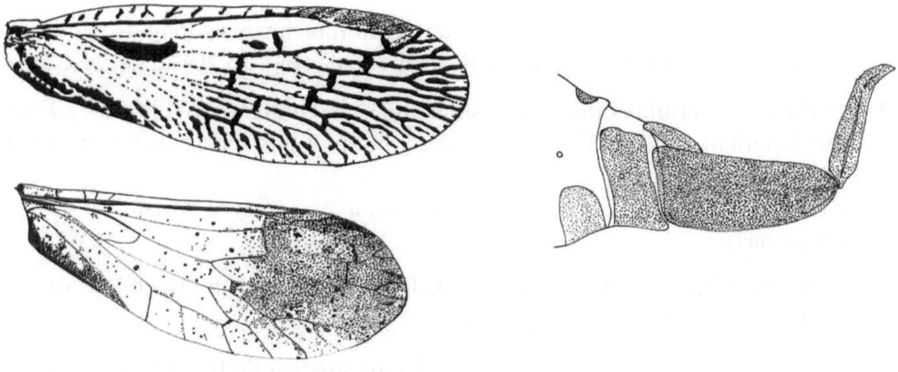

Fig. 15.19 *Climacia basalis* female: fore and hind wing (left) and apex of the abdomen in lateral view (right). Based on Parfin and Gurney (1956).

(Fig. 15.18). Fore-wing length: slightly longer than 4 mm; hind wing: c. 4.0 mm. The male has not been described.
....*Climacia nota* Parfin and Gurney, 1956
(Venezuela, Guyana, Bolivia, Pará).

– The radiomedial streak of the female is long and somewhat crescent-shaped; it clearly reaches the first r cross vein. The darkened bases of the setae on the fore-wing are pronounced. The subcostal vein is pale. The antepterostigmal spot is either absent or faint (Fig. 15.19). Fore-wing length: c. 4.5 mm; hind wing: c. 4.3 mm. The male has not been described.

....*Climacia basalis* Navás, 1933
(Guyana, Mato Grosso do Sul).

15.3 Key to the Species of Adult *Sisyra* in South America

The key was prepared with information found in Parfin and Gurney (1956), Penny (1981), Penny and Rafael (1983), Flint (2006), and Hamada et al. (2014).

1. There are no light streaks on any of the longitudinal veins in the fore-wing. The two basal antenna segments are fuscous, and the other segments are yellowish. All femurs are fuscous. The tenth sternite of the male is beak-like, being sharply curved dorso-ventrally and ending in an acute tip (Fig. 1.7). Fore-wing length: 2.9–3.7 mm.
   ....*Sisyra minuta* Esben-Petersen, 1935
   (Amazonas, Pará).
   - All longitudinal veins in the fore-wing are bordered by a broad field of light streaks. At least the fore-femur is pale yellow (Fig. 15.2) 2
2. (1) The tenth sternite of the male is about twice as long as wide. The antenna is entirely dark (Fig. 15.2)..3
   - The tenth sternite of the male is more than four times as long as wide. The base or apex of the antenna is yellowish or pale (Fig. 15.20) 4
3. (2) The tenth sternite of the male is stout, thick, and mitten-like (Fig. 15.2). Fore-wing length of male: c. 3.7 mm. The description of this species is based on one male specimen.
   ....*Sisyra ariasi* Penny 1981
   (Amazonas).
   - The tenth sternite of the male is quadrangular without a thumb-like ventral lobe (Fig. 15.20). Fore-wing length: 3.7–4.0 mm.

   ....*Sisyra panama* Parfin and Gurney 1956
   (Panama, Peru, Bolivia, Amazonas, Rondônia, Rio de Janeiro).
4. (2) The antenna consists of about 33 segments; the apical three are pale. The coxopodite appears strongly constricted in the middle in both dorsal and lateral view, and it bears a thick coating of setae on its apicomesal surface (Fig. 15.21). Fore-wing length: 3.7–4.0 mm.

   ....*Sisyra elongata* Penny and Rafael 1983
   (Peru, Amazonas, Bahía).
   - The antenna is yellowish only on the two basal segments, or it is yellow with only the 12 basal segments of the flagellum fuscous. The apical third of the coxopodite is not considerably wider than at midlength, and there is not an especially large tuft of setae on the apicomesal segment (Fig. 15.22)................5

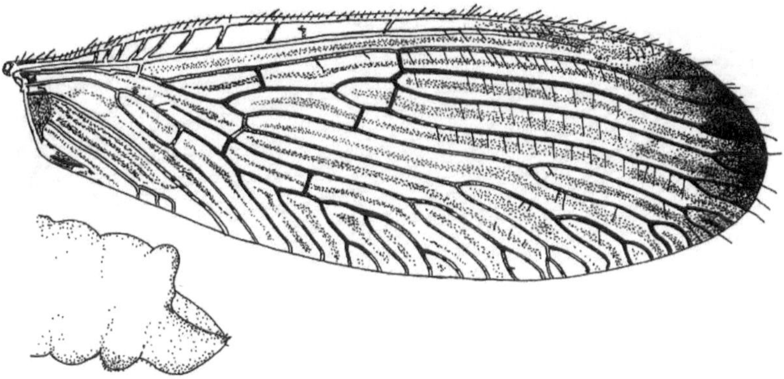

Fig. 15.20 *Sisyra panama* male: fore-wing (above) and the apex of the abdomen in lateral view (lower left). Based on Penny (1981).

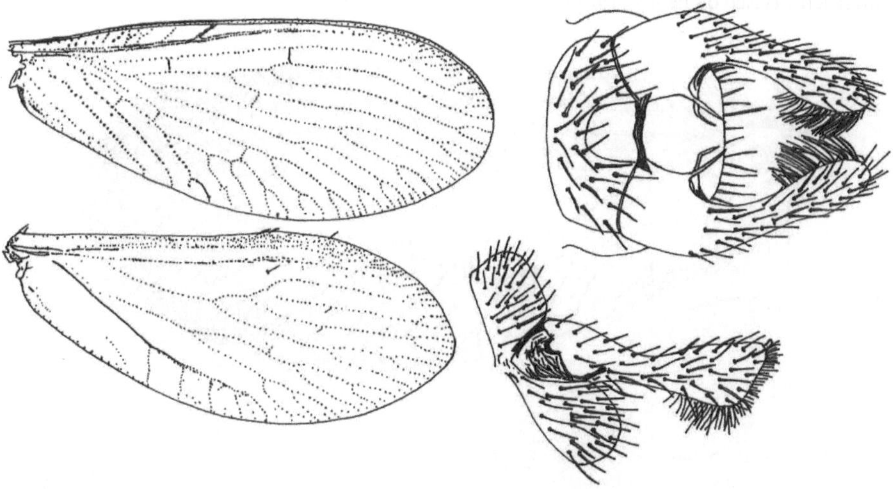

Fig. 15.21 *Sisyra elongata* male: fore and hind wing (left), and the genitalia in dorsal (upper right) and lateral view (lower right). Based on Flint (2006).

5. (4) The tenth sternite of the male lacks a ventral projection. The antenna is yellowish with the two basal segments somewhat darker (Fig. 15.22). Fore-wing length: 3.0–3.2 mm.
   ....*Sisyra amazonica* Penny, 1981 (Guyana, Paraguay, Amazonas).

   – The tenth sternite of the male bears a ventral projection. The first two antennal segments are yellowish, the next 12 are fuscous, and the terminal segments are yellowish (Figs. 2.21 and 15.23). Fore-wing length: 3.2–3.8 mm.

   ....*Sisyra apicalis* Banks, 1908 (U.S.A., West Indies, Central America, Surinam, Peru, Argentina, Uruguay, Amazonas, Minas Gerais). Syn: *Sisyra nocturna* Navás, 1932.

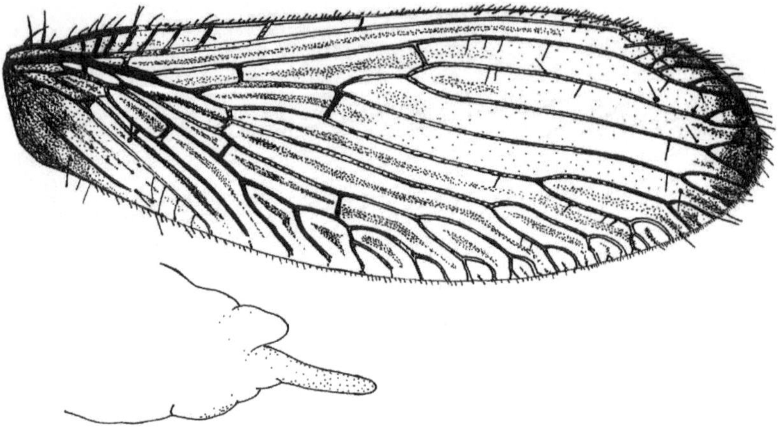

Fig. 15.22 *Sisyra amazonica* male: fore-wing (above) and the apex of the abdomen in lateral view (lower left). Based on Penny (1981).

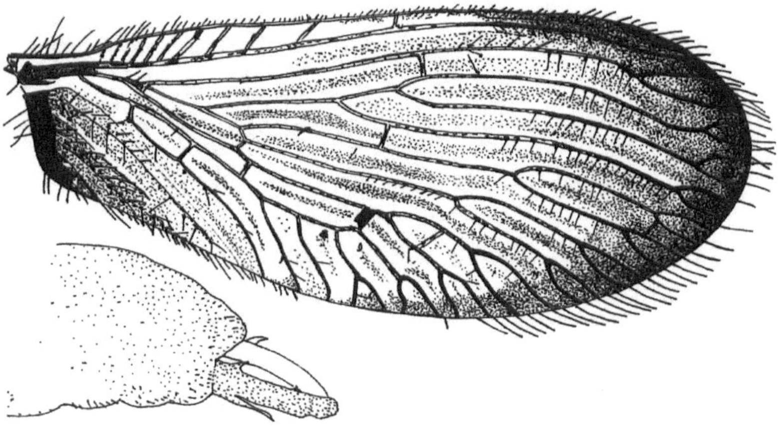

Fig. 15.23 *Sisyra apicalis* male: fore-wing (above) and the apex of the abdomen in lateral view (lower left). Based on Penny (1981).

Literature

Banks N (1908) New tropical American Neuroptera. Proceedings of the Entomological Society of Washington 9(1–4):30–33

Banks N (1913a) Synopses and descriptions of exotic Neuroptera. Transactions of the American Entomological Society 39:201–242 + pl. 23–26

Banks N (1913b) New exotic neuropteroid insects. Proceedings of the Entomological Society of Washington 15(3):137–143

Bowles DE (2015) New distributional records for Neotropical spongillaflies (Neuroptera: Sisyridae). Insecta Mundi 0400:1–7

Burmeister H (1839) Handbuch der Entomologie. Neuroptera 2(2–3):397–1050. Theodor Friedrich Enslin, Berlin

Chandler HP (1956) Megaloptera. In: Usinger RL (ed) Aquatic insects of California. University of California Press, Los Angeles, pp 229–236

Clavier S, Guillemet L, Rhone M, Thomas A (2011 [2010]) First report of the aquatic genus *Climacia* McLachlan, 1869 in French Guiana (Neuroptera, Sisyridae). Ephemera 12(1):31–36

Cover MR, Resh VH (2008) Global diversity of dobsonflies, fishflies, and alderflies (Megaloptera; Insecta) and spongillaflies, nevrorthids, and osmylids (Neuroptera; Insecta) in freshwater. Hydrobiologia 595:409–417

Flint OS Jr (1998) New species and records of *Climacia* from the Neotropics (Neuroptera, Sisyridae). Acta Zoologica Fennica 209:107–117

Flint OS Jr (2006) New species and records of Neotropical Sisyridae with special reference to *Sisyra* (Insecta: Neuroptera). Proceedings of the Biological Society of Washington 119(2):279–286

González Olazo EV (1983) Contribucion al conocimiento de *Climacia carpenteri* Parfin & Gurney, 1956, (Neuroptera, Sisyridae), nueva cita de la Republica Argentina. Acta Zoológica Lilloana 37(1):115–117

Hagan HA (1861) Synopsis of the Neuroptera of North America, with a list of South American species. Smithsonian Miscellaneous Collections 4(1):i–xx + 1–347

Hamada N, Pes AMO, Fusari LM (2014) First record of Sisyridae (Neuroptera) in Rio de Janeiro state, Brazil, with bionomic notes on *Sisyra panama*. Florida Entomologist 97(1):281–284

McLachlan R (1869) Considerations on the neuropterous genus *Chauliodes* and its allies; with notes and descriptions. Annals and Magazine of Natural History, Series 4 (4):35–46

Monserrat VJ (2005) Nuevos datos sobre angunas pequeñas familias de neurópteros (Insecta: Neuroptera: Nevrothidae, Osmylidae, Sisyridae, Dilaridae). Heteropterus Revista de Entomologia 5:1–26

Navás L (1933) Insectos Neotropicos. 8ª Serie. Revista Chilena de Historia Natural 37:194–197

Parfin SI, Gurney AB (1956) The spongilla-flies with special reference to those in the Western Hemisphere (Sisyridae, Neuroptera). Proceedings of the U. S. National Museum 105(3360):421–529 + 3 pl

Penny ND (1981) Neuroptera of the Amazon Basin, Part 1. Sisyridae. Acta Amazonica 11(1):157–169

Penny ND (2002) Family Sisyridae. In: Penny ND (ed) A guide to the lacewings (Neuroptera) of Costa Rica. Proceedings of the California Academy of Sciences 53(12):290–291

Penny ND, Rafael JA (1983 [1982]) Two new species of Sisyridae (Neuroptera) of the Amazon Basin. Neuroptera International 2:53–58

Pupedis RJ (1981 [1980]) Generic differences among new world *Spongilla*-fly larvae and a description of the female of *Climacia striata* (Neuroptera: Sisyridae). Psyche 87:305–314

Chapter 16
Chrysopidae

Abstract An illustrated key is provided for identification of the subfamilies and genera of adult chrysopids known to occur in South America. Illustrated keys are also provided to identify the species of each genus native to South America. Tentative keys for known larvae are also shown. The countries or states of Brazil in which specimens of each species have been found and invalid synonyms are reported. Problems recognizing technically valid species which lack adequate descriptions and complete type specimens are discussed.

Keywords Green lacewing • Key to species • Nothochrysinae • Apochrysinae • Chrysopinae

Members of this family are called green lacewings. All known South American species behave like terrestrial insects during all life stages. However, unidentified green lacewings have been seen on the emergent parts of floating plants, which are sometimes infested with aphids and other small herbivorous insects. These plants form floating islands that cover large areas of standing water in many parts of tropical and subtropical South America (Heckman, 1994), some permanently and others only during an annual rainy season.

Chrysopidae has a nearly cosmopolitan distribution and encompasses a great many species. Many recent studies of the individual species have been undertaken because of their potential for biologically controlling small insect pests in tropical agriculture. Unfortunately, most of the original descriptions are insufficient for identifying the genera according to features now used to distinguish them. A great many of the nominal species were placed in the genus *Chrysopa* when they were first described. Presently, no species of *Chrysopa* is known to inhabit South America, so the type specimens must be assigned to their correct genera before they can be identified. As the keys show, many of those which have been examined again turned out to be conspecific, so a great many of the older names are now considered junior synonyms of other species named even earlier. Many type specimens have been lost, and without better descriptions, species still in obsolete taxa cannot be assigned to modern genera. Most such nominal species cannot be placed in the keys, and they are regarded, at least temporarily, as *nomina dubia*. Many such names appear in old lists, but even such lists are not nearly complete. Much more revision will be

C.W. Heckman, *Neuroptera (Including Megaloptera)*, Encyclopedia of South American Aquatic Insects, DOI 10.1007/978-3-319-35125-4_16

required before it will be possible to reliably identify and classify many of the chrysopid species.

Morphological information on the larvae of South American chrysopids is still fragmentary. Although descriptions of the larvae and pupae of common genera from other continents are available, it is inadvisable to present a general key to the South American genera because it is not unlikely that some of the larvae yet to be described deviate in important features from those already described. The systematic classification of the green lacewings in South America is based entirely on the imagos, so the only fully reliable way to identify larvae is to raise them until their metamorphosis to adults. A few notes on the known larvae will appear in the keys to the adults, but reliable identifications will only be possible after a much larger percentage of the larvae have been described.

Because of their great potential importance as predators capable of controlling the small insect pests of crops raised in tropical and subtropical climatic zones, efforts have been made to list the plant species on which predatory neuropterans have been found. It is generally assumed that aphids and other small insects feeding on these plants are suitable prey for the neuropterans. This knowledge will assist greatly in choosing native species for biological control of pests.

16.1 Key to the South American Genera in the Family Chrysopidae

Information for the key was provided by Navás (1913, 1916, 1917a, b); Banks (1944); Adams (1982a, b); Adams and Penny (1987, 1992); Brooks and Bernard (1990); Oswald and Penny (1991); Winterton (1995); Penny (1996); Winterton and Brooks (2002), and Tauber et al. (2005). The key is intended mainly for male specimens, and problems will be encountered because descriptions of many species were provided for only one sex. It is not yet known whether or not the larval characteristics of genera determined from species on other continents hold true for South American species. Notes on larvae are placed in the keys below.

1. Vein Psm in the fore-wing is continuous with the inner row of gradate veins. The jugal lobe is well developed (Fig. 16.1). The tympanal organ is absent ....Subfamily Nothochrysinae.. 2
   A junior synonym of this subfamily is Dictyochrysinae.
   - Vein Psm in the fore-wing joins the outer row of gradate veins. The jugal lobe is vestigial or absent (Fig. 16.2). A tympanal organ is present............. 3
2. (1) The fore-wing is much less than four times as long as its maximum width. Between the gonosaccus and genital port of the male, there is a heavily sclerotized plate bearing adpressed spines directed posteriad. The subgenitale of the female can be everted to reveal a complex internal structure. The second and third anal veins in the fore-wing are not fused (Fig. 16.1). The color of the only

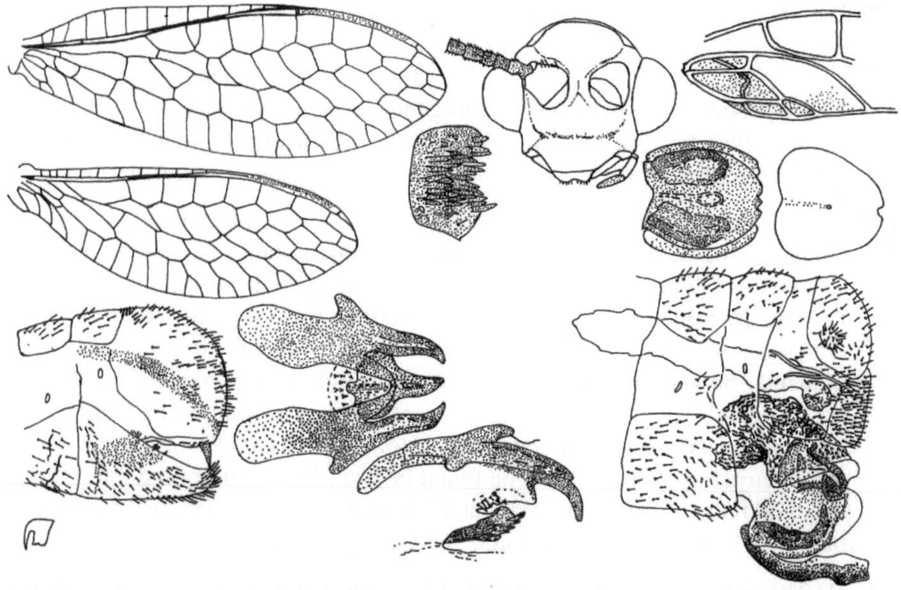

Fig. 16.1 *Asthenochrysa viridula:* fore and hind wing (upper and middle left), head in anterior view with segments of appendages removed (upper center), enlarged axial area of the fore-wing (upper right), hind tarsal claw (lower left), apex of the male abdomen in lateral view (lower middle left), gonosaccal plate in ventral view (above center), internal genitalia of the male in dorsal (lower left center) and lateral view (lower right center), apex of the female abdomen with the subgenitale and internal structures everted (lower right), and the subgenitale of two varying specimens in ventral view (upper middle left, left and right). Based on Adams and Penny (1992).

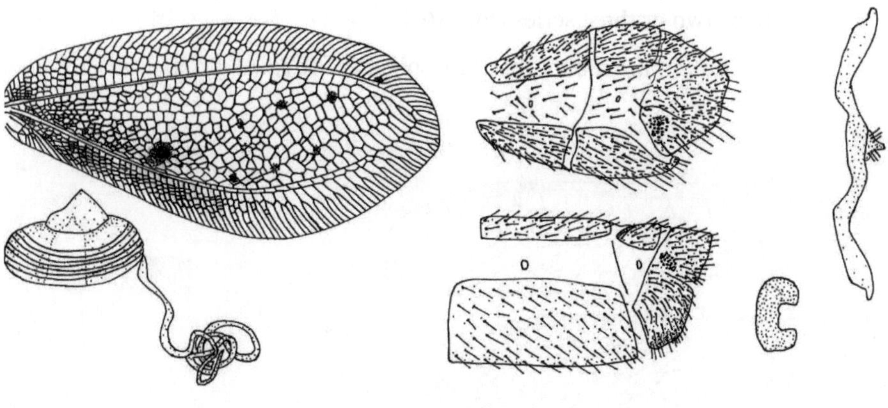

Fig. 16.2 *Joguina constellatus:* fore-wing (upper left), apex of the male abdomen in lateral view (upper center), male genitalia in dorsal view (right), apex of the female abdomen in lateral view (lower center), spermatheca (lower left), and female subgenitale in ventral view (lower right center). Based on Brooks and Bernard (1990), who referred to the species as *Lainius constellatus.*

known South American species is predominantly green. The length of the fore-wing is c. 9.2 mm; hind wing: c. 8.2 mm.
....*Asthenochrysa* Adams and Penny, 1992
The only species yet described in this genus is *Asthenochrysa viridula* (Adams, 1978), which has been found in Argentina, Espirito Santo, and Minas Gerais and was first called by its synonym: *Hypochrysa viridula* Adams, 1978.

– The fore-wing is about four times as long as its maximum width. The subgenitale of the female is longer than wide, tapers somewhat toward the apex, which is bilobed with an emargination in the middle. The second anal veins in the fore-wing is forked, but the posterior branch does not reach the wing margin (Fig. 16.3). The color of the only known South American species is mainly brown. The fore-wing length of the only known species is c. 20 mm.
....*Leptochrysa* Adams and Penny, 1992
The only species yet described in this subfamily is *Leptochrysa prisca* Adams and Penny, 1992, which has been found in Peru. The length of the fore-wing is about 20 mm, and its width is about 4.7 mm. All the veins are dark brown with pale brown setae arranged in single rows along the veins.

3. (1) The third cubital cell is undivided. The costal area is conspicuously broad. Gradate veinlets, if present, never form a regular series. The radius branches fire or six times apical to the subcosta. The radial veins are often connected by crossveins (Fig. 16.2). The wings usually contain bullae.
.............Subfamily Apochrysinae, p. 464.. 4

– The third cubital cell is divided. The costal area is considerably narrowed near the pterostigma. The radius only rarely branches more than three times apical to the subcosta. The cross veins near the mid-length of the wing usually form two or three series (Fig. 16.4).

.................................Subfamily Chrysopinae, p. 467....................................... 6

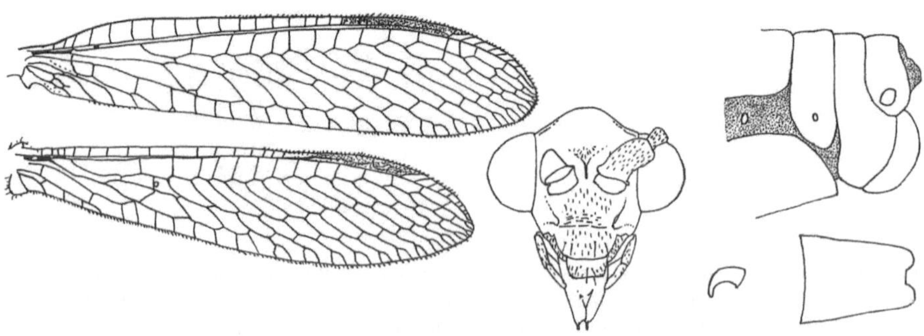

Fig. 16.3 *Leptochrysa prisca* female: fore and hind wing (left), head in anterior view (center), apex of the abdomen in lateral view (upper right), hind tarsal claw (lower right center), and subgenitale (lower right). Based on Adams and Penny (1992).

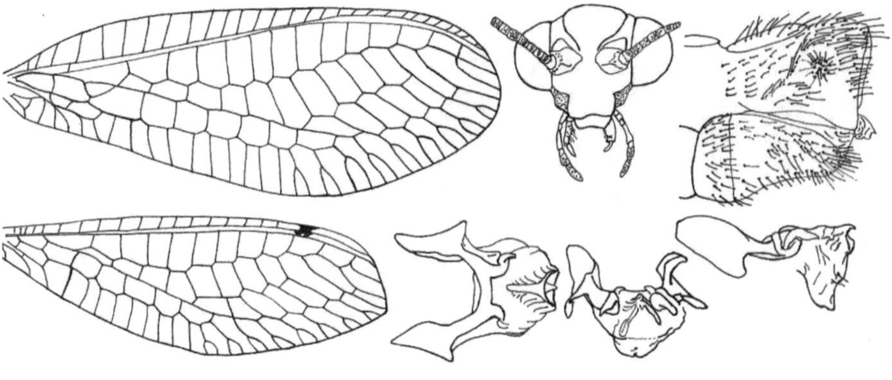

Fig. 16.4 *Leucochrysa incognita:* fore and hind wing (left), head in anterior view (upper right center), apex of the male abdomen in lateral view (upper right), male genitalia in dorsal (lower center), posterior (lower right center), and lateral view (right). Based on de Freitas and Penny (2001).

4. (3) The costal area of the wing is filled by two or more rows of cells, and the entire costal area, as well as the posterior marginal area, are densely reticulated. Rs in the fore-wing is irregular and not distinct from the area of reticulation. There is no accessory row of cross veins in the marginal area of the apical half of either wing (Fig. 16.2).
   ....*Joguina* Navás, 1912..p. 464
   – The costal and posterior marginal areas of the fore-wing have only a single row of cells (Fig. 16.5). 5
5. (4) The wing cells in the posterior marginal area are simple, that is, not irregularly subdivided by cross veins. Vein 1A in the fore-wing may be forked, but 2A is always unforked (Fig. 16.5). The spermatheca lack lateral striations.
   ....*Loyola* Navás, 1913..p. 465
   – The wing cells in the posterior marginal area are irregularly subdivided by cross veins. Veins 1A and 2A in the fore-wing are both forked, but vein 3A is not (Fig. 16.6). The spermatheca have lateral striations.
   ....*Domenechus* Navás, 1913..p. 466
   Of the two species assigned to this genus, one occurs in South America: *Domenechus marianellus* (Guérin, 1853), found in Brazil. The other is known as far south as Panama, close enough to be included in a key.
6. (4) The pterostigma of the fore-wing has a black spot at the base. The cross-veins in the outer gradate series form an angle of less than 45° with the pseudomedial vein. The wings are shorter than the antenna (Fig. 16.4).................... 7
   – The pterostigmal area of the fore-wing is entirely pale, but it may be heavily sclerotized. The crossveins in the outer gradate series form an angle of more than 60° and usually nearly 90° with the pseudomedial vein. The wings or most species are longer than the antennae or only slightly shorter (Fig. 16.7) 14

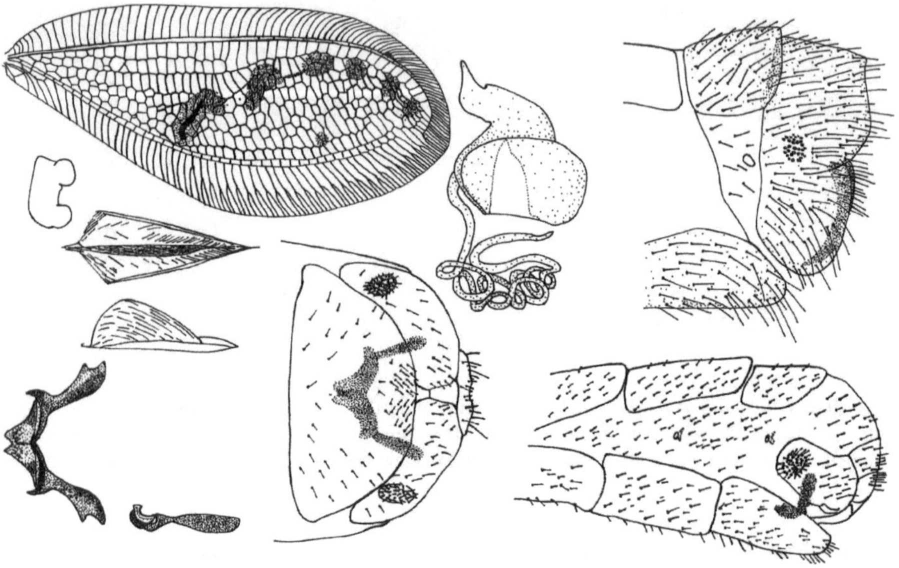

Fig. 16.5 *Loyola croesus:* fore-wing (upper left), apex of the male abdomen in ventral (lower center) and lateral view (lower right), female abdomen in lateral view (upper right), spermatheca of a female (upper center), female subgenitale (upper middle left), male hypandrium internum in dorsal (right of subgenitale) and lateral view (lower middle left), and the male gonarcus in dorsal and lateral view (lower left, left and right, respectively). Based on Brooks and Bernard (1990) and Tauber et al. (2005).

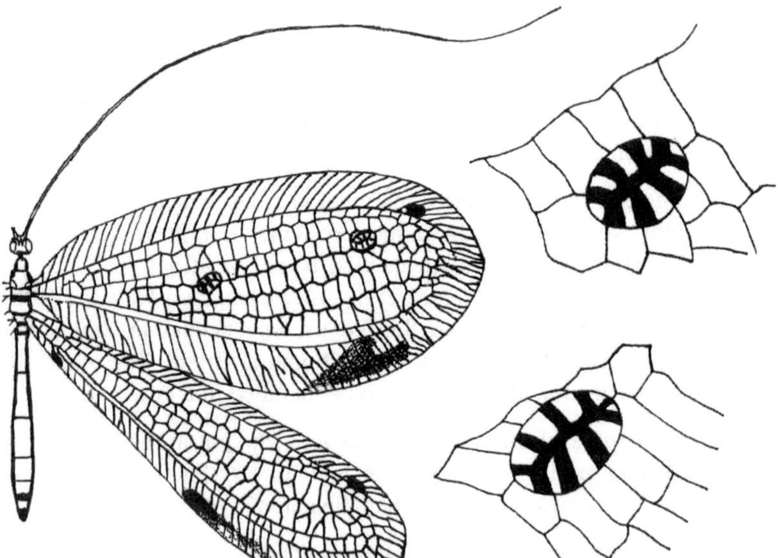

Fig. 16.6 *Domenechus marianellus:* habitus with appendages on the left side removed (left) and the structure of two blister-like structures on the fore-wing, called bullae. Based on Guérin-Méneville (1853).

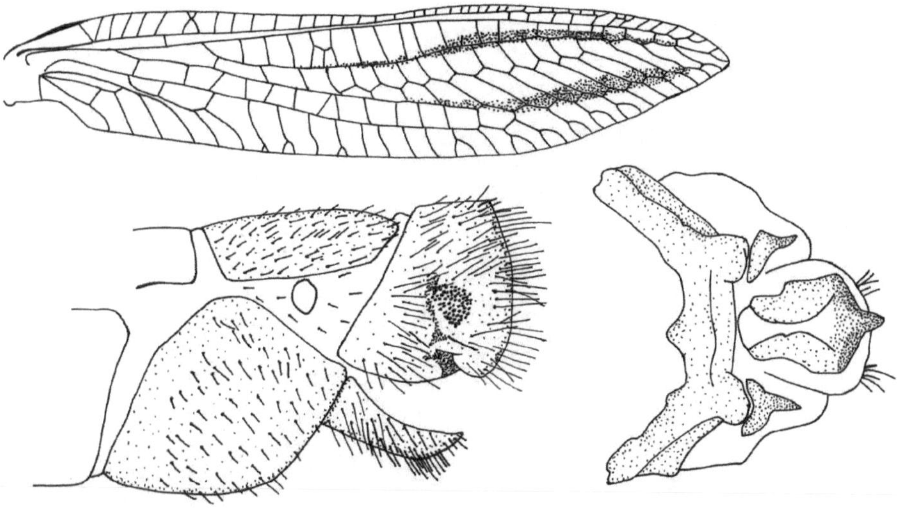

Fig. 16.7 *Belonopteryx arteriosa* male: fore-wing (above), apex of male abdomen in lateral view (lower left), and male genitalia in dorsal view (lower right). Based on Brooks and Bernard (1990).

7. (6) In both wings, there are two series of gradate veins, with occasional isolated third gradates. The first radial cross vein originates at or very near the origin of vein Rs. There are prominent dark markings on the wing and body, including a dark frons, although these markings may fade in preservative. The abdomen of the male is slender and bears gonocristae at the apex of the fused eighth and ninth sternites or on subapical tubercles. The male has an elongate, tubular mediuncus bearing dorsal rods or apodemes. The female has a round spermatheca with a velum shaped like a sail and a shallow invagination (Fig. 16.8). The antenna is longer than the fore-wing but shorter than twice the length of the fore-wing.
   .... *Santocellus* Tauber and Albuquerque, 2008..p. 467
   - There are fewer than two or three or more series of gradate veins, and the antenna is shorter or more than twice as long as the fore-wing, and the origins of the first radial cross vein and vein Rs are not very close together (Fig. 16.4). In case of doubt, there are no prominent dark markings on the wings and body, and the genitalia differ from the descriptions above.......... 8
8. (7) At least the fore-wing has at least three series of gradate veins. The antenna is longer than the fore-wing (Fig. 16.9)... 9
   - Each wing has fewer than two series of gradate veins, although exceptional cells may have isolated extra cross veins. (Fig. 16.4)............................... 10
9. (8) Both the fore and hind wings have three series of gradate veins. Aside from the small black mark at the base of the pterostigma, the fore-wings are unmarked

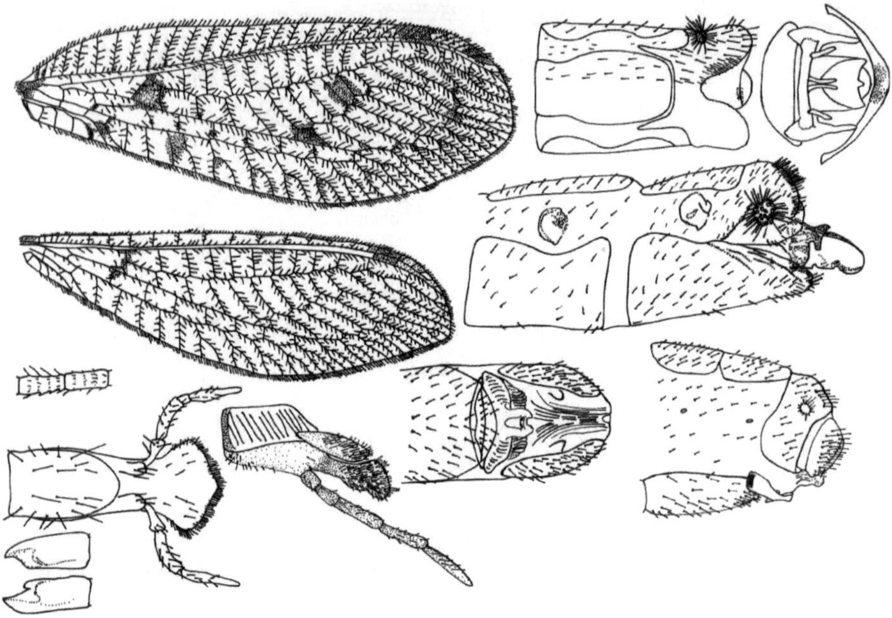

Fig. 16.8 *Santocellus riodoce:* fore and hind wing (upper and upper middle left), sixth and seventh segments of the antenna (middle left), labrum in ventral view (lower middle left), mandibles in dorsal view (lower left), maxilla in ventral view (lower left center), apex of the male abdomen in dorsal (upper right center) and lateral view (middle right), gonarcal complex in posterodorsal view (upper right), apex of the female abdomen in ventral (lower right center) and lateral view (lower right). Based on Tauber (2007), illustrated under its synonym: *Vieira riodoce.*

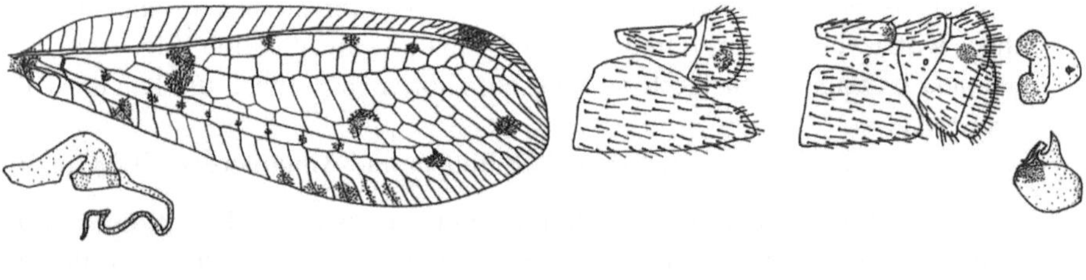

Fig. 16.9 *Cacarulla maculipennis:* fore-wing (upper left), apex of the male (upper center) and female abdomen in lateral view (upper right center), male genitalia in lateral view (lower right), female subgenitale in ventral view (upper right), and spermatheca (lower left). Based on Brooks and Bernard (1990).

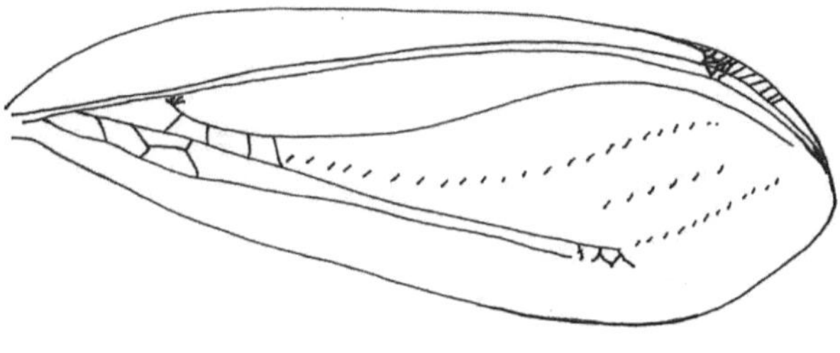

Fig. 16.10 Fore-wing of *Neula mesana* without most veins. Based on Navás (1917a).

(Fig. 16.10). The head has a red stripe on the vertex, and the thorax has a red lateral stripe. The ground color is yellowish green. Length of fore-wing: c. 25 mm.

....*Neula* Navás 1917

Only one species in this genus is known from South America: *Neula mesana* Navás, 1917, from Colombia. Unfortunately, Brooks and Bernard (1990) could not find a specimen in any collection. However, three series of gradate veins in the hind wing will serve to distinguish this genus from any other member of the Leucochrysini.

– There are three series of gradate veins in the fore-wing, but only two in the hind wing. Wing cell *im* is quadrangular. The fore-wing has numerous black spots (Fig. 16.9). The antenna is 1½ times as long as the fore-wing. The fore-wing of the only knows species is 20–21 mm long.

....*Cacarulla* Navás 1910

The only known species in this genus is *Cacarulla maculipennis* (Banks, 1910). It was originally described under its synonym, *Allochrysa maculipennis* Banks, 1910.

10. (8) Either the fore-wing of the male lacks markings, or markings are limited to narrow dark shading on the membrane bordering the crossveins (Fig. 16.4). In addition, the fore-wing of the female may also have tiny black spots and a blackish vein Sc. The species are usually characterized as having a very narrow pronotum and thread-like palps. Two subgenera are recognized, *Leucochrysa* and *Nodita*, as shown in the key to the species.

....*Leucochrysa* MacLachlan 1868..p. 469

– There are distinct black, brown, or yellowish brown spots on the fore-wing (Fig. 16.11)... 11

11. (10) In addition to a small blackish spot at the base of the stigma, there are somewhat faint yellowish brown transverse stripes on the fore-wing. The radial vein makes a characteristic sigmoid curve posteriad near the wing apex to parallel vein Sc. It then curves back to reach the wing margin near the apex of the wing. The gradate veins are arranged in two parallel series (Fig. 16.11). Fore-wing length of the only known species: c. 17 mm.

....*Nuvol* Navás 1916

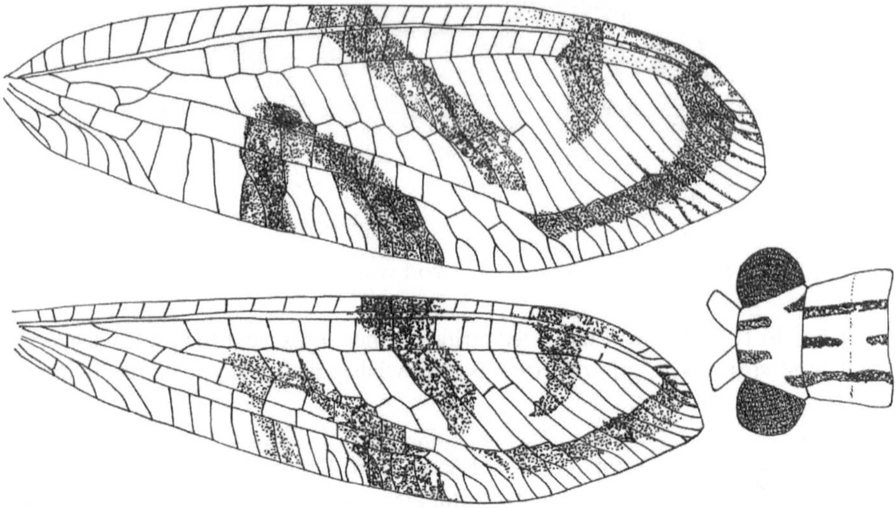

Fig. 16.11 *Nuvol umbrosus:* fore (upper left) and hind wing (lower left), and the head and anterior part of the thorax in dorsal view (lower right). Based on Navás (1916).

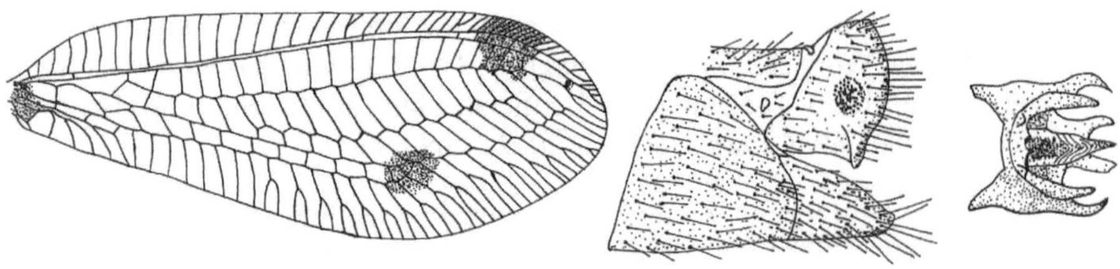

Fig. 16.12 *Gonzaga torquatus* male (left to right): fore-wing, apex of the abdomen in lateral view, and genitalia in dorsal view. Based on Brooks and Bernard (1990).

The only species in this genus is *Nuvol umbrosus* Navás 1916, from Rio de Janeiro. The holotype seems to have been lost, and Brooks and Bernard (1990) could only locate a damaged specimen, which they were not able to examine. The unusual curvature of the radial vein is a characteristic of the genus.

– There are more than one black or dark brown spots on the fore-wing (Fig. 16.12) 12

12. (11) Wing cell *im* is quadrangular. The fore-wing always has black or dark brown markings (Fig. 16.12). .... *Gonzaga* Navás 1913..p. 511

– Wing cell *im* is triangular or ovate (Fig. 16.13) 13

13. (12) The basal radial cross vein originates proximal to the origin of vein Rs and meets the apex of wing cell *im* (Fig. 16.13). An unbroken antenna is at least twice as long as the fore-wing. .... *Berchmansus* Navás 1913..p. 515

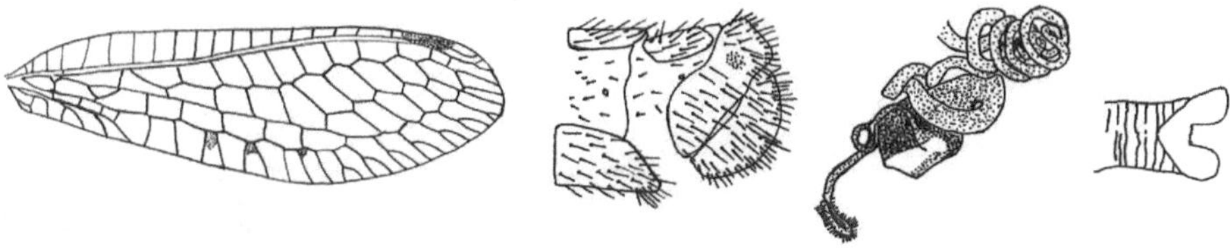

Fig. 16.13 *Berchmansus adumbratus* female (left to right): fore-wing, apex of the abdomen in lateral view, spermatheca in lateral view, and subgenitale in ventral view. Based on Brooks and Bernard (1990).

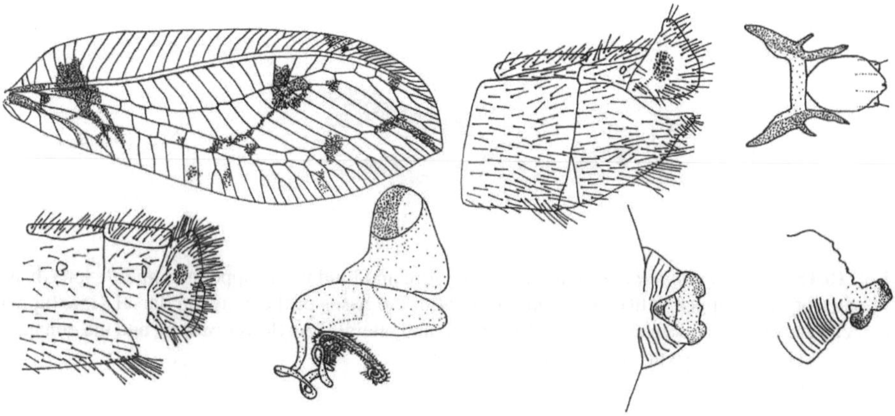

Fig. 16.14 *Vieira leschenaulti* (above, left to right): fore-wing, apex of the male abdomen in lateral view, male genitalia in dorsal view, and (below, left to right): apex of the female abdomen in lateral view, lateral view of the spermatheca, and the female subgenitale in ventral and lateral view. Based on Brooks and Bernard (1990).

– There is no radial cross vein originating proximal to vein Rs (Fig. 16.14). The wings have extensive markings. The basal costal cross veins are sinuate. The antenna is shorter than the fore-wing.

.... *Vieira* Navás 1911..p. 516

14. (6) The prothorax is wider than long, making the insect appear thick-bodied. The abdominal setae of the male are usually short (Fig. 16.7)...................... 15

– The prothorax is usually at least as long as wide. Beyond the first few segments, the antennal segments are longer than wide (Fig. 16.15). The larvae inhabit plants, sometimes covering themselves with debris or the skins of the insects on which they prey.. 17

15. (14) In the fore-wing, cell *im* is absent (Fig. 16.7). Beyond the first few segments, the antenna segments are usually wider than long. The larvae inhabit ant nests.

.... *Belonopteryx* Gerstaecker, 1863

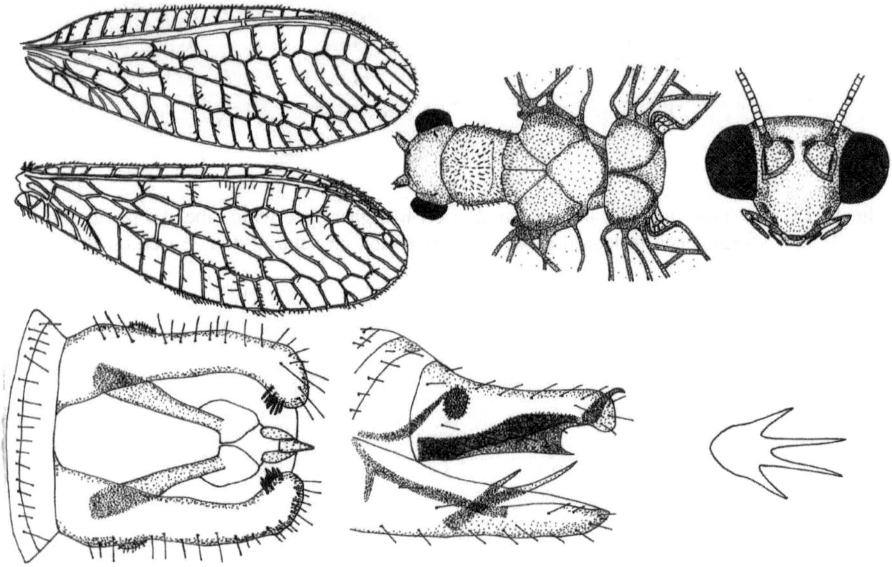

Fig. 16.15 *Parachrysopiella pallidicornis* male: fore and hind wing (upper and middle left), head in anterior view (upper right), head and thorax with the bases of the wings in dorsal view (upper center), apex of the abdomen in dorsal (lower left) and lateral view (lower center), and the gonapsis in dorsal view (lower right). Based on Penny (1996).

The only species in this genus is *Belonopteryx arteriosa* Gerstaecker, 1863, which is known from Argentina and Brazil.

– In the fore-wing, cell *im* is present (Fig. 16.16). Males lack parameres. ... 16

16. (15) The intermedial cell of the wing is generally triangular. Gonocornua are absent from the gonarcus. The subgenitale is cordate, with a lobate dorsal part (Fig. 16.16). The fore-wing is usually shorter than 15 mm. The larvae hunt actively near the floors of forests, camouflaging themselves with bits of debris. .... *Titanochrysa* Sosa and Freitas, 2012...p. 518

– The intramedial cell (*im*) of each wing forms a short rectangle, and the posterior marginal cells are forked. Long lateral gonocornua are present on the gonarcus. The diameters and lengths of the middle segments of the antennal flagella are subequal. There is a praegenitale close to the base of the subgenitale of the female (Fig. 16.17). The fore-wing is usually longer than 16 mm. The larvae inhabit ant nests. .... *Nacarina* Navás, 1915...p. 519

17. (14) There is only one series of gradate cross veins in the hind wing. Inner gradate crossveins are absent from all wings. The claws are dilated. The pedicel is elongate. The ectoprocts of the male extend apicad and bear strong apical teeth.

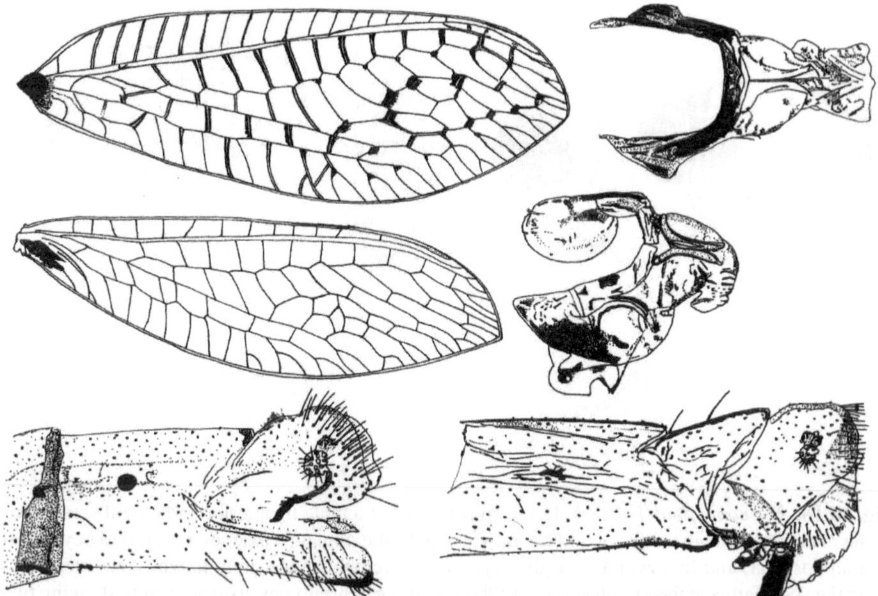

Fig. 16.16 *Titanochrysa circumfusa:* fore and hind wing (upper and middle left), apex of the male abdomen in lateral view (lower left), apex of the female abdomen in lateral view (lower right center), gonarcal complex in dorsal (upper right) and lateral view (middle right), and the subgenitale in anterior view (lower right). Based on Sosa and de Freitas (2012a).

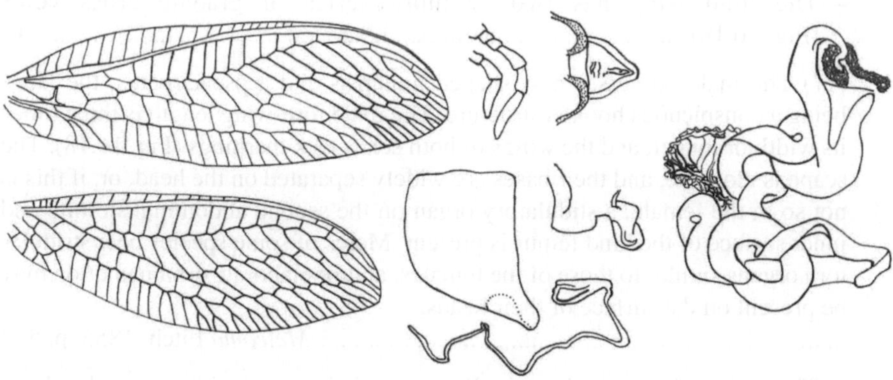

Fig. 16.17 *Nacarina wagneri:* fore and hind wing (left), labial and maxillary palp (upper center, left and right, respectively), seventh sternite (center), colleterial gland (lower center), subgenitalia in ventral (upper right center) and lateral view (right center), and the spermatheca (right). Based on de Freitas and Penny (2001).

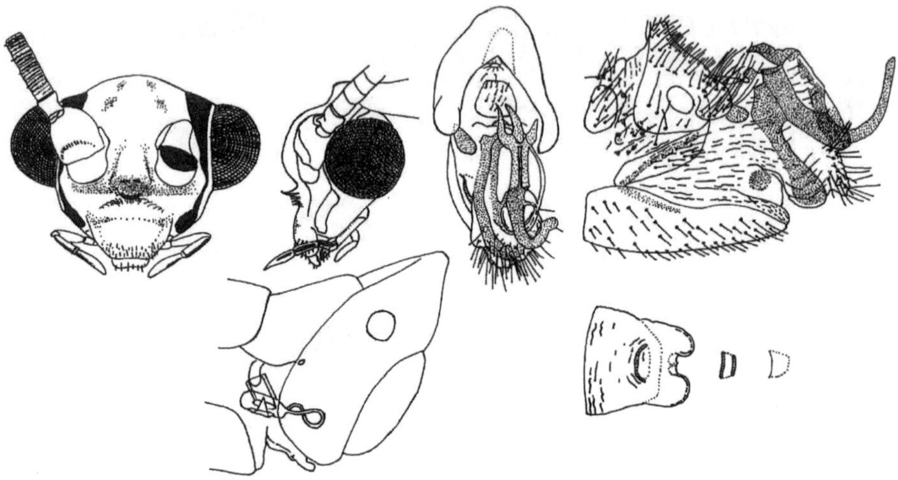

Fig. 16.18 *Meleoma poolei* (above: left to right): head of a male in anterior and lateral view; apex of the male abdomen in posterior and lateral view, with the gonarcus extended posteriad beyond its normal position; and (below, left to right): apex of the female abdomen in lateral view, cleared to show the spermatheca; the eighth sternite of the female in ventral view, its pocket area showing two lips, and a fully developed pocket. Based on Adams (1969).

The gonapsis is short, broad, and bifurcated. The subgenitale of the female is extended basally (Fig. 16.15). ....*Parachrysopiella* Brooks and Barnard, 1990..p. 525

– The hind wing has two or more series of gradate cross veins (Fig. 16.18). 18

18. (17) The male possesses a W-shaped gonapsis and a pseudopenis, the latter being a conspicuous hooked structure. The male fore-wing length is three times its width or longer, and the wings of both sexes lack markings (Fig. 16.18). The scape is elongate, and their bases are widely separated on the head, or, if this is not so in the female, a stidulatory organ on the second abdominal sternite and inner surface of the hind femur is present. Males of some species bear stridulatory organs similar to those of the females, and/or elaborate ornamentation may be present on the surface of their heads. ....*Meleoma* Fitch, 1855..p. 527

– The male lacks a pseudopenis. If a gonapsis is present, it is not in the shape of the letter W. Stridulatory organs are absent, and there is no elaborate ornamentation of the head of the male (Fig. 16.19)... 19

19. (18) The ninth segment of the male has a subapical constriction to isolate a small apical node. There is no pseudopenis. Reddish markings on the head are confined exclusively to the genae. There are no stripes on the scape. The pronotum is not notably enlarged and lacks red markings (Fig. 16.19). .... *Chrysoperla* Steinmann, 1964..p. 528

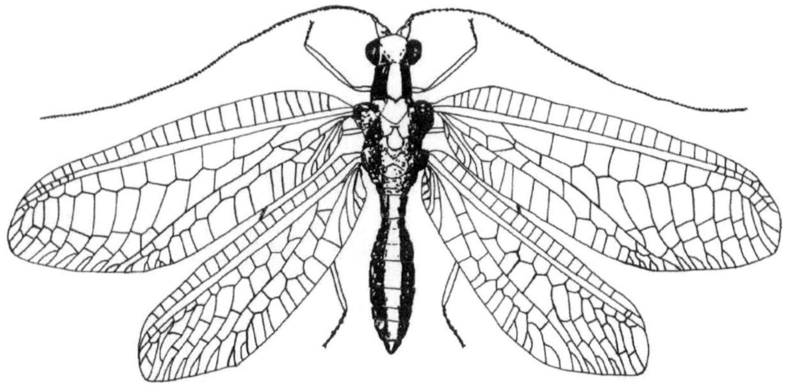

Fig. 16.19 Habitus of *Chrysoperla externa*. Based on Nuñez (1988).

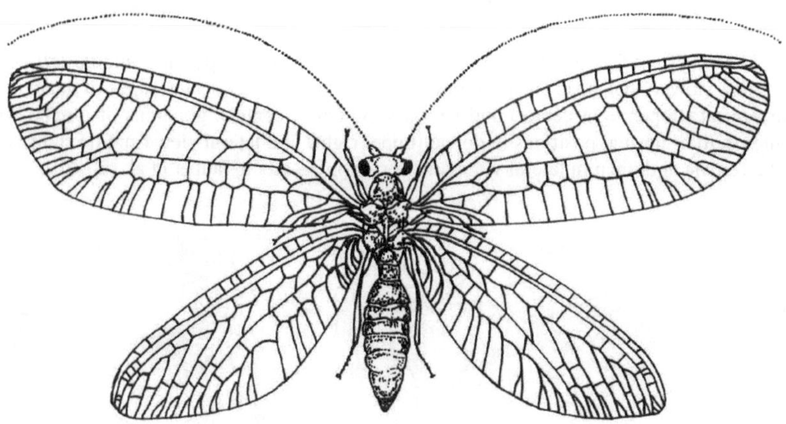

Fig. 16.20 Habitus of *Ceraeochrysa cincta*. Based on Núñez (1988).

– Even if the ninth sternite of the male has lateral lobes or curves dorsad toward the apex, it tapers evenly and lacks a subapical constriction. Dark rings around the base of the antenna or red or dark transverse bands may be present on the frons (Fig. 16.20)... 20

20. (19) The male genitalia include an elongate gonapsis. The gonarcus or arcessus bears two horn-like structures (Fig. 16.20). .... *Ceraeochrysa* Adams, 1982..p. 533

– The gonapsis is not elongate or is completely absent (Fig. 16.21)............. 21

21. (20) There is no pseudopenis. A tignum is absent (Fig. 16.21). The frons is pale with no markings, or it has two transverse red bands. There is no lateral stripe on the scape, and the hind wing has at least five inner gradate veins. There is no pseudopenis.. 22

– A pseudopenis is present. A tignum is present in South American species but absent from species inhabiting other regions of the world (Fig. 16.22). On

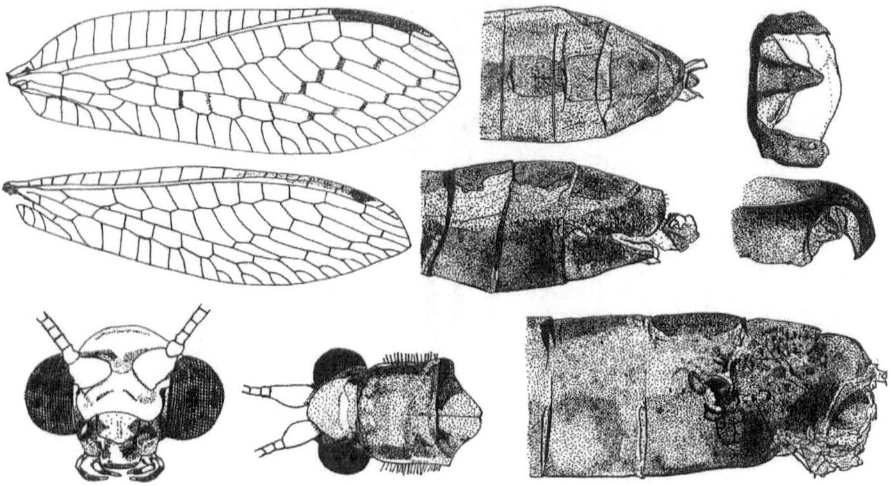

Fig. 16.21 *Chrysopodes serrabonitensis:* fore and hind wing (upper and middle left), head in anterior view (lower left), head and pronotum in dorsal view (lower center), apex of the male abdomen in ventral (upper right center) and lateral view (right center), gonarcus of a male with the apodemes in their normal positions in dorsal (upper right) and lateral view (middle right), and the apex of a female abdomen in lateral view (lower right). Based on Tauber et al. (2012b).

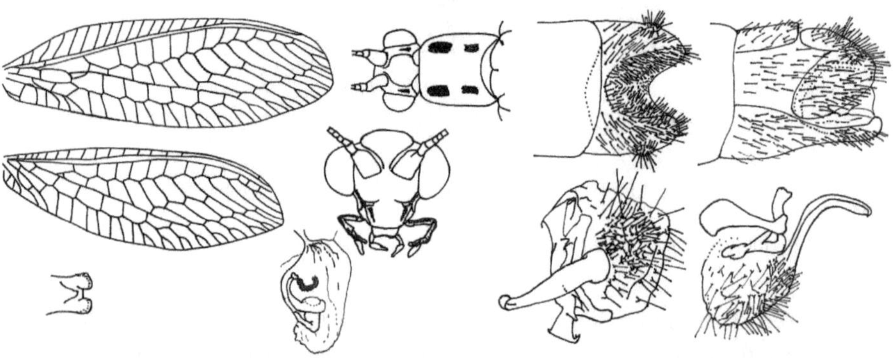

Fig. 16.22 *Plesiochrysa brasiliensis:* fore (upper left) and hind wing (middle left), head and pronotum in dorsal view (upper left center), head in anterior view (left center), apex of male abdomen in dorsal (upper right center) and lateral view (upper right), male genitalia in dorsal (lower right center) and lateral view (lower right), female genitalia (lower left center), and the subgenitale (lower left). Based on de Freitas and Penny (2001).

the species outside of South America which lack a tignum, the color pattern on the frons does not include two transverse red bands 24

22. (21) The apices of the mandibles are fang-like. In only a few species is a gonapsis present. The costal space on the wing is usually wide (Fig. 16.21). .... *Chrysopodes* Navás, 1913. pars p. 560

– The apices of the mandibles are blunt. A gonapsis is never present. The costal space on the wing is usually narrow (Fig. 16.23) 23

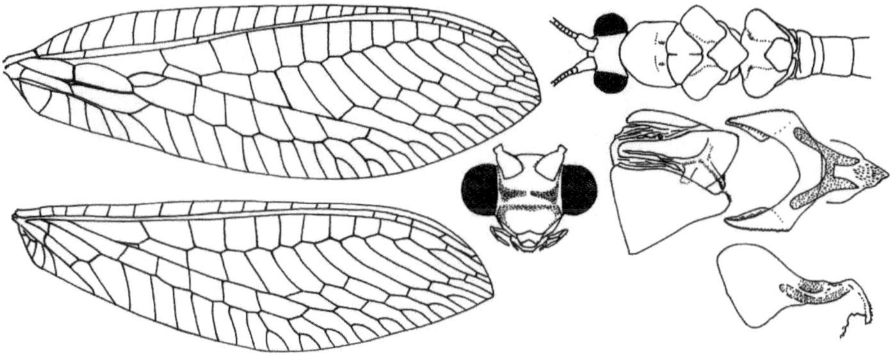

Fig. 16.23 *Chrysopodes collaris* male: fore and hind wing (upper and lower left); head in anterior view (center); head, thorax, and anterior abdominal segments in dorsal view (upper right); apex of the male abdomen in lateral view (right of center); male genitalia in dorsal (middle right) and lateral view (lower right). Based on Adams and Penny (1987) and Penny (2002).

23. (22) The apodeme of the male ninth abdominal tergite protrudes beyond the apex of the abdomen. The ectoprocts of the male are elongated basally. The fused eighth and ninth abdominal sternites are short. The gonosaccus is short, and gonosetae are usually absent. The arcessus is weakly sclerotized and bears dorsal microsetae and lateral rods. The spermatheca of the female is large and barely constricted at its junction with the vela (Fig. 16.23). The mandibles are wide for their entire lengths. The left mandible has a rounded mesal knob, while the right lacks such a knob. The prominent frons and the vertex of the head are smooth and lack pits and folds. The anterior margin of the fore-wing is almost straight along its midlength.
.... *Chrysopodes (Neosuarius)* Adams and Penny, 1987..p. 560

– The ninth abdominal tergite lacks an apodeme. The ectoproct is rounded at the base. The fused eighth and ninth abdominal sternite is elongate. The gonosaccus is large and armed with numerous setae. The arcessus is strongly sclerotized. The spermatheca of the female is short and shaped like a pill box (Fig. 16.24).
.... *Ungla* Navás, 1914..p. 584

24. (21) The pronotum lacks spots, or, if four spots are present, they are orange.
.... *Chrysopa* Leach, 1815
Many South American species were described under the generic name, *Chrysopa*. No South American species are presently included in this genus, which encompasses only Old World and Nearctic species. There are many poorly known species which remain to be reassigned to their correct genera. See p. 38, *et seq*.

– The pronotum is enlarged and has a pattern of four red spots or lateral stripes. The radial cross veins are darkened in the middle (Fig. 16.22).
.... *Plesiochrysa* Adams, 1982..p. 587

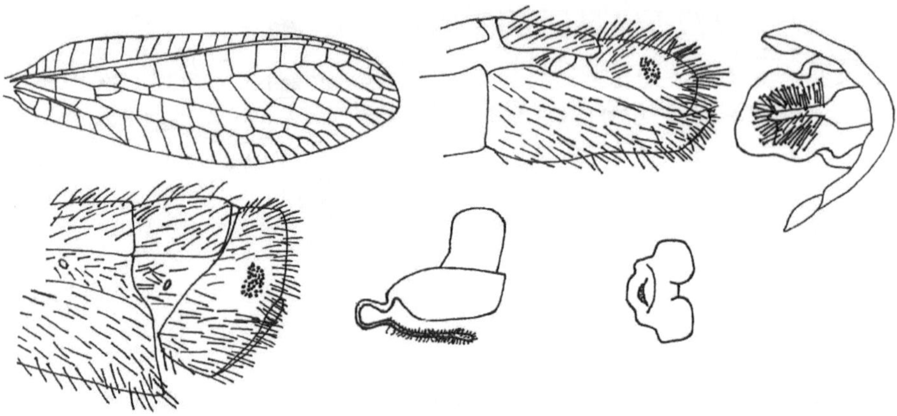

Fig. 16.24 *Ungla binaria* (upper row, left to right): fore-wing, lateral view of the apex of the male abdomen, and male genitalia in dorsal view, and (lower row, left to right): apex of the female abdomen in lateral view, spermatheca, and subgenitale. Based on Brooks and Bernard (1990).

16.2 Subfamily Apochrysinae

Key to the Species of *Joguina* in South America

Information for the key was provided by Navás (1913, 1930); Winterton and Brooks (2002).

1. The maximum width of the fore-wing is located in the middle 1/3 of the wing, and it is more than half as long as the wing. There are eight bullae, blister-like structures, on the fore-wing (Fig. 16.2). Total length: c. 15 mm. Length of fore-wing: 24 mm; hind wing: 24 mm.
   ....*Joguina constellatus* (Navás, 1914)
   (Guatamala). Syn: *Lainius constellatus* Navás, 1914.

   – The fore-wing is widest in its apical third, and it is less than 1/3 as long as the wing. There are 13 bullae on the for-wing (Fig. 16.25). Total length: c. 17 mm. Length of fore-wing: c. 27.5 mm; hind wing: 27.5 mm.
   ....*Joguina decoratus* (Navás, 1930)
   (Northern South America). Syn: *Lainius decoratus* Navás, 1930.

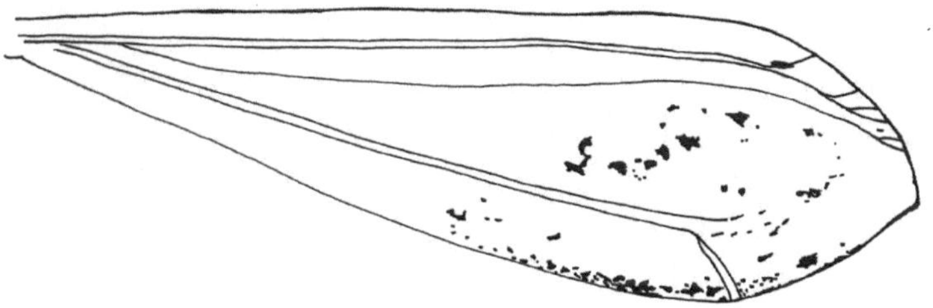

Fig. 16.25 Fore-wing of *Joguina decoratus*. Based on Navás (1930), who named the species *Lainius decoratus*.

Key to the Species of Adult Female *Loyola* in South America

Information for the key was provided by Navás (1913, 1930); Winterton and Brooks (2002), and Tauber et al. (2005). These species can only be tentatively distinguished from one another based on the general coloration. The male anatomy had been unknown until Tauber et al. (2005) provided a description of a male *Loyola croesus*. More detailed descriptions of the species in this genus would be helpful.

1. Dorsally and laterally, the insect is green with yellow tergites. The first three abdmonal sternites, the anterior part of the fourth, and the sixth to the tenth are also yellowish; the posterior part of the fourth and the fifth are brown. The first six segments of the male abdomen are relatively thin, while the abdomen appears swollen from the seventh to the apex (Fig. 16.5).
   .... *Loyola croesus* (Gerstaecker, 1893) (Mexico, Central America, Venezuela, Rio de Janeiro). Syn: *Apochrysa croesus* Gerstaecker, 1893.

   – The entire female is whitish and covered by fine, whitish setae. The expanse of the fore-wings is about 48 mm.. 2

2. The fore-wing has a basal bulla with nine black-tipped veins, a discal bulla with six such veins, and a smaller, round bulla with four dark-tipped veins distal to the discal bulla. Each bulla appears to be an elevated, pustule-like swelling on the wing. Reddish markings are found on the head anterior to the bases of the antennae and along the lateral margin of the pronotum. There is a tiny sigmoidal mark on each wing. The male has not been described.
   .... *Loyola tripunctata* Banks, 1924 (Amazonas).

   – The fore-wing has a basal and discal bulla but no additional bulla. There is one bulla on the hind wing. Veins 1A and 2A in the fore-wing are forked, but vein 3A is not (Fig. 16.26).
   .... *Loyola beata* (Walker, 1858) (Central America, Venezuela, Amazonas). Syn: *Apochrysa beata* Walker, 1858; *Claverina beata* (Walker, 1858).

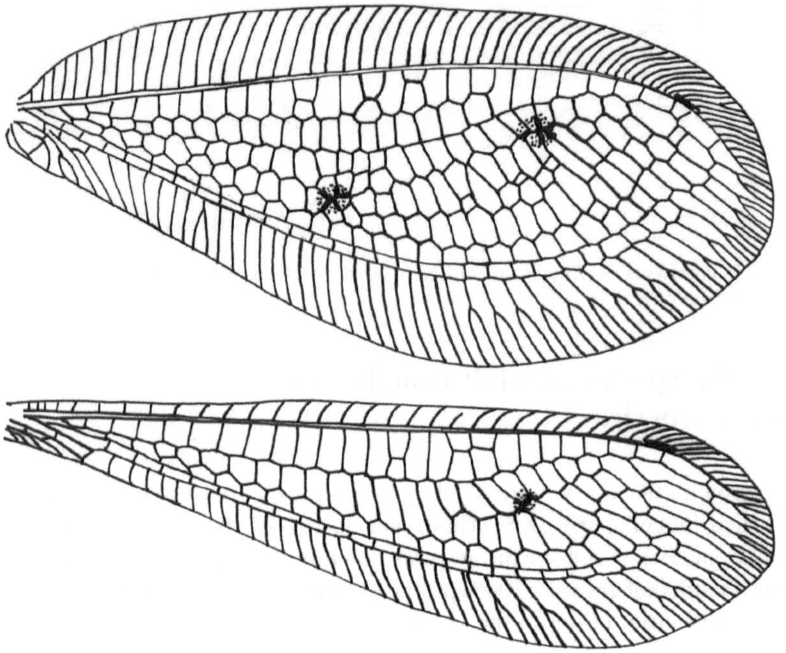

Fig. 16.26 Fore and hind wing of *Loyola beata*. Based on Kimmins (1952), who referred to it by its synonym: *Claverina beata*.

Key to the Species of *Domenechus* in South America

Information for the key was provided by Guérin-Méneville (1853); Gerstaecker (1888), Navás (1913); Winterton and Brooks (2002), and Penny (2002). The two species are similar in size, and most of the differences enumerated by Gerstaecker (1888) involve coloration of the body and minor differences in the coloration and bullae on the wings. More detailed descriptions of these two species would be useful.

1. The bulla in the apical part of the fore-wing is at least three times as large as that in the basal part, and it has five blackish bands, compared to two in the basal bulla. The apex of the fore-wing is not evenly rounded, and the wing appears elongate. The hind wing does not appear to be much narrower and more elongate than the fore-wing (Fig. 16.27). Total length: c. 20 mm. Fore-wing length: c. 33 mm. Color: light reddish brown, somewhat darker on the head, basal segment of the antenna, and the prothorax.
   .... *Domenechus mirificus* (Gerstaecker, 1888) (Central America). Syn: *Apochrysa mirifica* Gerstaecker, 1888; *Domenechus sigillatus* Navás, 1913.

   – The basal and apical bullae in the fore-wing are subequal. Both bullae have four blackish bands connected in the middle by a dark midlength stripe. The apex of the fore-wing is more evenly rounded, and the wing does not appear

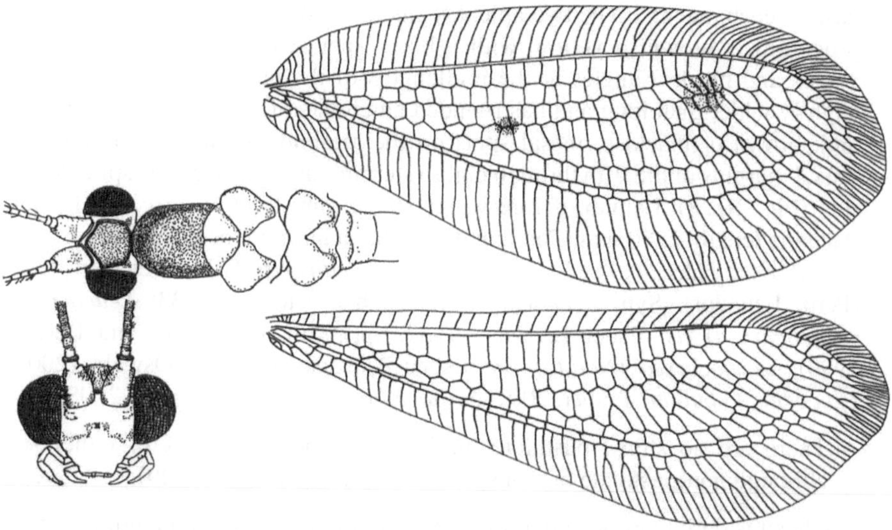

Fig. 16.27 *Domenechus mirificus:* fore and hind wing (above and below right), head and thorax in dorsal view (middle left), and the head in anterior view (lower left). Based on Kimmins (1952) and Penny (2002).

elongated. The hind wing appears considerably more elongate and narrower at the apex than the fore-wing (Fig. 16.6). Total length: c. 20 mm. Wing span: c. 65 mm. Color: pale yellow, including the head and antennae.

....*Domenechus marianellus* (Guérin-Méneville, 1853) (Pará, Guyana). Syn: *Chrysopa Marianella* Guérin-Méneville, 1853; *Apochrysa Marianella* (Guérin-Méneville, 1853); *Loyola Marianella* (Guérin-Méneville, 1853); *Hemerobius marianellus* (Guérin-Méneville, 1853).

16.3 Subfamily Chrysopinae

Key to the Species of Adult *Santocellus* in South American with Notes on the Larvae

Information for the key was provided by Tauber (2007, 2012) and (Tauber et al. 2008a, b, c). After analyzing larval characters of this genus, Tauber et al. (2008a, b, c) concluded that it is distinctly different from *Leucochrysa.*

1. The fore-wing of the male has shiny, raised, pustulate markings surrounding many cross veins and an opaque pterostigma with a slight brownish tinge. The hind wing is unspotted, and most of its veins are pale, except for a few of the costal cross veins. Its pterostigma is opaque with a yellowish tinge. The head is yellow with an oblong black spot on the vertex and a brown spot ventral to the

base of each antenna. The two apical segments of the labial palp are black. The pronotum is brownish yellow with a black spot at each of the anterolateran angles and black triangles at the midlengths of the dorsolateral margins; it is slightly wider than long and has rounded anterolateral corners. There are large, dark brown or blackish, mesal spots on the mesonotum and metanotum. The dorsal margins of the fifth through eighth sternites are concave. The spiracles are tiny (Fig. 16.28). Length of fore-wing: 16–20 mm; hind wing: c. 13–18 mm. Antenna length: c. 40 mm. The female and larva have not been described. .... *Santocellus risi* (Esben-Petersen, 1933) (Peru, Ecuador). Syn: *Leucochrysa risi* Esben-Petersen, 1933; *Leucochrysa (Leucochrysa) risi* Esben-Petersen, 1933, appearing in (Brooks and Bernard 1990); *Leucochrysa (Nodita) risi* Esben-Petersen, 1933, in Oswald (2007); *Leucochrysa bullata* Tauber, 2007; *Santocellus bullata* (Tauber et al. 2008).

– There are no shiny, pustulate markings on the fore-wings or either sex, although non-pustulate brown spots may be present (Fig. 16.29).................. 2

2. The head is dark maroon and amber with a light green spot in the middle of the frontoclypeal region of the head. The dorsal margins of the fifth through eighth sternites are straight or slightly concave. The pterostigma of the fore-wing is a dark brown spot, and there is a second brown spot in the subcostal area distal to the pterostigma. The abdomens of neither sex have enlarged spiracles (Fig. 16.29). Length of fore-wing: 14.0–16.8 mm; hind wing: c. 12.4–15.1 mm. The third larval instar is about 6.7–6.8 mm long, whitish or cream in color, and with two pairs of brown stripes and a brown marking roughly in the shape of a triangle on the dorsal side of the head. ....*Santocellus atlanticus* Tauber and Albuquerque, 2008 (Rio de Janeiro, Rio Grande do Sul).

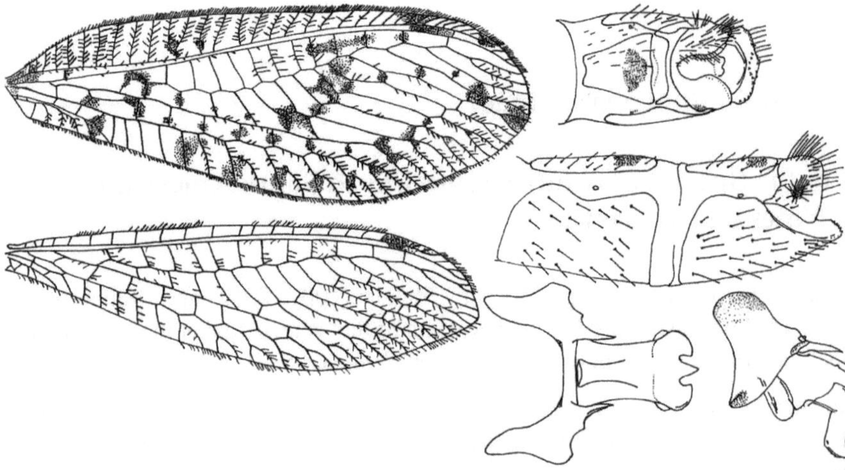

Fig. 16.28 *Santocellus risi* male: fore and hind wing (left), apex of the abdomen in ventral (upper right) and lateral view (middle right), and the gonarcal complex in dorsal (lower right center) and lateral view (lower right). Based on Tauber (2007), who named the species *Leucochrysa bullata*.

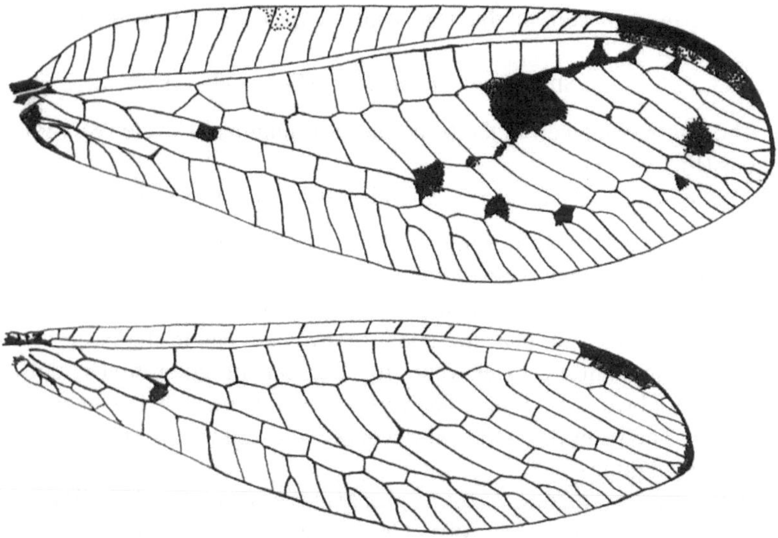

Fig. 16.29 *Santocellus atlanticus:* fore and hind wings (left) and the mandibles and palps of a larva (right). Based on Tauber et al. (2008a, b, c).

– The basal radial crossvein is located at or very near the origin of vein Rs. There is a dark marking at each end of the pterostigma (Fig. 16.8). The spiracles in the fifth through eighth abdominal segments or the male are enlarged and obvious, while those of the female are typically small. Length of forewing: 16.3–16.9 mm; hind wing: c. 14.5–16.6 mm. The first larval instar is c. 2.5–2.6 mm long and has reddish brown to purple markings on all segments. On the dorsal part of head, there are three pairs of brown stripes and a brown isosceles triangle.
.... *Santocellus riodoce* (Tauber, 2007) (Espirito Santo). Syn: *Vieira riodoce* Tauber, 2007.

Key to the Species of *Leucochrysa* in South America

Information for the key was provided by Navás (1916, 1922a, b, 1923a, b, 1925a, 1929a, 1932b, 1933a, b); Banks (1910, 1920); Adams (1987); de Freitas and Penny (2001); Penny (2002), and (Tauber et al. 2011). The intermedian cells of exceptional specimens sometimes vary enough to cause misidentifications of the subgenus. Species described by Navás after 1916 are assumed to have been placed in the correct subgenus. A revision of the genus would be desirable, and identifications of species not re-examined since described by Navás must be regarded as doubtful.

1. The intermedian cell in the fore-wing is quadrilateral. Vein Rs is straight (Fig. 16.30).
   .... Subgenus *Leucochrysa* MacLachlan, 1868.........................2

   – The intermedian cell in the fore-wing is triangular or ovate. Vein Rs is sinuous (Fig. 16.31).
   ....Subgenus *Nodita* Navás, 1916...................11

2. (1) There are bright white spots on the vertex of the head and mesoscutellum, which turn to a creamy yellow color on preserved specimens. The scape of the antenna has one dark, transverse band on the dorsal side at the apex and a dark spot on the inner side near the apex. Many segments in the basal half of the antennal flagellum are black ventrally. The general coloration is dull yellow with reddish brown markings. The vertex is mainly reddish, but an elevated part is yellow. The wing is hyaline with a dark pterostigma (Fig. 16.32). Fore-wing length: c. 23 mm.
   .... *Leucochrysa magnifica* (Banks, 1920) (French Guiana, Argentina, Mato Grosso). Syn: *Allochrysa magnifica* Banks, 1920.

   – The vertex and mesoscutellum never have white or creamy yellow spots (Fig. 16.33)..3

3. (2) The thorax is testaceous with a ferrugineous tinge on the pronotum dorsally. The pronotum is slightly longer than wide. The head is yellowish with darker

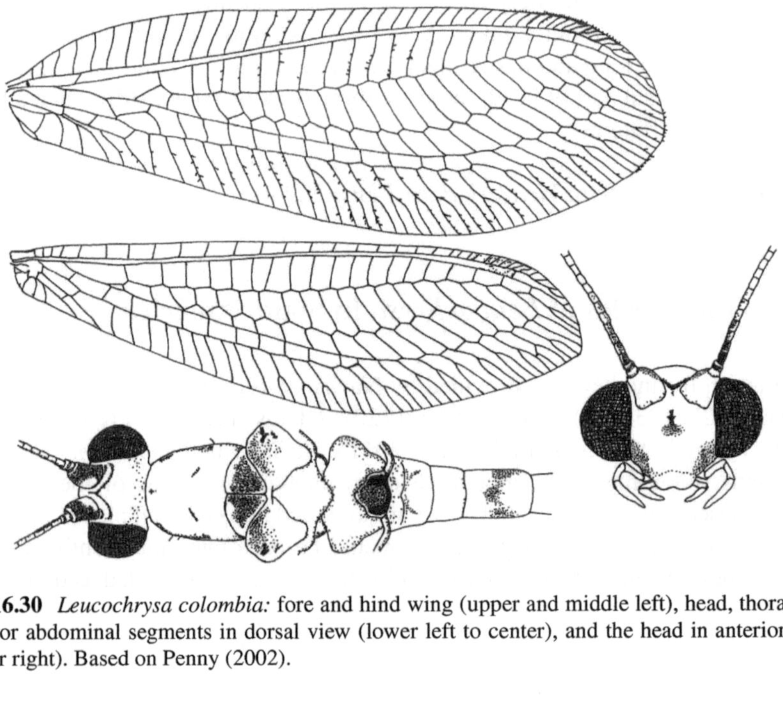

Fig. 16.30 *Leucochrysa colombia:* fore and hind wing (upper and middle left), head, thorax and anterior abdominal segments in dorsal view (lower left to center), and the head in anterior view (lower right). Based on Penny (2002).

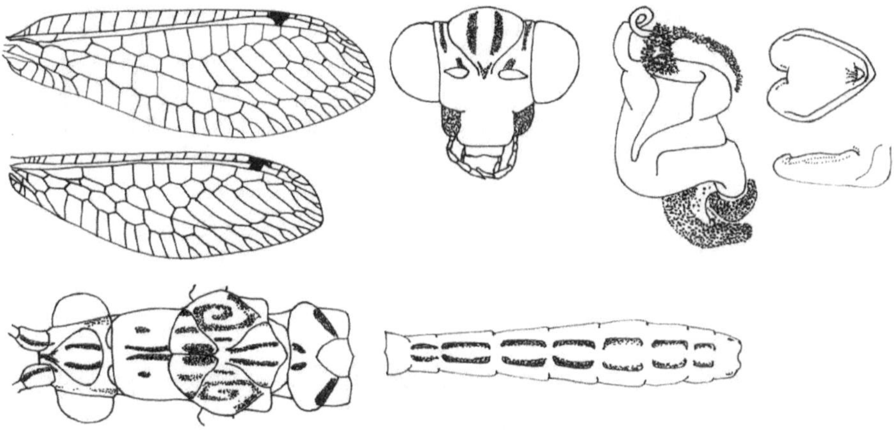

Fig. 16.31 *Leucochrysa vittata:* fore (upper left) and hind wing (middle left), head in anterior view (upper center), head and thorax in dorsal view (lower left), abdomen in dorsal view (lower right), spermatheca (upper right center), and the subgenitale in dorsal (upper right) and lateral view (middle right). Based on de Freitas and Penny (2001).

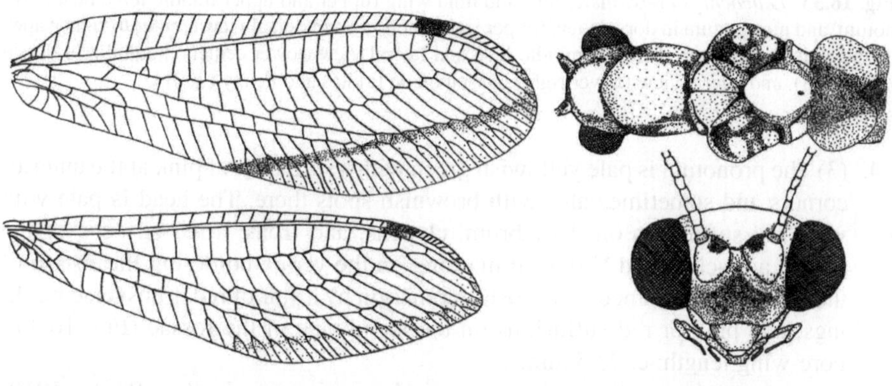

Fig. 16.32 *Leucochrysa magnifica:* fore and hind wing (left), head and thorax in dorsal view (upper right), and head in anterior view (lower right). Based on Penny (2002).

markings, and the antenna is yellow. There is a blackish longitudinal stripe divided in the middle on the mesonotum and metanotum. The abdomen is ferrugineous dorsally and brownish yellow ventrally. The wings have yellow veins and a dark marking on the proximal part of the pterostigma. There is a dark oblique cloud running from the posterior margin near the base of the fore-wing toward the rounded apex. Total length: c. 12.5 mm. Length of fore-wing: 24.5 mm; hind wing: 21.5 mm.
....*Leucochrysa ignatii* (Navás, 1923) (Argentina). Syn: *Leucochrysa Ignatii* Navás, 1923 (capitalization).

– The pronotum is pale green, green, or brownish yellow with black markings, reddish markings, or suffused with pink (Fig. 16.33)................................ 4

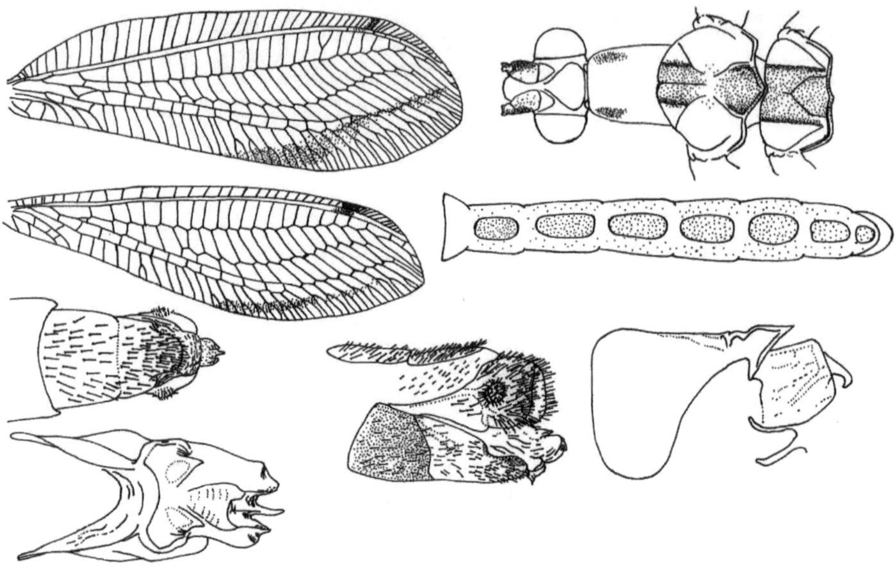

Fig. 16.33 *Leucochrysa boxi* male: fore and hind wing (upper and upper middle left); head, pronotum, and mesonotum in dorsal view (upper right); abdomen in dorsal view (center to right); apex of the abdomen in ventral (lower middle left) and lateral view (lower center); genitalia in dorsal (lower left), and lateral view (lower right). Based on de Freitas and Penny (2001).

4. (3) The pronotum is pale yellowish green with a suffusion of pink at the anterior corners and sometimes also with brownish spots there. The head is pale with wine red suffusions on the labrum, clypeus, and frons; a red stripe over the genae; a blackish red V-shaped marking on the vertex bordering the fossae of the antennae and joined at its posterior margin by a pair of red, transverse markings; and pink or red suffusions on the membrane of the fossae (Fig. 16.30). Fore-wing length: c. 23.5 mm.
.... *Leucochrysa colombia* (Banks, 1910) (Colombia). Syn: *Allochrysa colombia* Banks, 1910; *Leucochrysa claveria* Navás, 1928; *Leucochrysa californica* Navás 1928; *Allochrysa californica* (Navás, 1928) Banks (1938). The synonym *L. californica* may have been given the species due to a misread label (Adams, 1977).

– The pronotum is either uniformly green and pale green, or there are red spots near the anterior margin of the pronotum and sometimes near its midlength, as well, or it is pale with dark spots at the anterolateral corners as seen in dorsal view (Fig. 16.33). The markings on the head differ from those described for *L. colombia* 5

5. (4) The mesonotum and metanotum both have two dark longitudinal stripes (Fig. 16.33). There are two red spots near the anterior margin of the pronotum. Length of fore-wing: c. 24.2 mm; hind wing: c. 22.1 mm.
   ....*Leucochrysa boxi* Navás, 1930 (Peru, Bolivia, Argentina, Espirito Santo, Santa Catarina, Rio Grande do Sul, Rio de Janeiro). Syn: *Hemerobius longicornis* G. Gray in Cuvier, 1832 (preoccupied); *Hemerobius longicollis* Gray in Cuvier, 1832 (for figure in Schneider, 1851); *Leucochrysa longicollis* (Gray) Navás, 1929; *Chrysopus longicornis* (Gray in Cuvier, 1832) Schneider, 1851; *Allochrysa longicornis* (Gray in Cuvier, 1832) Navás, 1932; *Leucochrysa (Leucochrysa) loretana* Navás, 1935, *Leucochrysa lorenata* Navás, 1935 (misspelling).
   - The mesonotum and metanotum lack longitudinal stripes (Fig. 16.34)...........6
6. (5) The head, thorax, and abdomen lack dark markings (Fig. 16.34). The apical half of the velum of the female spermatheca is an elongated and flattened structure. Fore-wing length: c. 18.1 mm; width: c. 5.6 mm. Hind wing length: c. 17.3 mm; width: c. 5.5 mm.
   ....*Leucochrysa catarinae* Freitas and Penny, 2001 (Santa Catarina).
   - There are thick red markings and suffusions on the head, or a black Y-shaped marking between the antennae, or two pairs of small, dark spots on the pronotum, seen in dorsal view. There are large dark markings on the fifth and sixth abdominal tergites, or the abdomen is green dorsally with a yellowish mid-dosal stripe and yellow ventrally with yellowish setae, or it is uniformly yellow dorsally or pale yellow with yellow setae ventrally (Fig. 16.35)...... 7

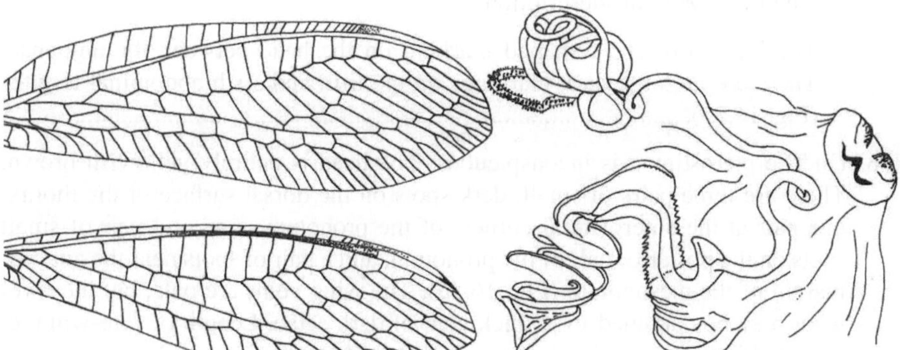

Fig. 16.34 *Leucochrysa catarinae* female: fore and hind wing (left), genitalia (right). Based on de Freitas and Penny (2001).

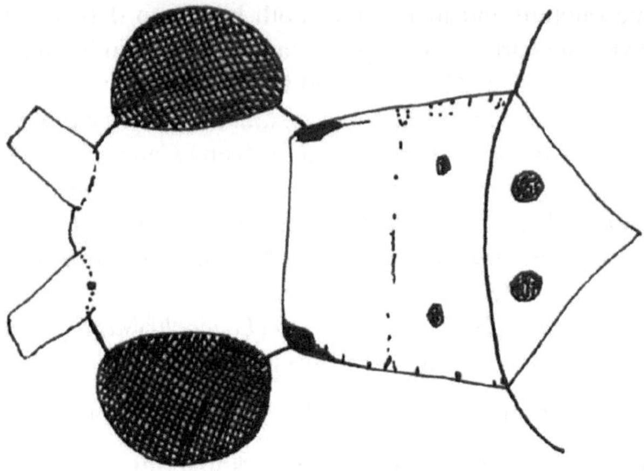

Fig. 16.35 Head and pronotum of *Leucochrysa ehrhardti*. Based on Navás (1929a).

7. (6) The head is green with some suffusions of red and a black Y-shaped marking between the antennae. The pronotum is green dorsally with a yellowish mid-dorsal stripe. The mesonotum and metanotum are dark with green posterior margins on the metanotum. The abdomen is green dorsally with a yellowish mid-dosal stripe and yellow ventrally with yellowish setae. Total length of male: 11 mm. Length of fore-wing: c. 19.5 mm; hind wing: c. 17.5 mm. ....*Leucochrysa reedi* Navás, 1919 (Argentina). This species is presumed to be in the subgenus *Leucochrysa* because Navás had described the other South American subgenus, *Nodita,* 3 years earlier but placed his type specimen of this species in *Leucochrysa.* However, Navás provided no illustration of the wing and did not mention the shape of cell *im* in his description.

   – There is no black, Y-shaped marking on the head between the antennae. There are no large dark markings on the fifth and sixth abdominal tergites (Fig. 16.35). 8

8. (7) The pterostigma is inconspicuous. The head is entirely yellowish brown. There are three pairs of small, dark spots on the dorsal surface of the thorax: one pair at the anterolateral corners of the pronotum, a second pair of small spots on the posterior half of the pronotum, and a pair of spots near the anterior margin of the mesonotum (Fig. 16.35). The wing veins are pale, but the fore-wing may be darkened by a thick coat of dark setae. Length of fore-wing: c. 25 mm; hind wing c. 22 mm. ....*Leucochrysa ehrhardti* Navás, 1929 (Santa Catarina). Syn: *Leucochrysa Ehrhardti* Navás, 1929 (capitalization). From the original description, it cannot be determined with certainty that the species was correctly assigned to this genus. Navás provided no illustration of

the wing and did not mention the shape of cell *im* in his description. A specimen fitting the description in the key should be preserved for further study.

– The location of the pterostigma is marked with a dark spot, at least a faint one (Fig. 16.36). Neither wing has a dense coat of dark setae 9

9. (8) At the base of the intramedial vein in the fore-wing, there is a dark spot. The frons is pale and unmarked or suffused with red. The gonocornua of the male are short and separated by more than five timenys the length of one (Fig. 16.36). Length of fore-wing: 20.8–21.3 mm; hind wing: 18.3–18.6 mm. .... *Leucochrysa varia* (Schneider, 1851) (Mexico, Central America, Guyana? Surinam? Bolivia? Paraguay? Peru, Ecuador, Argentina, Bahia, Espirito Santo? Minas Gerais, Rio de Janeiro, São Paulo, Mato Grosso, Rondônia, Pará, Distrito Federal, Santa Catarina). Syn: *Chrysopa varia* Schneider, 1851; *Allochrysa varia* (Schneider, 1851); *Leucochrysa ampla* (Walker, 1853); *Chrysopa internata* Walker, 1853; *Allochrysa internata* (Walker, 1853); *Nodita internata* (Walker, 1853); *Leucochrysa internata* (Walker, 1853); *Leucochrysa walkeriana* Navás, 1913; *Leucochrysa walkerina* Navás, 1913, spelling error in Freitas and Penny (2001) and (Tauber et al. 2012a, b); possibly *Leucochrysa vegana* Navás, 1917 *nomen dubium*.

– At the base of the intramedial cell in the fore-wing, there is no dark spot. The gonocornua of the male are moderately long and separated by no more than twice the length of one (Fig. 16.37)... 10

10. (9) The arms of the male gonocoxite are oriented about 45° from the vertical axis. The inner margins of the gonocornua are parallel (Fig. 16.37). .... *Leucochrysa bruneola* Freitas and Penny, 2001 (Mato Grosso).

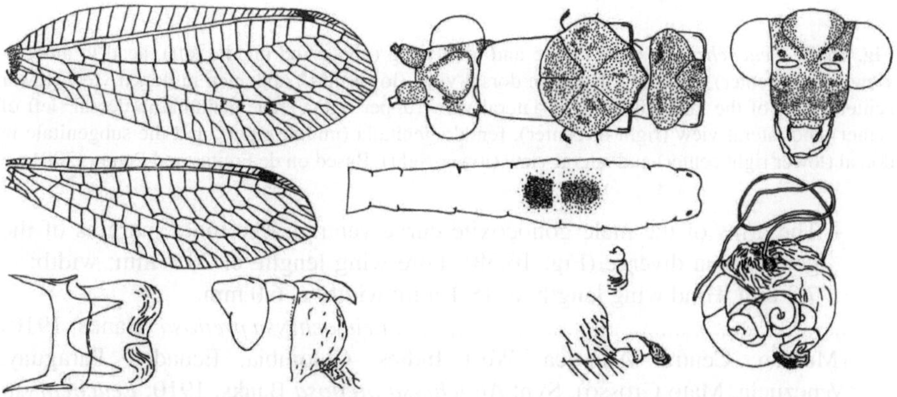

Fig. 16.36 *Leucochrysa varia:* fore and hind wing (upper and middle left); head, pronotum, and mesonotum in dorsal view (upper center); head in anterior view (upper right); abdomen in dorsal view (center); male genitalia in dorsal (lower left) and lateral view (lower left center); female genitalia (lower right); subgenitale in ventral and lateral view (lower right center, above and below, respectively). Based on de Freitas and Penny (2001).

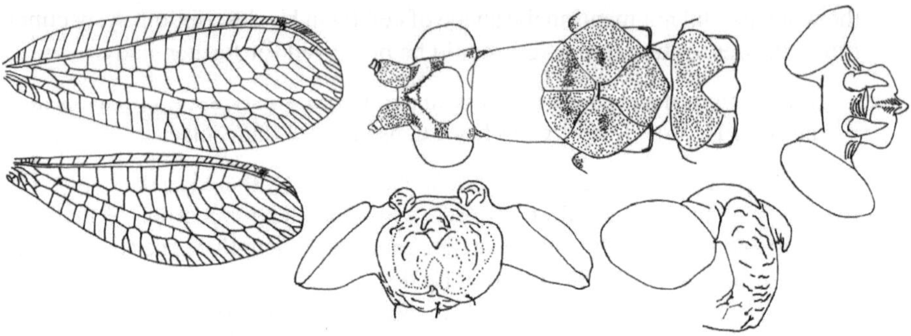

Fig. 16.37 *Leucochrysa bruneola* male: fore and hind wing (left); head, prothorax, and mesothorax in dorsal view (upper center); and the genitalia in dorsal (right), lateral (lower right center), and posterior view (lower left center). Based on de Freitas and Penny (2001).

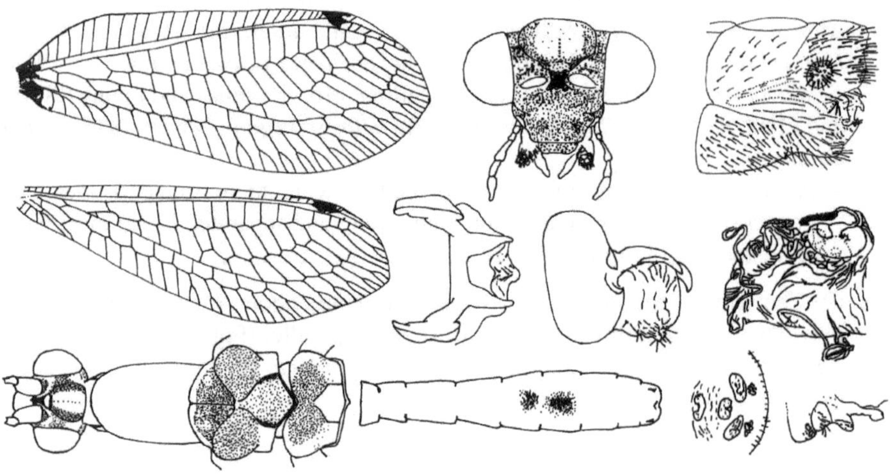

Fig. 16.38 *Leucochrysa pretiosa:* fore and hind wing (upper and middle left), head in anterior view (upper center), head and thorax in dorsal view (lower left), abdomen in dorsal view (lower center), apex of the male abdomen in lateral view (upper right), male genitalia in dorsal (left of center) and lateral view (right of center), female genitalia (middle right), and the subgenitale in dorsal (lower right center) and lateral view (lower right). Based on de Freitas and Penny (2001).

– The arms of the male gonocoxite curve ventrad. The inner margins of the gonocornua diverge (Fig. 16.38). Fore-wing length: c. 20.3 mm; width: c. 7.2 mm. Hind wing length: c. 18.1 mm; width: c. 6.0 mm.
....*Leucochrysa pretiosa* (Banks, 1910) (Mexico, Central America, West Indies, Colombia, Ecuador, Paraguay, Venezuela, Mato Grosso). Syn: *Allochrysa pretiosa* Banks, 1910; *Leucochrysa delicata* Navás, 1925; *Leucochrysa erminea* Banks, 1946. Tauber et al. (2013) resurrected two names long regarded as synonyms of *Leucochrysa pretiosa* but provided no thorough descriptions of the specimens examined. Damaged specimens were used to restore the validity of *Allochrysa angrandi* (Navás, 1911). A

description was promised in a subsequent publication. A variety of questionably identified specimens were used to restore the name *Allochrysa variata* Navás, 1913, which had been correctly declared to be a junior synonym of *L. pretiosa*. However, a second specimen was described by Navás (1917a, b) under the same name, and this is allegedly a distinct species. Descriptions for these two species were promised in a subsequent publication, but until then, neither can be identified using a key with any degree of confidence.

11. (1) The scape, which is the enlarged basal section of the antenna, is light in color and lacks a dark longitudinal stripe, and the male arcessus has a truncate subapical plate. The metanotum and abdominal segments lack dark dorsal markings (Fig. 16.39). At its mid-length, vein Rs is dark. There is a crescent-shape marking on the vertex. The length of the fore-wing is more than 20 mm. ....*Leucochrysa amazonica* (Navás, 1913) (Central America, Guyana, Amapa, Pará, Amazonas). Syn: *Nodita amazonica* (Navás, 1913); *Leucochrysa egregia* Navás, 1913; *Leucochrysa luctuosa* Banks, 1914.

   – The scape has one or more longitudinal stripes, other markings, or is entirely dark in color. If the male arcessus has a truncate subapical plate, the metanotum and abdominal segments have dark dorsal markings (Fig. 16.40) 12

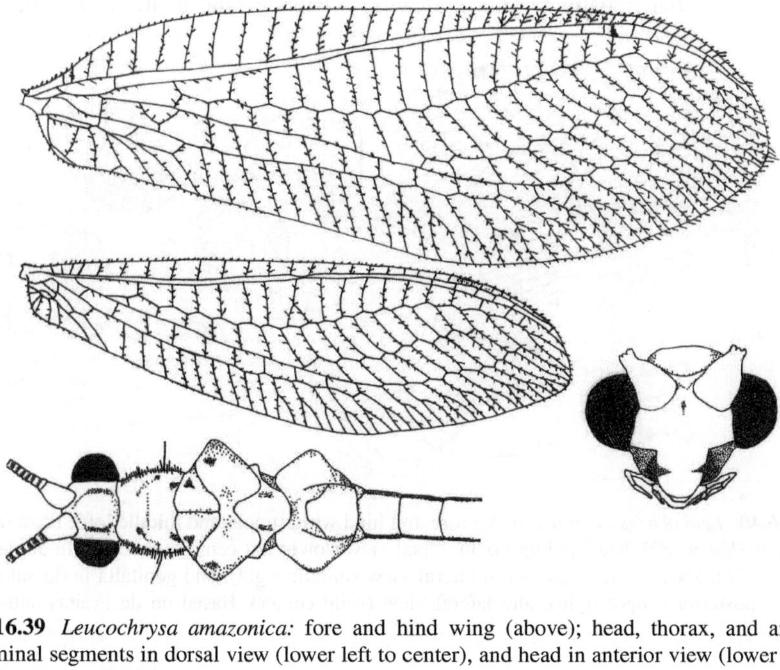

Fig. 16.39 *Leucochrysa amazonica:* fore and hind wing (above); head, thorax, and anterior abdominal segments in dorsal view (lower left to center), and head in anterior view (lower right). Based on Penny (2002).

12. (11) There is one longitudinal stripe on the scape, which is the enlarged basal segment of the antenna (Fig. 16.40), or, in faded specimens, the scape is light, and no stripe is apparent 13

 – The scape is completely dark, has two dark stripes, or has other markings (Fig. 16.41)..36

13. (12) The dark longitudinal stripe on the scape is on the dorsal surface (Fig. 16.31). The gradate cross veins are darker than the longitudinal veins.......14

 – The dark stripe is on the dorsolateral or lateral surface of the scape (Fig. 16.40)...15

14. (13) The head is pale green with a brown mid-dorsal stripe. The pedicel and flagellum of the antenna is pale. The frons is pale and unmarked, and the genae are brown. The pronotum has brown submedial stripes and a small lateral spot at mid-length on each side. The pterostigma is black at its base (Fig. 16.31). Length of fore-wing: c. 15.9 mm; hind wing c. 14.3 mm.
 ....*Leucochrysa vittata* Freitas and Penny, 2001 (São Paulo). Syn: *Leucochrysa vittatus* Freitas and Penny, 2001 (alternate spelling with mistake in Latin grammar).

 – The scape is pale yellowish with a red stripe dorsally. There is a broad red transverse marking above the base of each antenna. The pronotum has red spots at each anterolateral corner and a large, red, transverse marking near its midlength, which is divided along the midline. There are also smaller reddish markings on the dorsal and dorsolateral surface. The pterostigma is somewhat indistinct. The hind wing is slender and acute at the apex. The

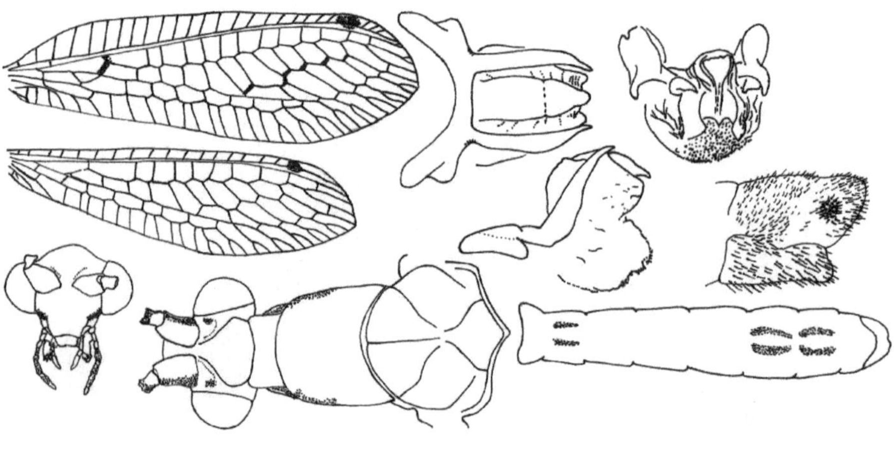

Fig. 16.40 *Leucochrysa confusa* male: fore and hind wing (upper and middle left), head in anterior view (lower left), head and thorax in dorsal view (lower left center), abdomen in dorsal view (lower right), apex of the abdomen in lateral view (middle right), and genitalia in dorsal (upper center), posterior (upper right), and lateral view (right center). Based on de Freitas and Penny (2001).

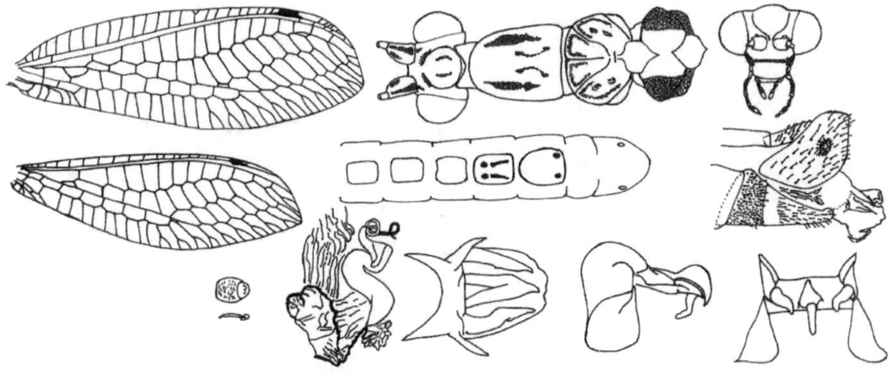

Fig. 16.41 *Leucochrysa lancala:* fore (upper left) and hind wing (middle left), head and thorax in dorsal view (upper center), head in anterior view (upper right), abdomen in dorsal view (center), apex of the male abdomen in lateral view (middle right), male genitalia in dorsal (lower center), lateral (lower right center), and posterior view (lower right); spermatheca (lower left center), and subgenitale in ventral and lateral view (lower left, above and below, respectively). Based on de Freitas and Penny (2001).

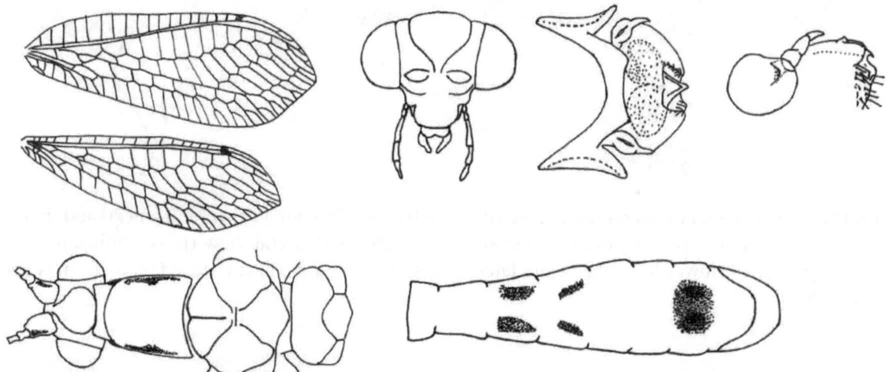

Fig. 16.42 *Leucochrysa vignisi* male: fore and hind wing (upper and middle left), head and thorax in dorsal view (lower left), head in anterior view (upper left center), abdomen in dorsal view (lower right), and male genitalia in dorsal (upper right center) and lateral view (upper right). Based on de Freitas and Penny (2001).

wings are longer, but the length was not precisely recorded. The description provides only 40 mm as the wing span. It is very sketchy, and the species needs to be redescribed.
....*Leucochrysa neuralis* Banks, 1910 (Colombia). Syn: *Nodita neuralis* (Banks, 1910).

15. (13) The dark stripe on the scape is wide and reaches both the anterior and posterior ends (Fig. 16.40).. 16

– The dark stripe on the scape is narrow, or it does not reach the anterior and posterior ends (Fig. 16.42).. 28

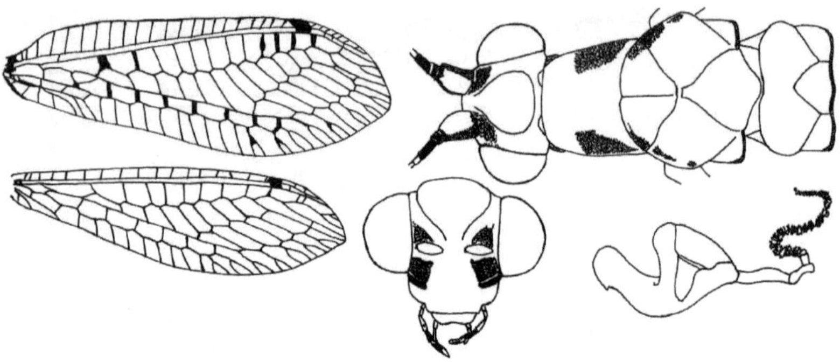

Fig. 16.43 *Leucochrysa maculosa* female: fore and hind wing (left), head and thorax in dorsal view (upper right), head in anterior view (lower center), and spermatheca (lower left). Based on de Freitas and Penny (2001).

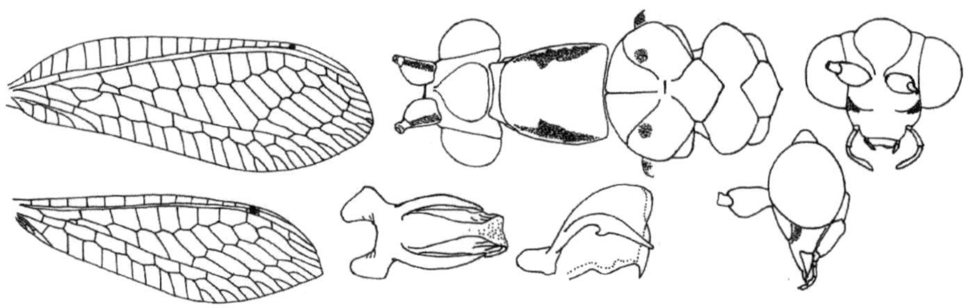

Fig. 16.44 *Leucochrysa barrei* male: fore (upper left) and hind wing (lower left), head and thorax in dorsal view (upper center), head in dorsal (upper right) and lateral view (lower right), and the genitalia in dorsal (lower left center) and lateral view (lower right center). Based on de Freitas and Penny (2001).

16. (15) The antennal flagellum is pale (Fig. 16.40).. 17

    – At least part of the antennal flagellum is dark (Fig. 16.43)........................ 19

17. (16) The base of Rs, MA, and the inner gradate veins of the fore-wing are dark and have darkened margins around them (Fig. 16.40). Length of fore-wing: c. 13 mm. Only the male has been described.
    ....*Leucochrysa confusa* Freitas and Penny, 2001 (Santa Catarina).

    – The base of Rs, MA, and the inner gradate veins of the fore-wing are dark but lack dark margins or shading around them (Fig. 16.44)............................ 18

18. (17) The gonarcus has a short, thick central arch. The gonocornua are long (Fig. 16.44). Fore-wing length: c. 27.5 mm; hind wing length: c. 23.0 mm. Only the male has been described.
    ....*Leucochrysa barrei* Freitas and Penny, 2001 (Argentina, Mato Grosso).

– The gonarcus has a thin, elongate central arch. The gonocornua are small (Fig. 16.45). Fore-wing length: c. 15.1 mm; hind wing length: c. 13.4 mm. Only the male has been described.
....*Leucochrysa tenuis* Freitas and Penny, 2001 (São Paulo).

19. (16) There are dark spots on the cross veins of the fore-wing. There are dark spots on the dorsal insertion cavity of the antenna. The first two segments of the antennal flagellum are dark. The dark stripes on the pronotum do not reach the anterior margin (Fig. 16.43). Fore-wing length: c. 14.7 mm; hind wing length: c. 12.4 mm. Only the female has been described.
....*Leucochrysa maculosa* Freitas and Penny, 2001 (São Paulo). Syn: *Leucochrysa (Nodita) maculata* Freitas and Penny, 2001, mistake in figure caption.

– There are no dark spots on the cross veins of the fore-wing. There is no dark spot on the dorsal insertion cavity of the antenna. The entire antennal flagellum or one side of it is dark. The dark stripes on the pronotum reach the anterior margin (Fig. 16.46).. 20

20. (19) Only one side of the antennal flagellum is darkened (Fig. 16.46) 21

– The antennal flagellum is entirely dark (Fig. 16.47) 24

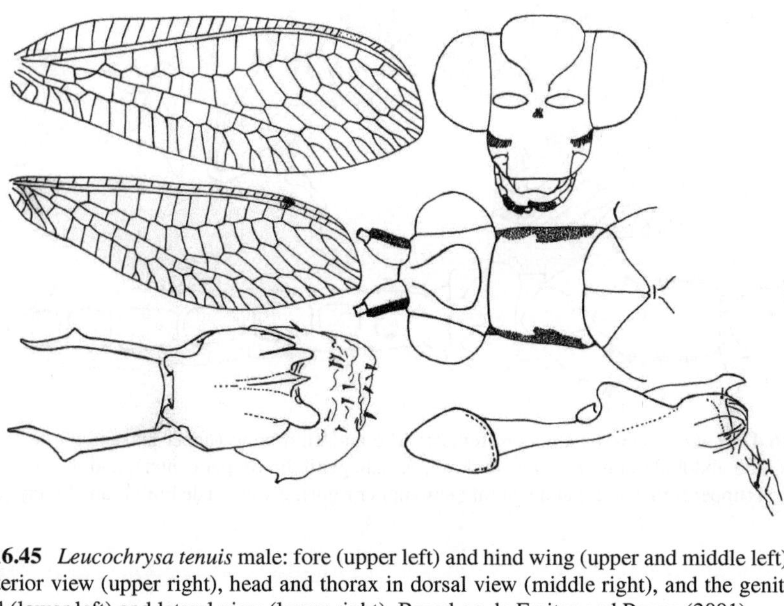

Fig. 16.45 *Leucochrysa tenuis* male: fore (upper left) and hind wing (upper and middle left), head in anterior view (upper right), head and thorax in dorsal view (middle right), and the genitalia in dorsal (lower left) and lateral view (lower right). Based on de Freitas and Penny (2001).

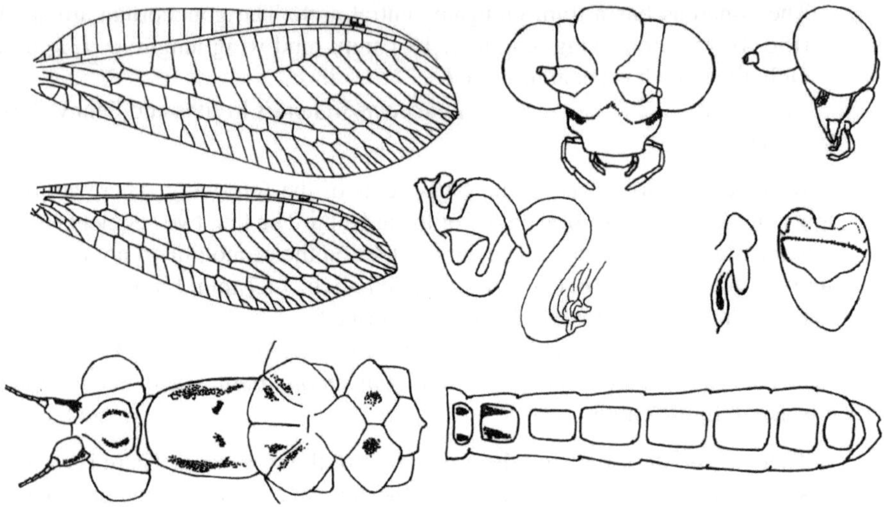

Fig. 16.46 *Leucochrysa interata:* fore and hind wing (upper and middle left), head and thorax in dorsal view (lower left), head in anterior (upper center) and lateral view (upper right), abdomen in dorsal view (lower right), spermatheca (center), and two views of the subgenitale (middle right). Based on de Freitas and Penny (2001).

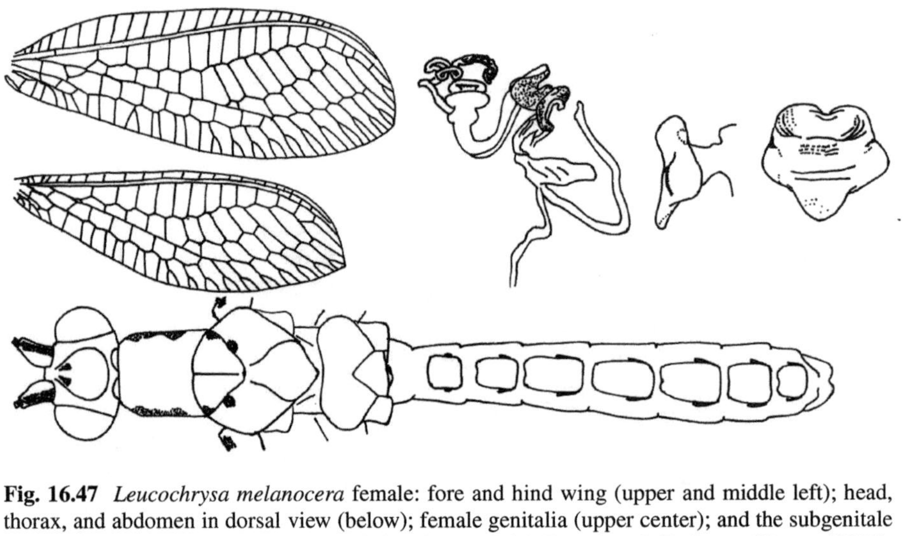

Fig. 16.47 *Leucochrysa melanocera* female: fore and hind wing (upper and middle left); head, thorax, and abdomen in dorsal view (below); female genitalia (upper center); and the subgenitale in lateral (upper right center) and ventral view (upper right). Based on de Freitas and Penny (2001).

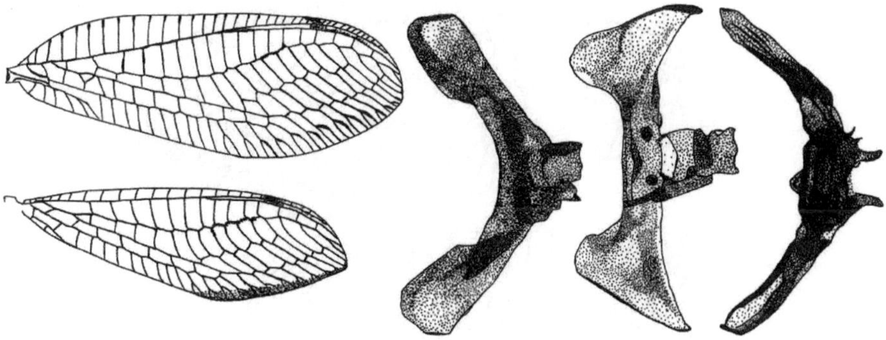

Fig. 16.48 *Leucochrysa azevedoi* male: fore (upper left) and hind wing (lower left), and the gonarcus in dorsal (center), dorsoposterior (right center), and ventral view (right). Based on Tauber et al. (2011a, b, c).

21. (20) Vein R_s in the fore-wing and its marginal forks are completely pale. The posterior margin of the fore-wing is pale (Fig. 16.46). Fore-wing length: c. 19.3 mm; hind wing length: c. 16.4 mm. Only the female has been described. .... *Leucochrysa interata* Freitas and Penny, 2001 (São Paulo).

 – The middle section of vein R_s in the fore-wing and its marginal forks are dark. The posterior margin of the fore-wing is darkened (Fig. 16.48) 22

22. (21) In males, the arms of the gonarcus diverge so that the structure nearly equals the total width of the segment. The gonocornua are located on the mesal part of the gonarcal bridge, well within the span of the mediuncus, and directed dorsad from the bridge. The spermatheca of the female is tubular, and the bursa is small with a small fluted duct (Fig. 16.48). Fore-wing length: 17.7–19.5 mm; hind wing length: 15.3–16.9 mm. .... *Leucochrysa azevedoi* Navás, 1913 (Rio de Janeiro, São Paulo, Mato Grosso, Pará?). Syn: *Leucochrysa camposi* nec Navás, 1933 (Freitas and Penny, 2001); *Nodita azevedoi* (Navás, 1913).

 – In males, the gonarcus is detached from the abdomen but connected to the gonarcal bridge by a clear, smooth membrane around the gonocornua, which protrude through the membrane. The gonocornua are broadly attached to the gonarcal bridge distal to the mediuncus, broad at the base, and tapering distally (Fig. 16.49). The two species in the next couplet are in need of better descriptions because the existing specimens are not in good condition. Tauber et al. (2011a, b, c) did not consider them well enough described to be placed in a key, so identifications are tentative. The female of only one of the two species in the couplet is known. Specimens appearing to belong to one of these two species should be retained for specialists to examine 23

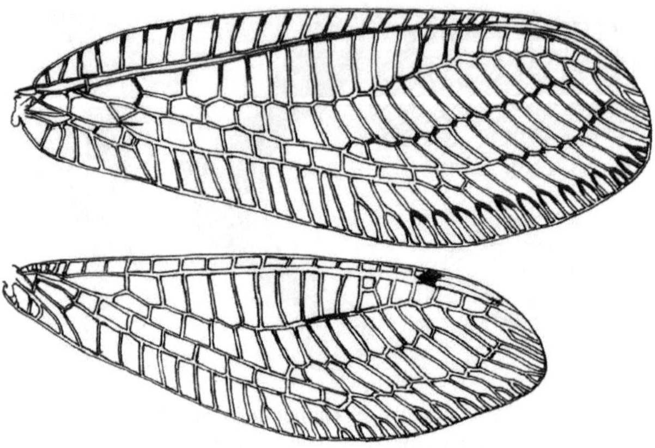

Fig. 16.49 *Leucochrysa morenoi:* fore and hind wing. Based on Tauber et al. (2011a, b, c).

23. (22) The posterior part of the gonarcus is rounded. The dorsal part of the mediuncus is also rounded. The gonocornua are flattened and laterally expanded (Fig. 16.49). The female has not been described.
....*Leucochrysa morenoi* (Navás, 1934) (Ecuador). Syn: *Nodita morenoi* Navás, 1934. The name of this species was considered a junior synonym of *Leucochrysa camposi* until recently (Tauber et al. 2011a, b, c).

  – In the male, the posterior part of the gonarcus and dorsal part of the mediuncus are not rounded, and the gonocornua are not flattened or laterally expanded. The spermatheca of the female is a broad, thick tube in the shape of a U, and the bursa is a lightly folded, shield-like, thick, membranous sac, which extends anteriorly over the spermatheca (Fig. 16.50). Fore-wing length: 21.3–21.6 mm; hind wing length: 18.5–18.8 mm.
....*Leucochrysa camposi* (Navás, 1933) (Ecuador). Syn: *Nodita camposi* Navás, 1933.

24. (20) There are one or two round spots on the mesonotum (Fig. 16.47).......... 25

  – There are no round spots on the mesonotum (Fig. 16.51)......................... 26

25. (24) There are two pairs of dark spots on the mesonotum and a dark spot at the base of the second axillary sclerite. The lateral surface of the scape is completely pale (Fig. 16.47). Fore-wing length: c. 17.7 mm; hind wing length: c. 15.2 mm.
....*Leucochrysa melanocera* Navás, 1916 (Mato Grosso).

  – There is only one pair of dark spots on the mesonotum and no dark spot on the second axillary sclerite. There is a small dark spot on lateral surface of

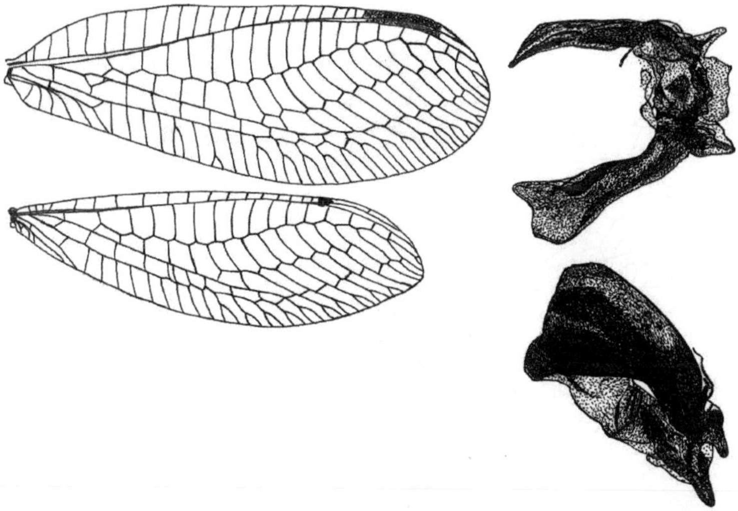

Fig. 16.50 *Leucochrysa camposi:* male: fore and hind wing (left), gonarcus in oblique ventral (*upper right*) and oblique lateral view. Based on Tauber, Albuquerque, and Tauber (1911).

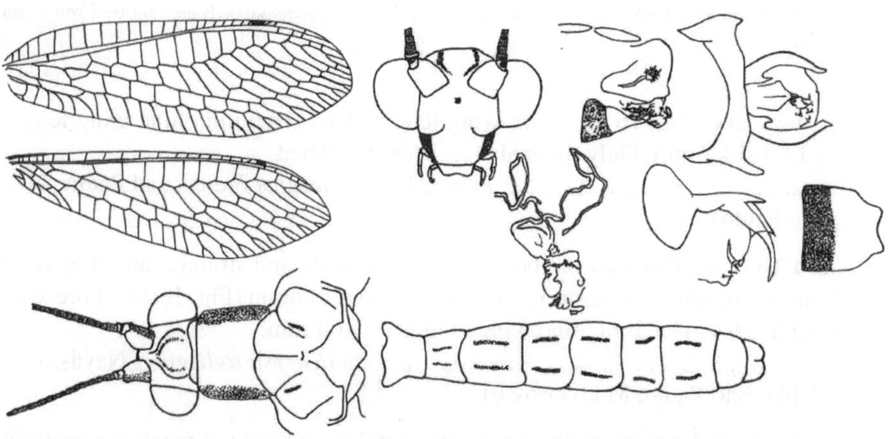

Fig. 16.51 *Leucochrysa rodriguezi:* fore (upper left) and hind wing (middle left), head and anterior part of thorax in dorsal view (lower left), head in anterior view (upper left center), abdomen in dorsal view (lower right), apex of the male abdomen in lateral view (upper right center), internal male genitalia in dorsal (upper right) and lateral view (right of center); spermatheca of a female (center), and the eighth and ninth sternites in ventral view (lower right). Based on de Freitas and Penny (2001).

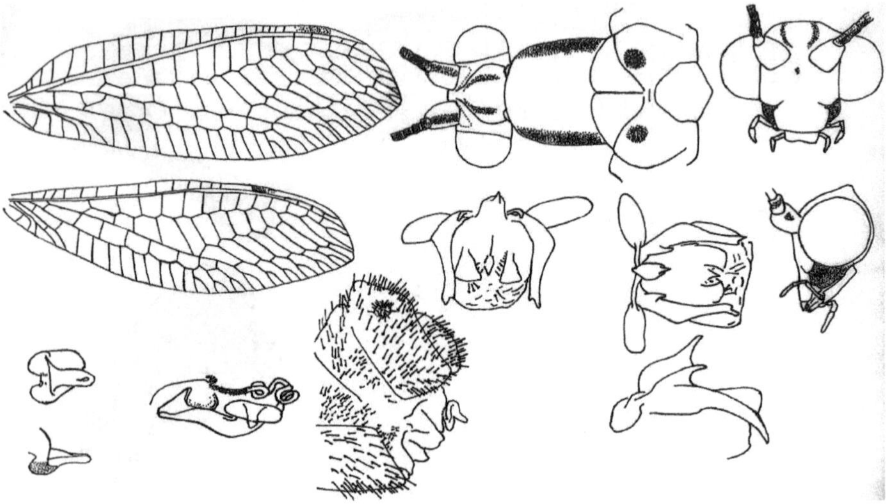

Fig. 16.52 *Leucochrysa icterica:* fore and hind wing (upper and upper middle left); head, pronotum, and mesonotum in dorsal view (upper right center); head in anterior (upper right) and lateral view (middle right); male genitalia in dorsal (right of center), posterior (center), and lateral view (lower right); apex of the female abdomen (lower center); spermatheca (lower left center), and the subgenitale in ventral (lower middle left), and lateral view (lower left). Based on de Freitas and Penny (2001).

the scape (Fig. 16.52). Fore-wing length: 15.5–16.9 mm; hind wing length: 13.3–14.7 mm. Only the male has been described.
....*Leucochrysa icterica* Freitas and Penny, 2001 (São Paulo).

26. (24) The lateral stripes on the pronotum are wide and straight, and they reach both the anterior and posterior margins of the pronotum (Fig. 16.51). Fore-wing length: 16.9–17.8 mm; hind wing length: c. 14.4 mm.
.... *Leucochrysa rodriguezi* Navás, 1913 (Bahia, São Paulo, Mato Grosso).

– The lateral stripes on the pronotum undulate and do not reach the posterior margins of the pronotum (Fig. 16.53). The inner and outer margins of the antenna are darkened to the same degree.. 27

27. (26) The genae are red. The hind wing has a darkened posterior margin. The spermatheca is long and coiled (Fig. 16.53). Fore-wing length: c. 16.3 mm; hind wing length: c. 14.1 mm.
....*Leucochrysa marginalis* Banks, 1915 (Bolivia, São Paulo).

– The genae are yellow. The hind wing has a pale posterior margin. The spermatheca is short and broad, and it has a small lateral projection opposite its

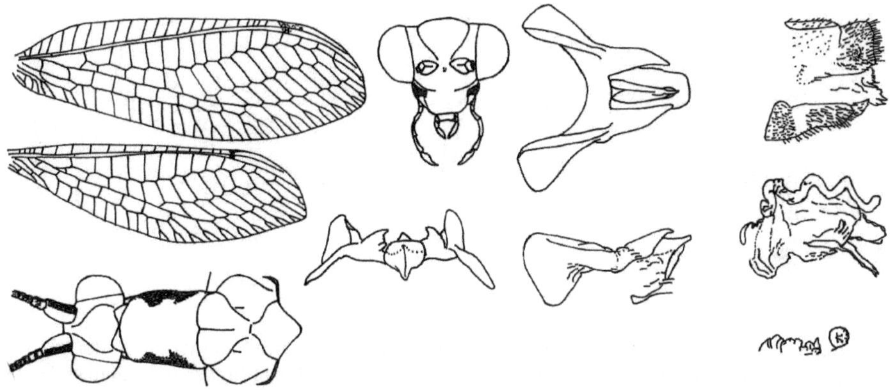

Fig. 16.53 *Leucochrysa marginalis:* fore (upper left) and hind wing (middle left), head and anterior part of the thorex in dorsal view (lower left), head in anterior view (upper left center), apex of the male abdomen in lateral view (upper right), male genitalia in dorsal (upper right center), lateral (lower right center), and posterior view (lower left center), female genitalia (middle right), and the subgenitale with the basal stalk shown to the left (lower right). Based on de Freitas and Penny (2001).

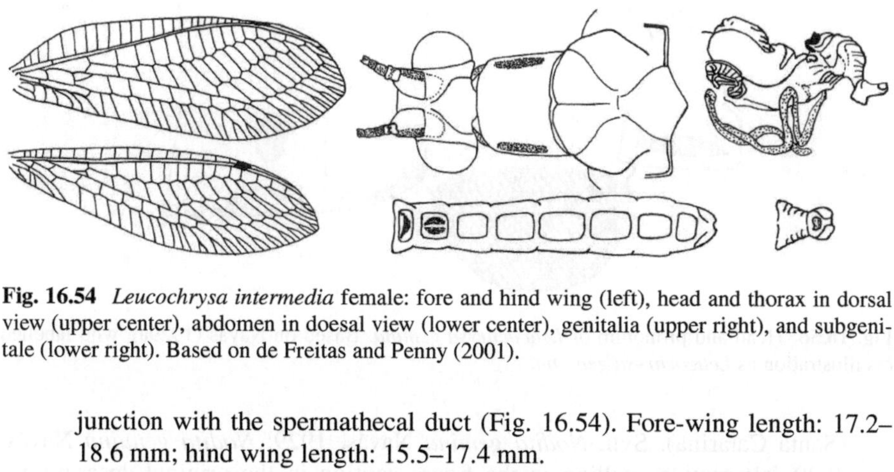

Fig. 16.54 *Leucochrysa intermedia* female: fore and hind wing (left), head and thorax in dorsal view (upper center), abdomen in doesal view (lower center), genitalia (upper right), and subgenitale (lower right). Based on de Freitas and Penny (2001).

junction with the spermathecal duct (Fig. 16.54). Fore-wing length: 17.2–18.6 mm; hind wing length: 15.5–17.4 mm.
....*Leucochrysa intermedia* (Schneider, 1951)
(São Paulo, Mato Grosso). Syn: *Chrysopa intermedia* Schneider, 1951.

28. (15) The genae are uniformly pale (Fig. 16.42)... 29
– There are reddish markings on the genae (Fig. 16.55) 31

29. (28) The pronotum is significantly wider than long, and it has a pair of small black spots on the posterior half. Its sides are subparallel and are somewhat darkened (Fig. 16.56). Length of fore-wing: c. 16 mm; hind wings: c. 14 mm.
....*Leucochrysa gemina* (Navás, 1929)

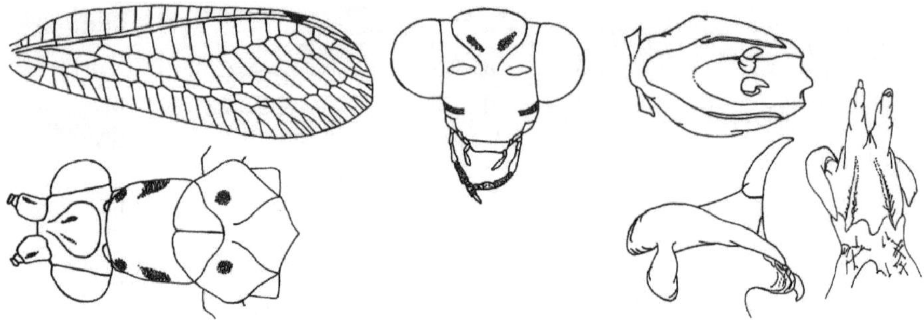

Fig. 16.55 *Leucochrysa clepsydra* male: fore-wing (upper left), head and thorax in dorsal view (lower left), head in anterior view (center), and the genitalia in dorsal (upper right center), lateral (lower right center), and posterior view (right). Based on de Freitas and Penny (2001).

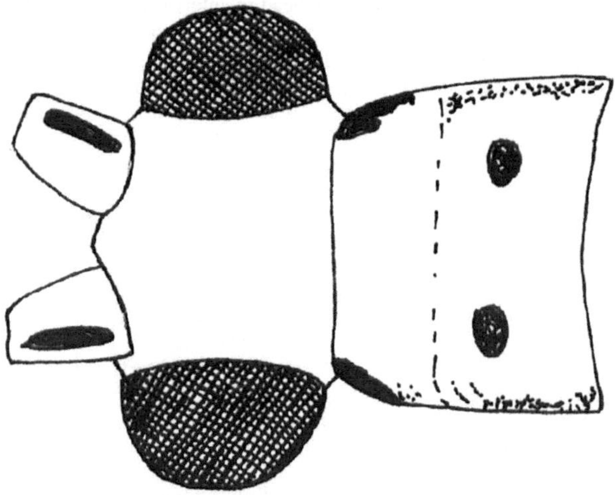

Fig. 16.56 Head and pronotum of *Leucochrysa gemina*. Based on Navás (1929a), who labelled his illustration as *Leucochrysa genuina*.

(Santa Catarina). Syn: *Nodita gemina* Navás, 1929; *Nodita genuina* Navás, 1929 (alternative spelling in the figure caption in the original description). The description does not permit a positive identification of this species, which has never been redescribed.

– The length and maximum width of the pronotum are subequal, or the pronotum is somewhat longer that wide (Fig. 16.57) 30

30. (29) The gonarcus is broad. The gonocornua are relatively narrow and are elongate near the apex (Fig. 16.42). Fore-wing length: c. 14.8 mm; hind wing length: c. 13.0 mm. Only the male has been described. .... *Leucochrysa vignisi* Freitas and Penny, 2001 (Argentina, Mato Grosso).

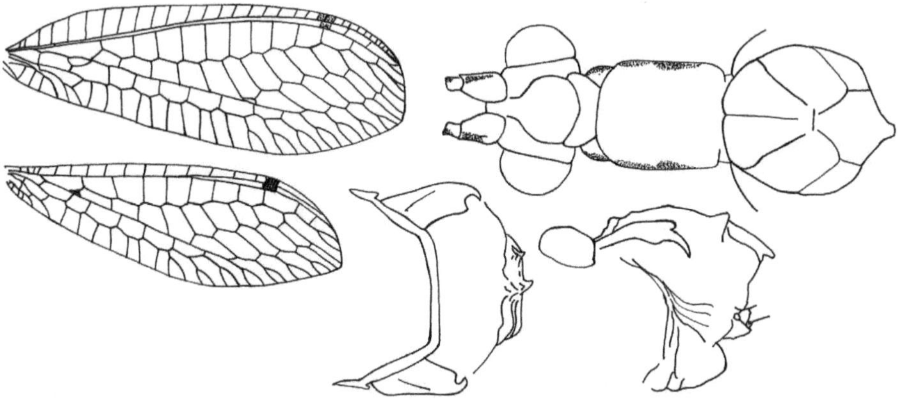

Fig. 16.57 *Leucochrysa retusa* male: fore and hind wing (left), head and thorax in dorsal view (upper right), and male genitalia in dorsal (lower center) and lateral view (lower right). Based on de Freitas and Penny (2001).

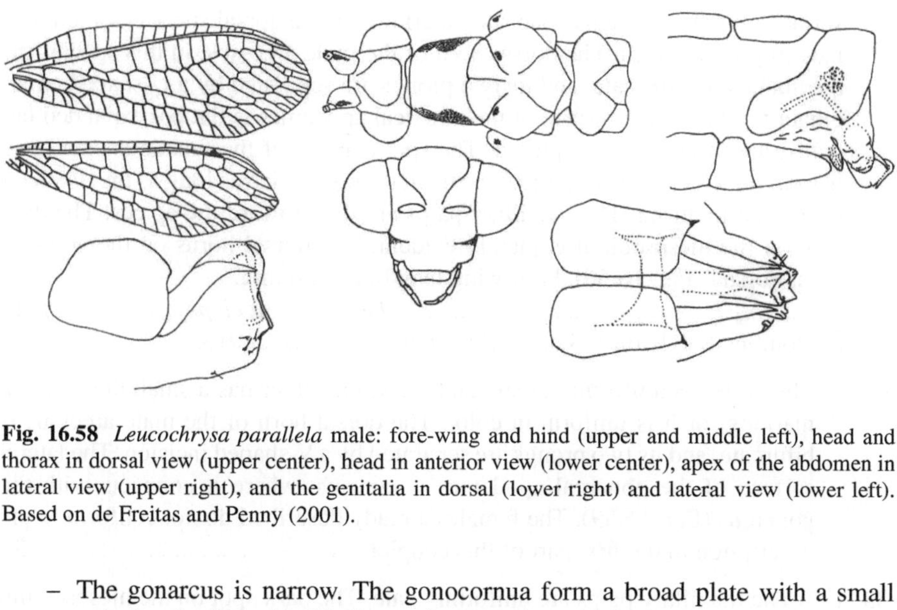

Fig. 16.58 *Leucochrysa parallela* male: fore-wing and hind (upper and middle left), head and thorax in dorsal view (upper center), head in anterior view (lower center), apex of the abdomen in lateral view (upper right), and the genitalia in dorsal (lower right) and lateral view (lower left). Based on de Freitas and Penny (2001).

– The gonarcus is narrow. The gonocornua form a broad plate with a small apical point (Fig. 16.57). Fore-wing length: c. 14.4 mm; hind wing length: c. 12.4 mm. Only the male has been described. ....*Leucochrysa retusa* Freitas and Penny, 2001 (São Paulo).

31. (28) The pronotal stripe is interrupted, leaving a small spot at the anterior margin and a longer segment across the mid-length. There is a pair of dorsal horns on the arcessus (Fig. 16.55) 32

– The pronotal stripe is continuous, although it may be constricted near the middle. There are no dorsal horns on the arcessus (Fig. 16.58) 34

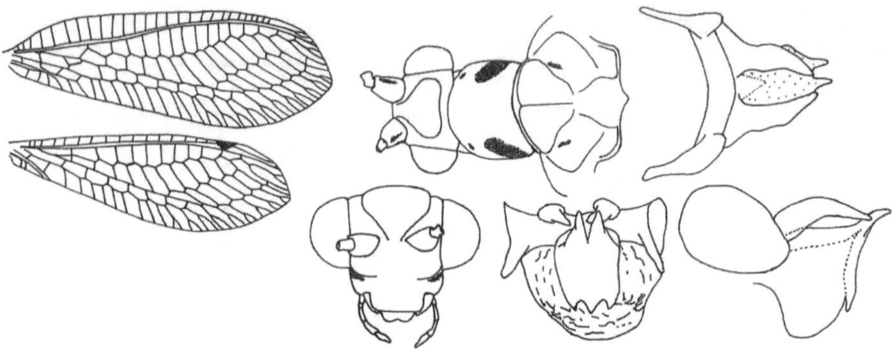

Fig. 16.59 *Leucochrysa cornuta* male: fore and hind wing (left), head and pronotum (upper center), head in anterior view (lower left center), and male genitalia in dorsal (upper right), lateral (lower right), and posterior view (lower right center). Based on de Freitas and Penny (2001).

32. (31) The vertex is green with a pair of red longitudinal markings on the dorsal surface, and the scape has one such marking on the dorsal surface, which may fade in preservatives. The dorsal horn of the male arcessus, in this species the mediuncus, is bifurcate, and its two prongs are separated by a U-shaped emargination. The lateral surface of the gonarcal apodemes is flat but expanded laterally, and it appears sculptured. The spermatheca of the female is small and has only one curve. The spermathecal duct is only twice as long as the diameter of the spermatheca. The maxillary palps are dark toward the apices. The dark spot on the mesoscutum is plainly evident. The dorsal horns on the arcessus point dorsad (Fig. 16.55). Fore-wing length: c. 14.6 mm. .... *Leucochrysa clepsydra* Banks, 1918 (Colombia, São Paulo). Syn: *Nodita clepsydra* Banks, 1918.

    – The vertex is uniformly green, and the scape either has a small dorsolateral marking, or it is uniform in color. The dorsal horn of the male arcessus is bifurcate, and its two prongs are separated by a V-shaped incision. The lateral surface of the gonarcal apodemes is concave where it connects with the gonarcus (Fig. 16.59). The females already described do not conform to the description in the first part of this couplet..33

33. (32) The maxillary palps are uniformly pale. The dark spot on the mesoscutum can barely be seen. The prothorax is approximately as long as the vertex. The dorsal margins of the male apodeme appear to flare slightly in dorsal view. Dorsal horns on the arcessus arise from the posterior part of the gonarcal bridge and point posteriad (Fig. 16.59). Fore-wing length of male: c. 15.4 mm; hind wing length of male: c. 13.0 mm. The female has not yet been described. ....*Leucochrysa cornuta* Freitas and Penny, 2001 (São Paulo).

    – The maxillary palps are dark at their apices. Dark brownish red spots are present on the mesoscutum. The prothorax is considerably longer than the

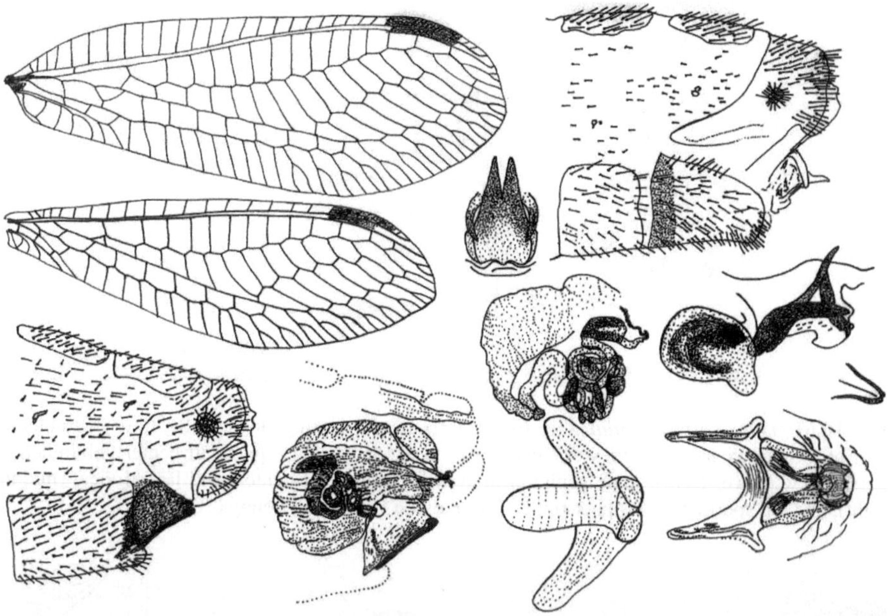

Fig. 16.60 *Leucochrysa digitiformis:* fore and hind wing (upper and middle left), apex of the male abdomen in lateral view (upper right), male gonarcal complex in lateral (upper middle right) and ventral view (lower right), internal hypandrium (lower middle right), frontal view of the male mediuncus (center), apex of the female abdomen in lateral view (lower left) and the apex cleared showing the internal female genitalia (lower left center), spermothecal complex in lateral view (below right center), and subgenitale in ventral view (lower right center). Based on Tauber et al. (2012a, b).

vertex. The dorsal margins of the male apodeme appear parallel for their entire lengths. The horns on the arcessus arise from the ventral side of the gonarcal bridge (Fig. 16.60). The ventral process on the on the exterior surface of the female subgenitale is oblong, elongate, and flat. The coiled bursal duct of the female is very long. The spermathecal duct is more than twice as long as the diameter of the spermatheca and thick at the base. Fore-wing length of male: 14.4–16.5 mm; hind wing length of male: 12.5–13.2 mm. ....*Leucochrysa digitiformis* Tauber and Albuquerque, 2008 (Rio de Janeiro, Rio Grande do Sul).

34. (31) A pair of prominent dark spots is present on the mesonotum (Fig. 16.58). Fore-wing length: c. 14.0 mm; hind wing length: c. 11.9 mm. ....*Leucochrysa parallela* Freitas and Penny, 2001 (São Paulo).

– The mesonotum is uniformly pale (Fig. 16.61) 35

35. (34) The maxillary palps are dark, and the genae are entirely red (Fig. 16.61). Fore-wing length: 12.8–15.8 mm; hind wing length: c. 12.1 mm. ....*Leucochrysa santini* Freitas and Penny, 2001 (São Paulo, Minas Gerais).

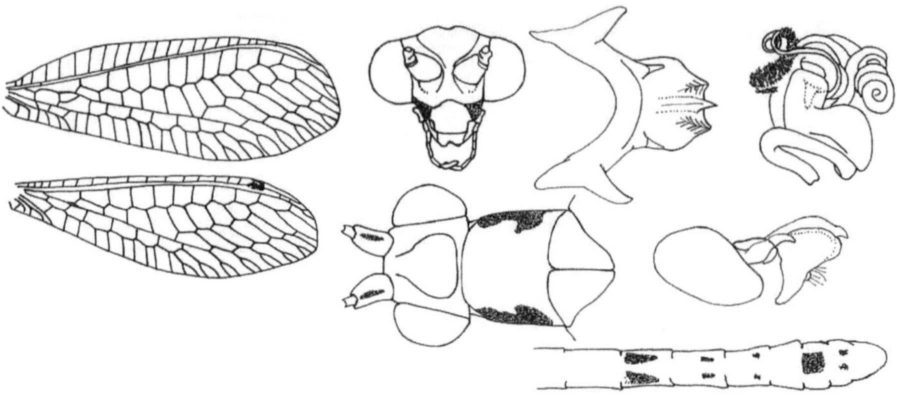

Fig. 16.61 *Leucochrysa santini:* fore (upper left) and hind wing (lower left), head and anterior part of the thorax in dorsal view (center), head in anterior view (upper left center), abdomen in dorsal view (lower right), male genitalia in dorsal (upper right center) and lateral view (middle right), and the spermatheca (upper right). Based on de Fretas and Penny (2001).

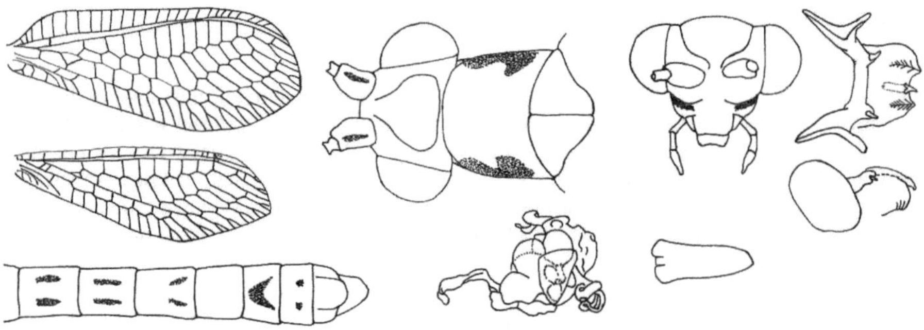

Fig. 16.62 *Leucochrysa robusta:* fore and hind wing (upper and middle left); head and prothorax in dorsal view (upper left center); head in anterior view (upper right center); abdomen in dorsal view (lower left); male genitalia in dorsal (upper right) and lateral view (lower right); female genitalia (lower center); subgenitale (lower right center). Based on de Freitas and Penny (2001).

– The maxillary palps are pale, and the red coloration on the genae is limited to the parts near the frons (Fig. 16.62). Fore-wing length: 13.3–15.1 mm; hind wing length: 11.7–12.5 mm.
.... *Leucochrysa robusta* Freitas and Penny, 2001 (São Paulo, Mato Grosso).

36. (12) The scape of the antenna, i.e., the basal segment, has two dark stripes (Fig. 16.41)... 37

– The dorsal surface of the scape is entirely dark (Fig. 16.63)...................... 40

37. (36) The pronotum has two pairs of stripes (Fig. 16.41) 38

– The pronotum has only one pair of lateral stripes (Fig. 16.64)...................... 39

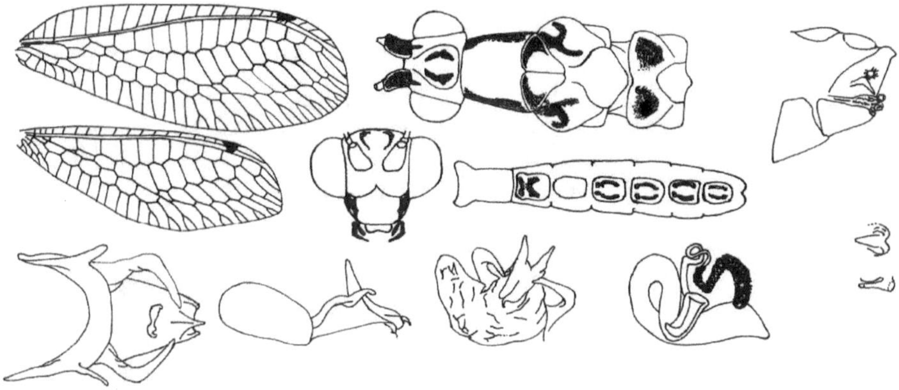

Fig. 16.63 *Leucochrysa cruentata:* fore (upper left) and hind wing (middle left); head and the pronotum and mesonotum in dorsal view (upper center); head in anterior view (left of center); abdomen in dorsal view (right of center); male genitalia in dorsal (lower left), lateral (lower left center), and posterior view (lower center); spermatheca (lower right center); subgenitale in ventral and lateral view (lower right, above and below, respectively). Based on de Freitas and Penny (2001).

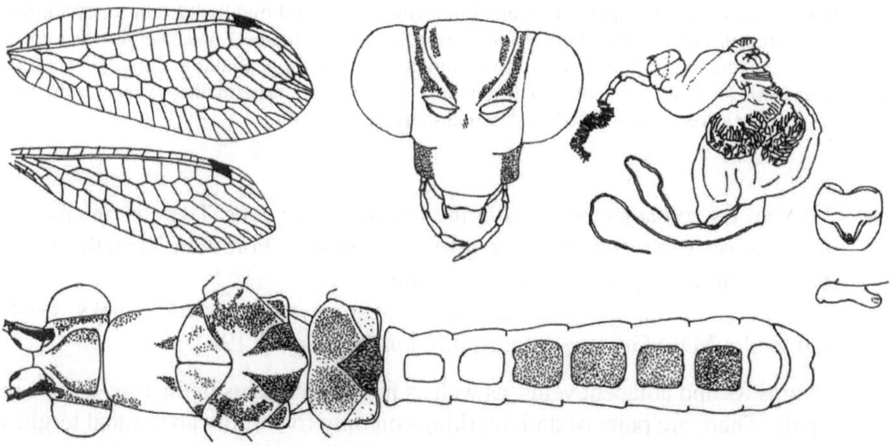

Fig. 16.64 *Leucochrysa scomparini* female: fore and hind wing (upper and middle left, respectively), habitus without appendages in dorsal view (below), head in anterior view (upper left center), genitalia (upper right center), and subgenitale in ventral (upper right) and lateral view (lower right). Based on de Freitas and Penny (2001).

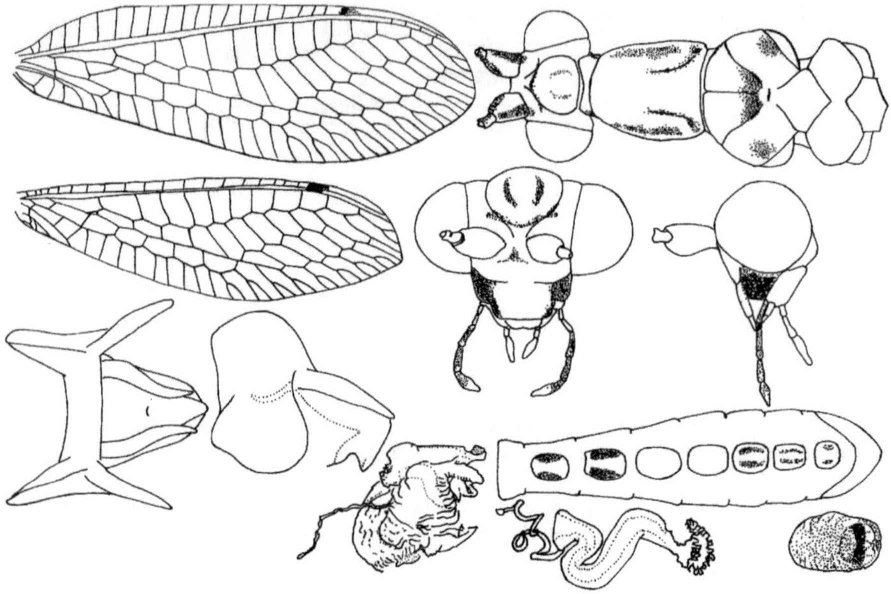

Fig. 16.65 *Leucochrysa marquezi:* fore and hind wing (upper and middle left); head, prothorax, and mesothorax in dorsal view (upper right); head in anterior (center) and lateral view (upper middle right); abdomen in dorsal view (lower middle right); male genitalia in dorsal (lower left) and lateral view (lower left center); female genitalia (lower center); spermatheca (lower right center); subgenitale (lower right). Based on de Freitas and Penny (2001).

38. (37) Vein Rs and adjacent veins in the fore-wing are dark. There is one pair of markings on each abdominal tergite (Fig. 16.41). Fore-wing length: 17.2–18.6 mm; hind wing length: 15.5–17.4 mm.
....*Leucochrysa lancala* (Banks, 1944) (São Paulo, Mato Grosso). Syn: *Nodita lancala* Banks, 1944.

– Veins Rs and adjacent veins, as well as most other veins in the fore-wing, are pale. There are pairs of dark markings on only five of the abdominal tergites (Fig. 16.65). Fore-wing length: 18–21 mm; hind wing length: c. 14.4 mm.
....*Leucochrysa marquezi* Navás, 1913 (São Paulo, Mato Grosso).

39. (37) The hind wing is pale toward the apex. On the vertex of the head there are dark lateral stripes and two crescent shaped marks in the middle. There are two pairs of small spots on the otherwise uniformly colored pronotum (Fig. 16.64). Fore-wing length: c. 12.0 mm; hind wing length: c. 10.8 mm. Only the female has been described.
....*Leucochrysa scomparini* Freitas and Penny, 2001 (Mato Grosso).

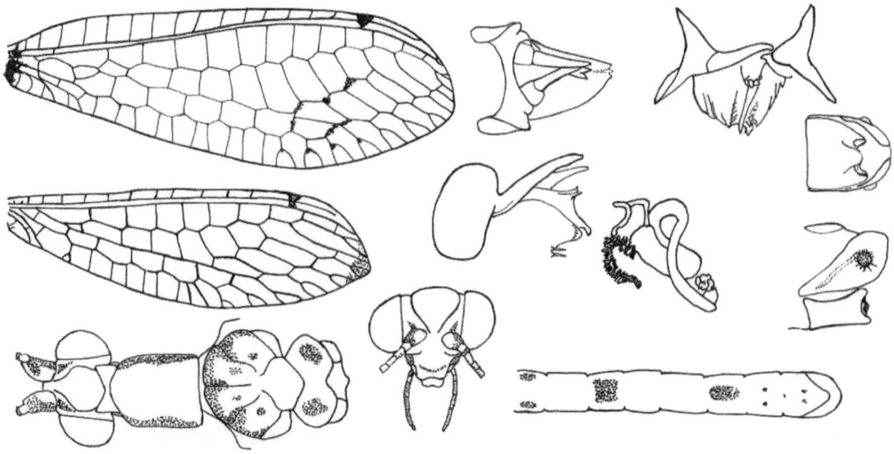

Fig. 16.66 *Leucochrysa heriocles:* fore (upper left) and hind wing (middle left), head and thorax in dorsal view (lower left), head in anterior view (lower center), abdomen in dorsal view (lower right), male genitalia in dorsal (upper center), lateral (center), and posterior view (upper right); spermatheca (right center), and apex of the male abdomen in ventral (upper middle right) and lateral view (lower middle right). Based on de Freitas and Penny (2001).

– The hind wing is dark toward the apex. On the vertex, the only markings are continuations of the stripes on the scapes. There pronotum has two complete lateral stripes (Fig. 16.66). Fore-wing length: c. 11.8 mm.

....*Leucochrysa heriocles* Banks, 1944 (Surinam, Mato Grosso).

40. (36) The gonocornua are elongate and spine-like (Fig. 16.63)......................41

– The gonocornua form short, flat plates (Fig. 16.67)...................................45

41. (40) The pterostigma in the fore-wing is only slightly darkened. The gonocornua are twisted (Fig. 16.4). Fore-wing length: c. 14.7 mm; hind wing length: c. 11.9 mm. Only the male has been described.

....*Leucochrysa incognita* Freitas and Penny, 2001 (Mato Grosso).

– The pterostigma in the fore-wing has a prominent dark marking. The gonocornua are straight and slender (Fig. 16.63) 42

42. (41) In the fore-wing, the base of vein R_s is darkened (Fig. 16.63)...............43

– In the fore-wing, the base of vein R_s is pale (Fg. 16.68) 44

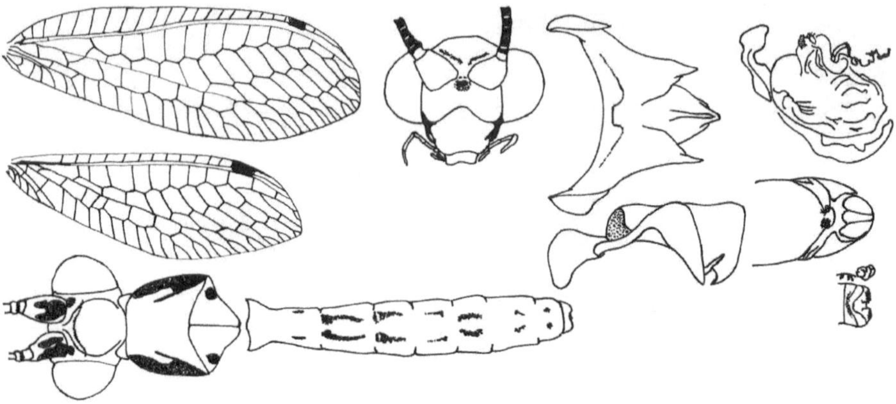

Fig. 16.67 *Leucochrysa guataparensis:* fore and hind wing (upper and middle left); head, prothorax, and mesothorax in dorsal view (lower left); head in anterior view (upper left center); abdomen in dorsal view (lower center); male genitalia in dorsal (upper right center) and lateral view (lower right center); female genitalia (upper right); subgenitale (lower right). Based on de Freitas and Penny (2001).

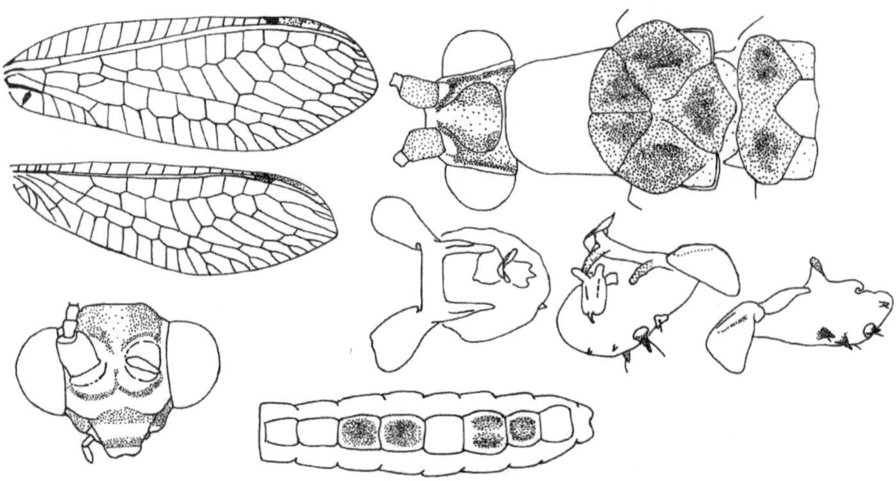

Fig. 16.68 *Leucochrysa squamisetosa* male: fore and hind wing (upper and middle left); head in anterior view (lower left); head, prothorax, and mesothorax in dorsal view (upper center to right); abdomen in dorsal view (lower center); and the male genitalia in dorsal (center), dorsoposterior (right center), and lateral view (lower right). Based on de Freitas and Penny (2001).

43. (42) The antennal flagellum is dark. The anterior part of the gonarcus forms a high arch. The gonocornua are forcipate. The apices of the dorsal horns on the arcessus are pointed (Figs. 2.22 and 16.63). Fore-wing length: c. 25.0 mm; hind wing length: c. 21.3 mm.
....*Leucochrysa cruentata* (Schneider, 1951) (Argentina, São Paulo).

– The antennal flagellum is pale. The anterior part of the gonarcus forms a weak arch. The gonocornua are straight. The apices of the dorsal horns on the arcessus are rounded (Fig. 16.69). Fore-wing length: c. 14.7 mm; hind wing length: c. 12.7 mm.
....*Leucochrysa gossei* Kimmins, 1940 (São Paulo, Mato Grosso).

44. (42) On the vertex of the head, there is a Y-shaped dark marking posterior to each antenna base. Cross veins a_2 and a_3 do not have darkened borders. The base of the fore-wing is darkened (Fig. 16.70).
....*Leucochrysa lateralis* Navás, 1913 (Mato Grosso).

– On the vertex of the head, there are dark lateral stripes beside the eyes and and two crescent-shaped marks on the medial surface. Cross veins a_2 and a_3 have darkened borders. The base of the fore-wing is pale (Fig. 16.68). Fore-wing length: c. 12.0 mm; hind wing length: c. 10.8 mm.
....*Leucochrysa squamisetosa* Freitas and Penny, 2001 (São Paulo).

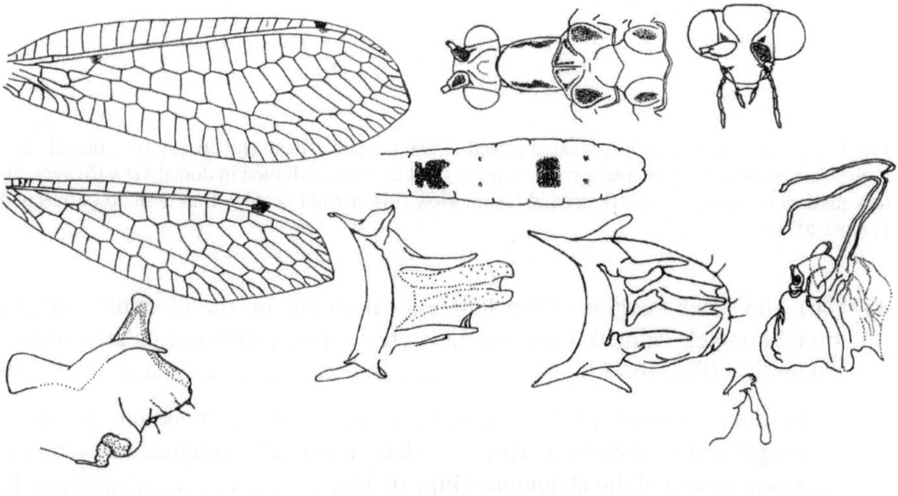

Fig. 16.69 *Leucochrysa gossei:* fore and hind wing (upper and middle left); head, pronotum, and mesonotum in dorsal view (upper right center); head in anterior view (upper right); abdomen in dorsal view (center); male genitalia in dorsal (lower center), posterodorsal (right center), and lateral view (lower left); spermatheca (middle right); subgenitale (lower right). Based on de Freitas and Penny (2001).

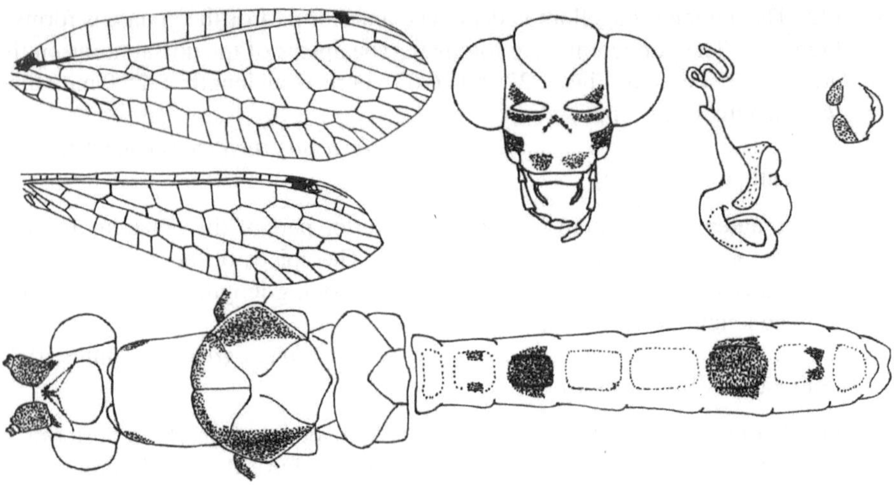

Fig. 16.70 *Leucochrysa lateralis* female: fore (upper left) and hind wing (middle left), head in anterior view (upper left center), habitus without appendages in dorsal view (below), spermatheca (upper right center), and subgenitale (upper right). Based on de Freitas and Penny (2001).

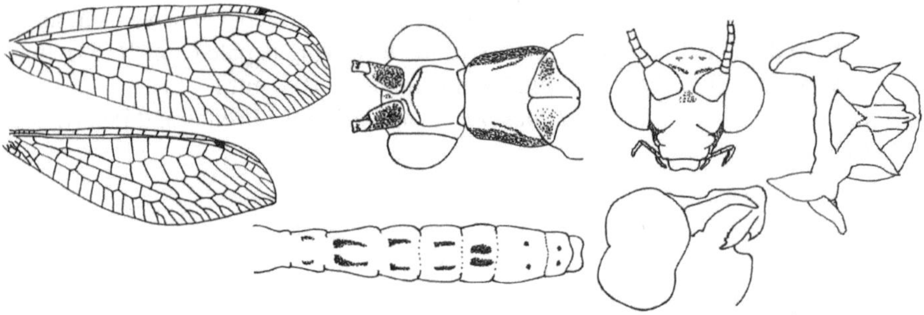

Fig. 16.71 *Leucochrysa affinis* male: fore and hind wing (left), head and pronotum in dorsal view (upper left center), head in anterior view (upper right center), abdomen in dorsal view (lower center), male genitalia in dorsal (right) and lateral view (lower right center). Based on de Freitas and Penny (2001).

45. (40) The inner margins of the longitudinal stripes on the pronotum are not straight or well-defined. There are dark, crescent-shaped marks on the abdominal tergites (Fig. 16.71) 46

– The inner margins of the longitudinal stripes on the pronotum are fairly straight and well-defined. There is a dark, medial, longitudinal stripe on the dorsal surface of the abdominal (Fig. 16.72) 47

46. (45) The gonocornua are directed apicolaterad and are widest at the mid-length (Fig. 16.67). Fore-wing length: 15.8–17.8 mm; hind wing length: c. 14.7 mm. .... *Leucochrysa guataparensis* Freitas and Penny, 2001 (São Paulo, Mato Grosso).

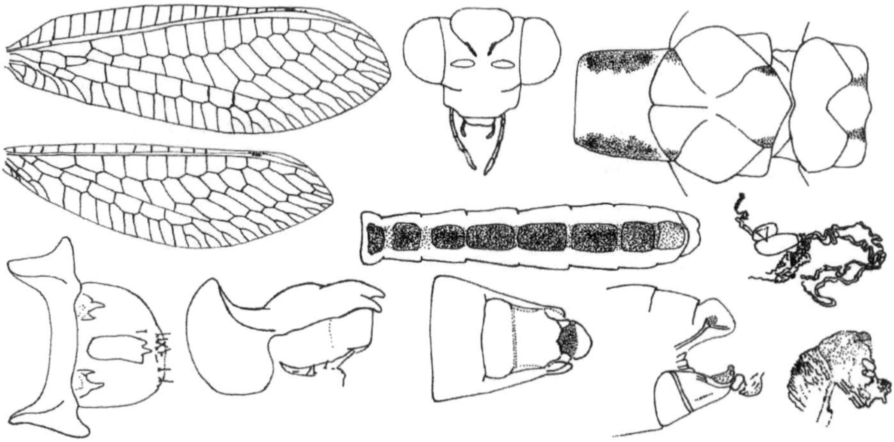

Fig. 16.72 *Leucochrysa furcata:* fore and hind wing (upper and middle left); head in anterior view (upper center); pronotum and mesothorax in dorsal view (upper right); abdomen in dorsal view (right center); apex of the male abdomen in ventral (lower center) and lateral view (lower right center); male genitalia in dorsal (lower right) and lateral view (lower right center); female genitalia (middle right); spermatheca (lower right). Based on de Freitas and Penny (2001).

– The gonocornua are directed apicad and are widest at the base (Fig. 16.71). Fore-wing length: c. 17.6 mm; hind wing length: c. 14.7 mm. ....*Leucochrysa affinis* Freitas and Penny, 2001 (São Paulo, Mato Grosso).

47. (45) The apex of the arcessus are much longer than the gonocornua (Fig. 16.72). Fore-wing length: c. 9.3 mm; hind wing length: c. 8.1 mm. ....*Leucochrysa furcata* Freitas and Penny, 2001 (Mato Grosso).

– The elongate gonocornua reach the apex of the arcessus (Fig. 16.73) 48

48. (47) The gonarcus is wide. The pterostigma is very dark (Fig. 16.73) 49

– The gonarcus is narrow. The pterostigma is only weakly pigmented (Fig. 16.74)... .50

49. (48) Posterior to the base of the antenna on the vertex, there is a complete, dark transverse band. The apices of the gonocornua are bifurcate (Fig. 16.73). Fore-wing length: c. 18.8 mm; hind wing length: c. 16.5 mm. ....*Leucochrysa forciformis* Freitas and Penny, 2001 (São Paulo, Mato Grosso).

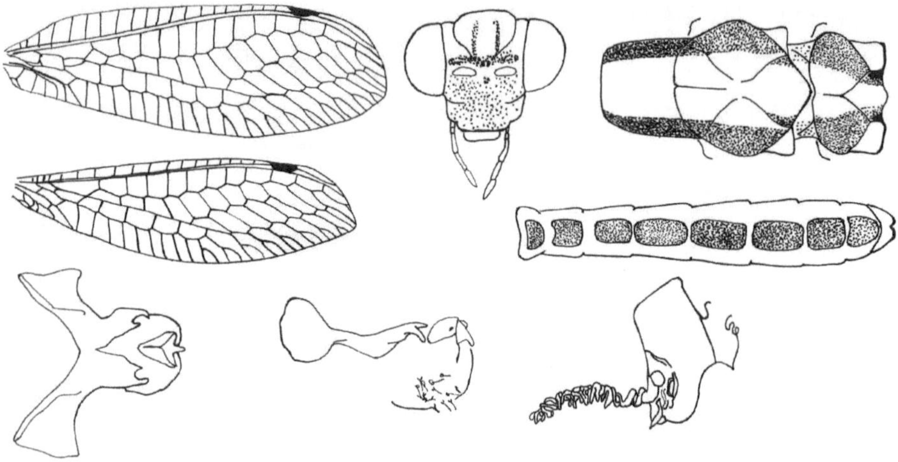

Fig. 16.73 *Leucochrysa forciformis:* fore (upper left) and hind wing (middle left), head in anterior view (upper center), thorax in dorsal view (upper right), abdomen in dorsal view (middle right), and male genitalia in dorsal (lower left) and lateral view (lower center), female genitalia (lower right). Based on de Freitas and Penny (2001).

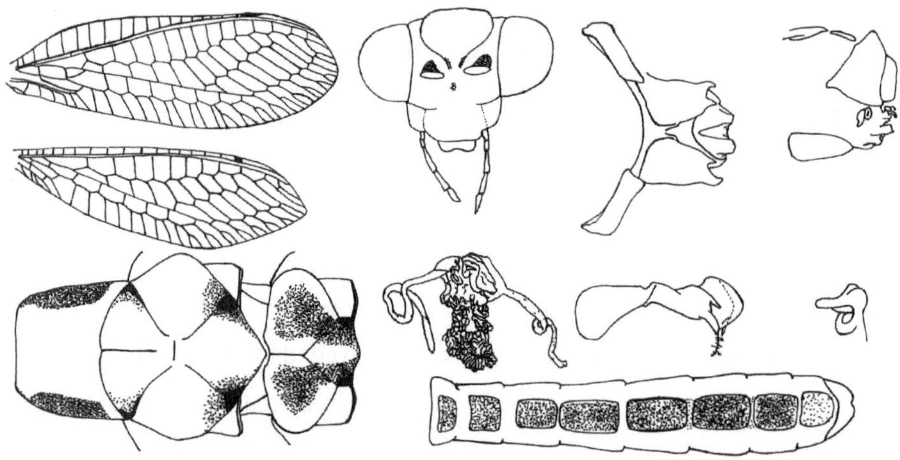

Fig. 16.74 *Leucochrysa tabacinus:* fore (upper left) and hind wing (middle left), head in anterior view (upper left center), thorax in dorsal view (lower left), abdomen in dorsal view (lower right), and male genitalia in dorsal (upper right center) and lateral view (right center), apex of the female abdomen in lateral view (upper right), female genitalia (left center), and subgenitale (middle right). Based on de Freitas and Penny (2001).

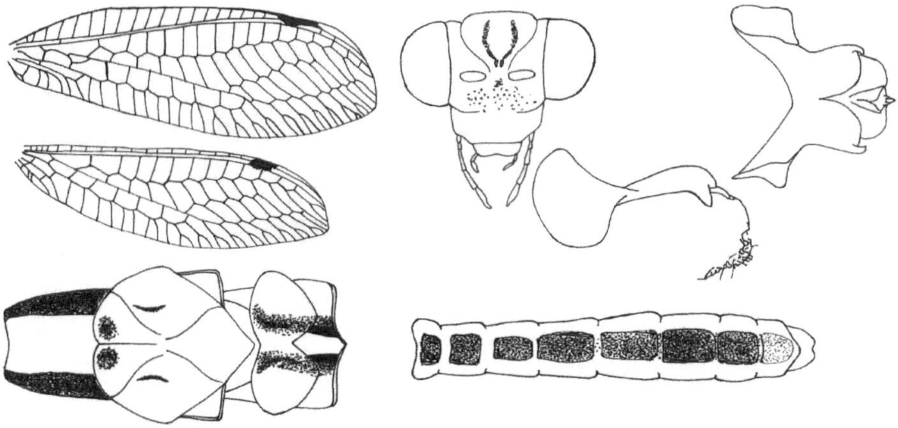

Fig. 16.75 *Leucochrysa michelini* male: fore (upper left) and hind wing (middle left), head in anterior view (upper center), thorax in dorsal view (lower left), abdomen in dorsal view (lower center to right), and genitalia in dorsal (upper right) and lateral view (right center). Based on de Freitas and Penny (2001).

– There is no transverse band on the vertex posterior to the antenna base. The apices of the gonocornua are not bifurcate (Fig. 16.75). Fore-wing length: c. 16.8 mm; hind wing length: c. 14.4 mm.

....*Leucochrysa michelini* Freitas and Penny, 2001 (São Paulo).

50. (48) Vein Rs in the fore-wing is pale. There is a dark marking on the dorsal part of the cavity for the antennal insertion. The apical intersections of the costal cross veins are pale. The apices of the gonocornua are bifurcate (Fig. 16.74). Fore-wing length: c. 30.6 mm; hind wing length: c. 26.6 mm.

....*Leucochrysa tabacinus* Freitas and Penny, 2001 (Mato Grosso).

– Vein Rs in the fore-wing and nears near it are dark. The dorsal part of the cavity for the antennal insertion has no dark markings. The apical intersections of the costal cross veins are darkened. The apices of the gonocornua are not bifurcate (Fig. 16.76). Fore-wing length: c. 15.8 mm; hind wing length: c. 14.1 mm. Only the male has been described.

....*Leucochrysa lineata* Freitas and Penny, 2001 (São Paulo).

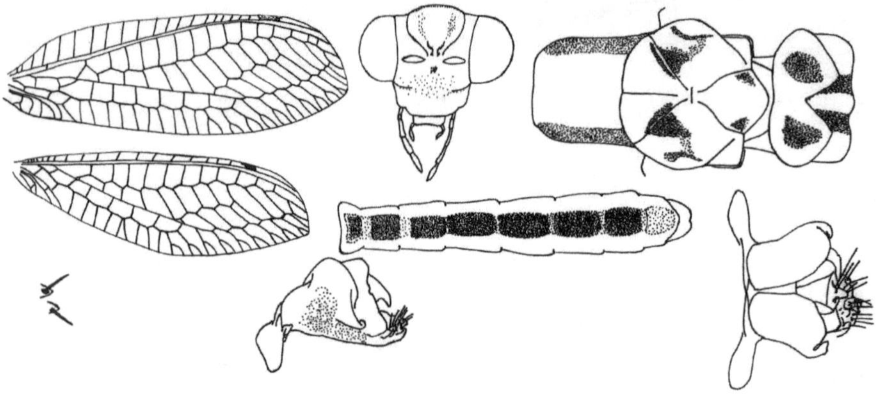

Fig. 16.76 *Leucochrysa lineata* male: fore and hind wing (upper and middle left), head in anterior view (upper center), thorax in dorsal view (upper right), abdomen in dorsal view (center), gonosetae (lower left), and genitalia in dorsal (lower right) and lateral view (lower left center). Based on de Freitas and Penny (2001).

Key to the First Instar Larvae of Known South American *Leucochrysa* Species

Information for the key was provided by Mantoanelli et al. (2006, 2011) and Tauber et al. (2011a, b, c)). The larvae of only a small fraction of the species in this genus have been described. The full descriptions appear to be good, but because chaetotaxy, which relies on the presence and location of specifically named or numbered setae has been found unsatisfactory for systematic work on other insect groups in the past, caution is advised. For example, after a complex system of classification to identify the species of mosquito larvae was developed several decades ago, it was quickly found that the development of setae was more influenced by environmental factors than by the genetic complements of the species. All known South American larvae of *Leucochrysa* species seem to paste debris on their body surfaces.

1. The markings on the dorsal surface of the head are paired bands that project anteriad, which do not come close to the eyes or the Y-shaped frontal marking. The epicranial markings lack a lateral part, or, if a lateral part is present, it does not extend very far posterior to the level of the posterior margin of the eye. The pronotal setae are hooked (Fig. 16.77)..2

   – The markings on the dorsal surface of the head are large, paired posterior patches that reach or nearly reach the eyes and the V-shaped frontal marking.

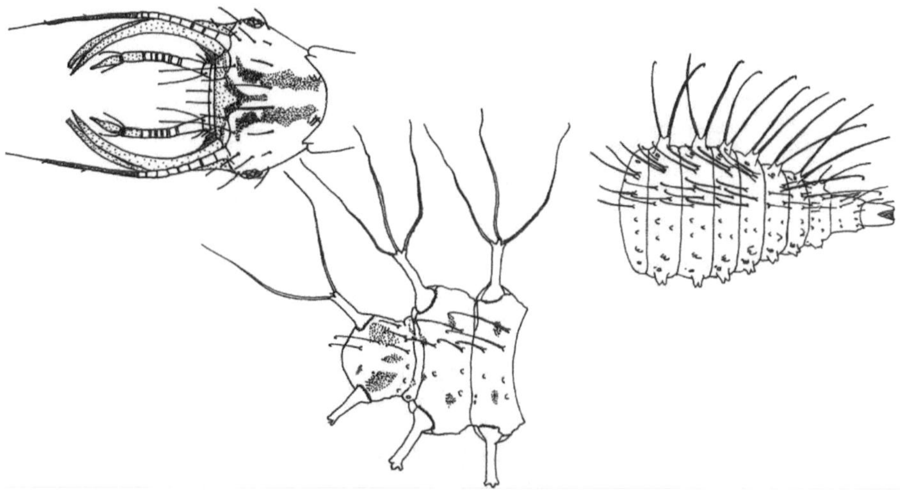

Fig. 16.77 First larval instar of *Leucochrysa rodriguezi:* head in dorsal view (left), thorax in dorsal view (lower center), abdomen in dorsal view (right). Based on Mantoanelli et al. (2011).

The epicranial markings have obvious mesal and lateral parts, both of which contact or almost contact the posterior border of the cranium. The pronotal setae are straight (Fig. 16.78)..3

2. (1) The elongate setae on the lateral tubercles of the thorax are straight and smooth. The thorax and abdomen lack dorsal markings (Fig. 16.77).
.... *Leucochrysa rodriguezi* Navás, 1913
(Bahia, São Paulo, Mato Grosso).

 – The elongate setae on the lateral tubercles of the thorax are not hooked at the apex, but the shortest one on each tubercle is. The thorax and abdomen have brown dorsal markings (Fig. 16.79).
....*Leucochrysa azevedoi* Navás, 1913
(Rio de Janeiro, São Paulo, Mato Grosso, Pará?). Syn: *Leucochrysa camposi* nec Navás, 1933; *Nodita azevedoi* (Navás, 1913).

3. (1) The thoracic spiracles are large, and their openings are located on tubercles elevated above the surface of the integument. Their chambers are bulbous (Fig. 16.78). 4

 – The thoracic spiracles are small, and their openings are not located on tubercles but rather on the same plane as the surface of the integument. Their chambers are narrow (Fig. 16.80). 5

4. (3). The anterior setae inserted on the lateral tubercles of the mesothoracic and metathoracic segments are shorter than the other two and hooked. The submesal segments on the same two thoracic segments are hooked at the apex (Fig. 16.78).
....*Leucochrysa boxi* Navás, 1930

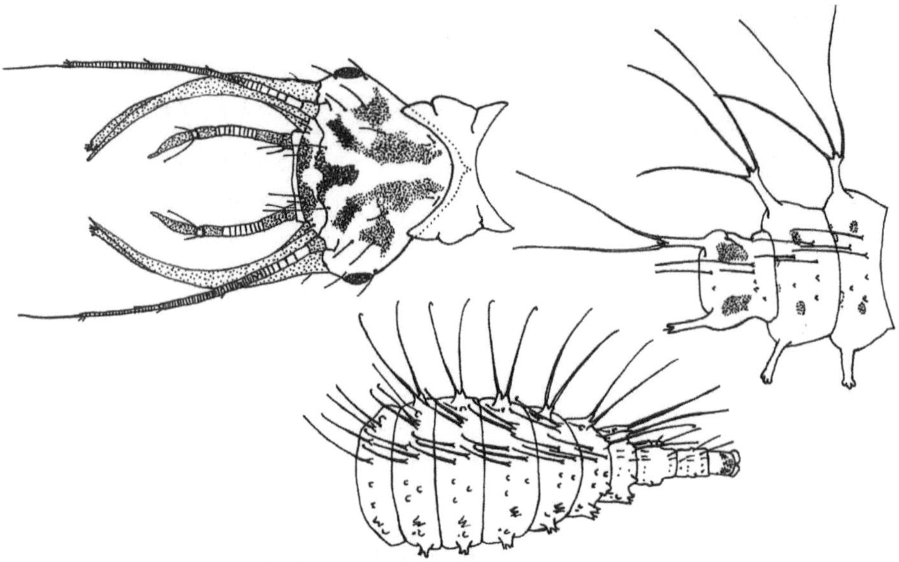

Fig. 16.78 First larval instar of *Leucochrysa boxi:* dorsal views of the head (upper left), thorax (upper right), and abdomen (lower center), showing thoracic and abdominal setae only on the right side. Based on Tauber et al. (2006).

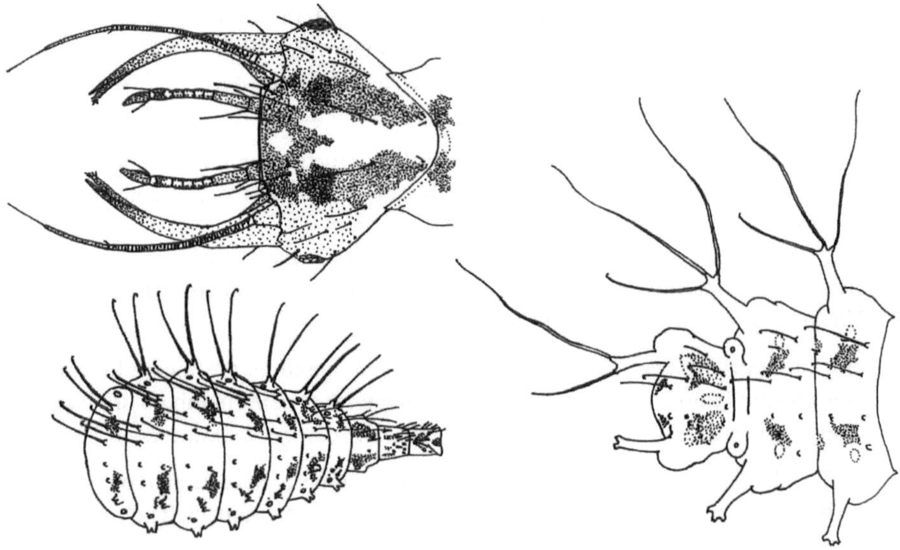

Fig. 16.79 First larval instar of *Leucochrysa azevedoi:* head in dorsal view (upper left), thorax in dorsal view (right), abdomen in dorsal view (lower left). Based on Mantoanelli et al. (2011).

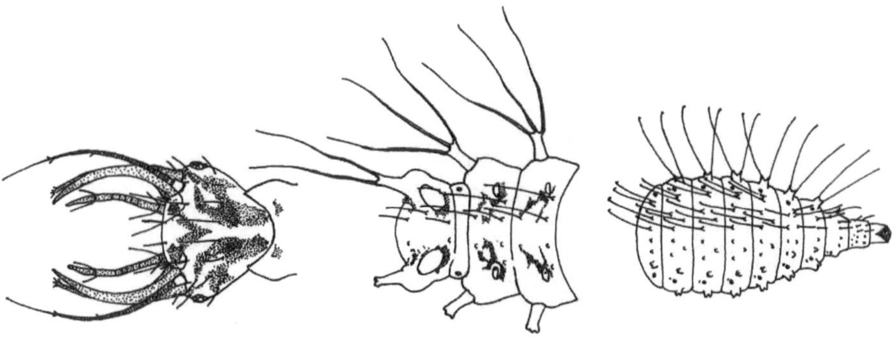

Fig. 16.80 First larval instar of *Leucochrysa marquezi* (left to right): dorsal views of the head, thorax, and abdomen. Based on Mantoanelli et al. (2011).

(Peru, Bolivia, Argentina, Espirito Santo, Santa Catarina, Rio Grande do Sul, Rio de Janeiro). Syn: *Hemerobius longicornis* G. Gray in Cuvier, 1832 (preoccupied); *Hemerobius longicollis* G. Gray in Cuvier, 1832 (error in figure); *Chrysopus longicornis* (Gray in Cuvier, 1832) Schneider, 1851; *Hemerobius longicollis* (for figure in Schneider, 1851); *Allochrysa longicornis* (Gray in Cuvier, 1832) Navás, 1932; *Leucochrysa (Leucochrysa) loretana* Navás, 1935, *Leucochrysa lorenata* Navás, 1935 (misspelling).

– The anterior and submesal setae inserted on the lateral tubercles of the mesothoracic and metathoracic segments are all of the same length, and all are unhooked and straight for their whole lengths. (Fig. 16.81).

....*Leucochrysa varia* (Schneider, 1851) (Mexico, Central America Guyana? Surinam? Bolivia? Paraguay? Peru, Ecuador, Argentina, Bahia, Espirito Santo? Minas Gerais, Rio de Janeiro, São Paulo, Mato Grosso, Rondônia, Pará, Distrito Federal Santa Catarina). Syn: *Chrysopa varia* Schneider, 1851; *Allochrysa varia* (Schneider, 1851); *Leucochrysa ampla* (Walker, 1853); *Chrysopa internata* Walker, 1853; *Leucochrysa internata* (Walker, 1853); *Leucochrysa walkeriana* Navás, 1913; *Leucochrysa walkerina* Navás, 1913, spelling error in de Freitas and Penny (2001) and Tauber et al. (2012a, b); possibly *Leucochrysa vegana* Navás, 1917 *nomen dubium. Leucochrysa phaeocephala* Navás, 1929, was long considered a synonym of *Leucochrysa varia,* but Tauber et al. 2013, restored its validity because it was found in Surinam, and *L. varia* was not previously reported from northern South America. With its sketchy description and without type specimens, *L. phaeocephala* will remain an undefined name on a list for the foreseeable futute and perhaps forever.

5. (3) Posterior to the dorsal sclerites, the pronotum has a scallop-shaped marking. Each leg is white with brown markings at the base of the coxa and on the basal half of the tibia (Fig. 16.80). There are hooked setae on the lateral tubercles of the fourth and fifth abdominal segments.

.... *Leucochrysa marquezi* Navás, 1913 (São Paulo, Mato Grosso).

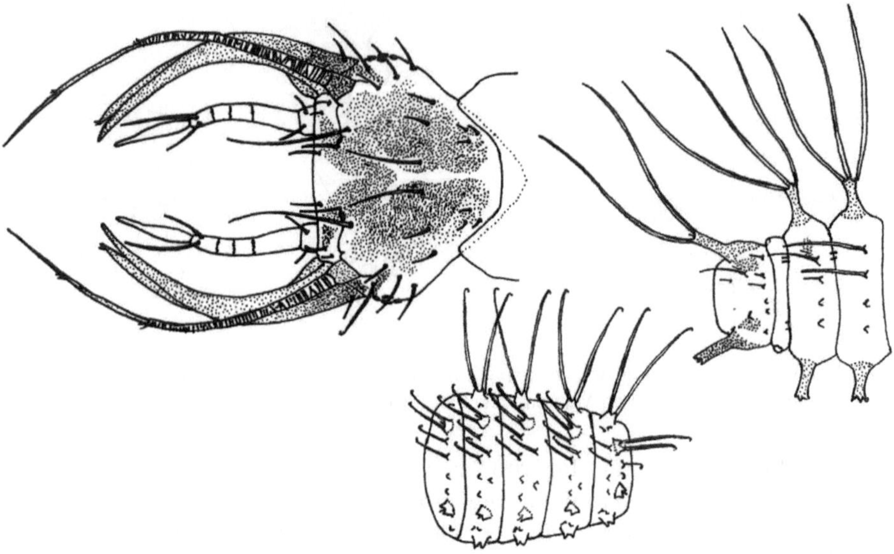

Fig. 16.81 First larval instar of *Leucochrysa varia:* dorsal views of the head (upper left), thorax (upper right), and the five anterior segments of the abdomen (lower center), showing thoracic and abdominal setae only of the right side. Based on Mantoanelli et al. (2006).

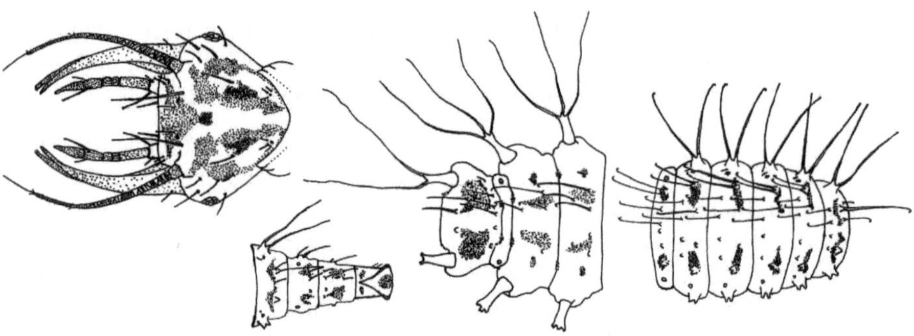

Fig. 16.82 First larval instar of *Leucochrysa digitiformis:* head in dorsal view (left), prothorax in dorsal view (upper center), anterior six segments of the abdomen in dorsal view (right), and posterior four segments of the abdomen in dorsal view (lower left center). The setae are shown only of the right sides of the thorax and abdomen. Based on Mantoanelli et al. (2011).

– The thorax is uniformly white, or it is white with small, light brown markings. Each leg is white with a brown marking at the base of the coxa (Fig. 16.82). All setae on the lateral tubercles of the fourth and fifth abdominal segments are straight.

......................................*Leucochrysa digitiformis* Tauber and Albuquerque, 2008 (Rio de Janeiro, Rio Grande do Sul).

Key to the Second and Third Instar Larvae of Known South American *Leucochrysa* Species

Information for the key was provided by Mantoanelli et al. (2006, 2011) and Tauber et al. (2011a, b, c). For comments, see the key to the first instar larvae, above.

1. The epicranial marking lacks a lateral part, or only a distal section of the lateral part near the eye is present. Only the mesal part of the epicranial marking come in contact with the posterior border of the head (Fig. 16.83)............................ 2

   – The epicranial marking has a mesal and a lateral part, both of which come into contact with the posterior margin of the head (Fig. 16.84) 3

2. (1) There is a dark brown marking adjacent to the eyes on the dorsal side of the head. The setae on the lateral tubercles of the thorax are serrated, with some straight and some hooked (Fig. 16.83). The second through sixth abdominal segments bear more than one seta beside some of the spiracles.
   ....*Leucochrysa azevedoi* Navás, 1913
   (Rio de Janeiro, São Paulo, Mato Grosso, Pará?). Syn: *Leucochrysa camposi* nec Navás, 1933; *Nodita azevedoi* (Navás, 1913).

   – There is no dark brown marking adjacent to the eyes on the dorsal side of the head. All setae on the lateral tubercles of the thorax are smooth and hooked

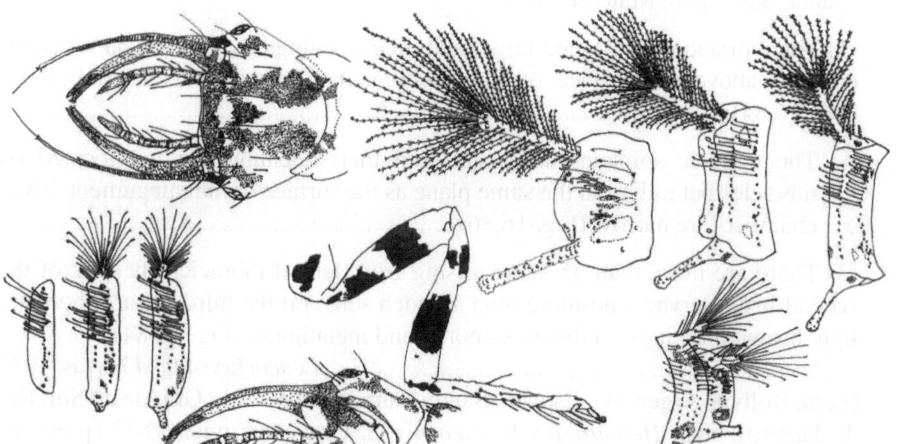

Fig. 16.83 Third larval instar of *Leucochrysa azevedoi:* head in dorsal view (upper left) and the left side of the head in ventral view (lower left center); the prothorax, mesothorax, and metathorax in dorsal view (upper center to right, left to right); the first, second, and third abdominal segments in dorsal view (lower left, left to right); hind leg (lower center); sixth and seventh abdominal segments (lower right). Based on Mantoanelli et al. (2011).

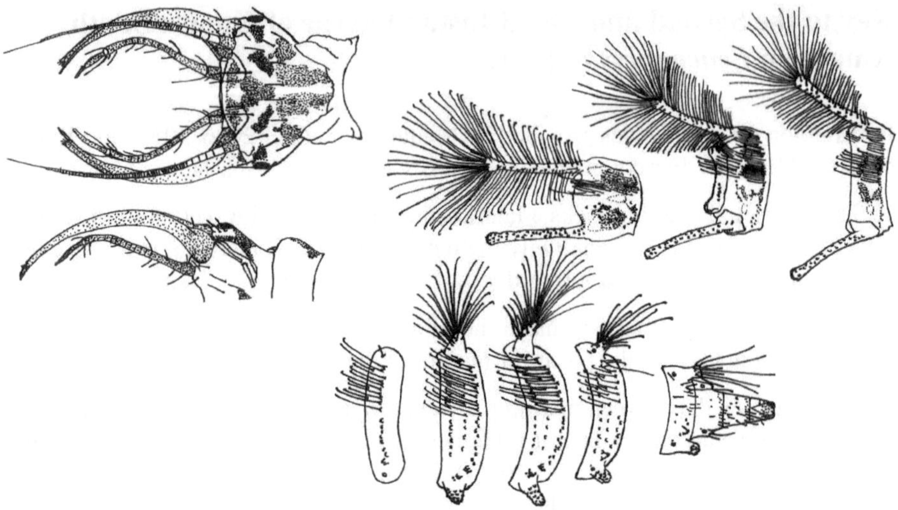

Fig. 16.84 Third larval instar of *Leucochrysa boxi:* head in dorsal view (upper left) and the left side of the head in ventral view (lower left); the prothorax, mesothorax, and metathorax in dorsal view (upper center to right, left to right); the first, second, third, sixth, seventh, and tenth abdominal segment in dorsal view. Based on Mantoanelli et al. (2011).

(Fig. 16.85). The second through sixth abdominal segments bear no more than one seta beside each the spiracle.
....*Leucochrysa rodriguezi* Navás, 1913 (Bahia, São Paulo, Mato Grosso).

3. (1) The thoracic spiracles are large, and their openings are located on tubercles elevated above the surface of the integument. Their chambers are bulbous (Fig. 16.84).. .4

   – The thoracic spiracles are small, and their openings are not located on tubercles but rather on the same plane as the surface of the integument. Their chambers are narrow (Fig. 16.86)...5

4. (3) There are more than 35 setae arising from lateral thoracic tubercles of the second instar larvae, and more than 70 such setae on the third instar. There are light brown markings on the mesothorax and metathorax (Fig. 16.84).
....*Leucochrysa boxi* Navás, 1930 (Peru, Bolivia, Argentina, Espirito Santo, Santa Catarina, Rio Grande do Sul, Rio de Janeiro). Syn: *Hemerobius longicornis* G. Gray in Cuvier, 1832 (preoccupied); *Hemerobius longicollis* G. Gray in Cuvier, 1832 (error in figure); *Chrysopus longicornis* (Gray in Cuvier, 1832) Schneider, 1851; *Hemerobius longicollis* (for figure in Schneider, 1851); *Allochrysa longicornis* (Gray in Cuvier, 1832) Navás, 1932; *Leucochrysa (Leucochrysa) loretana* Navás, 1935, *Leucochrysa lorenata* Navás, 1935 (misspelling).

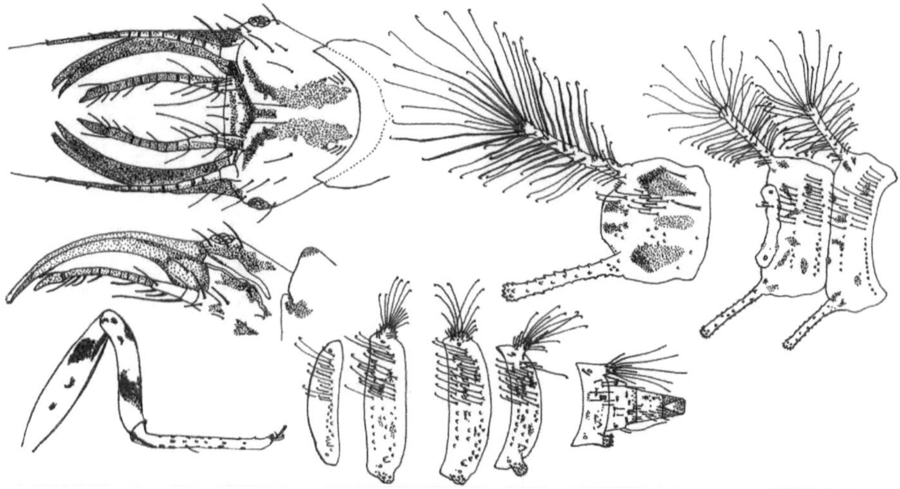

Fig. 16.85 Third larval instar of *Leucochrysa rodriguezi:* head in dorsal view (upper left) and the left side of the head in ventral view (middle left), dorsal views of prothorax (upper center) and the mesothorax and metathorax (right), first through fourth and seventh through tenth abdominal segments in dorsal view (lower center, left to right), and hind leg (lower left). Setae are shown only on the right sides of the segments. Based on Mantoanelli et al. (2011).

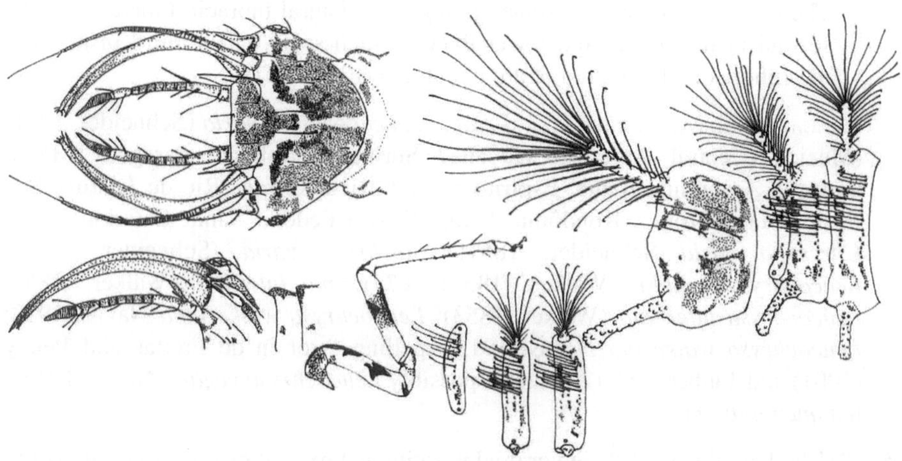

Fig. 16.86 Third larval instar of *Leucochrysa marquezi:* dorsal view of the head (upper left) and left half of the head in ventral view (lower left), pronotum (right center) and mesonotum and metanotum in dorsal view (right), first three segments of the abdomen in dorsal view (lower right center, left to right), and hind leg (lower left center). Based on Mantoanelli et al. (2011).

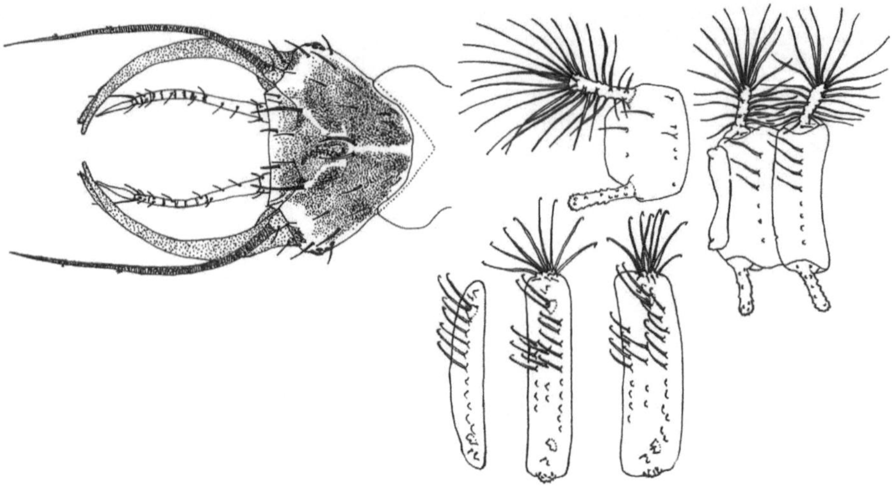

Fig. 16.87 Third larval instar of *Leucochrysa varia:* dorsal views of the head (upper left), pronotum in dorsal view (upper right center), mesonotum and metanotum in dorsal view (upper right), and anterior three abdominal segments in dorsal view (lower center, left to right). The thoracic and abdominal setae are shown only of the right side. Based on Mantoanelli et al. (2006).

– There are fewer than 25 setae arising from lateral thoracic tubercles of the second instar larvae, and fewer than 30 such setae on the third instar. The mesothorax and metathorax are unmarked (Fig. 16.87).

....*Leucochrysa varia* (Schneider, 1851) (Mexico, Central America, Guyana? Surinam? Bolivia? Paraguay? Peru, Ecuador, Argentina, Bahia, Espirito Santo? Minas Gerais, Rio de Janeiro, São Paulo, Mato Grosso, Rondônia, Pará, Distrito Federal Santa Catarina). Syn: *Chrysopa varia* Schneider, 1851; *Allochrysa varia* (Schneider, 1851); *Leucochrysa ampla* (Walker, 1853); *Chrysopa internata* Walker, 1853; *Leucochrysa internata* (Walker, 1853); *Leucochrysa walkeriana* Navás, 1913; *Leucochrysa walkerina* Navás, 1913, spelling error in de Freitas and Penny (2001) and Tauber et al. (2012a, b); possibly *Leucochrysa vegana* Navás, 1917a, b *nomen dubium.*

5. (3) The lateral part of the epicranial marking is broad; it extends to or almost to the eyes (Fig. 16.86). The dorsal setae on the prothorax are smooth and straight, and they are acutely pointed at their apices. The dorsal surfaces of the mesothorax and metathorax have dark brown markings.
....*Leucochrysa marquezi* Navás, 1913 (São Paulo, Mato Grosso).

– The lateral part of the epicranial marking is narrow; it extends to the postfrontal marking but not toward the eyes. The prothorax bears hooked setae on the dorsal surface, with a patch of about six or seven secondary setae on each side

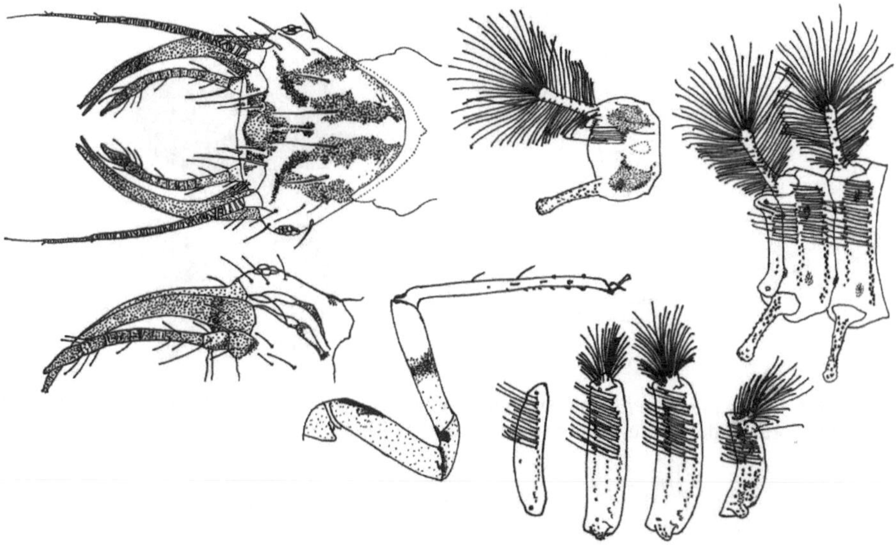

Fig. 16.88 *Leucochrysa digitiformis* third larval instar: head in dorsal view (upper left) and one side of the head in ventral view (lower left), pronotum (upper center) and mesonotum and metanotum in dorsal view without setae on the left side (upper right), hind leg (lower center), and the first, second, third, and sixth abdominal segments in dorsal view without the setae on the left side (lower right, left to right, respectively). Based on Mantoanelli et al. (2006).

of the midline mesad from the prothoracic sclerite, which is not heavily marked along its lower or middle margin. The mesothorax and metathorax are unmarked (Fig. 16.88).
....................................*Leucochrysa digitiformis* Tauber and Albuquerque, 2008 (Rio de Janeiro, Rio Grande do Sul).

Key to the Species of *Gonzaga* in South America

Information for the key was provided by Banks (1944) and Penny (2002).

1. Neither wing has dark markings, except in the pterostigmal area (Fig. 16.89). .... *Gonzaga palliceps* (McLachlan, 1867) (Central America, Guyana, Brazil). Syn: *Leucochrysa palliceps* (McLachlan, 1867).
   – The wings have other dark markings besides those in the pterostigmal area (Fig. 16.90)... 2
2. (1) The hind wing has a large dark spot near the center, and its pterostigma is yellowish. On the fore-wing, there is a small dark spot near the stigma, a larger one distal to the middle of the radial sector, a broad stripe curving from the distal

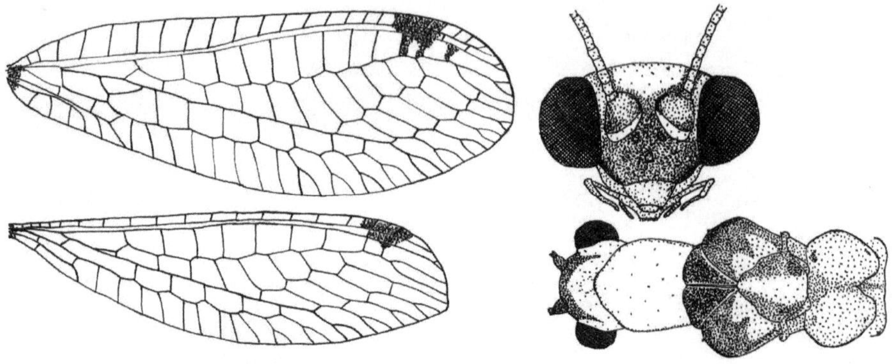

Fig. 16.89 *Gonzaga palliceps:* fore and hind wing (left), head in anterior view (upper right), and head and thorax in dorsal view (lower right). Based on Penny (2002).

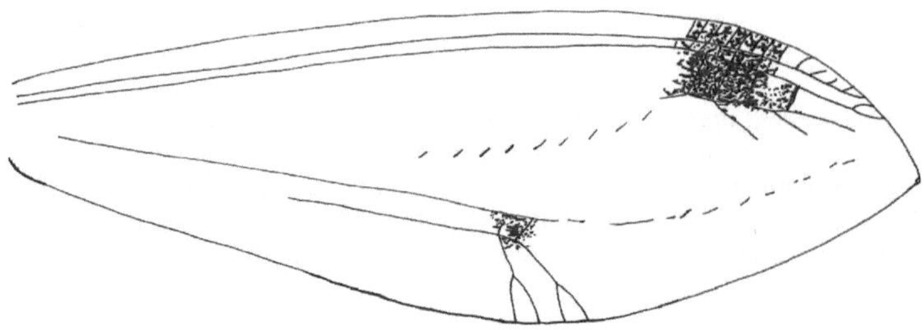

Fig. 16.90 Fore-wing of *Gonzaga notatus* showing only a few veins and the locations of the dark markings. Based on Navás (1929b).

cubital cross vein toward the anterior part of the wing, and several other small dark spots. There are about six cubital cross veins distal to the divisory cell. Length of fore-wing: c. 14 mm.
.... *Gonzaga calliptera* Banks 1944 (Guyana, Amazonas).

– There is no dark spot near the middle of the hind wing, and/or its pterostigma is not yellowish. Distal to the divisory cell, there are about eight to ten cubital cross veins. There is one large spot extanding distad from the stigma on the fore-wing to the radial sector (Fig. 16.12)...3

3. (2) The pronotum is pale in color. The elongate marking distal to the cubital cross veins does no extend far or approach the apex of the wing. There are 9–11 cubital cross veins; the two or three nearest the apex of the wing are usually close together (Fig. 16.12).
....*Gonzaga torquatus* Navás 1913 (Central America, Venezuela, Argentina).

– The pronotum and mesonotum are black or dark brown. The fore-wing has a large dark marking posterior to the pterostigma (Fig. 16.90).............................4

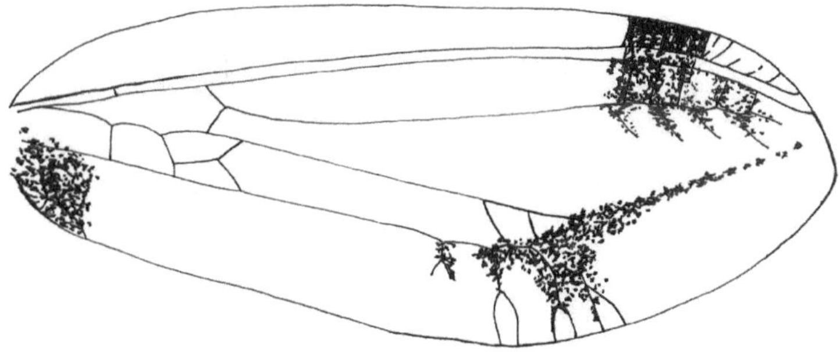

Fig. 16.91 Fore-wing of *Gonzaga palliatus* showing only a few veins and the locations of the dark markings. Based on Navás (1929b).

4. (3) The costal space, that is, the space between veins C and Sc, remains almost the same width from near the base to near the apex. There is a dark marking posterior to the pterostigma of the fore-wing but no blackish or dark brown marking at the posterior angle at the base of the wing (Fig. 16.90). A small, dark ferruginous mark may be present, however. Fore-wing length: c. 19.4 mm; hind wing: c. 17.5 mm.
   .... *Gonzaga notatus* Navás 1929
   (Peru).

   – The costal space is clearly widened a short distance from the wing base but then narrows considerably toward the pterostigma. There are large, dark markings posterior to the pterostigma, another in the middle of the apical half of the wing, and a third another at the posterior angle at the base of the wing (Fig. 16.91).. 5

5. (4) Almost the entire dorsal surface from the head to abdomen is black, although the pronotum may appear slightly lighter (Fig. 16.91). Length of fore-wing: c. 21 mm; hind wing: c. 18.5 mm.
   .... *Gonzaga palliatus* Navás 1929
   (Espirito Santo).

   – The dorsal surface is marked by contrasting colors (Fig. 16.92)...................... 6

6. (5) The first three abdominal segments are pale yellow dorsally, while the fourth through ninth are dark (Fig. 16.92). The pronotum is yellowish, while the rest of the thorax is mainly dark brown dorsally. Length of fore-wing: c. 18.4 mm; hind wing: 17.4 mm.
   .... *Gonzaga amabilis* Navás 1932
   (Ecuador).

   – The first four abdominal segments are mainly whitish dorsally, and the apical segments are cream. The fifth through eighth segments are dark brownish, and there is a brown mark in the middle of the whitish third tergite. The elon-

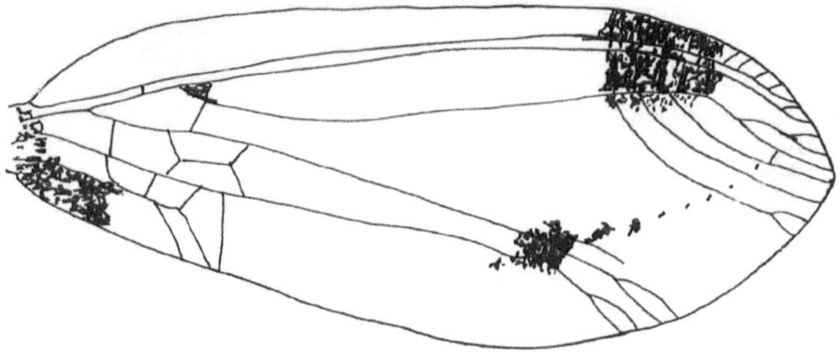

Fig. 16.92 Fore-wing of *Gonzaga amabilis* showing only a few veins and the locations of the dark markings. Based on Navás (1932b).

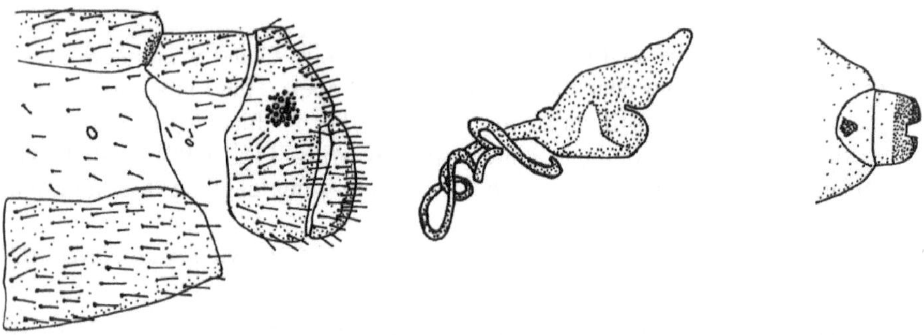

Fig. 16.93 *Gonzaga nigriceps* female (left to right): apex of the abdomen in lateral view, spermatheca in lateral view, and the subgenitale in ventral view. Based on Brooks and Bernard (1990).

gate marking distal to the cubital cross veins extends distad and anteriad toward the apex of the wing. The pterostigmata and area of the membrane posterior to them are dark brown. The pronotum and mesonotum are black. There are about 7 or 8 cubital cross veins. The median emargination in the female subgenitale is approximately trapezoidal (Fig. 16.93). Fore-wing length: 14.9–19.2 mm; width: 5.4–7.0 mm. Hind wing length: 13.1–17.6 mm; width: 3.8–5.1 mm.

....*Gonzaga nigriceps* (McLachlan, 1867) (Ecuador, Peru, Bolivia, Venezuela, Guyana, Surinam, Amazonas, Rio de Janeiro, Santa Catarina). Syn: *Chrysopa nigriceps* MacLachlan, 1867; *Leucochrysa nigriceps* (MacLachlan, 1867).

Key to the Species of *Berchmansus* in South America

Information for the key was provided by Tauber (2007). Only one of the two species is known from South America.

1. The unbroken antenna is more than twice as long as a wing. The fore-wing is unmarked, and the hind wing has a dark cloud along vein Rs (Fig. 16.13). Fore-wing length: 11.9–13.0 mm; width: 3.6–4.5 mm. Hind wing length: 10.5–11.9 mm; width: 3.2–3.6 mm.
 ....*Berchmansus adumbratus* Navás, 1913 (Bolivia, Pará, Amazonas).

 – The antenna is only about twice as long as a wing. The fore-wing is unmarked, and the hind wing has slight, variable brownish shading but no distinct cloud along vein Rs (Fig. 16.94). Fore-wing length: 13.6–14.2 mm; width: 4.3–4.6 mm. Hind wing lenth: 12.6–12.9 mm; width: 3.6–3.9 mm.
 ....*Berchmansus cinctipes* (Banks, 1915) (Costa Rica, Panama). Syn: *Leucochrysa cinctipes* Banks, 1915; *Berkmansus cinctipes* Banks, 1945 (misspelled).

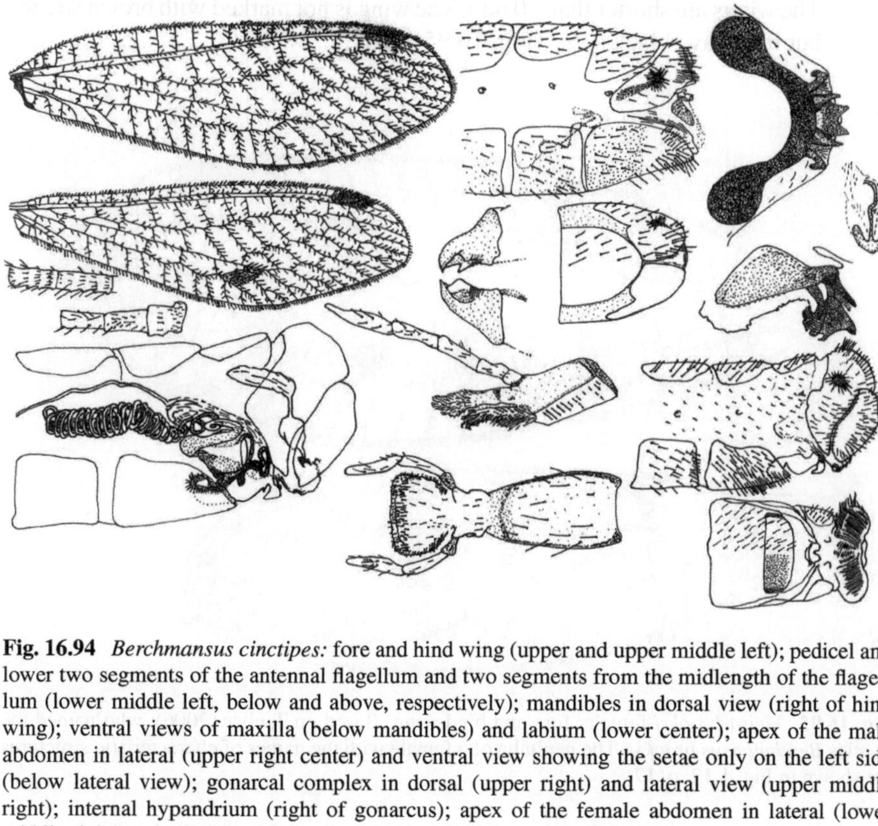

Fig. 16.94 *Berchmansus cinctipes:* fore and hind wing (upper and upper middle left); pedicel and lower two segments of the antennal flagellum and two segments from the midlength of the flagellum (lower middle left, below and above, respectively); mandibles in dorsal view (right of hind wing); ventral views of maxilla (below mandibles) and labium (lower center); apex of the male abdomen in lateral (upper right center) and ventral view showing the setae only on the left side (below lateral view); gonarcal complex in dorsal (upper right) and lateral view (upper middle right); internal hypandrium (right of gonarcus); apex of the female abdomen in lateral (lower middle right) and ventral view (lower right); cleared female abdomen in lateral view showing the internal structures (lower left). Based on Tauber (2007).

Key to the Species of *Vieira* in South America

Information for the key was provided by Tauber (2007). Two nominal species are distinguished in the first couplet. There are no features in the literature to distinguish them with certainty, and Tauber (2007) noted that earlier authors who examined type specimens could find no differences between them. They are considered probable synonyms, but nobody has yet declared them to be such. It appears that if they are identical, the older name, *Vieira iridea* (Olivier, 1792), would be valid.

1. Length of fore-wing: 22.4–30.5 mm; hind wing: 21.9–24.5 mm. The wings are hyaline with distinctive brown spots and streaks, most abundant on the basal parts of the wing (Fig. 16.14). A pair of probable synonyms, not positively distinguishable, key out here.
   .... *Vieira iridea* (Olivier, 1792) (Surinam). Syn: *Hemerobius iridea* Olivier, 1792.
   .... *Vieira leschenaulti* (Navás, 1911) (Central America, Colombia, French Guiana, Surinam, Peru, Bolivia, Mato Grosso, Rondônia). Syn: *Leucochrysa Leschenaulti* Navás, 1911; *Leucochrysa Lechesnaulti* Navás, 1911 (misspelled).
   - The wings are shorter than 20 mm. The wing is not marked with brown streaks, but may have black spots (Fig. 16.95) 2

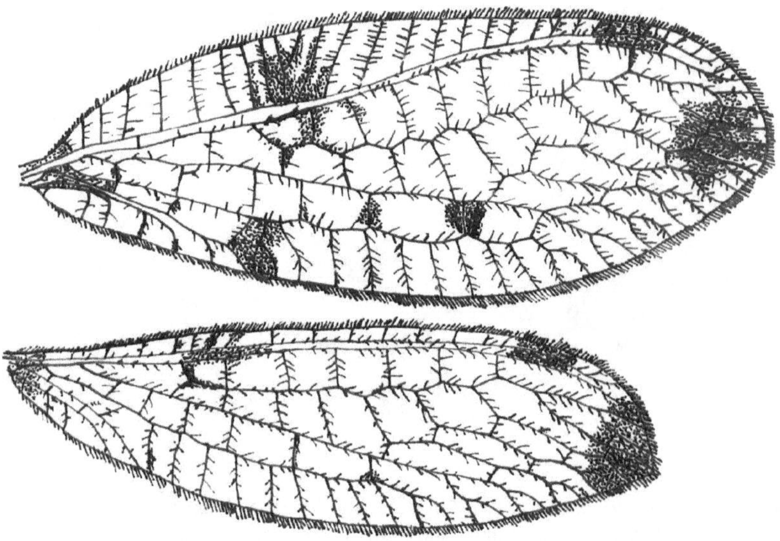

Fig. 16.95 *Vieira brooksi* female: fore and hind wing. Based on Tauber (2006), who named the species *Berchmansus brooksi*. The genitalia of a female with the names of characteristic structures are shown in Fig. 1.12, p. 17.

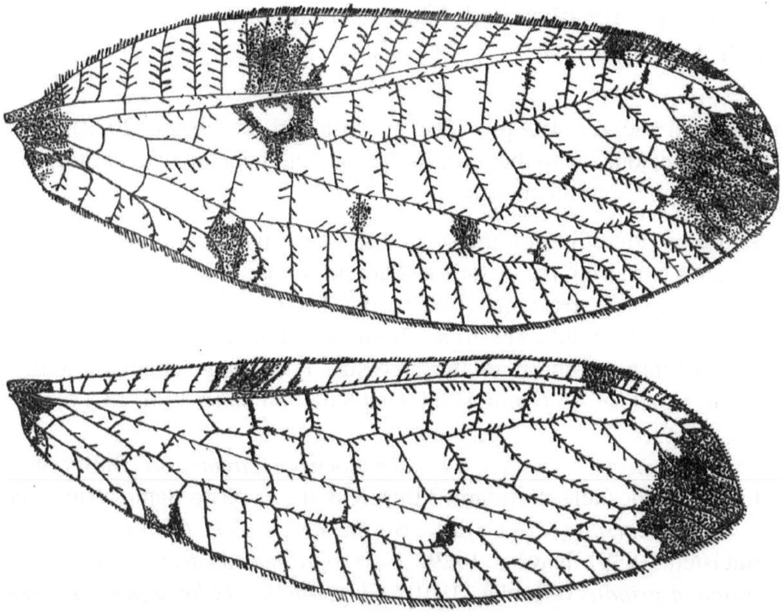

Fig. 16.96 Fore and hind wing of *Vieira elegans*. Based on Tauber (2007).

2. Length of fore-wing: 11–12 mm. The dark marking in the costal area covers six crossveins; one of the middle cross veins is bifurcated (Fig. 16.95). The male has a broad and rather flat mediuncus, with a complex gonarcal hood located ventral to it (Fig. 1.6). The base of the posterior part of the female subgenitale is deeply folded, but there are no ventral processes extending from the subgenitale. The bursal duct of the female is fluted, and there are spines on the bursal membrane (Fig. 1.12).
.... *Vieira brooksi* (Tauber, 2006) (Central America, Bolivia, Brazil). Syn: *Berchmansus brooksi* Tauber, 2006.

– Length of fore-wing: 13.5–15 mm. The dark marking in the costal area covers four to six crossveins, none of which is bifurcated. The male gonarcal hood is simple and lacks processes on its central arm; it is located dorsal to the mediuncus. The base of the posterior part of the female subgenitale is deeply folded, and there are a large ventral spine extending from the subgenitale. The membrane of the female is folded but not spiny (Fig. 16.96).

.... *Vieira elegans* (Guérin-Méneville, 1844) (Trinidad, Guyana, Surinam?, Peru, Rio de Janeiro, Espirito Santo). Syn: *Hemerobius elegans* Guérin-Méneville, 1844 (nec Stephens, 1936); *Chrysopa elegans* auctt.; *Nodita elegans* Navás, 1928; *Berkmansus* (sic) *elegans* Banks, 1944; *Berchmansus elegans* auctt.

Key to the Species of Adult *Titanochrysa* in South America

Information for the key was provided by Sosa and de Freitas (2012) and Tauber et al. (2012a). In addition to the three in the couplet, there are two additional species known only from Central America.

1. There are no microtholi on the second through eighth abdominal sternites. The membranes bordering the cross veins are not infuscated, except those bordering some of the costal cross veins (Fig. 16.16). The mandibles are symmetrical. The dorsal rods on the male arcessus are X-shaped; there are two mesal lines of setae on the membrane ventral to the arcessus. Fore-wing length of male: 14.0–14.7 mm; female: c. 15.6 mm. Hind wing length of male: 12.6–13.1 mm; female: c. 14.3 mm.
   .... *Titanochrysa circumfusa* (Burmeister, 1839) (Colombia, Venezuela, Surinam, Minas Gerais, Espirito Santo, Rio Grande do Sul). Syn: *Chrysopa circumfusa* Burmeister, 1839; *Cintomeva circumfusa* (Burmeister, 1839) Navás, 1929; *Chrysopa burmeisteri* Burmeister, 1929; *Chrysopa nigripalpis* Banks, 1910; *Chrysopodes (Chrysopodes) circumfusa* (Burmeister¸ 1839) Brooks and Bernard, 1990.
   – There are microtholi on the second through eighth abdominal sternites. The dorsal rods on the male arcessus run parallel to each other (Fig. 16.97)...... 2

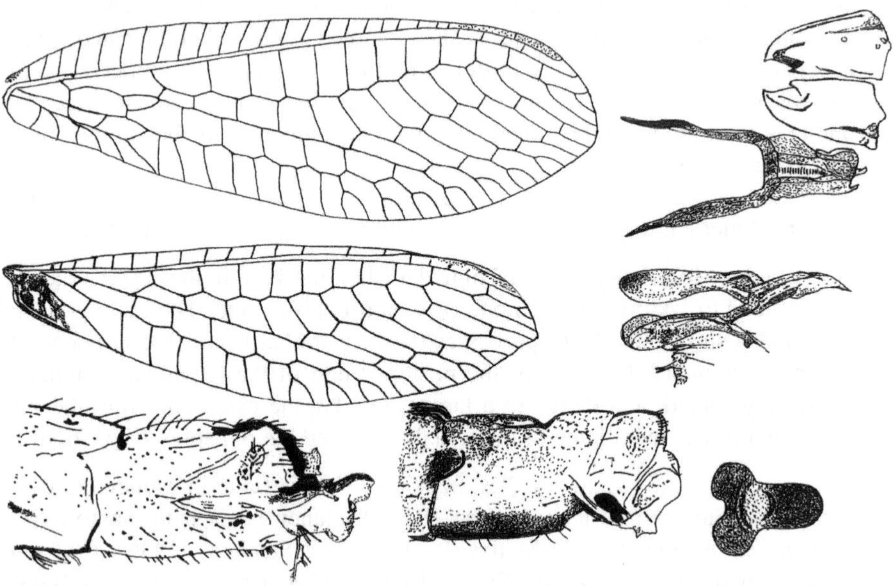

Fig. 16.97 *Titanochrysa trespuntensis:* fore and hind wing (upper and middle left), mandibles (upper right), apex of the male abdomen in lateral view (lower left), apex of the female abdomen in lateral view (lower right center), gonarcal complex in dorsal (upper middle right) and lateral view (middle right), and the subgenitale in anterior view (lower right). Based on Sosa and de Freitas (2012).

2. There is no infuscation at all on the fore-wing membrane (Fig. 16.97). The mandibles are symmetrical. There are two mesal lines of small setae on the membrane ventral to the male arcessus. Fore-wing length of male: 10.1–10.8 mm; female: 11.0–11.3 mm. Hind wing length of male: 9.3–9.7 mm; female: 9.6–10.3 mm. The apical part of the fore-wing is densely covered with black microtrichia.
   ....*Titanochrysa trespuntensis* Sosa and Freitas, 2012 (Minas Gerais).

   – The membrane bordering the inner gradate cross veins is infuscated in places (Fig. 1.8). The mandibles are asymmetral. Mesal lines of setae are completely absent from the membrane ventral to the arcessus of the male. Fore-wing length of male: 11.4–11.8 mm; female: 12.6–13.8 mm. Hind wing length of male: 9.9–10.4 mm; female: 11.2–11.9 mm.
   .... *Titanochrysa ferreirai* Sosa and Freitas, 2012 (Minas Gerais).

Key to the Species of *Nacarina* in South America

Information for the key was provided by Navás (1920, 1930), Brooks and Bernard (1990), de Freitas and Penny (2001), and Penny (2002). Identification of specimens remains tentative because the available descriptions are sketchy, and both sexes of most have not yet been described. A revision is needed to confirm that the apparent specific differences are not due to sexual dimorphism or dichromatism and thereby establish the validity of the species. Sexual dimorphism is charactistic of the genus, which will require separate descriptions of males and females.

Two species could not be placed in the key because their descriptions, published in 1848 and 1915, did not contain descriptions of features now considered important for distinguishing species. Specialists have not yet been able to find type specimens suitable for examination to revise the descriptions. Nevertheless, each of these specimens has been found to have several junior synonyms. The two species needing new descriptions are at the end of this key.

1. The predominant color on the antenna, dorsal surface of the head, thorax, wing veins, pterostigma, legs, and abdomen is yellowish green with yellow, longitudinal stripes along the dorsal midline of the thorax and abdomen. Light red markings are present along the genae, and reddish suffusions are present on the thorax. The apex of the hind wing is eliptically rounded. Length of fore-wing: c. 21.9 mm; hind wing: c. 18.6 mm.
   .... *Nacarina egena* Navás, 1930 (Guyana, Venezuela). Syn: *Nadiva egena* Navás, 1930. The species has not been well described, and undamaged specimens have not been found for redesctiption. Specimens fitting the description of this species should be preserved for examination.

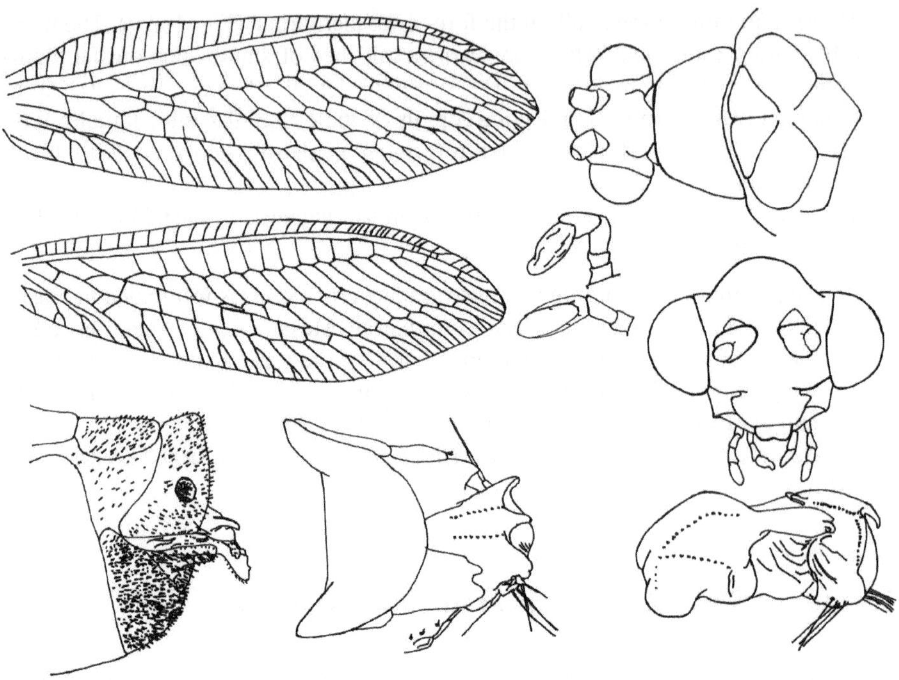

Fig. 16.98 *Nacarina pletorica* male: fore and hind wing (upper and middle left), head and pronotum in dorsal view (upper right), head in anterior view (middle right), two views of a maxillary palp (center), apex of the abdomen in lateral view (lower left), and the genitalia in dorsal (lower center) and lateral view (lower right). Based on de Freitas and Penny (2001).

– Either the specimen is not predominantly greenish in color with a yellow mid-line stripe on the thorax and abdomen, or it has distinct reddish markings on the head, thorax, and abdomen. If in doubt, the fore-wing length is less than 20 mm, or the pterostigma is pale without markings rather than yellowish green 2

2. (1) The maxillary palp has an apical segment that is swollen and bulbous apical segment (Fig. 16.98) 3

– The maxillary palp has a tubular apical segment, which tapers toward the apex (Fig. 16.99) 4

3. (2) The head and pronotum are uniformly pale without dark markings Dark markings on the wings are limited to the intersections of the costal vein and some of the cross veins meeting it at the anterior margin of the wing (Fig. 16.98). The pterostigma is pale without markings. Fore-wing length: c. 17.8 mm; hind wing: c. 16 mm.
.... *Nacarina pletorica* (Navás, 1919) (Colombia, Argentina, São Paulo). Syn: *Nadiva pletorica* Navás, 1919.

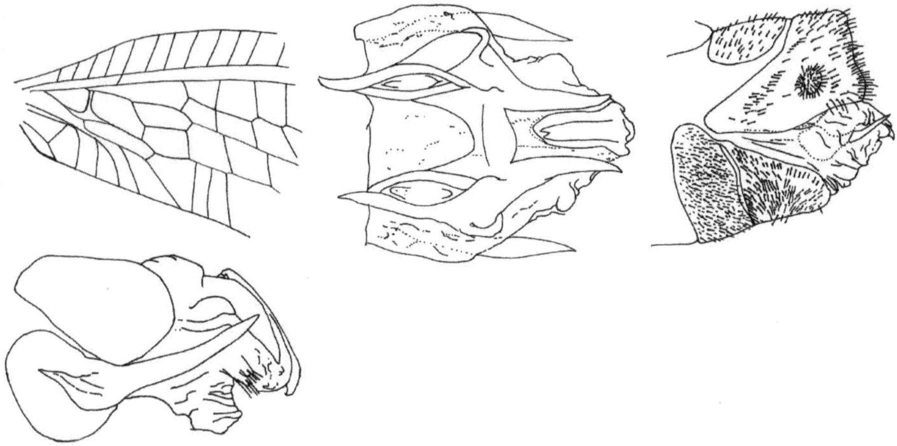

Fig. 16.99 *Nacarina sagitta* male: basal cells of a wing (upper left), apex of the male abdomen in lateral view (right), and the male genitalia in dorsal (center) and lateral view (lower left). Based on de Freitas and Penny (2001).

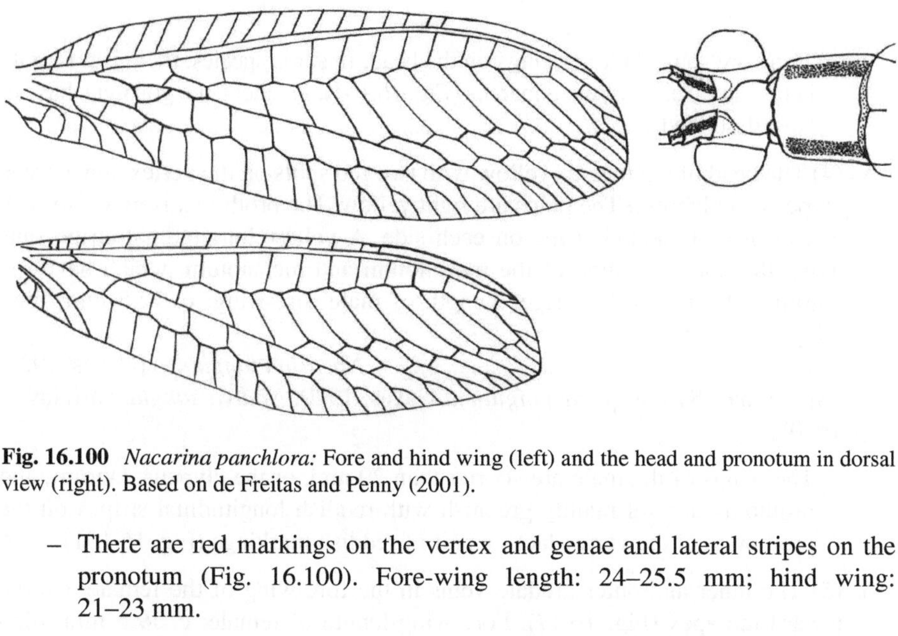

Fig. 16.100 *Nacarina panchlora:* Fore and hind wing (left) and the head and pronotum in dorsal view (right). Based on de Freitas and Penny (2001).

- There are red markings on the vertex and genae and lateral stripes on the pronotum (Fig. 16.100). Fore-wing length: 24–25.5 mm; hind wing: 21–23 mm.

  ....*Nacarina panchlora* (Gerstaecker, 1888) (Santa Catarina, São Paulo). Syn: *Nothochrysa panchlora* Gerstaecker, 1888.

4. (2) The head lacks dark markings. There is a funnel-shaped cavity in the lateral arms of the gonarcus (Fig. 16.99).

   ....*Nacarina sagitta* Freitas and Penny, 2001 (São Paulo).

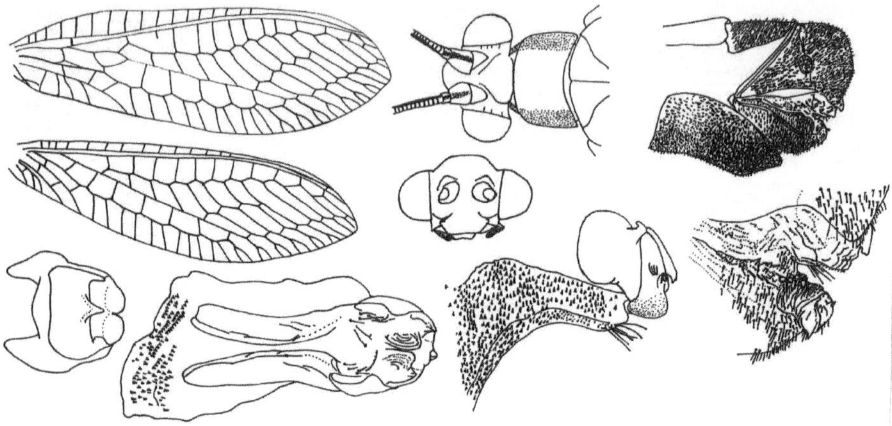

Fig. 16.101 *Nacarina aculeata* male: fore (upper left) and hind wing (middle left); head and anterior part of the thorax in dorsal view (upper center); head in anterior view (center); apex of the abdomen in lateral view (upper right); denticles on the ventral part of the abdominal apex in lateral view (lower right); the genitalia in dorsal (lower left), ventral (lower left center), and lateral view (lower right center). Based on de Freitas and Penny (2001).

– There are some dark markings on the head. In some species, there is no cavity in the lateral arms of the gonarcus (Fig. 16.101), in others, the gonarcus has not been described 5

5. (4) The head of the male is yellow with two red spots on the vertex and a green clypeus and labrum. The palps are light yellow. The prothorax is greenish with a reddish longitudinal stripe on each side. A yellow longitudinal stripe runs along the dorsal midline of the mesonotum and metanotum, which have red markings lateral to the stripe. Length of male fore-wing: c. 22.5 mm; hind wing: c. 20.4 mm.
.... *Nacarina sanguinea* (Navás, 1920) (Argentina). Syn: *Rameta sanguinea* Navás, 1920; *Nadiva sanguinea* (Navás, 1920).

– The wings of the male are shorter than 20 mm, or the clypeus, labrum, and prothorax are not mainly greenish with reddish longitudinal stripes on the prothorax, or the vertex lacks a pair of reddish markings (Fig. 16.101) 6

6. (5) The inner and outer gradate veins in the fore-wing of the female diverge toward the apex (Fig. 16.17). Fore-wing length of female: c. 19.5 mm; hind wing length of female: c. 16.7 mm. Only the female has been described.
....*Nacarina wagneri* (Navás, 1924) (Argentina, Brazil). Syn: *Nadiva wagneri* (Navás, 1924).
(Argentina, Mato Grosso).

– The inner and outer gradate veins in the male fore-wing are parallel (Fig. 16.101). Descriptions of the females are not available 7

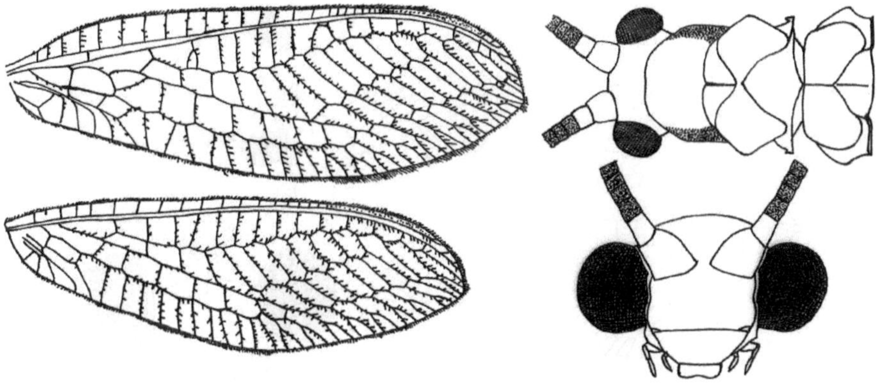

Fig. 16.102 *Nacarina megaptera:* fore and hind wing (left), head and thorax in dorsal view (upper right), and head in anterior view (lower right). Based on Penny (2002).

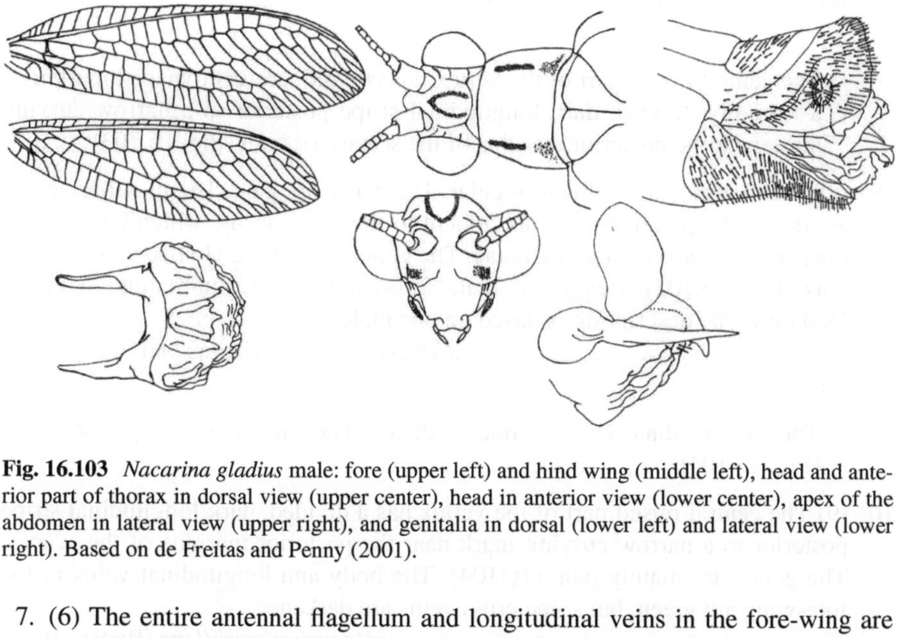

Fig. 16.103 *Nacarina gladius* male: fore (upper left) and hind wing (middle left), head and anterior part of thorax in dorsal view (upper center), head in anterior view (lower center), apex of the abdomen in lateral view (upper right), and genitalia in dorsal (lower left) and lateral view (lower right). Based on de Freitas and Penny (2001).

7. (6) The entire antennal flagellum and longitudinal veins in the fore-wing are very dark. The body is dark. The intermedial cell is triangular (Fig. 16.102). ....*Nacarina megaptera* (Navás, 1927) (Central America, Guyana, Brazil).

   – The antennal flagellum is at least partially pale (Fig. 16.103)..................... 8

8. (7) The central raised part of the vertex lacks red markings. The parameres adhere to the gonosaccus, which is covered with numerous small gonocristae (Fig. 16.101). Length of male fore-wing: c. 16.0 mm: hind wing: c. 13.8 mm. The description is based on one male. ....*Nacarina aculeata* Freitas and Penny, 2001 (São Paulo).

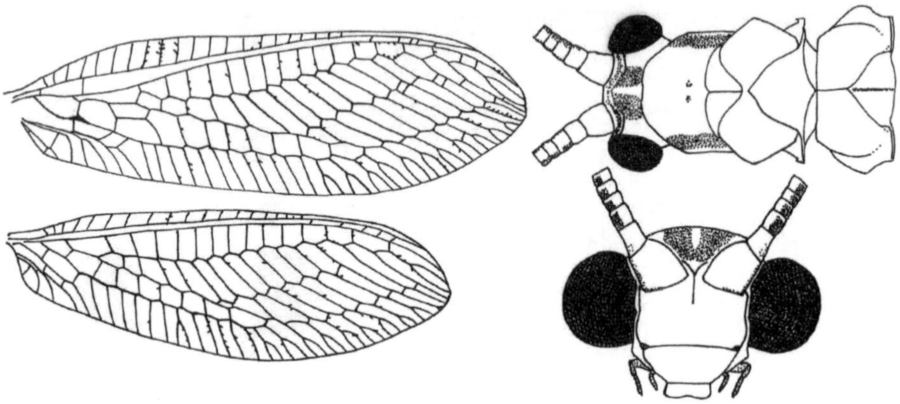

Fig. 16.104 *Nacarina cordillera:* fore and hind wing (left), head and thorax in dorsal view (upper right), and head in anterior view (lower right). Based on Penny (2002). The synonym, *Nacarina balboana,* was given to the female of this species.

– The central raised part of the vertex has two red, crescent-shaped markings, a medially divided, dark longitudinal stripe posterior to a narrow curving mark near the posterior margins of the scapes, or both (Fig. 16.103) 9

9. (8) The intermedian cell is triangular. The genae are pale. The maxillary palps are dark. The parameres do not adhere to the gonosaccus, which bears a few long setae but no small gonocristae. The gonocornua have U-shaped inner margins (Fig. 16.103). Length of male fore-wing: c. 22.3 mm: hind wing: c. 18.0 mm. The description is based on one male. .... *Nacarina gladius* Freitas and Penny, 2001 (São Paulo).

– The intermedian cell is quadrilateral. The maxillary palps are pale (Fig. 16.104). .... .10

10. (9) The central raised part of the vertex has a divided, dark longitudinal stripe posterior to a narrow curving mark near the posterior margins of the scapes. The genae are mainly pale (16.104). The body and longitudinal veins in the fore-wing are green, but some cross veins are darkened. .... *Nacarina cordillera* (Banks, 1910) (Mexico, Central America, Colombia, Venezuela, Rondônia). Syn: *Allochrysa cordillera* Banks, 1910; *Nacarina balboana* Banks, 1941.

– The central raised part of the vertex has two red, crescent-shaped markings. The genae are red. The parameres do not adhere to the gonosaccus, which bears a few long setae but no small gonocristae (Fig. 16.105). The gonocornua have V-shaped inner margins. Length of male fore-wing: c. 22.8 mm: hind wing: c. 20.2 mm. The description is based on one male, which seems also to fit the sketchy description of *Nacarina sanguinea* by Navás (1920).
.... *Nacarina lavrasana* Freitas and Penny, 2001 (São Paulo).

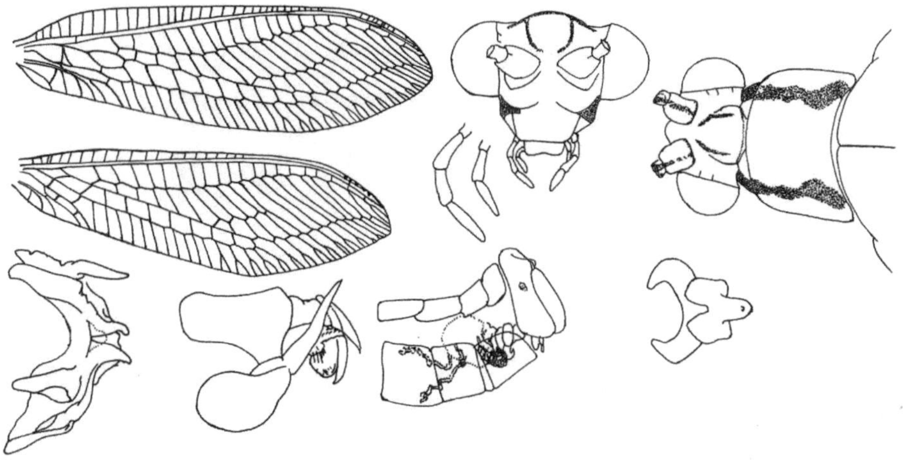

Fig. 16.105 *Nacarina lavrasana:* fore and hind wing (upper and middle left), head in anterior view showing the mandibular and labial palps to its lower left (upper center), head and anterior part of the thorax in dorsal view (upper right), internal male genitalia in dorsal (lower left) and lateral view (lower left center), apex of the female abdomen in lateral view (lower center), and the female subgenitale (lower right). Based on de Freitas and Penny (2001).

Species Inquirendae

1. *Nacarina valida* (Erichson, 1848): found only in Guyana. Junior synonyms: *Hemerobius validus* Erichson, 1848; *Chrysopa valida* (Erichson, 1848); *Berchmansus valida* (Erichson, 1848); *Nadiva valida* (Erichson, 1848).
2. *Nacarina furcata* Navás, 1915: found in Argentina and Bolivia. Junior synonyms: *Goliva deletangi* Navás, 1920; *Nadiva deletangi* (Navás, 1920), *Nacarina deletangi* (Navás, 1920), *Cintameva anomala* Navás, 1929; *Chrysopa anomala* (Navás, 1929).

An additional three Neotropical species of *Nacarina* have not been found in South America.

Key to the Species of *Parachrysopiella* in South America

Information for the key was provided by Banks (1910), Navás (1918a), and Penny (1996).

1. The antenna is pale, even on the pedicel, and there is no marking on the inner surface of the scape. The mediuncus bears a small dorsal hook (Fig. 16.15). Fore-wing length of male: 9.0–10.7 mm; female: 9.5–11.7 mm. .... *Parachrysopiella pallidicornis* Penny, 1996 (Argentina, Chile).

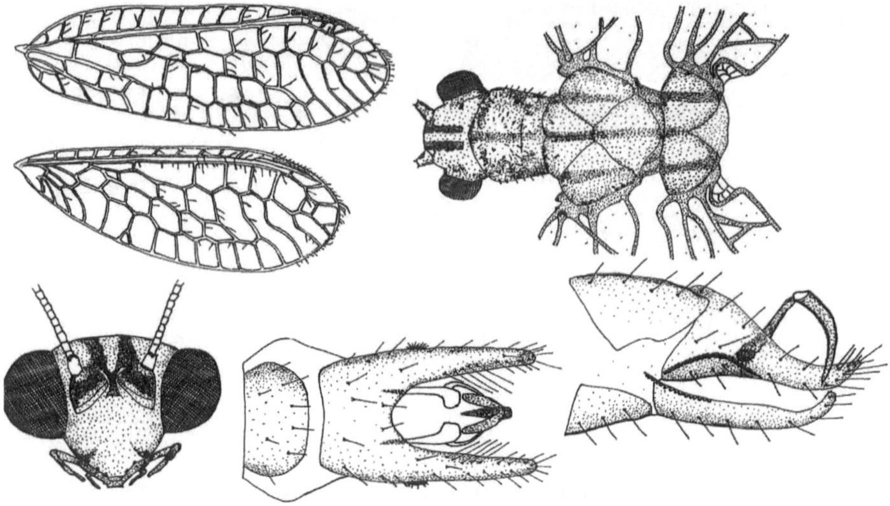

Fig. 16.106 *Parachrysopiella argentina* male: fore and hind wing (upper and middle left), head in anterior view (lower left), head and thorax with the bases of the wings in dorsal view (upper right), and apex of the abdomen in dorsal (lower center) and lateral view (lower right). Based on Penny (1996).

– The antenna is dark with its pedicel also darkened. The scape has a dark stripe on the medial surface. There is a long narrow projection from the mediuncus (Fig. 16.106) 2

2. There is no gonapsis on the male genitalia. The gonarcus lacks flattened, lateral plates. The subgenitale of the female is nearly ovoid and lacks membraneous extensions. The pterostigma on the fore-wing is conspicuously narrow and reaches only as far as the subcostal vein (Fig. 16.106). Fore-wing length of male: 11.5–12.0 mm; female: 12.1–13.0 mm.
.... *Parachrysopiella argentina* (Banks, 1910) (Argentina, Bolivia, Chile). Syn: *Eremochrysa argentina* Banks, 1910; *Chrysopiella argentina acysta* Navás, 1918; *Chrysopiella argentina extranea* Navás, 1918; *Chrysopiella argentina trapezia* Navás, 1918.

– There is a long, forked gonapsis on the male genitalia. The gonarcus has large flattened plates surrounding the mediuncus. The subgenitale of the female is elongate and has narrow membranous extensions. The pterostigma on the fore-wing is broad and extends as far as the radial vein (Fig. 16.107). Fore-wing length of male: 7.7–8.0 mm; female: 7.1–9.0 mm.

.... *Parachrysopiella talquensis* Penny, 1996 (Chile).

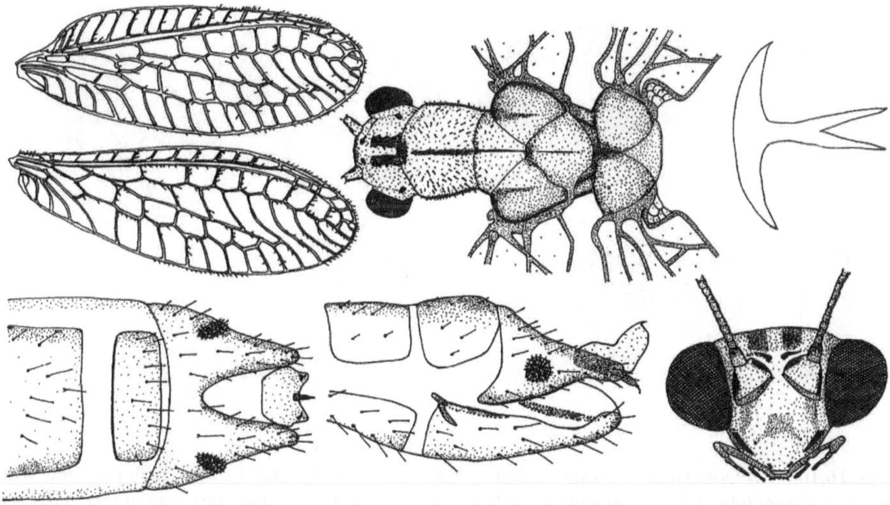

Fig. 16.107 *Parachrysopiella talquensis* male: fore and hind wing (upper and middle left), head in anterior view (lower right), head and thorax with the bases of the wings in dorsal view (upper center), apex of the abdomen in dorsal (lower left) and lateral view (lower center), and the gonapsis in dorsal view (upper right). Based on Penny (1996).

Key to the Species of *Meleoma* in South America

Information for the key was provided by Adams (1969). This genus is well represented in North America, but only two species have been reported from South America.

1. The face is green with an ochre yellow vertex, which becomes paler laterally. The clypeus is either unmarked, or it is reddish laterally. The dorsal surface of the pronotum is green with a broad yellow or white median stripe and broad reddish stripes along the lateral margins of the green protion. The apical portion of the mediuncus is straight (Fig. 16.18). Fore-wing length of the male: 15.0–15.5 mm; female: 15.5–16.5 mm.
   ....*Meleoma poolei* Adams, 1969
   (Central America, Venezuela).
   – The face is pale with the ventral part of the head greenish and the vertex orange. There are red stripes on the genae and between the eyes and the vertex. The thorax and abdomen are green laterally and have a mid-dorsal stripe, which is yellow on most of the thorax and ivory bordered by red on the abdomen. The apical portion of the mediuncus is slightly sinuous (Fig. 16.108). Fore-wing length of the male: c. 12.8 mm.
   ....*Meleoma festivata* Adams, 1969
   (Colombia).

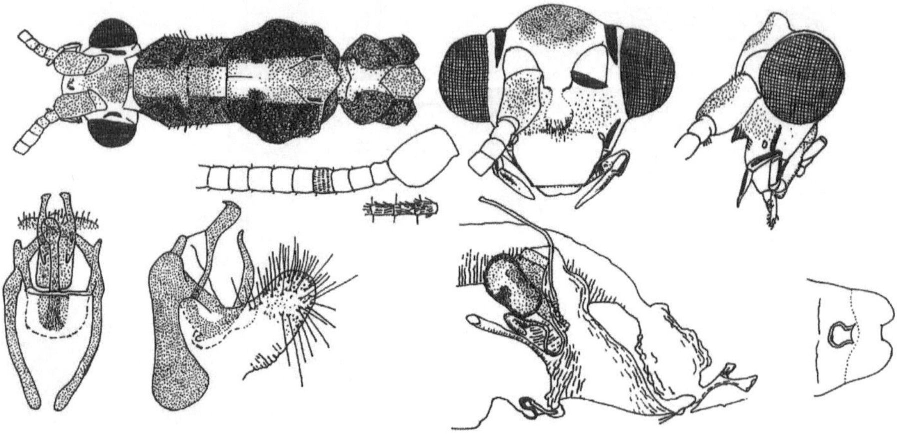

Fig. 16.108 *Meleoma festivata* (above: left to right): head and thorax of a male in dorsal view with the basal segments of its antenna enlarged below and segments near the apex shown just under the enlarged scape; the head in anterior and lateral view; and (below, left to right): the male genitalia in posterior and lateral view; the female genital organs in lateral view; the eighth sternite of the female in ventral view. Based on Adams (1969).

Key to the Species of *Chrysoperla* in South America

Information for the key was provided by Navás (1917b, 1920, 1922b), Adams and Penny (1987), Brooks (1994), de Freitas (2003), and Haramboure et al. (2014).

1. The head is mainly yellow or greenish yellow, but the genae are black (Fig. 16.109). The wing veins are green. Length of fore-wing: c. 11.9 mm, hind wing: c. 11.7 mm.
   ....*Chrysoperla genanigra* Freitas, 2003 (Rio Grande do Norte).

   – The genae are reddish or dark brown (Fig. 16.19). 2
2. (1) The gradate veins in the fore-wing are black (Fig. 16.110). Length of fore-wing: c. 13.0 mm: hind wing: c. 11.9 mm.
   ....*Chrysoperla defreitasi* Brooks, 1994 (São Paulo, Mato Grosso).

   – The gradate veins in the fore-wing are green (Fig. 16.19) 3
3. (2) The genae are dark brown (Fig. 16.111).
   ....*Chrysoperla argentina* González Olazo and Reguillón, 2002 (Argentina).

   – The genae are reddish (Fig. 16.112) 4
4. (3) There is a reddish postocular spot (Fig. 16.112).
   ....*Chrysoperla asoralis* (Banks, 1915) (Argentina).

   – There is no reddish postocular spot (Fig. 16.113)..5

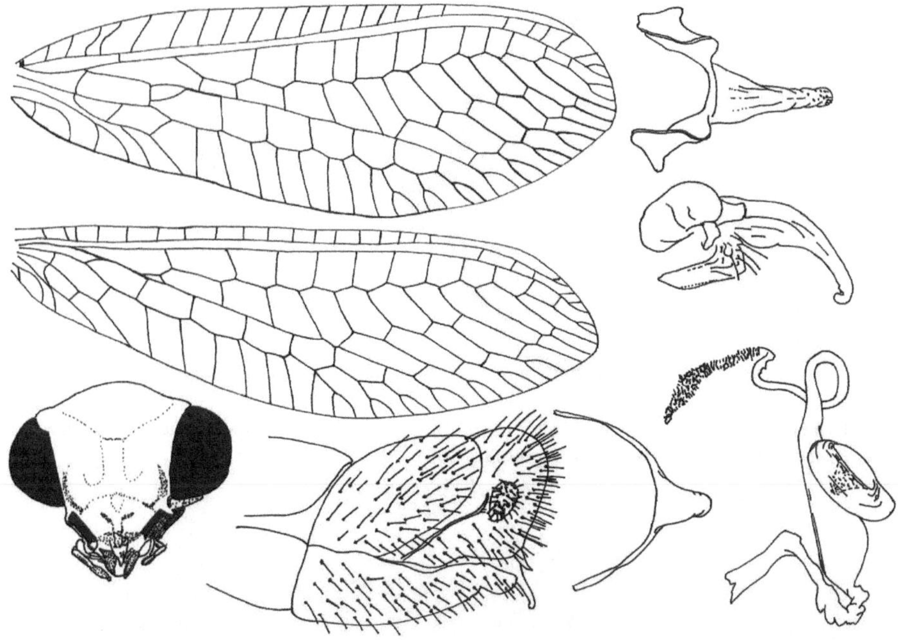

Fig. 16.109 *Chrysoperla genanigra;* fore and hind wing (upper and middle left), head in anterior view (lower left), apex of the male abdomen in lateral view (lower left center), male gonarcal complex in dorsal (upper right) and lateral view (middle right), tignum in dorsal view (lower right center), and the female spermatheca in lateral view (lower right). Based on de Freitas (2003).

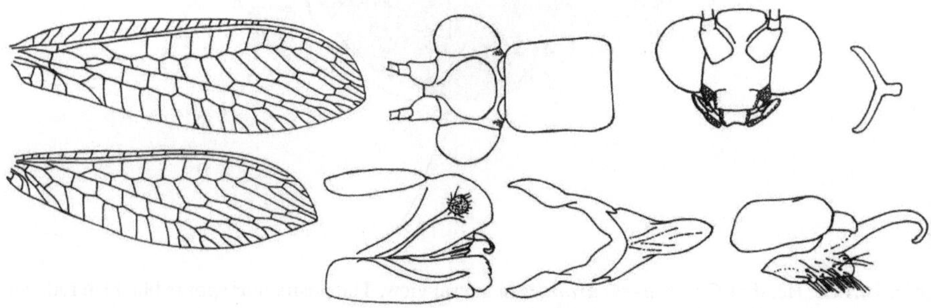

Fig. 16.110 *Chrysoperla defreitasi* male: fore and hind wing (left), head and pronotum (upper left center), head in anterior view (upper right center), apex of the abdomen in lateral view (lower left center), male genitalia in dorsal (lower right center) and lateral view (lower right), and the tignum (upper right). Based on de Freitas and Penny (2001).

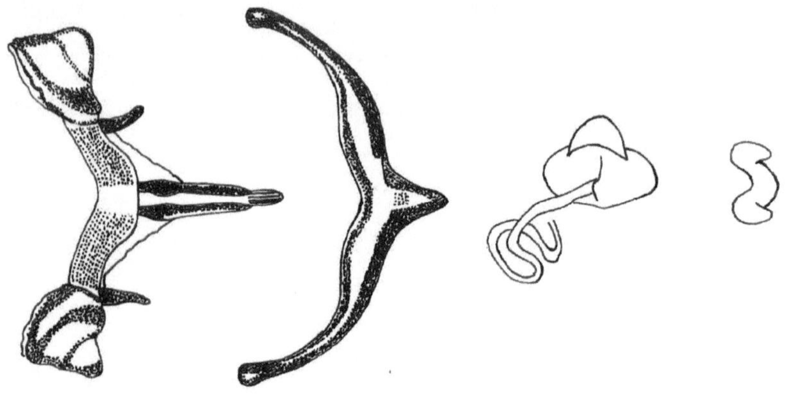

Fig. 16.111 *Chrysoperla argentina* (left to right): gonarcus, tignum, spermathecal, and subgenitale. Based on González Olazo and Reguilón (2002).

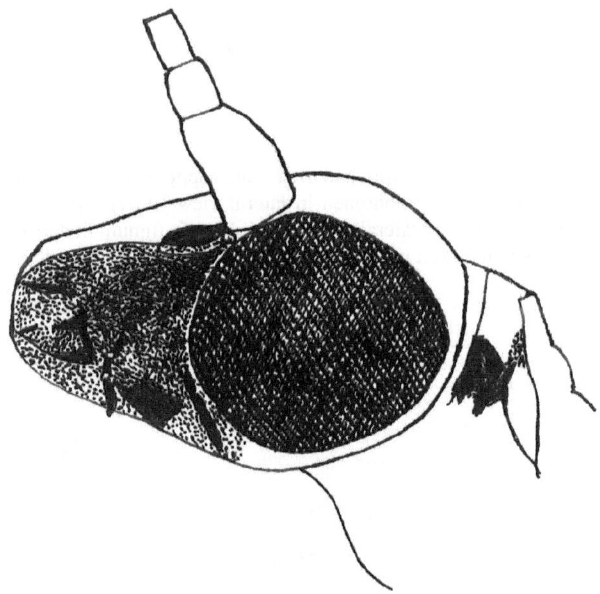

Fig. 16.112 Head of *Chrysoperla asoralis* in lateral view. The postocular spot is blackish red, and the genae are bright red. Based on Haramboure et al. (2014).

5. (4) The pronotum has reddish lateral stripes. The arcessus of the male is only slightly decurved at the apex (Figs. 16.19 and 16.113). Length of fore-wing: c. 12.3 mm: hind wing: c. 11.9 mm.
   .... *Chrysoperla externa* (Hagen, 1861) (North and Central America, West Indies, Cocos Islands, Easter Island, Colombia, Venezuela, Ecuador, Peru, Chile, Paraguay, Uruguay, Argentina, Amazonas, Rio Grande do Norte, Rio de Janeiro). Syn: *Chrysopa externa* Hagen, 1861; *Chrysopa plorabunda* Fitch, 1856; *Chrysopa lanata* Banks, 1910; *Chrysopa graciana*

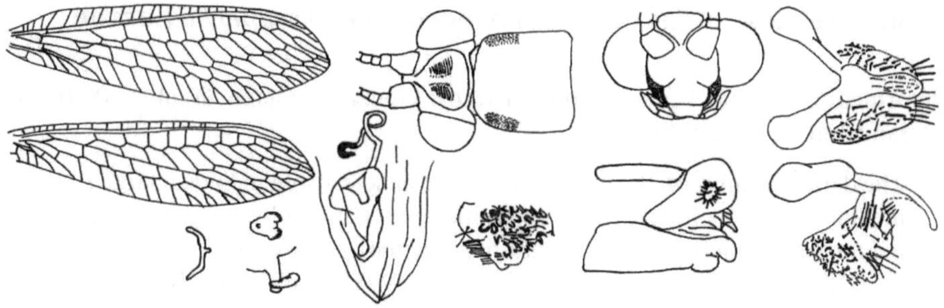

Fig. 16.113 *Chrysoperla externa:* fore (upper left) and hind wing (middle left), head and pronotum in dorsal view (upper left center), head in anterior view (upper right center), the apex of the male abdomen in lateral view (lower right center), male genitalia in dorsal (upper right) and lateral view (lower right), female genitalia (lower left center), gonosaccus (lower center), tignum (lower left), and subgenitale in ventral and lateral view (between tignum and female genitalia, above and below, respectively). Based on de Freitas and Penny (2001).

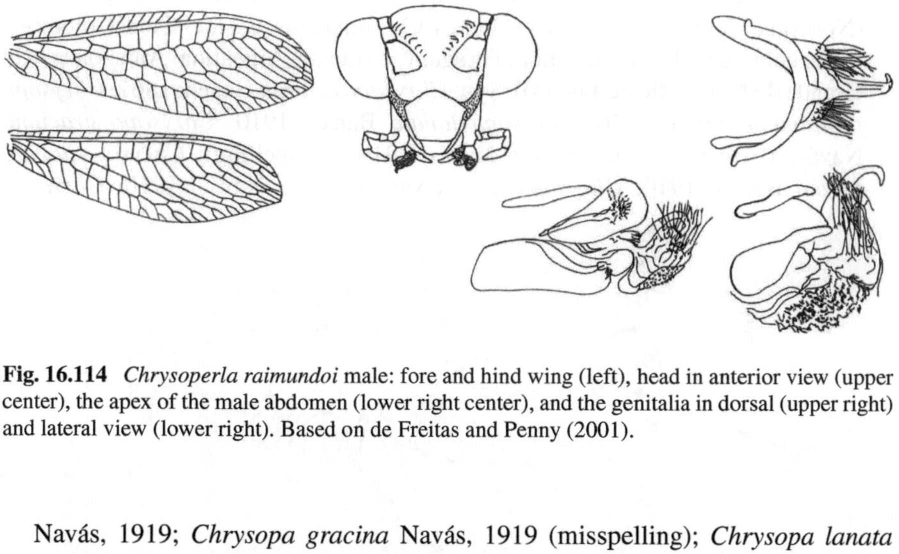

Fig. 16.114 *Chrysoperla raimundoi* male: fore and hind wing (left), head in anterior view (upper center), the apex of the male abdomen (lower right center), and the genitalia in dorsal (upper right) and lateral view (lower right). Based on de Freitas and Penny (2001).

Navás, 1919; *Chrysopa gracina* Navás, 1919 (misspelling); *Chrysopa lanata lanata* Banks, 1910; *Chrysopa lanata* var. *platensis* Navás, 1916; *Chrysopa lanata* var. *climacia* Navás, 1917; *Chrysopa lanata* var. *gradata* Navás, 1919; *Chrysopa lanata* var. *basalis* Navás, 1920; *Chrysopa lanata* var. *unita* Navás, 1920; *Chrysopa externa cocosensis* Adams, 1983.

– The arcessus of the male is strongly decurved at the apex (Fig. 16.114). Length of fore-wing: c. 12.2 mm: hind wing: c. 10.2 mm.
.... *Chrysoperla raimundoi* Freitas and Penny, 2001 (São Paulo).

Tentative Key to Know South American *Chrysoperla* Larvae

Information of this key was provided by Gonzalez Olazo et al. (2009) and Haramboure et al. (2014).

1. The dark markings on the dorsal side of the head capsule include a large rounded median mark touching two lateral markings separated by a lighter stripe along the midline, which ends at the dark median marking (Fig. 16.115). ....*Chrysoperla asoralis* (Banks, 1915) (Argentina).
   – If a dark median marking is present on the head capsule, it is separated from the lateral dark markings (Fig. 16.116) 2
2. (1) There is a small dark markings near the anterior margin of the head capsule, which is well separated from the dark, lateral, somewhat divergent markings (Fig. 16.116). .... *Chrysoperla externa* (Hagen, 1861) (North and Central America, West Indies, Cocos Islands, Easter Island, Colombia, Venezuela, Ecuador, Peru, Chile, Paraguay, Uruguay, Argentina, Amazonas, Rio Grande do Norte, Rio de Janeiro). Syn: *Chrysopa externa* Hagen, 1861; *Chrysopa plorabunda* Fitch, 1856; *Chrysopa lanata* Banks, 1910; *Chrysopa graciana* Navás, 1919; *Chrysopa gracina* Navás, 1919 (misspelling); *Chrysopa lanata lanata* Banks, 1910; *Chrysopa lanata* var. *platensis* Navás, 1916; *Chrysopa*

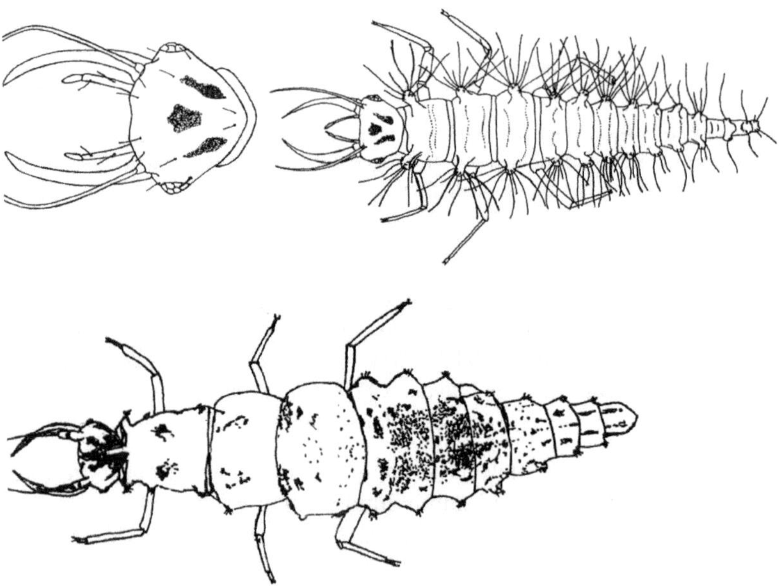

Fig. 16.115 First instar larva of *Chrysoperla asoralis:* head (left) and habitus (right), both in dorsal view. Based on González Olazo et al. (2009).

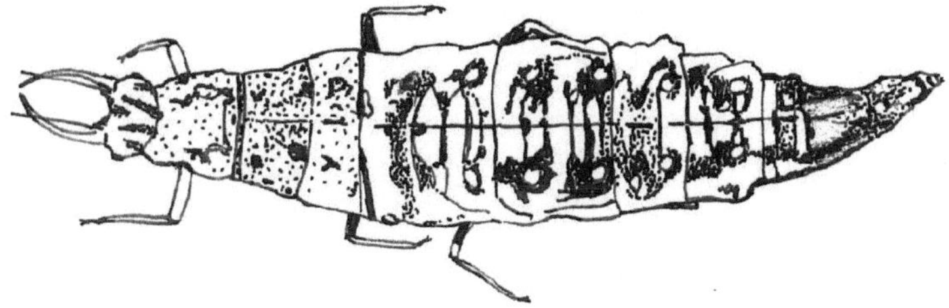

Fig. 16.116 *Chrysoperla externa* third instar larva: habitus in dorsal view. Note markings on the head. Based on González Olazo et al. (2009).

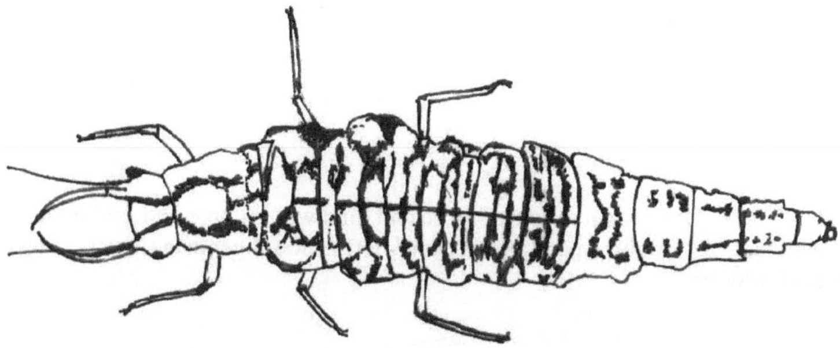

Fig. 16.117 *Chrysoperla argentina:* third instar larva: habitus in dorsal view. Note markings on the head. Based on González Olazo et al. (2009).

lanata var. *climacia* Navás, 1917; *Chrysopa lanata* var. *gradata* Navás, 1919; *Chrysopa lanata* var. *basalis* Navás, 1920; *Chrysopa lanata* var. *unita* Navás, 1920; *Chrysopa externa cocosensis* Adams, 1983.

– There is no dark median marking. There are only two curving, lateral stripes, which diverge toward the anterior margin and do not reach that mergin (Fig. 16.117)

..............................*Chrysoperla argentina* González Olazo and Reguillón, 2002 (Argentina).

Key to the Species of *Ceraeochrysa* in South America

Information for the key was provided by Navás (1925a), Adams and Penny (1987), de Freitas and Penny (2001), de Freitas et al. (2009), Sosa and de Freitas (2010), and Tauber and Flint (2010a, b). A few species are known only from female specimens, and those in this key can only be identified tentatively until males become available for examination. Even in the case of female specimens belonging to species

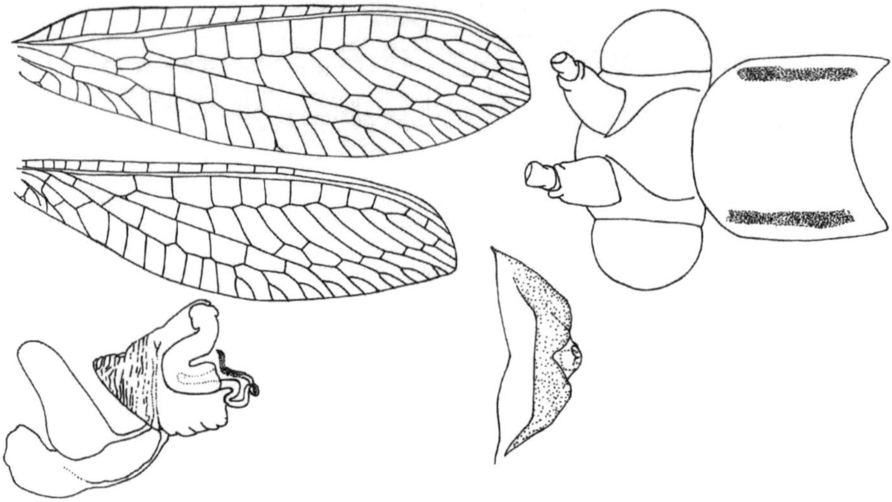

Fig. 16.118 *Ceraeochrysa reducta* female: fore and hind wing (upper and middle left), head and pronotum in dorsal view (right), spermatheca (lower left), and subgenitale (lower center). Based on de Freitas et al. (2009).

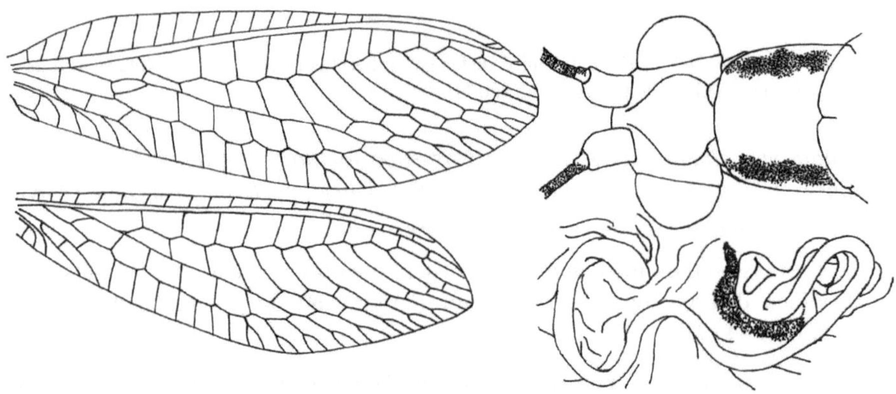

Fig. 16.119 *Ceraeochrysa dolichosvela* female: fore and hind wing (left), head and pronotum in dorsal view (upper right), and spermatheca (lower right). Based on de Freitas et al. (2009).

described from both sexes, idetification can only be tentative because the females of many other species remain to be described.

1. The body is pale green, as is the entire scape, which lacks dark stripes and a lateral patch. The flagellum is pale and densely covered by setae. There is a red-orange or yellow-orange lateral stripe on each side of the pronotum (Fig. 16.118). The fore-wing length is c. 13–15 mm. .2

   – The scape has dark or red markings, or, if not, the basal segments of the antennal flagellum are black, or all segments of the flagellum are longer than their width (Fig. 16.119). The spermatheca of the female is not U-shaped if the head is completely unmarked or it is otherwise not C-shaped. 3

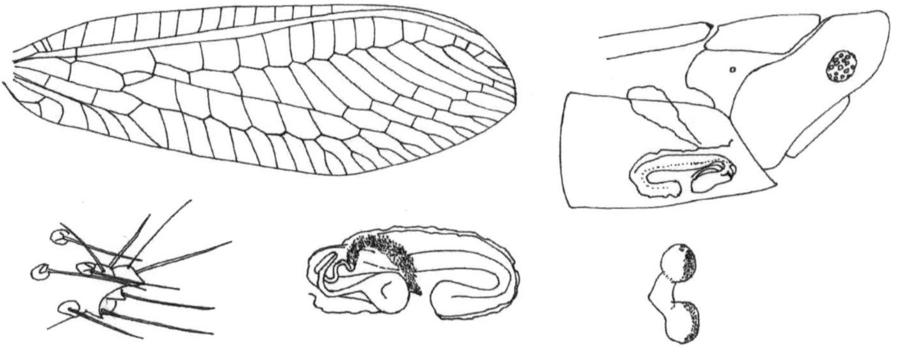

Fig. 16.120 *Ceraeochrysa fairchildi* female (above, left and right, respectively): wing and the apex of the abdomen in lateral view, and (below, left to right): apex of the seventh abdominal sternite, spermatheca, and subgenitale. Based on Adams and Penny (1987), who called the species *Ceraeochrysa acutipuppis*.

2. (1) The spermatheca of the female is U-shaped, and the copulatory bursa is wrinkled. There are no markings whatsoever on the head, including the antennae. The general coloration is pale greenish, with red-orange lateral stripes on the pronotum. The wings of the female have three inner and seven outer gradate crossveins, which are black (Fig. 16.118). The middle of the costal and the R-Rs crossveins are blackened at midlength. Fore-wing length of female: c. 14 mm. The male has not been described, making identification of this species tentative.
   .... *Ceraeochrysa reducta* (Banks, 1944) (Surinam, Colombia). Syn: *Chrysopa reducta* Banks, 1944.

   – The spermotheca of the female is in the shape of a C (Fig. 16.120). The fifth segment of the flagellum is about 1.25 times as wide as its length. The ninth sternite and fused ectoproct has a strongly sclerotized apodeme, which is forked dorsally and ventrally at the level of the callus cerci. Fore-wing length: c. 13.3–14.4 mm.

   ....*Ceraeochrysa fairchildi* (Banks, 1946) (Panama, Trinidad, Venezuela, Roraima). Syn: *Chrysopa fairchildi* Banks, 1946; *Ceraeochrysa acutipuppis* Adams and Penny, 1987.

3. (1) The scape of the female is uniformly pale in color, while the basal flagellar segments are almost entirely black. The spermatheca is long and uncoiled for most of its length (Fig. 16.119). Fore-wing length: c. 14.9 mm; hind wing: c. 12.6 mm. The male has not been described.
   .... *Ceraeochrysa dolichosvela* Freitas and Penny, 2001 (Mato Grosso do Sul, São Paulo).

   – The scape usually has dark or red markings. The spermathecas of all females known from South America are obviously coiled (Fig. 16.121).................. 4

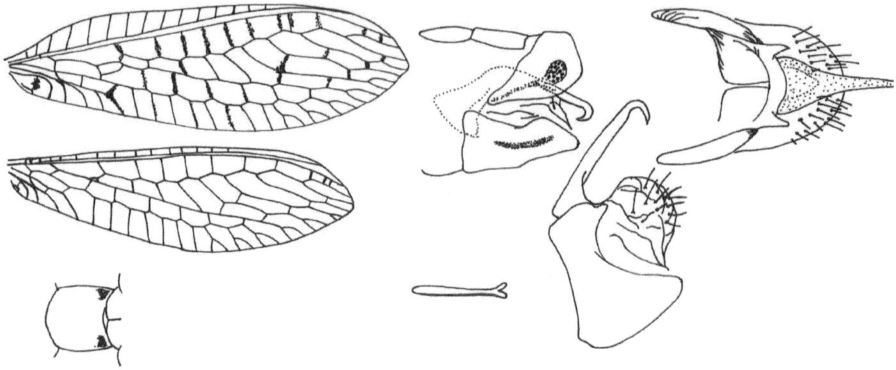

Fig. 16.121 *Ceraeochrysa tauberae* male: fore and hind wing (upper and middle left), pronotum showing the pair of posterolateral spots (lower left), apex of the male abdomen in lateral view (upper center), male genitalia in dorsal (upper right) and lateral view (lower right center), and the gonapsis (lower center). Based on de Freitas et al. (2009).

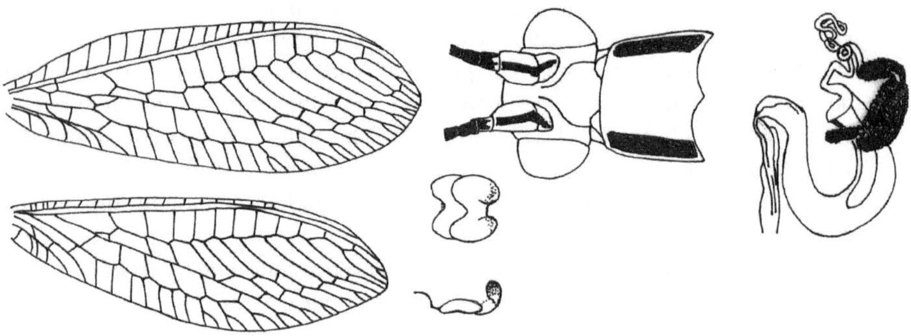

Fig. 16.122 *Ceraeochrysa montoyana* female: fore and hind wing (left), head and pronotum in dorsal view (upper center), spermatheca (right), and the subgenitale in ventral (left center) and lateral view (lower left center). Based on de Freitas and Penny (2001).

4. (3) The pronotum is marked with one or two pairs of lateral spots or thin submedian stripes (Fig. 16.121) 5

   – The pronotum has a pair of red or brown lateral stripes or no stripes at all (Fig. 16.122) 7

5. (4) The pronotum has one pair of lateral spots. There are dark margins along the gradate cross veins, the distal cubital cell, and cross vein 2a-3a. The gonocornua of the male is simple and tapers to a point (Fig. 16.121).
   .... *Ceraeochrysa tauberae* Penny, 1977 (Central America, Venezuela).

   – The pronotum has two pairs of dark lateral spots (Fig. 16.123)......................6

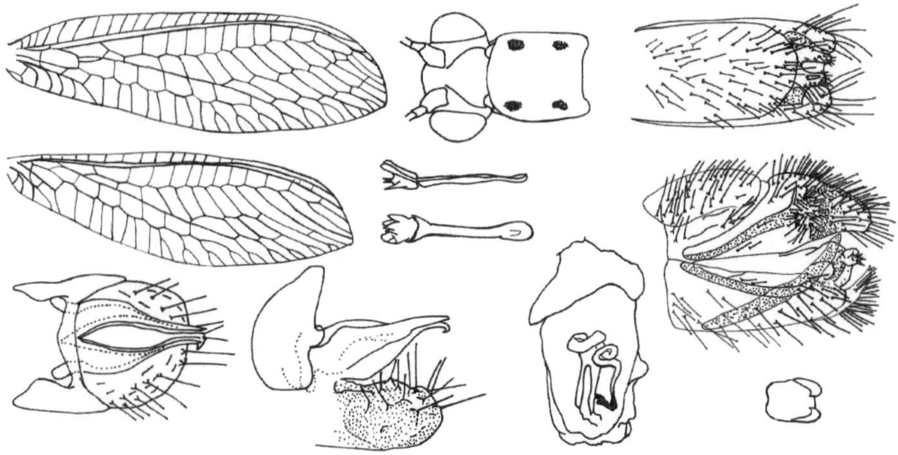

Fig. 16.123 *Ceraeochrysa angulata* male: fore and hind wing (upper and middle left); head and pronotum in dorsal view (upper center); apex of the male abdomen in lateral view (lower center); male genitalia in dorsal (upper right), lateral (middle right), and posterior view (lower left); gonapsis (lower right). Based on de Freitas et al. (2009).

6. (5) The base of the distal cubital cell is darkly shaded and sometimes swollen with a short triangular cell at the fork in CuP. The mesoscutum lacks dark spots. The arcessus of the male has a bifid apex and two basal horns, which are widely separated (Fig. 16.123). Length of fore-wing of female: c. 14 mm. ....*Ceraeochrysa angulata* Navás, 1929 (Central America, Colombia, Venezuela, Ecuador). Syn: *Chrysopa josephina* Navás, 1926, *nomen dubium.*

   – The base of the distal cubital cell is neither darkly shaded nor swollen, and there is never a short triangular cell at the fork in CuP. The two pairs of dark circular markings on the pronotum each have discrete borders. The eighth and ninth sternites of the male bear a pair of forcipate projections, and there are two spiny knobs between the gonapsis and the eighth and nonth sternites (Fig. 16.124). Fore-wing length: c. 12.5–12.8 mm; hind wing: c. 10.8–11.0 mm.

   ....*Ceraeochrysa paraguaria* (Navás, 1920) (Paraguay, Argentina, São Paulo). Syn: *Chrysopa paraguaria* Navás, 1920; *Chrysopa paraquaria* Navás, 1924; *Plesiochrysa paraguaria* (Navás, 1920) Brooks and Barnard, 1990.

7. (4) On the dorsal surface of the antennal scape, there are two dark or red stripes (Fig. 16.122). In addition to the two South American species in Couplet 8, a color variety of *Ceraeochrysa cincta,* known only from Florida, fits the description in that couplet. Its antennal flagellum is entirely pale, and the apex of the ninth abdominal sternite of the male is truncate at the apex, distinguishing it

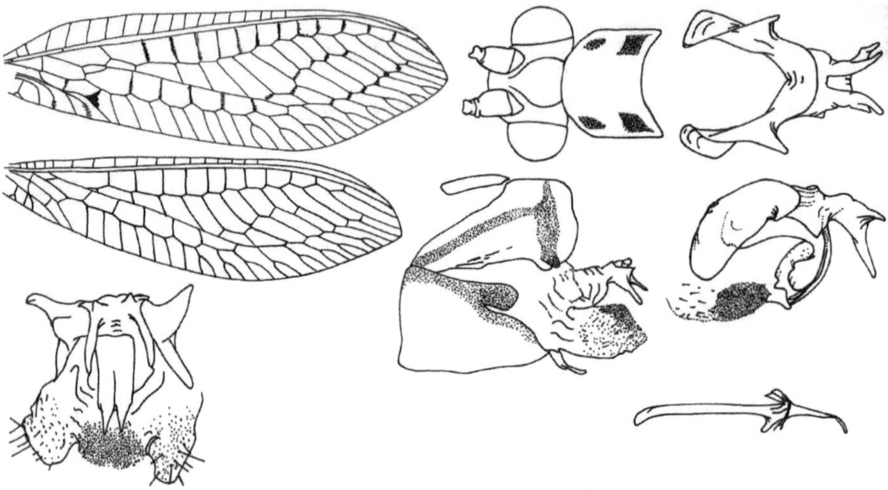

Fig. 16.124 *Ceraeochrysa paraguaria:* fore and hind wing (upper and middle left), head with only bases of the antennae and pronotum in dorsal view (upper center), apex of the male abdomen in ventral (upper right) and lateral view (middle right), subgenitale (lower right), male genitalia in dorsal (lower left) and lateral view (lower left center), and female genitalia (lower right center). Based on de Freitas and Penny (2001).

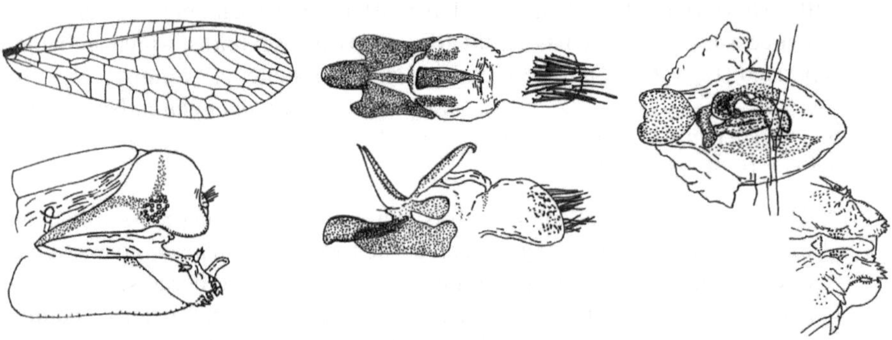

Fig. 16.125 *Ceraeochrysa cubana* (above, left to right): fore-wing, gonarcus in dorsal view, female genital system in ventral view, and (below, left to right): male abdomen in lateral view, gonarcus apparatus in lateral view, and apex of the ninth sternite in dorsal view. Based on Adams and Penny (1987).

from the two species in Couplet 8. Only specimens with color patterns of *C. cincta* known to occur in South America are included in this key 8

– The ground color on the dorsal surface of the antennal scape is either uniformly pale, or it is red with a single dark stripe (Fig. 16.125) 9

8. (7) The basal segments of the antennal flagellum of the female are dark brown or black. The scape of the antenna has two dark stripes. The spermatheca is U-shaped. The dorsal apodeme of the ectoproct does not reach ventrally farther

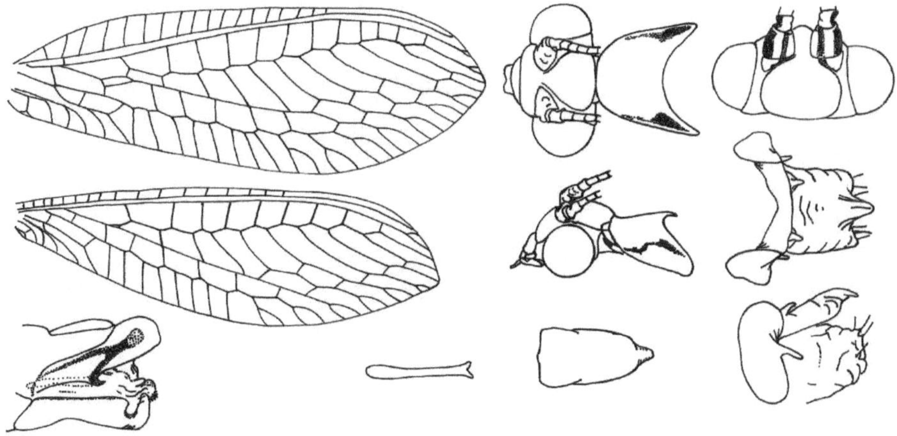

Fig. 16.126 *Ceraeochrysa arioles* male: fore and hind wing (upper and middle left), head and pronotum in dorsal (upper right center) and lateral view (right center), head in dorsal view (upper right), apex of the abdomen in lateral view (lower left), genitalia in dorsal (middle right) and lateral view (lower right), gonapsis (lower center), and the fused eighth and ninth sternites in ventral view (lower right center). Based on de Freitas et al. (2009).

than the ectoproct itself. The gonapsis not forked (Fig. 16.122). Fore-wing length of female: c. 15 mm. The male of this species has not been described. .... *Ceraeochrysa montoyana* (Navás, 1913) (Venezuela, Paraguay, Mato Grosso, Minas Gerais, São Paulo). Syn: *Chrysopa montoyana* Navás, 1913.

– The basal segments of the antennal flagellum are pale. The ninth abdominal segment of the male is strongly recurved (Fig. 16.126). Length of male fore-wing: c. 13 mm; hind wing: c. 11 mm. The female has not been described.

.... *Ceraeochrysa arioles* (Banks, 1944) (Mexico, Central America, Guyana). Syn: *Chrysopa adornata* LaCroix, 1926, treated as a nomen dubium by Freitas et al. (2009) because its type is so badly damaged; *Chrysopa binaria* Navás, 1928 preoccupied by *Chrysopa binaria* Navás, 1923; *Chrysopa arioles* Banks, 1944.

9. (7) The proximal segments of the antennal flagellum are predominantly pale (Fig. 16.125)... 10

– The proximal segments of the antennal flagellum are dark (Fig. 16.127)......23

10. (9). The maxillary palp is black or dark brown (Fig. 16.125)........................ 11

– The maxillary palp is amber yellow or paler (Fig. 16.128)...................... 12

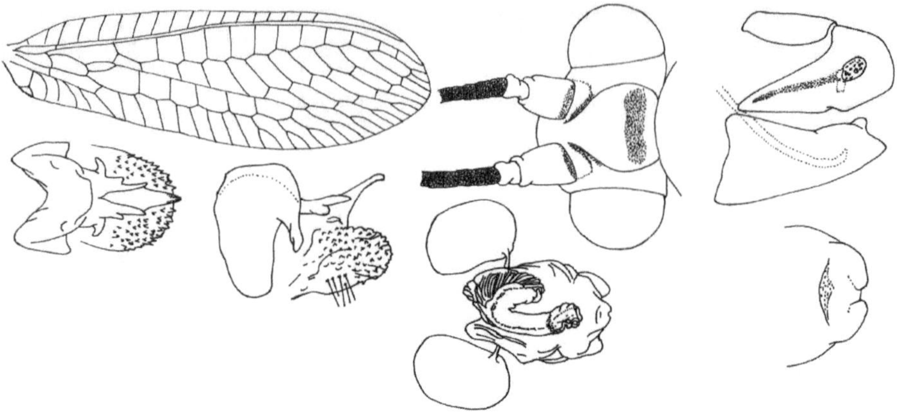

Fig. 16.127 *Ceraeochrysa nigripes* (above, left to right): wing, head in dorsal view, apex of the male abdomen, and (below, left to right): male genitalia in ventral and lateral view, female spermatheca, and subgenitale. Based on de Freitas et al. (2009).

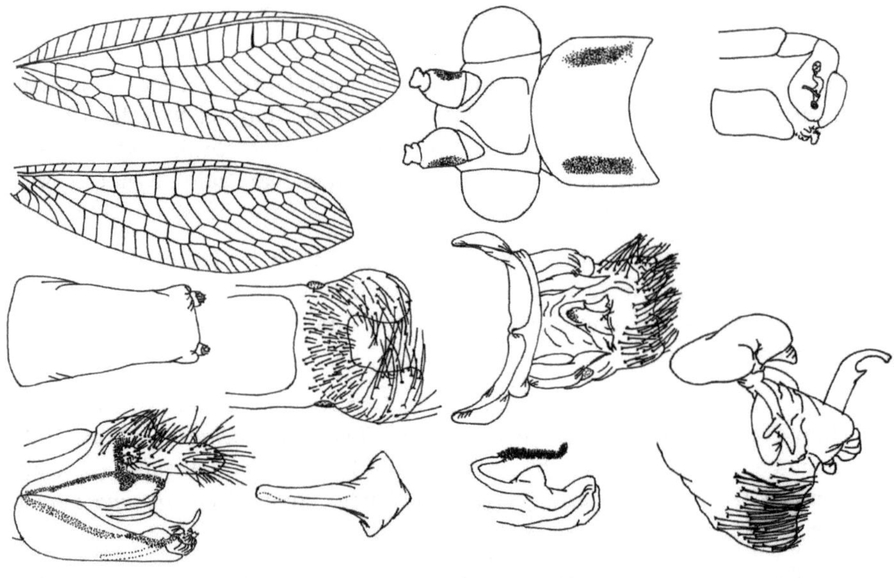

Fig. 16.128 *Ceraeochrysa forcipata:* fore and hind wing (upper and upper middle left), head and pronotum in dorsal view (upper right center), eighth and ninth sternites in ventral view (lower middle left), apex of the male abdomen in dorsal (left of center) and lateral view (lower left), male genitalia in dorsal (right of center) and lateral view (lower right), apex of the female abdomen in lateral view (upper right), gonapsis (lower left center), and spermatheca (lower right center). Based on de Freitas et al. (2009).

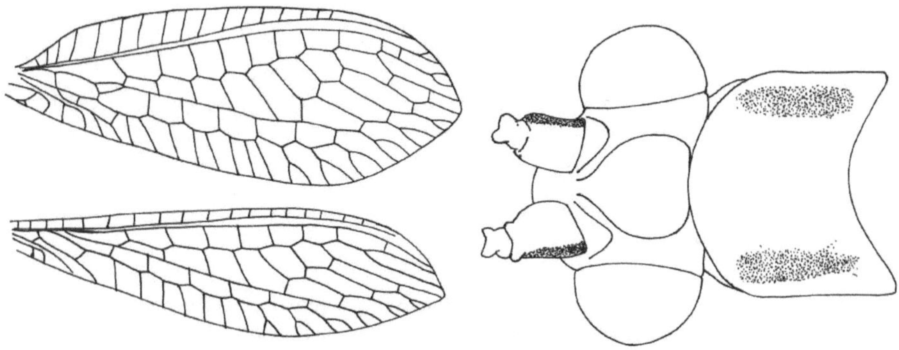

Fig. 16.129 *Ceraeochrysa adornata:* fore and hind wing (left) and the head and pronotum in dorsal view (right). Based on de Freitas et al. (2009).

11. (10) The genae are uniformly pale. The flagellum is entirely pale. The stripe on the scape is red (Fig. 16.125). There is a dorsomedial projection on the gonarcus. Several large gonocristae are present on the gonosaccus. The mesonotum lacks markings.
   .... *Ceraeochrysa cubana* (Hagen, 1861) (North America, West Indies, Venezuela, Guyana, Surinam, Argentina, Pará, Roraima, Amazonas, Rondônia). Syn: *Chrysopa cubana* Hagen, 1861; *Chrysopa tolteca* Banks, 1901; *Chrysopa albatala* Banks, 1913; *Chrysopa venularis* Navás, 1913; *Chrysopa imbecilla* Navás, 1935; *Chrysopa epheba* Navás, 1924; *Chrysopa seminole* Banks, 1924; *Chrysopa freemani* Smith, 1931; *Chrysopa damiensis jamaicensis* Banks, 1941.
   – There is a dark spot on each gena. There is a dark lateral stripe on the scape of the antenna. There are four inner gradate veins in the fore-wing (Fig. 16.129). Fore-wing length: c. 10 mm.

   .... *Ceraeochrysa adornata* (LaCroix, 1926) (Minas Gerais). Syn: *Chrysopa adornata* LaCroix, 1926. This name was treated as a *nomen dubium* by Freitas et al. (2009) because the type specimen is too badly damaged to identify. In addition to the two species in Couplet 10, a species not known south of Panama, *Ceraeochrysa panamensis* Freitas and Penny (2009) fits this description. It can be distinguished by the very dark margins of the cross veins on its fore-wings.
12. (10) There are eight inner gradate veins in the fore-wing. The length of the male ectoproct is three times its width (Fig. 16.128). Fore-wing length: c. 11 mm.
   .... *Ceraeochrysa forcipata* Penny and Freitas, 2009 (Bolivia).
   – There are no more than five inner gradate veins in the fore-wing. The length of the male ectoproct is no longer than twice its width (Fig. 16.130) 13
13. (12) There is a stripe on the scape of the antenna (Fig. 16.130) 14
   – The scape of the antenna lacks a stripe (Fig. 16.131) 19

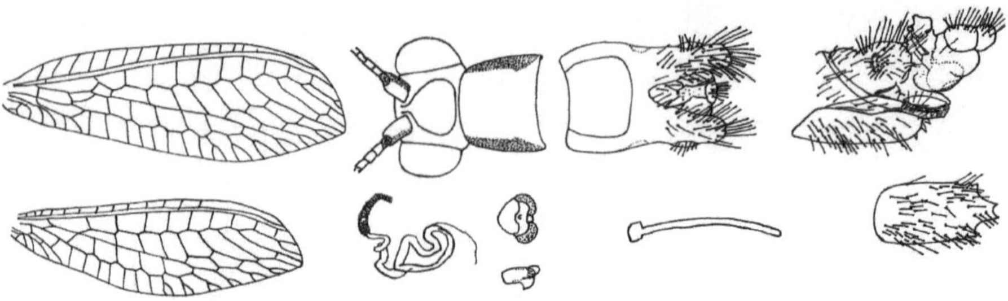

Fig. 16.130 *Ceraeochrysa tucumana* (above, left to right): fore-wing, head in pronotum in dorsal view, apex of the male abdomen in dorsal and lateral view, and (below, left to right): hind wing, spermatheca, subgenitale in ventral (above) and lateral view (below), gonapsis, and the fused sternite 8+9. Based on de Freitas and Penny (2001).

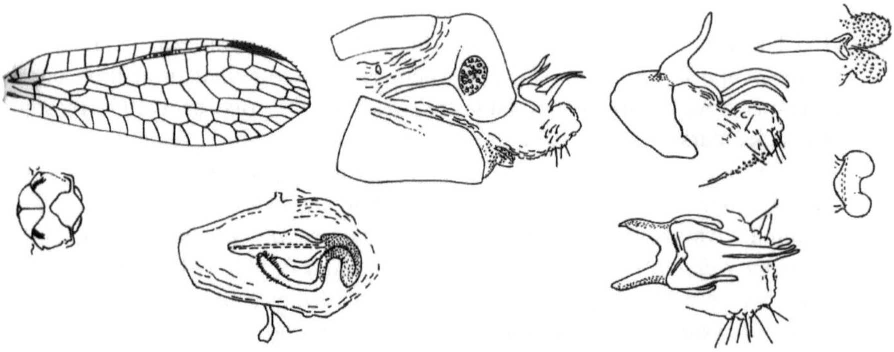

Fig. 16.131 *Ceraeochrysa michaelmuris:* fore-wing (upper left), metanotum (lower left), apex of the male abdomen in lateral view (upper center), gonarcus apparatus in lateral (upper right center) and dorsal view (lower right center), gonapsis and gonocristae in dorsal view (upper right), subgenitalia (lower right), and the female genital system (lower left center). Based on Adams and Penny (1987).

14. (13) The stripe on the scape of the antenna is dorsolateral (Fig. 16.130)....... 15
    – The stripe on the scape of the antenna is mid-dorsal (Fig. 16.132)............ 16
15. (14) The ventral apodeme of the male is straight, and it lacks a dorsal lobe. The fused eighth and ninth sternites of the male have two lateral and two median points and scattered setae, which are not arranged in distinct groups. The gradate veins of the fore-wings are not heavily shaded (Fig. 16.130). Most of the wing veins are green. Fore-wing length: c. 11 mm. ....*Ceraeochrysa tucumana* (Navás, 1919) (Argentina, São Paulo). Syn: *Chrysopa tucumana* Navás, 1919.
    – The ventral apodeme of the male has a subapical dorsal lobe. The fused eighth and ninth sternites of the male are simply rounded at the apex (Fig. 16.133). Many of the cross veins in the fore-wing are dark; most of those in the hind wing are pale. Fore-wing length: c. 12 mm. ....*Ceraeochrysa valida* (Banks, 1895)

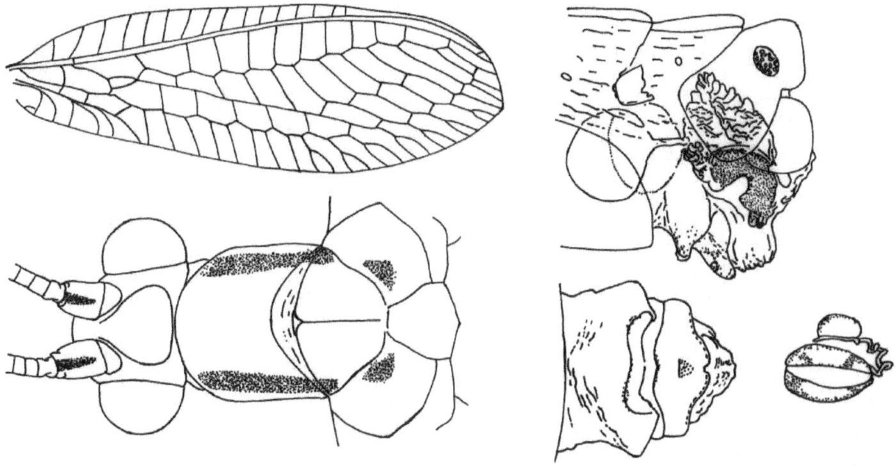

Fig. 16.132 *Ceraeochrysa rafaeli* female: fore-wing (upper left), head with basal antennal segments and thorax in dorsal view (lower left), apex of the abdomen in lateral view (upper right), subgenitale (lower right center), and spematheca (lower right). Based on de Freitas et al. (2009).

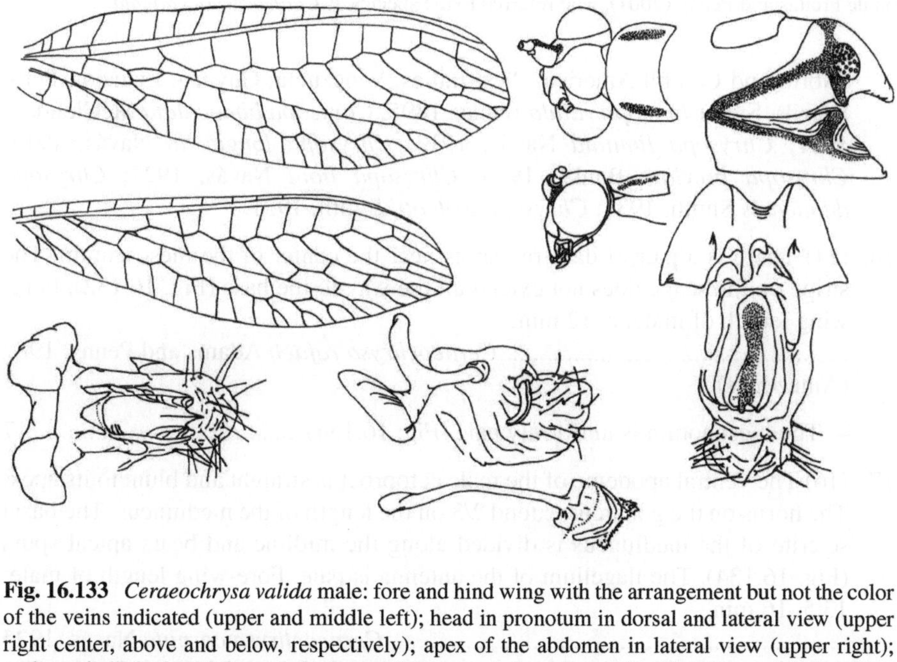

Fig. 16.133 *Ceraeochrysa valida* male: fore and hind wing with the arrangement but not the color of the veins indicated (upper and middle left); head in pronotum in dorsal and lateral view (upper right center, above and below, respectively); apex of the abdomen in lateral view (upper right); male genitalia in dorsal (lower left), lateral (below center), and posterior view (lower right); gonapsis (lower center). Based on de Freitas et al. (2009).

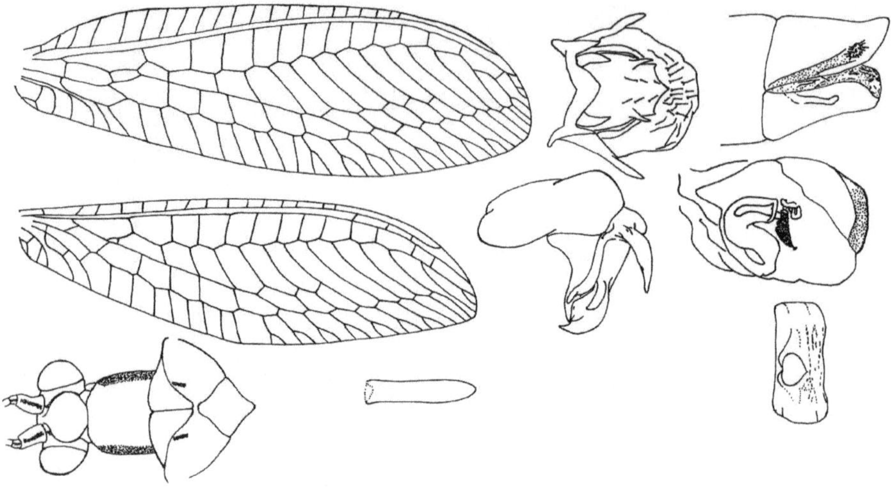

Fig. 16.134 *Ceraeochrysa cornuta:* fore and hind wing (upper and middle left), head and anterior part of the thorax in dorsal view (lower left), male genitalia in dorsal (upper right center) and lateral view (right of center), apex of the male abdomen in lateral view (upper right), gonapsis (lower center), female genitalia in a cleared specimen (middle right), and subgenitale (lower right). Based on de Freitas and Penny (2001), who referred to the species as *Ceraeochrysa caligata.*

(North and Central America, West Indies, Venezuela, Guyana, Ecuador, Peru, Brazil). Syn: *Chrysopa valida* Banks, 1895; *Chrysopa bimaculata* McClendon, 1901; *Chrysopa limitata* Navás, 1913; *Chrysopa longicella* Navás, 1914; *Chrysopa breviata* Banks, 1915; *Chrysopa lioni* Navás, 1927; *Chrysopa damiensis* Smith, 1931; *Chrysopa wolcotti* Smith, 1931.

16. (14) There is a pair of dark red spots near the center of the mesoscutum. The stripe on the scape does not extend all the way to the base (Fig. 16.132). Fore-wing length of male: c. 12 mm.
....*Ceraeochrysa rafaeli* Adams and Penny, 1987 (Amazonas).

– The mesonotum is uniformly pale (Fig. 16.134) 17

17. (16) The ventral apodeme of the male ectoproct is straight and blunt to its apex. The horns on the gonarcus extend 2/3 on the length of the mediuncus. The basal sclerite of the mediuncus is divided along the midline and bears apical spurs (Fig. 16.134). The flagellum of the antenna is pale. Fore-wing length of male: 13.5–16 mm.
....*Ceraeochrysa cornuta* Navás, 1925 (Mexico, Central America, Ecuador, Amazonas, Pará, Minas Gerais, São Paulo). Syn: *Chrysopa cornuta* Navás, 1926; *Chrysopa caligata* Banks, 1946; *Ceraeochrysa cincta* Adams, 1982 (*nec* Schneider, 1851); *Ceraeochrysa caligata* (Banks, 1946) auctt.

– The ventral apodeme of the male ectoproct is recurved (Fig. 16.135) 18

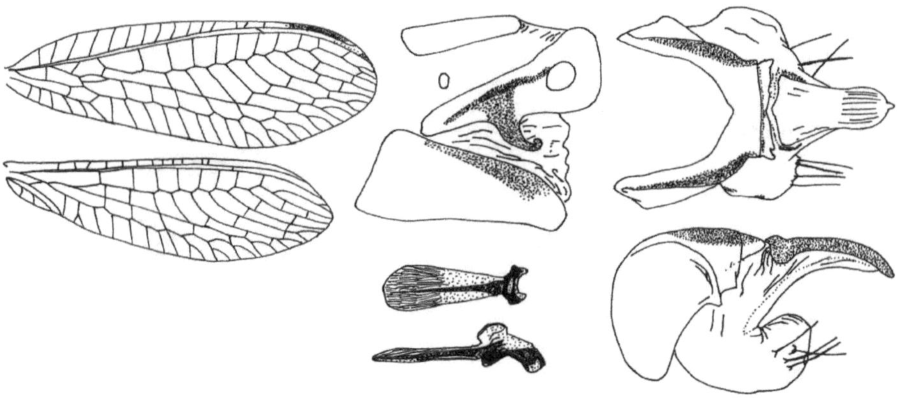

Fig. 16.135 *Ceraeochrysa pittieri* male: fore and hind wing (left), the ninth sternite and ectoproct (upper center), gonarcus in dorsal (upper right) and lateral view (lower right), the gonapsis in dorsal (below center) and lateral view (lower center). Based on Sosa and de Freitas (2010).

18. (17) The recurved ventral apodeme of the male ectoproct comes to a point at its apex. The dark stripe on the scape does not extend into the fossa at the base of the antenna. The mediuncus is divided into separate basal sclerites bearing blunt dorsal spines. The gonapsis lacks apical horns (Figs. 16.20 and 16.136). The female subgenital plate has a vertical invagination. Fore-wing length of male: 13.1–15.2 mm; female: 13.7–14.2 mm.
....*Ceraeochrysa cincta* (Schneider, 1851) (North and Central America, West Indies, Galapagos Islands, Guyana, Surinam, Peru, Bolivia, Paraguay, Argentina, Uruguay, Rio de Janeiro, São Paulo, Ceará). Syn: *Chrysopa cincta* Schneider, 1851; *Chrysopa bilineata* Navás, 1914; *Chrysopa bruchi* Navás, 1914; *Chrysopa lafonei* Navás, 1914; *Chrysopa incalis* Banks, 1915; *Chrysopa rochina* Navás, 1915; *Chrysopa thallina* Navás, 1919; *Chrysopa bicarnea* Banks, 1920; *Chrysopa advena* Navás, 1922; *Chrysopa habana* Navás, 1922; *Chrysopa bessona* Navás, 1922; *Chrysopa mestiza* Navás, 1924; *Cintameva bina* Navás, 1924; *Nodita bina* (Navás, 1924) Alayo, 1968; *Chrysopa villosula* Navás, 1924; *Chrysopa silvestrina* Navás, 1929; *Chrysopa alternans* Navás, 1933; *Chrysopa wollebaeki* Esben-Petersen, 1934; *Chrysopa iona* Banks, 1944; *Chrysopodes sallei* Banks, 1946; *Ceraeochrysa rochina* (Navás, 1915) auctt.

– The recurved ventral apodeme of the male ectoproct is twisted at its apex (Fig. 16.135). Fore-wing length of male: c. 12.0 mm; hind wing: c. 10.7 mm. ....*Ceraeochrysa pittieri* Sosa and Freitas, 2010 (Venezuela).

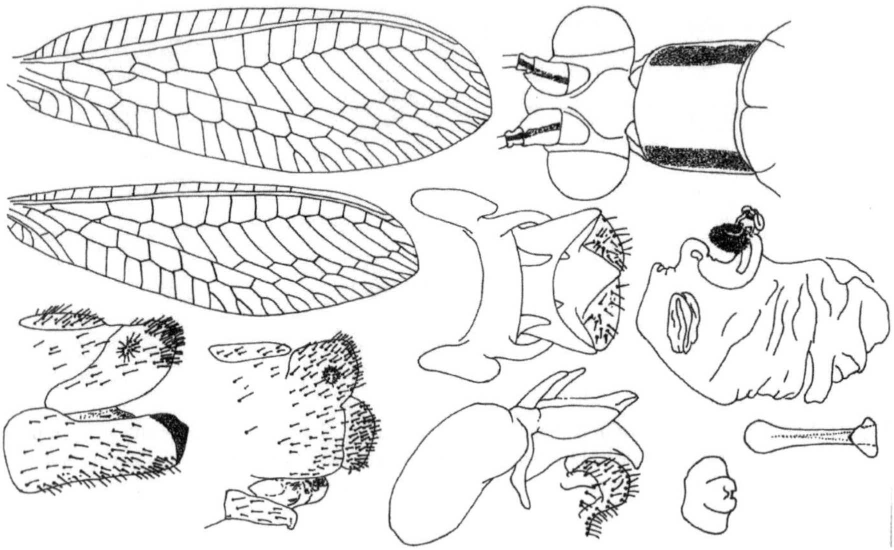

Fig. 16.136 *Ceraeochrysa cincta:* fore and hind wing (upper and middle left); head and pronotum showing the scape and pedicel of the antenna in dorsal view (upper right); apex of the male (lower left) and female abdomen (lower left center), both in lateral view; male genitalia in dorsal (center) and lateral view (lower center); female genitalia (middle right); gonapsis (lower right); subgenitale (lower right center). Based on de Freitas et al. (2009). Caution: some earlier illustrations were of misidentified specimens of other species.

19. (13) The mesoscutum has a pair of dark markings at its anteriolateral border with the prescutum. The gonocristae of the male are located on conspicuous, protruding lobes (Fig. 16.131). Fore-wing length of male: c. 8.8 mm; female: c. 9.9 mm.
 .... *Ceraeochrysa michaelmuris* Adams and Penny, 1987 (Amazonas).

 – The mesoscutum in completely uniformly pale in color. The gonocristae of the male are not located on protruding, spiny lobes (Fig. 16.137) 20

20. (19) On dorsal surface of each gena, there is a dark spot from the eye to the anterior tentorial pit. There is dark shading along the margins of the gradate veins (Fig. 16.137) 21

 – There is no dark marking of the dorsal surface of the gena. There is no dark shading along the margins of the gradate veins (Fig. 16.138) 22

21. (20) The gonocornua of the male are longer than the two medial horns of the mediuncus (Fig. 16.139). The gradate veins of the fore-wings are heavily shaded. The apical hook of the arcessus is about five times as long as wide. There are two apical fields of long setae at the apex of the eighth and ninth sternite. Fore-wing length of male: c. 9.1 mm; female: c. 10.2 mm.
 .... *Ceraeochrysa tenuicornis* Adams and Penny, 1987 (Amazonas, Pará).

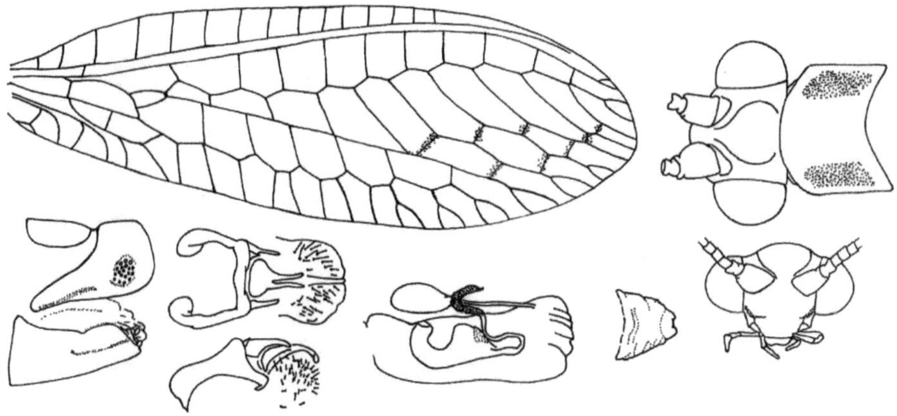

Fig. 16.137 *Ceraeochrysa ariasi:* fore-wing (upper left), head and pronotum in dorsal view (upper right), and (below, left to right) apex of the male abdomen in lateral view, male genitalia in dorsal and lateral view (above and below, respectively), female genitalia in lateral view, female subgenitale, and the head in anterior view. Based on de Freitas et al. (2009).

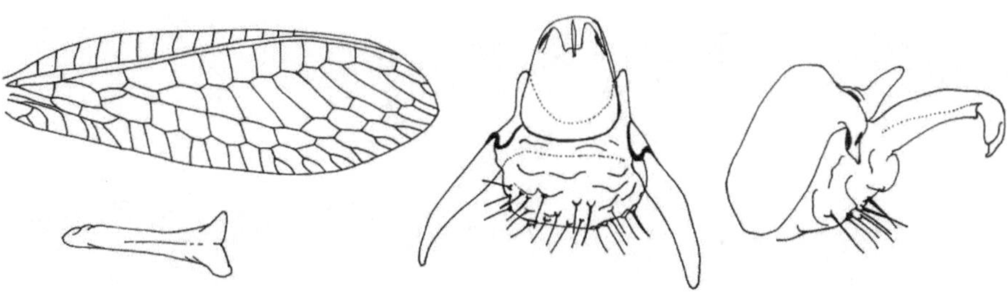

Fig. 16.138 *Ceraeochrysa reddyi* male: wing (upper left), genitalia in posterior (center) and lateral view (right), and gonapsis (lower left). Based on de Freitas et al. (2009).

– The gonocornua of the male appear vestigial and are shorter than the two medial horns of the mediuncus (Fig. 16.137). Fore-wing length: c. 10.0 mm for both sexes.

.... *Ceraeochrysa ariasi* Adams and Penny, 1987 (Rondônia, Amazonas).

22. (20) All cross veins in the fore-wing are dark, as are the branches of vein Rs. The ninth sternite of the male bulges at the apex and has dentate apicolateral lobes (Fig. 16.138). A vestige of prenotal stripes may be evident. Fore-wing length of male: 10.0 mm.

.... *Ceraeochrysa reddyi* Adams and Penny, 1987 (Argentina, Pará).

– Most cross veins in the fore-wing are green (Fig. 16.140). Fore-wing length: c. 11 mm.

.... *Ceraeochrysa silvanoi* (Navás, 1916) (Rio de Janeiro). Syn: *Chrysopa silvanoi* Navas, 1916.

Fig. 16.139 *Ceraeochrysa tenuicornis* male: fore and hind wing (upper and middle left), head and pronotum in dorsal view (upper center), male genitalia in dorsal (upper right) and lateral view (lower right), apex of the abdomen in lateral view (lower center), and gonapsis (lower left). Based on de Freitas and Penny (2001).

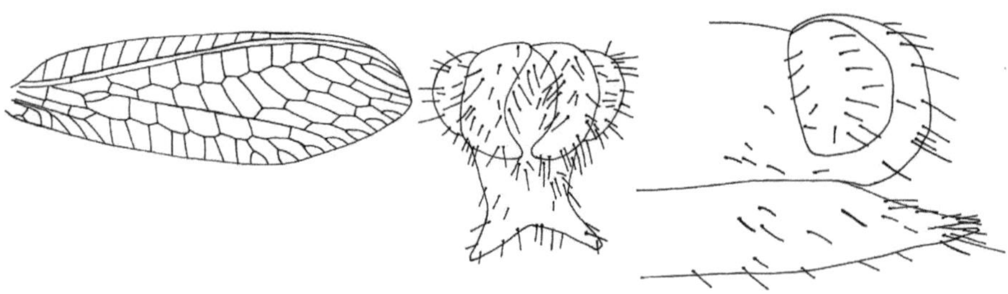

Fig. 16.140 *Ceraeochrysa silvanoi* male (left to right): wing and apex of the abdomen in posterior and lateral view. Based on de Freitas et al. (2009).

23. (9) The legs have dark brown or black markings. The scape has a basodorsal band (Fig. 16.127). Fore-wing length of male: 11.8–13.0 mm; female: c. 12.3 mm. .... *Ceraeochrysa nigripes* Adams and Penny, 1987 (Pará, Amazonas, Rondônia, Acre).

   – The legs are pale (Fig. 16.141) 24

24. (23) The scape of the antenna is uniform in color (Fig. 16.141) 25

   – There is a dark stripe on the dorsal or dorsolateral surface of the antennal scape (Fig. 16.142) 28

25. (24) Each gena has a red marking (Fig. 16.141) 26

   – The genae are uniformly pale in color (Fig. 16.143) 27

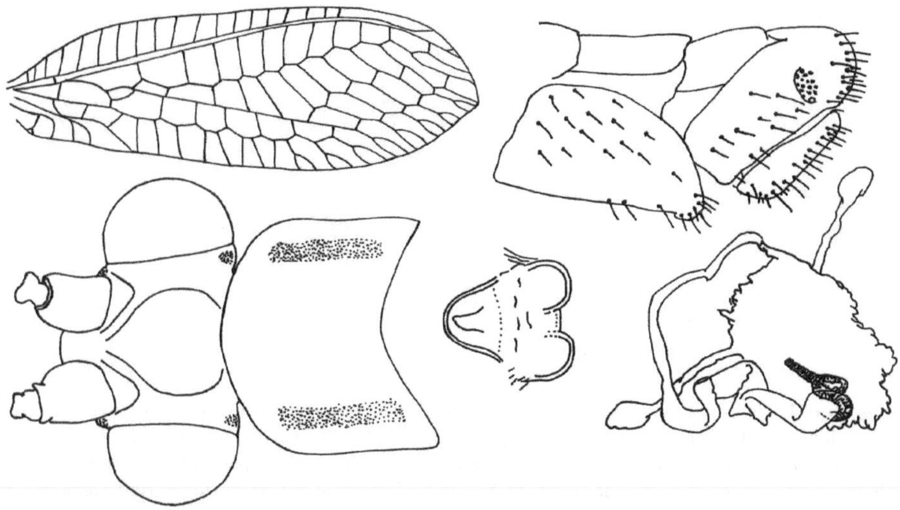

Fig. 16.141 *Ceraeochrysa anceps* female: fore-wing (upper left), head without the antennal flagellum and pronotum in dorsal view (lower left), apex of the abdomen in lateral view (upper right), subgenitale (lower center), and the genitalia with the spermatheca (lower right). Based on de Freitas et al. (2009).

Fig. 16.142 *Ceraeochrysa fiebrigi* female: fore and hind wing (left), head and pronotum in dorsal view (upper right), spermatheca (lower right center), and the subgenitale in lateral view (lower right). Based on de Freitas et al. (2009).

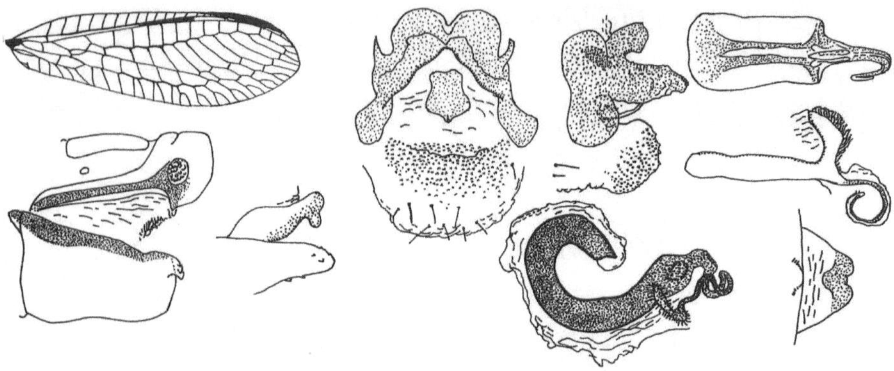

Fig. 16.143 *Ceraeochrysa squalidens:* fore-wing (upper left), apex of the male abdomen in lateral view (lower left), gonarcus apparatus in lateral (upper right center) and posterior view (upper center), gonapsis in dorsal (upper right) and lateral view (middle right), spermatheca in ventral view (lower right center), subgenitalia (lower right), and subgenitale in ventral (lower right) and lateral view (lower left center). Based on Adams and Penny (1987).

26. (25) The frons is entirely pale. The maxillary palps are amber. The vertex has a pair of small red spots along its posterior margin (Fig. 16.141). .... *Ceraeochrysa anceps* (Navás, 1926) (Rio de Janeiro).

   – The frons has a pair of red spots along its midline. The maxillary palps are pale. The vertex is uniformly pale (Fig. 16.144). Fore-wing length of female: c. 15 mm. The male has not been described. Whether the species is correctly assigned to *Ceraeochrysa* is uncertain (de Freitas et al. 2009).

   .... *Ceraeochrysa caucana* (Banks, 1910) (Colombia, Brazil). Syn: *Chrysopa caucana* Banks, 1910.

27. (25) At the dorsolateral corners of the fused eighth and ninth abdominal sternites of the male, there are ligulate projections bearing stout setae with conical bases. The gonarcus has a bilobed dorsal hood. The gonapsis has a double crest bearing adpressed scales (Fig. 16.143). The branches of vein Rs in the fore-wing are pale. Fore-wing length of male: 10.8–11.6 mm; female: c. 13.1 mm. Length of male antenna: 17.5 mm. .... *Ceraeochrysa squalidens* Adams and Penny, 1987 (Amazonas).

   – The fused seventh and eighth abdominal sternites are evenly rounded at the along the posterior margin. The gonapsis has a scythe-shaped process (Fig. 16.145). Wing length: c. 13.5 mm. The female has not been described.

   .... *Ceraeochrysa falcifera* Adams and Penny, 1987 (Pará).

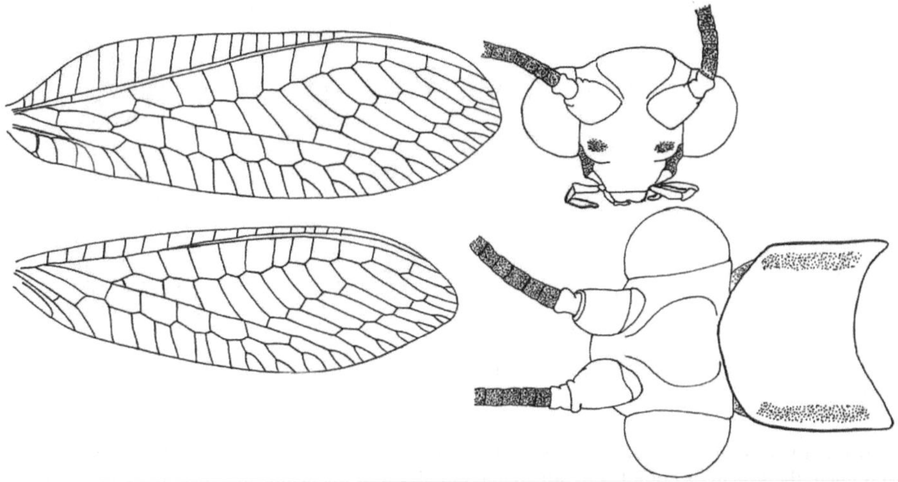

Fig. 16.144 *Ceraeochrysa caucana* female: fore and hind wing (left), head and pronotum in dorsal view (lower right), and the head in anterior view (upper right). Based on de Freitas et al. (2009).

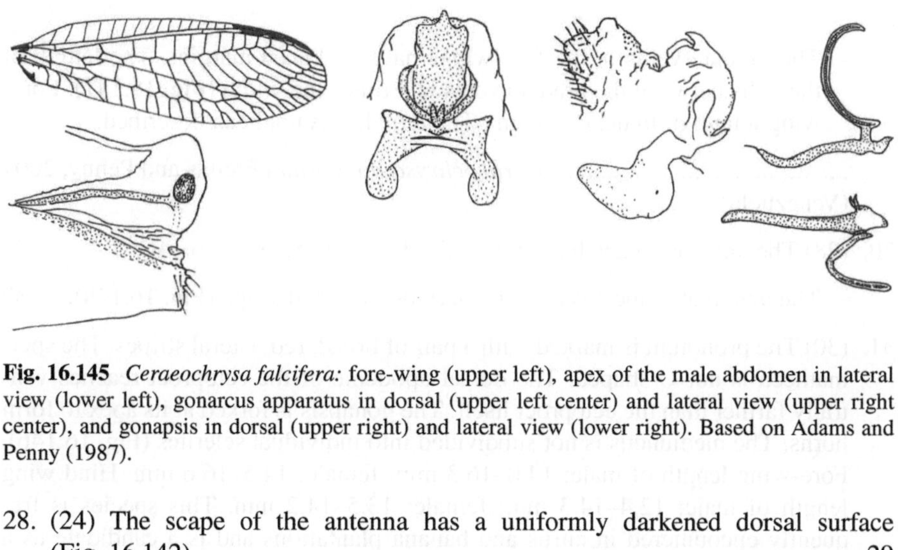

Fig. 16.145 *Ceraeochrysa falcifera:* fore-wing (upper left), apex of the male abdomen in lateral view (lower left), gonarcus apparatus in dorsal (upper left center) and lateral view (upper right center), and gonapsis in dorsal (upper right) and lateral view (lower right). Based on Adams and Penny (1987).

28. (24) The scape of the antenna has a uniformly darkened dorsal surface (Fig. 16.142).. 29
 – The dorsal surface of the antennal scape is divided into discrete dorsal and dorsolateral stripes (Fig. 16.146)... 30
29. (28) The gradate veins in the fore-wing do not have darkened margins. The length of the cells between the gradate veins is 1½ times the width (Fig. 16.142). Fore-wing length of female: c. 14 mm. The male has not been described. .... *Ceraeochrysa fiebrigi* (Navás, 1913) (Paraguay). Syn: *Chrysopa fiebrigi* Navás, 1913.

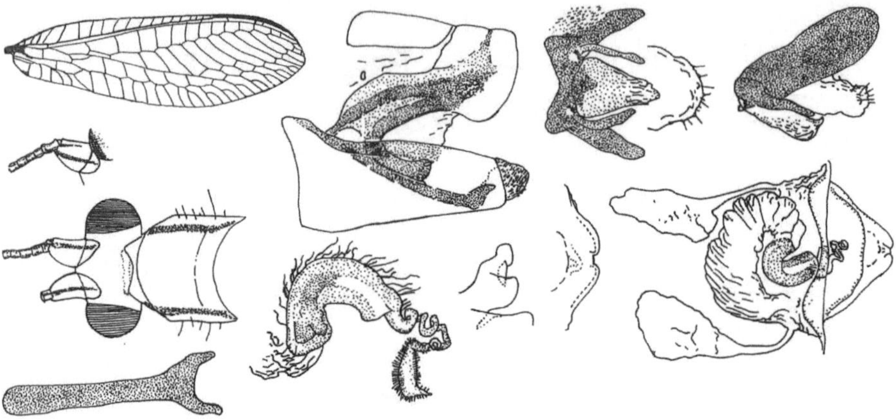

Fig. 16.146 *Ceraeochrysa claveri* (above, left to right): wing, lateral view of the apex of the male abdomen, the gonarcus complex in dorsal and lateral view, the scape and basal flagellar segments in dorsal view (upper middle left), head and pronotum in dorsal view (lower middle left), gonapsis in dorsal view (lower left), right side of the spermatheca (lower left center), subgenitale in lateral (lower center) and ventral view (lower right center), and the female genital system in ventral view. Based on Adams and Penny (1987).

– The gradate veins in the fore-wings have darkened margins. The length of the cells between the gradate veins is 3 times the width (Fig. 16.147). Fore-wing length of male: c. 10 mm. The female has not been described.

.... *Ceraeochrysa bitacornua* Freitas and Penny, 2009 (Venezuela).

30. (28) The antennal scape has a thin, mid-dorsal stripe (Fig. 16.146)............... 31

– The antennal scape has a broad, dark dorsolateral stripe (Fig. 16.148)...... 32

31. (30) The pronotum is marked with a pair of broad, red, lateral stripes. The spermatheca is not U-shaped. The dorsal apodeme of the ectoproct reaches ventrally farther than the ectoproct itself. The gonapsis is forked at its apex to form horns. The mediuncus is not subdivided into individual sclerites (Fig. 16.146). Fore-wing length of male: 14.0–16.3 mm; female: 14.5–16.6 mm. Hind wing length of male: 12.4–14.3 mm; female: 13.5–14.2 mm. This species is frequently encountered in citrus and banana plantations and is a candidate as a predator for pest control.

.... *Ceraeochrysa claveri* (Navás, 1911) (Central America, West Indies, Colombia, Venezuela, French Guiana, Guyana, Surinam, Argentina, Amazonas, Pará, Ceará, Bahia, Pernambuco, Minas Gerais, Mato Grosso, São Paulo). Syn: *Chrysopa claveri* Navás, 1911; *Chrysopa silvana* Navás, 1913; *Chrysopa deficiens* Navás, 1930; *Citameva sibirica* Navás, 1930; *Chrysopa haitiensis* Smith, 1931; *Chrysopa adoina* Banks, 1945; *Chrysopa inexpectata* Alayo, 1968.

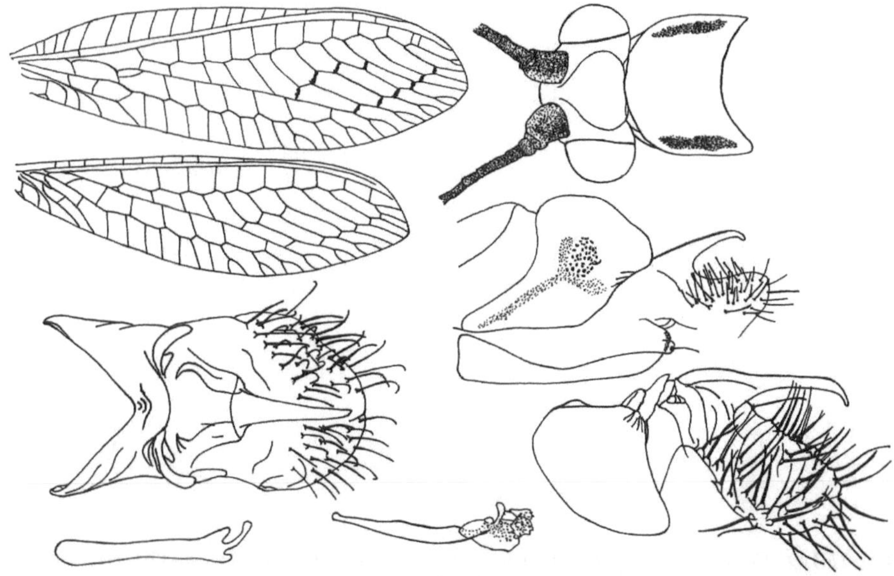

Fig. 16.147 *Ceraeochrysa bitacornua* male: fore and hind wing (upper left), head and pronotum in dorsal view (upper right), apex of the abdomen in lateral view (middle right), genitalia in dorsal (lower middle left) and lateral view (lower right), and the gonapsis in dorsal (lower left) and lateral view (lower center). Based on de Freitas et al. (2009).

Fig. 16.148 *Ceraeochrysa achillea* male: fore and hind wing (upper and middle left), head in anterior view (upper center), head and pronotum in lateral view (upper right), apex of the abdomen in lateral view (lower left), genitalia in dorsal (center) and lateral view (lower right), and gonapsis (lower right center). Based on de Freitas et al. (2009).

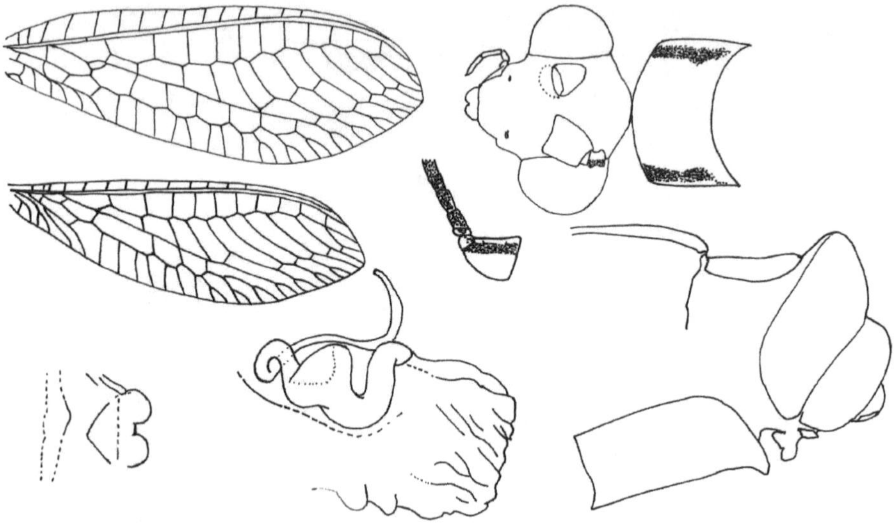

Fig. 16.149 *Ceraeochrysa castilloi* female: fore and hind wing (upper and middle left), marking on the base segments of the antenna (center), head and pronotum in dorsal view (upper right), apex of the abdomen in lateral view (lower right), spermatheca (lower center), and subgenitale (lower left). Based on de Freitas et al. (2009).

– The pronotum is marked with a pair of narrow, brown, lateral stripes (Fig. 16.149). Fore-wing length of female: c. 14 mm. The male has not been described.

....*Ceraeochrysa castilloi* (Navás, 1913) (Paraguay, Rio de Janeiro). Syn: *Chrysopa Castilloi* Navás, 1913.

32. (30) The genae have dark markings. The gonapsis of the male is forked apically (Fig. 16.148). Fore-wing length: c. 11 mm. ....*Ceraeochrysa achillea* Freitas and Penny, 2009 (Panama, Venezuela).

– The genae are entirely pale in color. The apex of the male gonapsis is unforked (Fig. 16.150)... 33

33. (32) The posterolateral lobe of the fused eighth and ninth abdominal sternite of the male curves dorsad. The gonocornua is massive and has teeth at its apex; its length is about twice its width. The ectoproct is produced ventrally so that it almost surrounds the abdomen, and it bears a ventrally elongated dorsal apodeme. The gonosaccus has very large basal tubercles bearing gonosetae (Fig. 16.150). The antennal flagellum is black. The mesonotum of living specimens is yellow. Fore-wing length: c. 11 mm. ....*Ceraeochrysa sanchezi* (Navás, 1924) (Mexico, Central America, Cuba, Amazonas, Rio Grande do Norte, Mato Grosso, São Paulo). Syn: *Chrysopa sanchezi* Navás, 1924.

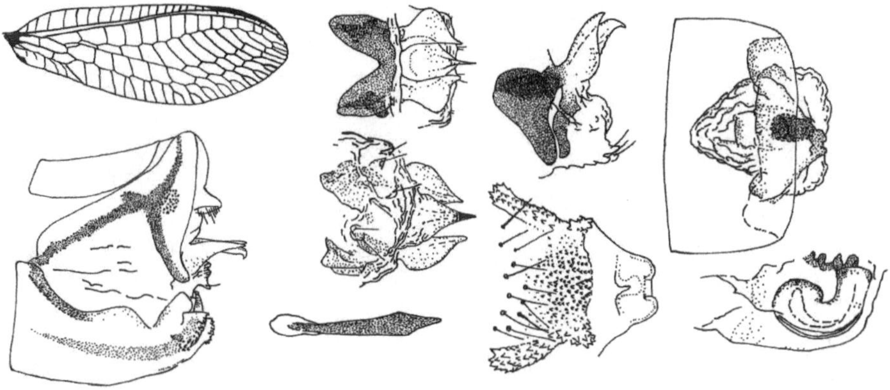

Fig. 16.150 *Ceraeochrysa sanchezi* (above, left to right): fore-wing, male gonarcus complex in dorsal and lateral view, female genitalia in ventral view, and apex of the male abdomen in lateral view (lower left), male gonarcus complex in ventral view (left center), male gonapsis (lower left center), gonocristae in posterior view (lower right center), and the spermatheca in right dorsolateral view (lower right). Based on Adams and Penny (1987).

– The posterolateral lobe of the fused eighth and ninth abdominal sternite of the male gradually tapers toward the apex and lacks a posterolateral lobe. The gonocornua is a relatively slender, digitiform process with only one rounded point (Fig. 16.151) 34

34. (33) The ventral branch of the fork in the dorsal apodeme of the male extends ventral to the margin of the ectoproct and forms a broad, anvil-shaped sclerite at its ventral apex (Fig. 16.151). Fore-wing length: 14–15 mm. .... *Ceraeochrysa acmon* Penny, 1998 (Central America, São Paulo, Mato Grosso do Sul).

– The ventral branch of the fork in the dorsal apodeme of the male does not extend ventral to the margin of the ectoproct (Fig. 16.152) 35

35. (34) The gonarcus of the male lacks a dorsomedial hood. A pair of dark spots is located on the mesoscutum at its suture with the premesoscutum. The fused eighth and ninth sternites of the male are forked at the apex. There are large gonocristae on the membrane ventral to the gonapsis. The ectoproct does not bear a ventrally elongated dorsal apodeme. The gonosaccus has small basal tubercles bearing long setae. The antennal flagellum is dark. The stripe on the scape is black (Fig. 16.152). Fore-wing length of male: 8.9–9.6 mm; female: 9.4–10.4 mm. .... *Ceraeochrysa scapularis* (Navás, 1914) (Colombia, Pará, Roraima, São Paulo). Syn: *Chrysopa scapularis* Navás, 1914.

– There is a dorsomedial hood on the male gonarcus, which arches above the arcessus (Fig. 16.153) 36

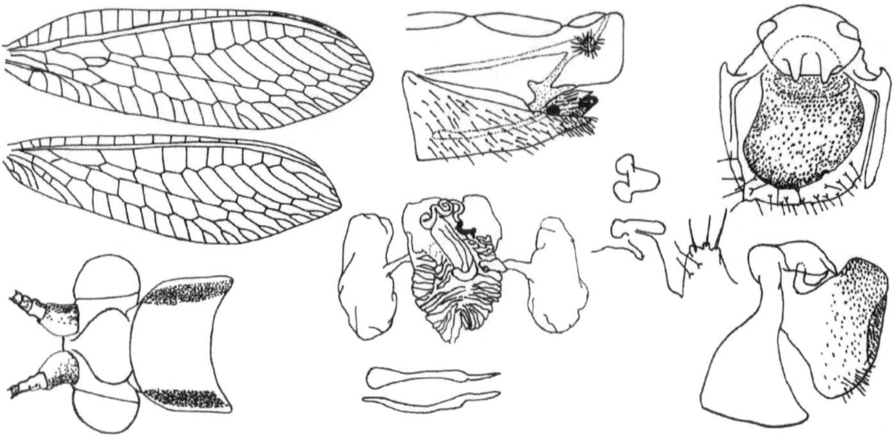

Fig. 16.151 *Ceraeochrysa acmon:* fore and hind wing (upper and middle left), head and pronotum in dorsal view (lower left), apex of the male abdomen in lateral view (upper center), male gentialia in posterior (upper right) and lateral view (lower right), female genitalia (center), subgenitale in posterior and lateral view (right of center, above and below, respectively), and two views of the gonapsis (lower center). Based on de Freitas and Penny (2001).

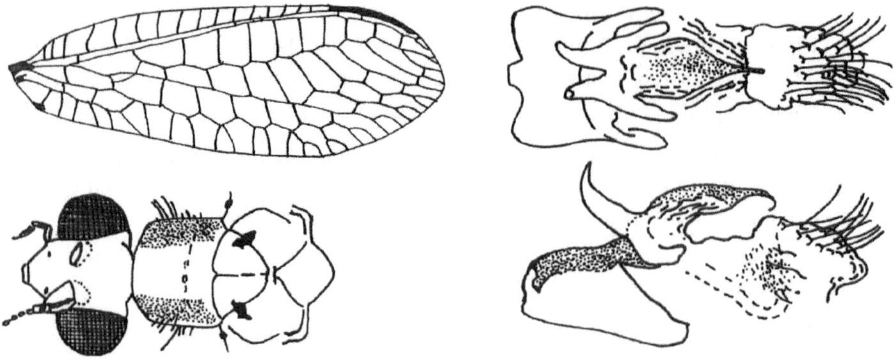

Fig. 16.152 *Ceraeochrysa scapularis:* fore-wing (upper left), head and thorax in dorsal view (lower left), and gonarcus apparatus in posterior (upper right) and lateral view (lower right). Based on Adams and Penny (1987).

36. (35) There are no scales present on the membrane below the ventral margin of the ectoproct (Fig. 16.153).. 37
    - There are scales on the membrane below the ventral margin of the ectoproct (Fig. 16.154) 38
37. (36) The dorsal hood of the male gonarcus is indented in the middle. The ectoproct lacks a ventrally elongated dorsal apodeme. There are teeth at the posterior apex of the gonapsis (Fig. 16.153). Fore-wing length: c. 11 mm. .... *Ceraeochrysa dislepis* Freitas and Penny, 2001 (São Paulo, Goiás, Mato Grosso).

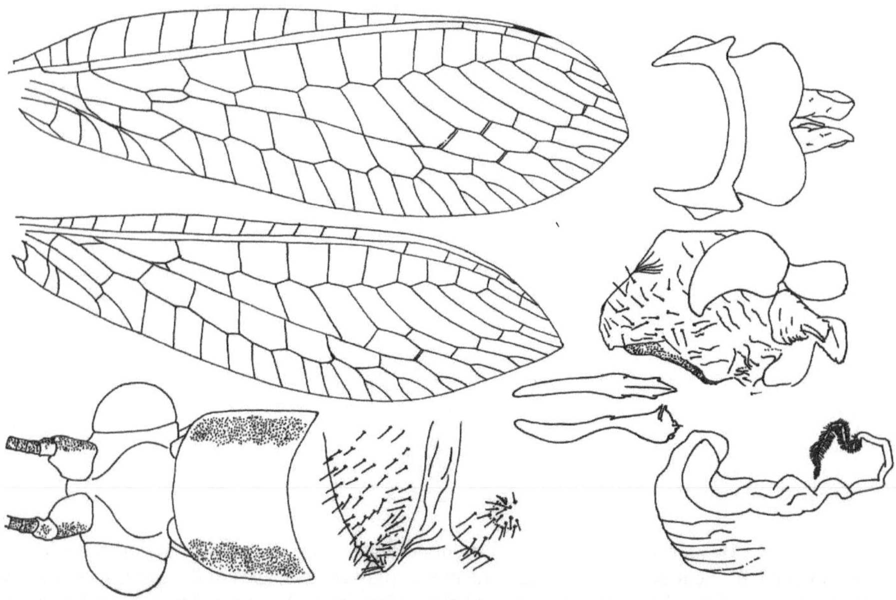

Fig. 16.153 *Ceraeochrysa dislepis:* fore and hind wing (upper and middle left), head and pronotum in dorsal view (lower left), male genitalia in dorsal (upper right) and lateral view (middle right), apex of the male abdomen with the dorsal part right and ventral part on the left (lower center), two views of the gonapsis (just below right center), and the female spermatheca (lower right). Based on de Freitas and Penny (2001).

– The dorsal hood of the male gonarcus is wider than long and not notably indented in the middle. The gonarcus is square. The gonapsis is dilated and forked at its apex (Fig. 16.155). Fore-wing length: 13.9–14.1 mm; hind wing: 12.4–12.7 mm.

.... *Ceraeochrysa melaenopareia* Sosa and Freitas, 2010 (Venezuela).

38. (36) The ectoproct of the male is simply rounded ventrally, and scales are scattered on the membrane below ectoproct. The gonapsis lacks lateral processes. The dark apodeme of the ectoproct does not project ventrad. The gonarcus has a plate-like, dorsomedial projection. The antennal flagellum is pale. The stripe on the scape is sometimes obscure. The male epiproct is not produced ventrally so that it nearly surrounds the abdomen (Fig. 16.154). The mesonotum of living specimens is green. Fore-wing length: c. 15.2 mm; hind wing: c. 13.5 mm. .... *Ceraeochrysa everes* (Banks, 1920) (Central America, West Indies, Colombia, French Guiana, Amazonas, Rondônia, São Paulo, Minas Gerais, Mato Grosso). Syn: *Chrysopa everes* Banks, 1920; *Chrysopa furcata* Navás, 1922 preoccupied; *Chrysopa furculata* Navás, 1923 (replacement); *Chrysopa gundlachi* Navás, 1924;

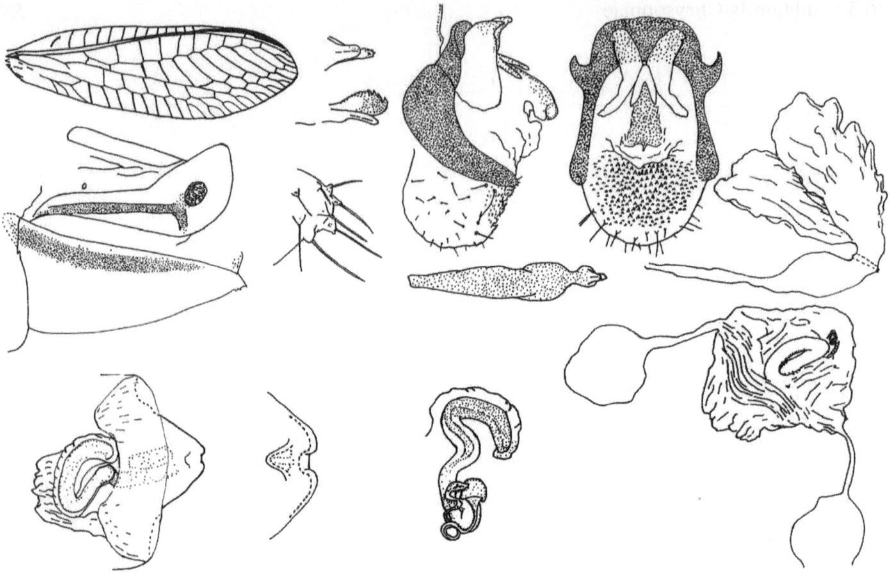

Fig. 16.154 *Ceraeochrysa everes:* fore-wing (upper left), apex of the male abdomen in lateral view (middle left), ventral view of gonapsis (center to right center), apex of the gonapsis in ventral and lateral view (upper left center, above and below, respectively), male gonarcus complex in lateral (upper center) and posterior view (upper right center), apex of the ninth sternite of a male (left center), the entire female genital system in dorsal view showing enlarged colleterial gland reservoirs (right), subgenitale and spermatheca in ventral view (lower left), enlarged spermatheca in ventral view (lower right center), and the apex of the subgenitale (lower left center). Based on Adams and Penny (1987).

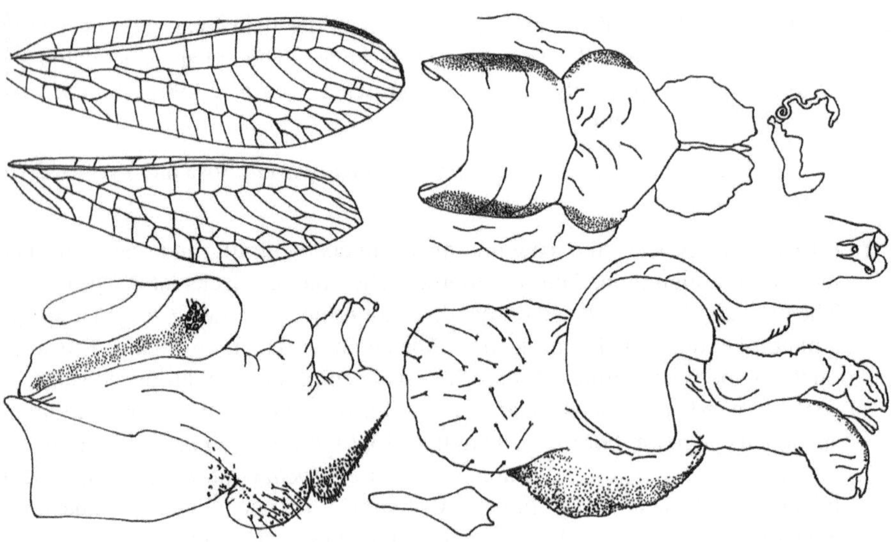

Fig. 16.155 *Ceraeochrysa melaenopareia:* fore and hind wing (upper and middle left), the ninth sternite and the ectoproct with the male genitalia (lower left), the gonarcus in dorsal (upper right center) and lateral view (lower right), the gonapsis in dorsal view (lower center), and the spermatheca (upper right) and subgenitale of a female (middle right). Based on Sosa and de Freitas (2010).

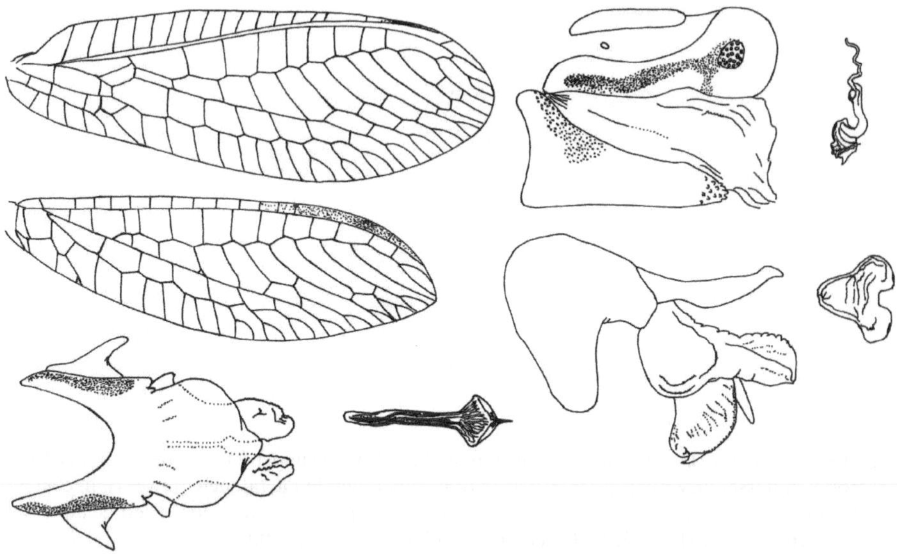

Fig. 16.156 *Ceraeochrysa torresi:* fore and hind wing (upper and middle left), the ninth sternite and ectoproct (upper right center), gonarcus in dorsal (lower left) and lateral view (lower right center), the gonapsis in dorsal view (lower center), the female spermatheca (upper right) and subgenitale (lower right). Based on Sosa and de Freitas (2010).

Chrysopa instabilis Navás, 1925; *Chrysopa jacobaea* Navás, 1925; *Chrysopa peterseni* Navás, 1929; *Chrysopa petersenia* Navás, 1931; *Chrysopa gloriae* Alayo, 1968.

– The ectoproct of the male has a ventral process. The gonapsis has lateral processes (Fig. 16.156) 39

39. (38) The ventral process on the male ectoproct is digitiform. The scales below the ectoproct are scattered on the membrane. The gonapsis has short lateral arms, and its apex is rounded (Fig. 16.156). Fore-wing length: c. 13.5 mm; hind wing: c. 11.8 mm.
.... *Ceraeochrysa torresi* Sosa and Freitas, 2010 (Venezuela).

– The ventral process on the male ectoproct is acutely pointed. The ectoproct bears a ventrally elongated dorsal apodeme. The scales on the ectoproct are attached to a plate near its ventral projection. The gonapsis has lateral processes, and its apex is smooth and lacks teeth (Fig. 16.157). Fore-wing length: c. 13 mm.
....*Ceraeochrysa squama* Freitas and Penny, 2001 (São Paulo).

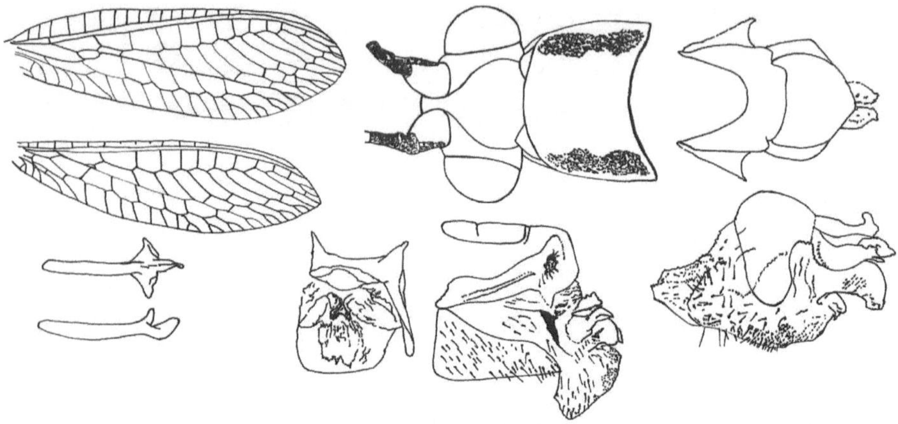

Fig. 16.157 *Ceraeochrysa squama* male: fore and hind wing (upper and middle left); head and prothorax in dorsal view (upper center); apex of the abdomen in lateral view (lower right center); genitalia in dorsal (upper right), lateral (lower right), and posterior view (lower left center), two views of the gonapsis (lower left). Based on de Freitas and Penny (2001).

Key to the Species of Adult *Chrysopodes* in South America

Information for the key was provided by Navás (1918b, 1925a), Adams and Penny (1987), de Freitas and Penny (2001), Penny (2002), Tauber (2010), and Tauber and Albuquerque in Tauber et al. (2012b). The key includes the subgenus *Neosuarius* Adams and Penny, 1987, which has been treated in some publications as an independent genus. The rest of the genus will probably be subject to a necessary revision in the near future (Tauber, 2010, Tauber et al. 2012b). Until then, the assignment to subgenera will continue to be uncertain, and no division of the genus into subgroups is proposed or employed in this key.

1. The features are known only from female specimens. The intermedian cell (im) in the fore-wing has four sides. The mandible is sickle-shaped (Fig. 16.158)...2
   – The intermedian cell (im) in the fore-wing is triangular or ovate (Fig. 16.159). The shape of the mandible varies. 3
2. (1) There are short setae and microtrichia on the copulatory bursa of the female, and these are arranged in diagonal rows on the basal part of the bursal duct, which is only of moderate length (Fig. 16.160). Fore-wing length: c. 14 mm. Only a single female specimen has been described. .... *Chrysopodes duckei* Adams and Penny, 1987 (Amazonas).

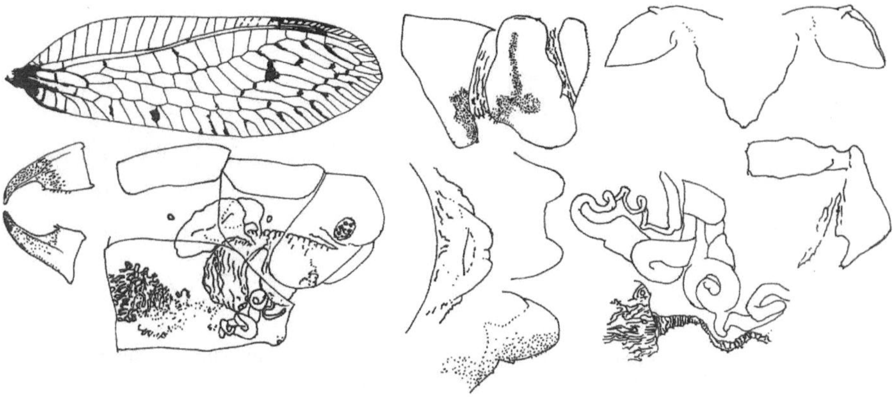

Fig. 16.158 *Chrysopodes pulchella* (upper row, left to right): fore-wing, apex of the male abdomen in lateral view, and gonarcal complex in dorsal view; (lower row, left to right): mandibles of a female, apex of a cleared female abdomen in lateral view, subgenitale, lobes from the gonapsis region bearing gonocristae in posterior and lateral view (above and below), bursal duct attached to spermathecal, and lateral view of male gonarcal complex. Based on Adams and Penny (1987).

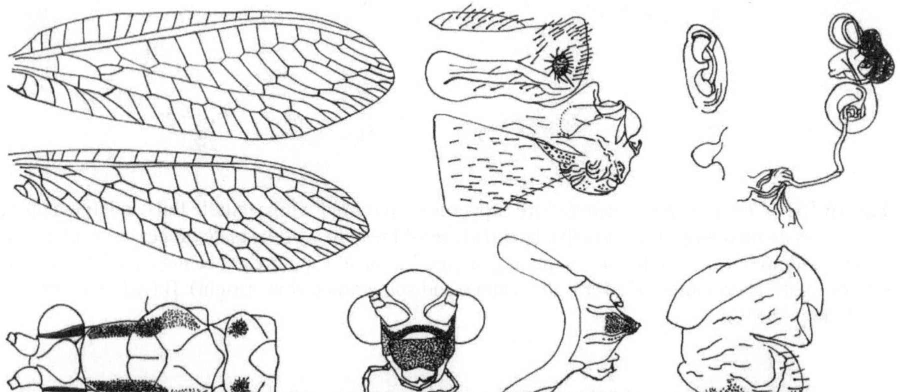

Fig. 16.159 *Chrysopodes copia:* fore and hind wing (upper and middle left), head and thorax in dorsal view (lower left), head in anterior view (lower left center), apex of the male abdomen in lateral view (upper center), male genitalia in dorsal (lower right center) and lateral view (lower right), spermatheca (upper right) and the subgenitale in posterior and lateral view (upper right center, above and below, respectively). Based on de Freitas and Penny (2001).

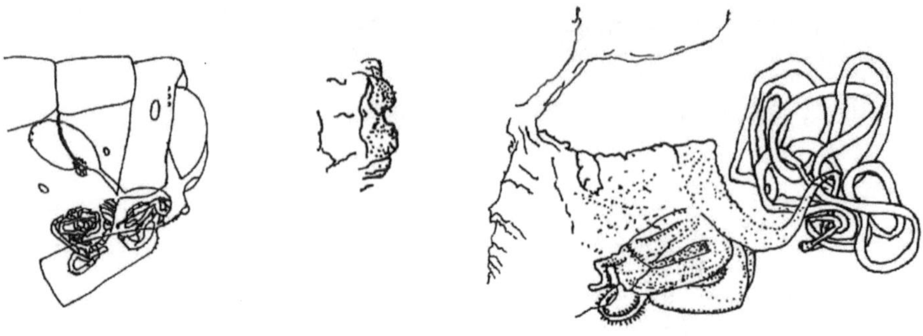

Fig. 16.160 *Chrysopodes duckei* female (left to right): apex of the abdomen in lateral view, subgenitale in ventral view, and genitalia. Based on Adams and Penny (1987).

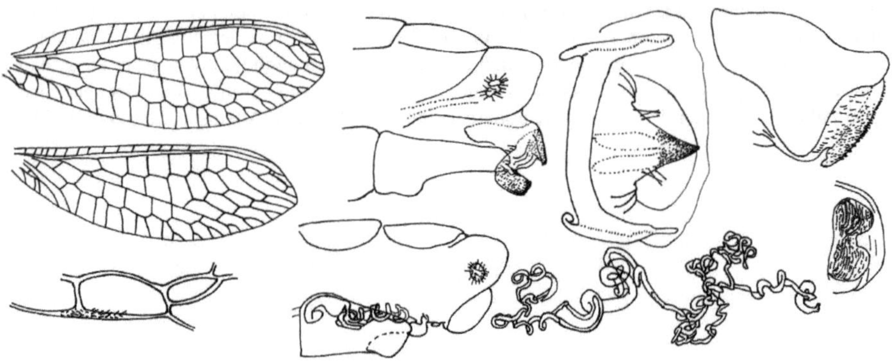

Fig. 16.161 *Chrysopodes crocinus:* fore (upper left) and hind wing (middle left), enlargement of the incrassate median cell (lower left), lateral views of the apices of the male (upper left center) and female abdomen (lower left center), male genitalia in dorsal (upper right center) and lateral view (upper right), spermatheca (lower right center), and subgenitale (lower right). Based on de Freitas and Penny (2001).

– There are no short setae or microtrichia on the copulatory bursa of the female, and the bursal duct is about 16 mm long (Fig. 16.158). Fore-wing length: c. 14.0–14.5 mm.

....*Chrysopodes pulchella* (Banks, 1910) (Surinam, Pará, Amazonas, Rondônia). Syn: *Allochrysa pulchella* Banks, 1910.

3. (1) There dark spots on the metanotum, while the mesonotum has irregular dark spots forming an interrupted stripe, a pair of round, dark spots, or is entirely pale (Fig. 16.159).. 4

– Neither the mesonotum nor the metanotum is marked with dark spots (Fig. 16.161) 6

4. (3) There is an interrupted stripe consisting of irregular dark spots on the mesonotum. Parts of the frons, the lateral margin of the vertex, and the genae

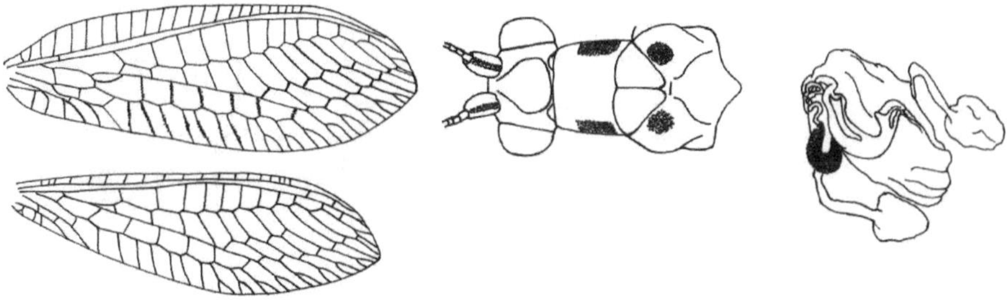

Fig. 16.162 *Chrysopodes nigropicta* female: fore and hind wing (left), head and anterior part of thorax (center), and genitalia (right). Based on de Freitas and Penny (2001).

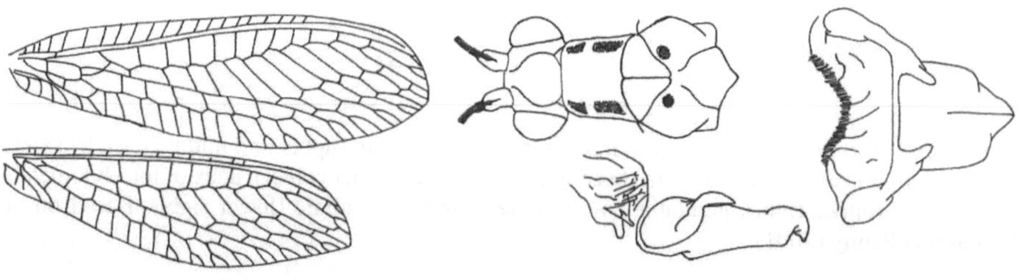

Fig. 16.163 *Chrysopodes adynatos* male: fore and hind wing (upper and lower left), head and thorax in dorsal view (upper center), and male genitalia in dorsal (right) and lateral view (lower right center). Based on de Freitas and Penny (2001).

and clypeus are red (Fig. 16.159). Length of fore-wing: c. 12.7 mm; hind wing: c. 11.8 mm. Width of fore-wing: c. 4.3 mm; hind wing: c. 3.6 mm. .... *Chrysopodes copia* Freitas and Penny, 2001 (São Paulo).

– There is a pair of dark, round spots on the mesootum, or it is completely pale. The frons is completely pale, or it has only one transverse, dark stripe (Fig. 16.162) 5

5. (4) The antennal flagellum is pale. The scape has a dorsomedial stripe that reaches to its proximal and distal ends. The pronotum has one pair of elongate dark spots, which extend onto the mesopraescutum. The fore-wing veins, including the gradate cross veins, are extensively clouded along their margins (Fig. 16.162). Length of fore-wing: c. 14.1 mm; hind wing: c. 12.4 mm. Width of fore-wing: c. 5.0 mm; hind wing: c. 4.0 mm. .... *Chrysopodes nigropicta* Freitas and Penny, 2001 (São Paulo, Minas Gerais).

– The antennal flagellum is black. The dark, dorsomedial stripe on the scape does not reach both ends. There are two pairs of lateral stripes on the pronotum. The mesopraescutum is completely pale. The margins of the cross veins on the fore-wing are not clouded (Fig. 16.163). Length of fore-wing: c. 13 mm; hind wing: c. 11.8 mm. Width of fore-wing: c. 4.5 mm; hind wing: c. 3.4 mm. .... *Chrysopodes adynatos* Freitas and Penny, 2001 (São Paulo).

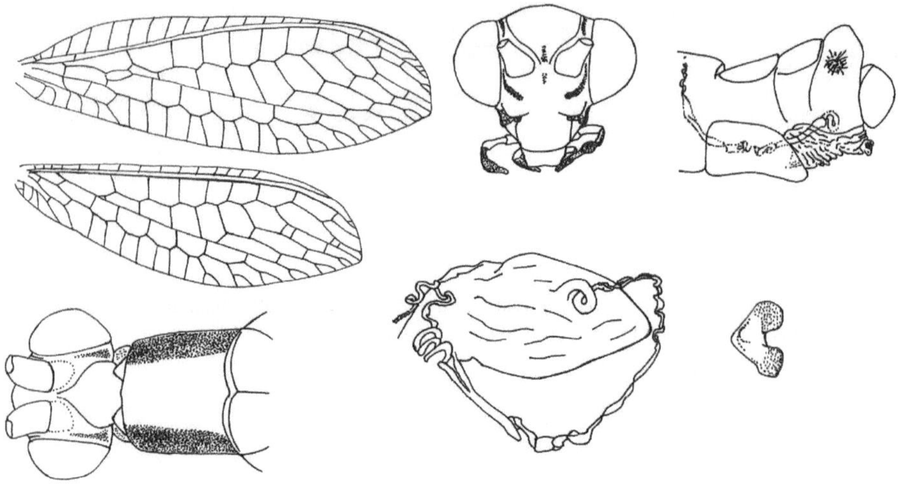

Fig. 16.164 *Chrysopodes elongata* female: fore and hind wing (upper and middle left), head and prothorax in dorsal view (lower left), head in anterior view (upper center), apex of the abdomen in lateral view (upper right), genitalia (lower center), and subgenitale (lower right). Based on de Freitas and Penny (2001).

6. (4) The mandible is fang-like at the apex (Fig. 16.161). Most species with this character were assigned to the sub-genus *Chrysopodes*. For more certainty, check both branches of this key 7

   – The mandible is blunt at the apex (Fig. 16.23). Most, but not all, species with this character have been assigned to the sub-genus *Neosuarius*........ 15

7. (6) There are no stripes on the frons (Fig. 16.161) 8

   – There are curving red stripes ventral to the origins of the antennae (Fig. 16.164). 12

8. (7) The genae are entirely red, and the clypeus and labrum are completely pale, although some may be amber along the margins (Fig. 16.161)........ 9

   – There are two longitudinal stripes in the genae. The clypeus and labrum are dark along their lateral margins (Fig. 16.165) 11

9. (8) The scape and pedicel lack dark markings, and the antennal flagella are cream or white. Some cross veins are margined by fumose membranes. The male and female genitalia are distinctive (Fig. 16.21). Length of fore-wing: 12.6–15.1 mm; hind wing: 11.3–13.2 mm.
   ....*Chrysopodes serrabonitensis* Tauber and Albuquerque in Tauber et al., 2012 (Bahia, Rio de Janeiro).

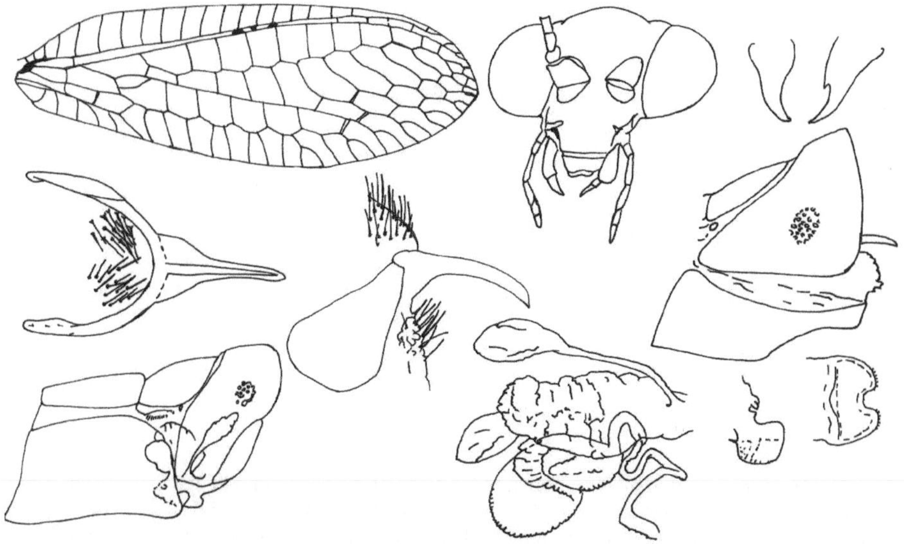

Fig. 16.165 *Chrysopodes polygonica* (above, left to right): fore-wing, head in anterior view, mandibles, and (middle row, left to right): male gonarcal complex in posterior and lateral view, apex of the male abdomen in lateral view, and (lower row, left to right): apex of the female abdomen in lateral view, bursa and spermatheca of a female, and the subgenitale in lateral and ventral view. Based on Adams and Penny (1987).

– The antennal flagellum is dark. No cross veins are bordered by darkly clouded membranes (Fig. 16.166) 10

10. (9) The subanal plate has many small setae and lateral pockets. The fore-wing has five outer and seven inner gradate cross veins. The predominant color is yellow, but the antennal flagella are dark, and the scapes each have a dark lateral marking. The maxillary and labial palps are pale. The length and width of the pronotum are subequal. The spermatheca of the female is relatively elongate and thin (Fig. 16.166). Length of fore-wing: c. 14.5 mm; hind wing: c. 13.3 mm. The male has not been described.
.... *Chrysopodes karinae* Freitas and Penny, 2001 (São Paulo). This species was formerly placed in the subgenus *Neosuarius,* but the male is unknown, and the mandible of the female type specimen has not been described. Its position in the genus remains uncertain.

– The subanal plate does not have many small setae and lateral pockets. There is no dark lateral markings on the scape (Fig. 16.161). The pterostigma is unmarked. Fore-wing length: c. 11.8 mm; width: c. 4.1 mm. Hind wing length: c. 11.2 mm; width: c. 3.5 mm.
.... *Chrysopodes crocinus* Freitas and Penny, 2001 (São Paulo).

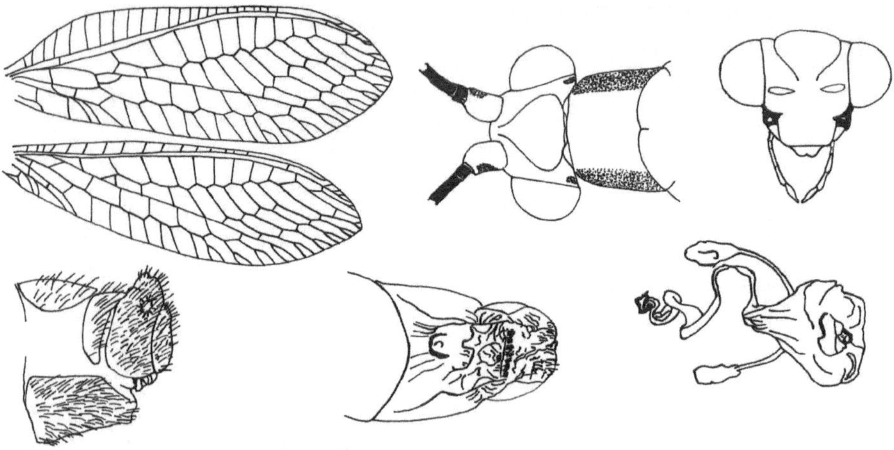

Fig. 16.166 *Chrysopodes karinae:* fore and hind wing (upper and middle left), head and pronotum in dorsal view (upper middle right), head in anterior view (upper right), apex of the male abdomen in lateral view (lower left), female abdomen in ventral view (lower center), and female genitalia (lower right). Based on de Freitas and Penny (2001).

11. (8) The apical veins in the male fore-wing appear swollen, but vein Rs is not. The apical processes on the male gonarcus are longer than the space between them. The inner gradate series is straight and converges with the outer series toward the wing apex. The genae have fuscous stripes (Fig. 16.165). Fore-wing length: 10.0–11.0 mm.
.... *Chrysopodes polygonica* Adams and Penny, 1987 (Argentina, Pará, Amazonas, Rondônia, Goiás).

    – The apical processes on the male gonarcus are reduced to small plates, shorter than the space between them (Fig. 16.167). Only the male has been described.
    .... *Chrysopodes delicata* Freitas and Penny, 2001 (Mato Grosso).

12. (7) The vertex and frons have dark markings between the bases of the antennae (Fig. 16.164). The seventh sternite of the female is elongate and bears no knob. The spermatheca is long and narrow.
.... *Chrysopodes elongata* Freitas and Penny, 2001 (São Paulo).

    – The vertex and frons lack dark markings between the bases of the antennae (Fig. 16.168)... 13

13. (12) The ectoproct has only one ventral projection, which is an elongation of the apodeme (Fig. 16.168). Fore-wing length: 11.1–12.8 mm.
.... *Chrysopodes lineafrons* Adams and Penny, 1987 (Amazonas, Santa Catarina, São Paulo).

    – The male ectoproct bears two ventral projections (Fig. 16.169).................. 14

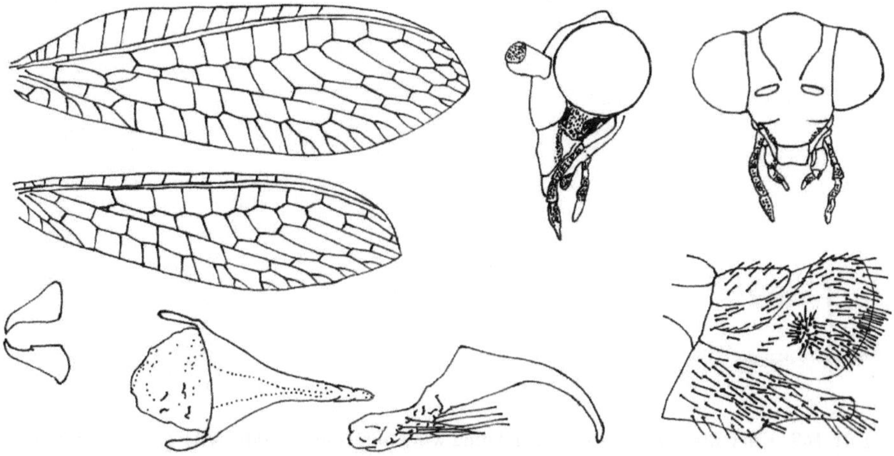

Fig. 16.167 *Chrysopodes delicata* male: fore and hind wing (upper and middle left), head in lateral (upper right center) and anterior view (upper right), and (below, left to right): mandibles, genitalia in dorsal and lateral view, and apex of the abdomen. Based on de Freitas and Penny (2001).

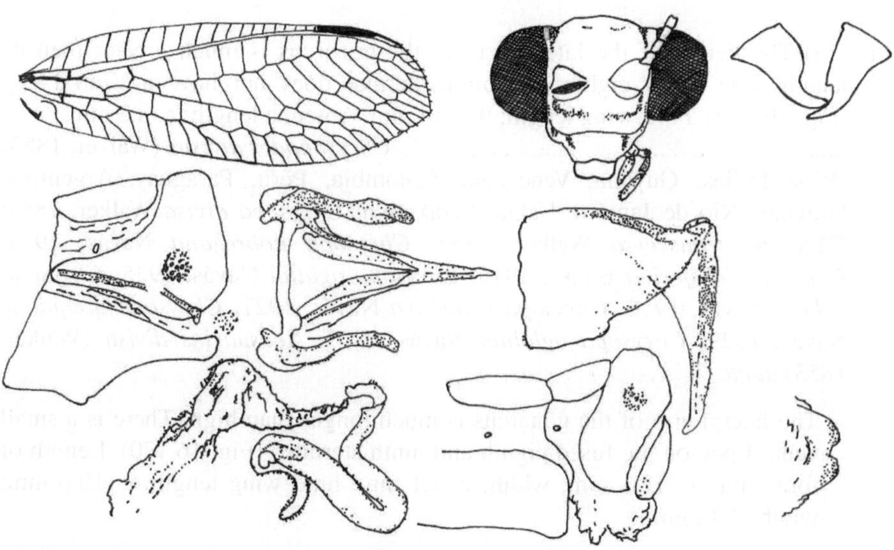

Fig. 16.168 *Chrysopodes lineafrons* (above, left to right): fore-wing, head in anterior view, mandibles; (middle row, left to right): apex of the male abdomen in lateral view and the gonarcal complex in dorsal and lateral view; (lower row, left to right): spermatheca, apex of the female abdomen in lateral view, and the subgenitale. Based on Adams and Penny (1987).

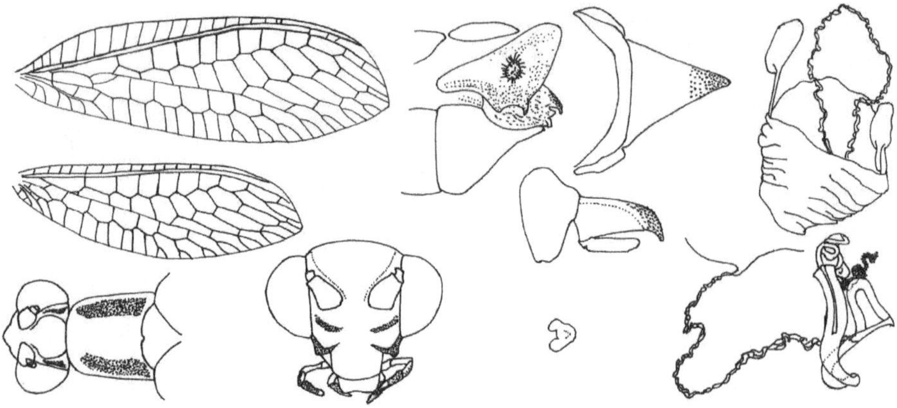

Fig. 16.169 *Chrysopodes divisus:* fore and hind wing (upper and middle left), head and pronotum in dorsal view (lower left), head in anterior view (lower left center), apex of the male abdomen in lateral view (upper center), male genitalia in dorsal (upper right center) and lateral view (right of center), female genitalia (upper right), spermatheca (lower right), and subgenitale (lower right center). Based on de Freitas and Penny (2001).

14. (13) The height of the lateral arm of the gonarcus is much greater than its length. The fused eighth and ninth sternite does not have a ventral peg (Fig. 16.169). Fore-wing length: 9.7–14 mm. Antenna length: c. 8.2 mm. .... *Chrysopodes divisus* (Walker, 1853) (West Indies, Guyana, Venezuela, Colombia, Peru, Paraguay, Argentina, Uruguay, Rio de Janeiro, Bahia, Pará). Syn: *Chrysopa divisa* Walker, 1853; *Chrysopa transversa* Walker, 1853; *Chrysopa nobregana* Navás, 1913; *Chrysopa hesperina* Banks, 1915; *Chrysopa agatha* Navás, 1925; *Chrysopa debilis* Navás, 1926; *Chrysopa uruguaya* Navás, 1927; *Chrysopa arequipae* Navás, 1929; *Chrysopa oglobini* Navás, 1931; *Neosuarius divisa* (Walker, 1853) *auctt.*

   – The lateral arm of the gonarcus is much longer than high. There is a small ventral peg on the fused eighth and ninth sternites (Fig. 16.170). Length of fore-wing: c. 12.1 mm; width: c. 4.1 mm; hind wing length: c. 10.6 mm; width: 3.4 mm.

   ....*Chrysopodes spinellus* Adams and Penny, 1987 (Argentina, Pará). Syn: *Chrysopodes spinella* Adams and Penny, 1987, mistaken gender.

15. (6) The mandibles are wide for their entire lengths. The left mandible has a rounded mesal knob, while the right mandible lacks such a knob. The prominent frons and the vertex of the head are smooth and lack pits and fields of folds. The middle part of the anterior margin of the fore-wing is almost straight, and the costal area is relatively narrow (Fig. 16.23). A gonapsis is never present.. 16

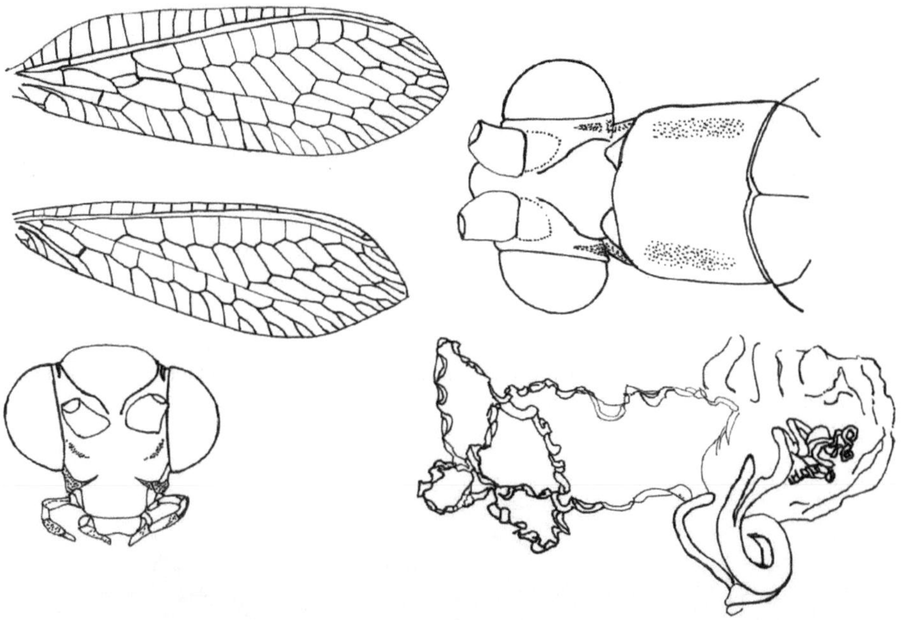

Fig. 16.170 *Chrysopodes spinellus* female: fore and hind wing (upper and middle left), head and pronotum in dorsal view (upper right), head in anterior view (lower left), and genitalia (lower right). Based on de Freitas and Penny (2001).

– The apices of the mandibles are sickle or fang-like. A gonapsis is rarely present. The costal space on the wing is usually wide (Fig. 16.171). The frons and vertex of the head are pitted and have fields of folds, except in the species *Chrysopodes hagenorum* (See Couplet 25)................................... 23

16. (15) The predominant coloration of the dorsal surface of the body is green. A median stripe is usually present, and the head is yellowish, often with reddish or brownish markings. Old preserved specimens may have faded colors and lack traces of the median stripe. The frons lacks brown spots ventral to the toruli. There are no transverse sclerotized bands across the posterior part of the abdomen, and no sclerotized invagination across the fused eighth and ninth tergites. The female has a stout, funnel-shaped spermatheca and a long, usually narrow bursal duct (Fig. 16.23).. 17

– The predominant coloration of the dorsal surface of the body is cream or tan with brown markings, and the frons has brown spots ventral to the toruli. There are transverse sclerotized bands across the posterior part of the male abdomen, and a sclerotized invagination across the fused eighth and ninth tergites of the male may also be present. The female has a robust, tubular spermatheca and a fluted bursal duct (Fig. 16.172)................................... 18

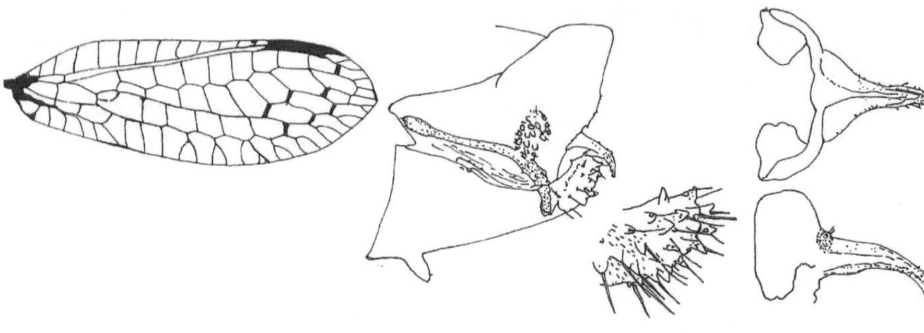

Fig. 16.171 *Chrysopodes tetifera* male: fore-wing (left), apex of abdomen in lateral view sketched from a damaged specimen (upper left center), apex of the ninth sternite enlarged (lower left center), and gonarcus in dorsal (upper right) and lateral view (lower right). Based on Adams and Penny (1987).

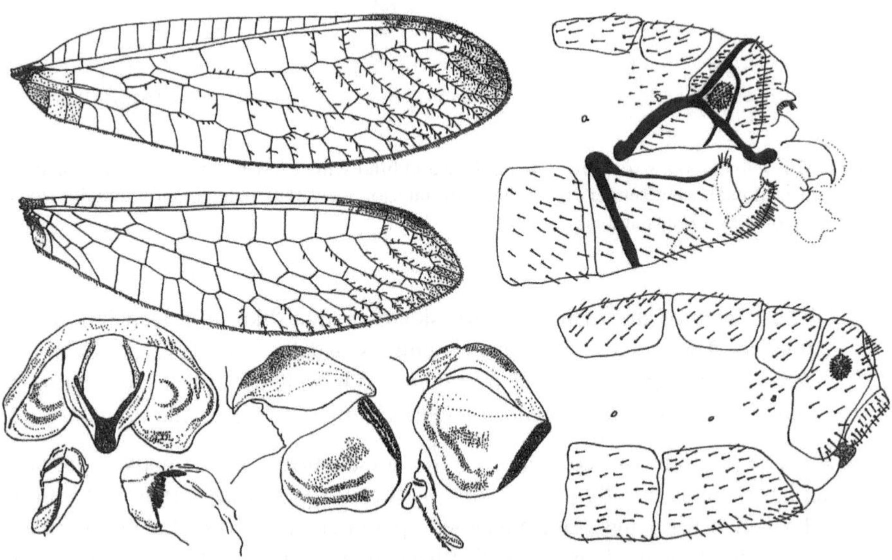

Fig. 16.172 *Chrysopodes escomeli:* fore and hind wing (upper and upper middle left), apex of the male abdomen (upper right), gonarcus in dorsal (lower middle left) and lateral view (lower left center) and shown unexpanded in lateral view (lower right center), hypandrium internum in dorsal (lower left) lateral (lower left center), and frontolateral view (lower center); apex of the female abdomen in lateral view (lower right). Based on Tauber (2010).

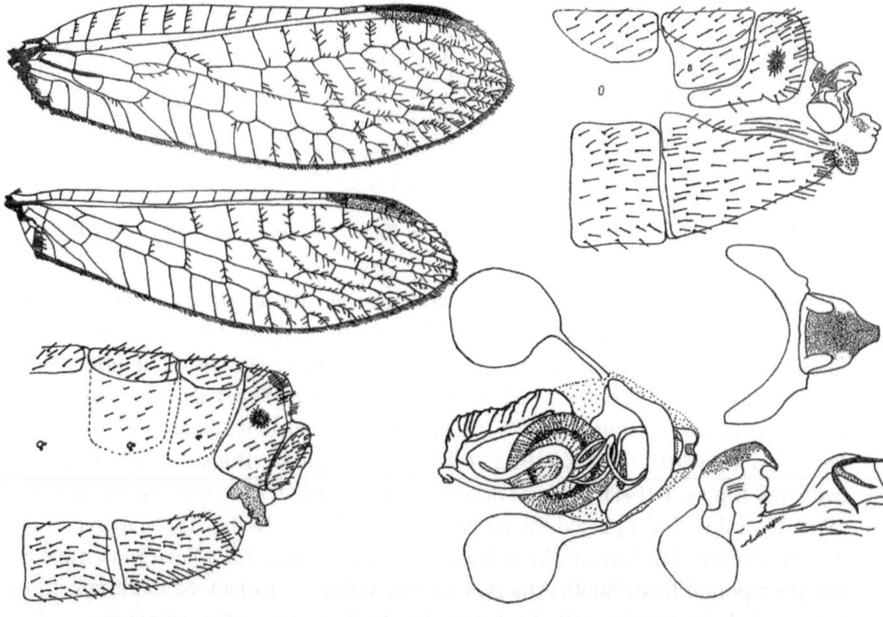

Fig. 16.173 *Chrysopodes figuralis:* fore and hind wing (upper and middle left), apex of the male abdomen in lateral view (upper right), male genitalia in dorsal (middle right) and lateral view (lower right), apex of the female abdomen in lateral view (lower left), and female genitalia (lower right center). Based on Tauber (2010).

17. (16) The prothorax is entirely green, or, if a mid-dorsal stripe is present, it is brown and not yellow. The vertex has a pair of broad reddish brown or brown markings. The apodeme on the fused ninth tergite and ectoproct of the male is slender, extends only as far as the callus cerci, and lacks well-developed ventral and caudal arms. The dorsal rods on the mediuncus arise from the anterior surface of the gonarcal bridge. The bursal duct of the female is not long and narrow; its length is approximately equal to or shorter than the length of the abdomen (Fig. 16.173). Fore-wing length: 11.6–15.3 mm. ....*Chrysopodes figuralis* (Banks, 1915) (Ecuador, Peru, Bolivia, Chile). Syn: *Chrysopa figuralis* Banks, 1915; *Mallada verticalis* Navás, 1929; *Chrysopa tacorensis* Navás, 1934; *Suarius figuralis* (Banks, 1915) Adams, 1975.

– The prothorax has a yellow mid-dorsal stripe. Reddish stripes on the vertex, if present at all, are light. The pronotum is much wider than long. The apodeme on the fused ninth tergite and ectoproct of the male are robust, extend beyond the callus cerci, and have well-developed ventral and caudal arms. Below the gonarcal bridge, dorsal rods arise from the mediuncus. The

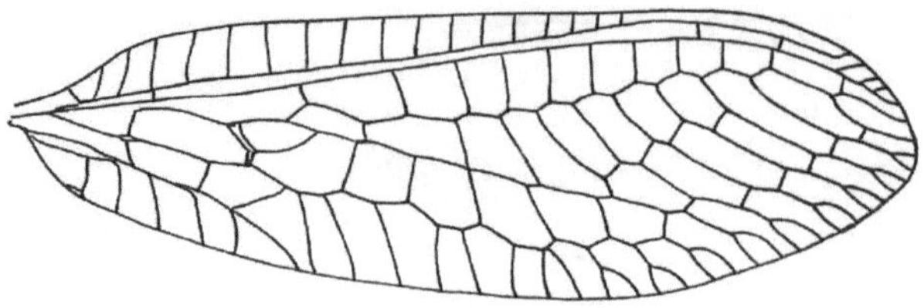

Fig. 16.174 Fore-wing of *Chrysopodes nigricubitus*. Based on Tauber and Tauber (2010).

bursal duct of the female is long and narrow; its length exceeds twice the length of the abdomen (Fig. 16.23). Fore-wing length: 10.9–14.3 mm.
....*Chrysopodes collaris* (Schneider, 1851) (North and Central America, West Indies, Colombia, Venezuela, Guyana). Syn: *Chrysopa collaris* Schneider, 1851; *Chrysopa thoracica* Walker, 1853; *Chrysopa thoracica cerverina* Navás, 1923; *Chrysopa signatalis* Banks, 1911, *Chrysopa rufolinea* Banks, 1914; *Chrysopa dampfina* Banks, 1927; *Chrysopa acolhua* Banks, 1948. Oswald (2007) placed the name *Chrysopa krugii* Kolbe, 1888, in synonymy with *N. collaris*, but Tauber (2010) has restored its validity for a West Indian species not yet reported from South America, distinguished by its lack of markings on the frons and clypeus and usually by a pair of reddish stripes on the pronotum.

18. (16) There are distinct brown or reddish markings on the dorsal surface of the vertex. There are distinct, jagged, median and sublateral longitudinal stripes on the pronotum. The male lacks large, eversible membranes on the fused eighth and ninth sternites or setose ducts on the gonosaccus (Fig. 16.172). Fore-wing length: 13.9–16.7 mm 19

   – Except for a brown stripe along the anterior margin, there are no markings on the dorsal surface of the vertex. The pronotum has light brown markings, which do not join to form distinct stripes. The male has large, eversible membranes on the fused eighth and ninth sternites and setose ducts originating in the gonosaccus (Fig. 16.174). Fore-wing length: 10.2–13.7 mm.....................................21

19. (18) The basal segments of the antennal flagellum are cream-colored. The scape has one dark dorsomedial and one mesolateral stripe. The cross veins on the fore-wing are white in the middle and brown or dark brown at each end (Fig. 16.172). Fore-wing length: 14.0–16.6 mm.
....*Chrysopodes escomeli* (Navás, 1922) (Ecuador, Peru, Chile). Syn: *Chrysopa escomeli* Navás, 1922; *Suarius escomeli* (Navás, 1922) in Núñez, 1989.

   – The basal segments of the antennal flagellum are black or dark brown. The scape has one dark stripe or none at all. The cross veins on the fore-wing are green or brown without white markings (Fig. 16.175)..................................20

20. (19) The long veins on the fore-wing are cream-colored with extensive black or dark brown markings. The crossveins are mainly dark brown. The apex of the female subgenitale is elongate, flat, and angled inward toward the base. The pronotum is yellow with five dark, longitudinal stripes evident in dorsal view.

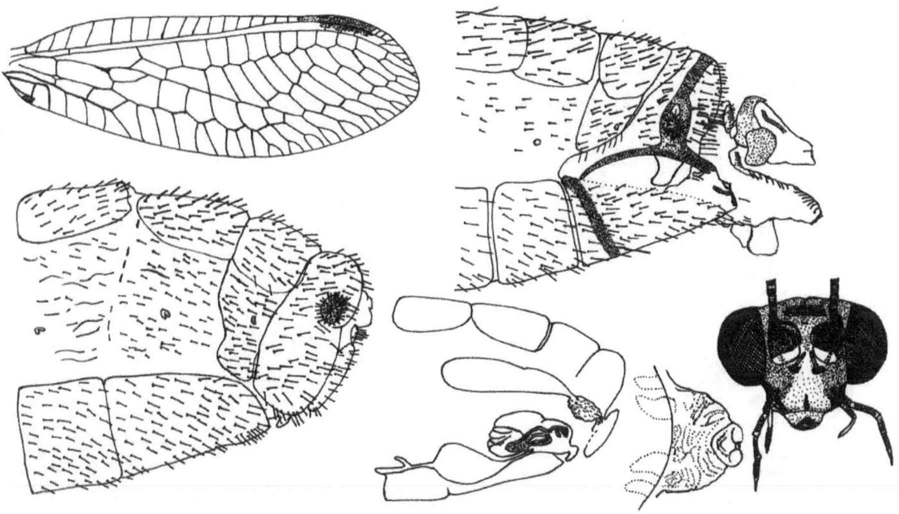

Fig. 16.175 *Chrysopodes nigripilosus:* fore-wing (upper left), apex of the male abdomen in lateral view showing the membranes on the eighth and ninth abdominal sternites everted (upper right), head in anterior view (lower right), apex of the female abdomen in lateral view (lower left), outline of the apex of the female abdomen in lateral view showing the location of the internal genitalia (lower center), and the subgenitale (lower right center). Based on Tauber and Tauber (2010).

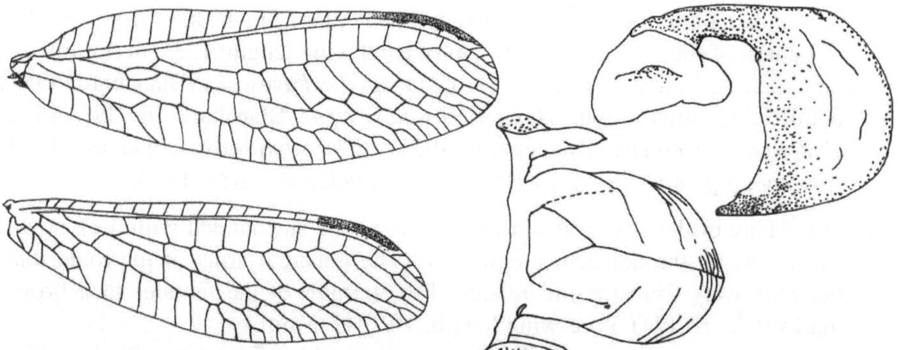

Fig. 16.176 *Chrysopodes porterinus* (male) fore and hind wing (left) and internal genitalia in dorsal (lower right center) and lateral view (upper right). Based on Monserrat and de Freitas (2005).

The median stripe is expanded greatly near mid-length (Fig. 16.176). The underside of the thorax and abdomen are mainly dark violet, but the posterior borders of the abdominal segments are pale. Length of fore-wing: 15.1–16.7 mm. Hind wing length: 12.7–14.7 mm.
.... *Chrysopodes porterinus* (Navás, 1910) (Chile, Argentina). Syn: *Chrysopa porterina* Navás, 1910; *Chrysopa porteri* Navás, 1928 (misspelling); *Chrysopa trisella* Navás, 1920; *Chrysopa ruizi* Navás, 1934.

– The long veins on the fore-wing are predominantly cream-colored with light brown markings only at the junctions of veins. The female subgenitale is

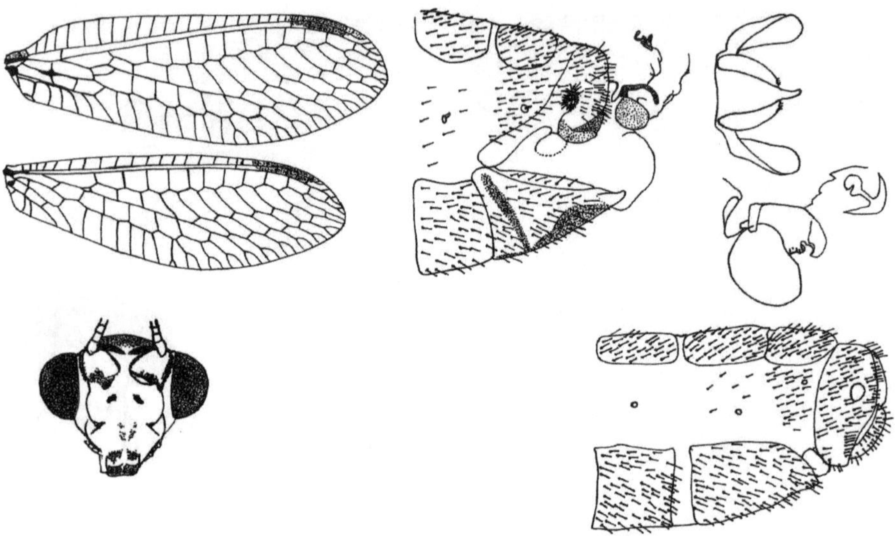

Fig. 16.177 *Chrysopodes flavescens:* fore and hind wing (upper and middle left), head in anterior view (lower left), apex of the male abdomen in lateral view (upper center), gonarcus in dorsal (upper right) and lateral view with the hypandrium internum to its right (middle right), and the apex of the female abdomen in lateral view (lower right). Based on Tauber (2010).

rounded, knob-like, and extended perpendicular from the base (Fig. 16.177). Length of fore-wing: 13.9–16.1 mm. Hind wing length: 12.7–14.4 mm. ....*Chrysopodes flavescens* (Blanchard, 1951) (Chile, Argentina). Syn: *Hemerobius flavescens* Blanchard in Gay, 1951; *Chrysopa flavescens* (Blanchard, 1951); *Chrysopa nosina* Navás, 1913; *Chrysopa jaffuelina* Navás, 1918; *Chrysopa bullocki* Navás, 1933.

21. (18) Many of the crossveins in the fore-wing are surrounded with dark brown suffusions on the adjacent membranes. The gonarcal arms of the male flare outward away from the mediuncus. The sternites of the females lack brown marks (Fig. 16.174). Fore-wing length: 10.2–13.7 mm. .... *Chrysopodes nigricubitus* Tauber and Tauber, 2010 (Santa Cruz and Pinta Island in the Galapagos Islands).

   – The fore-wing membrane lacks dark suffusions. The gonarcal arms of the male extend inward adjacent to the mediuncus. The sternites on the fifth and sixth or fifth and seventh abdominal segments of the female have brown marks (Fig. 16.175)..22

22. (6) The wing length does not exceed 13 mm. The length and width of the pronotum are approximately equal. There are no brown spots just posterior to the vertex. The second intracubital crossvein is not bent or swollen. In the male, the horizontal apodeme on the ventral part of the fused ninth tergite and ectoproct

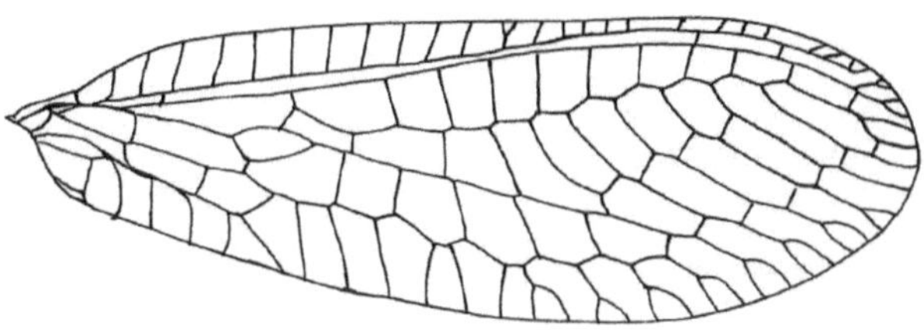

Fig. 16.178 Fore-wing of *Chrysopodes pecki*. Based on Tauber and Tauber (2010).

arches mesally and has a rounded, forked apex (Fig. 16.175). The fifth, sixth, and seventh sternites of the female have brown markings.
.... *Chrysopodes nigripilosus* (Banks, 1924) (Galapagos Islands). Syn: *Chrysopa nigripilosa* Banks, 1924.

– The wing length exceeds 13 mm. The length of the pronotum exceeds the width. There are two brown spots just posterior to the vertex. The second intracubital crossvein is slightly bent and has a dark brown swelling mesally. In the male, the horizontal apodeme on the ventral part of the fused ninth tergite and ectoproct is straight with an acute, unforked apex (Fig. 16.178). The fifth sternite of the female is dark brown; the sixth is brown only at the base, and the seventh is cream-colored. Fore-wing length: 15.1–16.7 mm.

.... Chrysopodes *pecki* Tauber and Tauber, 2010 (Isabela Island in the Galapagos Islands).

23. (15) There is a brown mid-dorsal stripe on the scape of the antenna (Fig. 16.171). Fore-wing length: c. 9 mm. Only the male has been described from a damaged specimen.
.... *Chrysopodes tetifera* Adams and Penny, 1987 (Pará).

– There is no brown mid-dorsal stripe on the scape, although small or narrow markings may be present on the ventral side (Fig. 16.179) 24

24. (23) The gradate cross-veins are pale but surrounded by brownish tinged membranes (Fig. 16.179). Fore-wing length: 10.5–14.5 mm.
.... *Chrysopodes nebulosa* Adams and Penny, 1987 (Amazonas, Rondônia).

– The gradate cross-veins are dark; the adjacent win membranes may or may not be darkened (Fig. 16.180) 25

25. (24) The veins in the gradate series are parallel, and the cells they enclose are about the same length. The frons and vertex of the head are smooth and shiny. The shape of the gonarcus is quadratic. The mediuncus is neither narrow and elongate nor simple. Most of the cross veins in the wings are brown, and none are clouded on the membranes (Fig. 16.180) Fore-wing length: c. 9.4–10.5 mm. Hind wing length: 8.3–9.4 mm.
.... *Chrysopodes hagenorum* Tauber and Albuquerque in Tauber et al., 2012 (Paraná).

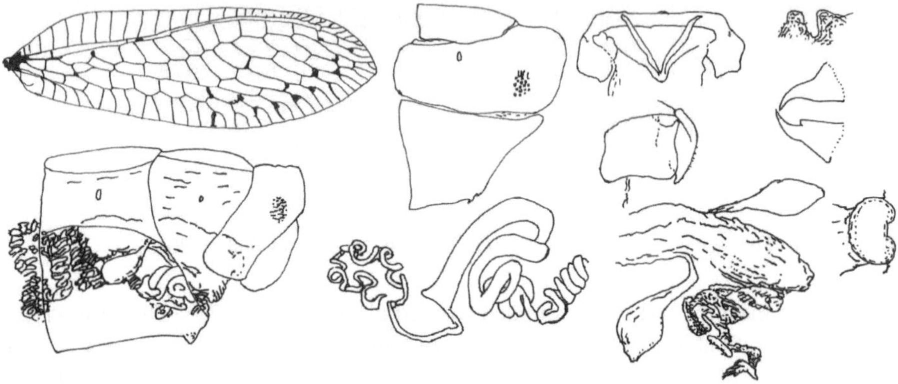

Fig. 16.179 *Chrysopodes nebulosa:* fore-wing (upper left); apex of male abdomen (upper left center); gonarcal complex in dorsal and lateral view (upper right center, above and below, respectively); membrane near the gonapsis in posterior view (upper right); mandibles (upper middle right); apex of female abdomen after clearing (lower left); bursa and bursal duct (lower right center); bursal duct and spermatheca (lower right center); subgenitale (lower middle right). Based on Adams and Penny (1987).

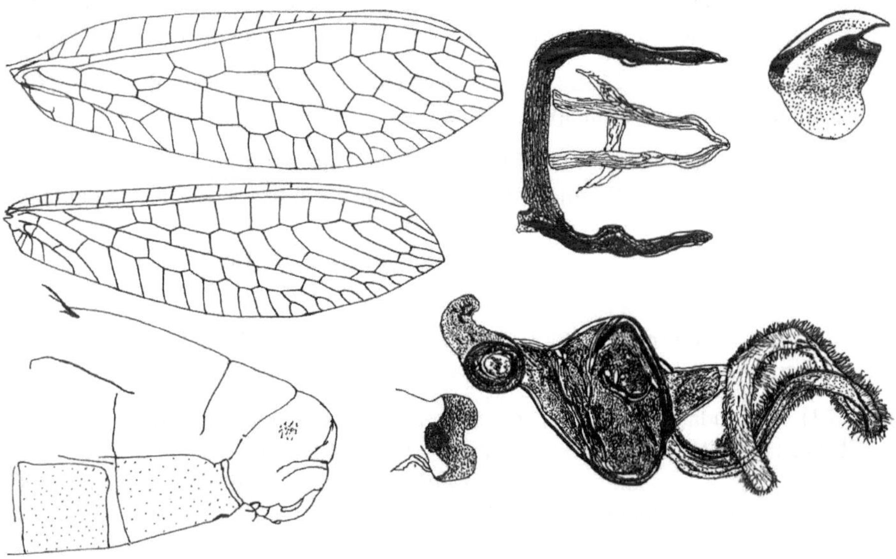

Fig. 16.180 *Chrysopodes hagenorum:* fore and hind wing (upper and middle left), male gonarcus in dorsal (upper right center) and lateral view (upper right), apex of the female abdomen in lateral view (lower left), bursa and spermatheca (lower right), and subgenitale (lower center). Based on Tauber et al. (2012b).

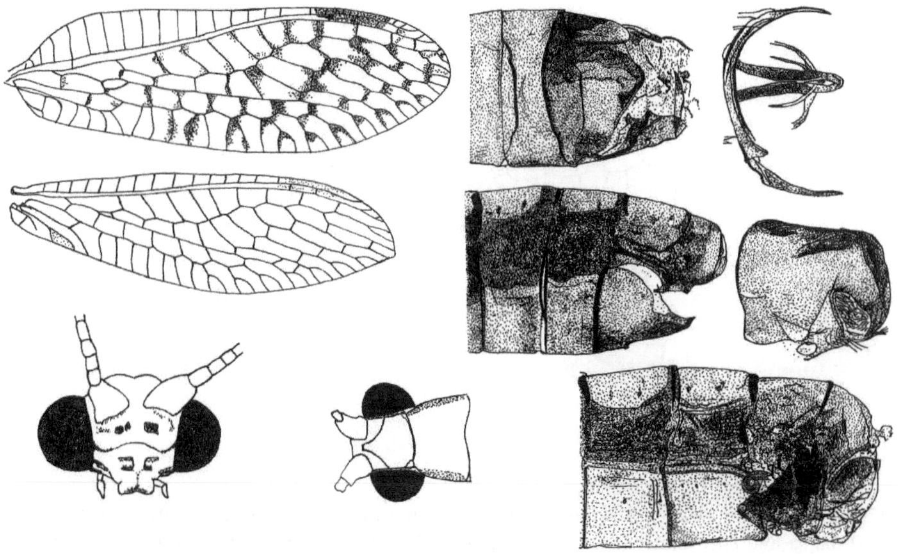

Fig. 16.181 *Chrysopodes fumosus:* fore and hind wing (upper and middle left), head in anterior view (lower left), head and pronotum in dorsal view (lower center), apex of the male abdomen with the apical membranes damaged in ventral (upper center) and lateral view (center), the gonarcus and adjacent sclerotized parts of the male genitalia in dorsal view (upper right), the gonarcus in lateral view showing the mediuncus raised (middle right), and the apex of the female abdomen in lateral view (lower right). Based on Tauber et al. (2012b).

– The veins in the gradate series are not parallel, and the cells they enclose vary greatly in length (Fig. 16.181) 26

26. (25) There is a deep medial incision in the labrum and a bulge in the clypeus, which projects in the middle and narrowly covers the basal midline of the labrum, in anterior view. The outer gradate veins are slightly darkened. Fore-wing length: c. 13.3 mm. Only the male has been described, from a single specimen.
....*Chrysopodes indentata* Adams and Penny, 1987 (Amazonas).

– The labrum lacks a deep incision, or the clypeus lacks a prominent median bulge (Fig. 16.181).. 27

27. (26) The fore-wing length is 14.0 – 15.5 mm. The wing has a wide costal area, which narrows rapidly toward the apex. The pronotum is clothed in both long and short fuscous setae. The radius curves foreward very strongly. Fore-wing length: c. 15.5 mm. Only one male specimen has been described.
....*Chrysopodes conisetosa* Adams and Penny, 1985 (Amazonas).

– The fore-wing length is 10.8 – 12.5 mm. The costal area of the wing narrows gradually toward the apex (Fig. 16.182).. 28

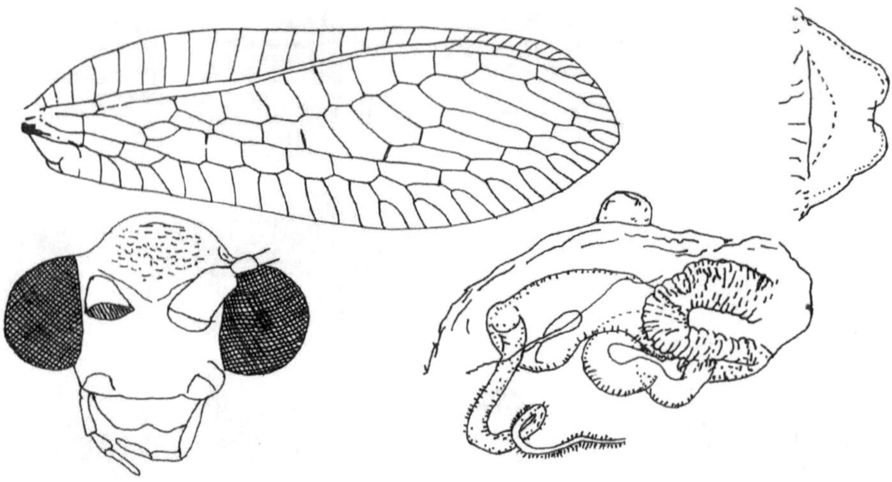

Fig. 16.182 *Chrysopodes breviata* female: fore-wing (upper left), head in anterior view (lower left), bursa and spermatheca (lower right), and subgenitale (upper right). Based on Adams and Penny (1987).

28. (27) There are only three or four inner gradate veins. The hind wing ends in an acute tip. The head has red marks on the frons and clypeus (Fig. 16.182). The fore-wing length is 10.8 – 11.5 mm. Only the female has been described. ....*Chrysopodes breviata* Adams and Penny, 1985 (Amazonas).

   – There are five or six inner gradate veins. The hind wing has a broad tip. The cross-veins sometimes have brown margins. The head lacks red markings on the frons and clypeus (Fig. 16.181). 29

29. (28) On the fore-wings, the gradate cross veins are brownish, and the membranes adjacent to most cross veins are strongly clouded with brown. The ectoproct of the male has two ventral projections (Fig. 16.181). The pronotum is green with a wide, yellow, median stripe and reddish lateral stripes; it is covered with many long, pale golden setae. Length of fore-wing: 12.3 – 12.5 mm; hind wing 10.8 – 11.1 mm. .... *Chrysopodes fumosus* Tauber and Albuquerque in Tauber et al., 2012 (Venezuela, Rio de Janeiro).

   – The membranes adjacent to the cross veins of the fore-wing are not clouded with brown, although there may be dark markings at the edges of the pterostigma. The fore-wing is c. 2.7 times as long as wide. The female bursa is simple (Fig. 16.183). Fore-wing length: 10.8 – 11.5 mm.

.... *Chrysopodes mediocris* Adams and Penny, 1987 (Amazonas, Rondônia).

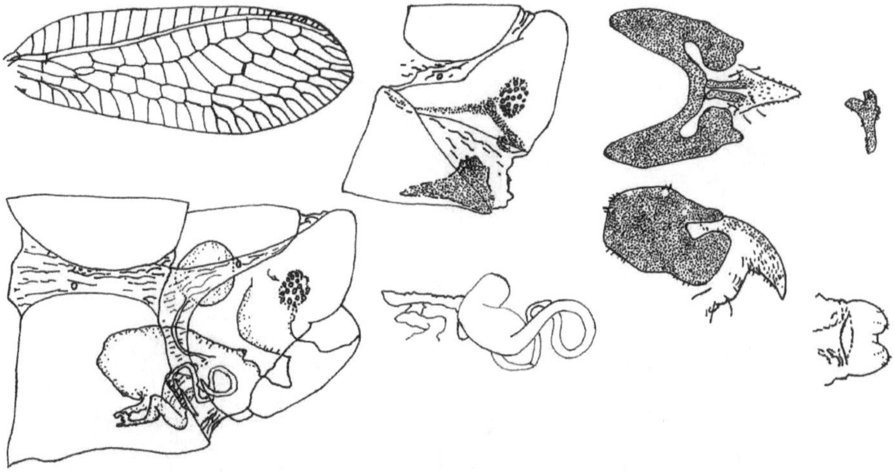

Fig. 16.183 *Chrysopodes mediocris* (above, left to right) fore-wing, apex of male abdomen showing the location of the gonapsis, gonarcus in posterior view, and (below, left to right): apex of female abdomen, spermatheca, gonapsis in lateral view, and subgenitale. Based on Adams and Penny (1987).

Key to the Species of *Chrysopodes* Larvae in South America

Information for the key was provided by Gonzales Olazo et al. (2010) and Tauber et al. (2013).

1. The head of the first instar is predominantly brown with a single marking consisting of a mesal and lateral arms, which are entirely of partially confluent along the midline. The first five abdominal segments bear a total of more than 14 long, hooked dorsal setae. The epicranial marking of the second and third instar are broad with two arms entirely or partially confluent; these instars also have paired frontal markings that are broadly fused mesally (Fig. 16.184)........................ .2

   – The head of the first instar is predominantly whitish or cream with a single marking consisting of two brown lateral arms, the mesal and lateral arms of which are completely separated. The first five abdominal segments bear a total of 14 long, hooked dorsal setae. The epicranial marking on the second and third instar consists of two paired longitudinal stripes. The frontal markings of these instars are close but separate, or their posterior edges are extended to meet mesally (Fig. 16.185)... 3

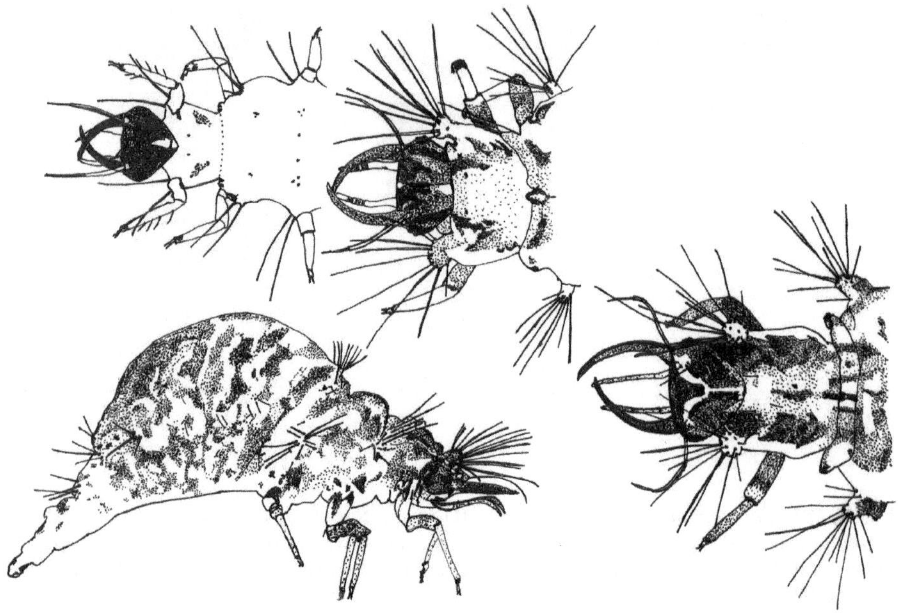

Fig. 16.184 *Chrysopodes divisus* larvae: dorsal views of the head and thorax of a first instar larva (upper left), heads and prothoraces of a second (upper right) and third instar larva (lower right); habitus of a third instar larva in lateral view (lower left). Based on Silva et al. (2013).

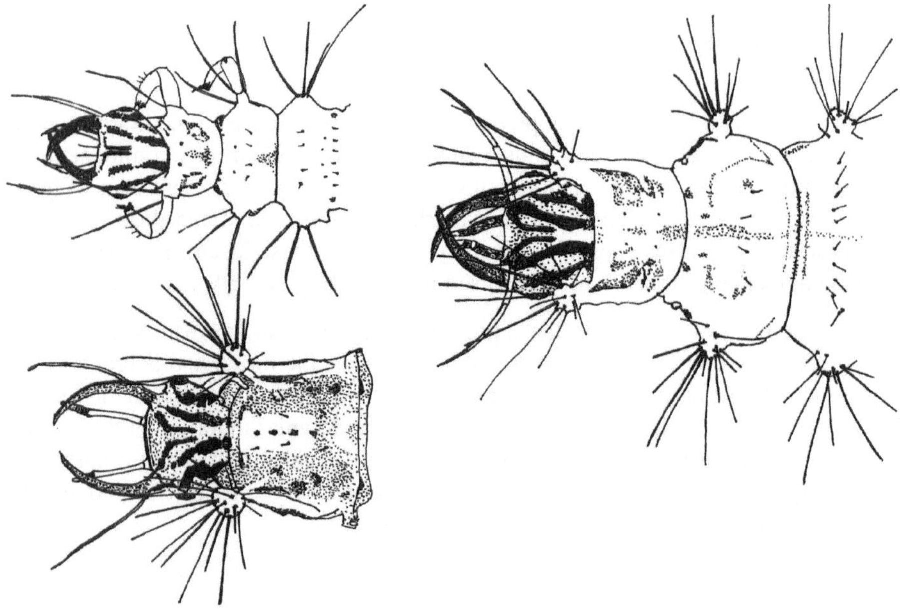

Fig. 16.185 *Chrysopodes fumosus* larvae: dorsal views of the heads and anterior parts of the thoraces of first instar (upper left), second instar (upper right), and third instar larva (center). Based on Silva et al. (2013).

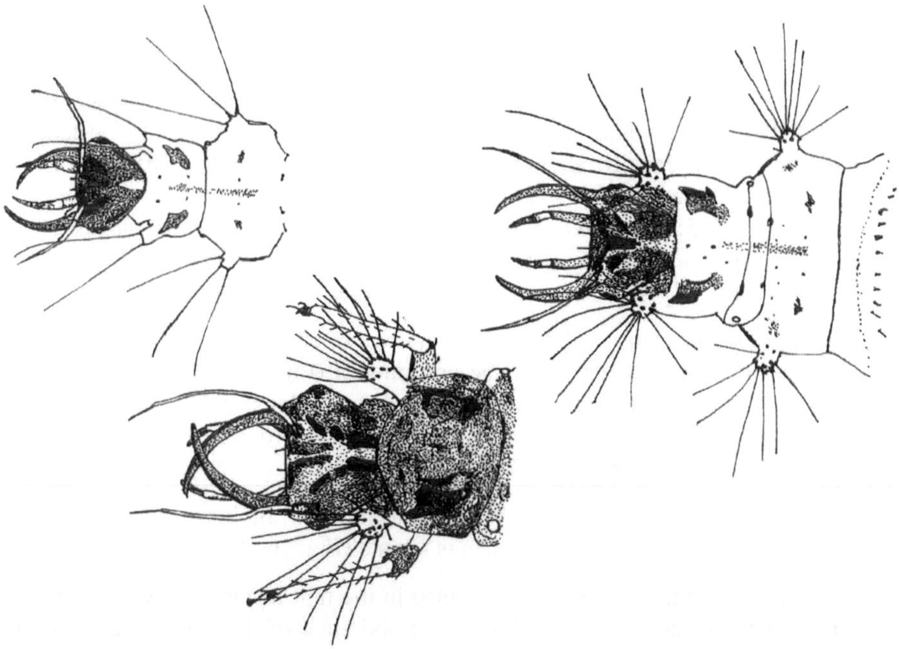

Fig. 16.186 *Chrysopodes lineafrons* larvae: dorsal views of the heads and anterior parts of the thoraces of first instar (upper left), second instar (upper right), and third instar larvae (center). Based on Silva et al. (2013).

2. (1) There are four lightly serrate dorsal setae on the cranium of the first instar, and the serrations become more pronounded on the second and third instars. The number of smooth, elongated, hooked setae on each of the first five abdominal segments of the first instar ranges from 16 to 20. The lateral and mesal markings on the epicranium of the second and third instars are confluent basally but separate toward their apices; at least one of the cranial setae is serrate. The pronota of the second and third instars are densely covered by fine, dark brown spinules. A posterior fold on the metathorax of these instars bears a transverse row of 12 or 13 long serrate setae inserted on strong, brown chalazae (Fig. 16.186). Length of first instar: 2.5–2.6 mm; second instar: 3.3–3.9 mm; third instar: 5.8–6.8 mm. .... *Chrysopodes lineafrons* Adams and Penny, 1987 (Argentina, Amazonas, Santa Catarina, São Paulo).

   – Only two dorsal setae on the cranium of the first instar are serrate. The number of smooth, elongated, hooked setae on each of the first five abdominal segments of the first instar exceeds 30. The lateral and mesal markings on the epicranium of the second and third instars are confluent from base to apex. The pronota of the second and third instars are sparsely covered by spinules, most of which are transparent. A posterior fold on the metathorax of these instars bears a transverse row of 14 or 15 long serrate setae inserted on strong, brown chalazae and one pair of smooth, lateral setae inserted on smaller chelazae (Fig. 16.184). Length of first instar: 2.5–2.9 mm; second instar: 3.3–3.4 mm; third instar: 5.7–6.3 mm.

.... *Chrysopodes divisus* (Walker, 1853) (West Indies, Guyana, Venezuela, Colombia, Peru, Paraguay, Argentina, Uruguay, Rio de Janeiro, Bahia). Syn: *Chrysopa divisa* Walker, 1853; *Chrysopa transversa* Walker, 1853; *Chrysopa nobregana* Navás, 1913; *Chrysopa hesperina* Banks, 1915; *Chrysopa agatha* Navás, 1925; *Chrysopa debilis* Navás, 1926; *Chrysopa uruguaya* Navás, 1927; *Chrysopa arequipae* Navás, 1929; *Chrysopa oglobini* Navás, 1931; *Neosuarius divisa* (Walker, 1853) *auctt.*

3. (1) The posterior ends of the frontal markings on all instars are straight and do not meet mesally. All cranial setae are smooth (Fig. 16.185). Length of first instar: 2.3–2.8 mm; second instar: 3.1–3.4 mm; third instar: 6.7–7.2 mm. .... *Chrysopodes fumosus* Tauber and Albuquerque in Tauber et al., 2012 (Venezuela, Rio de Janeiro).

   – The posterior ends of the frontal markings on all instars curve to meet mesally. At lease one cranial seta is serrate (Fig. 16.187) 4

4. (3) Only two or three of the cranial setae in the first instar are coarsely serrate. On the lateral margins of the thorax and abdomen of the first instar there are large, dark brown setae. There are no secondary serrae on the crania of the second and third instars. The pronotum has paired, dark brown sclerites, which fork

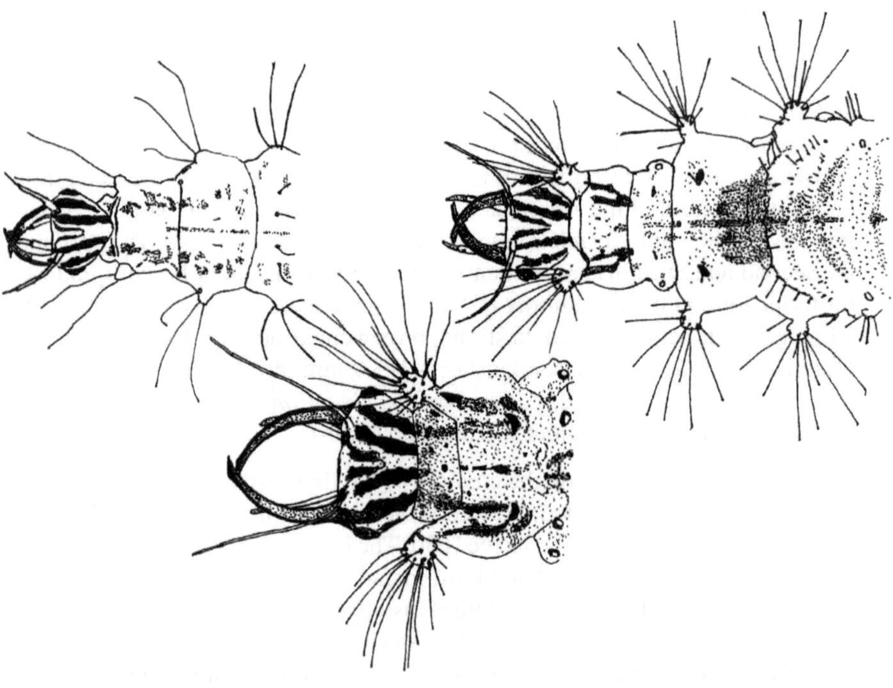

Fig. 16.187 *Chrysopodes geayi* larvae: dorsal views of the heads and anterior parts of the thoraces of first instar (upper left), second instar (upper right), and third instar larvae (center). Based on Silva et al. (2013).

at the base to form distinct mesal and lateral arms, which curve around the bases of lateral tubercles. There are several secondary sclerites along the midline of the pronotum (Fig. 16.187). Length of first instar: 2.7–3.1 mm; second instar: 4.6–5.1 mm; third instar: 7.5–7.9 mm.

.... *Chrysopodes geayi* (Navás, 1910) (French Guiana, Surinam?, Rio de Janeiro). Syn: *Nothochrysa geayi* Navás, 1910; *Allochrysa pulchella* Banks, 1944.

– More than three and usually six of the cranial setae in the first instar are coarsely serrate. On the lateral margins of the thorax and abdomen of the first instar, there are large, pale amber to light brown setae. There are three or four pairs of small secondary serrae on the crania of the second and third instars. The pronotum has paired, light brown or transparent sclerites, which fork at the base to form mesal and very indistinct lateral arms, which are light brown or transparent. There is only one sclerite and no secondary sclerites along the midline of the pronotum (Fig. 16.188). Length of first instar: 2.7–2.8 mm; second instar: 4.1–4.1 mm; third instar: 5.8–7.2 mm.

.... *Chrysopodes spinellus* Adams and Penny, 1987 (Argentina, Rio de Janeiro). Syn: *Chrysopodes spinella* Adams and Penny, 1987, mistaken gender. According to Silva et al. (2013), the illustration reported to be of this larva, published by González Olazo et al. (2010), is not of this species and remains unidentified.

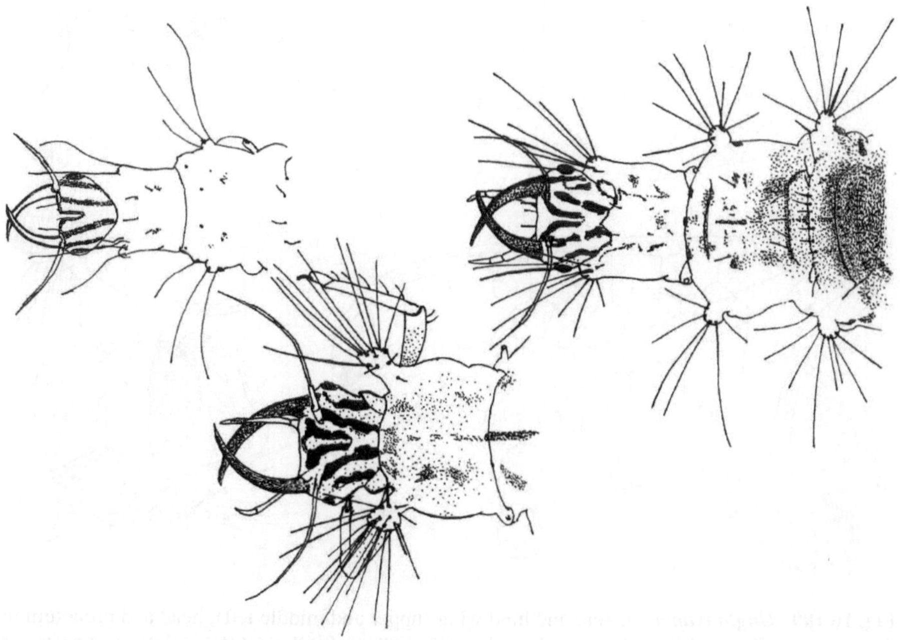

Fig. 16.188 *Chrysopodes spinellus* larvae: dorsal views of the heads and anterior parts of the thoraces of first instar (upper left), second instar (upper right), and third instar larvae (center). Based on Silva et al. (2013).

Key to the Species of Adult *Ungla* in South America

Information for the key was provided by Navás (1933c), de Freitas (2007a, b), de Freitas et al. (2009), von Ellenrieder (2009), and Tauber and Flint (2010a, b). An unplaced species under the name *Chrysopa reboredina* Navás, 1933, has been tentatively assigned to the genus *Ungla*. The only specimen is from Argentina, and it is badly faded and lacks an abdomen (von Ellenrieder, 2009), so the correct assignment to a genus may not be possible until undamaged specimens are found.

1. The markings on the head, pronotum, and mesonotum are vermillion to bright red. The long wing veins are green, and the transverse veins are black. The gonarcus of the male is strongly arched. The process on the gonarcus does not have a frontal projection. The arcessus is wide (Fig. 16.189). Length of forewing: 7.9–10.2 mm; hind wing: 7.2–9.2 mm.
.... *Ungla ivancruzi* Freitas, 2007
(Minas Gerais).

   – Most of the markings on the head, pronotum, and mesonotum are black or brownish black. The gonarcus of the known males is only slightly arched, and the process on the gonarcus has a frontal projection. The arcessus of all known males is narrow (Fig. 16.190). The male genitalia of *Ungla laufferi* remain unknown 2

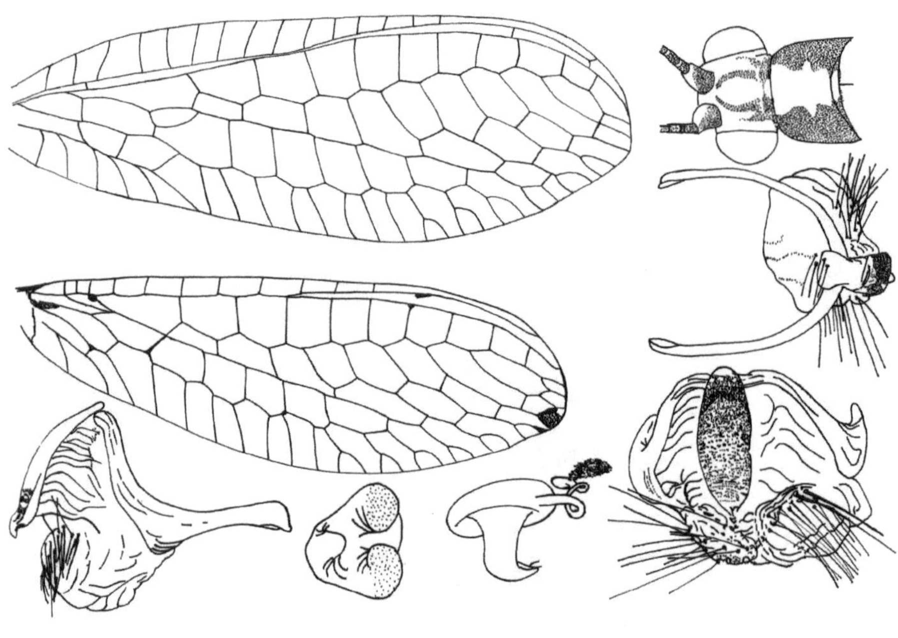

Fig. 16.189 *Ungla ivancruzi:* fore and hind wing (upper and middle left); head and pronotum in dorsal view (upper right); male genitalia in dorsal (middle right), lateral (lower left), and posterior view (lower right); spermatheca (lower right center) and subgenitale of a female (lower left center). Based on de Freitas (2007a, b).

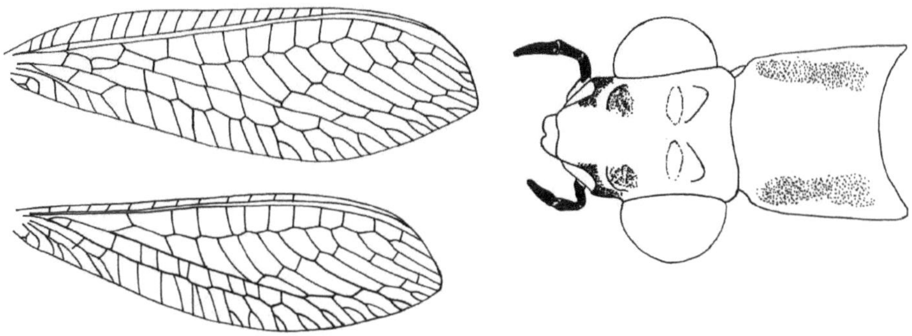

Fig. 16.190 *Ungla laufferi:* fore and hind wing (left) and the head and pronotum in dorsal view (right). Based on de Freitas et al. (2009).

2. (1) The antennae are pale, and the maxillary palps are black. The genae and lateral parts of the clypeus are blackish brown. There are faint red markings on the frons and brown spots on the vertex. There is a broad ivory longitudinal stripe on the pronotum bordered laterally by areas of small brown spots. The veins of the fore-wing are mainly pale but darkened at the ends of the costal and R-Rs cross-veins. The gradate crossveins are darkened and have narrow dark borders. The costal crossveins and the anterior end of crossvein R-Rs in the hind wing are darkened. Both wings have eight inner and seven outer gradate crossveins (Fig. 16.190). Fore-wing length: 13–15 mm.
   .... *Ungla laufferi* (Navás, 1922)
   (Colombia, Ecuador). Syn: *Chrysopa laufferi* Navás, 1922; *Chrysopa aroguesina* Navás, 1929; *Ceraeochrysa laufferi* (Navás, 1922) auctt. Because the abdomen of the male type specimen has been lost, there has been much disagreement about the genus to which this species belongs. According to Tauber and Flint ((2010a, b)), the pillbox shape of the female spermatheca, however, indicates that it belongs to *Ungla*.
   – The wing veins are predominantly uniform in color, and the coloration of the parts of the head show less distinct contrast (Fig. 16.191) 3
3. (2) The wing veins are blackish 4
   – The wing veins are green 5
4. (3) The gonarcus is narrow and has a long frontal projection on its process. The pronotum has a dense covering of black setae (Fig. 16.24). Fore-wing length: c. 11.mm.
   .... *Ungla binaria* (Navás, 1922)
   (Argentina). Syn: *Chrysopa binaria* Navás, 1922 nec 1928.

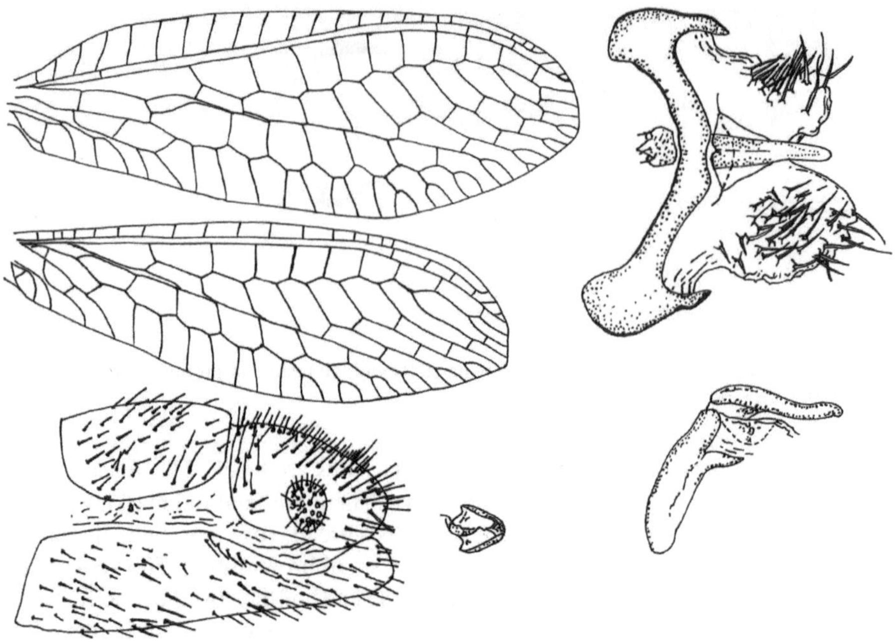

Fig. 16.191 *Ungla argentina:* fore and hind wing (upper and middle left), apex of the male abdomen in lateral view (lower left), male genitalia in dorsal (upper right) and lateral view (lower right), and the hypandrium (lower center). Based on Adams (1975), who referred to it by its synonym, *Suarius argentinus*.

– The gonarcus is wide with a short, expanded frontal projection on its process (Fig. 16.191). Fore-wing length: 12.6–15.0 mm.

.... *Ungla argentina* (Navás, 1911) (Argentina). Syn: *Hypochrysa argentina* Navás, 1911; *Ungla annulata* Navás, 1914; probably *Chrysopa graciana* Navás, 1919; *Chrysopa coronata* Navás, 1930; *Chrysopa lanata* v. *graciana* Bruch; *Chrysopa argentina* (Navás, 1911) Tjeder 1971; *Suarius argentinus* (Navás, 1911) Adams (1975).

5. (3) The process on the gonarcus has a frontal projection. .... *Ungla nesotala* (Banks, 1944) (Colombia). Syn: *Chrysopa nesotala* Banks, 1944.

– The process on the gonarcus lacks a frontal projection. .... *Ungla confraterna* (Banks, 1913)

(Argentina). Syn: *Chrysopa confraterna* Banks, 1913.

Preliminary Key to the Species of Known *Ungla* Larvae in South America

Information for the key was provided by Monserrat and Freitas (2005) and Reguilón (2010).

1. The anteromedial marking on the head is blackish and subrhomboidal, and the blackish markings lateral to it are subtriangular with the vertex of the triangle pointing at the prothorax. The mesothorax has two round, chestnut brown dorsolateral markings (Fig. 2.25).
   .... *Ungla argentina* (Navás, 1911) (Argentina). Syn: *Hypochrysa argentina* Navás, 1911; *Ungla annulata* Navás, 1914; probably *Chrysopa graciana* Navás, 1919; *Chrysopa lanata* v. *graciana* Bruch, 1917; *Chrysopa argentina* (Navás, 1911) Tjeder 1971; *Suarius argentinus* (Navás, 1911) Adams (1975).
   - The anteromedial marking on the head is subrhomboidal, and there are stripes lateral to it, which reach posteriad to the pronotum. The mesothorax has two subtriangular, dark chestnut brown dorsolateral markings (Figs. 1.19, 1.20, and 1.21).

   .... *Ungla binaria* (Navás, 1923) (Argentina).

Key to the Species of *Plesiochrysa* in South America

Information for the key was provided by Navás (1929c, 1930); Adams (1982a), and de Freitas and Penny (2001).

1. The antenna is dark (Fig. 16.192). Length of fore-wing: c. 15.6 mm; hind wing: c. 14.0 mm.
   .... *Plesiochrysa alytos* Freitas and Penny, 2001 (São Paulo, Mato Grosso).
   - The antenna is pale (Fig. 16.193) 2

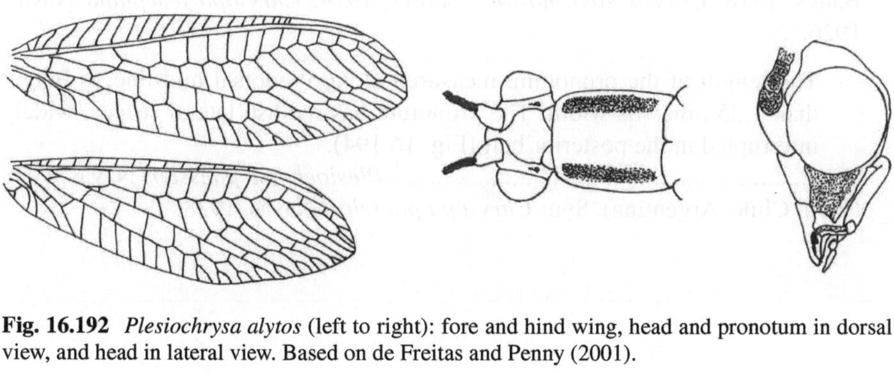

Fig. 16.192 *Plesiochrysa alytos* (left to right): fore and hind wing, head and pronotum in dorsal view, and head in lateral view. Based on de Freitas and Penny (2001).

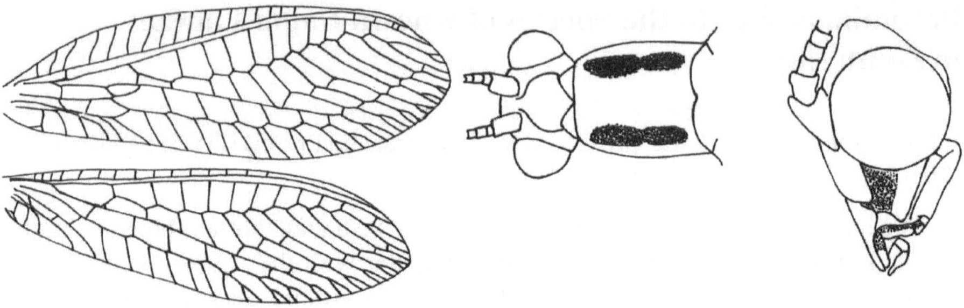

Fig. 16.193 *Plesiochrysa elongata:* fore and hind wing (left), head and pronotum in dorsal view (center), and head in lateral view (right). Based on de Freitas and Penny (2001).

2. (1) The length of the pronotum measured along its dorsal mid-line is shorter than its width. The pronotum has widely separated lateral spots. The graduate series of wing veins is divergent (Fig. 16.22).
....*Plesiochrysa brasiliensis* (Schneider, 1851) (North and Central America, West Indies, Colombia, Venezuela, Guyana, Argentina, Amazonas, Pará, Maranhão). Syn: *Chrysopa brasiliensis* Schneider, 1851; *Chrysopa bouvieri* Navás, 1923; *Chrysopa antillana* Navás, 1924; *Chrysopa rata* Lacroix, 1926; *Chrysopa uribei* Navás, 1927; *Chrysopa scalaris* Navás, 1929; *Chrysopa yucatanensis* Navás, 1929; *Chrysopa divergens* Navás, 1931.

   – The length of the pronotum, measured along its dorsal mid-line, is subequal to or greater than its width (Fig. 16.193).. 3

3. (2) The length of the pronotum, measured along its dorsal mid-line, is subequal to or slightly greater than its width. The pronotum has reddish lateral stripes, which are continuous or with only a small interruption. The wing veins in the graduate series are parallel (Fig. 16.193). Fore-wing length of male: 13 mm; female: 15 mm.
....*Plesiochrysa elongata* (Navás, 1913) (Central America, West Indies, Peru, Argentina, Amazonas). Syn: *Chrysopa elongata* Navás, 1913; *Chrysopa angusta* Navás, 1914; *Chrysopa submarginata* Banks, 1918; *Chrysocerca mariae* Lacroix, 1919; *Chrysopa josephina* Navás, 1926.

   – The length of the pronotum, measured along its dorsal mid-line, is longer than 1.25 times its width. The pronotum has reddish lateral stripes, widely interrupted in the posterior half (Fig. 16.194).
....*Plesiochrysa paessleri* Navás, 1928 (Peru, Chile, Argentina). Syn: *Chrysopa paessleri* Navás, 1928.

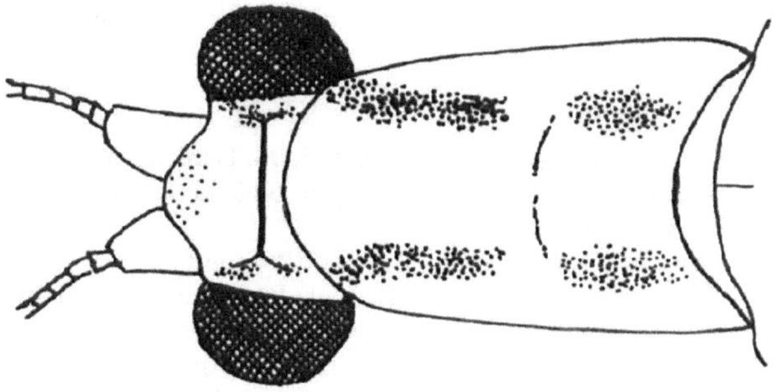

Fig. 16.194 Head and pronotum of *Plesiochrysa paessleri* in dorsal view. Based on Adams (1982a).

Literature

Adams PA (1969) A new species and synonymy in the genus *Meleoma* (Neuroptera, Chrysopidae), with a discussion of genitalic homologies. Postilla 136:1–18

Adams PA (1975) Status of the genera *Ungla* and *Mallada* Navás (Neuroptera: Chrysopidae). Psyche 82:167–173

Adams PA (1977) Taxonomy of United States *Leucochrysa* (Neuroptera: Chrysopidae). Psyche 84:92–102

Adams PA (1978a) Zoogeography of new world Chrysopidae, a progress report. Folia Entomologica Mexicana 30–40:210–211

Adams PA (1978b) A new species of *Hypochrysa* and a new subgenus and species of *Mallada* (Neuroptera: Chrysopidae). Pan-Pacific Entomologist 54(4):292–296

Adams PA (1982a) *Plesiochrysa*, a new subgenus of *Chrysopa* (Neuroptera) (Studies in new world Chrysopidae, Part 1). Neuroptera International 2(1):27–32

Adams PA (1982b) *Ceraeochrysa*, a new genus of Chrysopinae (Neuroptera) (Studies in new world Chrysopidae, Part 2). Neuroptera International 2(1):69–75

Adams PA (1987) Studies in neotropical Chrysopidae (Neuroptera) III. Notes on *Nodita amazonica* Navas and *N. oenops*, n. sp. Neuroptera International 4(4):287–294

Adams PA, Penny ND (1986) Faunal relations of Amazonian Chrysopidae. In: Gepp J, Aspöck H, Hölzel H (eds.) Recent research in neuropterology. In: Proceedings of the 2nd international symposium on neuropterology, Graz, pp 119–124

Adams PA, Penny ND (1987 [1985]) Neuroptera of the Amazon Basin. Part 11a. Introduction and Chrysopini..Acta Amazonica 15(3–4):413–479

Adams PA, Penny ND (1992) New genera of Nothochrysinae from South America (Neuroptera: Chrysopidae). Pan-Pacific Entomologist 68(3):216–221

Albertino SAS (2009) Biologia de três espécies de *Leucochrysa* (Insecta, Neuroptera, Chrysopidae) abundantes na Mata Atlântica e áreas agrícolas da região Norte Fluminense. – Anais do IX Congresso da Ecologia do Brasil, 13–17 Setembro de 2009, São Lourenço, Minas Gerais, pp 1–2

Arias JR, Penny ND (1980) A entomofauna da Bacia Amazónica. Acta Amazonica (Supplemento) 9(4):103–107

Banks N (1910) New South American neuropteroid insects. Proceedings of the Entomological Society of Washington 12:146–160

Banks N (1914) New American Chrysopidae. The Canadian Entomologist 2(4):24–27

Banks N (1918) New neuropteroid insects. Bulletin of the Museum of Comparative Zoology 62(1):1–22+pl. 1–2.

Banks N (1944) Neuroptera of northern South America. Part III. Boletin de Entomologia Venezolana 3(1):1–34
Banks N (1946 [1945]) A review of the Chrysopidae (Notochrysidae) of Central America. Psyche 52(3–4):139–174
Bezerra CES, Nogueira CHF, Sombra KDS, Demartelaere ACF, de Araujo EL (2009) Crisopídeos (Neuroptera: Chrysopidae): aspectos biológicos, potencial de utilização e perspectivas futuras. Revista Caatinga 22(3):1–5
Bezerra CES, Tavares PKA, Macedo LPM, de Freitas S, de Araujo EL (2010) Green lacewings (Neuroptera: Chrysopidae) associated with melon crop in Mossoró, Rio Grande do Norte State, Brazil. Neotropical Entomology 39(3):454–455
Brooks SJ (1994) A taxonomic review of the common green lacewing genus *Chrysoperla* (Neuroptera: Chrysoperla). Bulletin of the Natural History Museum, Entomology Series 63(2):137–210
Brooks SJ (1997) An overview of the current status of Chrysodidae (Neuroptera) systematics. Deutsche Entomologische Zeitschrift (NF) 44:267–275
Brooks SJ, Bernard PC (1990) The green lacewings of the world: a generic review (Neuroptera: Chrysopidae). Bulletin of the British Museum, Natural History (Entomology) 59(2):117–286
Bruch C (1917) Desarrollo de "*Chrysops lanata*" Banks (Neuróptero). Physis 3(15):361–369
Cadena P, Ángel F, Gómez LA, González R (2007) Diferenciación morfológica y molecular de especies de crisópidos (Neuroptera: Chrysopidae). –. Revista Colombiana de Entomología 33(2):171–177
de Freitas S (2003) *Chrysoperla* Steinman, 1864 (Neuroptera, Chrysopidae): descrição uma nova espécie no Brasil. Revista Brasileira de Entomologia 47(3):385–387
de Freitas S (2007a) New species of Brazilian green lacewings genus *Leucochrysa* McLachlan, 1868 (Neuroptera Chrysopidae). In: Pantaleoni RA, Letardi A, Corazza C (eds) Proceedings of the IX international symposium on neuropterology, Ferrara, Italy, 20–23 June, 2005. Annali del Museo civico di Storia naturale di Ferrara 8:I–XX:49–54
de Freitas S (2007b) Ocorrência de *Ungla* Navás (Neuroptera, Chrysopidae) no Brasil e descrição de nova espécie. Revista Brasileira de Entomologia 51(4):413–415
de Freitas S, Penny ND (2000) Two new genera of Neotropical Chrysopini (Neuroptera: Chrysopidae). Journal of the Kansas Entomological Society 73(3):164–170
de Freitas S, Penny ND (2001) The green lacewings (Neuroptera: Chrysopidae) of Brazilian agro-ecosystems. Proceedings of the California Academy of Sciences 52(19):245–395
de Freitas S, Penny ND, Adams PA (2009) A revision of the world genus *Ceraeochrysa* (Neuroptera: Chrysopidae). Proceedings of the California Academy of Sciences 60(16):503–610
Esben-Petersen P (1933) New and little-known Neuroptera. Meddelelser fra Dansk Naturhistorisk Forening København 94:109–123
Fitch A (1855) Report of Asa Fitch, M.D., on the noxious, beneficial, and other insects of the State of New York. Transactions of the New York State Agricultural Society 14:705–880
González EV, Heredia JF, Cichón L, Fernández D, Garrido S (2011) Chrysopidos (Neuroptera, Chrysopidae) asociados a frutales de pepita en el Alto Valle de Río Negro y Neuquén (región Patagonia Norte Argentina). Horticultura Argentina 30(73):5–8
González Olazo EV, Heredia JF, Federico J (2010) Morfología de los estadios larvales y datos biológicos de *Chrysopodes (Chrysopodes) spinellus* (Neuroptera, Chrysopidae), nueva cita para la Argentina. Acta Zoológica Lilloana 54(1–2):87–92
González Olazo EV, Lanati SJ, Heredia JF (2009) Morfología y datos biológicos de los estados preimaginales de *Chrysoperla asoralis* (Neuroptera: Chrysopidae). Acta Zoológica Lilloana 53(1–2):1–28
González Olazo EV, Reguilón C (2002) Una nueva especie de *Chrysoperla* (Neuroptera: Chrysopidae) para la Argentina. Revista de la Sociedad Entomológica Argentina 61(1–2):47–50
González Olazo EV, Toledo SN, Zaia G (1999) Nuevas citas de Chrysopidae (Neuroptera: Planipennia) para la Argentina. Acta Zoológica Lilloana 45:151–152

Guérin-Méneville FE (1853) Description d'une nouvelle espèce de Névroptères. Revue et Magasin de Zoologie Pure et Appliquée (2)5:261–263 + 1 pl

Haramboure M, Reguilón C, Alzogaray RA, Schneider MI (2014) First record of *Chrysoperla asoralis* and *C. argentina* (Neuroptera: Chrysoperlidae) in horticultural fields of La Plata associated with the sweet pepper (*Capsicum annuum* L.). Revista de la Sociedad Entomologica Argentina 73(3–4):187–190

Heckman CW (1994) New limnological nomenclature to describe ecosystem structure in the tropical wet-and-dry climatic zone. Archiv für Hydrobiologie 130:385–407

Kimmins DE (1952) XCVII. A revision of the genera of the Apochrysinae (Fam. Chrysopidae). Annals and Magazine of Natural History (12) 5:929–944

Legrande J, Tauber CA, Albuquerque GS, Tauber MJ (2008 [2009]) Navás' type and non-type specimens of Chrysopidae in the MNHN Paris (Neuroptera). Revue française d'Entomologie (N.S.) 30:103–183

Mantoanelli E, Albuquerque GS (2007) Desenvolvimento e comportamento larval de *Leucochrysa (Leucochrysa) varia* (Schneider) (Neuroptera: Chrysopidae) em laboratorio. Revista Brasileira de Zoologia 24(2):302–311

Mantoanelli E, Albuquerque GS, Tauber CA, Tauber MJ (2006) *Leucochrysa (Leucochrysa) varia* (Neuroptera: Chrysopidae): larval descriptions, developmental rates, and adult color variation. Annals of the Entomological Society of America 99(1):7–18

Mantoanelli E, Tauber CA, Tauber MJ, Albuquerque GS (2011) Larvae of four *Leucochrysa (Nodita)* species (Neuroptera: Chrysopidae: Leucochrysini) from Brazil's Atlantic Coast. Annals of the Entomological Society of America 104(6):1233–1259

Monserrat VJ, de Freitas S (2005) Contribución al conocimiento de los crisópidos de Coquimbo, Patagonis y Tierra del Fuego (Argentina, Chile) (Insecta, Neuroptera, Chrysopidae). Graellsia 61:163–179

Navás L (1910a) Crisópidos nuevos ó poco conocidos.Revista de la Real Academia de Ciencias Exactas Fisicas y Naturales de Madrid 9:473–480

Navás L (1910b) Crisópidos nuevos. Brotéria, Série Zoológica 9(1):38–59

Navás L (1912) Crisópidos y Hemeróbidos (Ins. Neur.) nuevos o críticos. Brotéria Série Zoológica 10:98–113

Navás L (1913) Chrysópidos Sudamericanos. Brotéria, Série Zoológica 11(2–3):73–104, 149–168 + pl. VII

Navás L (1914a [1913]) Les Chrysopides (Ins. Névr.) du Musée de Londres. Annales de la Société scientifique de Bruxelles 37(2):292–330

Navás L (1914b) Les Chrysopides (Ins. Névr.) du Musée de Londres. Conclusion. Annales de la Société scientifique de Bruxelles 38(1):73–114

Navás L (1915) Neue Neuropteren. Zweite (II) serie. Entomologische Mitteilungen 4:194–202

Navás L (1916) Neurópteros Sudamericanos. Tercera Serie. Brotéria, Série Zoológica 14(1):14–35

Navás L (1917a) Neue Neuropteren. Dritte Serie. Entomologische Mitteilungen 6(7/9):274–282

Navás L (1917b) Algunos insectos Neurópteros de la Argentina. Serie I. Physis 3(13–15):186–196

Navás L (1918a [1917]) Insecta Nova. IV Series. Neuroptera. Memorie della'Pontifica Accademia Romana dei Nuovi Lincei, Serie Seconda 4:4–23, Fig. 10–19

Navás L (1918b) Insectos chilenos. Boletin Sociedad Aragonesa de Ciencias Naturales 17:212–230

Navás L (1919 [1918]) Algunos insectos Neurópteros de la República Argentina. Serie Tercera (III). Revista de la Real Academia de Ciencias Exactas Fisicas y Naturales de Madrid 17:287–305

Navás L (1920 [1919]) Insecta Nova. VII Series. Neuroptera. Memorie della'Pontifica Accademia Romana dei Nuovi Lincei, Serie Seconda 5:21–29

Navás L (1922a [1921]) Insectos Sudamericanos Cuarta Serie (1). Revista de la Real Academia de Ciencias Exactas Fisicas y Naturales de Madrid 19:255–267

Navás L (1922b) Insectos nuevos o poco conocidos. Memorias de la Real Academia de Ciencias y Artes de Barcelona. Tercera Época 17(15):383–400

Navás L (1923a) Insecta Nova. X Series. Neuroptera. Memorie della'Accademia Pontifica dei Nuovi Lincei, Rome, Serie Seconda 6:19–27

Navás L (1923b) *Chrysopa furculata* nom. nov. for *Chrysopa furcata* Navás, 1922. Boletin de la Sociedad Entomológica de España 6:39

Navás L (1924a) Chrysópidos (Ins. Neur.) Neotrópicos. 1.a Serie. Revista Chilena de Historia Natural 27:110–116

Navás L (1924b) Algunos insectos de Cuba. Secunda Serie. Revista de la Academia de Cíencias Exactas Físicas y Naturales de Madrid 21:333–340

Navás L (1925a) Neuropteren aus Brasilien. Mitteilungen der Münchener Entomologischen Gesellschaft 15:64–68

Navás L (1925b) Insectos Neotrópicos. 1.a Serie. Revista Chilena de Historia Natural 29:305–313

Navás L (1926a) Insectos exóticos Neurópteros y afines. Brotéria, Série Zoológica 23:79–93

Navás L (1926b) Insecta Nova. XI Series. Neuroptera. Memorie della'Pontifica Accademia Romana dei Nuovi Lincei, Roma (2)9:101–110

Navás L (1927) Insecta Nova. Serie XII. Neuroptera. Memorie della'Accademia Pontifica dei Nuovi Lincei, Rome, Serie Seconda 10:1–10

Navás L (1929a [1928]) Insectos neotropicos, 4.ª Serie. Revista Chilena de Historia Natural 32:106–128

Navás L (1929b) Insectos del Brasil. 3.ª serie. Revista do Museu Paulista 16:857–864

Navás L (1929c) Insectos Neurópteros del Museo de Hamburgo. Memorias de la Real Sociedad España de Historia Natrual 15:315–322

Navás L (1930) Insectos neotropicos, (Sexta Serie). Revista Chilena de Historia Natural 34:62–75, 299–307

Navás L (1931) Decadas de insectos nuevos. Decada 2.ª. Revista de la Real Academia de Ciencias Exactas, Fisicas y Naturales de Madrid 26:79–86

Navás L (1933a) Insectos Suramericanos, Septima Serie. Revista de la Academia de Cíencias Exactas Físicas y Naturales de Madrid 31:303–314

Navás L (1933b) Insectos Neotropicos. 8ª Serie. Revista Chilena de Historia Natural 37:194–197

Navás L (1933c [1932]) Insectos de la Argentina. Revista de la Academia de ciencias exactas, fisico-quimicas y Naturales de Zaragosa 16:87–120

Navás L (1934) Insectos Suramericanos, Octava Serie. Revista de la Academia de Cíencias Exactas Físicas y Naturales de Madrid 31:9–28

Navás L (1935 [1934]) Insectos Suramericanos. Décima Serie. Revista de la Academia de Cíencias Exactas Físicas y Químicas y Naturales de Madrid 32:360–375

Núñez E (1988) Chrysopidae (Neuroptera) del Perú y sus especies más comunes. Revista Peruana de Entomologia 31:69–75

Olivier GA (1792) Encyclopédie méthodique. Histoire naturelle. Insectes. Tome septième. Panckoucke, Paris. 827 pp

Oswald JD (1987) *Hypochrysa* Hagen, 1866 (Chrysopidae) and *Stenorrhachus* Mel., 1886 (Nemopteridae) are valid names in the Neuroptera. Neuroptera International 4:225–229

Penny ND (1996) A revision of *Parachrysopiella* (Neuroptera: Chrysopidae). Journal of the Kansas Entomological Society 69(3):222–233

Penny ND (2002) Family Chrysopidae. In: Penny ND (ed) A guide to the lacewings (Neuroptera) of Costa Rica. Proceedings of the California Academy of Sciences 53(12):187–227

Reguilón C, González Olazo EV, Núñez Campero SR (2006) Morfología de los estados inmaduros de *Chrysoperla argentina* (Neuroptera: Chrysopidae). Acta Zoológica Lilloana 50(1–2):31–39

Schneider GT (1851) Symbolae ad monographiam generis Chrysopae, Leach. Apud Ferdinandum Hirt, Vratislaviae. 178 pp

Silva PS, Albuquerque GS, Tauber CA, Tauber MJ (2007) Life history of a widespread Neotropical predator, *Chrysopodes (Chrysopodes) lineafrons* (Neuroptera: Chrysopidae). Biological Control 41(1):33–41

Silva PS, Tauber CA, Albuquerque GS, Tauber MJ (2013) Larva of five horticulturally important species of *Chrysopodes* (Neuroptera, Chrysopidae): shared generic features, descriptions and keys. ZooKeys 262:39–92

Sosa F, de Freitas S (2010) New Neotropical species of *Ceraeochrysa* Adams (Neuroptera: Chrysopidae). Zootaxa 2562:57–65

Sosa F, de Freitas S (2011) A new synonym, a new male description and new geographical records for three *Ceraeochrysa* species (Neuroptera: Chrysopidae). Zootaxa 2913:47–58

Sosa F, de Freitas S (2012) A new genus of Neotropical Chrysopini (Neuroptera: Chrysopidae). Zootaxa 3351:1–14

Tauber CA (2003) Generic characteristics of *Chrysopodes* (Neuroptera: Chrysopidae), with new larval descriptions and a review of the species from the United States and Canada. Annals of the Entomological Society of America 96(4):472–490

Tauber CA (2006) A new species of *Berchmansus* (Neuroptera: Chrysopidae) and nomelclatural notes on *B. elegans* (Guérin Méneville). Proceedings of the Entomological Society of Washington 108(4):830–841

Tauber CA (2007) Review of *Berchmansus* and *Viera* and description of two new species of *Leucochrysa* (Neuroptera: Chrysopidae). Annals of the Entomological Society of America 100(2):110–138

Tauber CA (2010) A revision of *Neosuarius,* a subgenus of *Chrysopodes* (Neuroptera: Chrysopidae). ZooKeys 44:1–104

Tauber CA, Albuquerque GS, Tauber MJ (2005) Characteristics of the *Loyola* Navás male (Neuroptera: Chrysopidae: Apochrysinae). Proceedings of the Entomological Society of Washington 107(3):543–547

Tauber CA, Albuquerque GS, Tauber MJ (2008a) A new species of *Leucochrysa* and a redescription of *Leucochrysa (Nodita) clepsydra* Banks (Neuroptera: Chrysopidae). Zootaxa 1781:1–19

Tauber CA, Albuquerque GS, Tauber MJ (2008b) *Gonzaga nigriceps* (MacLachlan) (Neuroptera: Chrysopidae): descriptions of larvae and adults, biological notes, and generic affiliation. Proceedings of the Entomological Society of Washington 112(3):417–438

Tauber CA, Albuquerque GS, Tauber MJ (2011a) Nomenclatorial changes and redescriptions of three of Navás's *Leucochrysa (Nodita)* species (Neuroptera, Chrysopidae). ZooKeys 92:9–33

Tauber CA, Albuquerque GS, Tauber MJ (2012a) The Neotropical genus *Titanochrysa* (Neuroptera, Chrysopidae): larval descriptions, biological notes, a new species, and taxonomic changes. Zootaxa 3514:1–26

Tauber CA, Albuquerque GS, Tauber MJ (2012b) Three new Brazilian species of *Chrysopodes* (Neuroptera, Chrysopidae). Annals of the Entomological Society of America 105(5):638–663

Tauber CA, Flint OS (2010a) Resolution of some taxonomic and nomenclatural issues in a recent revision of *Ceraeochrysa* (Neuroptera: Chrysopidae). Zootaxa 2565:55–67

Tauber CA, Flint OS (2010b) Erratum. Resolution of some taxonomic and nomenclatural issues in a recent revision of *Ceraeochrysa* (Neuroptera: Chrysopidae). Zootaxa 2565:55–67. Zootaxa 2572:68

Tauber CA, Mantoanelli E, Albuquerque GS, Reguilón C, Gonzalez Olazo E, Tauber MJ (2011b) A taxonimcally significant polymorphism in *Leucochrysa* (Neuroptera: Chrysopidae): nomenclature, larval and adult descriptions, and biological notes. Zootaxa 3130:1–29

Tauber CA, Sosa F, Albuquerque GS (2013) Two common and problematic leucochrysine species – *Leucochrysa (Leucochrysa) varia* (Schneider) and *L. (L.) pretiosa* (Banks): redescriptions and synonymies. ZooKeys 310:57–101

Tauber CA, Tauber MJ (2010) Two new endemic species of *Chrysopodes (Neosuarius)* (Neuroptera: Chrysopidae) from the Galapagos Islands. ZooKeys 42:47–78

Tauber CA, Tauber MJ, Albuquerque GS (2001) *Plesiochrysa brasiliensis* (Neuroptera: Chrysopidae): larval stages, biology, and taxonomic relationships. Annals of the Entomological Society of America 94:858–865

Tauber CA, Tauber MJ, Albuquerque GS (2006) *Berchmansus elegans* (Neuroptera: Chrysopidae): larval and adult characteristics and new tribal affiliation. European Journal of Entomology 103:221–231

Tauber CA, Tauber MJ, Albuquerque GS (2008c) A new genus and species of green lacewings from Brazil (Neuroptera: Chrysopidae: Leucochrysini). Annals of the Entomological Society of America 101(2):314–326
Tauber CA, Tauber MJ, Albuquerque GS (2011c) Nomenclatorial changes and redescriptions of three of Navás' *Leucochrysa (Nodita)* species (Neuroptera, Chrysopidae). ZooKeys 92:9–33
Tauber CA, Tauber MJ, Albuquerque GS (2014) Debris-carrying in larval Chrysopidae. Unraveling its evolutionary history. Annals of the Entomological Society of America 107(2):295–314
Thouvenot M (2008) Névroptères de Guyane (première note). L'Entomologiste 64(3):177–178
von Ellenrieder N (2009) Type specimens of Insecta housed at the Museo de Ciencias Naturales de Salta, Argentina. Revista de la Sociedad Entomológica Argentina 68(3–4):253–262
Weber NA (1942) A neuropterous myrmecophile, *Nadiva valida* Erichs. Psyche 49(1–2):1–3
Winterton S (1995) A new genus and species of Apochrysinae (Neuroptera: Chrysopidae) from Australia, with a checklist of Australian Chrysopidae. Journal of the Australian Entomological Society 34:134–145
Winterton SL, Brooks SJ (2002) Phylogeny of the apochrysine green lacewings (Neuroptera: Chrysopidae: Apochrysinae). Annals of the Entomological Society of America 95(1):16–28
Winterton S, de Freitas S (2006) Molecular phylogeny of the green lacewings (Neuroptera: Chrysopidae). Australian Journal of Entomology 45:235–243

Chapter 17
Berothidae

Abstract An illustrated key is provided for identification of the genera of adult berothids known to occur in South America. Illustrated keys are also provided to identify the species of each genus native to South America. The countries or states of Brazil in which specimens of each species have been found and invalid synonyms are reported.

Keywords Beaded lacewing • Key to species • Cyrenoberothinae • Berothinae • Chile

Members of this family are called beaded lacewings. All known South American species are exclusively terrestrial in all life stages. The family is very small and includes only a few species worldwide; five of them are known from South America. The first South American berothid described, *Ormiscocerus nitidipennis,* was discovered in Chile during the nineteenth century and then rediscovered more than a century later (Penny and Winterton 2007). All species seem to be terrestrial throughout their lives, but details of the development of all larvae are not yet known.

17.1 Key to the South American Genera of Berothidae

Information for the key was provided by McLeod and Adams (1968), Monserrat (2006), Penny and Winterton (2007), and Aspöck and Randolph (2014). The species in this family are few and not well known. Two subfamilies were proposed for its South American species by McLeod and Adams (1968): Cyrenoberothinae only for the genus *Cyrenoberotha,* and Berothinae for the rest. Later, specimens of an unusual species, the type specimen of which could not be found, were discovered and described by Penny and Winterton (2007). Although it was tentatively suggested that the genus, *Ormiscocerus,* should belong to the subfamily Cyrenoberothinae, a final assignment of the South American species to subfamilies is not attempted here, and more studies will be required to elucidate the systematics of this family.

C.W. Heckman, *Neuroptera (Including Megaloptera)*, Encyclopedia of South American Aquatic Insects, DOI 10.1007/978-3-319-35125-4_17

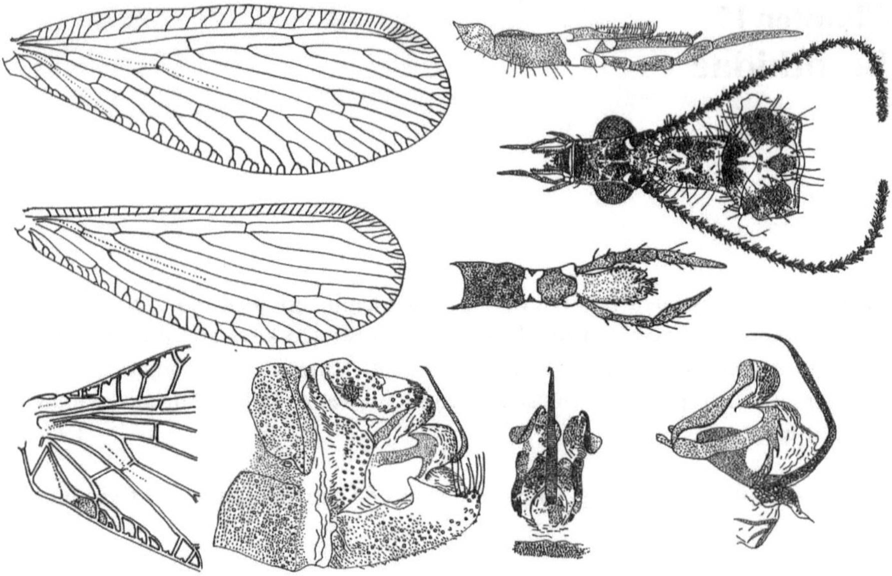

Fig. 17.1 *Cyrenoberotha penai* male: fore and hind wing (upper and middle left), base of the fore-wing (lower left), head and anterior part of the thorax in dorsal view (right center to upper right), right maxilla in anterior view (upper right center), labium in posterior view (right center), apex of the abdomen in lateral view (lower left center), and the internal genitalia in posterior (lower right center) and lateral view (lower right). Based on McLeod and Adams (1968).

1. Either the antennal flagellum consists of approximately 18 segments, or the face is considerably elongated ventrad below the ventral margins of the eyes. The ninth abdominal sternite is not fused with the ectoprocts (Fig. 17.1)................. 2

   – The antennal flagellum consists of at least c. 30 segments, and the face is short and does not extend very much below the ventral margins of the eyes. The fore-wing is usually subfalcate or falcate. The ninth abdominal sternite is fused with the ectoprocts (Fig. 17.2) 3

2. (1) The antennal flagellum consists of at least c. 30 segments. The face extends conspicuously ventrad from the ventral margins of the eyes (Fig. 17.1). ....*Cyrenoberotha* McLeod and Adams, 1968
   The only species in this genus is *Cyrenoberotha penai* McLeod and Adams, 1968, found in Chile.

   – The antennal flagellum consists of about 18 segments (Fig. 2.23). Fore-wing length of the only known species: c. 3.8 mm.
   ....*Ormiscocerus* Blanchard, 1851
   The only known species in this genus is *Ormiscocerus nitidipennis* Blanchard, 1851, from Chile. *Berothimerobius reticulatus* Monserrat and Deretsky, 1999, appears to be a junior synonym of this species (Penny and Winterton 2007).

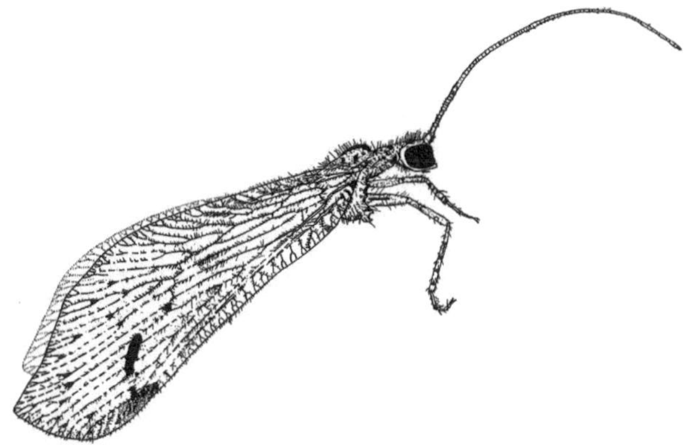

Fig. 17.2 Habitus of *Lomamyia trombetensis* in lateral view. Based on Penny (1983).

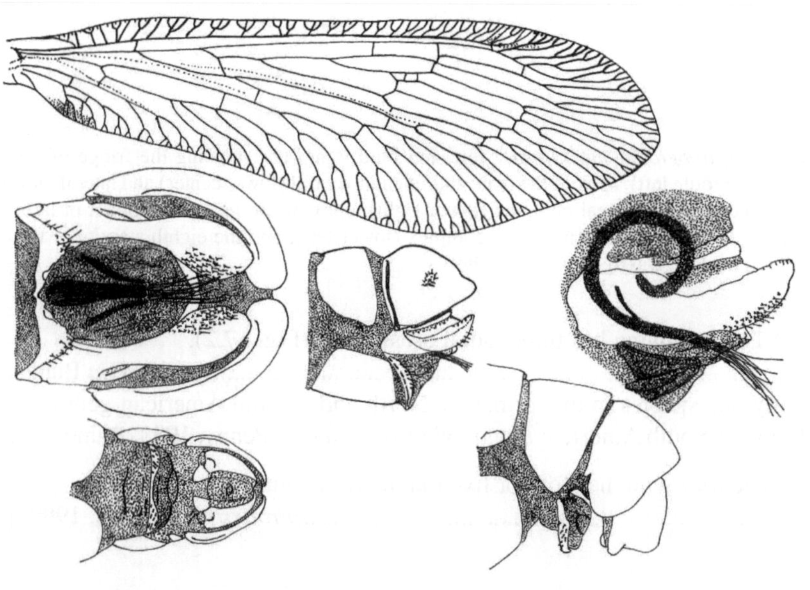

Fig. 17.3 *Naizema mendozina:* fore-wing of a male (above), apex of the male abdomen in posteroventral (middle left) and lateral view (center), male internal genitalia in lateral view (middle right), and the apex of the female abdomen in ventral (lower left) and lateral view (lower right). Based on McLeod and Adams (1968).

3. (1) The mandibles appear vestigial, and the mouthparts are sunken into a ventral cavity of the head. The pronotum is transverse, arched, and slightly sclerotized; it has two transverse furrows (Fig. 17.3). The spermatheca of the female is short. .... *Naizema* Navás, 1919..p. 598

   – The mandibles are well developed and easily visible. The pronotum is quadratic, sometimes slightly constricted, and not flexed laterally (Fig. 17.2). The spermatheca of the female is long. 4

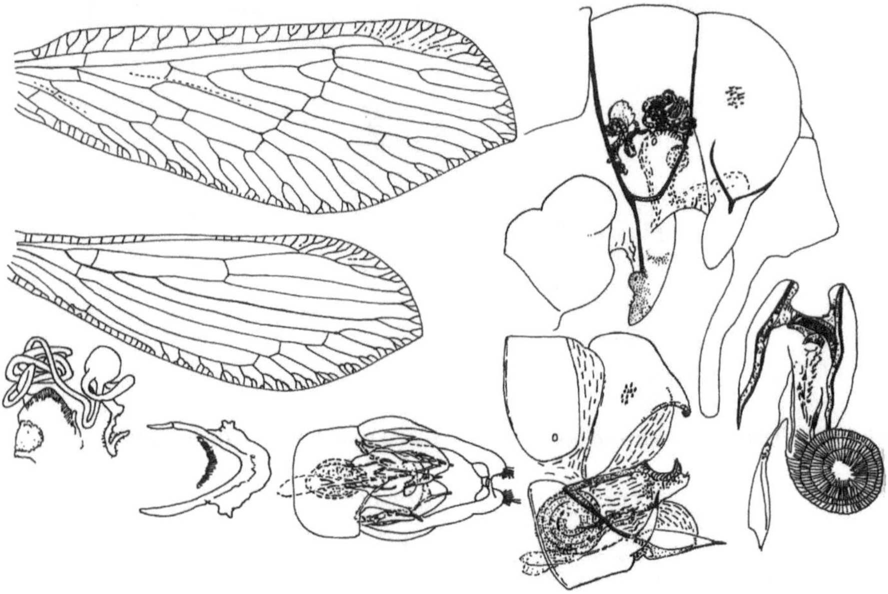

Fig. 17.4 *Spiroberotha sanctarosae:* fore and hind wing, not showing the fringe of long setae (upper and middle left); apex of the male abdomen in ventral (lower center) and lateral view (lower right center), male gonarcal complex (lower right), apex of the female abdomen in lateral view (upper right), bursa and spermathecal complex (lower left), and the eighth sternite of the female (lower left center). Based on Adams (1989).

4. (3) The fore-wing has three radial cross veins (Fig. 17.2).
   .... *Lomamyia* Banks, 1904
   Only one species in this primarily North and Central American genus has been found in South America: *Lomamyia trombetensis* Penny, 1983, found in Pará.

   – The fore-wing has four or five radial cross veins (Fig. 17.4).
   .... *Spiroberotha* Adams, 1989..p. 599

17.2 Subfamily Berothinae

Key to the Species of Adult *Naizema* in South America

Information for the key was provided by Esben-Petersen (1912), Navás (1919), and McLeod and Adams (1968).

1. The pterostigma has a reddish tinge (Fig. 17.3). There are many small brown spots on the legs, which are otherwise pale with brown on the knees, at the apices of the tibiae, and on rings around the tarsal segments. Length: c. 8 mm. Fore-wing length: c. 10 mm. The dorsal surface from the head to the apex of the

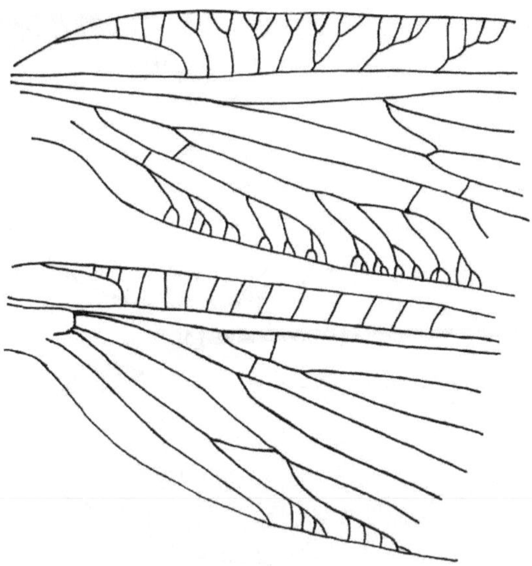

Fig. 17.5 Basal portions of the fore and hind wings of *Naizema patagonica*. Based on Navás (1919).

abdomen is yellowish brown with a narrow, unbroken mid-dorsal stripe. There are also broad brown stripes on the thorax.
....*Naizema mendozina* (Esben-Petersen, 1912) (Argentina). Syn: *Berotha mendozina* Esben-Petersen, 1912.

– The pterostigma is pale. Most cross veins in the costal space are forked once or twice near the apex (Fig. 17.5). The legs are dark ferrugineous with yellow tibiae. The tarsi are dark at the apices. There is a long white polosity on the upper parts of the leg, while near the apices, long dark and light setae are intermixed. Length: c. 7.4 mm. Fore-wing length: c. 10.4 mm; hind wing: c. 9.3 mm. The dorsal surface is mainly dark yellow with a narrow midline stripe, sometimes narrowly divided along the midline if the scutellum, and with wide, lateral stripes on parts of the thorax.
....*Naizema patagonica* Navás, 1919 (Argentina).

Key to the Species of *Spiroberotha* in South America

Information for the key was provided by Adams (1989).

1. The male has two strong, black setae at the apex of each posterolateral projection on the sternites. The wing is strongly falcate and has only short setae along the posterior margin (Fig. 17.6). Fore-wing length of the male: 5.5–7.2 mm; female: 5.7–7.8 mm.
....*Spiroberotha fernandezi* Adams, 1989 (Venezuela).

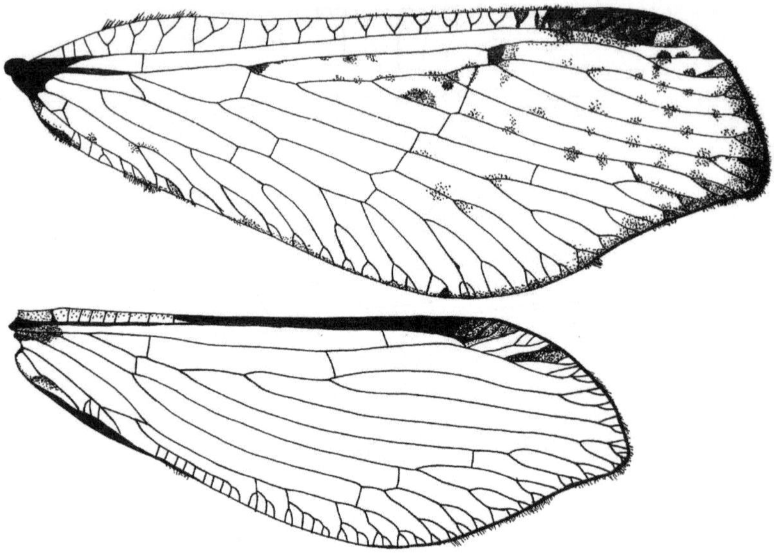

Fig. 17.6 Fore and hind wing of *Spiroberotha fernandezi*. Based on Adams (1989).

– The male has one strong seta at the apex of each posterolateral projection on the sternites. The wing is weakly falcate and has a fringe of long setae along the posterior margin (Fig. 17.4). Fore-wing length of the male: 5.0–6.1 mm; female: 6.5–6.8 mm.

.... *Spiroberotha sanctarosae* Adams, 1989 (Central America, Colombia, Venezuela).

Literature

Adams PA (1989) A new genus of Berothidae from tropical America, with two new species. Psyche 96:187–194

Ardila-Camacho A (2013) First record of beaded lacewings (Neuroptera, Berothidae) from Colombia. Zootaxa 3669(2):159–164

Aspöck U (1986) The present state of knowledge of the family Berothidae (Neuropteroidea: Planipennia). In: Gepp J, Aspöck H, Hölzel H (eds) Recent research in Neuropterology. Proceedings of the 2nd international symposium on neuropterology, Graz, pp 87–101

Aspöck U, Randolph S (2014) Beaded lacewings – a pictorial identification key to the genera, their biogeographics and a phylogenetic analysis (Insecta: Neuroptera: Berothidae). Deutsche Entomologische Zeitschrift (Neue Folge) 61(2):155–172

Blanchard CÉ (1851) Mirmeleonianos. In: Gay C (ed) Historia Fisica y Politica de Chile. Zoologia. E. Thunot, Paris, pp 119–129

Carpenter FM (1940) A revision of the Nearctic Hemerobiidae, Berothidae, Sisyridae, Polystoechotidae and Dilaridae (Neuroptera). Proceedings of the American Academy of Arts and Sciences 74(7):193–280

Esben-Petersen P (1912) A few new Neuroptera Planipennia. Entomologische Mitteilungen 1(9):267–273

Gurney AB (1947) Notes on Dilaridae and Berothidae, with special references to immature stages of the Nearctic genera (Neuroptera). Psyche 54:145–169

McLeod EG, Adams PA (1968 [1967]) A review of the taxonomy and morphology of the Berothidae, with the description of a new subfamily from Chile (Neuroptera). Psyche 74:237–265

Monserrat VJ (2006) Nuevos datos sobre angunas especies de la familia Berothidae (Insecta: Neuroptera). Heteropterus Revista de Entomologia 6:173–207

Monserrat VJ, Deretsky Z (1999) New faunistical, systematic and taxonomic data on brown lacewings (Neuroptera: Hemerobiidae). Journal of Neuropterology 2:45–66

Navás L (1919 [1918]) Algunos insectos Neurópteros de la República Argentina. Serie Tercera (III). Revista de la Real Academia de Ciencias Exactas Fisicas y Naturales de Madrid 17:287–305

Penny ND (1983) Neuroptera of the Amazon Basin. Part 8. – Berothidae. Acta Amazonica 13(3–4):689–695

Penny ND (2002) Family Berothidae. In: Penny ND (ed) A guide to the lacewings (Neuroptera) of Costa Rica. Proceedings of the California Academy of Sciences 53(12):186–187

Penny ND, Winterton SL (2007) Rediscovery of the unusual genus *Ormiscocerus* (Neuroptera: Berothidae: Cyrenoberothinae). Proceedings of the California Academy of Sciences, Fourth Series 58(1):1–6

Tauber CA, Tauber MJ (1968) *Lomamyia latipennis* (Neuroptera: Berothidae) life history and larval descriptions. The Canadian Entomologist 100(6):623–629

Index

Valid names of genera and species are printed in italics. Invalid names and the names of higher taxa are printed in Roman type. Page numbers on which illustrations of the species appear are printed in bold type.

C.W. Heckman, *Neuroptera (Including Megaloptera)*, Encyclopedia of South American Aquatic Insects, DOI 10.1007/978-3-319-35125-4

MIX
Papier aus verantwortungsvollen Quellen
Paper from responsible sources
FSC® C105338

If you have any concerns about our products,
you can contact us on
ProductSafety@springernature.com

In case Publisher is established outside the EU,
the EU authorized representative is:
Springer Nature Customer Service Center GmbH
Europaplatz 3, 69115 Heidelberg, Germany

Printed by Libri Plureos GmbH
in Hamburg, Germany